FERROELECTRICS

Edited by **Indrani Coondoo**

Ferroelectrics
Edited by Indrani Coondoo

Published by InTech

Janeza Trdine 9, 51000 Rijeka, Croatia

Publishing Process Manager Jelena Marusic

Technical Editor Teodora Smiljanic

Cover Designer InTech Design Team

Additional hard copies can be obtained from orders@intechopen.com

Ferroelectrics, Edited by Indrani Coondoo
p. cm.
ISBN 978-953-307-439-9

Meet the editor

Indrani Coondoo is currently a research staff member at the Department of Glass & Engineering Ceramics and CICECO in University of Aveiro, Portugal. She was awarded Ph. D. in Physics from the University of Delhi, India. She was invited for a Japan-India joint - research on ferroelectric materials at the Institute of Materials Science, University of Tsukuba, Japan. She has worked as a research associate in the field of ferroelectric liquid crystals at National Physical Laboratory, India. She then worked as a volunteer researcher at the University of Puerto Rico, USA. She serves as a member of the Materials Research Society and is a life member of The Indian Science Congress Association (ISCA). She is a recipient of the national scholarship conducted jointly by Council of Scientific and Industrial Research & University Grants Commission, India. Her current research interests are in the field of ferroelectric and multiferroic nanostructures and thin films, composite materials, nanosacle phenomena in Pb-free piezoelectrics for energy harvesting, nonvolatile information storage technologies and liquid crystals. She has authored several scientific papers in refereed journals and presented several research papers at international conferences.

Contents

Preface

This book reviews a wide range of diverse topics related to the phenomenon of ferroelectricity (in the bulk as well as thin film form) and provides a forum for scientists, engineers, and students working in this field. It contains records of original research that provide or lead to fundamental principles in the science of ferroelectric materials. The present book is a result of contributions of experts from international scientific community working in different aspects of ferroelectricity related to experimental and theoretical work aimed at the understanding of ferroelectricity and their utilization in devices. It provides an up-to-date insightful coverage to the synthesis, characterization, functional properties and potential device applications in specialized areas. The readers will perceive a trend analysis and examine recent developments in different fields of applications of ferroelectric materials. Based on the thematic topics, it contains the following chapters:

Chapter 1 gives a detailed description of new operation principle of memory devices based on piezoacousto properties of ferroelectric films – the Acousto-FeRAM (AFeRAMs). It is based on the two physical properties of a ferroelectric film (two-in-one), the intrinsic memory phenomenon for information storage and piezoacousto property for effective read-out info. It allows the design and implementation of 3-D memory devices with universal characteristics of the inherent ferroelectric memory.

Chapter 2 deals with the studies on the tunneling effect and the magnetoelectric coupling in ferroelectric-based junctions. This chapter suggests connecting conventional electronics with spintronics at the nanoscale and several low-power approaches for spin-based information control.

Chapter 3 reports the observance of photoluminescence in low-dimensional oxide ferroelectric materials in lanthanide-doped $Bi_4Ti_3O_{12}$ thin films. Co-doping of rare earth ions such as Eu/Gd, Pr/La, Er/Yb in bismuth titanate thin films is found to be an effective way to improve photoluminescence and electrical properties of the thin films.

Chapter 4 examines the optical properties and electronic band structures of perovskite-type ferroelectric and conductive metallic oxide films. Temperature influence on the optical properties of the BLT film prepared by chemical solution method on quartz substrate has been studied by the transmittance measurements. The Adachi's model has been employed to calculate the dielectric functions of high-quality LNO and LSCO conductive oxide films.

Chapter 5 reports on $SrRuO_3$ bottom electrodes grown with an epitaxial relationship with SrO buffered-Si(001) substrates by pulsed laser deposition. The structural and electrical properties of the $SrRuO_3$ films have been studied with deposition parameters of $SrRuO_3$ on the optimized SrO buffer layered Si (001) substrates. 100nm thick-$Pb(Zr_{0.2}Ti_{0.8})O_3$ thin films deposited at 575°C on $SrRuO_3$/SrO/Si substrates showed a (00*l*) preferred orientation and exhibited a $2P_r$ of 40 μC/cm² and a E_c of 100 kV/cm.

In **Chapter 6** the experimental results on Electro Caloric Effect (ECE) in ferroelectric polymer films. The relaxor ferroelectric P(VDF-TrFE-CFE) terpolymer has been studied which reveal a very large ECE at ambient condition in the relaxor terpolymers.

Chapter 7 studies the subtitutional effect in $Bi_4Ti_3O_{12}$ ferroelectric thin films. BIT, BLT, BTV and BLTV thin films were fabricated on the $Pt/TiO_2/SiO_2$/p-Si(100) substrates by sol-gel processes. The BLTV thin film shows the largest $2P_r$ of 50.8 μC/cm² with small $2E_c$ of 194 kV/cm among these films. For the BLTV thin film, the fatigue test showed the strongest fatigue endurance up to 10^{10} cycles with leakage current density in the order of 10^{-9}-10^{-8} A/cm² .

Chapter 8 reports the successful fabrication of uniaxially aligned semiconducting, conducting, and carbon nanofiber yarns by electrospinning and subsequent thermal conversion or carbonization of PXTC, which would open up a new field of applications in organic nanoelectronics.

Chapter 9 reports on the introduction of carbon materials where in the excellent electrical properties of carbon materials enable them to assist the applications of ferroelectric and related materials.

In **Chapter 10**, the dielectric relaxation phenomenon is discussed in ferroelectric perovskite-related structures considering the relaxation mechanisms and the influence of the vacancies on them.

Chapter 11 reports on the ferroelectric-ferromagnetic composite ceramics with high permittivity and high permeability in hyper-frequency. The co-fired composite ceramics of $0.8Pb(Ni_{1/3}Nb_{2/3})O_3$-$0.2PbTiO_3/Ba_2Zn_{1.2}Cu_{0.8}Fe_{12}O_{22}$ are discussed in this chapter, which has excellent co-firing behavior, dense microstructure and good electromagnetic properties.

Chapter 12 investigates the aging-induced double ferroelectric hysteresis (P–E) loops and recoverable electrostrain (S–E) curves in an Mn-doped orthorhombic $KNbO_3$-based $[K(Nb_{0.90}Ta_{0.10})O_3]$ lead-free ceramic: $K[(Nb_{0.90}Ta_{0.10})_{0.99}Mn_{0.01}]O_3$. Double P–E loops and large recoverable S–E curves with amplitudes in excess of 0.13% at 5 kV/mm have been observed in the aged samples over a wide temperature range of 25–140 °C.

Chapter 13 deals with the co-relation between piezoelectric properties and grain size affected by defects in Nb^{5+} and Mn^{3+} doping in Pb-free $(Bi_{0.5}Na_{0.5})(Ti_{1-x}D_x)O_3$ (D = Nb or Mn) ceramics.

Chapter 14 shows enhanced dielectric and ferroelectric properties of donor (W^{6+}, Eu^{3+}) substituted $SrBi_2Ta_2O_9$ ferroelectric ceramics. Also, W and Eu doping in SBT results in reduced dielectric loss and conductivity.

Chapter 15 talks about non-equilibrium thermodynamics of ferroelectric phase transitions. The authors provide both experimental and theoretical evidence for the existence of stationary states to a ferroelectric phase transition. Moreover they have given the non-equilibrium (irreversible) thermodynamic description of phase transitions and explanation of the irreversibility of ferroelectric phase transitions.

Chapter 16 gives a detailed description of the theories and methods of first order ferroelectric phase transitions Heavy-computer relying methods, such as first-principle calculations and molecular dynamic simulations etc, have been applied to investigate the physical properties of ferroelectrics.

Chapter 17 deals with electroacoustic waves in a ferroelectric crystal with moving domain walls (DW). The effect of the uniform motion of ferroelectric DWs that form a dynamic superlattice on the spectral properties of EWs has been analyzed for the first time. It is put forth that the velocity of domain-wall motion serves as a new parameter that is convenient for controlling the reflection and transmission of waves in combination with their frequency shifts and this induced acoustic nonreciprocity and the Doppler frequency conversion can be used for the development of sensors and acousto-electronic devices for data conversion with the frequency output.

Chapter 18 deals with an theoretical approach to ferroelectric Optics: Optical Bistability in Nonlinear Kerr Ferroelectric Materials wherein the Maxwell-Duffing analysis has been employed to study the optical bistability of a ferroelectric slab as well as a Fabry-Perot resonator coated with two identical partially-reflecting dielectric mirrors.

Chapter 19 exposes the use of nonlinear treatments for energy harvesting enhancement. It has been demonstrated that the conversion enhancement by application of switching approach (allowing both a voltage increase and a reduction of the time shift between voltage and velocity), concept to energy harvesting (SSHI) allows a significant gain in terms of harvested power (7-8 times greater than the standard interface) or significantly reduce the amount of piezoelectric material required to harvest a given amount of energy.

Chapter 20 deals with quantum chemical investigations of the energetics and structures corresponding to the different structural conformations for the PVDF, P(VDF-TrFE) and P(VDF-CTFE) units.

Chapter 21 provides a solution for an exact impedance control for electro-mechanical systems. The authors give a new model of DC motor with dynamics between the torque and control input. They propose a new impedance control which is based on Casimir function.

In **Chapter 22** the problem of designing an H_∞ control for networked control systems (NCSs) with the effects of both the input saturation as well as the network-induced delay have been addressed.

In **Chapter 23**, robust sampled-data control and observer design for uncertain fuzzy systems with discrete, neutral and distributed delays has been considered. A controller design method has been proposed via LMI conditions.

Chapter 24 proposes a new method of numerical solution to solve a system of linear equations. The presented method can be applied on any system of equations with LR fuzzy number coefficients.

We hope that this book will prove to be timely and thought provoking and will serve as a valuable reference for researchers working in different areas of ferroelectric materials. Special thanks go to the authors for their valuable contributions.

Indrani Coondoo

Departamento de Engenharia Cerâmica e do Vidro (DECV) & CICECO
Universidade de Aveiro, 3810-193 Aveiro,
Portugal

Part 1

Ferroelectrics and Its Application: Bulk and Thin Films

1

Principle Operation of 3–D Memory Device based on Piezoacousto Properties of Ferroelectric Films

Ju. H. Krieger
Academgorodoc
Russia

1. Introduction

At present, the main driving forces for rapidly growing the market of nonvolatile memory devices and nonvolatile memory technology are portable electronics. Technical innovations will continue to drive the increase of memory density and speed in the future. For traditional nonvolatile memory devices like NOR flash, scaling could stop even before the 32nm node, while scaling of mass storage devices like NAND flash, is going to be very difficult beyond the 20nm node, too.

In recent years, more research efforts have been devoted to finding a new memory technology able to overcome performance and scalability limits of currently memory devices. However as lithographic scaling becomes more challenging. Novel architectures are needed to improve memory device performances in embedded as well as stand-alone memory applications. It is now generally recognized that 3–D integration and vertically stacking are possible alternatives to scaling. In order to realize stacked memory cells a novel physical principle operation of a memory element is needed.

Many new ideas have been proposed to find out the solution of such problems. Recently, it were developed a rewriteable 3–D NAND flash memory chip called Bit Cost Memory (BiCS) in which it stacked memory arrays vertically (Fukuzumi & et al., 2007) and 3–D double-stacked multi-level NAND flash memory (Park & et al., 2009). However, there is a cross talk between neighbouring memory cells in the same plane and between vertical neighbours in adjacent planes, too. It is a new challenge in addition to scaling issues. The major limitation, however, is likely to be power dissipation where 3–D designs are not efficient at dissipating heat. More over 3–D architectures are required to overcome a tyranny of interconnects.

From physical point of view, a memory device based on ferroelectric phenomena is more suitable for creating 3–D architectures of universal memory devices. Ferroelectric materials can be electrically polarized in certain direction. Then, the polarization is retained after the polarizing field is removed. Therefore, these materials have intrinsic memory. Ferroelectric random access memory (FeRAM) is widely considered as an ideal nonvolatile memory with high write speeds, low-power operation and high endurance (Scott, 2000; Pinnow & Mikolajick, 2004). FeRAM also have an added significant advantage, which is high radiation hardness. FeRAM are inherently radiation-hard making it very suitable for space applications. The FeRAM with lowest active power is a voltage-controlled device as opposed to a magnetic, a phase change and a switchable resistor type of memory, which are current-controlled devices with high power consumption. In scaled memory device, current-driven devices have not survived. There is needed a low resistance material for a nanoscale signal-line.

In order to realize all good performance of ferroelectric memory devices and to simplify a problem of creating 3–D memory devices a novel physical principle of memory device operations are needed. The one of the advantages a physical principle of ferroelectric memory device operation is Acousto-Ferroelectric Memory Device (AFeRAM) has been published recently (Krieger, 2008, 2009). The new universal memory device is called Acousto-Ferroelectric RAM (AFeRAM) makes use of acoustic method of detecting the direction of the spontaneous polarization of the ferroelectric memory cells. The physical principle of AFeRAM device operations is based on a polarization dependent of an acoustic response from a ferroelectric memory cell under the action of applied electrical field pulse. In this chapter, it will be presents physical principle, 3–D architectures and operation modes of the new type of universal memory devices based on piezoacusto properties of ferroelectric materials.

2. Ferroelectric random access memory (FeRAM)

Ferroelectric Random Access Memory (FeRAM) has been studied for over fifty years. The FeRAM devices are divided into two categories based on the readout technique: destructive read-out (DRO) FeRAM and non-destructive read-out (NDRO) FeRAM (Scott, 2000). The first one is capacitor-type, where ferroelectric capacitors are used to store the data and FeRAM with 1T/1C (or 2T/2C, etc.) configurations. This structure is composed of a ferroelectric capacitor (C) to store data and a transistor (T) to access it similar to a DRAM cell. The read operation is based on reading current, which results from changes in polarization when voltage is applied to the cell. Accordingly, the data stored in the cell are destroyed in each read cycle. In other words, the read operation is destructive. Therefore, the data needs to be re-written in each cycle. Another main issue of FeRAM memories with 1T/1C or 2T/2C configuration is that it cannot be easily scaled down. The absolute value of the ferroelectric polarization is important for capacitor-type FeRAM. It is necessary to look for new ferroelectric materials with the high intrinsic remanent polarization and using high program voltage. At the same time, ferroelectric memory cells based on ferroelectric materials with the high remanent polarization are usually characterised by low endurance as a consequence of fatigue phenomena in ferroelectric films. Moreover this FeRAM device based on ferroelectric materials with a high remanent polarization, as a rule, have bad endurance and fatigue property, which don't allow to use this type of FeRAM as DRAM device. Therefore, many companies have suspended development of this type of memory.

The second type of memory is transistor-type (like Flash), in which ferroelectric-gate field effect transistors (FeFET-type) are used to store the data (Miller, 1992; Scott, 2000). Ferroelectric-gate field effect transistors, in which the gate insulator film is composed of a ferroelectric material, have attracted much attention because the ferroelectric gate area can be scaled down in proportion to the FeFET size. Thus, FeFET has high potential for use in large-scale FeRAM with 1 Gigabit or higher density. In other words, charge density (charge per unit area) of the ferroelectric polarization is an important parameter to achieve non-volatility in transistor-type FeRAM, whereas the absolute value of the ferroelectric polarization is important for capacitor-type FeRAM.

Typical values of remanent polarization for the ferroelectric gate material are extremely different from those for FeRAM with 1T/1C configuration. The latter require remanent polarization of 20–40 μC/cm², whereas the ferroelectric-gated FeFETs can function well with 100-200 times lower remanent polarization (0.1–0.2 μC/cm²) (Miller, 1992; Dawber, 2005). However, FeFET memory devices generally have short data retention times due to leakage, depolarization, or both.

Another main issue of FeFET memory is high program voltage. In order to prevent the chemical reaction between the Si substrate and the ferroelectric film, an insulator buffer layer must be inserted between them. For this reason, the program voltage becomes large due to voltage drop on the buffer layer between Si and ferroelectric films. Nonetheless, the FeFET type memory continues to attract much interest. It is well recognized that only non-destructive read-out (NDRO) of ferroelectric memory cells should realize all good performance of ferroelectric memory devices. By the way another important feature of FeFET is its high sensitivity to intrinsic remanent polarization of the ferroelectric gate that can be used for detecting mechanical stress or acoustic waves (Greeneich & Muller, 1972; Greeneich & Muller, 1975). At the same time, the information about ferroelectric gate polarization is not lost and can be extracted by heating (pyroelectric approach) or mechanical deformation (piezoacousto approach) of ferroelectric memory cell (Krieger, 2009).

Ferroelectric materials display a number of physical phenomena (pyroelectricity, piezoelectricity), which allow us to obtain information about the intrinsic remanent polarization by generating additional charge in the ferroelectric film, which is proportional to the value and polarity of the intrinsic remanent polarization. In general, all ferroelectric materials in their ferroelectric phase are pyroelectrics and piezoelectrics. Table 1 lists based properties of some of the more commonly used ferroelectric materials (Settera, 2006).

Materials	$LiNbO_3$	SBT	$BiFeO_3$	PZT	PMN-PT
$\varepsilon/\varepsilon_0$	28	200–300		400–1700	5000–7000
$\mathbf{2P_r}$ (μC/cm²)	1.2	10–30	80–120	30–60	42
$\mathbf{d_{33}} \cdot 10^{-12}$ (m/V⁻¹)	6–23	15–20	60–80	300–600	2000–3000
p C/(Kcm²)10⁻¹⁰	83	71	14,7	260	300

Table 1. The properties of the commonly used ferroelectric materials: $\varepsilon/\varepsilon_0$ – dielectric constant, $\mathbf{2P_r}$ – remanent polarization, and $\mathbf{d_{33}}$ – piezoelectric charge constant, p-pyroelectric charge constant

There are three general methods for determining a value and polarity of the intrinsic remanent polarization of ferroelectric film which is associated with intrinsic memory. They are a charge generation by electrical fields, charge generation by heating or cooling (pyroelectricity), and charge generation by mechanical stress (piezoelectricity). The electrical field approach is used in conventional FeRAM devices.

One of the advantageous methods to obtain information about the intrinsic remanent polarization of ferroelectric films is piezoacousto approach. Below we present piezoacousto approach which should allow us to extract information about ferroelectric polarization of memory cells by detecting an acoustic response from a ferroelectric memory cells under applied reading electrical pulse.

3. Physical principle AFeRAM device operation

Ferroelectric crystals are a subgroup of piezoelectric materials. It means that all ferroelectric materials are piezoelectric. Some ferroelectrics are excellent piezoelectrics (see Table 1). It is known that piezoelectric materials can generate useful output under simple input signal. For example, piezoelectric material will generate an electric field with the input of stress or vice versa.

The piezoelectric effect is a natural ability of piezoelectric materials to produce an electric charge, which is proportional to an applied mechanical stress. This is termed the direct piezoelectric effect. By reversing the direction (from tension to compression) of the stress applied to piezoelectric material, a sign of the created electric charge is reversed as well. The piezoelectric effect is also reversible. When an electric charge or voltage is applied, a mechanical strain is created. This property is called the inverse piezoelectric effect. In other words, when piezoelectric film is stressed electrically by voltage, its dimensions change and generate acoustic wave. When it is stressed mechanically by force or acoustic wave, it generates an electric charge. If electrodes are not short-circuited, voltage associated with the charge appears. Both piezoelectric effects are used in AFeRAM devices.

A basic characteristic of piezoelectric materials is the piezoelectric charge constant (**d**). The piezoelectric charge constant, **d** (C/N), is the polarization generated per unit of mechanical stress applied to a piezoelectric material or, alternatively, the piezoelectric charge constant, **d** (m/V), is the mechanical strain experienced by a piezoelectric material per unit of applied electric field (1).

$$\mathbf{d} = \frac{\text{Charge density}}{\text{Applied mechanical stress}}\left(\mathrm{C/N}\right) = \frac{\text{Strain developed}}{\text{Applied electric field}}\left(\mathrm{m/V}\right) \qquad (1)$$

In general, since piezoelectric charge constant is a second-rank tensor value, it is correct to write **d** in the equations as the tensor components; in the case of polycrystalline thin films we deal with the effective value of piezoelectric charge constant, $\mathbf{d}_{eff}$. Sometimes notation d_{33} is also used to represent piezoelectric charge constant measured in the direction perpendicular to film surface.

Piezoelectric charge constant ($\mathbf{d}_{33}$) is an important characteristic of the material's suitability for AFeRAM device. Others are Young's (or elastic) modulus (**Y**), electromechanical coupling factor (**K**), and dielectric constant (ε). Y is an indicator of the stiffness (elasticity) of a material. Typical value of the bulk piezoelectric charge constant, $\mathbf{d}_{33}$ (C/N) for several fully poled conventional ferroelectric materials are tabulated in Table 1. At the same time, it is known that piezoelectric charge constants of ferroelectric thin films (100 nm or less) increase in several times for feature sizes of cells with lateral dimensions below 300 nm (Buhlmann, & et al, 2002).

There are two methods for determining a value and polarity of the intrinsic remanent polarization of memory cells by piezoacousto approach. The first one, the ferroelectric memory cell is stressed mechanically by pulse force or acoustic solitary wave and then the momentarily generated electric charge in the ferroelectric memory cell is measured by conventional methods or with the use of FeFET as memory element.

The second method is to measure the generated acoustic elastic solitary wave, when the ferroelectric memory cell is stressed electrically by the reading voltage pulse with special waveform. As it was mentioned before, the field effect transistor with piezoelectric gate can be used for detecting the acoustic solitary wave, which is generated by the ferroelectric memory cell under the action of the reading voltage pulse.

The most obvious advantage of the second method is that the memory cells are not directly connected to the access and sensitive and measuring transistor by conductor wires, but exchange data with the access transistor by acoustic solitary waves. It means that there is an acoustic interconnection between the ferroelectric memory cells and the access transistor, which allows creating 3-D structure of memory device without using additional metal interconnections.

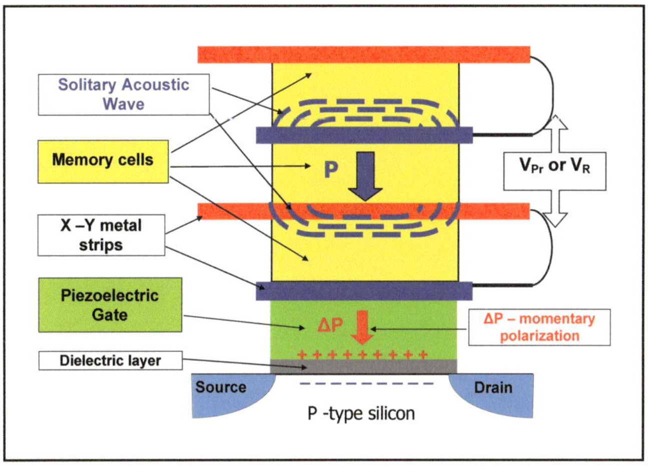

Fig. 1. Simplified structure and principle operation of AFeRAM element

In simple case, the AFeRAM element has a columnar structure (Fig. 1). The structure of the AFeRAM element can be implemented using two well know devices: a) ferroelectric capacitor or capacitors which acts as ferroelectric memory cell where two binary logic states "0" and "1" are represented by the direction of the spontaneous polarization as well as piezoacoustic transmitter; b) field effect transistor with piezoelectric or polarized ferroelectric gate which acts as acoustic sensor and access transistor. The memory cell or cells are stacked above the transistor with the piezoelectric-ferroelectric gate. The memory cells are merged with the access transistor with ferroelectric/ piezoelectric gate to form AFeRAM element. At the same time, memory cells have no electrical link to the access transistor. Ferroelectric memory cell and the access transistor exchange the information by an acoustic elastic solitary deformation wave (Fig. 1).

There is vague similarity between the AFeRAM element and piezoelectric transformer. The direct and converse piezoelectric effects are used in AFeRAM memory element as well as in the piezoelectric transformer, where voltage from one unit (in our case, from memory cell) to another (in our case, to transistor gate) is transformed acoustically.

The memory cell can be programmed in accordance with the conventional method by applying high negative electrical field (state of "0") or positive electrical field (state of "1") to the memory cell (Fig. 1).

In order to read the stored information, reading pulse voltage (V_R^{MC}) weaker than the coercive voltage is applied to the memory cell, and two types of acoustic elastic solitary deformation waves can be generated. For example, the reading electrical field, additive to the ferroelectric memory cell polarization (state "1"), will produce a tensile stress. The reading electrical field, opposite to the ferroelectric memory cell polarization (state "0"), will produce a compressive stress. In turn, these stresses produce the acoustic solitary wave, which produces stress on the piezoelectric-ferroelectric gate of the access transistor. This stress produces a momentary "polarization" (ΔP) of the transistor piezoelectric-ferroelectric gate and controls channel conductivity of the transistor. Positive piezoelectric charges increase the effective donor concentration, reducing thickness of the surface depletion layer. Therefore, the conductivity of the semiconductor channel can be altered by the momentary piezoelectric charge due to acoustic solitary wave when it goes through the ferroelectric gate. The reading operation is thus carried out by identifying whether the current flows from the source to the drain of the transistor (state "1") or not (state "0") or vice versa. The memory cells exchange data with the access transistor by means of acoustic solitary waves. The momentary generated charges or polarization can be relatively easily calculated in terms of piezoelectric and elastic constants. Equation for the value of the momentary generated charges or polarization is as follows:

$$\mathbf{\Delta P} = \mathbf{d_{33}}^{TG} \cdot \mathbf{d_{33}}^{MC} \cdot \mathbf{Y}^{MC} \cdot \mathbf{V_R}^{MC}; \quad (2)$$

Here $\mathbf{\Delta P}$ (μC/cm²) is the momentarily generated charge or polarization within the transistor gate; $\mathbf{d_{33}}^{TG}$ and $\mathbf{d_{33}}^{MC}$ are the piezoelectric charge constants for the transistor gate (TG) and memory cell materials (MC), respectively; $\mathbf{Y}^{MC}$ is the Young modulus for the memory cell materials and $\mathbf{V_R}^{MC}$ is the reading voltage applied to the memory cell. The momentary polarization calculated within the transistor gate for different materials of the fixed memory cell and the transistor gate are presented in Table 2.

Typical thicknesses of the memory cell and the transistor gate film are 100 nm, and the reading voltage ($\mathbf{V_R}^{MC}$) is about 0.5 V. As shown in Table 2, the momentary polarization of the piezoelectric gate can exceed several times the critical value of polarization (0.1–0.2 μC/cm²) required for the transistor functioning (Miller, 1992; Dawber, 2005).

Materials for ferroelectric transistor gates	Materials for ferroelectric memory cells			
	$LiNbO_3$	SBT	$BiFeO_3$	PZT
PZT	0.25–1.72	0.47–1.0	1.22–2.25	2.25–9.0
PMN-PT	1.50–2.25	2.80–4.30	7.50–11.25	15.0–45.0

Table 2. Momentary polarization $\mathbf{\Delta P}$(μC/cm²) generated within the ferroelectric-piezoelectric transistor gate by solitary acoustic wave

It shows a very high acoustic wave sensitivity of FeFET with ferroelectric/piezoelectric gate which allows using a broad assortment of the ferroelectric materials with low value of an intrinsic remanent polarization of ferroelectric memory cells and low temperature deposition for memory cells as well as for ferroelectric/piezoelectric gate; to create 3-D structure of memory device with many memory cell layers stacked above the access transistor and to operate on a multi-bit mode to store more than one bit per memory cell (Kimura H., Hanyu T. & Kameyama M. 2003).

Another advantageous way to detect the acoustic solitary wave is to use an access and acoustic sensing transistor based on ferroelectric or piezoelectric semiconductors (e.g. GaAs,

GaN, AlN) or their heterostructure (e.g. GaAs/AlGaAs, GaN/AlN) which characterised fast the channel transit time (Kang Y., Fan Q., Xiao B. & et al., 2006; Lu S. S. & Huang C.L., 1994; Wu & Singh, 2004). In both cases the transistor channel charge and current are controlled by the gate voltage and/or by acoustic solitary wave.

Ferroelectric materials with high piezoelectric charge constants for transistor gate, like PZT or PMN-PT; and "fatigue-free" ferroelectric materials with relatively small piezoelectric charge and dielectric constant for memory cell, like $LiNbO_3$ or SBT, can be recommended.

The most obvious advantage of detecting direction and value of the polarization of ferroelectric memory cells by acoustic solitary wave is that the memory cells are not directly connected to access and acoustic sensing transistor by conductor wires. It allows easy creating 3–D structure of memory device by stacking the several layers of memory cells above array of the access transistors without using additional metal interconnections.

One of the main issues of AFeRAM devices is the echo phenomena. This problem can be eliminated and neglected by using composite packed materials with high acoustic absorption factor and acoustical impedance value equal to the silicon acoustical impedance.

4. 3-D architectures of Acousto-Ferroelectric RAM

There are several types of the AFeRAM memory storage element structures and operation modes. Basically, the AFeRAM element consists of two parts: saving (memory) area (ferroelectric memory cell or cells) and reading area (a ferroelectric field transistor) (Fig. 2, 3). There is perfect analogy between reading area of AFeRAM element and FeFET memory devices. It means that FeFET can be used as a parent element of AFeRAM memory devices. Moreover design problems for AFeRAM are already known very well and have been solved by FeFET and other FeRAM device solutions (Wang S. & et al. 2009). Furthermore, the new principle for ferroelectric memory operation does not have great demands to the remanent polarization and conductivity of the submicron ferroelectric film. It allows easy finding of new candidate ferroelectric materials for AFeRAM devices (Settera, 2006).

One of promising structures of memory elements (resembling the NAND Flash configuration) are shown in Fig. 2. In these structures, the memory area consists of cross point memory ferroelectric cells and the memory cell size is determined by crossing array between the top and bottom electrodes (Fig. 2). This memory area is relatively simple to manufacture, as one involve only three layers of metal and two ferroelectric layers sandwiched between metal strips. In this case, patterning of a submicron ferroelectric film is not needed for the memory cells. It is easy to create two levels and inexpensive memory device. Fig. 2a shows an equivalent electrical circuit of the two-layer memory area. The reading area of this type of AFeRAM element consists of FeFET with an elongated conductance channel and multi metal gate.

Another promising structure of a three-layer memory element is shown in Fig. 3. Fig. 3a shows an equivalent electrical circuit of the three-layer memory area. In these structures, the memory area also consists of cross point ferroelectric memory cells which are separated by dielectrics with different values of acoustical impedance and dielectric constant. It allows creating acoustic channels, improving anisotropy of solitary acoustic wave propagation in the stratified structure and eliminates acoustic cross talk. In these cases the reading area of AFeRAM element also consists of the piezoelectric/ferroelectric field transistor with the elongated conductance channel and the multi-metal gate.

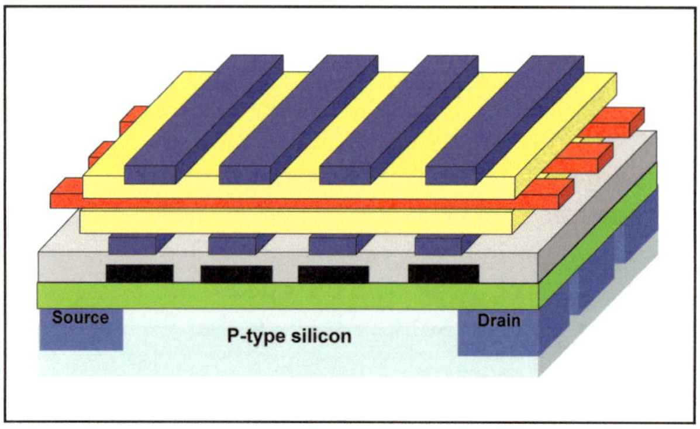

Fig. 2. Two-layer AFeRAM structure with elongated conductance channel. The memory cell size is defined by crossing strips

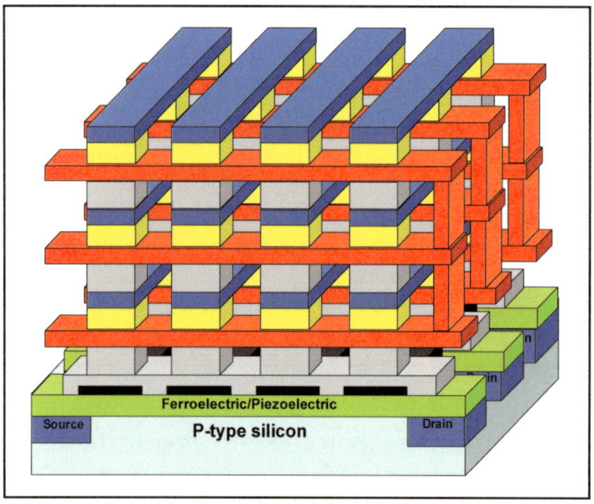

Fig. 3. Three-layer AFeRAM structure with elongated a conductance channel of access transistors. The memory cells are separated by different type of dielectrics and form acoustic channels

This type of memory element allows creating practically unlimited number of memory cell layers and can be easily scaled down to tens of nanometers electrode stripe width and use nano-ferroelectric materials (Scott , 2006).

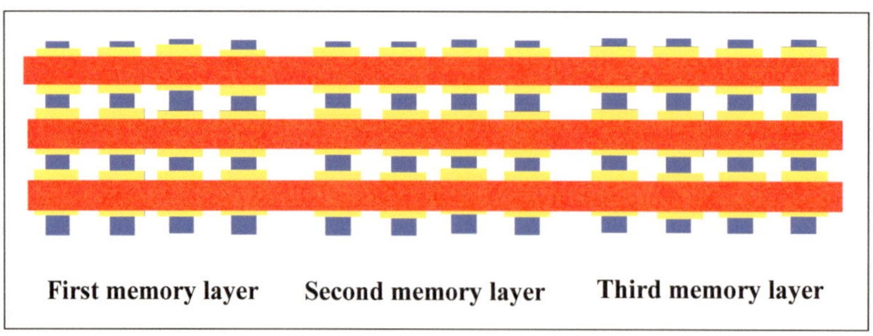

Fig. 3a. Equivalent electrical circuit for the three-layer memory area

The AFeRAM element architecture allows stacking more than one layer (four or six layers) of memory arrays (3-dimensional stacking), offers scalability, the simplest and lowest cost technology and the highest integration density (~10 Gbit/cm^2 for F=90 nm or ~100 Gbit/cm^2 for F=32 nm and four layers of memory arrays, where F denotes the smallest feature size limited by lithography resolution). The additional increase in info density can be reached by means of a multi-bit mode to store more than one bit per memory cell (Kimura, Hanyu & Kameyama, 2003).

5. Operation modes of Acousto-Ferroelectric RAM

A memory area of all type of AFeRAM elements consist of cross point ferroelectric memory arrays. The cross point memory arrays are relatively simple to manufacture, as they only involve two layers of metal and some ferroelectric layer sandwiched between. They typically have high density. The electrodes are typically arranged in an X and Y conductive strips, with each cell of the cross point array being located at the points in the ferroelectric materials where the X and Y strips cross over each other. The data bit stored in each cell has a value determined by the polarity of the ferroelectric material at that point. The polarity is controlled by application of voltages on the X and Y strips. Typically, the X strips are referred to as word lines and the Y strips are referred to as bit lines. Write operations and read operations are normally accomplished by applying voltages to the word line and bit line of the cell. In this case, ferroelectric capacitors can form a ferroelectric memory array in the same way as magnetic cores form a core memory.

5.1 Writing operation modes

Writing operation modes will be illustrated by the example of the two-layer AFeRAM structure with elongated conductance channel (Fig. 2). As it was mentioned above, the ferroelectric memory cells of AFeRAM element can be programmed with conventional method (Fig. 2a) by applying a high negative electrical field (state of "0") or positive electrical field (state of "1") to the memory cell. In order to program the ferroelectric memory cell, the program voltage (V_{Pr}) across the ferroelectric capacitor must exceed the coercive voltage (V_C), hence, $V_{Pr} >= V_C$.

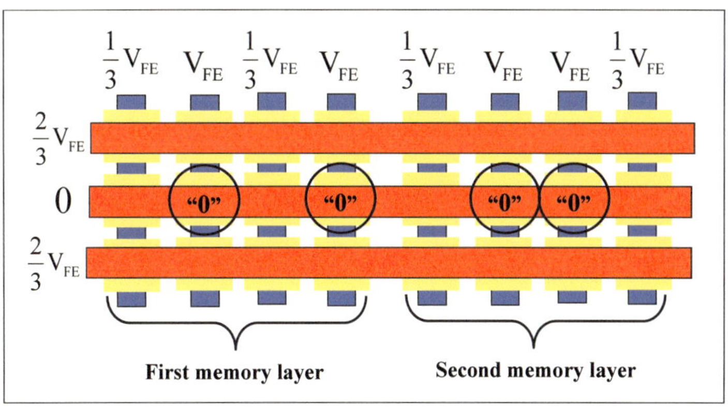

Fig. 2a. Equivalent electrical circuit for the two layer memory area and writing mode (program "0" state)

There are two voltage pulse protocols to write information into the ferroelectric memory cells which are the 1/2V protocol and the 1/3V protocol. Despite the fact that passive memory array is relatively small, a voltage pulse protocol based on the 1/3 voltage selection rule (1/3V protocol) should be used in order to keep disturb voltages at minimum (Fig. 2a). It is possible to write data to multiple cells in both schemes; word line by word line (Fig. 2a) or bit line by bit line, but this cannot be performed simultaneously for the "1" state and "0"states.

To write a "0" state by using the 1/3V protocol, the corresponding bit line of the cells to be written is connected to V_{Pr} and the corresponding word line to V=0. All other bit lines are connected to $1/3V_{Pr}$ and all other word lines to $2/3V_{Pr}$ as can be seen in Fig. 2a. The voltage drop on all non accessed cells is $1/3V_{Pr}$ or $-1/3V_{Pr}$, which means, that the distortion voltage is reduced in comparison to the 1/2V protocol. Writing a "1" is similar but this time the voltages of the word lines and bit lines are exchanged (Fig. 2b). The three-layer as well as multi-layers AFeRAM devices would operate in much the same manner as analogously to the two-layer AFeRAM elements.

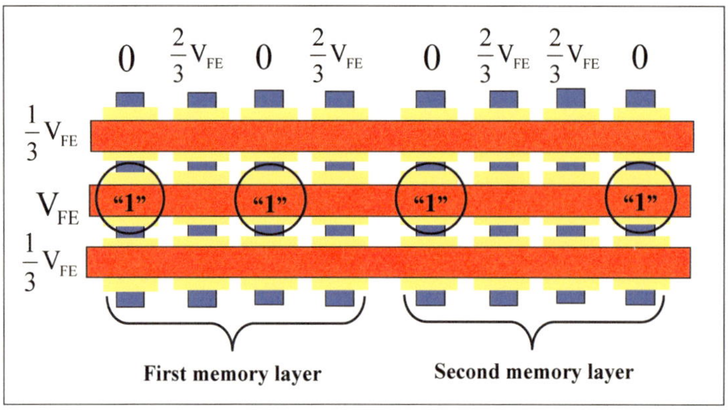

Fig. 2b. Equivalent electrical circuit for the two-layer memory area and writing mode (program "1" state)

As it was mentioned before, the one of important features of FeFET which uses as reader of the polarization memory cell is its high sensitivity to a mechanical stress or acoustic waves (Greeneich & Muller, 1975). It allows using a low voltage to program memory cell. Therefore, it should be possible to use value of the program voltage very closely to the coercive voltage V_C as soon as possible. As result endurance can be increase to up 10^{15-17} (Joshi, 2000). It allows using AFeRAM element as DRAM device. Moreover, the low switching voltages for programming is a definitive advantage for low-power applications, system management issues, and scalability.

5.2 Reading operation modes

Reading operation modes will be illustrated by the example of the three-layer AFeRAM structure with elongated conductance channel and multi-gate field effect transistor (Fig. 3). The reading operation mode of for the three-layer memory area of AFeRAM memory element is shown in Fig. 3b, 3c. In order to information read-out from AFeRAM element; the reading pulse voltage (V_R) weaker than the coercive voltage is applied sequentially to the memory cells (a, b, c, and d) (Fig. 3b, 3c).

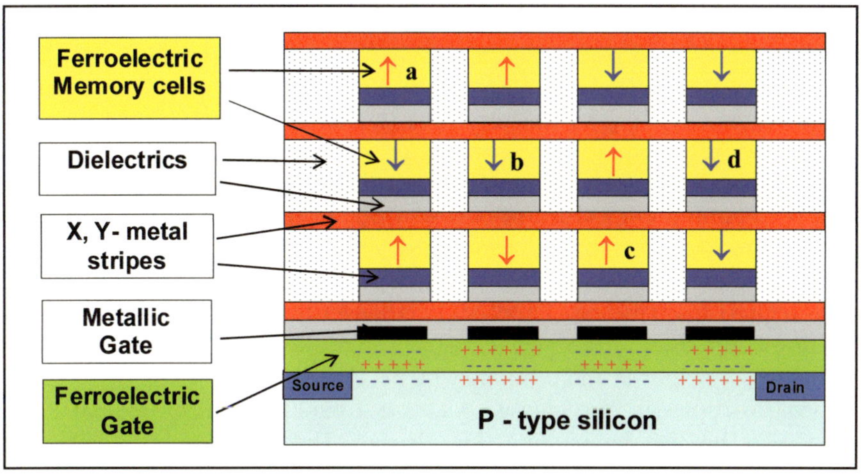

Fig. 3b. Cross section of three-layer 3-D structure memory element and reading mode of memory cells (**a**, **b**, **c** and **d**)

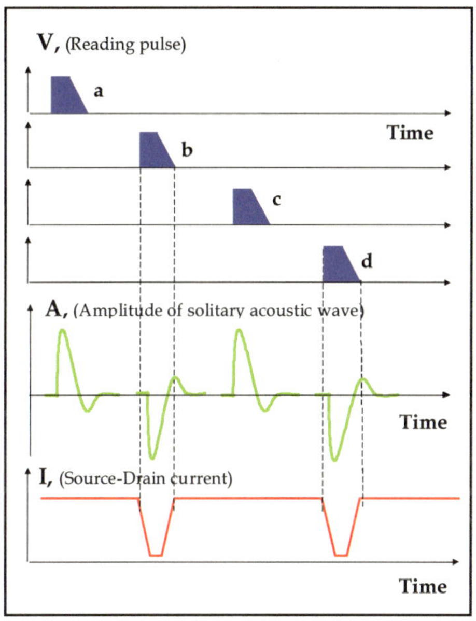

Fig. 3c. Timing diagram of the reading pulse voltages for each of the memory cells (a, d, c and d on Fig. 3b), amplitude of solitary acoustic wave generated by memory cells and the corresponding source-drain current through the semiconductor channel of the access transistor

Fig. 3c shows a timing diagram of the reading pulse voltages with special waveform for each of the memory cells, its acoustic response under the action of the reading pulse voltages and the corresponding current through the semiconductor channel of the access transistor. Initially, the access transistor is opened by metal gates and the maximum current flows easy from the source to the drain of the access transistor (Fig. 3c). Two types of acoustic elastic solitary deformation waves (or shock waves) can be generated (Fig. 3c) by

memory cells. Memory cell (state "1") will produce a tensile acoustic solitary wave and memory cell (state "0") will produce a compressive acoustic solitary wave under the action of the positive reading pulse (Fig. 3c). In turn, only compressive acoustic solitary wave which are generated (Fig. 3c) by memory cells (b and d) (state "0") produce the positive momentary "polarization" (**ΔP**) of the transistor ferroelectric gate and interrupts current through the semiconductor channel of the access transistor. The reading operation is thus carried out by identifying interrupt of current through the semiconductor channel of the access transistor by check out state "0" of memory cells (Fig. 3c).
The memory cells exchange data with the access transistor by means of acoustic solitary waves. The highest frequency exchange data between the memory cell and the sensitive transistor can extend to the GHz range (Greeneich & Muller, 1975) and is determined by the transistor frequency response and ultimately by the channel transit time.
The piezoacousto approach also allows us to use a parallel reading without distortion of nonselected cells, bit line by bit line (Fig. 2c, 3c). These modes effectively reduce the programming and reading time per byte by factor of 16 or 64 or more. This factor is determined by size of the memory array. It is also important to notice that the reading and programming periphery circuits of the passive memory array are the same.

6. Conclusion

In order to improve the performance and scalability of today's memory devices, a lot of new memory technologies are under investigation now. The goal these investigations to find the ultimate universal memory that combines fast read, fast write, non-volatility, low-power, unlimited endurance and high info density.
Forecasting the future of solid state memory leads us to the inescapable conclusion that lithographic scaling becomes more challenging and memory companies should be turning their sights to investigation of 3-D memory technology based on stackable cross-point memory arrays. At the same time, in 3-D memory architecture, current-driven memory devices with high power dissipation which is developing now can not be used.
Today it is known only one physical principle operation of a voltage-controlled memory device with low power consumption and dissipation; it is ferroelectric memory with intrinsic memory phenomenon. Unfortunately, reading operation mode of conventional FeRAM devices which are developed and manufactured now do not use all good performances of the ferroelectric memory phenomenon and its lithographic scaling becomes more challenging beyond the 90 nm node.
The AFeRAM concept uses the two physical properties of a ferroelectric film (two-in-one), the intrinsic memory phenomenon for information storage and piezoacousto property for effective read-out info. It allows the design and implementation of 3-D memory devices with universal characteristics of the inherent ferroelectric memory. Furthermore, there is no limitation in future AFeRAM miniaturization in the sense of the operational principle and can be shrunk down into the nanometer range that is a great advantage compared to current-controlled memory devices. Acoustic principle of ferroelectric memory operation does not require high remanent polarization value and value of conductivity of a submicron ferroelectric film. It allows us to easily find a new candidate among ferroelectric materials. Very high acoustic wave sensitivity of the FeFET with ferroelectric gate allows us to use a wide range of ferroelectric materials or use an access and acoustic sensing transistor based on ferroelectric or piezoelectric semiconductors, which characterised fast the channel transit

time. More over, the new type of memory device can be based on very well investigated and developed a ferroelectric or piezoelectric transistor which can be used as a parent element of AFeRAM device and a memory array with passive cross-point structure. Due to the many analogies between AFeRAM and 1T/1C FeRAM, FeFET FeRAM and Flash, several design issues for AFeRAM are already known and have been solved by applying prior Flash, FeRAM solutions.

The new operation principle of memory devices based on piezoacousto properties of ferroelectric films will help to realize all good performance of ferroelectric memory devices and to simplify a problem of creating 3–D memory devices for portable electronics with universal characteristics and low power consumption. The acoustical interconnection allows creating not only 3–D stand-alone or embedded ferroelectric memory device, but other implementations have been proposed. Specifically, it can be use for the fast exchange (GHz frequency range) info between memory zones and a microprocessor.

New electrically accessible and voltage-controlled AFeRAM device with universal characteristics, high density and low power dissipation should involve a revolution in computer architectures. The simplicity of having the ability to use a single memory type and having all computer memory in a nonvolatile system brings us back to the architectures used by system designers forty years ago, when simple memory cells based on magnetic cores was used. Acousto-Ferroelectric RAM would become the ultimate universal memory device compatible with the CMOS technology and can replace DRAM, Flash as well as HDD in modern portable electronic devices such as mobile phones and notebook computers.

7. References

Buhlmann S., Dwir, B., Baborowski J., & Muralt P. (2002). Size effect in mesoscopic epitaxial ferroelectric structures: Increase of piezoelectric response with decreasing feature size, *Applied Physics Letters,* Vol. 80, No. 17, pp. 3195-3197.

Dawber M., Rabe K. M., & Scott J. F. (2005). Physics of thin-film ferroelectric oxides, *Reviews Modern Physics,* Vol. 77, No. 4, pp. 1083–1496.

Fukuzumi Y., Matsuoka Y., Kito M., Sato M., Tanaka H., Nagata Y., & et al. (2007). Optimal integration and characteristics of vertical array devices for ultra-high density, bit-cost scalable flash memory. - *International Electronics Devices Meeting Tech. Dig,* (Dec 2007), pp. 449-52.

Greeneich E. W. & Muller R. S. (1972). Acoustic-wave detection via a piezoelectric field-effect transducer, *Applied Physics Letters*, Vol. 20, pp.156–158.

Greeneich E. W. & Muller R. S. (1975). Theoretical transducer properties of piezoelectric insulator FET transducers, *Journal Applied Physics*, Vol. 46, No. 11, pp. 4631–4640.

Hoffman J., Pan X., Reiner J. W. & et al. (2010). Ferroelectric field effect transistors for memory applications. *Advanced Materials*, Vol. 22, No. 26-27, pp. 2957–2961.

Joshi V., Solayappan N. & Araujo C. A. (2000). FeRAM reliability studies for 10^{15} switching cycle regime. *12th International Symposium on Integrated Ferroelectrics,* March 2000; Aachen: Germany.

Kang Y., Fan Q., Xiao B. & et al. (2006). Fabrication and current-voltage characterization of a ferroelectric lead zirconate titanate/AlGaN/GaN field effect transistor. *Applied Physics Letters,* Vol. 88, No. 12, pp.123508 -10.

Kimura H., Hanyu T. & Kameyama M. (2003). Multiple-Valued Logic-in-Memory VLSI Using MFSFETs and its Applications. *Journal of Multiple-Valued Logic and Soft Computing* Vol. 9, pp. 23-42.

Krieger Ju. H. (2008). Acousto-ferroelectric nonvolatile RAM, *Integrated Ferroelectrics,* Vol. 96, No. 1, pp. 120–128.

Krieger Ju. H. (2009). Physical concepts of FeRAM device operation based on piezoacousto and pyroelectric properties of ferroelectric films. *Journal Applied Physics,* Vol. 106, No. 6, pp. 0616291–0616296.

Lu S. S. & Huang C.L. (1994). Piezoelectric field effect transistor (PEFET) using In0.2Ga0.8As/Al0.35Ga0.65As/In0.2Ga0.8As/GaAs strained layer structure on (111)B GaAs substrate. *Electronics Letters,* Vol. 30, No. 10, p. 823-825.

Miller S. L. & McWhorter P. J. (1992). Physics of the ferroelectric nonvolatile memory field effect transistor, *Journal Applied Physics,* Vol. 72, No. 12, pp. 5999–6010.

Park K-T., Kang M., Hwang S. & et al. (2009). A fully performance compatible 45 nm 4-gigabit 3–D double-stacked multi-level NAND flash memory with shared bit-line structure. *IEEE Journal of Solid-State Circuits,* Vol. 44, No. 1, pp. 208–216.

Park S. O., Bae B. J., Yoo D. C. & Chung U. I. (2010). Ferroelectric random access memory. *Nanotechnology,* Vol. 9, pp.397–418.

Pinnow C.-U., & Mikolajick T. (2004). Material aspects in emerging nonvolatile memories. *Journal Electrochemical Society,* Vol. 151, No. 6, pp. K13-K19.

Scott J. F. (2000). *Ferroelectric Memories,* Berlin: Springer.

Scott J. F. (2006). Nanoferroelectrics: statics and dynamics, *Journal of physics: Condensed Matter,* Vol.18, pp. R361–R386.

Settera N., Damjanovic D., Eng L. & et al. (2006). Ferroelectric thin films: Review of materials, properties, and applications, *Journal Applied Physics,* Vol. 100, No. 5, pp. 51606–51646.

Wang S., Takahashi M., Li Q-H., Takeuchi K., Sakai S. (2009) Operational method of a ferroelectric (Fe)-NAND flash memory array. *Semiconductor Science and Technology,* Vol. 24, No. 10, pp. 5029–5037.

Wu Y.-R. & Singh J. (2004). Metal piezoelectric semiconductor field effect transistors for piezoelectric strain sensors. *Applied Physics Letters,* Vol. 85, No. 7, pp. 1223-1225.

2

Ferroelectric and Multiferroic Tunnel Junctions

Tianyi Cai[1], Sheng Ju[1], Jian Wang[2] and Zhen-Ya Li[1]
[1]*Department of Physics and Jiangsu Key Laboratory of Thin Films*
Soochow University, Suzhou, 215006
[2]*Department of Teaching Affairs,*
Wuxi City College of Vocational Technology, Wuxi, 214153
China

1. Introduction

The phenomenon of electron tunneling has been known since the advent of quantum mechanics, but it continues to enrich our understanding of many fields of physics, as well as offering a route toward useful devices. A tunnel junction consists of two metal electrodes separated by a nanometer-thick insulating barrier layer, in which an electron is allowed to transverse a potential barrier exceeding the electron's energy. The electron therefore has a finite probability of being found on the opposite side of the barrier. In the 1970's, spin-dependent electron tunneling from ferromagnetic metal electrodes across an amorphous Al_2O_3 film was observed by Tedrow and Meservey(1)(2). Based on this discovery, Jullière proposed and demonstrated that in *a magnetic tunnel junction* tunnel current depends on the relative magnetization orientation of the two ferromagnetic electrodes(3). Such a phenomenon nowadays is known as *tunneling magnetoresistance*(TMR)(4). Magnetic tunnel junctions may be very useful for various technological applications in spintronics devices such as magnetic field sensors and magnetic random access memories. Other insulators are also used for tunnel barriers. For example, epitaxial perovskite $SrTiO_3$ barriers were studied by De Teresa *et al.* to demonstrate the importance of interfaces in spin-dependent tunneling(5). In tunnel junctions with MgO barriers, Ikeda *et al.* found large magnetoresitance as high as 604% at room temperature and 1144% at 5 K(6), which approaches the theoretical predictions of Butler *et al.*(7) and Mathon *et al.*(8). Despite the diversity of materials used as the barrier of the tunnel junctions, the common feature is that almost all the barriers are nonpolar dielectrics.

On the other hand, magnetic insulators, i.e, EuO, EuS and EuSe, are used for tunnel barriers. Spin filtering has been observed in these junctions as were first discussed by Moodera *et al.*(9). in 1988. They observed that the tunneling current in Au/EuS/Al junction has a spin polarization with the magnitude as high as 80%. and attributed it to the electron tunneling across the spin-dependent barriers (Fig.1). Later, they reported that the tunneling current across Ag/EuSe/Al junctions has an enhanced spin-polarization reaching 97%(10). Recently, using EuO with a higher Curier temperature (69 K) than EuS (16.7 K) and EuSe (4.6 K), Santos et al. obtained 29% spin-polarized tunneling current(11). Naturally, if electrodes are not normal metals, but ferromagnetic materials, both TMR and spin filter effects can be observed(Fig.2)(12).

Another important concept is the *ferroelectric tunnel junction* (FTJ)(13)(14)(15), which take advantage of a ferroelectric as the barrier material. Ferroelectrics possess a spontaneous

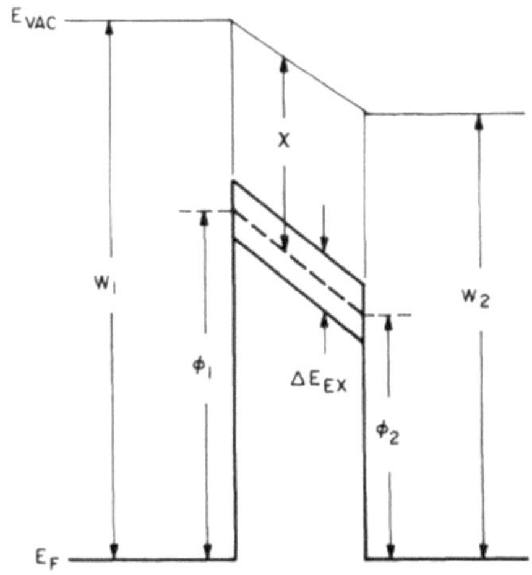

Fig. 1. Schematic illustration of tunnel barrier of a Au/EuS/Al junction. W_1 and W_2 are the work functions of Au and Al, respectively. χ is the electron affinity of EuS. The barrier heights at the Au and Al interfaces are shown as Φ_1 and Φ_2 at the bottom of the EuS conduction band (dashed line) at T>16.7 K. The bottom of the two bands shown at $T \leq T_C$ by the solid lines separated by ΔE_{ex} are the barriers seen by the two spin directions.(9)

electric polarization that can be switched by an applied electric field. This adds a new functional property to a tunnel junction. Nowadays, there are worldwide efforts to include FTJs into various nanoscale devices such as Gbit nonvolatile semiconductor memories. This

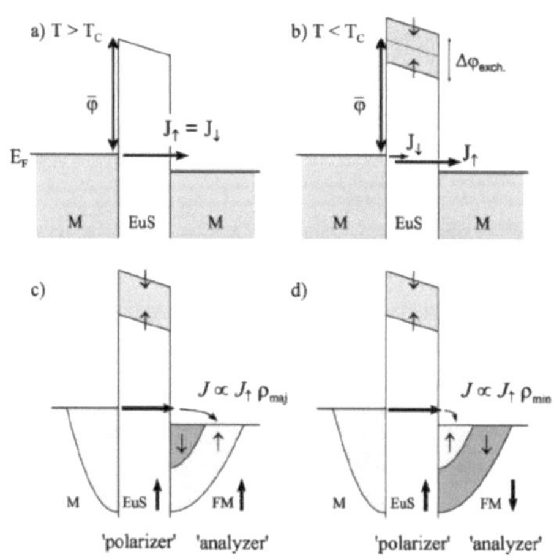

Fig. 2. Schematic illustration of spin filtering and the MR effect. (a) above T_C of the EuS filter the two spin currents are equal. (b) below the T_C of EuS, the tunnel barrier is spin split, resulting in a highly spin polarized tunnel current. With a ferromagnetic (FM) electrode, the tunnel current depends on the relative magnetization orientation. For parallel alignment (P), (c) a large current results, while for antiparallel alignment (AP), (d) a small current results.(12)

may open new exciting perspectives but also give rise to important fundamental questions. For example, *can ferroelectricity exist in a nanometer-thick barrier film?* As is well-known, ferroelectricity is a collective phenomenon (as magnetism or superconductivity) and it results from a delicate balance between long-range Coulomb forces (dipole-dipole interaction) which are responsible for the ferroelectric state and short-range repulsion which favor the paraelectric cubic state. When the size of a ferroelectric sample is reduced, both Coulomb and short-range forces are modified. This leads to a behavior at very small size that cannot be trivially predicted and causes eventually a suppression of the interesting functional properties below what is referred to the correlation volume. The ferroelectric instability of ultra-thin films and ultra-small particles has been an open question for several decades. Recently, experimental and theoretical investigations showed that ferroelectricity may persist even at a film thickness of a few unit cells under appropriate mechanical (lattice strain) and electric boundary (screening) conditions. In particular, it was discovered that, in organic ferroelectrics, ferroelectricity can be sustained in thin films of a few monolayer thickness (16). In perovskite ferroelectric oxides, ferroelectricity was observed down to a nanometer scale (17). This fact is consistent with first-principle calculations that predict a nanometer critical thickness for a FTJ (18). As a result, the existence of ferroelectricity at such a small film thickness makes it possible to use ferroelectrics as tunnel barriers.

Let us now turn to a general outline of this chapter. We will begin with a discussion of the concept of a ferroelectric tunnel junction, then show that the reversal of the electric polarization in the ferroelectric produces a change in the electrostatic potential profile across the junction. This leads to the resistance change which can reach a few orders of magnitude, namely, the giant *tunneling electroresistance* (TER) effect. Interface effect, strain effect and composite barrier are also discussed. Next, we will show that functional properties of FTJs can be extended by adding the spin degree of freedom to FTJs. This makes the junctions multiferroic (that is, simultaneously ferromagnetic and ferroelectric). The interplay between ferroelectric and magnetic properties in a multiferroic tunnel junction (MFTJ) may affect the electric polarization of the ferroelectric barrier, the electronic and magnetic properties of the interface, and the spin polarization of the tunneling current. Therefore, TMR and spin filtering effect observed in MTJs can also be observed in MFTJs. Such a new kind of tunnel junction may be very useful for future technological applications. Several ways to obtain MFTJs are introduced, such as (1) replacing one normal metal electrode with ferromagnetic one, (2) replacing ferromagnetic barriers with multiferroic materials, and (3) using a composite of ferromagnets and ferroelectrics as the barrier. These studies open an avenue for the development of novel electronic devices in which the control of magnetization can be achieved by the electric field via magnetoelectric coupling. Finally, we look at the magnetoelectric coupling effect in the ferroelectric-based junctions, which is independent of particular chemical or physical bonding.

2. Ferroelectric tunnel junction

The concept of a FTJ is illustrated in Fig.3(15), which shows the simplified band structure of a tunnel junction with a ferroelectric barrier. If the ferroelectric film is sufficiently thin but still maintains its ferroelectric properties, the surface charges in the ferroelectric are not completely screened by the adjacent metals [Fig.4(a)] and therefore the depolarizing electric field E in the ferroelectric is not zero. The electrostatic potential associated with this field depends on the direction of the electric polarization [Fig.4(b)]. If a FTJ is made of metal electrodes which have different screening lengths, this leads to the asymmetry in the potential profile for the

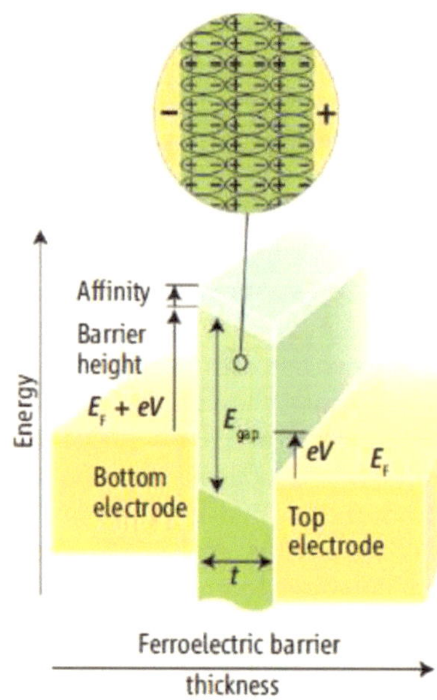

Fig. 3. (Color online). Schematic diagram of a tunnel junction, which consists of two electrodes separated by a nanometer-thick ferroelectric barrier layer(15). (E_{gap} is the energy gap. E_F is the Fermi energy, V is the applying voltage, t is the barrier thickness.)

opposite polarization directions. Thus, the potential seen by transport electrons changes with the polarization reversal which leads to the TER effect.

Electrostatic effect. The above arguments can be made quantitative by applying a Thomas-Fermi model. The screening potential within metal 1 and metal 2 electrode is given by(19)

$$\varphi(z) = \begin{cases} \frac{\sigma_S \delta_L e^{-|z|/\delta_L}}{\varepsilon_0}, & z \leq 0 \\ -\frac{\sigma_S \delta_R e^{-|z-d|/\delta_R}}{\varepsilon_0}, & z \geq d \end{cases} \tag{1}$$

Here δ_L and δ_R are the Thomas-Fermi screening lengths in the M_1 and M_2 electrodes. σ_S is the magnitude of the screening charges and can be found from the continuity of the electrostatic potential:

$$\varphi(0) - \varphi(d) = \frac{d(P - \sigma_S)}{\varepsilon_F} \tag{2}$$

Here P is considered to be the absolute value of the spontaneous polarization, and the introduction of the dielectric permittivity ε_F is required to account for the induced component of polarization resulting from the presence of an electric field in the ferroelectric. Using Eqs.(1)-(2) and introducing the dielectric constant $\varepsilon = \varepsilon_F/\varepsilon_0$, σ_S can be expressed as $\sigma_S = dP/[\varepsilon(\delta_L + \delta_R) + d]$.

Figure 4(b) shows the electrostatic potential in a M_1-FE-M_2 junction assuming that metals M_1 and M_2 have different screening lengths, such that $\delta_L > \delta_R$. It follows from Eq.(1) that different screening lengths result in different absolute values of the electrostatic potential at the interfaces, so that $\varphi_1 \equiv |\varphi(0)| \neq \varphi_2 \equiv |\varphi(d)|$, which makes the potential profile highly asymmetric. The switching of the polarization in the ferroelectric layer leads to the change in the potential which transforms to the one shown in Fig.4(b) by the dashed line, thus, inevitably leading to the change in the resistance of the junction.

Tunneling electroresistance effect. If the thickness of the ferroelectric barrier is so small that the dominant transport mechanism across the FTJ is the direct quantum-mechanical electron tunneling. The conductance of FTJ can be calculated, for example, at a small applied bias voltage the conductance of a tunnel junction per area A is(20)

$$\frac{G}{A} = \frac{2e^2}{h} \int \frac{d^2 k_{\parallel}}{(2\pi)^2} T\left(E_F, k_{\parallel}\right) \tag{3}$$

Here, T is the transmission coefficient evaluated at the Fermi energy E_F for a given value of the transverse wave vector $k_{\parallel}$, which can be obtained from the Schrödinger equation for an electron moving in the potential.

The overall potential profile seen by the transport electrons is a superposition of the electrostatic potential shown in Fig.4(b), the electronic potential which determines the bottom of the bands in the two electrodes with respect to the Fermi energy E_F, and the potential barrier created by the ferroelectric insulator. The resulting potential for the two opposite orientations of polarization in the ferroelectric barrier is shown schematically in Fig.5 for

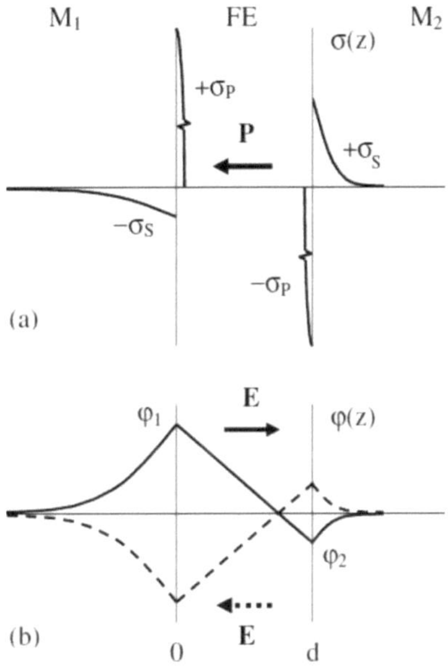

Fig. 4. Electrostatics of a M_1-FE-M_2 junction: (a) charge distribution and (b) the respective electrostatic potential profile (solid line)(14). The polarization P creates surface charge densities, $\pm\sigma_P = \pm|P|$, on the two surfaces of the ferroelectric film. These polarization charges $\pm\sigma_P$, are screened by the screening charge per unit area, $\mp\sigma_S$, which is induced in the two metal electrodes. It is assumed that metal 1 (M_1) and metal 2 (M_2) electrodes have different screening lengths ($\delta_L > \delta_R$) which lead to the asymmetry in the potential profile. The dashed line in (b) shows the potential when the polarization P in the ferroelectric is switched, resulting in the reversal of the depolarizing field E. The following assumptions are made: (1) The ferroelectric is assumed to be uniformly polarized in the direction perpendicular to the plane. (2) The ferroelectric is assumed perfectly insulating so that all the compensating charges resides in the electrodes. (3) The short-circuited FTJs are discussed.

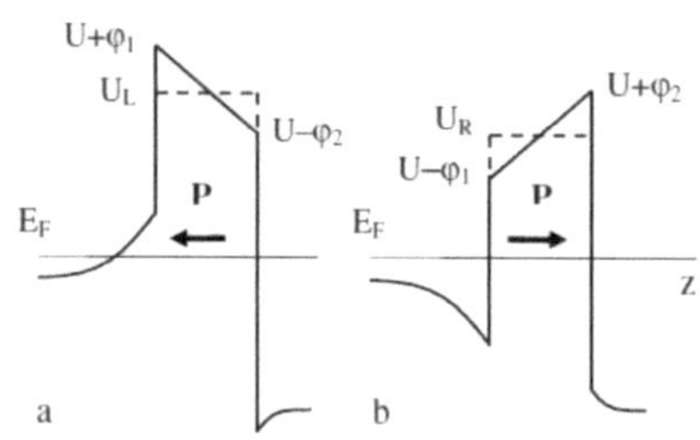

Fig. 5. Schematic representation of the potential profile $V(z)$ in a M_1-FE-M_2 junction for polarization pointing to the left (a) and for polarization pointing to the right (b), assuming that $\delta_1 > \delta_2$. The dashed lines show the average potential seen by transport electrons tunneling across the ferroelectric barrier(14). The horizontal solid line denotes the Fermi energy, E_F.

$\delta_L > \delta_R$. Indeed, the average potential barrier height seen by the transport electrons travelling across the ferroelectric layer for polarization pointing to the left, $U_L = U + (\varphi_1 - \varphi_2)/2$, is not equal to the average potential barrier height for polarization pointing to the right, $U_R = U + (\varphi_1 - \varphi_2)/2$, as is seen from Figs. 5(a) and 5(b). This make the conductance G_L for polarization pointing to the left much smaller than the conductance G_R for polarization pointing to the right (Fig.6(a)), thereby resulting in the TER effect(Fig. 6(b))(14).
Experimentally, ferroelectric tunnel junctions with different ferroelectric barriers have been fabricated successfully and giant tunneling electroresistance effects have been observed. Garcia et al.(21) and Gruverman et al.(22) reported that giant TER effects reached 75000% through 3 nm-thick $BaTiO_3$ barrier at room temperature. Crassous et al. observed that the TER reached values of 50000% through a 3.6 nm $PbTiO_3$(23). Maksymovych et al. found

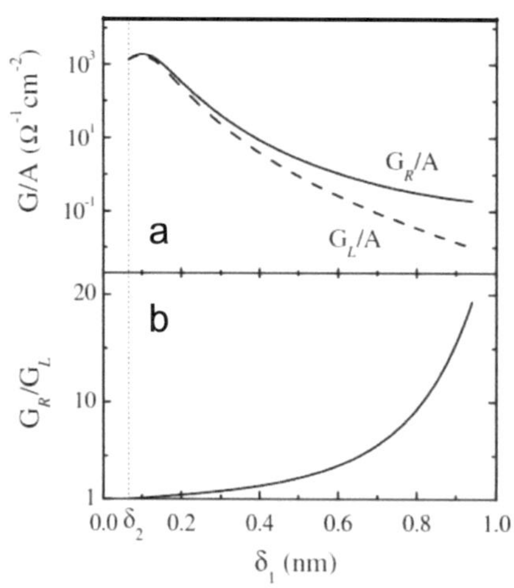

Fig. 6. (a) conductance per unit area for polarization oriented to the right, G_R/A (solid line) and for polarization oriented to the left, G_L/A (dashed line); (b) conductance change, $G_R = G_L$, associated with the polarization switching in the ferroelectric barrier (14). The vertical dotted line indicates the value of $\delta_1 = \delta_2$ at which no asymmetry in the potential profile and, hence, no conductance difference is predicted.

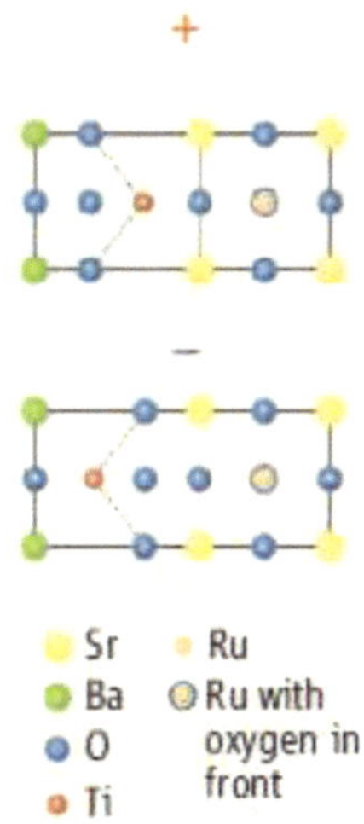

Fig. 7. Interface effect of a FTJ (15).

that the large spontaneous polarization of the $Pb(Zr_{0.2}Ti_{0.8})O_3$ film resulted in up to 500-fold amplification of the tunneling current upon ferroelectric switching(24).

Interface and strain effect. Interestingly, recent experimental(25) and theoretical(26) studies indicate that ionic displacements within the electrodes, in a few atomic monolayers adjacent to the ferroelectric, may affect the electron screening. The polarization switching alters positions of ions at the interfaces that influences the atomic orbital hybridizations at the interface and hence the transmission probability (see Fig.7). On the other hand, the piezoelectricity of a ferroelectric barrier under an applied voltage produces a strain (see Fig.8) that changes transport characteristic of the barrier such as the barrier width and the attenuation constant(27).

FTJs with ferroelectric/dielectric composite barriers. It is an efficient way to enhance the TER by using a layered composite barrier combing a functional ferroelectric film (FE) and a thin film of a nonpolar dielectric material (DI)(28). Due to the change in the electrostatic potential

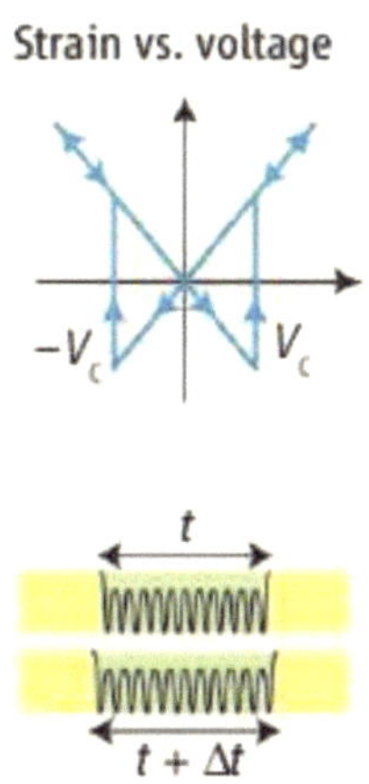

Fig. 8. Strain effect of a FTJ (15).

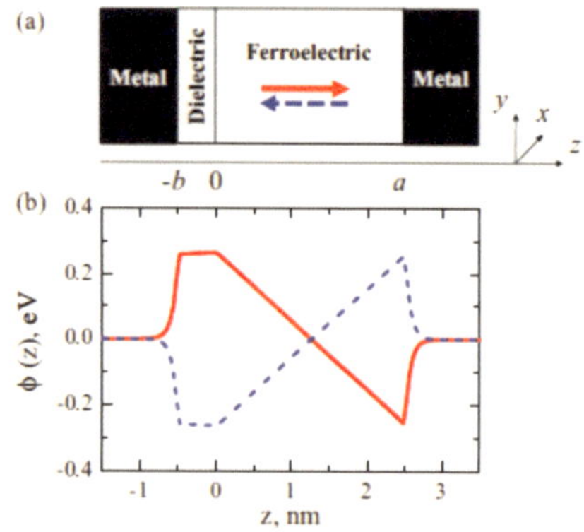

Fig. 9. (Color online) Geometry (a) and the electrostatic potential profile for the two opposite polarization orientations (b) of a FTJ: a=25 Å, b=5 Å, ε_d=300, ε_f =90, δ =1 Å, and P=20 $\mu C/cm^2$(28).

induced by polarization reversal, the nonpolar dielectric film adjacent to one of the interfaces acts as a switching changing its barrier height from a low to high value (Fig.9), resulting in a dramatic change in the transmission across the FTJ. The predicted values of TER are giant, indicating that the resistance ratio between the two polarization-orientation states in such FTJs may reach hundred thousands and even higher as shown in Fig.10. Furthermore, Wu et al. proposed that if the interface between the FE and the dielectric layer is very sharp and space charges exist at this interface, the TER will be enhanced strongly(29).

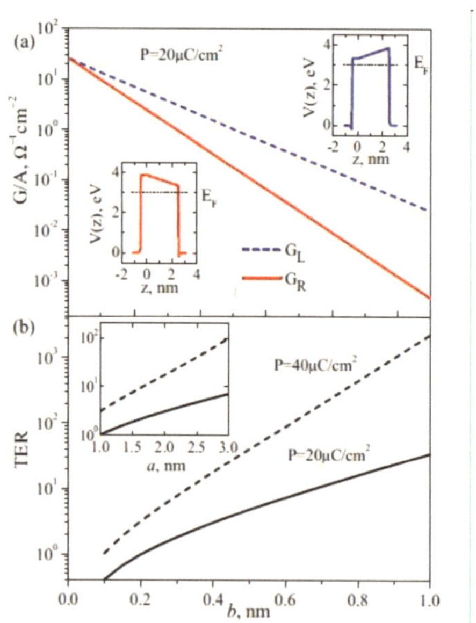

Fig. 10. (Color online) (a) Conductance of a FTJ for two opposite polarization orientations: Left (solid line) and right (dashed line), as a function of dielectric layer thickness. The insets show the corresponding tunneling barrier profiles. (b) TER as the function of dielectric layer thickness for two polarizations P=20 $\mu C/cm^2$ (solid line) and P=40 $\mu C/cm^2$ (dashed line). The inset shows TER as the function of the ferroelectric film thickness. U_d=0.6 eV, ε_d=300(28).

It should be noted that FTJs with composite barriers (M/FE/DI/M) does not require different electrodes. This may be more practical for device application than the conventional FTJs (M_1/FE/M_2). For FTJs as M_1/FE/M_2, asymmetry is necessary for the TER effect, which may be intrinsic (e.g., due to nonequivalent interfaces) or intentionally introduced in the system (e.g., by using different electrodes).

3. Multiferroic tunnel junctions (MFTJ)

A MFTJ is simultaneously ferromagnetic and ferroelectric. The transport behaviors of MFTJs can be controlled by the magnetic and electric field. Furthermore, the interplay between ferromagnetic and ferroelectric properties may affect the electric polarization of the ferroelectric barrier, the electronic and magnetic properties of the interface, and the spin polarization of the tunneling current. This indicates that not only TER effect, TMR and spin filtering effects may also be observed in MFTJs.

3.1 Ferromagnet/ferroelectric/normal metal junctions

By replacing normal metal electrode with a highly spin-polarized (ferromagnetic) material, such as diluted magnetic semiconductor(30), doped manganite(31), double perovskite manganites, CrO_2 and Heussler alloys, spin degrees of freedom can be incorporated into existing FTJs. In such MFTJs, the spin-polarized electrons from a ferromagnetic metallic electrode tunnel through a ferroelectric thin film which serves tunneling barrier. The reversal of the electric polarization of the ferroelectric film leads to a sizable change in the spin polarization of the tunneling current. This provides a two-state electric control of the spin polarization, including the possibility of switching from zero to nonzero or from negative to positive spin polarization and vice versa.

Electrostaticeffect. As is discussed on FTJs, the switching of the electric polarization changes the potential profile of the whole junction. Then, how does this change affect the conductance of the minority- and majority-spin carriers? As shown in Fig.11, for the electric polarization of the FE barrier pointing to the left (i.e., towards the FM electrode), majority-spin carriers experience an additional barrier compared to minority-spin carriers [compare the solid and dashed lines in Fig. 11(a)], since the spin dependent potential in FM electrode is $V_1^\sigma = V_1 \pm 1/2\Delta_{ex}$, σ is the spin index $\sigma =\downarrow,\uparrow$, Δ_{ex} is the exchange splitting strength. This occurs if the magnitude of the electrostatic potential at the FM/FE interface, $\varphi_1 \equiv \varphi(0)$, is larger than the Fermi energy with respect to the bottom of the minority-spin band, i.e., $E_F - V_1^\downarrow - \varphi_1 < 0$. If this condition is met, the spin polarization of the tunneling current is positive and weakly dependent on the potential barrier height. On the other hand, for the electric polarization pointing to the right (Fig. 11(b)), i.e., towards the NM electrode, the tunneling barrier is the same for majority and minority spins [compare the solid and dashed lines in Fig. 11(b)]. In this case, the magnitude of the spin polarization of the tunneling current is largely controlled by the exchange splitting of the bands and the potential profile across the structure. When $E_F - V_1^\downarrow - \varphi_1 > 0$, the asymmetry between R and L is due to the different barrier transparencies as a result of the different band structures of the two electrodes. Thus, by reversing the electric polarization of the FE barrier it is possible to switch the spin polarization of the injected carriers between two different values, thereby providing a two-state spin-polarization control of the device.

Spin filtering effect. The spin polarization of the conductance can be defined by $\Pi = G_\uparrow - G_\downarrow / G_\uparrow + G_\downarrow$, where the conductance can be calculated from Eq. (3). Figs. 12(a) and

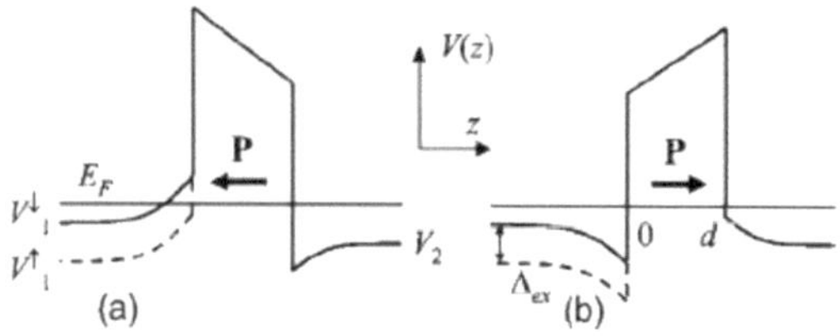

Fig. 11. (Color online) (a) Conductance of a FTJ for two opposite polarization orientations: Left (solid line) and right (dashed line), as a function of dielectric layer thickness(30). The insets show the corresponding tunneling barrier profiles. (b) TER as the function of dielectric layer thickness for two polarizations P=20 $\mu C/cm^2$ (solid line) and P=40 $\mu C/cm^2$ (dashed line). The inset shows TER as the function of the ferroelectric film thickness. U_d=0.6 eV, ε_d=300.

12(b) show the calculated conductance and spin polarization of the conductance as a function of the potential barrier height, U, in the ferroelectric barrier. It is seen that, for P pointing towards the ferromagnetic electrode, the spin polarization, Π_L, is positive and is weakly dependent on U, reflecting an additional tunneling barrier for minority spins (Fig. 11(a)). On the other hand, for the P pointing towards the NM electrode, the spin polarization, Π_R, is slightly negative at not too large values of U and becomes positive when U is larger than a certain value. The latter result can be understood in terms of spin-dependent tunneling across a rectangular barrier. Thus, using an appropriate FE barrier, it is possible to change the spin polarization of injected carriers from positive to negative and vice versa by reversing the electric polarization of the FE barrier. The degree of the spin polarization change in response to the electric polarization reversal depends on the carrier density in the semiconductors. This is illustrated in Fig. 12(c), which shows the dependence of the Π_L and Π_R on the Fermi energy with respect to the bottom of the minority-spin band in the FM electrode. When $E_F \leq V_1^\downarrow$ the DMS is fully spin polarized and hence Π_L=Π_R=1. With increasing the carrier concentration

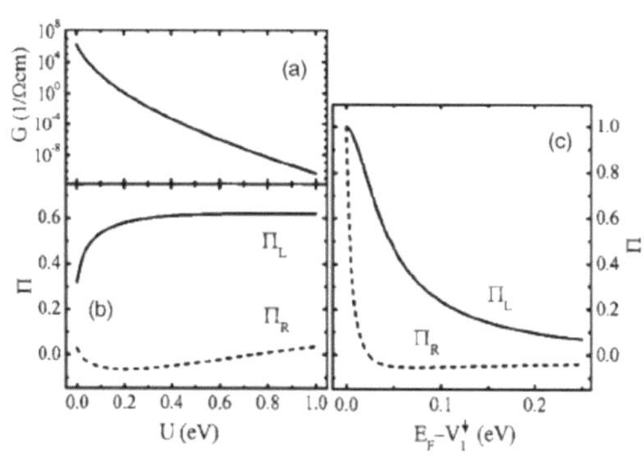

Fig. 12. (Color online)Total conductance, $G = G_\uparrow + G_\downarrow$ (a) and spin polarization (b,c) of injected current in a FM/FE/NM tunnel junction as a function of potential barrier height (a,b) and the Fermi energy (c) for the polarization of the ferroelectric barrier pointing to the left (solid lines) and pointing to the right (dashed lines) for d=3 nm. In (a) and (b) E_F-$V_1^\uparrow$=0.06 eV and $V_1 = V_2$; in (c) U=0.5 eV and $V_2 - V_1^\uparrow$=0.025 eV(30).

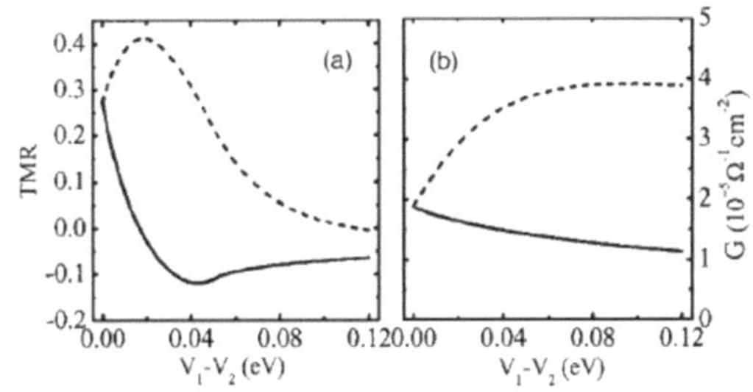

Fig. 13. (Color online) Tunneling magnetoresistance (a) and conductance, G, for parallel magnetization of the electrodes (b) in a FM/FE/FM tunnel junction versus potential difference in the two magnetic semiconductors for the electric polarization of the ferroelectric barrier pointing to the left (solid lines) and pointing to the right (dashed lines) for d=3 nm and U=0.5 eV(30).

and hence E_F, the spin polarization drops down much faster for the P pointing to the right than for the P pointing to the left, resulting in a sizable difference in the spin polarizations Π_R and Π_L. Therefore, by changing the density of carriers in the semiconductors it is possible to tune values of the spin polarization for a two-state control of the electronic device.

TMR effect. Such multiferroic tunnel junctions (MFTJ) have not yet been realized experimentally but might be promising in providing an additional degree of freedom in controlling TMR. Fig. 13(a) shows the calculated TMR in a tunnel junction with two FM electrodes separated by a FE barrier. The TMR ratio was defined by TMR = $G_P - G_{AP}/G_{AP}$, where G_P and G_{AP} are the conductances for the parallel and antiparallel magnetization, respectively. As is seen from Fig. 13(a), for $V_1 = V_2$ the TMR is independent of the orientation of P. The increasing potential difference in the two FM electrodes results in the enhancement of TMR for the P pointing to the right, whereas for the P pointing to the left the TMR drops down and becomes negative. At these conditions the MFTJ works as a device which allows switching the TMR between positive and negative values. As follows from Fig. 13(b), there is a sizable difference in the overall conductance of the junction for the two orientations of polarization, namely, the TER effect. Therefore, there is a coexistence of TMR and TER effects in such MFTJs.

3.2 MFTJs with a single-phase multiferroic barrier

Another type of MFTJ is feasible in which the barrier itself is made of a material that exhibits MF properties in the bulk, such as $BiFeO_3$ and $BiMnO_3$. In multiferroic materials, the coexistence of ferroelectric and ferromagnetic orders will provide a unique opportunity for encoding information independently in electric polarization and magnetization. Consequently, it will open new applications of multiferroic tunnel junctions on logic programming. Nowadays, several multiferroic tunneling junctions have been successfully fabricated. For example, Gajek et al. showed that $BiMnO_3$ tunnel barriers may serve as spin filters in magnetic tunnel junctions(32). This work was further advanced to demonstrate the presence of ferroelectricity in ultrathin $BiFeO_3$ films grown epitaxially on a half-metallic $La_{2/3}Sr_{1/3}MnO_3$ electrode(33)(34) and $La_{0.1}Bi_{0.9}MnO_3$(35).

In this section, two kinds of single-phase MFTJs will be discussed. One is normal metal/multiferroic/ferromagnetic metal (NM/MF/FM) junction(36). The other is FM_1/MF/FM_2 junctions(37), in which both electrodes and the barrier are ferromagnetic. TMR

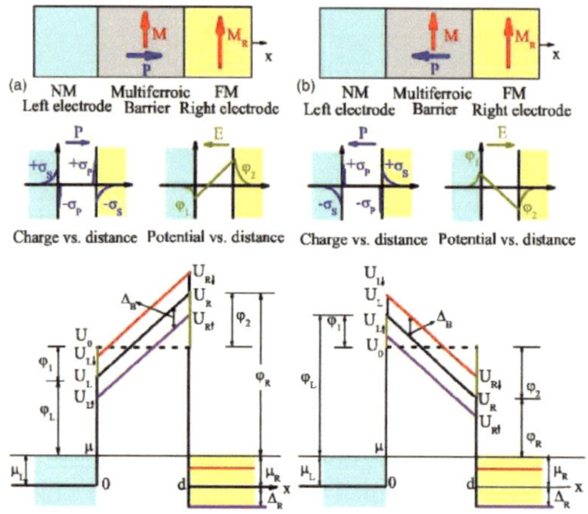

Fig. 14. (Color online) Schematic illustration of our multiferroic spin-filter tunneling junction, the charge distribution and corresponding electrostatic potential, and the overall potential profile (from top to bottom)(36). A NM electrode is placed in the left half-space $z < 0$, a multiferroic barrier of thickness t, and a semifinite FM electrode placed in the the right half-space $z > d$. m_L, m_B, and m_R are the effective masses in three regions. μ_L and μ_R are the Fermi energies of the left and right electrodes, respectively. Δ_R and Δ_B represent the exchange splitting of the spin-up and spin-down bands in FM electrode and the multiferroic barrier, respectively. φ_L and φ_R are, respectively, the electrostatic potentials at two interfaces relative to the Fermi level μ of the system.(a) The electric polarization P points to the right (positive). (b) P points to the left (negative). Here, it is assumed that two electrodes have different screening lengths and $\delta_1 < \delta_2$.

and spin filtering effects in these MFTJs are discussed. We also introduce the progress of the theoretical studies on these single-phase MFTJs.

Structure of NM/MF/FM Junctions. A NM/MF/FM MFTJ is illustrated in Fig.14. Similar to a FTJ, the screening potential $\varphi(z)$ of a MFTJ is $\sigma_S \delta_L e^{-|x|/\delta_L}/\varepsilon_L$ $(x \leq 0)$, and $-\sigma_S \delta_R e^{-|z-t|/\delta_R}/\varepsilon_R$ $(x \geq d)$. Here, δ_L and δ_R are the Thomas-Fermi screening lengths in the NM and FM electrodes, respectively. ε_L and ε_R are the dielectric permittivities of the NM and FM electrodes. The screening charge σ_S can be found from the continuity of the electrostatic potential: $\sigma_S = (dP/\varepsilon_B)/(\delta_L/\varepsilon_L + \delta_R/\varepsilon_R + d/\varepsilon_B)$ and ε_B is the dielectric permittivity of the tunneling barrier. The overall potential profile is asymmetric, as shown in Fig.14, because it is the sum of the electrostatic potential $\varphi(x)$, the electronic potential in the electrodes, and the rectangular potential profile U_0. Under the applied bias voltage V, the difference of the interfacial barrier heights is $\delta U = U_L - U_R = \delta\varphi + eV$, where $U_L = \mu + \varphi_L$ and $U_R = \mu + \varphi_R - eV$.

TMR and Spin filtering effects in NM/MF/FM Junctions. The model Hamiltonian for such MFTJs can be given by

$$\hat{H}_\sigma = -\left(\hbar^2/2m_v\right)\nabla^2 + U(z) - \sigma_v \Delta_v, \tag{4}$$

with the z-dependent potential $U(z) = 0$ when $z \leq 0$, $U(z) = U_L - (\delta U/d)z$ when $0 \leq x \leq d$, and $U(z) = \delta\mu - eV$ when $z > d$, and the spin indice is $\sigma_v = \theta_v \sigma$, where σ is the conserved spin orientation in three regions and v indicates L, B, or R. $\sigma = +1(\uparrow)$ or $-1(\downarrow)$ means up spin or down spin with respect to z. $\theta_v = +1(\Uparrow)$ or $-1(\Downarrow)$ denotes the magnetization orientation in the region v, parallel or antiparallel to the positive z direction. Then, $\sigma_v = +1(\uparrow)$ or $-1(\downarrow)$ is

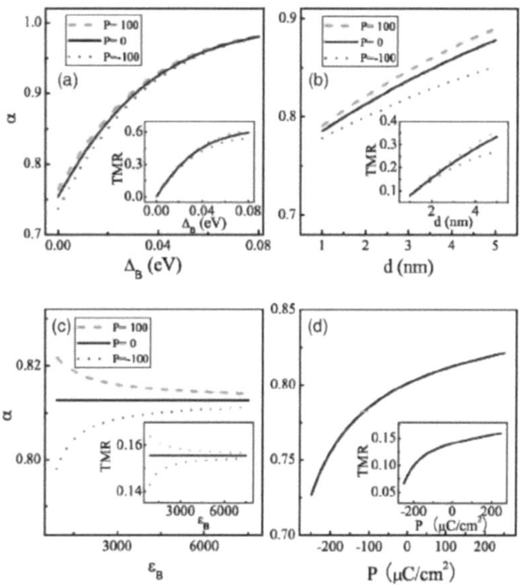

Fig. 15. (Color online) (a) The exchange splitting dependence of spin-filtering efficiency with B=1000 and d=2 nm. (b) Dielectric constant dependence of spin-filtering efficiency with Δ_B=0.01 eV and d=2 nm. (c) The thickness dependence of spin-filtering efficiency of the barrier with B=1000 and Δ_B=0.01 eV. (d) The P dependence of spin-filtering efficiency with ε_B=1000, Δ_B=0.01 eV, and d=2 nm(36). In the calculation, we set δ_L=0.07 nm, δ_R=0.08 nm, Δ_R=0.09 eV, E_F=0.1 eV, m_L=m_R =0.9m_e, m_B=1.1m_e, and U_0=0.5 eV.

the relative spin orientation, parallel or antiparallel to the given magnetization in the v region. Without loss of generality, we fix $\theta_B = +1(\Uparrow)$ and let θ_R vary. If the eigenenergy E and the transverse momentum q are conserved in this structure, the asymptotic expansion of Airys functions gives a good approximation for the transmission coefficient of each spin channel σ under a certain magnetization $\theta_R = \Uparrow or \Downarrow$(38),

$$T_{\sigma}^{\Uparrow\theta_R}(E,q) = \frac{16k_L k_{R\sigma_B}\kappa_{L\sigma_B}\kappa_{R\sigma_B}e^{-2\xi_{\sigma_B}d}}{\left(k_L^2+\kappa_{L\sigma_B}^2\right)\left(k_{R\sigma_R}^2+\kappa_{R\sigma_B}^2\right)} \tag{5}$$

where the reduced wave vectors are $k_L = \lambda_L/(E_L)^{1/2}$, $k_{R\sigma_R} = \lambda_R(E_R - \delta\mu + eV + \sigma_R\Delta_R)^{1/2}$,$\kappa_{L\sigma_B} = \lambda_B(U_{L\sigma_B} - E_B)^{1/2}$, and $\kappa_{R\sigma_B} = \lambda_B(U_{R\sigma_B} - E_B)^{1/2}$, and ξ is the decaying WKB wave vector. Here, $\lambda_v = \sqrt{2m_e^2/m_v\hbar^2}$ (v=L,B or R) with m_e the free-electron mass. When the applied bias voltage V is small and the barrier width t is large, the transmission at the Fermi level μ with $q = 0$ contributes predominantly. The zero-temperature conductance can be obtained as

$$G_{\sigma}^{\Uparrow\theta_R} = \frac{e^2}{8\pi^2\hbar}\frac{\xi_{\sigma_B}}{d}T_{\sigma}^{\Uparrow\theta_R}(\mu,0) \tag{6}$$

For a certain magnetization configuration, the total conductance is given by the sum of two channels (up spin and down spin): $G^{\Uparrow R} = G_{\uparrow}^{\Uparrow R} + G_{\downarrow}^{\Uparrow R}$. The TMR ratio is defined as TMR=1 $- G^{\Uparrow\Downarrow}/G^{\Uparrow\Uparrow}$. To show the spin-filtering effect, two sub-TMR ratio are defined as $\text{TMR}_{\sigma} = 1 - G^{\Uparrow\Downarrow_\sigma}/G_{\sigma}^{\Uparrow\Uparrow}$, corresponding to the higher barrier ($\sigma = \downarrow$) and lower barrier ($\sigma = \uparrow$), respectively. Thus, a ratio can be defined as $\alpha = G_{\uparrow}^{\Uparrow\Uparrow}/(G_{\uparrow}^{\Uparrow\Uparrow} + G_{\downarrow}^{\Uparrow\Uparrow})$ to represent the ratio between conductances through the higher and lower barriers. The TMR can be reformulated as TMR=$\alpha\times$ TMR$_{\uparrow}$+(1-α)TMR$_{\downarrow}$. Clearly, when the spin-filtering effect is very strong, $\alpha \to 1(0)$, then TMR$\approx$TMR$_{\uparrow}$ (TMR$_{\downarrow}$), which means that tunneling electrons are fully spin polarized and TMR is dominated by one of the spin channels. The sub-TMRs can be written as

$$\mathrm{TMR}_{\sigma} = \frac{k_{R\sigma} - k_{R\bar{\sigma}}}{k_{R\sigma}} \frac{\kappa_{R\sigma}^2 - k_{R\sigma}k_{R\bar{\sigma}}}{\kappa_{R\sigma}^2 + k_{R\bar{\sigma}}^2} \tag{7}$$

where $\bar{\sigma} = -\sigma$. If spin-filtering effect is very strong, TMR$\approx$ TMR$_{\uparrow} = P_{R\uparrow}P_{B\uparrow}$, where $P_{R\uparrow} = (k_{R\uparrow} - k_{R\downarrow})/k_{R\uparrow}$ and $P_{B\uparrow} = (\kappa_{R\uparrow}^2 - k_{R\uparrow}k_{R\downarrow})/(\kappa_{R\uparrow}^2 + k_{R\downarrow}^2)$ are effective spin polarization, respectively. Because of the positive $P_{R\uparrow}$, the sign of TMR is dominated by the term in $P_{B\uparrow}$,

$$\kappa_{R\sigma}^2 - k_{R\sigma}k_{R\bar{\sigma}} \approx \lambda_B\left(\varphi_{R\uparrow} - eV\right) - \lambda_R\sqrt{(\mu_R + eV)^2 - \Delta_R^2} \tag{8}$$

which is related with the barrier height $\varphi_{R\uparrow}$ induced by electric polarization. Figs. 15 and 16 shows the effect of the barrier's properties on the spin filtering coefficient and TMR ratio.

Four Logic states. As displayed in Fig. 17 (upper panel), there are overall eight resistive states with four independent pairs (A, A'; B, B'; C, C'; and D, D'), depending on the relative orientation of neighboring magnetizations and the sign of P. Because of the magnetoelectric coupling in the multiferroics, the P and the M can be reversed by an electric field separately or simultaneously. Thus, electric-field controlled functionality can be realized, including normal electroresistance (the transition from A to B or the transition from C to D) and more significant change in resistance, i.e., electromagnetoresistance (the transition from A to D). The difference between these states is complex but important for practical application. In the lower panel of Fig. 17, we show the resistance (normalized to its value at P_c) as a function of electric polarization, and the inset displays the exchange splitting dependence of P_c, where states B and C cross. Compared with conventional TMR elements which have been applied in magnetic random access memory, and also with FTJs, the present multiferroic structure possesses both electric controllable switching and large contrast between resistive states.

TMR effect in FM$_1$/MF/FM$_2$ junctions. In the following sections, another kind of MFTJs is discussed. As shown in Fig. 18, a multiferroic barrier is separated by **two ferromagnetic** metallic electrodes, for example, half-metallic $La_{2/3}Sr_{1/3}MnO_3$ and ferromagnetic metal Co(33)(34). Evidently, it is also a kind of magnetic tunnel junction, which will show

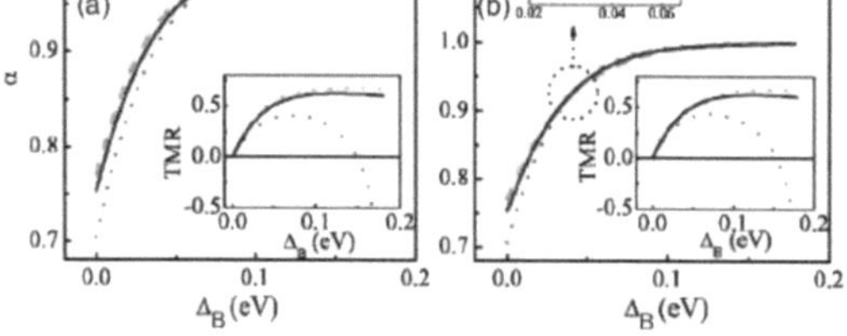

Fig. 16. (Color online) (a) The exchange splitting dependence of spin filtering efficiency for different P. δ_L=0.07 nm and δ_R =0.08 nm. The inset shows the corresponding TMR. (b) The same with (a) but with a stronger contrast between δ: δ_L=0.07 nm and δ_R=1 nm(36).

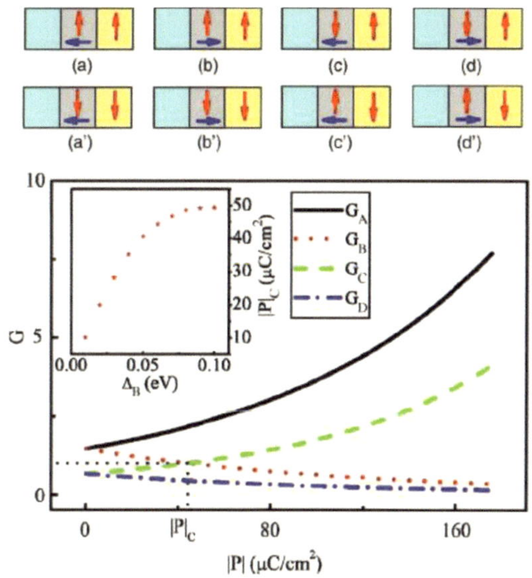

Fig. 17. (Color online) (Upper panel) Schematic illustration of multiple resistive states, depending on the orientation of P, M and MR. (Lower panel) The electric polarization dependence of normalized conductance for the four resistive states. δ_L=0.07 nm, δ_R =1 nm, ε_B=2000, and Δ_B=0.06 eV. The inset shows the Δ_B dependence of P_c(36).

significant TMR behaviors. Two theoretical models have been proposed to explain TMR effects in MTJs. One is Jullieres formula with TMR = $\frac{2P^2}{1+P^2}$(3), where the spin polarization

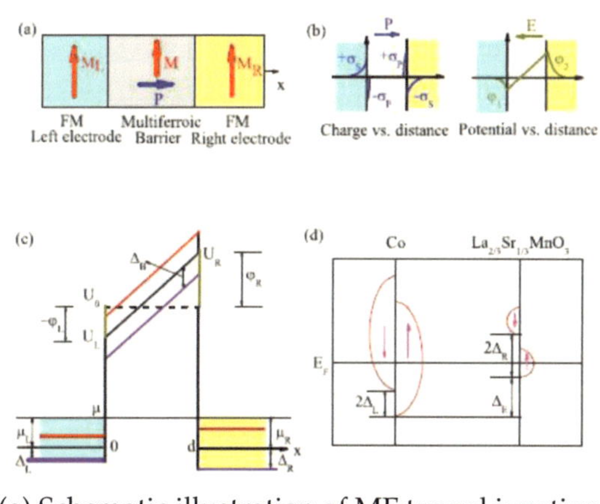

Fig. 18. (Color online) (a) Schematic illustration of MF tunnel junction. (b) Charge distribution and corresponding electrostatic potential. (c) The overall potential profile. (d) Schematic band structures of Co and half-metallic $La_{2/3}Sr_{1/3}MnO_3$(37). The exchange splittings are Δ_L, Δ_R, and Δ_B, respectively, for the left and right FM electrodes and the MF barrier. The electric polarization P induces surface charge densities, $\pm\sigma_P = \pm|P \cdot x| = P\cos\alpha$, on the two surfaces of the barrier, where α is the relative orientation between the electric polarization P and the x axis. $\theta_R = +1(\Uparrow)$ is fixed, while let θ_B and θ_L vary.

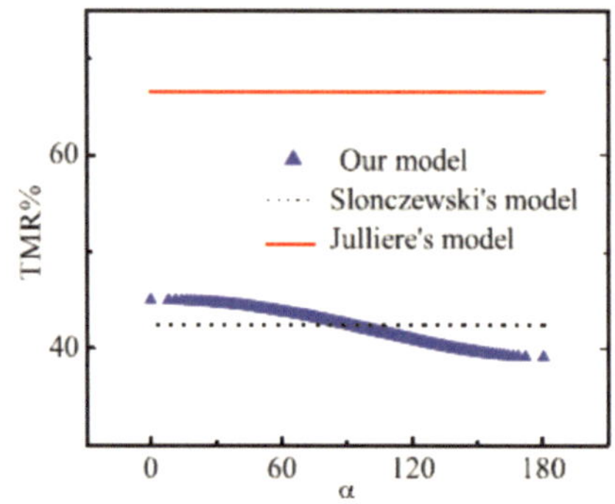

Fig. 19. (Color online) The electric polarization orientation (α) dependence of the TMR from Eq. (22)(37). The TMR from both Jullieres and Slonczewskis models is α independent. The parameters used are the same as that used in the calculations in Fig. 20. Here a 5 nm thick nonmagnetic barrier is adopted.

$P = (N_\uparrow - N_\downarrow)/(N_\uparrow + N_\downarrow)$ and N is the density of states at Fermi level. On the other hand, in the model by Slonczewski(39), the barrier height on the tunneling is considered and $P = [(k_\uparrow - k_\downarrow)/(k_\uparrow + k_\downarrow)][(\kappa^2 - k_\uparrow k_\downarrow)/(\kappa^2 + k_\uparrow k_\downarrow)]$, where $\kappa = \sqrt{(2m/\hbar^2)/(U - E_F)}$ and U is the height of barrier. However, in both models of Julliere and Slonczewski, neither the electric polarization nor the magnetism of the barrier was considered. For MFTJs, Ju *et al.* extended the previous TMR models and firstly pointed out the TMR is a function of the orientation of the electric polarization (Fig. 19). They also proved that Slonczewskis model is actually a special case of their model with P=0 (or α =0) and $\Delta_B = 0$. Their calculations show that the TMR of MFTJs are strongly influenced by the orientation of the electric polarization and the barrier properties, i.e., effective barrier height $\bar{U}$; the exchange splitting of the barrier, Δ_B; and the electric polarization in the barrier, P, which is shown in Fig.19 and Fig.20.

Tunneling electroresistance effect (TER). Since both electrodes and barrier are ferromagnetic, such tunnel electroresistance (TER), $TER = G(\pi)/G(\alpha = 0)$, is a little different from that in FTJs. In Fig. 20(e), we show such TER as for junctions with various barrier thicknesses. It is found that TER increases with the increase of d and when the magnetization of the barrier is parallel to the magnetization of right electrode, the presence of weak ferromagnetism in $BiFeO_3$ will make TER more significant (Fig. 20(f)). However, it is also noted that TER is almost independent of the magnetic configuration of two electrodes, i.e., parallel or antiparallel. These TER effects will be studied experimentally by Bea *et al.*(33)(34).

Converse piezoelectric effect. Converse piezoelectric effect may also have an important influence on the tunneling across a multiferroic(40). When the junction is applied with a bias voltage V, the converse piezoelectric property causes the strain in the barrier, which hence induces changes in the barrier thickness d, electron effective mass m_B, and position of the conduction band edge E_c. There are (27)

$$\begin{aligned} d &= d^0 + d_{33}V \\ m_B &= m_B^0(1 + \mu_{33}\Delta S_3) \\ E_C &= E_C^0 + \kappa_3 \Delta S_3 \end{aligned} \tag{9}$$

where $\Delta S_3 = d_{33}V/d^0$ is the lattice strain and d_{33}, μ_{33}, and κ_3 are, respectively, the out-of-plane piezoelectric coefficient, strain sensitivity of the effective mass, and relevant deformation potential of the conduction band in the barrier. d^0, m_B^0 , and E_c^0 are their values at $V = 0$.

If the applied voltage is positive, the barrier thickness d will be compressed and both the electron effective mass m_B and barrier height U will increase. Obviously, each of these strain-induced changes will change the electron tunneling probabilities, and therefore the spin filtering efficiency and TMR ratio (Figs. 21-22).

4. MFTJs with ferromagnet/ferroelectric composite barriers

For practical applications, the single-phase MF barrier junctions are limited by the scarcity of existing single-phase multiferroics, and none of which combine large and robust electric and magnetic polarizations at room temperature. It might be a good idea to use a FE/FM composite barrier to substitute the single-phase MF barrier(41). Such a composite barrier junction may be thought as an addition of a conventional spin filter and a FTJ. Here, FM insulator (FI) barrier acts as a SF generator, and FE barrier acts as a SF adjustor through the interplay between ferroelectricity and ferromagnetism at the interface. The large SF effect, the TMR and TER effects, can be achieved in this two-phase composite barrier based tunnel junction. The eight resistive states with large difference can also be realized.

Electrostatic effect. Figure 23 shows a junction in which a FE/FI composite barrier is sandwiched by two metallic electrodes(41). The electrostatic potential induced by the electric polarization in FE layer is obtained, as shown in Fig. 23(b). The overall potential profile $U(x)$ across the junction is the superposition of the electrostatic potential $\varphi(x)$, the electronic potential in the electrodes, and the rectangular potential in FE barrier and FI barrier, as shown in Fig. 23(c).

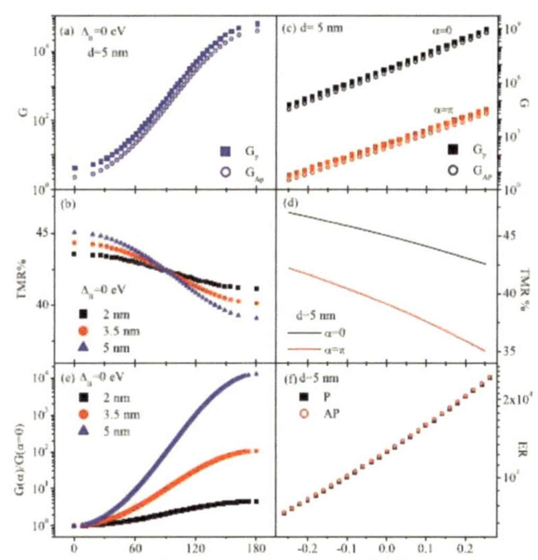

Fig. 20. (Color online) (a) The orientation of electric polarization α dependence of conductance. (b) dependence of TMR. (c) The exchange splitting of the barrier Δ_B vs conductance. (d) Δ_B vs TMR. (e) Δ_B dependence of the normalized conductance $G\alpha/G(\alpha=0)$ with parallel magnetization in two electrodes. (f) Δ_B dependence of TER with parallel (P) and antiparallel (AP) magnetizations in two electrodes(37).

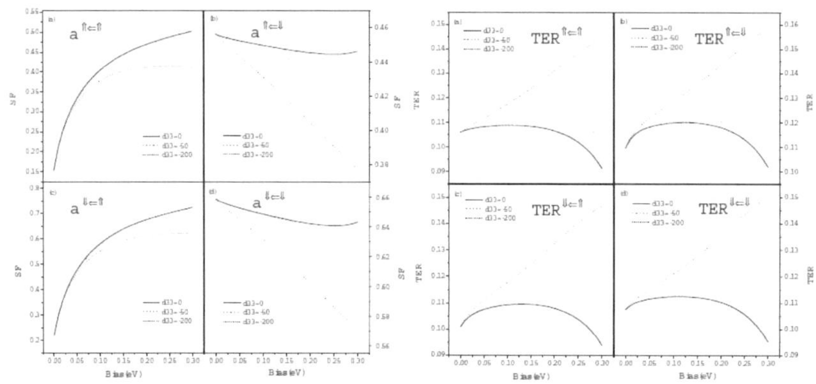

Fig. 21. (left panel)Piezoelectric effect on the SF and (right panel) TER(40) with Δ_B=0.015 eV, P=0.5 C/m^2, κ_3=-4.5 eV, μ_{33}=10, and Δ_L=Δ_R=0 eV.

Spin filtering effect, TMR and TER effect. Choosing the nonmagnetic metals (NMs) as two electrodes, the spin filtering efficiency can be defined as $\alpha = (G_\uparrow - G_\downarrow)/(G_\uparrow + G_\downarrow)$, while the TMR is defined by TMR=$(G_P - G_{AP})/(G_P + G_{AP})$ with changing the two metallic electrodes

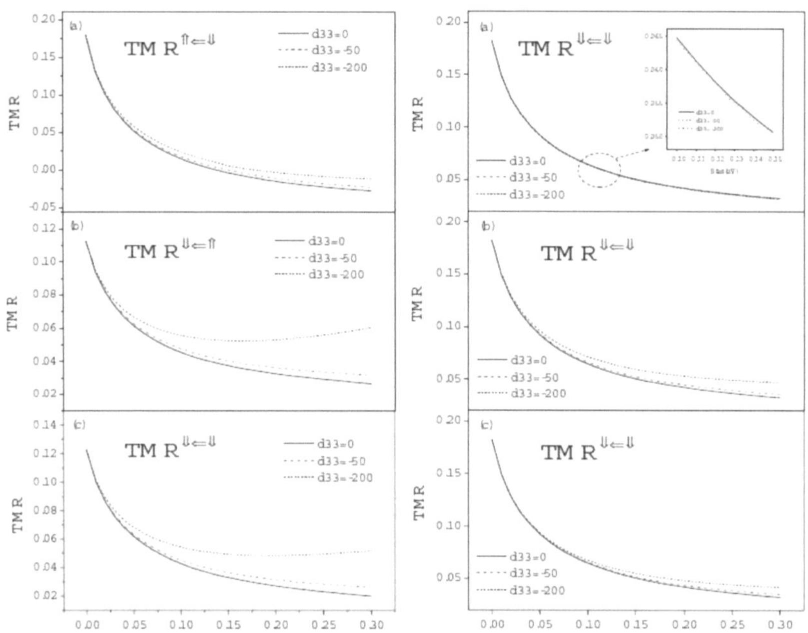

Fig. 22. (left panel)Piezoelectric effect on the TMR when Δ_B=0.015 eV, P=0.5 C/m^2, κ_3=-4.5 eV, μ_{33}=10, and Δ_L=0.05 eV, and Δ_R =0.09 eV. (right panel) Influence of only the (a) strain-induced barrier thickness change, (b) electron effective mass change, or (c) barrier height change on TMR. The inset in (a) shows the magnified curve around V=0.12 eV. The parameters are Δ_B=0.015 eV, P=0.5 C/m^2, Δ_L=0.05 eV, and Δ_R =0.09 eV(40).

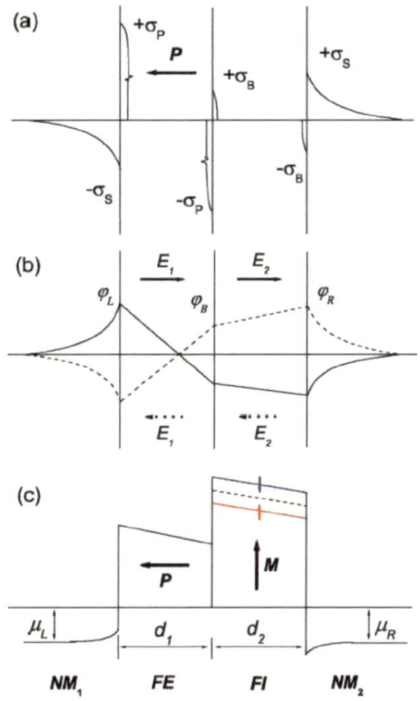

Fig. 23. (Color online) Schematic illustration of the charge distribution (a), corresponding electrostatic potential (b), and potential profile in a NM1/FE/ FI/NM2 tunnel junction C(41). The solid and dashed lines in (b) show the potential for the polarization P in FE barrier pointing to the left and to the right, respectively. The blue and red lines in FI barrier in (c) show the potential seen by the spin-down and spin-up electrons, respectively. Here d_1 and d_2 are the thicknesses of the FE and FI barriers, respectively.

from NMs to FM metals. To calculate TER, the direction of polarization in FE layer is reversed from pointing to the right to pointing to the left, while the magnetization in FI layer remains unchanged.

It can be seen from Fig. 24 that the SF and the TMR effect is mainly determined by the exchange splitting Δ_B in FI barrier, and is enhanced (reduced) by the polarization in FE barrier when P points the left (right) electrode. The results here are similar to those for the single-phase MF barrier junction, indicating that the physical mechanism responsible for SF effect is the same in both structures, i.e., the ferromagnetism in the barrier, making the barrier height spin dependent, acts as a SF generator, and the ferroelectricity, changing the profile of the potential across the junction, acts as a SF adjustor. A large TER effect can be found in MFTJs with composite barriers. The ferroelectricity in FE layer is the dominant factor to determine the magnitude of the TER effect.

Eight Logic states. By using one or two FM metal electrodes, eight independent logic states of tunneling conductance can be realized in such MFTJs. The conductances as a function of the exchange splitting Δ_B and the electric polarization P are shown in Fig. 24. For practical applications, the large contrast between eight states is very important. We find that when the carriers tunneling into the right electrode can be highly spin-polarized, the eight states are differentiated evidently.

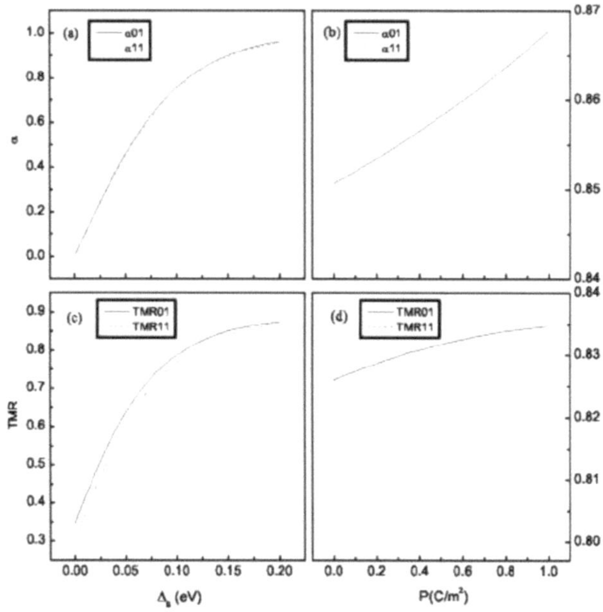

Fig. 24. (Color online) (a) The exchange splitting dependence of SF with P=0.4 C/m^2. (b) The electric polarization dependence of SF with Δ_B =0.13 eV. In (a) and (b), U_{01}=0.5 eV, U_{02}=1.1 eV, ε_{B1}=2000, ε_{B2}=1000, and d_1=d_2=2 nm. (c) and (d) show the corresponding TMR(41).

5. Magnetoelectric coupling at ferroelectrics/ferromagnetic metal interfaces

MFTJs have a great potential on the practical applications especially on the multistate logical elements. Due to the magnetoelectric coupling in MFTJ, it is perspective that the magnetization can be controled by the applied electric field, and vice versa. For single-phase multiferroic barrier, how the ferromagnetic order and ferroelectric order coupling is still a unsolved question. Therefore, we switch our attention to the ME coupling of the FM/FE/NM structures(Fig.26)(42).

Electrostatic effect at Ferromagnetic Metal / Ferroelectrics interfaces. For a FM/FE structure, when the FE layer is polarized, surface charges are created. These bound charges are compensated by the screening charge in both FM and NM electrodes. In the FM metal, the screening charges are spin polarized due to the ferromagnetic exchange interaction. The spin dependence of screening leads to additional magnetization in the FM electrode as illustrated in Fig. 26(b). If the density of screening charges is denoted as η and the spin polarization of screening charges is denoted as ζ, we can directly express the induced magnetization per unit area as

$$\Delta M = \frac{\eta}{e}\zeta\mu_B. \qquad (10)$$

As this effect depends on the orientation of the electric polarization in FE, the ME coupling is expected.

Two simple cases can be considered. (1) In an ideal capacitor where all the surface charges reside at the metal (FM or NM)/FE interfaces, the density of screening charge η reaches its maximum value $\eta = P_0$, where P_0 is the spontaneous polarization of the FE. This results in a large induced magnetization $[(P_0/e)\zeta\mu_B]$. (2) In half metals, there is only one type of carriers that can provide the screening. If a half metal is chosen to be the FM electrode, the screening electrons will be completely spin polarized. In this case, a large induced magnetization is also expected, $\Delta M = \frac{\eta}{e}\mu_B$.

Induced magnetization from screening charges. For FM/FE/NM junctions (Fig.26(a)), the additional magnetization, caused by spin-dependent screening(43)(44), will accumulate at each FM/FE interface. Due to the broken inversion symmetry between the FM/FE and the NM/FE interfaces, there would be a net additional magnetization in each FM/FE/NM unit cell, unlike the symmetric structures discussed in the previous work. The addition of magnetization in this superlattice will result in a large global magnetization. In the case of zero bias in Fig. 26(c), the local induced magnetization, defined as $\delta M(x) = [\delta n^{\uparrow}(x) - \delta n^{\downarrow}(x)]\mu_B$, is a function of distance from the interface x. Here, $\delta n^{\sigma}(x)$ is the density of the induced screening charges with spin σ. The total induced magnetization ΔM can be calculated by integrating $\delta M(x)$ over the FM layer, and

$$\Delta M = \int_{FM \quad layer} \delta M(x) = -\frac{\eta M_0/e}{N_0 + JN_0^2 - J(M_0/\mu_B)^2} \tag{11}$$

where $N_0 = N_{\uparrow} + N_{\downarrow}$ is the total density of states, $M_0 = (N_{\uparrow} - N_{\downarrow})\mu_B$ can be thought of as the spontaneous magnetization, ε_0 is the vacuum dielectric constant, J is the

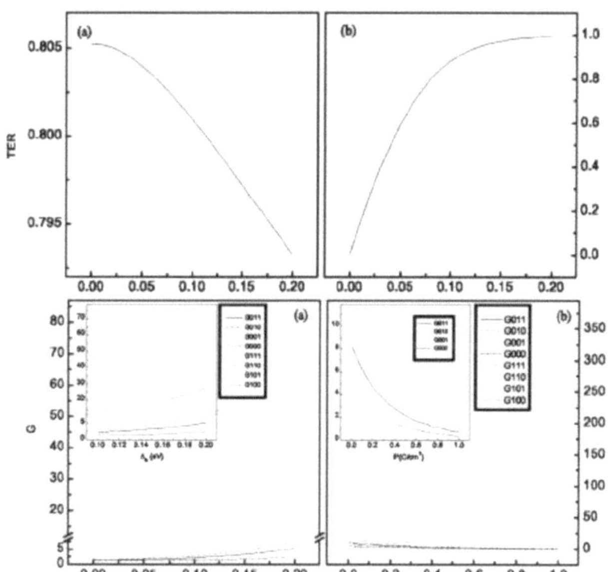

Fig. 25. (up panel) (a)The exchange splitting dependence of TER with P=0.4 C/m^2. (b) The electric polarization dependence of TER with Δ_B =0.13 eV. In (a) and (b), U_{01}=0.5 eV, U_{02}=1.1 eV, ε_{B1}=2000, ε_{B2}=1000, and d_1=d_2=2 nm. (down panel) (a) The exchange splitting and electric polarization dependence of tunneling conductances with P=0.4 C/m^2 in (a) and Δ_B =0.13 eV in (b)(41). In (a) and (b), the rectangular potential in FE barrier and FI barrier is 0.5 eV and 1.1 eV, respectively. ε_{B1}=2000, ε_{B2}=1000, and d_1=d_2=2 nm. The inset in (a) shows the enlarged part with Δ_B=0.1 eV. The inset in (b) shows the tunneling conductances with the polarization in FE barrier pointing to the left.

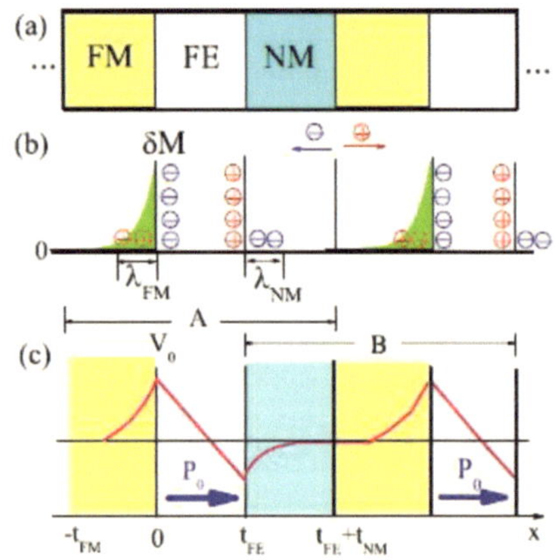

Fig. 26. (Color online) (a) Schematic illustration of FM/FE/NM tricomponent superlattice. (b) The distribution of charges and induced magnetization (green shaded area) calculated by our theoretical model. A and B are two different choices of the unit cell. The directions of arrows indicate the motions of positive and negative charges across the boundary of the unit cell A. (c) Electrostatic potential profile(42) . Here, the following assumptions are made. (1) The difference in the work function between FM and NM is ignored. (2) To screen the bound charges in FE, the charges in metal electrodes will accumulate at the FM/FE side, and there is a depletion at the NM/FE side. In this process, the total amount of charge is conserved; however, the spin density is not conserved because of the ferromagnetic exchange interaction in the FM metal.

strength of the ferromagnetic exchange coupling in the FM layer, η is the density of screening charges, $\lambda_{FM(NM)}$ is the screening length of FM (NM) electrode with $\lambda_{FM} = (e^2N_0/\varepsilon_0)[N_0 + JN_0^2 - J(M_0/\mu_B)^2]/(1+JN_0)^{-1/2}$, and t_{FM}, t_{FE}, and t_{NM} are the thicknesses of FM, FE, and NM layers, respectively. It is seen that the local induced magnetization $\delta M(x)$ decays exponentially away from the FM/FE interface.

The induced magnetization in FM/FE/NM tricomponent superlattice with several FM electrodes, i.e., Fe, Co, Ni, and CrO_2, are calculated.(42) Detailed parameters and calculated values of Δ_M are listed in Fig.27. The magnitude of Δ_M is found to depend strongly on the choice of the FM and FE. Among the normal FM metals (Ni, Co, and Fe), the largest Δ_M is observed in Ni for its smallest J and highest spontaneous spin polarization $M_0/\mu_B N_0$. On the other hand, we also predict a large Δ_M for the 100% spontaneous spin polarization in half-metallic CrO_2.

FM	J (eV nm^3)	N_0 (eV^{-1} nm^{-3})	$M_0/\mu_B N_0$ (%)	λ_{FM} (Å)	ΔM (μ_B nm^{-2})	τ (G cm/V)
Ni	0.65	1.74	−79.3	0.9	−0.280	0.015
Co	1.25	0.89	−58.4	1.5	−0.126	0.004
Fe	2.40	1.11	56.8	1.3	0.078	0.003
CrO_2	1.8	0.69	100	1.7	0.323	0.010

Fig. 27. Calculated induced magnetization(42). Here, ΔM is the value at V_a=V_C, where V_a is the applied bias and V_C is the coercive bias.

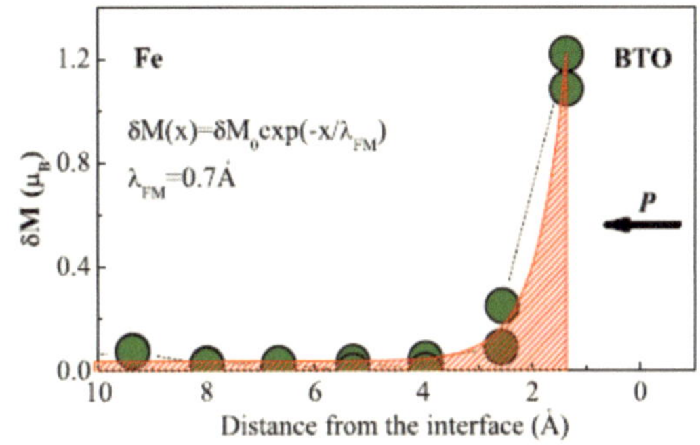

Fig. 28. (Color online) Layer-resolved induced magnetic moment of Fe near the interface between Fe and $BaTiO_3$ in the Fe/$BaTiO_3$ / Pt superlattice(42). Solid line is the fitted exponential function for the induced moment as a function of the distance from the interface.

First-principle calculations. The first-principle calculations are consistent with the results of the theoretical model. For example, the first-principle calculation of the Fe/FE/Pt superlattice will be shown.(45) The calculations are within the local-density approximation to density-functional theory and are carried out with VASP. We choose $BaTiO_3$ (BTO) and $PbTiO_3$ (PTO) for the FE layer. Starting from the ferroelectric P4mm phase of BTO and PTO with polarization pointing along the superlattice stacking direction, we perform a structural optimization of the multilayer structures by minimizing their total energies. The in-plane lattice constants are fixed to those of the tetragonal phase of bulk FEs. Figure 19 shows the calculated induced magnetic moment relative to that of bulk Fe near the Fe/BTO interface when the polarization in $BaTiO_3$ points toward the Fe/BTO interface. It is evident that the induced moments decay exponentially as the distance from the interface increases. This result is in line with our model for the magnetization accumulation in the FM at the FM/FE interface. A numerical fitting of the exponential function yields a screening length of 0.7 Åfor the Fe/$BaTiO_3$ / Pt structure. This value is comparable to the screening length parameters calculated using the theoretical model as shown in Fig.28.

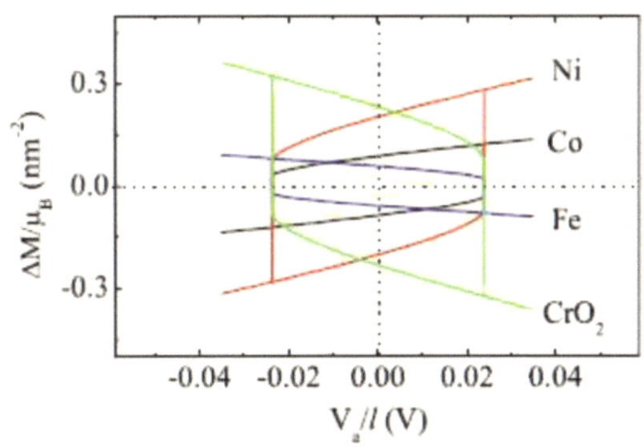

Fig. 29. (Color online) ΔM versus V_a/l for different ferromagnetic metal electrodes. V_a is the applied bias and l is the number of the unit cell(42). Here, the thickness of FE layer is 3 nm. However, a thicker FE layer can be used to avoid the possible electron tunneling effect.

Electric Control of Magnetization. A natural question is what happens to Δ_M when an external bias V is applied. In this case, the electric polarization P will have two parts: the spontaneous polarization P_0 and the induced polarization. The equation determining P is obtained by minimizing the free energy. From the continuity of the normal component of the electric displacement, we find equation relating η and P:$\eta = [(Pt_{FE}/\varepsilon_{FE}) + (V_a/l)]/[(\lambda_{FM} + \lambda_{NM})/\varepsilon_0] + (t_{FE}/\varepsilon_B)$. Here, ε_{FE} is the dielectric constant of the FE layer. These two equations need to be solved self-consistently. The value of η at a given bias can then be calculated and the induced magnetization ΔM is given by Eq. (11). The free-energy density F includes contributions from the FE layer, FM layer, and FM/FE interface and takes the form

$$F = \frac{t_{FE}F(P) + t_{FM}F(M) + F_I(F,M)}{t_{FE} + t_{FM} + t_{NM}} \tag{12}$$

M is the magnetization of the bulk ferromagnet and here $M = M_0$ because of zero external magnetic field. The interface energy $F_I(P,M)$ is the sum of the electrostatic energy and magnetic exchange energy of the screening charges,

$$F_I(F,M) = \frac{(\lambda_{FM} + \lambda_{NM})}{2\varepsilon_0}\eta^2 + \frac{J}{2\mu_B^2}(M + \Delta M)\Delta M \tag{13}$$

For FE, the free-energy density $F(P)$ can be expressed as $F(P) = F_P + \alpha_P P^2 + \beta_P P^4 + \int_0^P E_B dP$, where F_P is the freeenergy density in the unpolarized state. α_P and β_P are the usual Landau parameters of bulk ferroelectric. E_B is the depolarization field in the FE film. Similarly, $F(M)$ can be expanded as a series in the order parameter M, i.e., $F(M) = F_M + \mu_M M^2 + \nu_M M^4$, where F_M is the free-energy density of bulk ferromagnet and μ_M and ν_M are the Landau parameters of bulk ferromagnet. The calculated induced magnetization as a function of the applied bias is shown in Fig. 29. Clearly, the electrically controllable magnetization reversal is realized.

Magnetoelectric coupling energy at the FM/FE interface. To discuss the macroscopic properties of the electric control of magnetization, we analyze the magnetoelectric coupling energy in our tricomponent superlattice. For the macroscopic average polarization to be represented by the electric polarization obtained for a unit cell, this cell needs to be chosen with special care. Therefore, in the following calculation of total free energy, unit cell B in Fig. 26(b) is chosen, and

$$\bar{P} = \frac{Pt_{FE} + \eta(t_{FM} + t_{NM})}{t_{FE} + t_{FM} + t_{NM}} \tag{14}$$

The macroscopic average magnetization

$$\bar{M} = \frac{Mt_{FM} + \Delta M}{t_{FE} + t_{FM} + t_{NM}} \tag{15}$$

Considering the lowest-order term of the magnetoelectric coupling, $\bar{P}$ and $\bar{M}$ can be expanded as $\bar{P} = c_p P + c_p' PM^2$, $\bar{M} = c_m M + c_m' PM$. Therefore, the total free energy [Eq. (12)] can be expressed as the power series of $\bar{P}$ and $\bar{M}$, $F(\bar{P},\bar{M}) = F_0 + \alpha\bar{P}^2 + \beta\bar{P}^4 + \mu\bar{M}^2 + \nu\bar{M}^4 + \chi\bar{P}\bar{M}^2 + \cdots$. We would like to point out that biquadratic ME coupling $\bar{P}^2\bar{M}^2$ is easily achievable, but is usually weak and is not electrically controllable. However, because of the naturally broken inversion symmetry, the large ME coupling $\bar{P}\bar{M}^2$ is possible in our tricomponent structure.

6. Summary

Ferroelectric and multiferroic tunnel junctions have shown great promise for practical applications, i.e., high density data storage. However, to realize these junctions, a number of questions are required to be answered. For example, how the electric polarization switches in nanoscale ferroelectrics? How the ultrathin ferroelectric barrier change the magnetic and electric properties of electrode/barrier interfaces? How the ferroelectric domain affect the tunneling across FTJs and MFTJs? Nowadays, achievements in the field of complex oxide epitaxy and newly developed nanoscale characterization techniques, promise that the realization of FTJs and MFTJs is just a matter of time. The diversity of interesting physical phenomena that control the characteristics of these tunnel junctions and their multifunctional properties makes the research in this field challenging and promising.

7. Acknowledgments

This project was supported by the National Natural Science Foundation of China by grants No. 10774107 and No. 10974140.

8. References

[1] P. M. Tedrow, R. Meservey, Phys. Rev. Lett. 26, 192 (1971).
[2] P. M. Tedrow, R. Meservey, Phys. Rev. B 7, 318 (1973).
[3] M. Jullire, Phys. Lett. A 54, 225 (1975).
[4] J. S. Moodera *et al.*, Phys. Rev. Lett. 74, 3273 (1995).
[5] J. M. De Teresa *et al.*, Science 286, 507 (1999).
[6] S. Ikeda, J. Hayakawa, Y. Ashiizawa *et al.*, Appl. Phys. Lett. 93, 082508 (2008).
[7] W. H. Butler, X. G. Zhang, T. C. Schulthess, Phys. Rev. B 63, 054416 (2001).
[8] J. Mathon and A. Umerski, Phys. Rev. B 63, 220403(R) (2001).
[9] J. S. Moodera *et al.*, Phys. Rev. Lett. 61, 637 (1988).
[10] J. S. Moodera, R. Meservey, and X. Hao, Phys. Rev. Lett. 70, 853 (1993).
[11] T. S. Santos *et al.*, Phys. Rev. B 69, 241203 (2004).
[12] P. LeClair *et al.*, Appl. Phys. Lett. 80, 625 (2002).
[13] L. Esaki, R. B. Laibowitz, P. J. Stiles, IBM Tech. Discl. Bull. 13, 2161 (1971).
[14] M. Ye. Zhuravlev *et al.*, Phys. Rev. Lett. 94, 246802 (2005).
[15] E. Y. Tsymbal and H. Kohlstedt, Science 313, 181 (2006).
[16] A.V. Bune, V. M. Fridkin, S. Ducharme, L. M. Blinov, S. P. Palto, A.V. Sorokin, S. G. Yudin, and A. Zlatkin, Nature (London) 391, 874 (1998).
[17] D. D. Fong, G. B. Stephenson, S. K. Streiffer, J.A. Eastman, O. Auciello, P. H. Fuoss, and C. Thompson, Science 304, 1650 (2004).
[18] J. Junquera and P. Ghosez, Nature (London) 422, 506 (2003).
[19] N.W. Ashcroft and N. D. Mermin, Solid State Physics (Saunders College Publishing, New York, 1976), p. 340.
[20] C. B. Duke, Tunneling in Solids (Academic, New York, 1969).
[21] V. Garcia *et al.*, Nature 460, 81 (2009)
[22] A. Gruverman, Nano Letters 9, 3539 (2009)
[23] A. Crassous *et al.*, Appl. Phys. Lett. 96, 042901 (2010)
[24] P. Maksymovych *et al.*, Science 324, 1421 (2009)
[25] D. D. Fong *et al.*, Phys. Rev. B 71, 144112 (2005).
[26] G. Gerra *et al.*, Phys. Rev. Lett. 96, 107603 (2006).

[27] H. Kohlstedt *et al.*, Phys. Rev. B 72, 125341 (2005).
[28] M. Ye. Zhuravlev *et al.*, Appl. Phys. Lett. 95, 052902 (2009).
[29] Y. Wu, S. Ju, and Z. Y. Li, Appl. Phys. Lett. 96, 252905 (2010)
[30] M. Ye. Zhuravlev *et al.*, Appl. Phys. Lett. 87, 222114 (2009).
[31] V. Garcia et al., Science 327, 1106 (2010)
[32] M. Gajek *et al.*, Phys. Rev. B 72, 020406 (2005).
[33] H. Bèa *et al.*, Appl. Phys. Lett. 88, 062502 (2006)
[34] H. Bèa *et al.*, J. Phys. Condens. Matter 20, 434221 (2008).
[35] M. Gajek *et al.*, Nat. Mater. 6, 296 (2007).
[36] S. Ju *et al.*, Phys. Rev. B 75, 064419 (2007).
[37] S. Ju *et al.*, J. Appl. Phys. 104, 093504 (2008).
[38] D. F. Jin, Y. Ren, Z. Z. Li, M.W. Xiao, G. J. Jin, and A. Hu, Phys. Rev. B 73, 012414 (2006).
[39] J. C. Slonczewski, Phys. Rev. B 39, 6995 (1989)
[40] J. Wang, S. Ju, and Z. Y. Li, J. Appl. Phys. 105, 093920 (2009).
[41] J. Wang and Z. Y. Li, Appl. Phys. Lett. 93, 112501 (2008).
[42] T. Cai *et al.*, Phys. Rev. B 80, 140415(R) (2009)
[43] S. F. Zhang, Phys. Rev. Lett. 83, 640 (1999).
[44] J. M. Rondinelli, M. Stengel, and N. A. Spaldin, Nat. Nanotechnol. 3, 46 (2008).
[45] J. K. Lee, N. Sai, T. Y. Cai, Q. Niu, and A. A. Demkov, Phys. Rev. B 81, 144425 (2010).

3

Photoluminescence in Low-dimensional Oxide Ferroelectric Materials

Dinghua Bao
State Key Laboratory of Optoelectronic Materials and Technologies, School of Physics and Engineering, Sun Yat-Sen University, Guangzhou, 510275
P.R. China

1. Introduction

Ferroelectric materials exhibit important multifunctional electrical properties such as ferroelectric, dielectric, piezoelectric, pyroelectric, and electrooptic properties. They can be used to fabricate various microelectronic and optoelectric devices including nonvolatile ferroelectric random access memories, microsensors and microactuators, integrated capacitors, and electrooptic modulators [1-3]. Widely studied ferroelectric materials include $Pb(Zr,Ti)O_3$ (PZT), lanthanide doped $Bi_4Ti_3O_{12}$ (BiT), and $BaTiO_3$.
PZT is the most important piezoelectric material which has been used in various electronic devices. $BaTiO_3$ and its solid solution with $SrTiO_3$ exhibit high dielectric constant and other advantageous properties. Lanthanide doped BiT gained great interest on searching new ferroelectric thin films with fatigue-free polarization properties for non-volatile random access memories. Among them, La doped bismuth titanate and Nd-doped bismuth titanate exhibit excellent electrical properties such as high fatigue resistance, good retention, fast switching speed, high Curie temperature, large spontaneous polarization, and small coercive field [2-4]. In addition, lanthanide doped BiT do not contain lead, thus, environmental pollution and harm to health of human due to lead volatility in PZT can be avoided. Studies in the past decade indicate that lanthanide doped bismuth titanate ferroelectric thin films are the most promising candidate materials for nonvolatile ferroelectric random access memory applications. Most efforts have been devoted to improving the electrical properties of $Bi_4Ti_3O_{12}$ (BIT) thin films by rare earth ion doping for the development of non-volatile ferroelectric random access memory applications [5-7]. It is worth pointing out that besides excellent ferroelectric polarization fatigue-free characteristics, the bismuth layered perovskite structure ferroelectric thin films also exhibit good piezoelectric properties and large optical nonlinearity [8-10].
Recently, photoluminescence (PL) properties originated from defects or rare earth ions in oxide ferroelectric materials have attracted much attention for possible integrated photoluminescent ferroelectric device applications. This paper briefly reviews the status and new progress on study of photoluminescence in low-dimensional oxide ferroelectric materials including ferroelectric thin films, nanopowders, nanorods or nanowires, and nanotubes, and some of our own research work in this field with an emphasis on the photoluminescence properties of lanthanide doped bismuth titanate thin films such as $(Bi,Pr)_4Ti_3O_{12}$, $(Bi,Eu)_4Ti_3O_{12}$, $(Bi,Er)_4Ti_3O_{12}$, and codoped bismuth titanate thin films will be presented in this paper, also.

2. Photoluminescence in ferroelectric nanopowders and nanowires

2.1 Photoluminescence in ferroelectric nanopowders

Photoluminescence from rare earth ions in ferroelectric nanopowders has been studied by some research groups. Badr, et al. studied the effect of Eu^{3+} contents on the photoluminescence of nanocrystalline $BaTiO_3$ powders [11]. The powders were prepared by sol-gel technique using $Ba(Ac)_2$, and $Ti(C_4H_9O)_4$ as raw materials, annealed at 750 °C in air for 0.5 h. The crystallite size of the doped sample with 4% Eu^{3+} ions in the powder was found to be equal to 32 nm. The photoluminescence of nanocrystalline powders at 488 nm were observed. The luminescence spectra of ultra fine Eu^{3+}:$BaTiO_3$ powders are dominated by the $^5D_0 \rightarrow {}^7F_J$ (J=0-4) transitions, indicating a strong distortion of the Eu^{3+} sites. The structure disorder and charge compensation were suggested to be responsible for the strong inhomogeneous broadening of the $^5D_0 \rightarrow 7F$ luminescence band of the Eu^{3+}.

Fu, et al. studied characterization and luminescent properties of nanocrystalline Pr^{3+}-doped $BaTiO_3$ powders synthesized by a solvothermal method using barium acetylacetonate hydrate, titanium (IV) butoxide, and praseodymium acetylacetonate hydrate as precursors [12]. They discussed the luminescence mechanism, the band gap change and the size dependence of their fluorescence properties of the powders.

Lemos and coworkers studied up-conversion luminescence properties in Er^{3+}/Yb^{3+}-codoped $PbTiO_3$ perovskite powders prepared by Pechini method [13]. Efficient infrared-to-visible conversion results in green (about 555 nm) and red (about 655 nm) emissions under 980 nm laser diode excitation. The main up-conversion mechanism is due to the energy transfer among Yb and Er ions in excited states.

On the other hand, photoluminescence from defects such as oxygen vacancies or structure disorder is worth studying. Lin, et al. prepared nanosized $Na_{0.5}Bi_{0.5}TiO_3$ powders with particle size of about 45-85 nm using a sol-gel method and studied their photoluminescence properties [14]. It was observed from photoluminescence spectra that the main emission peak of the nanosized powder exhibited a blue shift from 440.2 to 445.3 nm with decreasing particle size, while the emission intensity increased. They explained that the visible emission band is due to self-trapped excitation, and this blue shift of the main emission was attributed to distortion of TiO_6 octahedra originated from the surface stress of $Na_{0.5}Bi_{0.5}TiO_3$ crystallite and the dielectric confinement effect.

Oliveira and coworkers studied synthesis and photoluminescence behavior of $Bi_4Ti_3O_{12}$ (BiT) powders obtained by the complex polymerization method [15]. The BiT powders annealed at 700 °C for 2 h under oxygen flow show an orthorhombic structure without impurity phases. UV-vis spectra indicated that there are intermediary energy levels within the band gap of the powders annealed at low temperatures. The maximum PL emissions were observed at about 598 nm, when excited by 488 nm wavelengths. Also, it was observed that there were two broad PL bands, which were attributed to the intermediary energy levels arising from alpha-Bi_2O_3 and BiT phases.

Pizani, et al. reported intense photoluminescence observed at room temperature in highly disordered (amorphous) $BaTiO_3$, $PbTiO_3$, and $SrTiO_3$ prepared by the polymeric precursor method [16]. The emission band maxima from the three materials are in the visible region and depend on the exciting wavelength. The authors thought that photoluminescence could be related to the disordered perovskite structure.

Highly intense violet-blue photoluminescence at room temperature was also observed with a 350.7 nm excitation line in structurally disordered $SrZrO_3$ powders by Longo, et al [17].

They discuss the role of structural order-disorder that favors the self-trapping of electrons and charge transference, as well as a model to elucidate the mechanism that triggers photoluminescence. In this model the wide band model, the most important events occur before excitation.

They also reported room temperature photoluminescence of $Ba_{0.8}Ca_{0.2}TiO_3$ powders prepared by complex polymerization method [18]. Inherent defects, linked to structural disorder, facilitate the photoluminescence emission. The photoluminescent emission peak maximum was around of 533 nm (2.33 eV) for the $Ba_{0.8}Ca_{0.2}TiO_3$. The photoluminescence process and the band emission energy photon showed dependence of both the structural order-disorder and the thermal treatment history.

Intense and broad photoluminescence (PL) emission at room temperature was observed on structurally disordered $Ba(Zr_{0.25}Ti_{0.75})O_3$ (BZT) and $Ca(Zr_{0.05}Ti_{0.95})O_3$ (CZT) powders synthesized by the polymeric precursor method [19,20]. The theoretical calculations and experimental measurements of ultraviolet-visible absorption spectroscopy indicate that the presence of intermediary energy levels in the band gap is favorable for the intense and broad PL emission at room temperature in disordered BZT and CZT powders. The PL behavior is probably due the existence of a charge gradient on the disordered structure, denoted by means of a charge transfer process from $[TiO_5]$-$[ZrO_6]$ or $[TiO_6]$-$[ZrO_5]$ clusters to $[TiO_6]$-$[ZrO_6]$ clusters.

Similar photoluminescence behavior was also observed in disordered $MgTiO_3$ and Mn-doped $BaTiO_3$ powders [21,22]. The experimental and theoretical results indicated that PL is related with the degree of disorder in the powders and also suggests the presence of localized states in the disordered structure.

Blue-green and red photoluminescence (PL) emission in structurally disordered $CaTiO_3$:Sm powders was observed at room temperature with laser excitation at 350.7 nm [23]. The generation of the broad PL band is related to order-disorder degree in the perovskitelike structure.

The photoluminescence from $SrZrO_3$ and $SrTiO_3$ crystalline, quasi-crystalline, and quasi-amorphous samples, prepared by the polymeric precursor method, was examined by ab initio quantum mechanical calculations [24]. It was used in the modeling the structural model consisting of one pyramidal TiO_5 or ZrO_5 unit piled upon the TiO_6 or ZrO_6, which are representative of disordered structures of quasi-crystalline structures such as ST and SZ. In quasi-crystalline powders, the photoluminescence in the visible region showed different peak positions and intensities in SZ and ST. The PL emission was linked to distinct distortions in perovskite lattices and the emission of two colors-violet-blue in SZ and green in ST-was also examined in the light of favorable structural and electronic conditions. First principles calculations on the origin of violet-blue and green light photoluminescence emission in $SrZrO_3$ and $SrTiO_3$ perovskites

Ma, et al. studied synthesis and luminescence of undoped and Eu^{3+}-activated Aurivillius-type Bi_3TiNbO_9 (BTNO) nanophosphors by sol-gel combustion method [25]. Photoluminescence measurements indicated that a broad blue emission was detected for BTNO nanoparticles, and the characteristic Eu^{3+} ions $^5D_0 \rightarrow {}^7F_J$ (J=1-4) transitions were observed for the doped samples. Further investigation illuminates that the Eu^{3+} ions substituted for Bi^{3+} ions at A site in the pseudo-perovskite layers. It can be confirmed from high bright fluorescence image and short decay time that the novel orange-red phosphor has the potential applications in luminescence devices.

2.2 Photoluminescence in ferroelectric nanowires, nanotubes, and nanosheets

Gu, et al. reported characterization of single-crystalline lead titanate nanowires using photoluminescence spectroscopy, and ultraviolet-visible spectroscopy [26,27]. The nanowires were synthesized by surfactant-free hydrothermal method at 200 °C. The nanowires with uniform diameters of about 12 nm and lengths up to 5 um, exhibit a tetragonal perovskite structure without impurity phases. A blue emission centered at about 471 nm (2.63 eV) is observed at room temperature. Oxygen vacancies are believed to be responsible for the luminescence in the $PbTiO_3$ nanowires.

Chen, et al. prepared ferroelectric $PbHPO_4$ nanowires by a hydrothermal method at 180 °C for 24 h using a single-source precursor, Pb(II)−IP6 (IP6, inositol hexakisphosphate acid) complex [28]. PL measurements indicate that the nanowires exhibit a broad emission peak at 460 nm under the excitation of 375 nm.

Yang et al. reported synthesis of high-aspect-ratio $PbTiO_3$ nanotube arrays by the hydrothermal method [29]. The $PbTiO_3$ nanotube arrays have a tetragonal perovskite structure without any other impurity phases. A strong green emission band centered at 550 nm (2.25 eV) was observed in $PbTiO_3$ nanotube arrays at room temperature. Local defects in $PbTiO_3$ nanotube arrays were thought to result in the photoluminescence behavior.

$PbTiO_3$ nanotube arrays have also been synthesized via sol–gel template method. These nanotubes have a diameter about 300 nm and a length 50 um, with a wall thickness of typically several tens of nanometers. The as-prepared PTO nanotubes possess polycrystalline perovskite structure. An intense and wide emission band centered at 505 nm was observed [30]. The photoluminescence of PTO nanotubes was attributed to the radiative recombination between trapped electrons and trapped holes in localized tail states due to structural disorder and gap states due to a large amount of surface defects and oxygen vacancies.

Ida, et al. successfully prepared layered perovskite $SrBi_2Ta_2O_9$ nanosheets with a thickness of about 1.3 nm. The nanosheets showed visible blue luminescence under excitation at 285 nm at room temperature [31]. The luminescence property of the nanosheets was found to be largely sensitive to the change in the surface environment such as adsorption of H+ and/or OH-.

3. Photoluminescence properties of $(Bi, Ln)_4Ti_3O_{12}$ thin films

As is known, lanthanide doped bismuth titanate thin films are the most promising candidate materials for ferroelectric nonvolatile random access memory applications due to their superior properties such as nonvolatility, long retention time, high fatigue resistance. Recently, photoluminescence properties originated from some rare earth ions in these kinds of the thin film materials have attracted much attention for possible integrated photoluminescent ferroelectric thin film devices.

3.1 Crystal structure of lanthanide doped bismuth titanate thin films

Lanthanide doped bismuth titanate, $Bi_{4-x}Ln_xTi_3O_{12}$ (BLnT) consists of a layered structure of $(Bi_2O_2)^{2+}$ and $(Bi_2Ti_3O_{10})^{2-}$ pseudo-perovskite layers. Its crystal structure can be formulated as $(Bi_{2-x}Ln_xTi_3O_{10})^{2-}(Bi_2O_2)^{2+}$, in which three perovskite-like unit cells are sandwiched between two bismuth oxide layers along the c axis of the pseudotetragonal structure. The lattice paramenters of $Bi_{4-x}Ln_xTi_3O_{12}$ (using rare element La as an example) are a=0.542 nm,b=0.541 nm, and c=3.289 nm, respectively. The lanthanide ions substitute for the Bi ions

in the pseudo-perovskite layer, and thus enhance the ferroelectric properties. Generally, high fatigue resistance of $Bi_{4-x}Ln_xTi_3O_{12}$ is attributed to the existence of net charge of $(Bi_2O_2)^{2+}$ layer which can compensate for the space charge in the ferroelectric/electrode interface [2]. In addition, good chemical stability of perovskite-like layer makes oxygen vacancies difficultly generate, also contributing the good polarization fatigue-free properties [32]. Due to the layered perovskite structure, $Bi_{4-x}Ln_xTi_3O_{12}$ exhibits strong anisotropic electrical properties, because the vector of the spontaneous polarization in the layered perovskite materials is almost along a axis. Therefore, the anisotropic electrical properties of $Bi_{4-x}Ln_xTi_3O_{12}$ have been widely studied [33-35]. On the other hand, BLnT thin films also show excellent optical properties. For example, $Bi_{3.25}La_{0.75}Ti_3O_{12}$ thin films exhibited remarkable optical nonlinearity [10], and $(Bi, Eu)_4Ti_3O_{12}$ thin films showed red photoluminescence (PL) properties of Eu^{3+} ions [36,37]. The photoluminescence properties of BLnT thin films might have potential applications for integrated photoluminescent ferroelectric thin film devices.
As mentioned before, some of BLnT thin films, e.g. $(Bi,Nd)_4Ti_3O_{12}$, have also been reported to exhibit excellent optical properties such as large optical nonlinearity, high optical transparency in the visible wavelength region, which are attractive for the potential uses in optic and optoelectronic devices. Some rare earth elements such as Eu, Pr, Er in BLnT thin films can, on the one hand, act as a structural modifier which greatly improves the electrical properties of BLnT thin films, on the other hand, these rare earth elements can also act as the activator ions of luminescent materials [36,37]. Besides, $Bi_4Ti_3O_{12}$ (BIT) thin films have a high Curie temperature of 675 ^oC and good chemical stability. In addition, the rare earth ions substitute for the Bi^{3+} sites in perovskite-like layer $(Bi_2Ti_3O_{10})^{2-}$ of BIT, therefore the doping content of the rare earth ions could be large, and no charge compensation is needed. Recent research indicates that these Eu, Pr, Er doped bismuth titanate thin films will possibly be potential luminescent ferroelectric materials.

3.2 $(Bi, Eu)_4Ti_3O_{12}$ luminescent ferroelectric thin films

Ruan, et al. first prepared europium-doped bismuth titanate, $(Bi_{4-x}Eu_x)Ti_3O_{12}$ (BEuT) thin films on indium-tin-oxide (ITO)-coated glass substrates, and studied their photoluminescent properties as well as ferroelectric properties [36,37]. The BEuT thin films prepared by chemical solution deposition had a polycrystalline bismuth-layered perovskite structure. Excellent optical transmittance of the BEuT thin films was confirmed as shown in Fig. 1 [36]. Figure 2 shows emission and excitation spectra of BEuT (x=0.85) thin films annealed at different temperatures [36]. The excitation spectra were monitored at 617 nm and the emission spectra were observed by excitation at 350 nm. Photoluminescence spectra of the thin films included two strong peaks which originated from two transitions of $^5D_0 \rightarrow {}^7F_1$ (594 nm) and $^5D_0 \rightarrow {}^7F_2$ (617 nm) of levels of Eu^{3+} ions [38,39]. The emission peak intensities of the two transitions increase with increasing annealing temperature, due to improved crystallinity of the thin films, resulting in higher oscillating strengths for optical transitions [40,41]. At higher temperature, Eu^{3+} ions can arrive at real lattice sites more easily, thus this leads to enhanced activation of Eu^{3+} ions [42]. Investigation to the effect of Eu^{3+} concentrations (x=0.25, 0.40, 0.55, 0.70, and 0.85, respectively) on the photoluminescence properties of BEuT thin films indicates that there is an unusual composition quenching effect of photoluminescence as shown in Fig.3 [36]. The quenching concentration of BEuT thin films is about 0.40. Similar unusual concentration quenching effect of photoluminescence has also been observed in some Eu^{3+}-ion-activated layered perovskite luminescent powder materials such as $Na_2Gd_{2(1-x)}Eu_{2x}Ti_3O_{10}$, $NaGd_{1-x}Eu_xTiO_4$, and $RbLa_{1-x}Eu_xTa_2O_7$ [43,44]. It was suggested that in these layered perovskite luminescent powder materials, the energy transfer was restricted in the Eu^{3+} sublattices. Similarly, in our case, BEuT has a bismuth layer perovskite structure, and the Eu^{3+} ions in BEuT thin films are mainly substituting for the Bi^{3+} ions in $(Bi_{2-x}Eu_xTi_3O_{10})^{2-}$ perovskite-like layers rather than in $(Bi_2O_2)^{2+}$ layers, therefore, the energy transfer might occur in the nearby Eu^{3+} ions in the perovskite-like layers, thus resulting in the unusual concentration quenching of BEuT thin films. When the concentration of Eu^{3+} ions exceeds the critical value, the radiative centers increase with increasing doping amount of Eu^{3+} ions, however, at the same time the quenching centers also increase, this would lead to increase of the non-radiative rate, consequently, the PL intensity decreases.

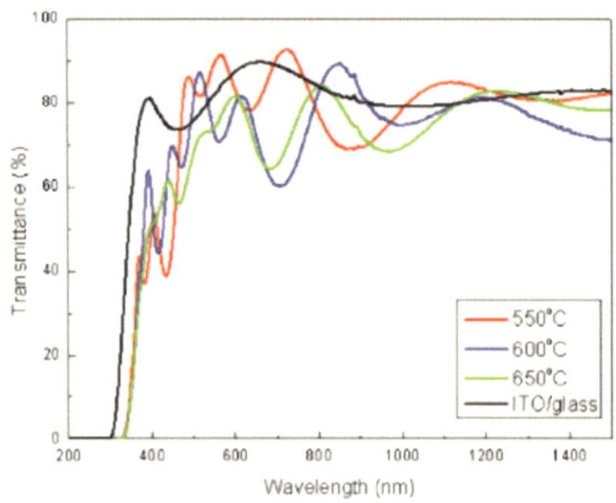

Fig. 1. Optic transmittance of BEuT (x=0.85) thin films annealed at different temperatures. From Ref. [36] Ruan, K.B., et al., *J. Appl. Phys., 103,* 074101 (2008). Copyright American Institute of Physics (2008)

Electrical measurements indicated that the BEuT thin films also showed ferroelectric properties comparable to those of BEuT thin films deposited on $Pt/Ti/SiO_2/Si$ substrates annealed at the same temperature for 1 h [45]. High fatigue resistance was also observed after 10^{10} switching cycles.

For most BLnT thin films, the doping amount of rare ions about 0.85 results in better electrical properties of the thin films [4, 46,47]. However, the quenching concentration x of $Bi_{4-x}Eu_xTi_3O_{12}$ thin films is about 0.40, which is much lower than 0.85. In order to obtain the BLnT thin films with both good electrical and photoluminescent properties at the same time, we proposed to prepare Eu- and Gd- codoped bismuth titanate ($Bi_{3.15}Eu_{0.425}Gd_{0.425}Ti_3O_{12}$, BEGT) thin films with both Eu^{3+} and Gd^{3+} doping amounts being 0.425, giving a total doping amount of rare earth ions of 0.85. It was expected that, on the one hand, the BEGT thin films might exhibit good photoluminescent properties since Eu^{3+} doping amount of 0.425 in the thin films is close to the quenching concentration about 0.40 of BEuT thin films, on the other hand, the BEGT thin films might also show good electrical properties comparable to those of BEuT thin films and of BGdT thin films because total doping amount of rare earth ions is maintained to be 0.85 in the thin films. Figure 7 shows the emission spectra of BEGT and BEuT thin films under 350 nm exciting wavelength [37]. The enhancement of emission intensities for two Eu^{3+} emission transitions of $^5D_0 \rightarrow ^7F_1$ (594 nm) and $^5D_0 \rightarrow ^7F_2$ (617 nm) was observed for the BEGT thin films as compared to BEuT thin films. This photoluminescence improvement can be attributed to Eu content of BEGT thin films close to quenching concentration of BEuT thin films and local distortion of crystal field surrounding the Eu^{3+} activator induced by different ionic radii of Eu^{3+} and Gd^{3+} ions. As mentioned before, the critical value of quenching concentration for $Bi_{4-x}Eu_xTi_3O_{12}$ thin films was near x=0.40. In the case of BEGT thin films, Eu^{3+} doping amount of 0.425 is close to the quenching concentration 0.40 of BEuT thin films, therefore, BEGT thin films exhibit better photoluminescence than $Bi_{3.15}Eu_{0.85}Ti_3O_{12}$ thin films. In addition, local distortion of crystal field surrounding the Eu^{3+} activator induced by different ionic radii of Eu^{3+} and Gd^{3+} ions in BEGT thin films also contributes to the enhancement of photoluminescence.

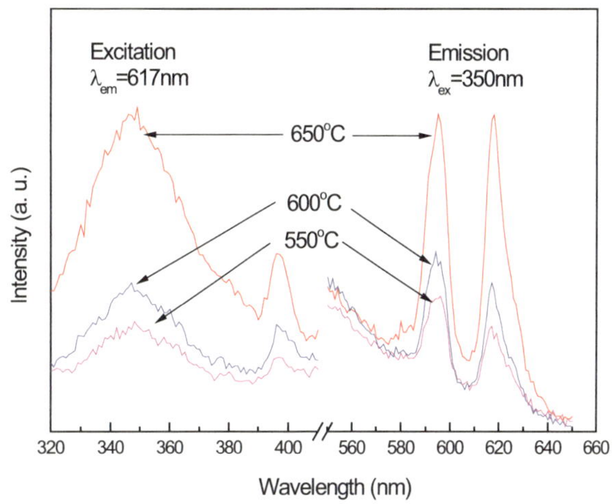

Fig. 2. Excitation and emission spectra of BEuT (x=0.85) thin films annealed at different temperatures. From Ref. [36] Ruan, K.B., et al., J. Appl. Phys., 103, 074101 (2008). Copyright American Institute of Physics (2008)

In addition, the BEGT thin films had larger remanent polarization and higher dielectric constant than BGdT and BEuT thin films prepared under the same experimental conditions. Therefore, co-doping of rare earth ions such as Eu and Gd in bismuth titanate thin films is an effective way to improve photoluminescence and electrical properties of the thin films.

As mentioned before, the bismuth layered perovskite structure BLnT exhibits strong structural anisotropy, which results in strong crystallographic orientation dependence of ferroelectric properties. Many studies indicate that c-axis-oriented (001) epitaxial $(Bi,La)_4Ti_3O_{12}$, $(Bi,Nd)_4Ti_3O_{12}$ thin films exhibit a very low remanent polarization along the film normal, because the vector of the spontaneous polarization in the layered perovskite materials is almost along a axis. Therefore, it is of interest to know whether there is similar orientation dependence of PL properties of BEuT thin films [48].

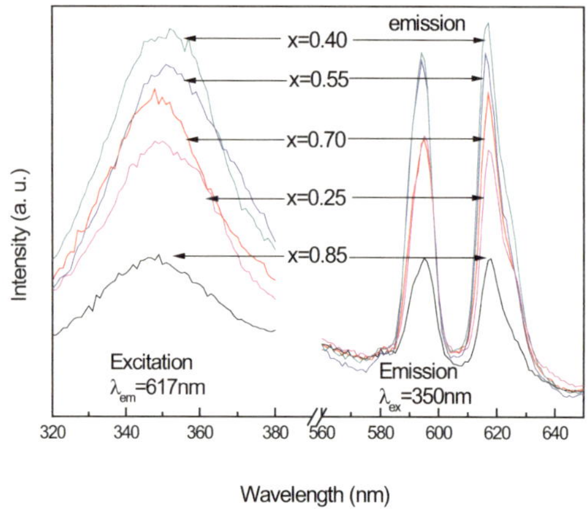

Fig. 3. Photoluminescence spectra of BEuT thin films with different Eu^{3+} concentrations. From Ref. [36] Ruan, K.B., et al., J. Appl. Phys., 103, 074101 (2008). Copyright American Institute of Physics (2008)

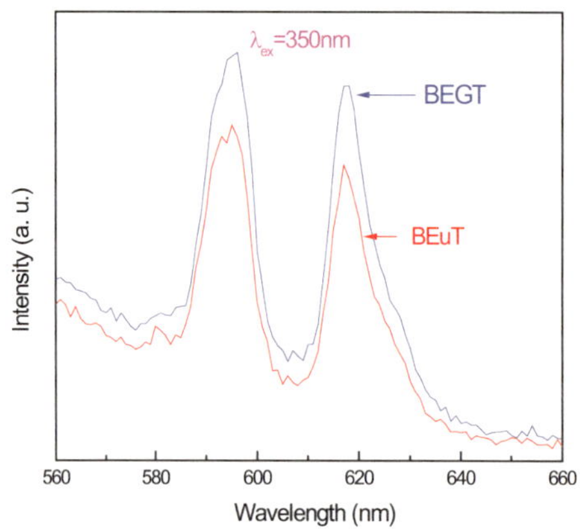

Fig. 4. Emission spectra of BEGT and BEuT thin films under exciting wavelength of 350 nm. From Ref. [37] Ruan, K.B., et al., J. Appl. Phys., 103, 086104 (2008). Copyright American Institute of Physics (2008)

The BEuT thin films were prepared on STO (100) and STO (111) substrates by using chemical solution deposition, and the effects of Eu ion concentration and crystallographic orientation on photoluminescent property of the thin films were investigated. The BEuT thin films prepared on STO (100) substrates grew with high c axis orientation due to very small film/substrate lattice mismatch, whereas BEuT thin films prepared on STO (111) substrates crystallized with random orientation. Figure 5 shows the PL emission spectra of $Bi_{4-x}Eu_xTi_3O_{12}$ thin films with different Eu ion concentrations on STO (100) substrates [48]. From Fig. 5, it can be observed that the PL intensity is obviously related to the doping amount of Eu^{3+} ions. The maximum PL intensity was obtained for the BEuT thin films with Eu^{3+}

concentration of x=0.55. It is worth noting that the quenching concentration of BEuT thin films prepared on STO (100) substrates is higher than that (x=0.4) for the thin films on ITO-coated glass. This may be ascribed to effects of the crystal orientation of BEuT thin films. The BEuT thin films prepared on STO (100) substrates show high c axis orientation, while BEuT thin films on ITO-coated glass and quartz substrates exhibit random orientation. For randomly oriented thin films, more non-radiative defects existed in the grain boundaries, resulting in decrease of quenching concentration.

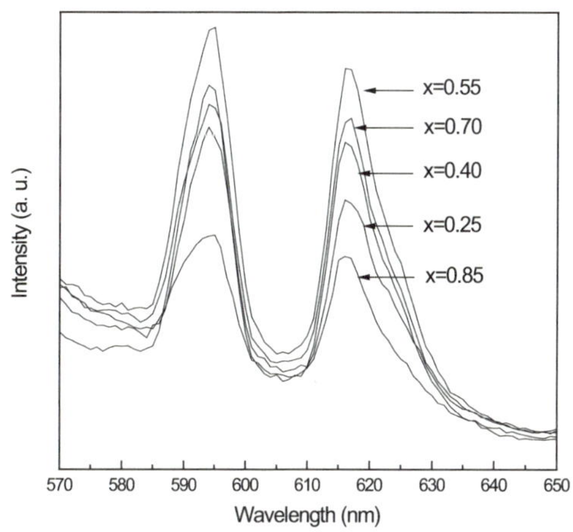

Fig. 5. PL emission spectra of $Bi_{4-x}Eu_xTi_3O_{12}$ thin films on STO (100) substrates (λ_{ex}=350 nm). Ref. [48] Ruan, K.B., et al., J. Appl. Phys., 104, 036101 (2008). Copyright American Institute of Physics (2008)

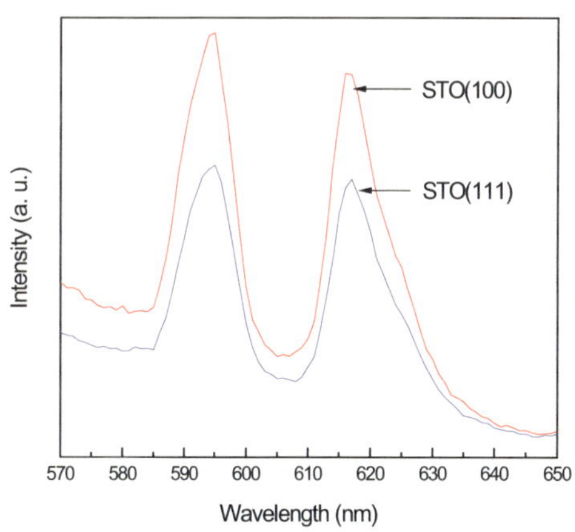

Fig. 6. PL emission spectra of $Bi_{3.45}Eu_{0.55}Ti_3O_{12}$ thin films prepared on STO (100) and STO (111) substrates. Ref. [48] Ruan, K.B., et al., J. Appl. Phys., 104, 036101 (2008). Copyright American Institute of Physics (2008)

Figure 6 shows the PL emission spectra of $Bi_{3.45}Eu_{0.55}Ti_3O_{12}$ thin films prepared on STO (100) and STO (111) substrates [48]. It can be observed that the emission intensity of BEuT thin films on STO (100) substrate is obviously stronger than that on STO (111) substrate. On the one hand, the orientation difference of the thin films might affect PL properties of BEuT thin films, specifically, well-aligned grains with c-axis oriented growth lead to low light scattering; on the other hand, the surface roughness may also affect detection of PL properties of thin film materials. It has been reported that rougher surfaces of photoluminescent thin films reduce internal reflections [49]. However, our AFM observation showed that the root mean square roughness values of both the BEuT thin films on the two different substrates are close (15.2 nm on STO (100) substrate and 13.8 nm on STO (111) substrate, respectively). This implies that different PL properties are mainly caused by the orientation difference of BEuT thin films. Note that different from orientation dependence of ferroelectric properties of the rare earth doped bismuth titanate thin films, the c-axis oriented BEuT thin films on STO (100) substrates exhibited stronger photoluminescence than the randomly oriented thin films on STO (111) substrates.

3.3 Nanocomposite films composed of ferroelectric $Bi_{3.6}Eu_{0.4}Ti_3O_{12}$ matrix and highly c-axis oriented ZnO nanorods

PL properties of BEuT thin films might have potential applications for integrated photoluminescent ferroelectric thin film devices. However, further enhancement of emission intensity of BEuT is desirable. For this purpose, Zhou, et al. developed a hybrid chemical solution method to prepare nanocomposite films composed of ferroelectric BEuT matrix and highly c-axis oriented ZnO nanorods with an attempt to achieve a more efficient energy transfer from ZnO nanorods to Eu^{3+} ions in BEuT, and thus, to enhance the PL intensity of BEuT [50].

Figure 7 shows emission spectra of (a) the nanocomposite film composed of BEuT matrix and highly c-axis oriented ZnO nanorods and (b) the BEuT thin film [50]. The excitation wavelength is chosen as 350 nm, as the internal band excitation is more efficient than the intrinsic excitation of Eu^{3+} in BEuT host. Compared with individual BEuT thin film, the nanocomposite film of BEuT matrix and ZnO nanorods exhibits much more intense emissions centered at 594 nm and 617 nm from Eu^{3+} ions. The emission intensities are about 10 times stronger than those for individual BEuT thin film. At the same time, in the emission spectrum, a sharp UV emission at about 380 nm and a broad band emission centered at 525 nm are also observed. The former can be attributed to near band edge emission of ZnO, and the latter is commonly believed to originate from defect-related deep-level emission in ZnO, such as the radiative recombination of photo-generated holes with electrons occupying the oxygen vacancies.

By comparing the emission spectrum of the nanocomposite film composed of BEuT matrix and highly c-axis oriented ZnO nanorods and excitation spectrum of the BEuT thin film, it was found that there are two spectral overlaps between the emission bands of ZnO nanorods and the absorption bands of Eu^{3+} ions in BEuT. One is located between sharp UV emission at 380 nm of ZnO and transition of $^7F_0 \rightarrow ^5L_6$ at 395 nm of Eu^{3+} ions, and the other one is between defect-related deep-level emission band centered at 525 nm of ZnO and transition of $^7F_0 \rightarrow ^5D_2$ at 465 nm of Eu^{3+} ions. It has been believed that energy transfer occurs only when the emission band of sensitizer (ZnO in this study) overlaps spectrally with the absorption band of activator (Eu^{3+} ions in this study). In our case, under the excitation of 350 nm radiation, ZnO nanorods firstly absorb the radiation energy, and promote electrons to move from the valence band to the conduction band, leading to band edge emission of ZnO. Then, due to the spectral overlap between the band edge emission of ZnO nanorods and the $^7F_0 \rightarrow ^5L_6$ excitation spectrum of Eu^{3+} ions centered at 395 nm in BEuT, an efficient energy transfer from the ZnO nanorods to Eu^{3+} ions occurs, promoting the Eu^{3+} ions from $^7F_0$ ground state to $^5L_6$ excited state [51]. The Eu^{3+} ions in the excited state $^5L_6$ undergo nonradiative decay to the $^5D_0$ state because the gaps of adjacent levels are small. Then, radiative transition takes place between the $^5D_0$ and the ^7F_J (J=0–6) states because of the larger gap [52,53]. This is one of the reasons that red PL of Eu^{3+} ions can be enhanced.

On the other hand, the defect states such as oxygen vacancies in the ZnO nanorods also capture electrons, and generate a broad band green emission centered at 525 nm. Due to the spectral overlap between the defect-related emission of ZnO and the absorption band of $^7F_0 \rightarrow ^5D_2$ transition at 465 nm of Eu^{3+} ions, an efficient energy transfer also occurs. Thus, some Eu^{3+} ions are promoted from the ground state $^7F_0$ to the excited state $^5D_2$. Similarly, this can also result in enhancement of red PL of Eu^{3+} ions. It has been reported that the defect states (trapping centers) in ZnO can temporarily store the excitation energy, then giving rise to efficient energy transfer from the traps in ZnO to Eu^{3+} ions [53].

For nonradiative energy transfer, the distance between the emission and absorption centers must be very close [51]. Therefore, in Eu^{3+}-doped ZnO materials the energy transfer is nonradiative energy transfer [54,55]. However, in our nanocomposite films, the energy transfer should be a radiative energy transfer because the BEuT matrix can only contact with the outside part of the ZnO nanorods. It has been reported that in some other materials with Eu^{3+} as activators combined with ZnO quantum dots or ZnO thin films the energy transfer is radiative energy transfer [51, 56].

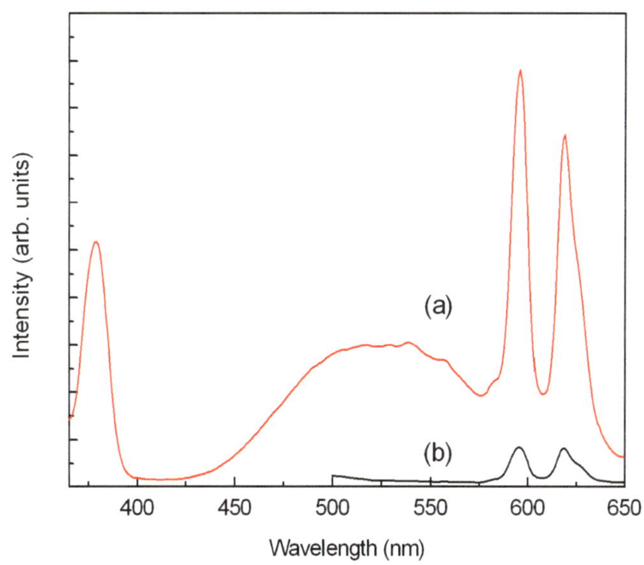

Fig. 7. Emission spectra of (a) the nanocomposite film composed of BEuT matrix and highly c-axis oriented ZnO nanorods and (b) the BEuT thin film. From Ref. [50] Zhou, H., et al., J. Am. Chem. Soc., 132, 1790 (2010). Copyright American Chemical Society (2010)

3.4 (Bi, Pr)$_4$Ti$_3$O$_{12}$ luminescent ferroelectric thin films

Recently, Pr doped $ATiO_3$ (A=Sr, Ca) titanate thin films have been widely studied for photoluminescence applications. Usually these thin films exhibit three emission peaks, which are located at 493 nm, 533 nm, and 612 nm, respectively [57]. Of them, the strongest peak is at about 612 nm. It is known that the lattice distortion of host materials and/or site symmetry of rare ions greatly affect the photoluminescence properties. Since Pr doped $SrTiO_3$ or $CaTiO_3$ thin films have simple perovskite structure, whereas Pr^{3+}-doped BIT (BPT) presents bismuth layered perovskite structure, and correspondingly, site symmetry of Pr ions is different, Pr^{3+}-doped BIT thin films are expected to show different photoluminescent properties. Our study confirmed that BPT thin films also exhibit three emission peaks, at 493 nm, 533 nm, and 612 nm, respectively, however, strongest peak is at 493 nm when Pr doping content is large than 0.01. It was observed that the $Bi_{3.91}Pr_{0.09}Ti_3O_{12}$ thin films have highest photoluminescence intensity, in other words, the quenching concentration for 493 nm emission is 0.09. Note that the doping amount of Pr^{3+} ions was so small that the ferroelectric properties of BPT thin films were not good. Taking into consideration that lanthanum doped BIT thin films exhibited good electrical properties, such as relatively high remanent polarization, low processing temperature, and good polarization fatigue-free properties, and that La^{3+} are usually used as host ions in rare earth luminescence materials because La^{3+} ions do not absorb ultraviolet emission, therefore, La^{3+} ions were selected to incorporate into $Bi_{3.91}Pr_{0.09}Ti_3O_{12}$ thin films in order to improve the photoluminescence and electrical properties of the thin films. The results confirmed that both the photoluminescence and ferroelectric polarization of BPT thin films have been enhanced by La^{3+} doping [58].

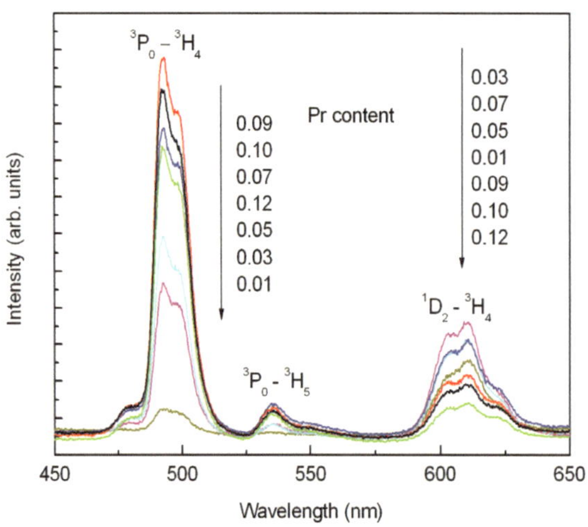

Fig. 8. Photoluminescence spectra of BPT thin films with different Pr contents. From Ref. [58] Zhou, H., et al., J. Am. Ceram. Soc., 93, 2109 (2010) Copyright American Ceramic Society (2010)

Figure 8 shows photoluminescence spectra of BPT thin films with different Pr^{3+} contents [58]. It can be observed that under the 350 nm UV excitation, the emission spectra include three peaks at around 493, 533 and 612 nm, which can be assigned to the transitions of $^3P_0 \rightarrow ^3H_4$, $^3P_0 \rightarrow ^3H_5$, and $^1D_2 \rightarrow ^3H_4$ of Pr^{3+} ions, respectively. As Pr^{3+} ion content increases, the intensities of the blue-green (493 nm) emissions increase first, and then decrease. When Pr^{3+}

ion content is 0.09, the blue-green emission reaches maximum, indicating that the quenching concentration of the blue-green emission is 0.09. We note that the red emission (612 nm) has been relatively weak for all Pr doping contents except Pr doping content of 0.01. This is very different from that of Pr^{3+} doped $SrTiO_3$ and $CaTiO_3$ thin films in which the red emission is strongest whereas the blue-green emission is very weak [59,60]. This is because the photoluminescence properties of Pr^{3+} are sensitive to host lattice symmetry. There is obvious difference of host lattice between Pr doped titanate with simple perovskite structure and BPLT with bismuth layered perovskite structure. It has been reported that, in Pr^{3+} doped $CaSnO_3$ perovskite with an orthorhombic symmetry, the emission from $^3P_0$ to $^3H_4$ (blue-green) is dominant over $^1D_2$ to $^3H_4$ red emission [61]. Okumura and coworkers confirmed that the $^3P_0 \rightarrow {}^3H_4$ transition (blue-green) in R_2O_3:Pr^{3+} (R=Y, La and Gd) became dominant instead of $^1D_2 \rightarrow {}^3H_4$ transition (red) when the crystal structure was distorted from cubic to monoclinic [62]. Besides the difference of their crystal structures, local structure or site symmetry of Pr ions should have a large influence on photoluminescence properties.

As is known, BIT layered perovskite has a monoclinic symmetry below Tc, but it can be represented as orthorhombic with the c-axis perpendicular to the $(Bi_2O_2)^{2+}$ layers. In BIT, a perovskite-like $(Bi_2Ti_3O_{10})^{2-}$ layer consists of three layers of TiO_6 octahedra, where Bi ions occupy the spaces in the framework of TiO_6 octahedra, and rare earth ions only substituted Bi^{3+} ions at $(Bi_2Ti_3O_{10})^{2-}$ perovskite layers. Owing to different lattice structures of BPT from $ATiO_3$ (A=Sr, Ca), BPT exhibits different photoluminescent properties. Since the transition from $^3P_0$ to $^3H_4$ is spin-allowed, it is understandable that the blue-green emission is dominant in the BPT thin films.

In order to improve the photolumincesence and ferroelectric properties, we fixed the Pr content at 0.09, and then added different content La ions into PBT thin films. Figure 9 shows photoluminescence spectra of BPLT-x thin films with various La contents [58]. Under the 350 nm UV excitation, the BPLT-x thin films also exhibit a strong blue-green emission peak at 493 nm, together with the other two weaker emission peaks at 533 nm and 612 nm, respectively. As can be seen, the peaks of emission spectra are similar for all the thin films, but their emission intensities vary with the La content. When La content x is 0.36, the emission reaches maximum.

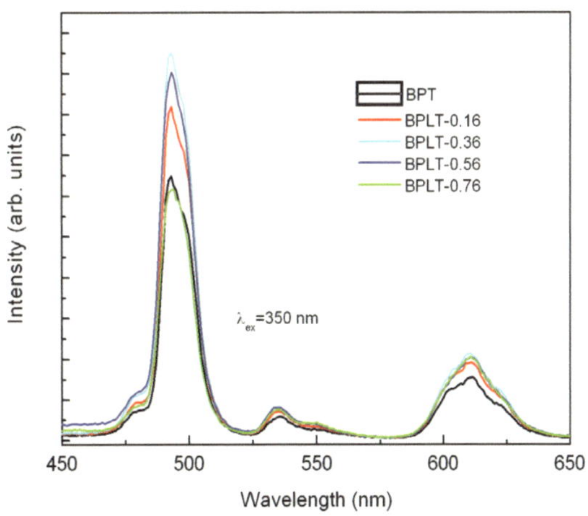

Fig. 9. Photoluminescence spectra of BPLT-x thin films. From Ref. [58] Zhou, H., et al., J. Am. Ceram. Soc., 93, 2109 (2010) Copyright American Ceramic Society (2010)

To understand the photoluminescence process in BPLT thin films excited under UV light, the excitation spectra by monitoring the emission at 493 nm and the transmission spectra of BPLT thin films were measured as shown in Fig.10 [58]. From Fig. 10 (a), it was observed that Pr doping or Pr-La codoping does not obviously influence the absorption spectra except the intensity of the broad band at around 350 nm. Meanwhile, the transmission spectra indicate that all the thin films with different La doping contents show a sharp absorption edge at around 350 nm as shown in Fig. 10 (b). The results showed that there is good agreement between excitation peak and absorption edge. This implies the possibility that the excitation energy from a UV light is first absorbed by the host lattice through the interband transition, and then the absorbed energy transfers from the host lattice to the activator Pr^{3+} ions.

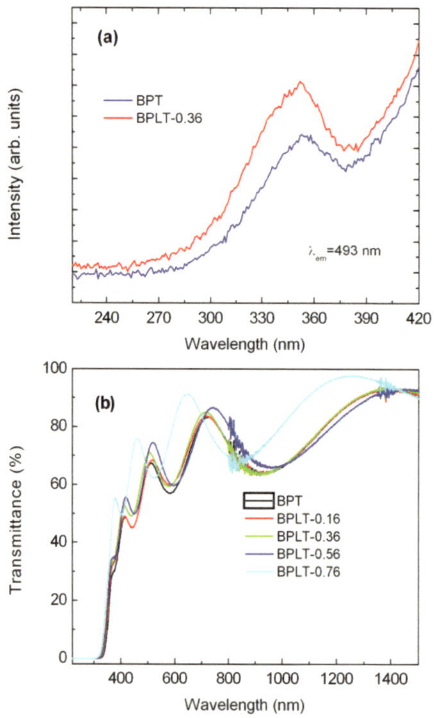

Fig. 10. Excitation and transmission spectra of BPLT-x thin films. From Ref. [58] Zhou, H., et al., J. Am. Ceram. Soc., 93, 2109 (2010) Copyright American Ceramic Society (2010)

The polarization-electric field (P-E) hysteresis loops of the BPLT-x thin films deposited on $Pt/TiO_2/SiO_2/Si$ substrates are measured. At an applied electric field of about 400 kV/cm, the remanent polarization (Pr) and coercive field (Ec) values of BPT and BPLT-0.36 films are 6.9 μC/cm^2, 147.3 kV/cm, and 19.8 μC/cm^2, 115.6 kV/cm, respectively. Obviously, the La^{3+} substitution raised Pr value of the BPT thin film from 6.9 to 19.8 μC/cm^2, and, at the same time, reduced the Ec value.

The polarization enhancement by La^{3+} substitution can be attributed to crystallization improvement of the thin films which has been confirmed by XRD and SEM. In addition, better crystallinity can also result in strong photoluminescence due to higher oscillating strengths for the optical transitions [41].

3.5 (Bi, Er)$_4$Ti$_3$O$_{12}$ luminescent ferroelectric thin films

As described before, BiT doped with Eu^{3+} or Pr^{3+} ions showed photoluminescence properties. However, they are only frequency down-conversion photoluminescence, up-conversion photoluminescence in BiT based thin films is also worth studying.

As is known, Er^{3+}/Yb^{3+} co-doped materials have been recognized as one of the most efficient systems for obtaining frequency up-conversion photoluminescence due to the efficient energy transfer from Yb^{3+} to Er^{3+} ions pumped by 980 nm [63,64], in which Er^{3+} ions acted as luminescence centers and Yb^{3+} ions as sensitizers. On the other hand, considering the lattice structure of BiT, Er^{3+}/Yb^{3+} ions can be accommodated in well defined sites of BiT lattice by substituting for Bi^{3+} ions without the need of charge compensation. Therefore, it is of much interest to study up-conversion luminescence properties of Er^{3+}/Yb^{3+} co-doped BiT thin films [65].

Figure 11 shows the up-conversion luminescence spectra for the BErT and BYET thin films under a 980 nm laser excitation at room temperature [65]. As can be seen from Fig.11, there are three distinct up-conversion emission bands, of which the two bands centered at 524 nm and 545 nm, correspond to strong green light emissions ascribed to $^2H_{11/2}$ to $^4I_{15/2}$ and $^4S_{3/2}$ to $^4I_{15/2}$ transitions of Er^{3+} ions, respectively, and another band centered at 667 nm to very weak red light emission originated from $^4F_{9/2}$ to $^4I_{15/2}$ transition of Er^{3+} ions. The green emission at 549 nm is 20 times as strong as the red emission at 667 nm. Meanwhile, BYET thin films exhibit higher green emission intensity by a factor of about 30 compared with BErT thin films. It is worth noting that the near pure green up conversion photoluminescence is bright enough to be clearly observed by the naked eyes. This indicates that the up-conversion photoluminescence efficiency of BiT thin films co-doped by Er^{3+} and Yb^{3+} has been greatly improved.

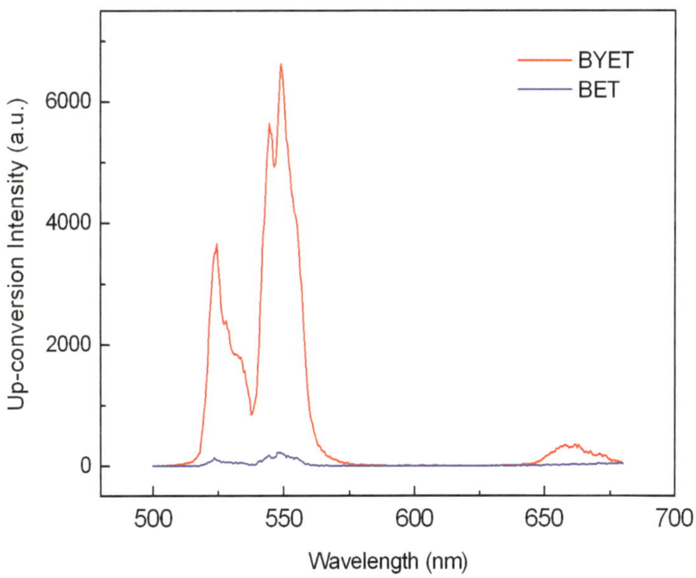

Fig. 11. Up-conversion luminescence spectra of BET and BYET thin films on fused silica substrates pumped by 980 nm. From Ref. [65] Gao, F., et al., J. Appl. Phys., 106, 126104 (2009). Copyright American Institute of Physics (2009)

4. Photoluminescence in other ferroelectric thin films

Reshmi, et al. studied photoluminescence of Eu^{3+}-doped $Ba_{0.7}Sr_{0.3}TiO_3$ thin films prepared by pulsed laser deposition [66]. The photoluminescence spectra show main transitions of Eu^{3+} ions at 550 nm ($^5D_1 \rightarrow {}^7F_2$), 615 nm ($^5D_0 \rightarrow {}^7F_2$), and 669 nm ($^5D_0 \rightarrow {}^7F_3$) upon excitation with 408 nm. It was confirmed that the photoluminescence of BST:Eu thin films was closely correlated to the crystallinity.

Kuo, et al. studied photoluminescent properties of Er^{3+}-doped $Ba_{0.7}Sr_{0.3}TiO_3$ thin films prepared by sol-gel method [67]. The excitation-dependent PL studies indicate that the green emission peaks do not shift with the change in excitation power, while the integrated intensity increases monotonically with increasing excitation power.

Recently, Garcia Hernandez, et al. [68] also investigated photoluminescence of Er^{3+}-doped $BaTiO_3$ thin films prepared by sol-gel method. Their films exhibited green emissions under near-infrared excitation.

Aizawa and Ohtani studied the effect of Eu/Sr ratios on ferroelectric and fluorescent properties of Eu-substituted strontium bismuth tantalate (Eu-SBT) films grown on Pt/Ti/SiO_2/Si substrates by spin-coating technique [69,70]. The remnant polarization values of the 260-nm-thick Eu-SBT films with Eu concentrations of 1.25 and 5 mol % were approximately 4.0 and 0.86 uC/cm^2, respectively. The PL intensity of the Eu-SBT films was significantly dependent on both annealing temperature and Eu concentrations. Emission peaks at approximately 600 nm were observed, which were associated with the $^5D_0 \rightarrow {}^7F$ transitions of Eu^{3+}.

Nakajima, et al. prepared $Sr_{1-x}Pr_xTiO_3$:Al^{3+} polycrystalline thin films using an excimer laser assisted metal-organic-deposition process at temperatures below 400 ^oC [71]. The Al^{3+}-added perovskite titanate thin films show strong red emission. The PL emission peaks were observed at 493, 533 and 612 nm, which are assigned to the transitions of $^3P_0 \rightarrow {}^3H_4$, $^3P_0 \rightarrow {}^3H_5$, $^1D_2 \rightarrow {}^3H_4$ of Pr^{3+}, respectively. Host-to-activator energy transfer due to the coupling of the band-edge recombination energy and an allowed transition of a rare-earth ion is essential to provide a high efficiency of Al-doped $SrTiO_3$:Pr^{3+} [72]. Nakajima, et al. also studied the substrate effect on excimer laser assisted crystal growth in $Ca_{0.997}Pr_{0.002}TiO_3$ polycrystalline thin films prepared by same method [73]. The thin films also show strong red luminescence. It was found that the substrate properties such as optical absorbance and thermal conductivity affected the crystal growth and the PL emission of the thin films in the excimer laser assisted metal organic deposition process.

Takashima, et al. observed intense red photoluminescence under ultraviolet excitation in epitaxial Pr-doped $Ca_{0.6}Sr_{0.4}TiO_3$ perovskite thin films prepared on $SrTiO_3$ (100) substrates by pulsed laser deposition [74]. The observed sharp PL peak centered at 610 nm was assigned to the transition of Pr^{3+} ions from the $^1D_2$ state to the $^3H_4$ state. It was suggested that the UV energy absorbed by the host lattice was transferred to the Pr ions, leading to the red luminescence.

Rho, et al. prepared $SrTiO_3$ thin films by rf-sputtering and studied the photoluminescence of the thin films after postannealing treatments [75]. The remarkable room temperature PL effects observed are contributed to both metastable and energetically stabilized defect states formed inside the band gap.

Moreira, et al. [76] studied photoluminescence of barium titanate and barium zirconate in multilayer disordered thin films at room temperature. The thin films were prepared by spin-coating and annealed at 350, 450, and 550 ^oC for 2 h. It was observed that the wide band emission showed a synergic effect on barium zirconate and barium titanate thin layers in alternate multilayer system at room temperature by 488 nm excitation. Visible and intense photoluminescence was governed by $BaZrO_3$ thin films in the multilayer system. A disordered model was used to explained photoluminescence emission results of these multilayered systems.

Cavalcante, et al. [77] investigated the photoluminescent behavior of $Ba(Ti_{0.75}Zr_{0.25})O_3$ (BTZ) thin films experimentally and theoretically. The thin films were prepared by the polymeric precursor method. They suggested that the presence of localized electronic levels and a charge gradient in the band gap due to a symmetry breaking, be responsible for the PL in disordered BTZ lattice.

Ueda, et al. [78] studied photoluminescence from epitaxial perovskite-type alkaline-earth stannate thin films. The epitaxial calcium and strontium stannate films with perovskite or its related structure were fabricated by pulsed laser deposition method. The highly transparent films in the visible region showed intense luminescence of several colors under ultraviolet excitation.

5. Conclusion

Photoluminescence in low-dimensional oxide ferroelectric materials is of much interest. Lanthanide-doped $Bi_4Ti_3O_{12}$ thin films are promising photoluminescence ferroelectric materials. Co-doping of rare earth ions such as Eu/Gd, Pr/La, Er/Yb in bismuth titanate thin films is found to be an effective way to improve photoluminescence and electrical properties of the thin films. In addition, constructing nanocomposite thin films composed of ferroelectric $Bi_{3.6}Eu_{0.4}Ti_3O_{12}$ matrix and highly c-axis oriented ZnO nanorods can effectively enhance photoluminescent properties of Eu^{3+} ions. Our own studies suggest that rare earth doped $Bi_4Ti_3O_{12}$ thin films can be considered as a promising multifunctional material which can find applications in integrated multifunctional ferroelectric devices as well as in optoelectronic devices. Further research should be focused on understanding the physical mechanism of photoluminescence in low-dimensional oxide ferroelectric materials.

6. Acknowledgments

The author is grateful for the contributions from Drs. K.B. Ruan, H. Zhou, F. Gao, G. H. Wu, and X.M. Chen. The financial support from NSFC (Nos. 50872156 and U0634006), the Natural Science Foundation of Guangdong Province, China (Grant No. 10251027501000007), and the Specialized Research Fund for the Doctoral Program of Higher Education of China (Grant No. 20090171110007), is acknowledged.

7. References

[1] Warusawithana, M.P., Cen, C., Sleasman, C.R., Woicik, J.C., Li, Y.L., Kourkoutis, L.F., Klug, J.A., Li, H., Ryan, P., Wang, L.P., Bedzyk, M., Muller, D.A., Chen, L.Q., Levy, J., Schlom, D.G. (2009) *Science*, 324, 367-370.

[2] Paz de Araujo, C.A., Cuchiaro, J.D., McMillan, L.D., Scott, M.C., & Scott, J.F. (1995) *Nature (London)*, *374*, 627.

[3] Park, B.H., Kang, B.S., Bu, S.D., Noh, T.W., Lee, J., & Jo, W. (1999) *Nature (London)*, *401*, 682.

[4] Chon, U., Jang, H.M., Kim, M.G., & Chang, C.H. (2002) *Phys. Rev. Lett., 89,* 087601.
[5] Bao, D.H., Chiu, T.W., Wakiya, N., Shinozaki, K. & Mizutani, N. (2003) *J. Appl. Phys., 93,* 497.
[6] Bao, D.H., Lee, S.K., Zhu, X.H., Alexe, M. & Hesse, D. (2005) *Appl. Phys. Lett., 86,* 082906.
[7] Lee, H. N., Hesse, D., Zakharov, N. & Gösele, U. (2002) *Science, 296,* 2006.
[8] Gu, B., Wang, Y.H., Peng, X.C., Ding, J.P., He, J.L. & Wang, H.T. (2004) *Appl. Phys. Lett., 85,* 3687.
[9] Hu, G. D., Fan, S. H. & Cheng, X. (2007) *J. Appl. Phys., 101,* 054111.
[10] Wang, Y. H., Gu, B., Xu, G. D. & Zhu, Y.Y. (2004) *Appl. Phys. Lett., 84,* 1686.
[11] Badr, Y., Salah, A., Battisha, I.K. (2005) *J. Sol-Gel Sci. & Technol.,* 34 (3): 219-226.
[12] Fu, Z., Moon, B.K., Yang, H.K., Jeong, J.H. (2008) *J. Phys. Chem. C,* 112 (15): 5724-5728 .
[13] Lemos, F.C.D., Melo, D.M.A., da Silva, J.E.C. (2005) *Mater. Res. Bull.,* 40 (1): 187-192.
[14] Lin, Y.H., Nan, C.W., Wang, J.F., He, H.C., Zhai, J.Y., Jiang, L. (2004) *Mater. Lett.,* 58 (5): 829-832.
[15] Oliveira, R. C., Cavalcante, L. S., Sczancoski, J. C., Aguiar, E. C., Espinosa, J. W. M., Varela, J. A., Pizani, P. S., Longo, E. (2009) *J. Alloys & Compds.,* 478 (1-2): 661-670.
[16] Pizani, P.S., Leite, E.R., Pontes, F.M., Paris, E.C., Rangel, J.H., Lee, E.J.H., Longo, E., Delega, P., Varela, J.A. (2000) *Appl. Phys. Lett.,* 77 (6): 824-826.
[17] Longo, V. M., Cavalcante, L. S., de Figueiredo, A. T., Santos, L. P. S., Longo, E., Varela, J. A., Sambrano, J. R., Paskocimas, C. A., De Vicente, F. S., Hernandes, A. C. (2007) *Appl. Phys. Lett.,* 90 (9): 091906.
[18] Motta, F. V., Marques, A. P. A., Espinosa, J. W. M., Pizani, P. S., Longo, E., Varela, J. A. (2010) *Curr. Appl. Phys.,* 10 : (1): 16-20.
[19] Cavalcante, L. S., Anicete-Santos, M., Sczancoski, J. C., Simoes, L. G. P., Santos, M. R. M. C., Varela, J. A., Pizani, P. S., Longo, E. (2008) *J. Phys. & Chem. Solids,* 69 (7): 1782-1789.
[20] Cavalcante, L. S., Simoes, A. Z., Espinosa, J. W. M., Santos, L. P. S., Longo, E., Varela, J. A., Pizani, P. S. (2008) *J. Alloys & Comp.,* 464 (1-2): 340-346.
[21] Ferri, E. A. V., Sczancoski, J. C., Cavalcante, L. S., Paris, E. C., Espinosa, J. W. M., de Figueiredo, A. T., Pizani, P. S., Mastelaro, V. R., Varela, J. A., Longo, E. (2009) *Mater. Chem. & Phys.,* 117 (1): 192-198.
[22] Gurgel, M. F. C., Espinosa, J. W. M., Campos, A. B., Rosa, I. L. V., Joya, M. R., Souza, A. G., Zaghete, M. A., Pizani, P. S., Leite, E. R., Varela, J. A., Longo, E. (2007) *J. Lumin.,* 126 (2): 771-778.
[23] de Figueiredo, A.T., Longo, V.M., de Lazaro, S., Mastelaro, V.R., De Vicente, F.S., Hernandes, A.C., Li, M.S., Varela, J.A., Longo, E. (2007) *J. Lumin.,* 126 (2): 403-407.
[24] Longo, V.M., Cavalcante, L.S., Costa, M.G.S., Moreira, M.L., de Figueiredo, A.T., Andres, J., Varela, J.A., Longo, E. (2009) *Theoretical Chemistry Accounts,* 124 (5-6): 385-394.
[25] Ma, Q., Zhou, Y.Y., Lu, M.K., Zhang, A.Y., Zhou, G.J. (2009) *Mater. Chem. & Phys.,* 116 (2-3): 315-318.
[26] Gu, H.S., Hu, Y.M., You, J., Hu, Z.L., Yuan, Y., Zhang, T.J. (2007) *J. Appl. Phys.,* 101 (2): 024319.
[27] Hu, Y.M., Gu, H.S., Sun, X.C., You, J., Wang, J. (2006) *Appl. Phys. Lett.,* 88 (19): 193120.
[28] Chen, G.Y., Qu,W.G., Ye, F., Zhang, W.X., Xu, A.W. (2008) *J. Phys. Chem. C,* 112 (43): 16818–16823.

[29] Yang, Y., Wang, X.H., Sun, C.K., Li, L.T. (2008) *J. Am. Ceram. Soc.*, 91, (11): 3820–3822.

[30] Liu L.F., Ning, T.Y., Ren, Y., Sun, Z.H., Wang, F.F., Zhou, W.Y., Xie, S.S., Song, L., Luo, S.D., Liu, D.F., Shen, J., Ma, W.J., Zhou, Y.L. (2008) *Mater. Sci. & Eng. B* 149: 41–46.

[31] Ida, S., Ogata, C., Unal, U., Izawa, K., Inoue, T., Altuntasoglu, O., Matsumoto, Y. (2007) *J. Am. Chem. Soc.*, 129 (29): 8956.

[32] Kang, B. S., Park, B. H., Bu, S. D., Kang, S. H. & Noh, T. W. (1999) *Appl. Phys. Lett., 75,* 2644.

[33] Lee, H. N. & Hesse, D. (2002) *Appl. Phys. Lett., 80,* 1040.

[34] Garg, A., Barber, Z. H., Dawber, M., Scott, J.F. & Barber, Z.H. (2003) *Appl. Phys. Lett., 83,* 2414.

[35] Lee, S.K., Hesse, D. & Gosele, U. (2006) *Appl. Phys. Lett., 88,* 062909.

[36] Ruan, K.B., Chen, X.M., Liang, T. Wu, G.H. & Bao, D.H. (2008) *J. Appl. Phys., 103,* 074101.

[37] Ruan, K.B., Chen, X.M., Liang, T. Wu, G.H. & Bao, D.H. (2008) *J. Appl. Phys., 103,* 086104.

[38] Pradhan, A. K., Zhang, K., Mohanty, S., Dadson, J. , Hunter, D., Loutts, G. B., Roy, U. N., Cui, Y., Burger, A. & Wilkerson, A. L. (2005) *J. Appl. Phys., 97,* 023513.

[39] Mercier, B., Dujardin, C., Ledoux, G., Louis, C., Tillement, O. & Perriat, P. (2004) *J. Appl. Phys., 96,* 650.

[40] Shim, K. S., Yang, H. K., Moon, B. K., Jeong, J. H., Yi, S. S. & Kim, K.H. (2007) *Appl. Phys. A, 88,* 623.

[41] Mishra, K. C., Berkowitz, J. K., Johnson, K. H. & Schmidt, P. C. (1992) *Phys. Rev., B 45,* 10902.

[42] Liu, Q.L., Bando, Y., Xu, F.F. & Tang, C.C. (2004) *Appl. Phys. Lett., 85,* 4890.

[43] Toda, K., Honma, T. & Sato, M. (1997) *J. Lumin., 71,* 71.

[44] Toda, K., Kameo, Y., Ohta, M. & Sato, M. (1995) *J. Alloys Comp., 218,* 228.

[45] Kim, K T., Kim, C. II., Kim, J. G. & Kim, G. H. (2007) *Thin Solid Films, 515,* 8082.

[46] Giridharan, N.V., Subramanian, M. & Jayavel, R. (2006) *Applied Physics, A83,* 123.

[47] Kim, S.S., Bae, J.C., & Kim, W.J. (2005) *J. Cryst. Growth, 274,* 394.

[48] Ruan, K.B., Gao, A.M., Deng, W. L. Chen, X.M. & Bao, D.H. (2008) *J. Appl. Phys., 104,* 036101.

[49] Bae, J.S., Shim, K.S., Kim, S.B., Jeong, J.H., Yi, S.S. & Park, J.C. (2004) *J. Cryst. Growth, 264,* 290.

[50] Zhou, H., Chen, X.M., Wu, G.H., Gao, F. Qin, N. & Bao, D.H. (2010) *J. Am. Chem. Soc.,* 132 (6):1790.

[51] Chong, M.K., Abiyasa, A.P., Pita, K. & Yu, S.F. (2008) *Appl. Phys. Lett., 93,* 151105.

[52] Shionoya, S. & Yen, W.H. (1998) Phosphor Handbook (CRC, Boca Raton, FL, USA), p.179.

[53] Jia, W., Monge, K. & Fernandez, F. (2003) *Opt. Mater., 23,* 27.

[54] Yu, Y. L., Wang, Y. S., Chen, D. Q., Huang, P., Ma, E. & Bao, F. (2008) *Nanotechnology, 19,* 055711.

[55] Du, Y.P., Zhang, Y.W., Sun L.D. & Yan, C.H. (2008) *J. Phys. Chem., C112,* 12234.

[56] Bang, J., Yang, H. & Holloway, P. H. (2005) *J. Chem. Phys., 123,* 084709.

[57] Tetsuka, H., Takashima, H., Ikegami, K., Nanjo, H., Ebina, T. & Mizukami, F. (2009) *Chem. Mater., 21,* 21.

[58] Zhou, H., Wu, G.H., Qin, N. & Bao, D.H. (2010) *J. Am. Ceram. Soc.*, 93, 2109.

[59] Takashima, H., Ueda, K. & Itoh, M. (2006) *Appl. Phys. Lett., 89,* 261915.
[60] Okamoto S. & Yamamoto, H. (2001) *Appl. Phys. Lett., 78,* 655.
[61] Goto, K., Nakachi, Y. & Ueda, K. (2008) *Thin Solid Films, 516,* 5885.
[62] Okumura, M., Tamatani, M., Albessard, A.K. & Matsuda, N. (1997) *Jpn. J. Appl. Phys, 36,* 6411.
[63] Wang, L. Y., Li, P. & Li, Y. D. (2007) *Adv. Mater., 19,* 3304.
[64] Lim, S. F., Riehn, R., Ryu, W. S., Khanarian, N., Tung, C. K., Tank, D. & Austin, R. H. (2006) *Nano Lett., 6,* 169.
[65] Gao, F., Wu, G.H., Zhou, H. & Bao, D.H. (2009) *J. Appl. Phys., 106,* 126104.
[66] Reshmi, R., Jayaraj, M. K., Jithesh, K., Sebastian, M. T. (2010) *J. Electrochem. Soc.*, 157 (7): H783-H786.
[67] Kuo, S.Y., Hsieh, W.F. (2005) *J. Vac. Sci. & Technol. A* , 23 (4): 768-772.
[68] Garcia Hernandez, M., Carrillo Romo, F. de J., Garcia Murillo, A., Jaramillo Vigueras, D., Meneses Nava, M. A., Bartolo Perez, P., Chadeyron, G. (2010) *J. Sol-Gel Sci. & Technol.*, 53 (2): 246-254.
[69] Aizawa, K., Ohtani, Y. (2007) *Jpn. J. Appl. Phys. Part 1*, 46 (10B): 6944-6947.
[70] Aizawa, K., Ohtani, Y. (2008) *Jpn. J. Appl. Phys. Part 2*, 47 (9): 7549-7552.
[71] Nakajima, T., Tsuchiya, T., Kumagai, T. (2008) *Curr. Appl. Phys.*, 8 (3-4): 404-407.
[72] Yamamoto, H., Okamoto, S., Kobayashi, H. (2002) *J. Lumin.*, 100 (1-4): 325-332.
[73] Nakajima, T., Tsuchiya, T., Kumagai, T. (2007) *Appl. Surf. Sci.*, 254 (4): 884-887.
[74] Takashima, H., Ueda, K., Itoh, M. (2006) *Appl. Phys. Lett.*, 89 (26): 261915.
[75] Rho, J., Jang, S., Ko, Y.D., Kang, S., Kim, D.W., Chung, J.S., Kim, M., Han, M., Choi, E. (2009) *Appl. Phys. Lett.*, 95 (24): 241906.
[76] Moreira, M. L., Gurgel, M. F. C., Mambrini, G. P., Leite, E. R., Pizani, P. S., Varela, J. A., Longo, E. (2008) *J. Phys. Chem. A*, 112 (38): 8938-8942 .
[77] Cavalcante, L. S., Gurgel, M. F. C., Paris, E. C., Simoes, A. Z., Joya, M. R., Varela, J. A., Pizani, P. S., Longo, E. (2007) *Acta Mater. A*, 55 (19): 6416-6426.
[78] Ueda, K., Maeda, T., Nakayashiki, K., Goto, K., Nakachi, Y., Takashima, H., Nomura, K., Kajihara, K., Hoson, H. (2008) *Appl. Phys. Exp.*, 1 (1): 015003 .

4

Optical Properties and Electronic Band Structures of Perovskite-Type Ferroelectric and Conductive Metallic Oxide Films

Zhigao Hu[1], Yawei Li[1], Wenwu Li[1], Jiajun Zhu[1], Min Zhu[2], Ziqiang Zhu[3] and Junhao Chu[3]
[1]*Key Laboratory of Polar Materials and Devices, Ministry of Education, Department of Electronic Engineering, East China Normal University, Shanghai 200241*
[2]*Department of Physics, Shanghai Jiao Tong University, Shanghai 200240*
[3]*Key Laboratory of Polar Materials and Devices, Ministry of Education, Department of Electronic Engineering, East China Normal University, Shanghai 200241*
People's Republic of China

1. Introduction

Ferroelectric (FE) materials have recently attracted considerable attention and intensive research due to their unique advantages in nonvolatile random-access memories, electro-optic devices, pyroelectric detectors, and optical mixers. (1; 2; 3; 4; 5; 6) It is well known that FE films can be deposited directly on diversified substrates and are expected to yield better sensitivity and faster response than the equivalent bulk single crystal. Nevertheless, the physical properties of FE films are strongly sensitive to the experimental conditions containing substrate, growth technique, crystalline quality, intrinsic defects, and doping elements, etc. Hence, it is significant to systematically investigate the physical properties of FE film materials, such as optical, electrical, and magnetic properties and their interactions. Among them, perovskite-type FE materials are the most promising compounds due to the polarization from the oxygen octahedra. Although there are lots of electrical properties reported on the FE bulk and film materials, optical and electronic properties are still scarce for their applications in the optoelectronic field. As we know, $Bi_4Ti_3O_{12}$ (BiT) is a very promising compound as a candidate of FE materials for integrated optics applications due to their combination of low processing temperatures, fatigue-free properties, high Curie temperature (T_c~675 °C) and large spontaneous polarization (P_s~50 μC/cm^2 along the a axis). (7; 8) Moreover, it was reported that the FE and optical properties of BiT materials can be evidently improved by the La doping, such as $Bi_{3.25}La_{0.75}Ti_3O_{12}$ (BLT). (9; 10; 11; 12) For Bi-site substitution, the coercive field usually becomes larger than that of pure BiT, while for Ti-site substitution, the coercive field becomes smaller. Theoretical calculation indicates that the La substitution strikingly decreases the T_c to about 400 °C. (13) Thus, BLT material can be widely accepted for an interesting compound in some optoelectronic applications owing to a lower T_c and larger coercive field.

Generally, the phase transitions of perovskite-based oxides represent a basilic class of structural phase transition that bear significant technological implications. (14) The high

FE phase transition temperature will be valuable for the electronic and optoelectronic device applications, which can be controlled at room temperature (RT). As we know, the phase transition of most FE materials can be derived by the soft mode, which can be studied by Raman scattering and/or far-infrared spectroscopy. (15; 16) Note that the ferroelectric-paraelectric phase transition of FE materials have been widely investigated to clarify the T_c with the aid of electrical measurements, such as the remnant polarization variations. However, there are some other structural transitions in FE materials except for the T_c temperature, which are also crucial to the polarization. (17; 18; 19) It can be expected that the physical properties of FE compounds, such as dielectric constants and loss tangent could be changed due to the structure variations at phase transition point. Therefore, the investigations on the optical properties during the phase transitions are necessary to further clarify its FE derivation and polarization.

For BiT nanocrystals and ceramics, some experimental results reveal a possible structural transformation at lower temperatures. (20; 21) Structural transitions are usually studied by various experimental techniques such as x-ray diffraction (XRD) and Raman scattering. (22; 23; 24) Although both of the two methods can provide the most direct information about the phase transitions, the variations of the optical band gap, optical constants, and band tail state behavior during phase transitions could not be derived due to the intrinsic technique limitations. Fortunately, far-infrared spectroscopy is commonly applied to observe the variation of soft mode and dielectric function during the phase transition. (15) Nevertheless, the band gap behavior and interband electronic transition are still keep as an open issue because of the wide-band-gap characteristics for most FE materials, which are located in the ultraviolet photon energy region. Therefore, current attention is focused on finding a new method to give more information about the phase transitions, especially for the band gap region. Motivated by the lack of experimental data about the dielectric functions and phase transitions for BLT film, here we focus on the optical properties of BLT nanocrystalline film with different temperatures to gain a more intensive understanding of the microscopic mechanism of phase transitions and novel physics in these materials.

On the other hand, top and bottom electrodes play an important role in optoelectronic devices in order to deal with electrical and/or optical signal. (25; 26; 27; 28; 29) For example, polarization charges, which can be screened by free carriers inside the electrode, are induced at the electrode/FE interface in a FE-based capacitor. As we know, $LaNiO_3$ (LNO) and $La_{0.5}Sr_{0.5}CoO_3$ (LSCO), which are the highly conductive metallic oxides, have a distorted perovskite structure with a cubic lattice parameter of 3.84 and 3.83 Å, respectively. (30; 31; 32) As the matched electrodes for perovskite FE-based devices, LNO and LSCO materials have been widely studied as alternatives for platinum (Pt) and Pt-based metals. (26; 29; 33) It is because these metallic oxides can significantly improve the physical properties of FE devices, as compared with noble metal electrodes. (33; 34) Therefore, understanding the intrinsic physical phenomena occurring within FE film and electrode/FE interface requires accurate knowledge of dielectric constants about the electrode. Owing to an increasing interest of nanostructured FE materials and devices, (35; 36) the physical properties of nanostructured electrodes should be further investigated in order to clarify the functionalities. In spite of the promising properties up to now, (27; 32; 29) there are no reports on dielectric function and optical conductivity of nanostructured LNO and LSCO materials, which can predicatively reflect the electrical transport properties and electronic band structure. (37; 38) It is well known that macroscopical dielectric function and/or optical conductivity can be directly related to the electronic band structure. (39) In particular, the optical and electronic properties could be remarkably different from the bulk crystal for LNO and LSCO materials with a low-dimensional structure. (38) Therefore, it is necessary to clarify the optical response behaviors of the nanostructured oxides.

Fortunately, optical transmittance, reflectance, and spectroscopic ellipsometry techniques, whi ch are the nondestructive and powerful technique to investigate the optical characteristics of materials, can directly provide electronic band energy and dielectric constants, etc. (40; 41; 42; 43) With the aid of the reasonable dielectric function model, one can reproduce the experimental spectra well and the dielectric functions can be easily extracted. (44) This makes it possible to investigate the optical properties of wide band gap FE films and conductive metallic oxide films in a wider photon energy range. In particular, Spectroscopic ellipsometry, which is very sensitive to ultrathin films and surfaces, is a nondestructive and powerful technique to investigate the optical characteristics of materials, and, in particular, to measure the thickness and the dielectric function of a multilayer system simultaneously. (44; 45; 46; 47; 48; 49) As compared to the traditional reflectance and/or transmittance spectroscopy, spectroscopic ellipsometry can determine exactly the ratio of linearly p- and s-polarized light intensity and the phase difference between p- and s-polarized light simultaneously. Therefore, spectroscopic ellipsometry can directly provide the optical constants of materials without Kramers-Krönig (K-K) analysis. (49; 50) The purpose of the present work is to point out that spectral technique is an effective tool to study the structural properties and phase transitions in FE materials and interband transitions in conductive metallic oxides. The results open a new vista for the experimental technique development of observing phase/structural transitions. Moreover, it indicates that the spectral techniques can be applied to study the optical response behavior of perovskite-type FE and conductive metallic oxides.

This chapter is arranged in the following way. In Sec. 2, the detailed growth processes of the BLT, LNO, and LSCO films and spectral setup are described; In Sec. 3, the crystal structures of the BLT, LNO and LSCO films have been presented; In Sec. 4, spectral transmittance, reflectance and ellipsometric theory is expressed; In Sec. 5, optical properties and phase transition characteristics of the BLT films have been discussed. In Sec. 6, the electronic band structures of the LNO and LSCO films on Si substrate are derived in detail; In Sec. 7, the thickness dependence of infrared optical constants in the LNO films on platinized silicon substrates ($Pt/Ti/SiO_2/Si$) is presented; In Sec. 8, the main results are summarized.

2. Experimental

2.1 The fabrications of BLT films

The BLT nanocrystalline film was prepared on quartz substrate by chemical solution deposition technique. Bismuth nitrate, lanthanum acetate, and titanium butoxide were used as starting materials. The BLT precursor solution with a stoichiometric molar ratio of Bi/La/Ti=3.25/0.75/3 was dissolved in heated glacial acetic acid instead of drastic toxicity 2-methoxyethanol, which makes the process of preparation of precursor solutions safer and simpler. Note that excess 8 mol% Bi precursor was added to compensate for Bi evaporation during the annealing process. Then an appropriate amount of acetylacetone was added to the solution as a stabilizing agent. Equimolar amounts of titanium butoxide were also added into solution. The concentration of the precursor solution was 0.05 M. In order to improve the hydrolysis and polymerization, the precursor solution was placed at atmosphere about 10 days before spin coating onto the substrate. The homogeneity and stability of the coating solution was greatly improved by the addition of acetylacetone. The BLT nanocrystalline film was deposited by spin coating of the solution onto quartz substrate at 3500 rpm for 30

s. Each layer was dried at 180 °C for 3 min, then calcined at 350 °C for 3 min. Finally, the samples were annealed at 675 °C for 3 min by rapid thermal annealing. By repeating the coating/calcining cycles about five times, the BLT film with the nominal thickness of about 150 nm can be obtained.

2.2 The growths of conductive metallic oxides

Nanocrystalline LNO and LSCO films were deposited on the single-side polished silicon (Si) wafers by radio frequency magnetron sputtering (RFMS) and pulsed laser deposition (PLD) methods, respectively. A 100-mm-diam LNO target was used for deposition. The distance between the target and Si(111) substrates was 70 mm. Before deposition, the target was pre-sputtered for 30 min with the substrate shutter closed to achieve stable conditions. The vacuum chamber pumped by a turbomolecular pump produced a base pressure of 2×10^{-4} Pa and was raised to 1.6 Pa by admitting argon (Ar) and oxygen (O_2). The Si(111) substrates were mounted with silver paste onto a resistively heated substrate holder. During the sputtering, the temperature of the substrate holder was kept at 265 °C and the radio frequency power on the target was 80 W, operating at 13.56 MHz, yielding a growth rate of about 2-3 nm/min. After deposition, a cooling was carried out in vacuum. Following with the above method, the LNO film with the nominal thickness of about 230 nm can be derived.

For the LSCO film, however, a KrF excimer laser with the pulse frequency of 5 Hz was used for the growth. The oxygen pressure was controlled at 25 Pa during the deposition process. The substrates were cleaned in pure ethanol with an ultrasonic bath to remove physisorbed organic molecules from the Si surfaces. Then the substrates were rinsed several times with de-ionized water. Finally the wafers were dried in a pure nitrogen stream before the deposition of the LSCO films. The ceramic LSCO target with a La:Sr stoichiometric ratio of 0.5:0.5 was sintered by solid state reaction. The substrate temperature was kept at 700 °C. In order to obtain the nominal thickness of about 1 μm, the LSCO layers have been deposited during the several periods.

On the other hand, the sol-gel technique is applied to fabricate the LNO with different thickness on platinized silicon ($Pt/Ti/SiO_2/Si$) substrates. Lanthanum nitrate [$La(NO_3)_3$] and nickel acetate [$Ni(CH_3COO)_2 \bullet 4H_2O$] were used as the start materials, and acetic acid and water were used as the solvents. Nickel acetate was dissolved in acetic acid and equimolar amounts of lanthanum nitrate dissolve in distilled water held at RT, respectively. Then the two solutions were mixed together with constant stirring. The concentration of the precursor solution was adjusted to 0.3 M by adding or distilling some acetic and water. The LNO films were deposited spin coating of the 0.3 M solution at the speed of 4000 rpm for 30 s. Each layer of the films was dried at 160 °C for 5 min, then pyrolyzed at 400 °C for 6 min to remove residual organic compounds, following annealed at 650 °C for 3 min in air by a rapid thermal annealing (RTA) process. The aforementioned coating, pyrolyzing and annealing were repeated different times in order to obtain the LNO films with different thickness.

2.3 The crystalline and optical characterizations

The crystallinity of the BLT, LNO, and LSCO films at RT was examined by XRD using a Ni filtered Cu Kα radiation source (D/MAX-2550V, Rigaku Co.). In the XRD measurement a vertical goniometer (Model RINT2000) was used, and continuous scanning mode (θ-2θ) was selected with a scanning rate of 10°/min and interval of 0.02°. The temperature dependent ultraviolet transmittance spectra were measured by a double beam ultraviolet-infrared spectrophotometer (PerkinElmer UV/VIS Lambda 2S) at the photon energy range of 1.1-6.5 eV

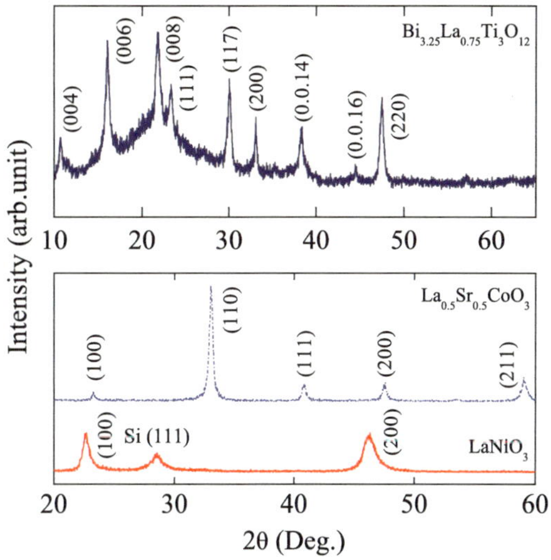

Fig. 1. The XRD patterns of $Bi_{3.25}La_{0.75}Ti_3O_{12}$, LNO and $La_{0.5}Sr_{0.5}CoO_3$ films, respectively.

(190-1100 nm). The BLT film was mounted into an optical cryostat (Optistat CF-V from Oxford Instruments) and the temperature was continuously varied from 80 to 480 K. Near-normal incident optical reflectance spectra ($\sim 8^\circ$) were recorded at room temperature (RT) with a double beam ultraviolet-infrared spectrophotometer (PerkinElmer Lambda 950) at the photon energy from 0.47 to 6.5 eV (190-2650 nm) with a spectral resolution of 2 nm. Aluminum (Al) mirror, whose absolute reflectance was directly measured, was taken as reference for the spectra in the photon energy region. The ellipsometric measurements were carried out at RT by a variable-angle infrared spectroscopic ellipsoetry (IRSE) synchronously rotating polarizer and analyzer. The system operations, including data acquisition and reduction, preamplifier gain control, incident angle, wavelength setting and scanning were fully and automatically controlled by a computer. In this study, the incident angles were 70°, 75° and 80° for the LNO films. Note that no mathematical smoothing has been performed for the experimental data.

3. The crystalline structures

Figure 1 (a) shows the XRD spectra of BLT, LNO, and LSCO nanocrystalline film. For the BLT film, there are the strong diffraction peaks (004), (006), (008), (117) and (220), which confirm that the present BLT nanocrystalline film has the tetragonal phase. Note that the broadening feature near 22° can be ascribed to the quartz substrate. On the other hand, the XRD patterns indicate that the LNO and LSCO films are crystallized with the single perovskite phase. It should be emphasized that the LNO film presents a highly (100)-preferential orientation. Besides the strongest (110) peak, some weaker peaks (100), (111), (200), and (211) appear, indicating that the LSCO film is polycrystalline. Generally, the grain size can be estimated by the well-known Scherrer's equation $r = K\lambda/\beta\cos\theta$, where r is the average grain size, β the full width at half maximum of the diffraction line, λ the x-ray wavelength, θ the Bragg angle, K the Scherrer's constant of the order of unity for usual crystals. For the BLT film on the quartz substrate, the average grain size from the (006) diffraction peak is estimated to about 16 nm. However, the grain size from the (200) and (110) peaks was evaluated to 78 and 27 nm for the LNO and LSCO films on Si substrates, respectively. Note that the grain size of the BLT and LSCO films is much less that that of the LNO film. The striking increment for the LNO film could be due to the better crystallization (i.e., highly preferential orientation). It should be emphasized that the LNO films on the $Pt/Ti/SiO_2/Si$ substrates are of the similar crystalline structure to that from the silicon substrate. Nevertheless, the grain size of the LNO films on the $Pt/Ti/SiO_2/Si$ substrates is estimated to 18.4, 17.5 nm and 27.8 nm for the 100-nm, 131-nm and 177-nm thick films, respectively. It indicates that the average grain size is similar for the LNO films with the thickness of about 200 nm on different substrates.

4. Transmittance, reflectance and SE theory

4.1 Transmittance and reflectance

Generally, transmittance and reflectance spectra can be reproduced by a three-phase layered structure (air/film/substrate) for the film materials with a finite thickness. (40; 41; 42) For the BLT film, the three-phase layered structure configuration is applied due to the thickness of about 150 nm. Nevertheless, it is noted that the nominal growth thickness is about 230 nm and 1 μm for the LNO and LSCO films on Si substrates, respectively. Therefore, considering the light penetration depth in the films due to the optical conductivity, the three-phase layered model and the semi-infinite medium approach are applied to calculate the reflectance spectra of the LNO and LSCO films, respectively. (27; 51) The optical component of each layer is expressed by a 2×2 matrix. Suppose the dielectric function of the film is $\tilde{\varepsilon}$, vacuum is unity, and the substrate is $\tilde{\varepsilon_s}$, respectively. The resultant matrix M_r is described by the following product form

$$M_r = M_{vf} M_f M_{fs}. \tag{1}$$

Here, the interface matrix between vacuum and film has the form

$$M_{vf} = \frac{1}{2\sqrt{\tilde{\varepsilon}}} \begin{bmatrix} (\sqrt{\tilde{\varepsilon}}+1) & (\sqrt{\tilde{\varepsilon}}-1) \\ (\sqrt{\tilde{\varepsilon}}-1) & (\sqrt{\tilde{\varepsilon}}+1) \end{bmatrix}. \tag{2}$$

and the propagation matrix for the film with thickness is described by the equation

$$M_f = \begin{bmatrix} \exp(i2\pi\sqrt{\tilde{\varepsilon}}d/\lambda) & 0 \\ 0 & \exp(-i2\pi\sqrt{\tilde{\varepsilon}}d/\lambda) \end{bmatrix}. \tag{3}$$

where λ is the incident wavelength, and correspondingly the interface matrix between film and substrate is

$$M_{fs} = \frac{1}{2\sqrt{\tilde{\varepsilon_s}}} \begin{bmatrix} (\sqrt{\tilde{\varepsilon_s}}+\sqrt{\tilde{\varepsilon}}) & (\sqrt{\tilde{\varepsilon_s}}-\sqrt{\tilde{\varepsilon}}) \\ (\sqrt{\tilde{\varepsilon_s}}-\sqrt{\tilde{\varepsilon}}) & (\sqrt{\tilde{\varepsilon_s}}+\sqrt{\tilde{\varepsilon}}) \end{bmatrix}. \tag{4}$$

thus, the transmittance T and reflectance R can be readily obtained from

$$T = \mathrm{Real}(\sqrt{\tilde{\varepsilon}}) \left| \frac{1}{M_{r1,1}} \right|^2, \quad R = \left| \frac{M_{r1,0}}{M_{r1,1}} \right|^2. \tag{5}$$

The multi-reflections from substrate are not considered in Eq (5). It should be emphasized that the absorption from the substrate must be taken into account to calculate the transmittance of the film-substrate system. (44)

The physical expression of the real and imaginary parts of the dielectric functions for semiconductor and insulator materials has been reported by Adachi. (52) The Adachi's model dielectric function is based upon the one electron interband transition approach, and relies upon the parabolic band approximation assuming energy independent momentum matrix elements. In our present spectral range, the dielectric functions of the nanocrystalline film are primarily ascribed to the fundamental optical transition. Moreover, for wide band gap FE materials, the dielectric response can be described by the contribution from typical critical point (CP). Therefore, the Adachi's model is employed to express the unknown dielectric functions of the BLT nanocrystalline film, and written as

$$\tilde{\varepsilon}(E) = \varepsilon_\infty + \frac{A_0\left[2 - \sqrt{1+\chi_0(E)} - \sqrt{1-\chi_0(E)}\right]}{\sqrt{E_0^3}\chi_0(E)^2}. \tag{6}$$

Here, $\chi_0 = (E + i\Gamma)/E_0$, A_0 and Γ are the strength and broadening values of the E_0 transition, respectively. (52) It should be emphasized that the dielectric functions of many semiconductor and dielectric materials have been successfully determined by fitting the Adachi's model to the measured data. (53; 54; 55; 56; 57; 58) On the other hand, the dielectric function of the metallic oxide films can be expressed using a Drude-Lorentz oscillator dispersion relation owing to the conductivity

$$\varepsilon(E) = \varepsilon_\infty - \frac{A_D}{E^2 + iEB_D} + \sum_{j=1}^{4} \frac{A_j}{E_j^2 - E^2 - iEB_j}. \tag{7}$$

Here ε_∞ is the high-frequency dielectric constant, A_j, E_j, B_j, and E is the amplitude, center energy, broadening of the jth oscillator, and the incident photon energy, respectively. (27; 29; 59) The refractive index n and extinction coefficient k can be calculated as follows

$$n = \frac{1}{\sqrt{2}}\sqrt{\sqrt{\varepsilon_1{}^2 + \varepsilon_2{}^2} + \varepsilon_1}, \quad k = \frac{1}{\sqrt{2}}\sqrt{\sqrt{\varepsilon_1{}^2 + \varepsilon_2{}^2} - \varepsilon_1}. \tag{8}$$

where ε_1 and ε_2 are the real part and imaginary part of the dielectric function, respectively. The root-mean-square fractional error σ, defined by

$$\sigma^2 = \frac{1}{N-M}\sum_{i=1}^{N}(T(R)_i^m - T(R)_i^c)^2. \tag{9}$$

where $T(R)_i^c$ and $T(R)_i^m$ are the calculated and measured values at the ith data point for the transmittance or reflectance spectra. N is the number of wavelength values and M is the number of free parameters. (11; 60; 61) A least squares-fitting procedure employing the modified Levenberg-Marquardt algorithm, the convergence of which is faster than that of the SIMPLEX algorithm, was used in the transmittance or reflectance spectral fitting.

4.2 SE technique

On other hand, SE, based on the reflectance configuration, provides a effective tool to extract simultaneously thickness and optical constants of a multilayer system. (48) It is a sensitive and nondestructive optical method that measures the relative changes in the amplitude and the phase of particular directions polarized lights upon oblique reflection from the sample surface. The experimental quantities measured by ellipsometry are the complex ratio $\tilde{\rho}(E)$ in terms of the angles $\Psi(E)$ and $\Delta(E)$, which are related to the structure and optical characterization of materials and defined as

$$\tilde{\rho}(E) \equiv \frac{\tilde{r}_p(E)}{\tilde{r}_s(E)} = \tan\Psi(E)e^{i\Delta(E)}. \tag{10}$$

here, $\tilde{r}_p(E)$ and $\tilde{r}_s(E)$ is the complex reflection coefficient of the light polarized parallel and perpendicular to the incident plane, respectively. (50) It should be noted that $\tilde{\rho}(E)$ is the function of the incident angle, the photon energy E, film thickness and optical constants $\tilde{n}(E)$, i.e., the refractive index n and extinction coefficient κ from the system studied. Generally, the pseudodielectric function $< \tilde{\varepsilon} >$ is a useful representation of the ellipsometric data Ψ and Δ by a two-phase (ambient/substrate) model (50)

$$< \tilde{\varepsilon} > = < \varepsilon_1 > + i < \varepsilon_2 > = \sin^2\varphi \left\{ 1 + \left[\frac{1-\tilde{\rho}}{1+\tilde{\rho}} \right]^2 \tan^2\varphi \right\}. \tag{11}$$

where φ is the incident angle. Although $\tilde{\rho}(E)$ and $\tilde{n}(E)$ may be transformed, there are no corresponding expressions for $\tilde{n}(E)$, which is distinct for different materials. (62) Therefore, the spectral dependencies of $\Psi(E)$ and $\Delta(E)$ have to be analyzed using an appropriate fitting model. Correspondingly, the film thickness d_f, optical constants and other basic physical parameters, such as optical band gap, the high frequency dielectric constant ε_∞, etc., can be extracted from the best fit between the experimental and fitted spectra. Similarly, the three-phase layered structure model cab be applied to reproduce the SE spectra of the LNO films on the $Pt/Ti/SiO_2/Si$ substrates due to a smaller thickness (the maximum value is about 180 nm). In SE fitting, the root-mean-square fractional error, which is defined as

$$\sigma^2 = \frac{1}{2J-K} \sum_{i=1}^{J} \left[\left(\frac{\Psi_i^m - \Psi_i^c}{\sigma_{\Psi,i}^m} \right)^2 + \left(\frac{\Delta_i^m - \Delta_i^c}{\sigma_{\Delta,i}^m} \right)^2 \right]. \tag{12}$$

has been used to judge the quality of the fit between the measured and model data. (11) Where, J is the number of data points and K is the number of unknown model parameters, has been used to judge the quality of the fit between the measured and model data. Note that Eq. 12 has $2J$ in the pre-factor because there are two measured values included in the calculation for each and pair. The standard deviations were calculated from the known error bars on the calibration parameters and the fluctuations of the measured data over averaged cycles of the rotating polarizer and analyzer. Note that the same Levenberg-Marquardt algorithm as the above T or R data was applied to reproduce SE spectra. Therefore, the infrared optical constants of the LNO films can be uniquely extracted.

5. Optical properties of the BLT films

Typical transmittance spectra of the BLT films at different temperatures are shown in Fig. 2 (a). The spectra can be roughly separated into three specific regions: a transparent oscillating one (labeled with "I"), a low transmittance one ("II"), and a strong absorption one ("III") at higher photon energies. The Fabry-Pérot interference behavior observed in the transparent region, which is due to the multi-reflectance between the film and substrate, is similar to those on silicon substrates. (10; 63) The transmittance differences with the temperature are ascribed to the thermal expansion and variations of optical constants in lower photon energies. Transmittance spectra ranging from 80 to 480 K were measured in order to analyze the shift of the absorption edge during the phase transition. As shown in Fig. 3 (a), the spectral transmittance value sharply decreases with the photon energy and down to zero in the ultraviolet region beyond 4.5 eV. This is due to the strong absorption from the fundamental band gap and high-energy CP transitions, which can not be detected by the present transmittance spectra.

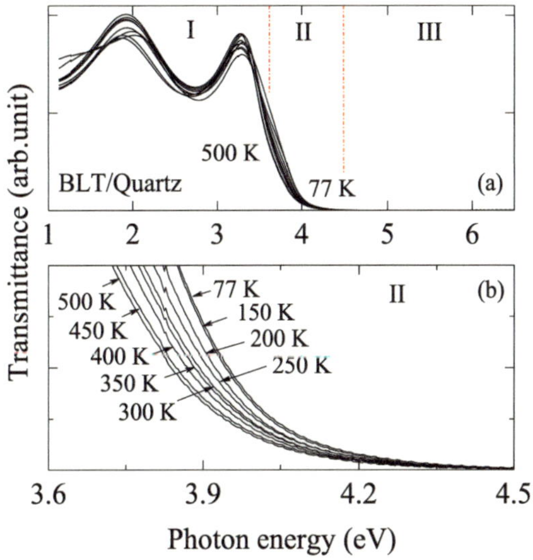

Fig. 2. Transmittance spectra of the BLT film at the temperature region from 77 to 500 K: (a) in the ultraviolet-infrared photon energy of 1.1-6.5 eV, and (b) near the OBG region of 3.6-4.5 eV. The arrows indicate the corresponding temperature values. (Figure reproduced with permission from (63). Copyright 2007, American Institute of Physics.)

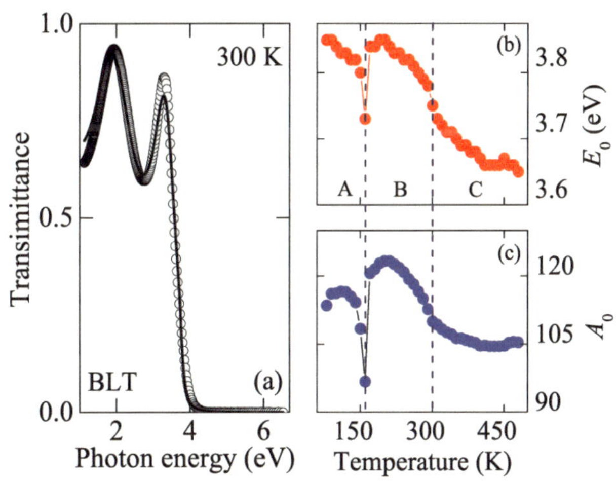

Fig. 3. (a) Experimental (dotted line) and fitting (solid line) transmittance spectra of the BLT film at 300 K. (b) The optical transition energy E_0 of the Adachi's model for the BLT film with different temperatures. It indicates that the optical transition energy E_0 decreases with increasing the temperature. Nevertheless, the anomalous behavior can be observed around 160 and 300 K. (c) The strength parameter A_0 of the E_0 transition with different temperatures.

Temperature (K)	A_0 ($eV^{3/2}$)	E_0 (eV)	Γ (eV)	S	E_c (eV)	E_d (eV)	$n(0)$
80	113.5	3.85	0.08	3.18	4.71	7.48	2.18
90	116.1	3.85	0.06	3.25	4.68	7.51	2.20
100	116.2	3.83	0.05	3.28	4.66	7.47	2.21
110	116.6	3.84	0.05	3.29	4.65	7.46	2.21
120	116.5	3.83	0.05	3.30	4.64	7.43	2.21
130	116.2	3.83	0.05	3.30	4.62	7.31	2.20
140	114.2	3.82	0.04	3.28	4.62	7.21	2.20
150	108.4	3.80	0.05	3.20	4.58	6.76	2.17
160	96.7	3.73	0.05	3.10	4.43	5.66	2.09
170	120.6	3.84	0.07	3.34	4.68	7.87	2.24
180	121.5	3.84	0.07	3.35	4.68	7.91	2.25
190	122.8	3.85	0.08	3.37	4.69	8.03	2.25
200	123.3	3.85	0.08	3.38	4.70	8.09	2.26
210	123.3	3.84	0.07	3.40	4.68	8.02	2.26
220	122.6	3.83	0.07	3.40	4.67	7.95	2.26
230	121.8	3.83	0.07	3.39	4.66	7.89	2.25
240	120.8	3.82	0.08	3.38	4.65	7.81	2.25
250	119.3	3.82	0.08	3.36	4.64	7.68	2.24
260	118.2	3.81	0.07	3.36	4.61	7.53	2.23
270	116.6	3.80	0.07	3.35	4.60	7.39	2.23
280	115.0	3.79	0.07	3.34	4.58	7.23	2.22
290	112.7	3.78	0.07	3.32	4.55	7.00	2.20
300	110.0	3.75	0.05	3.33	4.49	6.62	2.19
310	109.3	3.73	0.04	3.37	4.44	6.41	2.19
320	108.3	3.72	0.03	3.39	4.41	6.21	2.19
330	107.6	3.71	0.03	3.40	4.38	6.09	2.19
340	107.2	3.71	0.03	3.41	4.36	5.99	2.19
350	106.3	3.70	0.03	3.41	4.35	5.88	2.18
360	106.3	3.69	0.02	3.44	4.33	5.80	2.19
370	105.9	3.69	0.02	3.45	4.31	5.71	2.19
380	105.6	3.68	0.02	3.46	4.29	5.63	2.18
390	105.3	3.68	0.02	3.38	4.28	5.53	2.18
400	104.6	3.67	0.02	3.48	4.27	5.45	2.18
410	104.7	3.66	0.02	3.50	4.25	5.37	2.18
420	104.5	3.66	0.02	3.52	4.23	5.30	2.18
430	104.5	3.66	0.02	3.52	4.23	5.30	2.18
440	104.5	3.66	0.02	3.52	4.23	5.31	2.19
450	104.6	3.67	0.02	3.50	4.25	5.41	2.18
460	105.3	3.66	0.04	3.50	4.26	5.51	2.19
470	105.5	3.66	0.04	3.51	4.25	5.51	2.19
480	105.4	3.66	0.04	3.51	4.25	5.51	2.19

Table 1. The Adachi's model parameters of the BLT film at the temperature range of 80-480 K are determined from the simulation of the transmittance spectra. Note that the thickness is estimated to 145 nm and the ε_∞ is fixed to 1 for the film measured at different temperature. In addition, S, E_c, E_d, and $n(0)$ is the contribution from high-energy CP transitions, the oscillator energy, the dispersion energy of the Sellmeier relation, and long wavelength refractive index. respectively.

The fitted parameter values in Eq. 6 are summarized in Table 1 and Fig. 3 (b) and (c). A good agreement is obtained between the measured and calculated data in the experimental range, especially for the fundamental band gap region. For all temperatures, the fitting standard deviations are less than 2×10^{-3}. As can be seen in Fig. 3 (b), the optical transition energy E_0 of the BLT film increases with decreasing temperature except for the temperatures of 160 and 300 K. Fig. 3 (b) can be roughly separated into three specific regions: the monoclinic phase (labeled with "A"), the orthorhombi phase (labeled with "B") and the tetragonal phase (labeled with "C"). An obvious dip at 160 K for the fundamental band gap can be observed due to the phase/structural transition. (20) The origin of the structural anomaly is due to strain energy and lattice strain change with the temperature. The variation of E_0 with temperature can be mainly ascribed to thermal expansion and electron-phonon (*e-p*) interaction. The BLT film emits or absorbs the phonon with increasing temperature. It will result in the band gap perturbation and shift. The valence band top mainly consists of the O_{2p} orbital, which is strongly hybridized with the Ti_{3d} orbital below the Fermi level. (13) The strong hybridization between the O_{2p} and Ti_{3d} orbitals changes with the distortion of the crystal structure during the phase transition. Therefore, the E_0 variation clearly indicates that the absorption edge is strongly related to the phase transition with decreasing temperature.

It is generally believed that the symmetry of BiT at RT is orthorhombic and changes to tetragonal at 675 °C. (64) As we know, physical properties of FE nanocrystals are strongly dependent on the grain size. (65) Recently, a size-driven phase transition was found at a critical size r_c of 44 nm for BiT. The high-temperature tetragonal phase stabilizes at RT when the grain size is smaller than the r_c. (17) According to the XRD analysis, the BLT film has the tetragonal crystal structure at RT. Various anomalies of BiT at low temperature are believed to be a possible indication of the structural transitions. For instance, the dielectric constant decreases with decreasing temperature except for a broad hump around about 150 K, the spontaneous polarization (P_s) changes little but for a gradual decrease in the 100-250 K range with some thermal hysteresis and the E_c gradually decreases to a minimum at 100 K except for the broad hump about 200 K. (20) The similar phenomena from the optical properties can be observed in the present BLT nanocrystalline film. Note that the parameter A_0 of the Adachi's model is anomalous at 160 and 300 K. Moreover, the E_0 continuously decreases with increasing temperature except for the values at the temperatures of 160 and 300 K. The dielectric function anomaly indicates that the BLT nanocrystalline film undergoes a tetragonal to orthorhombic phase transition in the temperature range of 200-250 K. (21) Thermodynamic analysis indicate that the energy separation of the orthorhombic and monoclinic states decreases with decreasing the temperature. (66) When the temperature further decreases, the BLT nanocrystalline film with orthorhombic structure undergoes monoclinic distortion around 160 K. (21) From the dielectric function model, one can safely conclude that the low temperature phase transition of the BLT film can be detected by the spectral transmittance.

The calculated dielectric functions are exhibited in Fig. 4. The real part ε_1 increases and reaches a maximum, beyond which it gradually falls with further increasing of the photon energy. The peak position of ε_1 corresponding approximately with the optical transition energy E_0 of the BLT film shifts to high energies with decreasing the temperature. The peaks may be assigned to the transitions between the CP or lines with high symmetry in the Brillouin zone, termed as Van Hove singularities. In the present case, the value of the photon energy corresponding to the peaks of ε_1 is about 3.7 eV, which agrees with the value of the FE band gap (3-4 eV). (67). Therefore, the CP may be associated with the interband transition between the valence and conduction bands of BLT nanocrystalline film. The sites of the dielectric

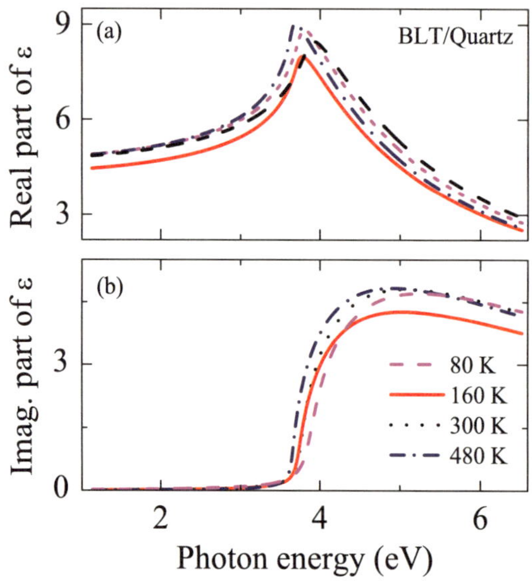

Fig. 4. Temperature dependence of dielectric function (a) real part and (b) imaginary part for the BLT nanocrystalline film in the photon energy range of 1.1-6.5 eV. The anomalies of dielectric function are observed at 160 and 300 K, respectively.

function or the refractive index peaks generally correspond to the band gap energies of the dielectrics and semiconductors. (68; 69) This can be understood from the K-K relation (70; 71)

$$n(v) = 1 + \frac{c}{\pi}\int \frac{d\alpha(v')}{dv'} \log(\frac{v'+v}{v'-v})dv'. \tag{13}$$

where c is the light velocity in vacuum, v and v' are light frequencies, $\alpha(v')$ is the absorption coefficient. In the frequency region near the absorption edge the value of $d\alpha(v')/dv'$ is very large; however, when the photon energy reaches the gap energy, the absorption curve changes its slope and becomes smoother, so the value of $d\alpha(v')/dv'$ decreases at the absorption edge. The imaginary part ε_2 is nearly zero in the interference region and increases rapidly with E, reaches a maximum, and falls slightly at higher energies. The ε_2 near the fundamental band gap energy is not zero due to defects and disorder in the nanocrystalline film. Moreover, the real part of ε decreases with the temperature and varies from 9.04 at 480 K to 8.22 at 80 K at the photon energy of 3.7 eV. The reduction is due to decreasing electron-phonon interactions at degraded temperatures, making direct transitions less probable. From Fig. 4, an anomaly of the dielectric function is observed at 160 and 300 K, which can be ascribed to the phase/strucutural transitions. Fig. 5 exhibits the temperature-dependent refractive index n at various photon energies. The anomalous variations occur at the temperature rang of 130-160 K and 290-330 K, indicating the subtle phase transitions in the corresponding temperature regions, as compared with lower-temperature Raman results. (20; 21) As previously discussed, the optical constants are strongly related to the crystalline structure, which can affect the valence and conduction band formation. Obviously, the structural variation of the BLT film can contribute to the optical response.

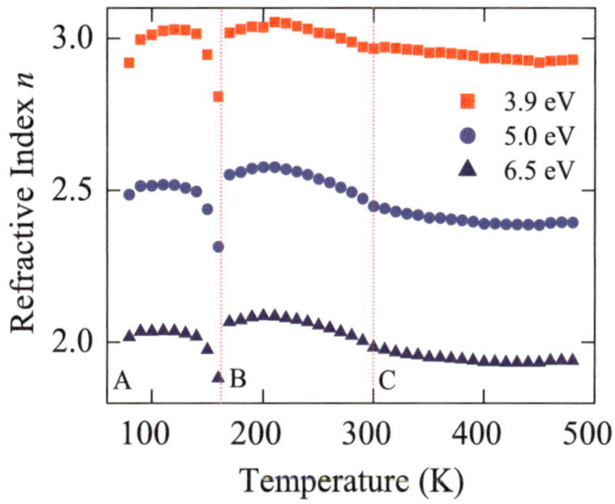

Fig. 5. Variation of the refractive index for the BLT nanocrystalline film with the temperature at the photon energies of 3.9, 5, and 6 eV, respectively.

In order to give an insight on the electronic band structure of the BLT film, we have fitted the refractive index n below the optical transition energy with a dispersion formula corresponding to an empirical Sellmeier equation

$$n^2 = S + \frac{E_c E_d}{E_c^2 - E^2}. \tag{14}$$

Here, S is the contribution from high-energy CP transitions, E_c the oscillator energy, and E_d the dispersion energy. The fitting quality from the BLT film at 80 K has been illustrated in Fig. 6 (a). The empirical Sellmeier equation gives a relatively good description to the optical

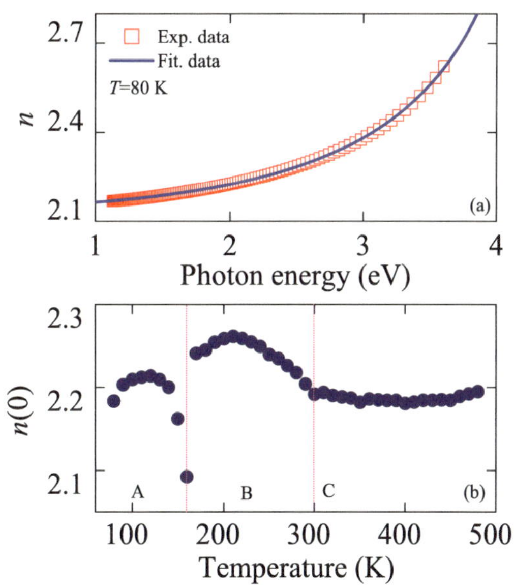

Fig. 6. (a) An empirical Sellmeier equation in the transparent region for BLT nanocrystalline film at 80 K. (b) The long wavelength refractive index $n(0)$ with different temperatures. The anomalous $n(0)$ values occur at 160 and 300 K, respectively.

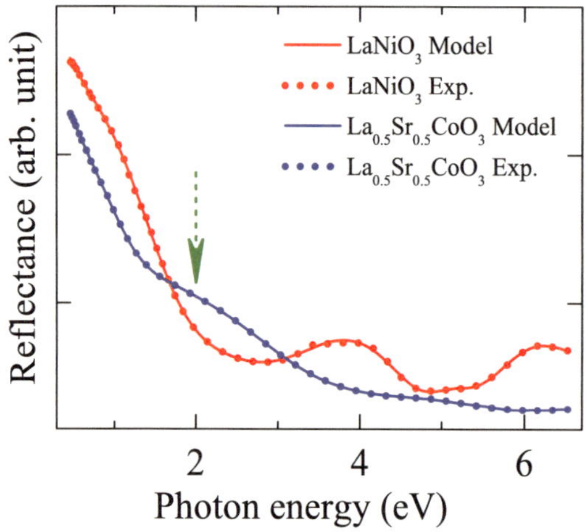

Fig. 7. Experimental (dotted lines) and best-fit (solid lines) near-normal incident reflectance spectra of LNO and $La_{0.5}Sr_{0.5}CoO_3$ films. The arrow is applied to approximately distinguish two different electronic transition regions. (Figure reproduced with permission from (25). Copyright 2009, American Institute of Physics.)

dispersion in the transparent range. The parameters of Sellmeier equation are summarized in Table 1. The S value generally increases with the temperature except for the phase transitions, indicating that the effects from high-frequency electronic transitions becomes stronger in the BLT film. It can be found that the maximum optical transition occurs near the energy range of 4.2-4.7 eV from the fitted oscillator energy, which agrees well with ε_2 observed (see Fig. 4). It indicates that the optical dispersion in the transparent region is mainly ascribed to the higher CP virtual transitions and not by the band gap energy. Note that the parameter S presents an opposite variation trend with the temperature, as compared with the oscillator energy E_c and the dispersion energy E_d apart from the temperature ranges around 160 and 300 K. Again, the anomaly can be ascribed to the phase transitions as previously discussed. The long wavelength refractive index $n(0)$, which can be calculated from $\sqrt{(E_d/E_c)+S}$, is shown in Fig. 6 (b). The $n(0)$ at 300 K is estimated to be about 2.19 at zero point, which indicates that the dielectric function is about 4.80. The $n(0)$ is related to the total effective number of valence electrons per atom in materials. For theoretically calculating $n(0)$, the f-sum-rule integral can be written as

$$n^2(0) - 1 = \frac{4}{\pi}\int_0^\infty \frac{n(\omega')k(\omega')}{\omega'}d\omega'. \tag{15}$$

It indicates that the long wavelength refractive index is smaller for materials with wider fundamental band gap because there is lower energy transition possibility, which results in less contributions to the imaginary part of dielectric function $[n(w)k(w)]$. In addition, the band gap of the present BLT nanocrystalline film could be larger due to the crystalline size effect, as compared to that of bulk crystal. (72; 73) Thus, the $n(0)$ will be further decreased. Although the phase transitions of the BLT film can be controlled by the crystalline/grain effect, the present transmittance spectra at variable temperature further indicate that the subtle phase transitions appear in the nanocrystalline film structure.

Samples	Lorentz (j)	A_j (eV)^2	E_j (eV)	B_j (eV)
LNO	1	1.97 (0.09)	2.32 (0.02)	2.28 (0.05)
	2	2.0 (0.1)	3.36 (0.01)	2.16 (0.06)
	3	0.19 (0.01)	5.10 (0.02)	0.9 (0.1)
	4	2.69 (1.07)	7.31 (0.41)	3.29 (0.93)
$La_{0.5}Sr_{0.5}CoO_3$	1	0.33 (0.06)	1.55 (0.01)	1.22 (0.06)
	2	2.31 (0.06)	2.11 (0.04)	3.06 (0.06)
	3	1.57 (0.07)	4.59 (0.01)	3.03 (0.07)
	4	0.10 (0.02)	6.01 (0.03)	0.73 (0.10)

Table 2. The Lorentz oscillator parameter values of LNO and $La_{0.5}Sr_{0.5}CoO_3$ films are determined from the simulation of reflectance spectra. The 90% reliability of the fitting parameters is given with the parentheses. (Table reproduced with permission from (25). Copyright 2009, American Institute of Physics.)

6. Electronic band structures of the LNO and LSCO films

The experimental reflectance spectra of the LNO and LSCO films are shown in Fig. 7 with the dotted lines. Generally, the reflectance spectra recorded show the similar dependence on the photon energy except for the amplitude, which is slightly higher in the LNO film. Note that the reflectance spectrum from the LSCO film is similar to that of the film grown on MgO substrate, where a slight dip appears from optical reflectance measurement. (74) The reflectance spectra can be roughly divided into two regions (see the dotted line in Fig. 7). The lower photon energy region below about 2 eV is assigned to a strong Drude response, which is derived from the intraband transition of free carrier. It was reported that the typical energy scale of the Jahn-Teller (JT) related excitations are varied between 0.5 and 3 eV. (75) It indicates that the Ni^{3+} and Co^{3+} ions related JT effect is important in the perovskite-type materials, whose lattice distortion can couple via an orbital exciton strongly in the $3d$ correlated metallic systems. (37) The results are similar to some high temperature superconductors. (75) The charge-transfer excitations are located above 2.0 eV for the LNO and LSCO films. The interband electronic transitions are mainly derived from O $2p$ and Ni/Co $3d$ energy states. (38) It was argued that the crystal field splitting and the JT effect in the metallic oxides can affect the charge-transfer transitions. (75)

In order to clarify the different transition energy bands, the Lorentz oscillator model expressing the respective contribution to the optical response is necessary. The reproduced optical reflectance results with the Drude-Lorentz model are also shown in Fig. 7 by the solid lines and the parameter values are given in Table 2. A good agreement is obtained between the experimental and calculated spectra in the entirely measured photon energy region. In particular, the oscillator broadening are generally less than the corresponding E_j values (see Table 2), indicating that the Drude-Lorentz dispersion is reliable and satisfying, similar to the results from some metal films. (28) It is testified that four Lorentz oscillators are requisite for the nanocrystalline films, as compared with some noble metal films. (28) These Lorentz oscillators can correspond to different interband electronic transitions, respectively. For the LNO material, the optical transition peaks are located at 2.32 ± 0.02, 3.36 ± 0.01, 5.10 ± 0.02, and 7.31 ± 0.41 eV. These center energy positions can be readily distinguished from the reflectance spectrum. For the LSCO film, however, the four center energy peaks are found at 1.55 ± 0.01, 2.11 ± 0.04, 4.59 ± 0.01, and 6.01 ± 0.03 eV, respectively. Compared with the result of the LNO film, these transition energy positions are overlapped on the reflectance

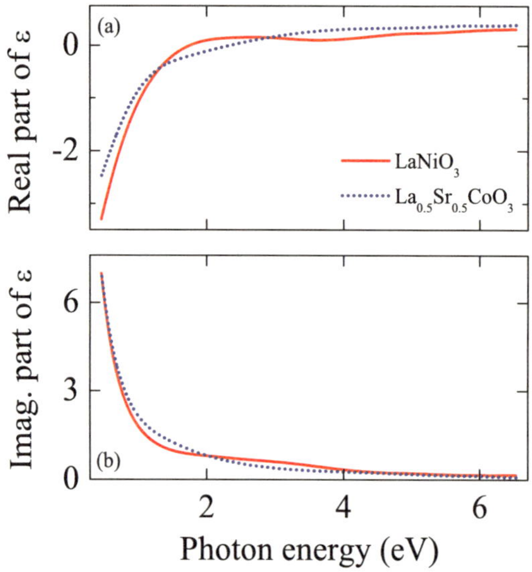

Fig. 8. The real part (a) and imaginary part (b) of dielectric function in $LaNiO_3$ and $La_{0.5}Sr_{0.5}CoO_3$ films from infrared to ultraviolet photon energy regions. (Figure reproduced with permission from (25). Copyright 2009, American Institute of Physics.)

spectrum due to the large broadening. As compared with the theoretical model of the band structure for the metallic oxides, (75; 76; 77) the four energy bands can be assigned to the following electronic transitions: (1) O 2p to Ni/Co 3d (t_{2g}-JT); (2) O 2p to Ni/Co 3d (t_{2g}+JT); (3) O 2p to Ni/Co 3d (e_g-JT); and (4) O 2p to Ni/Co 3d (e_g+JT), respectively. The results are in good agreement with the data from the photoemission spectroscopy, in which four prominent band structures can be observed above the Fermi level (E_F). (38) Based on the Drude model parameters, it can be found that the plasma frequency of the LSCO film (A_D is 3.47 $\pm$ 0.04 eV) is slightly larger than that from the LNO film (A_D is 3.34 $\pm$ 0.03 eV). Note that the parameter B_D is 0.74 $\pm$ 0.01 eV and 0.84 $\pm$ 0.01 eV for the LNO and LSCO films, respectively. The phenomena can be confirmed from the reflectance spectra.

From the model parameters, the dielectric constants can be readily obtained. The evolution of $\tilde{\varepsilon}(E)$ with the photon energy for the LNO and LSCO films is shown in Fig. 8 (a) and (b). Generally, the real part ε_r increases with the photon energy and the values for LNO and LSCO films are evaluated to be about -3.3 and -2.5 at 0.47 eV, respectively. Note that the ε_r of the LSCO film is slightly larger than that of the LNO film in the mid-infrared region. According to the Drude model, the free carrier results in the fact that the dielectric constant decreases to a high negative value as photon energy approaches zero. (28) It indicates that the contribution from the Drude response becomes more prominent for the LSCO material, which agrees well with the above analysis on the plasma frequency. On the other hand, the imaginary part ε_i strikingly decreases with the photon energy and approaches zero towards the ultraviolet energy region, indicating the contributions from the strong interband transitions. Note that the ε_i is slightly smaller that those of nobel metal films. The ε_i discrepancy between the LNO and LSCO films is not obvious, as compared with the ε_r. It can be attributed to the fact that they have the similar perovskite crystalline structure.

In order to give a further insight on the electronic structure of the LNO and LSCO films, the real part of optical conductivity σ can be calculated by $\sigma_r = \varepsilon_0 \omega \varepsilon_i$, here ε_0 and ω is the

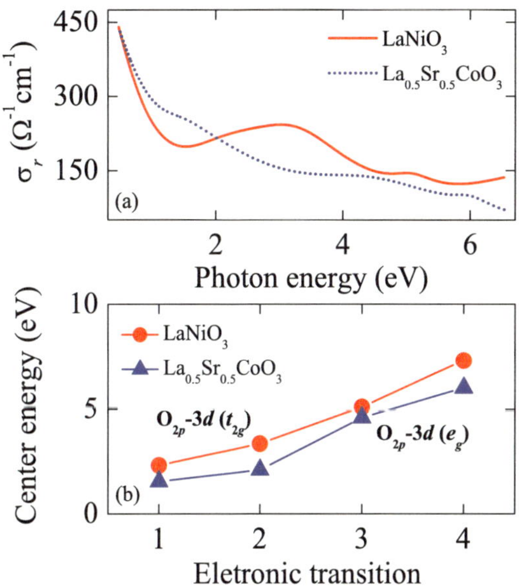

Fig. 9. The comparison of the optical conductivity between two films (a) and (b) the Lorentz oscillator E_j value corresponds to the electronic transitions from O $2p$ to Ni/Co $3d$, respectively. (Figure reproduced with permission from (25). Copyright 2009, American Institute of Physics.)

vacuum dielectric constant and the light frequency, respectively. Fig. 9 (a) gives a comparison of the σ_r for the LNO and LSCO films. Although the σ_r is approximately varied from 100 to 450 Ω^{-1}cm^{-1}, the optical conductivity presents a different behavior for the perovskite-type oxide materials. In the mid-infrared region, the σ_r rapidly increases with decreasing photon energy because of the contribution from the Drude response. Then the optical conductivity rapidly decreases with further increasing the photon energy. From the visible to ultraviolet energy regions, the absorption peaks owing to the interband electronic transitions can be easily distinguished for the LNO film, compared with the LSCO film. Note that the conductive behavior for the two samples is similar to those of high-temperature superconductivity. (74) Fig. 9 (b) shows the four energy bands for the charge-transfer excitations. Considering the corresponding electronic structure, the transition energy of the LNO film is slightly larger than those of the LSCO film. The small discrepancy is mainly due to Ni or Co $3d$ electronic states for the LNO and LSCO materials, respectively. Owing to the distinct $3d$ orbital energy, the t_{2g} and e_g states can be located at different level in the energy space, which induces the center energy variation for two perovskite-type oxides. Note that the La and/or Sr composition ratio can affect the O_{2p} and $3d$ orbital distributions as well. (75) It was reported that the structural distortions can affect the electronic band structures of the perovskite material. (77; 63) Therefore, the highly preferential orientation LNO and polycrystalline LSCO layers can contribute to the changed electronic structure. In addition, the LNO and LSCO films have different grain size, which results in the grain boundary discrepancy and maybe further affect the electronic state.

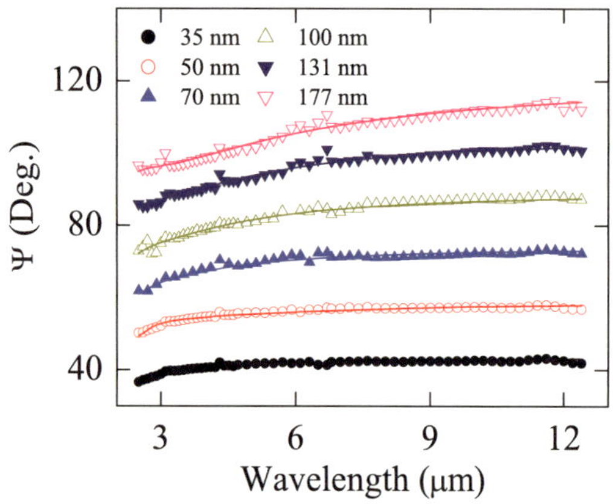

Fig. 10. Experimental (dotted lines) and fitted ellipsometric (solid lines) Ψ spectra of the LNO films with different film thickness at the incident angle of 75°. (Figure reproduced with permission from (47). Copyright 2003, The Japan Society of Applied Physics.)

7. Infrared optical constants of the LNO films

As an example, the measured Ψ and Δ spectra at the incident angle of 75° for the LNO films on $Pt/Ti/SiO_2/Si$ substrates in the wavelength range of 2.5-12.5 μm are shown by the dotted lines in Figs. 10 and 11, respectively. In order to estimate the infrared optical constants of the LNO films, the ellipsometric spectra were analyzed by a three-phase (Air/LNO/Pt) model system as the Pt layers were thick enough that the infrared light could not propagate through them. (50) The unknown infrared optical constants of the LNO films were described using the Drude model, i.e., the first term of Eq. 7. The optical constants of Pt in the fitting were taken from Ref. (62). The infrared optical constants of the LNO films were determined by fitting the model function to the measured data. Note that the fitted film thicknesses are in good agreement with the measured values by scanning electron microscopy within the experimental error bars. The Ψ and Δ spectra are shown in Figs. 10 and 11 by the solid lines, respectively. The best fit is obtained between the measured data and the model fit in the entirely measured wavelength range.

The evaluated optical constants n and κ of the LNO films are shown in Fig. 12 (a) and (b), respectively. The refractive index and extinction coefficient of the LNO films increase with increasing the wavelength, which indicates that $LaNiO_3$ is of some metallic characteristics. In addition, the refractive index of the LNO films decreases with increasing the thickness in the entirely measured wavelength range except for two thinner samples in the short wavelengths. However, the extinction coefficient of the LNO films monotonously decreases with increasing the thickness in the entirely measured wavelength range. In particular, the difference of the infrared optical constants decreases with increasing the thickness. It indicates that the optical properties of films are close to those of bulk materials with increasing the thickness. The value of the refractive index decreases from 10.4 to 1.41, and the extinction coefficient decreases from 14.5 to 1.15 with increasing the thickness in the wavelength of 12.5 μm. It is believed that the crystallinity of the films results in the thickness dependence of the infrared optical properties. From the XRD patterns of the LNO films (not shown), the crystallinity of the LNO films is improved with increasing the thickness. (47) It is possible that the thinner films contained a small fraction of amorphous phase, which remained undetected by XRD. To explain this deviation, the assumption that the change in the structure and texture sensitively reflects the refractive index is presented. In this case, the shift of the optical constants should be limited by the difference between the optical constants for the crystalline and amorphous phases, leading to a large shift of the infrared optical properties. (78; 79) The XRD measurements show that the full width at half maximum (FWHM) of the (200) peak in the LNO films are 0.341°, 0.358° and 0.346° for the 100-nm, 131-nm and 177-nm thick films, respectively. However, the peak positions do not shift with decreasing thickness except for the 177-nm thick films, the (200) diffraction peak of which shifts the smaller angle side.

The interaction between the grain boundaries and/or morphologies and the infrared light increases with increasing grain size, which can be changed as the thickness. It can induce the difference of the optical constants with increasing the thickness. On the other hand, the influence of the interface layer may be responsible for the observed deviation. Although there is the sharp interface feature between LNO and Pt substrates, the stress may lie in the interface because of the difference from the coefficient of thermal expansion between LNO and Pt. The stress or strain can affect not only the crystallinity of the LNO films, but also the infrared optical properties. Because the LNO films are semi-transparent in the measured wavelength region, the reflected light has the interaction to the interface layer. With increasing thickness, the effect of the interface layer could increase and induce the change of the detected light, and finally result in the difference of the infrared optical properties. Therefore, we believe that the difference of the infrared optical properties is due mainly to the crystalline size effect, the influence of the interface layer and existence of some amorphous phases in the films. In addition, the IRSE data do not reveal sensitivity to surface roughness, because scattering becomes inefficient for long wavelengths. (80) The change of the surface roughness with different film thickness can affect the infrared optical properties of the LNO thin films because the influence (such as lower refractive index and extinction coefficient) of the surface roughness has been added in the optical properties. Therefore, the LNO films approach the bulk behavior and show the metallic properties with increasing the thickness from the IRSE experiments.

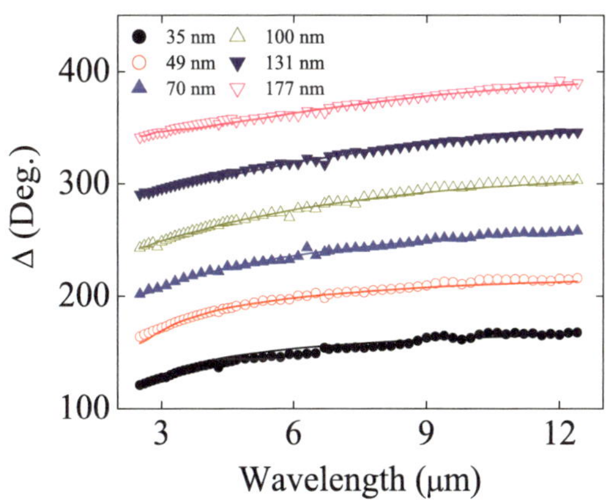

Fig. 11. Experimental (dotted lines) and fitted ellipsometric (solid lines) Δ spectra of the LNO films with different film thickness at the incident angle of 75°. (Figure reproduced with permission from (47). Copyright 2003, The Japan Society of Applied Physics.)

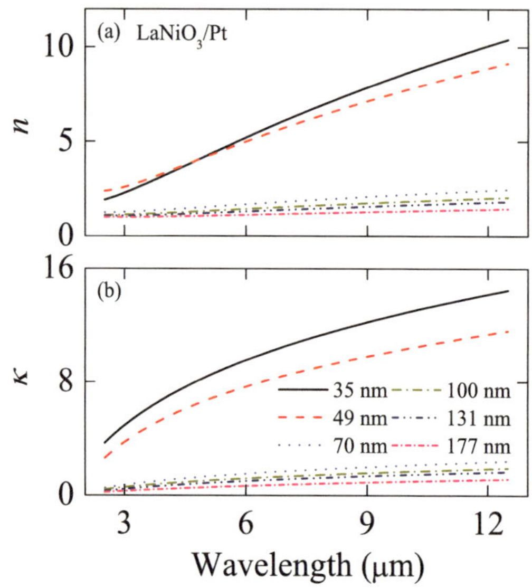

Fig. 12. The fitted spectroscopic dispersion of the optical constants (a) n and (b) κ of the LNO films with different film thickness. (Figure reproduced with permission from (47). Copyright 2003, The Japan Society of Applied Physics.)

8. Conclusion

In summary, the temperature influence on the optical properties of the BLT film prepared by chemical solution method on quartz substrate has been studied by the transmittance measurements. The Adachi's model has been employed to calculate the dielectric functions. The temperature dependence of the refractive index dispersion below the transition energy is found to obey an empirical Sellmeier equation. The results indicate that the BLT film transforms from tetragonal to orthorhombic structure in the temperature range 300-160 K, and then undergoes the orthorhombic to monoclinic transition with further decreasing the temperature. The traditional spectral transmittance technique provide a novel solution to observing the phase transitions of FE materials. On the other hand, the dielectric functions of high-quality LNO and LSCO conductive oxide films have been investigated by fitting the reflectance spectra and ellipsometric spectra with the Drude-Lorentz model. The optical conductivity presents a different behavior for two perovskite-type oxides due to the contributions from the charge-transfer excitations. Moreover, the crystallinity of the films results in the thickness dependence of the infrared optical properties. The present results indicate that the spectral techniques can be successfully used to study the optical properties of the perovskite-type FE and conductive metal oxide films. The experimental data could be crucial for the applications from the FE-based optoelectronic devices.

Although the optical properties, such as optical constants, optical band gap and electronic band structures of the perovskite-type FE and conductive metallic oxides have been discussed with the aid of the transmittance, reflectance and spectroscopic ellipsometry methods in this chapter, the external field (i.e., electrical and/or magnetic fields) effects on the optical properties of these oxide materials are still kept as an open issue. In order to easily operating the optoelectronic devices, the influences must be addressed before the "actual" devices can be applied. The projects will be developed in the near future.

9. Acknowledgments

This work is financially sponsored in part by Natural Science Foundation of China (**Grant No.** 60906046), National Basic Research Program of China (Grant Nos. 2007CB924901, 2007CB9249 02 and 2011CB922202), Program of New Century Excellent Talents, MOE (Grant No. NCET-08-0192), and Shanghai Municipal Commission of Science and Technology Project (Grant Nos. 10DJ1400201, **10ZR1409800**, 08JC1409000, 08520706100 and 09ZZ42), and the Fundamental Research Funds for the Central Universities.

10. References

[1] Lee, H. N.; Hesse, D.; Zakharov, N. & Gösele, U. (2002). Ferroelectric $Bi_{3.25}La_{0.75}Ti_3O_{12}$ films of uniform *a*-axis orientation on silicon substrates. *Science*, 296, 5575, 2006-2009

[2] Ahn, C. H.; Rabe, K. M. & Triscone, J. -M. (2004). Ferroelectricity at the nanoscale: Local polarization in oxide thin films and heterostructures. *Science*, 303, 5657, 488-491

[3] Yeh, C. S. & Wu, J. M. (2008). Characterization of Pt/multiferroic $BiFeO_3$/(Ba,Sr)TiO_3/Si stacks for nonvolatile memory applications. *Appl. Phys. Lett.*, 93, 15, 154101(1-3)

[4] Scott, J. F. (2007). Applications of modern ferroelectrics. *Science*, 315, 5814, 954-959

[5] Catalan, G. & Scott, J. F. (2009). Physics and applications of bismuth ferrite. *Adv. Mater.*, 21, 24, 2463-2485

[6] Kan, Y.; Liu, Y. F.; Mieth, O.; Bo, H. F.; Wu, X. M.; Lu, X. M.; Eng, L. M. & Zhu, J. S. (2009). Mechanical stress induced polarization reorientation in polycrystalline $Bi_{3.25}La_{0.75}Ti_3O_{12}$ films. *Phys. Lett. A*, 374, 2, 360-365

[7] Cummins, S. E. & Cross, L. E. (1968). Electrical and optical properties of ferroelectric $Bi_4Ti_3O_{12}$ single crystals. *J. Appl. Phys.*, 39, 5, 2268-2274

[8] Scott, J. F. & Paz de Araujo, C. A. (1989). Ferroelectric memories. *Science*, 246, 4936, 1400-1405

[9] Park, B. H.; Kang, B. S.; Bu, S. D.; Noh, T. W.; Lee, J. & Jo, W. (1999). Lanthanum-substituted bismuth titanate for use in non-volatile memories. *Nature*, 401, 6754, 682-684

[10] Hu, Z. G.; Ma, J. H.; Huang, Z. M.; Wu, Y. N.; Wang, G. S. & Chu, J. H. (2003). Dielectric functions of ferroelectric $Bi_{3.25}La_{0.75}Ti_3O_{12}$ thin films on Si(100) substrates. *Appl. Phy. Lett.*, 83, 18, 3686-3688

[11] Hu, Z. G.; Wang, G. S.; Huang, Z. M. & Chu, J. H. (2003). Optical properties of $Bi_{3.25}La_{0.75}Ti_3O_{12}$ thin films using spectroscopic ellipsometry. *J. Appl. Phys.*, 93, 7, 3811-3815

[12] Du, H. C.; Tang, L. J. & Kaskel, S. (2009). Preparation, microstructure, and ferroelectric properties of $Bi_{3.25}La_{0.75}Ti_{3-x}M_xO_{12}$ (M = Mo, W, Nb, V) ceramics. *J. Phys. Chem. C*, 113, 4, 1329-1339

[13] Shimakawa, Y.; Kubo, Y.; Tauchi, Y.; Asano, H.; Kamiyama, T.; Izumi, F. & Hiroi, Z. (2001). Crystal and electronic structures of $Bi_{4-x}La_xTi_3O_{12}$ ferroelectric materials. *Appl. Phys. Lett.*, 79, 17, 2791-2793

[14] Spanier, J. E.; Kolpak, A. M.; Urban, J. J.; Grinberg, I.; Lian, Q. Y.; Yun, W. S.; Rappe A. M. & Park, H. (2006). Ferroelectric phase transition in individual single-crystalline $BaTiO_3$ nanowires. *Nano Lett.*, 6, 4, 735-739

[15] Sirenko, A. A.; Bernhard, C.; Golnik, A.; Clark, A. M.; Hao, J. H.; Si, W. D. & Xi, X. X. (2000). Soft-mode hardening in $SrTiO_3$ thin films. *Nature*, 404, 6776, 373-376

[16] Tenne, D. A.; Bruchhausen, A.; Lanzillotti-Kimura, N. D.; Fainstein, A.; Katiyar, R. S.;

Cantarero, A.; Soukiassian, A.; Vaithyanathan, V.; Haeni, J. H.; Tian, W.; Schlom, D. G.; Choi, K. J.; Kim, D. M.; Eom, C. B.; Sun, H. P.; Pan, X. Q.; Li, Y. L.; Chen, L. Q.; Jia, Q. X.; Nakhmanson, S. M.; Rabe, K. M. & Xi, X. X. (2006). Probing nanoscale ferroelectricity by ultraviolet Raman spectroscopy. *Science*, 313, 5793, 1614-1616

[17] Du, Y. L.; Zhang, M. S.; Chen, Q.; Yuan, Z. R.; Yin, Z. & Zhang, Q. A. (2002). Size effect and evidence of a size-driven phase transition in $Bi_4Ti_3O_{12}$ nanocrystals. *Solid State Commun.*, 124, 3, 113-118

[18] Ríos, S.; Scott, J. F.; Lookman, A.; McAneney, J.; Bowman, R. M.; & Gregg, J. M. (2006). Phase transitions in epitaxial $Ba_{0.5}Sr_{0.5}TiO_3$ thin films. *J. Appl. Phys.*, 99, 2, 024107(1-8)

[19] Hoshina, T.; Kakemoto, H.; Tsurumi, T.; Wada, S. & Yashima, M. (2006). Size and temperature induced phase transition behaviors of barium titanate nanoparticles. *J. Appl. Phys.*, 99, 5, 054311(1-8)

[20] Idink, H.; Srkanth, V.; White, W. B. & Subbarao, E. C. (1994). Raman study of low temperature phase transitions in bismuth titanate, $Bi_4Ti_3O_{12}$. *J. Appl. Phys.*, 76, 3, 1819-1823

[21] Du, Y. L.; Chen, G. & Zhang, M. S. (2004). Grain size effects in $Bi_4Ti_3O_{12}$ nanocrystals investigated by Raman spectroscopy. *Solid State Commun.*, 132, 3-4, 175-179

[22] Sohn, J. I.; Joo, H. J.; Ahn, D.; Lee, H. H.; Porter, A. E.; Kim, K.; Kang, D. J. & Welland, M. E. (2009). Surface-stress-induced Mott transition and nature of associated spatial phase transition in single crystalline VO_2 nanowires. *Nano Lett.*, 9, 10, 3392-3397

[23] Tenne, D. A.; Turner, P.; Schmidt, J. D.; Biegalski, M.; Li, Y. L.; Chen, L. Q.; Soukiassian, A.; Trolier-McKinstry, S.; Schlom, D. G.; Xi, X. X.; Fong, D. D.; Fuoss, P. H.; Eastman, J. A.; Stephenson, G. B.; Thompson, C. & Streiffer, S. K. (2009). Ferroelectricity in ultrathin $BaTiO_3$ films: Probing the size effect by ultraviolet Raman spectroscopy. *Phy. Rev. Lett.*, 103, 17, 177601(1-4)

[24] Zhang, S. X.; Chou, J. Y. & Lauhon, L. J. (2009). Direct correlation of structural domain formation with the metal insulator transition in a VO_2 nanobeam. *Nano Lett.*, 9, 12, 4527-4532

[25] Hu, Z. G.; Li, W. W.; Li, Y. W.; Zhu, M.; Zhu, Z. Q. & Chu, J. H. (2009). Electronic properties of nanocrystalline $LaNiO_3$ and $La_{0.5}Sr_{0.5}CoO_3$ conductive films grown on silicon substrates determined by infrared to ultraviolet reflectance spectra. *Appl. Phys. Lett.*, 94, 22, 221104(1-4)

[26] Yang, B.; Aggarwal, S.; Dhote, A. M.; Song, T. K.; Ramesh, R. & Lee, J. S. (1997). $La_{0.5}Sr_{0.5}CoO_3/Pb(Nb_{0.04}Zr_{0.28}Ti_{0.68})O_3/La_{0.5}Sr_{0.5}CoO_3$ thin film heterostructures on Si using TiN/Pt conducting barrier [capacitors]. *Appl. Phys. Lett.*, 71, 3, 356-358

[27] Hu, Z. G.; Huang, Z. M.; Wu, Y. N.; Zhao, Q.; Wang, G. S. & Chu, J. H. (2004). Ellipsometric characterization of $LaNiO_{3-x}$ films grown on Si(111) substrates: Effects of oxygen partial pressure. *J. Appl. Phys.*, 95, 8, 4036-4041

[28] Choi, W. S.; Seo, S. S. A.; Kim, K. W.; Noh, T. W.; Kim, M. Y. & Shin, S. (2006). Dielectric constants of Ir, Ru, Pt, and IrO2: Contributions from bound charges. *Phys. Rev. B*, 74, 20, 205117(1-8)

[29] Berini, B.; Keller, N.; Dumont, Y.; Popova, E.; Noun, W.; Guyot, M.; Vigneron, J.; Etcheberry, A.; Franco, N. & M. C. da Silva, R. (2007). Reversible phase transformation of $LaNiO_{3-x}$ thin films studied in situ by spectroscopic ellipsometry. *Phys. Rev. B*, 76, 20, 205417(1-9)

[30] Rossel, C.; Rosová, A.; Hušeková, K.; Machajdík, D. & Fröhlich, K. (2006). Phase stability of $La_{0.5}Sr_{0.5}CoO_{3-y}$ films upon annealing in hydrogen atmosphere. *J. Appl. Phys.*, 100,

4, 044501(1-6)

[31] Xie, C. K.; Budnick, J. I.; Wells, B. O. & Woicik, J. C. (2007). Separation of the strain and finite size effect on the ferromagnetic properties of $La_{0.5}Sr_{0.5}CoO_3$ thin films. *Appl. Phys. Lett.*, 91, 17, 172509(1-3)

[32] Noun, W.; Berini, B.; Dumont, Y.; Dahoo, P. R. & Keller, N. (2007). Correlation between electrical and ellipsometric properties on high-quality epitaxial thin films of the conductive oxide $LaNiO_3$ on STO (001). *J. Appl. Phys.*, 102, 6, 063709(1-7)

[33] Mambrini, G. P.; Leite, E. R.; Escote, M. T.; Chiquito, A. J.; Longo, E.; Varela, J. A. & Jardim, R. F. (2007). Structural, microstructural, and transport properties of highly oriented $LaNiO_3$ thin films deposited on $SrTiO_3$(100) single crystal. *J. Appl. Phys.*, 102, 4, 043708(1-6)

[34] Angadi, M.; Auciello, O.; Krauss, A. R. & Gundel, H. W. (2000). The role of electrode material and polarization fatigue on electron emission from ferroelectric $Pb(Zr_xTi_{1-x})O_3$ cathodes. *Appl. Phys. Lett.*, 77, 17, 2659-2661

[35] Naumov, I. I.; Bellaiche, L. & Fu, H. X. (2004). Unusual phase transitions in ferroelectric nanodisks and nanorods. *Nature*, 432, 7018, 737-740

[36] Kim, J.; Yang, S. A.; Choi, Y. C.; Han, J. K.; Jeong, K. O.; Yun, Y. J.; Kim, D. J.; Yang, S. M.; Yoon, D.; Cheong, H.; Chang, K. -S.; Noh, T. W. & Bu, S. D. (2008). Ferroelectricity in highly ordered arrays of ultra-thin-walled $Pb(Zr,Ti)O_3$ nanotubes composed of nanometer-sized perovskite crystallites. *Nano Lett.*, 8, 7, 1813-1818

[37] Sreedhar, K.; Honig, J. M.; Darwin, M.; McElfresh, M.; Shand, P. M.; Xu, J.; Crooker, B. C. & Spalek, J. (1992). Electronic properties of the metallic perovskite $LaNiO_3$: Correlated behavior of 3*d* electrons. *Phys. Rev. B*, 46, 10, 6382-6386

[38] Horiba, K.; Eguchi, R.; Taguchi, M.; Chainani, A.; Kikkawa, A.; Senba, Y.; Ohashi, H. & Shin, S. (2007). Electronic structure of $LaNiO_{3-x}$: An in situ soft x-ray photoemission and absorption study. *Phys. Rev. B*, 76, 15, 155104(1-6)

[39] Kuhns, P. L.; Hoch, M. J. R.; Moulton, W. G.; Reyes, A. P.; Wu, J. & Leighton, C. (2003). Magnetic phase separation in $La_{1-x}Sr_xCoO_3$ by ^{59}Co nuclear magnetic resonance. *Phys. Rev. Lett.*, 91, 12, 127202(1-4)

[40] Heaven, O. S. (1991). *Optical Properties of Thin Solid Films*, Dover, NewYork

[41] Hu, Z. G.; Huang, Z. M.; Wu, Y. N.; Hu, S. H.; Wang, G. S.; Ma, J. H. & Chu, J. H. (2004). Optical characterization of ferroelectric $Bi_{3.25}La_{0.75}Ti_3O_{12}$ thin films. *Eur. Phys. J. B*, 38, 3, 431-436

[42] Hu, Z. G.; Strassburg, M.; Dietz, N.; Perera, A. G. U.; Asghar, A. & Ferguson, I. T. (2005). Composition dependence of the infrared dielectric functions in Si-doped hexagonal $Al_xGa_{1-x}N$ films on *c*-plane sapphire substrates. *Phys. Rev. B*, 72, 24, 245326(1-10)

[43] Hu, Z. G.; Weerasekara, A. B.; Dietz, N.; Perera, A. G. U.; Strassburg, M.; Kane, M. H.; Asghar, A. & Ferguson, I. T. (2007). Infrared optical anisotropy of diluted magnetic $Ga_{1-x}Mn_xN/c$-sapphire epilayers grown with a GaN buffer layer by metalorganic chemical vapor deposition. *Phys. Rev. B*, 75, 20, 205320(1-8)

[44] Djurišić, A. B.; Chan, Y. & Li, E. H. *Mater. Sci. Eng. R*, (2002). Progress in the room-temperature optical functions of semiconductors. 38, 6, 237-293

[45] Chen, C.; An, I. & Collins, R. W. (2003). Multichannel mueller matrix ellipsometry for simultaneous real-time measurement of bulk isotropic and surface anisotropic complex dielectric functions of semiconductors. *Phys. Rev. Lett.*, 90, 21, 217402(1-4)

[46] Kasic, A.; Schubert, M.; Einfeldt, S.; Hommel, D. & Tiwald, T. E. (2000). Free-carrier and phonon properties of *n*- and *p*-type hexagonal GaN films measured by infrared

ellipsometry. *Phys. Rev. B.*, 62, 11, 7365-7377

[47] Hu, Z. G.; Meng, X. J.; Huang, Z. M.; Wang, G. S.; Zhao, Q. & Chu, J. H. (2003). Thickness dependence of infrared optical properties of $LaNiO_3$ thin films prepared on platinized silicon substrates. *Jpn. J. Appl. Phys.*, 42, 11, 7045-7049

[48] Hu, Z. G.; Li, Y. W.; Zhu, M.; Zhu, Z. Q. & Chu, J. H. (2008). Structure and optical properties of ferroelectric $PbZr_{0.40}Ti_{0.60}O_3$ films grown on $LaNiO_3$-coated platinized silicon determined by infrared spectroscopic ellipsometry. *J. Phys. Chem. C*, 112, 26, 9737-9743

[49] Hu, Z. G.; Huang, Z. M. & Chu, J. H. (2005). *Trends in Semiconductor Research*, edited by Thomas B. Elliot, Chapter 5, page: 111-136, ISBN: 1-59454-414-X, Nova Science Publishers, Inc. NY

[50] Azzam, R. M. A. & Bashara, N. M. (1999). *Ellipsometry and Polarized Light*, Chapter 3, Page: 153-267, ISBN: 0-444-87016-4, North-Holland, Amsterdam

[51] Li, W. W.; Hu, Z. G.; Li, Y. W.; Zhu, M.; Zhu, Z. Q. & Chu, J. H. (2010). Growth, microstructure, and infrared-ultraviolet optical conductivity of $La_{0.5}Sr_{0.5}CoO_3$ nanocrystalline films on silicon substrates by pulsed laser deposition. *ACS Appl. Mater. Interfaces*, 02, 3, 896-902

[52] Adachi, S. (1987). Model dielectric constants of GaP, GaAs, GaSb, InP, InAs, and InSb. *Phys. Rev. B*, 35, 14, 7454-7463

[53] Hu, Z. G.; Wang, G. S.; Huang, Z. M.; Meng, X. J.; Shi, F. W. & Chu, J. H. (2003). Investigations of the optical properties of $Ba_{0.9}Sr_{0.1}TiO_3$ ferroelectric thin films by spectroscopic ellipsometry. *Jpn. J. Appl. Phys.*, 42, 3, 1400-1404

[54] Hu, Z. G.; Li, Y. W.; Zhu, M.; Zhu, Z. Q. & Chu, J. H. (2008). Composition dependence of dielectric function in ferroelectric $BaCo_xTi_{1-x}O_3$ films grown on quartz substrates by transmittance spectra. *Appl. Phys. Lett.*, 92, 08, 081904(1-3)

[55] Hu, Z. G.; Li, W. W.; Wu, J. D.; Sun, J.; Shu, Q. W.; Zhong, X. X.; Zhu, Z. Q. & Chu, J. H. (2008). Optical properties of pulsed laser deposited rutile titanium dioxide films on quartz substrates determined by Raman scattering and transmittance spectra. *Appl. Phys. Lett.*, 93, 18, 181910(1-3)

[56] Li, W. W.; Hu, Z. G.; Wu, J. D.; Sun, J.; Zhu, M.; Zhu, Z. Q. & Chu, J. H. (2009). Concentration dependence of optical properties in arsenic-doped ZnO nanocrystalline films grown on silicon (100) substrates by pulsed laser deposition. *J. Phys. Chem. C*, 113, 42, 18347-18352

[57] Kim, T. J.; Yoon, J. J.; Hwang, S. Y.; Aspnes, D. E.; Kim, Y. D.; Kim, H. J.; Chang, Y. C. & Song, J. D. (2009). Interband transitions of $InAs_xSb_{1-x}$ alloy films. *Appl. Phys. Lett.*, 95, 11, 111902(1-3)

[58] Yu, W. L.; Li, W. W.; Wu, J. D.; Sun, J.; Zhu, J. J.; Zhu, M.; Hu, Z. G. & Chu, J. H. (2010). Far-infrared-ultraviolet dielectric function, lattice vibration, and photoluminescence properties of diluted magnetic semiconductor $Sn_{1-x}Mn_xO_2/c$-sapphire nanocrystalline films. *J. Phys. Chem. C*, 114, 18, 8593-8600

[59] Qazilbash, M. M.; Schafgans, A. A.; Burch, K. S.; Yun, S. J.; Chae, B. G.; Kim, B. J.; Kim, H. T. & Basov, D. N. (2008). Electrodynamics of the vanadium oxides VO_2 and V_2O_3. *Phys. Rev. B*, 77, 11, 115121(1-10)

[60] Hu, Z. G.; Wang, G. S.; Huang, Z. M.; Meng, X. J. & Chu, J. H. (2002). Infrared optical properties of $Bi_{3.25}La_{0.75}Ti_3O_{12}$ ferroelectric thin films using spectroscopic ellipsometry. *J. Phys. D: Appl. Phys.*, 35, 24, 3221-3224

[61] Hu, Z. G.; Wang, G. S.; Huang, Z. M.; Meng, X. J. & Chu, J. H. (2003). Investigations

on the infrared optical properties of $BaTiO_3$ ferroelectric thin films by spectroscopic ellipsometry. *Semicond. Sci. Technol.*, 18, 6, 449-453

[62] Palik, E. D. (1985). *Handbook of Optical Constants of Solids*, Page: 341, Academic, Orlando. FL

[63] Hu, Z. G.; Li, Y. W.; Yue, F. Y.; Zhu, Z. Q. & Chu, J. H. (2007). Temperature dependence of optical band gap in ferroelectric $Bi_{3.25}La_{0.75}Ti_3O_{12}$ films determined by ultraviolet transmittance measurements. *Appl. Phys. Lett.*, 91, 22, 221903(1-3)

[64] Subbarao, E. C. (1961). Ferroelectricity in $Bi_4Ti_3O_{12}$ and its solid solutions. *Phys. Rev.*, 122, 3, 804-807

[65] Tsunekawa, S.; Ito, S.; Kawazoe, Y. & Wang, J. T. (2003). Critical size of the phase transition from cubic to tetragonal in pure zirconia nanoparticles. *Nano Lett.*, 3, 7, 871-875

[66] Cross, L. E. & Pohanka, R. C. (1968). A thermodynamic analysis of ferroelectricity in bismuth titanate. *J. Appl. Phys.*, 39, 8, 3992-3995

[67] Bennett, J. W.; Grinberg, L. & Rappe, A. M. (2008). New highly polar semiconductor ferroelectrics through d^8 cation-O vacancy substitution into $PbTiO_3$: A theoretical study. *J. Am. Chem. Soc.*, 130, 51, 17409-17412

[68] Arslan, M.; Duymus, H. & Yakuphanoglu, F. (2006). Optical properties of the poly(N-benzylaniline) thin film. *J. Phys. Chem. B*, 110, 1, 276-280

[69] Hu, Z. G.; Li, Y. W.; Zhu, M.; Yue, F. Y.; Zhu, Z. Q. & Chu, J. H. (2008). Growth and ellipsometric characterizations of highly (111) oriented $Bi_2Ti_2O_7$ films on platinized silicon by metal organic decomposition method. *J. Vac. Sci. Technol. A*, 26, 5, 1287-1292

[70] Myhre, C. E. L.; Christensen, D. H.; Nicolaisen, F. M. & Nielsen, C. J. (2003). Spectroscopic study of aqueous H_2SO_4 at different temperatures and compositions: Variations in dissociation and optical Properties. *J. Phys. Chem. A*, 107, 12, 1979-1991

[71] Hu, Z. G.; Li, Y. W.; Zhu, M.; Zhu, Z. Q.; & Chu, J. H. (2008). Microstructural and optical investigations of sol-gel derived ferroelectric $BaTiO_3$ nanocrystalline films determined by spectroscopic ellipsometry. *Phys. Lett. A*, 372, 24, 4521-4526

[72] Alivisatos, A. P. (1996). Semiconductor clusters, nanocrystals, and quantum dots. *Science*, 271, 5251, 933-937

[73] Frey, M. H. & Payne, D. A. (1996). Grain-size effect on structure and phase transformations for barium titanate. *Phys. Rev. B*, 54, 5, 3158-3168

[74] Bozovic, I.; Kim, J. H.; Harris, J. S.; Eom, C. B.; Phillips, J. M. & Cheung, J. T. (1994). Reflectance and Raman spectra of metallic oxides, LaSrCoO and CaSrRuO: Resemblance to superconducting cuprates. *Phys. Rev. Lett.*, 73, 10, 1436-1439

[75] Rusydi, A.; Rauer, R.; Neuber, G.; Bastjan, M.; Mahns, I.; Müller, S.; Saichu, P.; Schulz, B.; Singer, S. G.; Lichtenstein, A. I.; Qi, D.; Gao, X.; Yu, X.; Wee, A. T. S.; Stryganyuk, G.; Dörr, K.; Sawatzky, G. A.; Cooper, S. L. & Rübhausen, M. (2008). Metal-insulator transition in manganites: Changes in optical conductivity up to 22 eV. *Phys. Rev. B*, 78, 12, 125110(1-5)

[76] Barman, S. R.; Chainani, A. & Sarma, D. D. (1994). Covalency-driven unusual metal-insulator transition in nickelates. *Phys. Rev. B*, 49, 12, 8475-8478

[77] Dobin, A. Yu.; Nikolaev, K. R.; Krivorotov, I. N.; Wentzcovitch, R. M.; Dahlberg, E. D. & Goldman, A. M. (2003). Electronic and crystal structure of fully strained $LaNiO_3$ films. *Phys. Rev. B*, 68, 11, 113408(1-4)

[78] Bao, D. H.; Yao, X.; Wakiya, N.; Shinozaki, K. & Mizutani, N. (2001). Band-gap energies of sol-gel-derived $SrTiO_3$ thin films. *Appl. Phys. Lett.*, 79, 23, 3767-3769

[79] Majumder, S. B.; Jain, M. & Katiyar, R. S. (2002). Investigations on the optical properties of sol-gel derived lanthanum doped lead titanate thin films. *Thin Solid Films*, 402, 1-2, 90-98

[80] Franke, E.; Trimble, C. L.; Devries, M. J.; Woollam, J. A.; Schubert, M. & Frost, F. (2000). Dielectric function of amorphous tantalum oxide from the far infrared to the deep ultraviolet spectral region measured by spectroscopic ellipsometry. *J. Appl. Phys.*, 88, 9, 5166-5174

5

Epitaxial $SrRuO_3$ Thin Films Deposited on SrO buffered-Si(001) Substrates for Ferroelectric $Pb(Zr_{0.2}Ti_{0.8})O_3$ Thin Films

Soon-Gil Yoon

School of Nano Science and Technology, Graduate of Analytical Science and Technology (GRAST), Chungnam National University, Daeduk Science Town, 305-764, Daejeon Korea

1. Introduction

At present there is considerable interest in utilizing ferroelectric thin films as a medium for non-volatile data storage [1]. In particular, much attention has been focused on investigating high density giga-bit data storage using scanning probe techniques [2]. It is known that the thickness of a 180° domain wall in ferroelectric thin film is 1-2 nm [3]. As such, these films have potential to serve as a medium for scanning probe microscopy (SPM)-based ultrahigh density (100 $Gbit/cm^2$ class) data storage [2]. It has been proposed that employing ferroelectric films as recording media inherently has several advantages: 1) the recorded data are non-volatile, 2) the recording density can be ultra high because of narrow domain wall thickness, 3) the ferroelectric domains can have fast switching speeds, and 4) information bits can be written and read electrically.

Among the various perovskite oxide materials, PZT films appear to be suitable as a storage medium, since their remanent polarizations are high. The highly tetragonal PZT films exhibit high remanent polarization showing a square-shaped hysteresis curve, compared with films having morphotropic and rhombohedral compositions. Therefore, $Pb(Zr_{0.2}Ti_{0.8})O_3$ films with a highly tetragonal structure are suitable for nano-data storage system applications. Ferroelectric $Pb(Zr_{0.2}Ti_{0.8})O_3$ thin films grown epitaxially on $SrRuO_3/SrTiO_3$ are reported to exhibit large P_r (remanent polarization) [4-5].

$SrRuO_3$ is a conductive oxide (a room temperature resistivity of 280 $\mu\Omega\cdot cm$ for single crystals), pseudo-cubic perovskite with a lattice constant of 0.55 nm, and has a low mismatch with $Pb(Zr_{0.2}Ti_{0.8})O_3$(~ 0.5 %) and SrO (~ 1.7 %), thus allowing high quality epitaxial growth. Single-crystal epitaxial thin films of $SrRuO_3$ and $Sr_{1-x}Ca_xRuO_3$[6] are frequently grown on $SrTiO_3$ substrates. In order to achieve epitaxial growth of $SrRuO_3$ on Si, yttria-stabilized zirconia (YSZ) [7-8] and MgO [9] have been employed as buffer layer materials. Another promising candidate for the buffer layer is SrO which has a NaCl-type cubic structure with a lattice parameter of a = 5.140 Å [10-11]. The SrO(110) dielectric layer also exhibits good compatibility with Si(001) showing a low lattice mismatch of about 0.4 % [12].

In this study, highly tetragonal $Pb(Zr_{0.2}Ti_{0.8})O_3$ films were grown on epitaxial $SrRuO_3$ thin film electrodes using a SrO buffer layer on Si(001) substrates by pulsed laser deposition. The effect of the SrO buffer layer on the surface morphologies and the electrical properties of the

$SrRuO_3$ thin film electrodes was investigated as a function of SrO buffer layer thickness. Ferroelectric properties in c-axis oriented $Pb(Zr_{0.2}Ti_{0.8})O_3$ thin films deposited on an epitaxial $SrRuO_3$ electrode were also investigated.

2. Experimental procedure

The SrO target for pulsed laser deposition is difficult to make due to the low sinterability of SrO materials. A SrO_2 target was used to deposit SrO on Si (001) substrates in a vacuum ambient. Before deposition of SrO layers on Si (001) substrates, Si wafers were etched using HF solution to remove the native oxide layers. SrO, $SrRuO_3$ and PZT films were deposited on etched-silicon substrates using a KrF excimer laser (λ = 248 nm) with a maximum repetition rate of 10 Hz. Laser pulse energy density used for SrO, $SrRuO_3$ and PZT films deposited using ceramic targets was approximately 1.5 J/cm^2. 30 mol % excess PbO in the PZT targets was added to compensate for the PbO loss during both the sintering and deposition processes. The distance between the target and substrate was varied depending on the deposited materials. The typical SrO, $SrRuO_3$, and PZT deposition conditions are summarized in Table 1. The SrO, $SrRuO_3$, and PZT were in-situ deposited at each temperature in order to ensure chemical stability of the SrO films with Si. The PZT films were cooled down to room temperature at an oxygen pressure of 300 Torr to preserve the oxygen content in the films during the cooling procedure.

Deposition parameters	SrO	$SrRuO_3$	$Pb(Zr_{0.2}Ti_{0.8})O_3$
Target	SrO_2	$SrRuO_3$	$Pb(Zr_{0.2}Ti_{0.8})O_3$
Deposition temperature	700°C	550~750°C	575~600°C
Film thickness	3~30nm	50~300nm	100nm
Deposition pressure	1x10^{-6} Torr	1x10^{-2} Torr	1x10^{-1} Torr
Energy density	1.5 J/cm^2	1.5 J/cm^2	1.5 J/cm^2
Repetition rate	1 Hz	10 Hz	10 Hz
Target-substrate distance	3 cm	4 cm	6 cm
Substrates	Si (001)	SrO/Si	$SrRuO_3$/SrO/Si

Table 1. Deposition conditions of $Pb(Zr_{0.2}Ti_{0.8})O_3$ thin films, $SrRuO_3$ electrodes, and SrO buffer layers on Si (001) substrates by pulsed laser deposition

The surface morphologies of $SrRuO_3$ and PZT were characterized by atomic force microscopy (AFM, AUTOPROBE CP, PSI). Crystalline properties of the films were investigated by θ–2θ, ω-scan, and Φ-scan using a high resolution X-ray diffraction (HRXRD, Rigaku RINT2000). The composition of the $SrRuO_3$ films was identified by Rutherford backscattering spectroscopy (RBS) using a beam energy of 2.236 MeV (4He^{++}) and an incident angle of 160°. The elemental distribution in the PZT/$SrRuO_3$/SrO/Si structure was investigated by secondary ion mass spectroscopy (SIMS) and Auger electron spectroscopy (AES). The resistivity of $SrRuO_3$ thin films was measured by an electrometer (CMT-SR 1000) using a four-point probe. The ferroelectric properties and leakage current characteristics were measured using a RT66A ferroelectric tester (Radiant Technology) operating in the virtual ground mode and a Keithley 617 electrometer, respectively. The measurements were carried out using a metal–insulator–metal (MIM) configuration. A Pt top electrode (Area = 7.85 × 10^{-5} cm^2) patterned using lift-off lithography was prepared at room temperature by dc sputtering.

3. Results and discussion

Figure 1(A) shows XRD patterns of 100 nm thick-$SrRuO_3$ thin films deposited on SrO buffer layers with various thicknesses. The XRD patterns were plotted using a log-scale to check the existence of the minor portion of SrO_2 phase within the $SrRuO_3$ films. The $SrRuO_3$ films were deposited at about 650°C in 1×10^{-2} Torr. As shown in Fig. 1(A), $SrRuO_3$ films deposited on 6 nm-thick-SrO buffered Si (001) substrates exhibit a c-axis preferred orientation, indicating the (001) and (002) planes alone. In the case of buffer layers above 12 nm thickness, SrO_2 phase was observed in the $SrRuO_3$ films. From the XRD results, the SrO buffer layer was found to play an important role for the preferred growth of $SrRuO_3$ films on Si substrates at below approximately 6 nm-thickness. In order to investigate the epitaxial relationship between $SrRuO_3$ and SrO(6 nm)/Si(001), a Φ-scan in the $SrRuO_3$/SrO/Si structure was performed and the results are shown in Fig. 1 (B). Peaks of $SrRuO_3$ {111} can be observed at every 90°, indicating that the $SrRuO_3$ films are epitaxially grown on a SrO/Si (001).

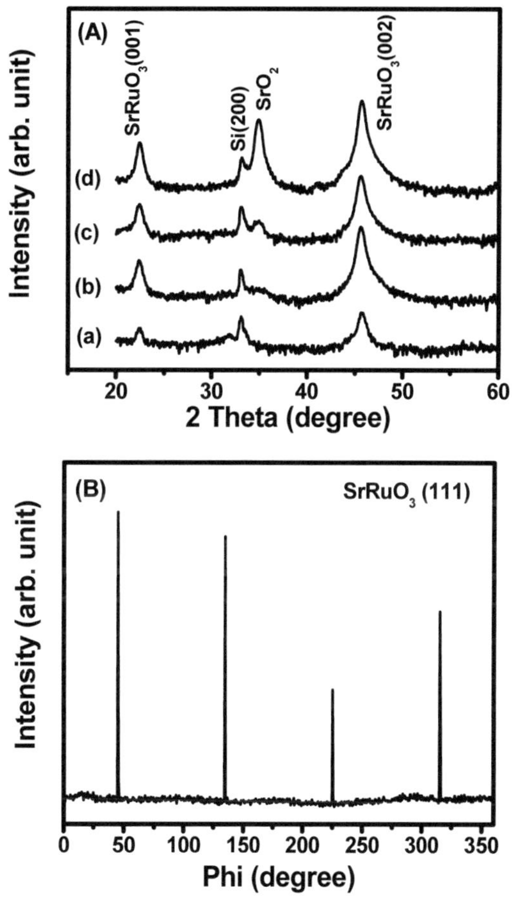

Fig. 1. (A) XRD patterns of 100nm thick-$SrRuO_3$ thin films deposited on SrO buffer layers of (a) 3 nm, (b) 6 nm, (c) 12 nm, and (d) 30 nm thickness. (B) Φ-scan profile of $SrRuO_3$ films deposited on SrO (6nm)/Si. (XRD patterns were plotted using a log-scale)

Figure 2 shows the variation in resistivity and rms roughness of 100nm thick-$SrRuO_3$ thin films as a function of SrO thickness. The $SrRuO_3$ films deposited on ultra-thin SrO films of about 3nm show the highest rms roughness and resistivity values because the SrO films do not play a role as a buffer layer. On the other hand, above 6nm thickness, $SrRuO_3$ films exhibit a low rms roughness of about 3-5 Å and a resistivity of 1700 - 1900 μΩ-cm. The resistivity values of the $SrRuO_3$ films deposited on 6nm-thick SrO buffer layers are approximately 1700 μΩ-cm, higher than that of $SrRuO_3$ films (~ 400 μΩ-cm) deposited on $SrTiO_3$ single crystals[13] and (100) $LaAlO_3$ single crystals[14]. The high resistivity of the $SrRuO_3$ films on Si substrates originates from the existence of silicon oxide formed from the diffusion of silicon, because the thin SrO layer does not prevent the diffusion of silicon during the deposition of $SrRuO_3$ at high temperature. The full-width-half-maximum (FWHM) values of the $SrRuO_3$ films deposited on SrO/Si and $SrTiO_3$ single crystal substrates are approximately 7.32° and 0.15°, measured by ω-scan, respectively. These results suggest that the $SrRuO_3$ films grown on Si substrates are inferior in terms of film qualities such as crystallinity and defects relative to those grown on $SrTiO_3$ single crystals. The inset in Fig. 2 shows three-dimensional AFM images of $SrRuO_3$ films deposited on (a) 3 nm thick- and (b) 30 nm thick-SrO buffer layers.

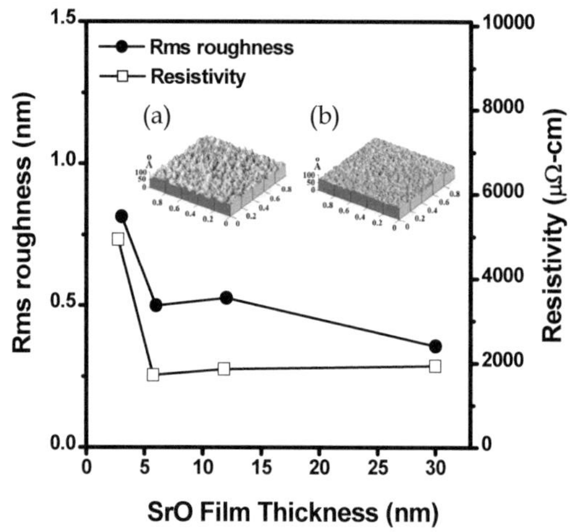

Fig. 2. Variation in resistivity and rms roughness of 100 nm thick-$SrRuO_3$ thin films as a function of SrO thickness. The inset shows three-dimensional AFM images of $SrRuO_3$ thin films deposited on SrO buffer layers of (a) 3 nm and (b) 30 nm thickness

Figures 3(a) and 3(b) show XRD patterns and rms roughness and resistivity of $SrRuO_3$ films, respectively, as a function of $SrRuO_3$ thickness. The $SrRuO_3$ films were deposited at 650°C on 6 nm thick-SrO buffer layers. As shown in Fig. 3(a), the peak intensity of $SrRuO_3$ {001} increases with increasing $SrRuO_3$ thickness, and the $SrRuO_3$ films of 50 nm thickness also exhibit a *c*-axis preferred relationship with Si(001) substrates. The resistivity of the $SrRuO_3$ films exhibits a constant value of about 1700 μΩ-cm irrespective of the film thickness above 50nm. The rms roughness of the $SrRuO_3$ films exhibits a similar tendency with the variation of resistivity as a function of $SrRuO_3$ thickness. The film roughness in conducting materials is inversely proportional to the mobility of the charge carriers. The resistivity in the conducting films is also inversely proportional to the mobility of the charge carriers if the concentration of the charge carriers is constant.

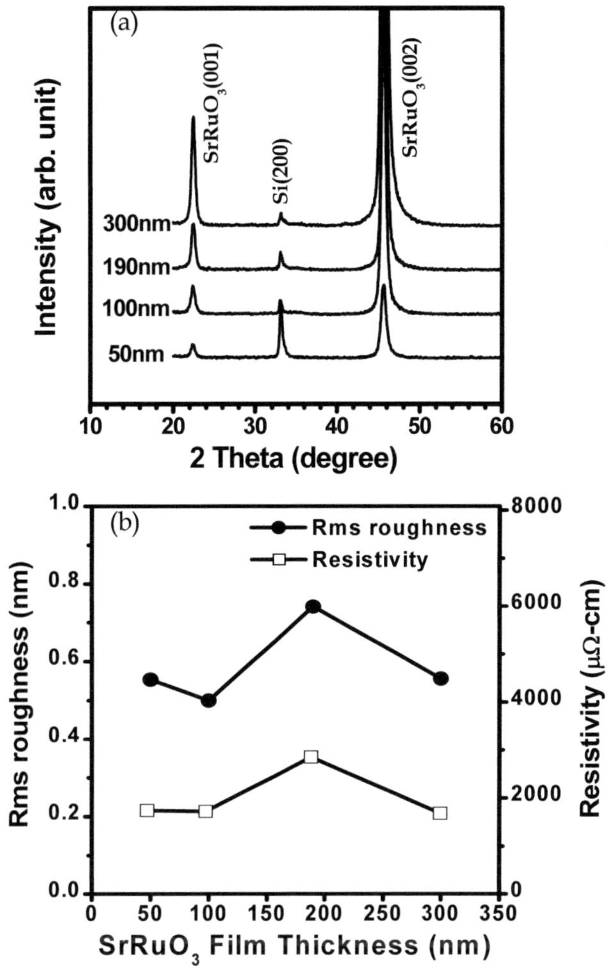

Fig. 3. (a) XRD patterns, (b) rms roughness and resistivity of the $SrRuO_3$ films deposited on 6nm-thick SrO buffer layers as a function of $SrRuO_3$ thickness

Figure 4(a) and 4(b) show XRD patterns and variation of rms roughness and resistivity, respectively, for 190nm-thick $SrRuO_3$ films as a function of $SrRuO_3$ deposition temperature. As shown in Fig. 4(a), $SrRuO_3$ films deposited at 550 and 600°C show $SrRuO_3$ (110) peaks in addition to the $SrRuO_3${001} peaks. This indicates that the films deposited at lower temperatures exhibit a polycrystalline nature rather than an epitaxial relationship. The $SrRuO_3$ (110) peak disappears in the films deposited above 650°C and the films are grown with an epitaxial relationship with Si (001) substrates. As shown in Fig. 4(b), the rms roughness of the films continuously increases with increasing deposition temperature, and a rms roughness of 7.4 Å is noted in the films deposited at 650°C. The rms roughness of the $SrRuO_3$ films deposited on SrO/Si substrates is higher than that of the films deposited on $SrTiO_3$ single crystals [13]. In the $SrRuO_3$/SrO/Si structure, even though the films deposited above 650°C were epitaxially grown, the high rms roughness of the $SrRuO_3$ films was attributed to the unstable interface structure of SrO/Si, compared with the $SrTiO_3$ single crystal substrate. The resistivity of the $SrRuO_3$ films abruptly decreases as the deposition temperature increases up to 650°C and maintains a constant value of about 1700 μΩ-cm at deposition temperatures above 650°C.The $SrRuO_3$ films grown with an epitaxial relationship exhibit lower resistivity than the polycrystalline films.

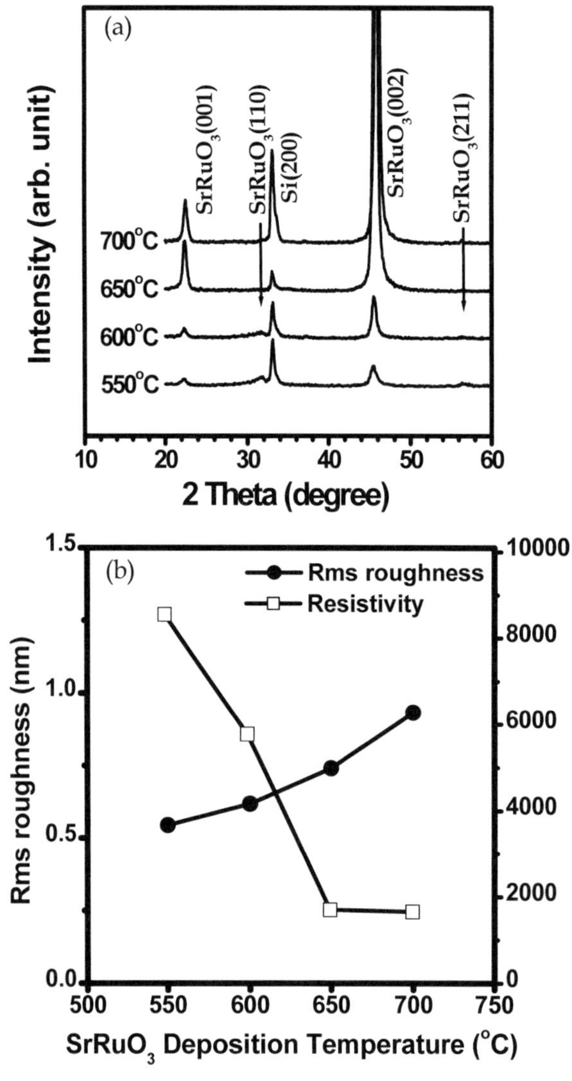

Fig. 4. (a) XRD patterns and (b) rms roughness and resistivity of the $SrRuO_3$ films deposited on 6nm-thick SrO buffer layers as a function of $SrRuO_3$ deposition temperature

Figure 5 shows the RBS spectrum of 190 nm thick-$SrRuO_3$ films deposited at 650°C on SrO/Si substrates. The compositions of Sr and Ru in the $SrRuO_3$ films deposited from stoichiometric $SrRuO_3$ targets are Sr/(Sr+Ru) = 50.52 % and Ru/(Sr+Ru) = 49.48 % from the RBS analysis. Because the Sr and Ru elements are overlapped, the composition of the films was fitted by repeated adjustments with the standard composition of $SrRuO_3$. The results suggest that the epitaxial $SrRuO_3$ films deposited by pulsed laser deposition show the same composition as $SrRuO_3$ targets.

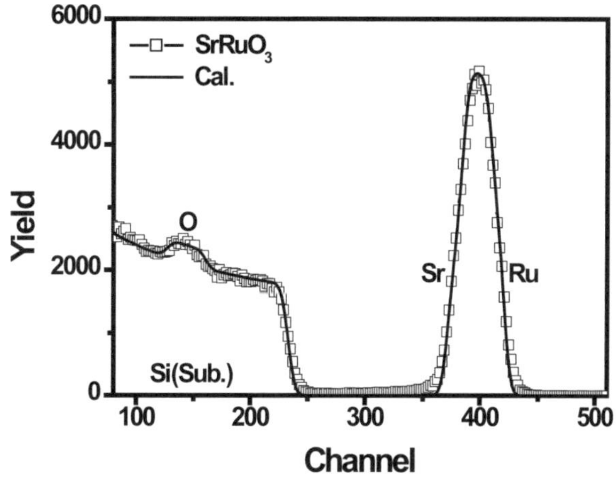

Fig. 5. RBS spectrum of 190nm-thick $SrRuO_3$ films deposited at 650°C on SrO/Si substrates

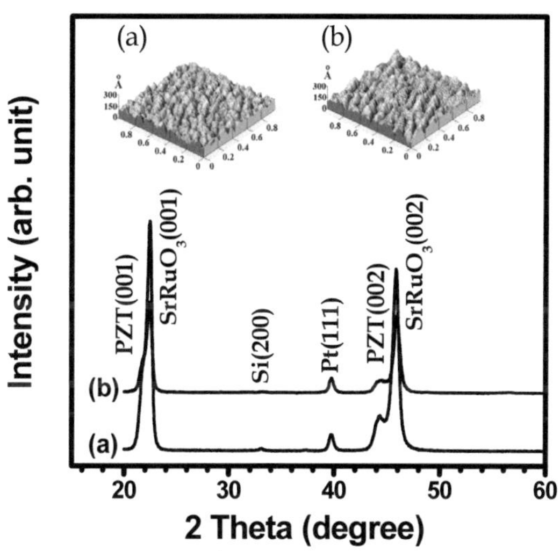

Fig. 6. XRD patterns of PZT thin films deposited at (a) 575°C and (b) 600°C on 190nm-thick $SrRuO_3$ films. The insets show three-dimensional AFM images of PZT films deposited at (a) 575°C and (b) 600°C

Figure 6 shows XRD patterns of 100 nm thick-$Pb(Zr_{0.2}Ti_{0.8})O_3$ (PZT) films deposited at 575 and 600°C on epitaxial $SrRuO_3$/SrO/Si substrates. The (a) and (b) in inset of Fig. 6 show three dimensional AFM images of PZT films deposited at 575°C and 600°C, respectively. The rms roughness of the PZT films deposited at 575 and 600°C are approximately 23Å and 26Å, respectively. The PZT films, as revealed in the XRD patterns of Fig. 6, exhibit a close relationship with $SrRuO_3$\{001\} at deposition temperatures of 575 and 600°C. The Pt (111) peak originates from the top electrode in the Pt/PZT/$SrRuO_3$/SrO/Si capacitor structures. However, the peak intensities of PZT\{001\} decrease with increasing deposition temperature. The PZT films react with the $SrRuO_3$ bottom electrode at high deposition temperature, resulting in Pb diffusion into the $SrRuO_3$/$SrTiO_3$ [15]. Even though Pb is diffused into the $SrRuO_3$ films, PZT films deposited on $SrRuO_3$/$SrTiO_3$ exhibit a good epitaxial relationship maintaining the predominating crystallinity. In order to investigate the distributions of each element in the PZT films deposited on $SrRuO_3$/SrO/Si, the elemental distribution in each layer was analyzed by secondary ion mass spectroscopy (SIMS), as shown in Fig. 7 (a). The Pb from the PZT films deposited at 600°C was clearly observed within the $SrRuO_3$ layer, indicating similar diffusion behavior of Pb into the $SrRuO_3$/$SrTiO_3$ [15]. In addition, silicon was also observed at the PZT layer as well as at the $SrRuO_3$ layer. The silicon existing in the PZT and $SrRuO_3$ films will present as silicon oxide, because silicon oxide phase is thermodynamically stable compared with Si element. The silicon oxide will exert a harmful influence upon the crystallinity and the morphologies of the PZT films. The existence of silicon in the PZT layer and Pb in the $SrRuO_3$ layer was also verified by the AES depth-profile as shown in Fig. 7(b). A phi-scan was performed to identify whether the PZT films are epitaxially grown on the $SrRuO_3$ films. From the results (not shown here), PZT films do not exhibit epitaxial growth on the $SrRuO_3$ bottom electrode. The PZT films were only grown with (00l) preferred orientation on $SrRuO_3$ electrodes. From the ω-scan of the PZT films deposited at 575°C, the FWHM value of the PZT (002) is approximately 7.08°. Thus, the crystalline quality of the PZT films deposited on $SrRuO_3$/SrO/Si was distinctly influenced by the silicon oxide within the PZT layers.

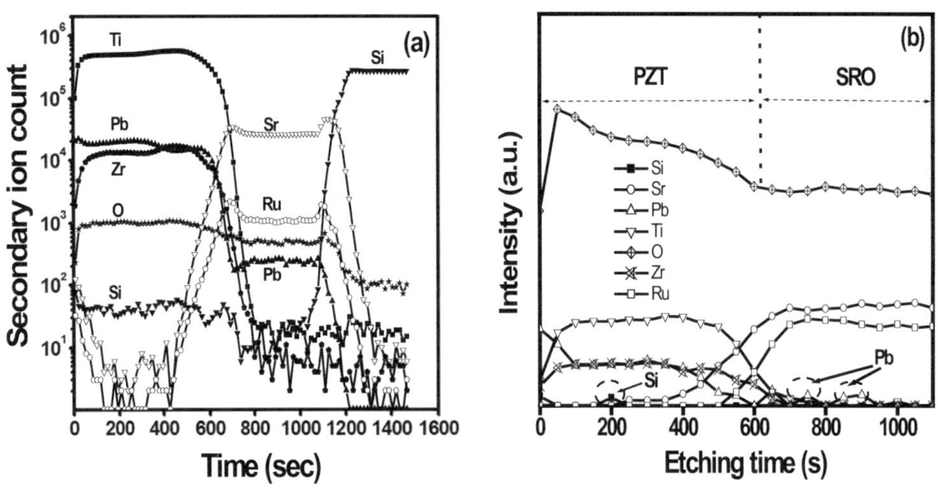

Fig. 7. (a) SIMS and (b) AES depth-profiles of PZT (100nm) films deposited at 600°C on $SrRuO_3$/SrO/Si

Figures 8(a) and 8(b) show P-E hysteresis loops and leakage current densities, respectively, as a function of applied voltage in 100 nm thick-PZT films deposited at 575 and 600°C. The PZT films deposited at both temperatures show similar P-E hysteresis loops, and were polycrystalline nature. The $2P_r$ (remanent polarization) and E_c (coercive field) of the PZT films are approximately 40 μC/cm² and 100 kV/cm, respectively. The lower P_r values relative to reported values may be due to the poor crystallization of the PZT films resulting from inclusion of silicon oxide phase. As shown in Fig. 8(b), leakage current densities of the PZT films deposited at 575°C are approximately 2×10^{-7} A/cm^2 at 1 V. The breakdown strength of the films was approximately 150kV/cm. The lower breakdown strength of the films may be due to the rough surface morphologies, as shown in the inset of Fig. 6.

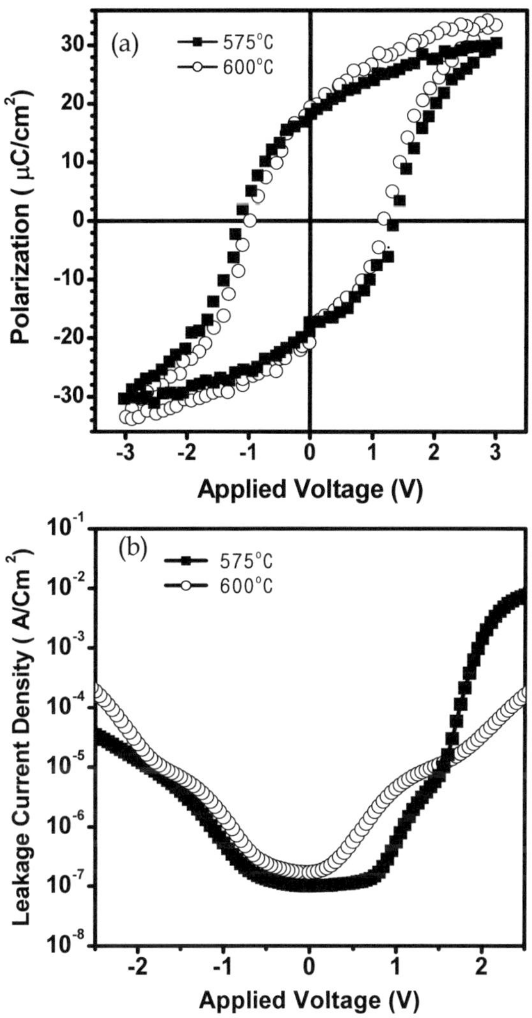

Fig. 8. (a) P-E hysteresis loops and (b) leakage current densities of 100nm thick-PZT thin films deposited at 575°C and 600°C

4. Conclusions

$SrRuO_3$ bottom electrodes were grown with an epitaxial relationship with SrO buffered-Si(001) substrates by pulsed laser deposition. The structural and electrical properties of the $SrRuO_3$ films were studied with deposition parameters of $SrRuO_3$ on the optimized SrO buffer layered Si (001) substrates. The optimum conditions of SrO buffer layers for $SrRuO_3$ preferred growth were a deposition temperature of 700°C, deposition pressure of 1×10^{-6} Torr, and thickness of 6 nm. 100nm thick-$SrRuO_3$ bottom electrodes deposited at 650°C on SrO buffered-Si(001) substrates showed a rms roughness of approximately 5.0 Å and a resistivity of 1700 μΩ-cm. 100nm thick-$Pb(Zr_{0.2}Ti_{0.8})O_3$ thin films deposited at 575°C on $SrRuO_3$/SrO/Si substrates showed a (00*l*) preferred orientation and exhibited a $2P_r$ of 40 μC/cm^2 and a E_c of 100 kV/cm. The leakage current density of the PZT films was approximately 1×10^{-7} A/cm^2 at 1 V. The silicon oxide phase, which presents within the PZT and $SrRuO_3$ films, influences the crystallinity of the PZT films and the resistivity of the $SrRuO_3$ electrodes.

5. Acknowledgments

This research was funded by the Center for Ultramicrochemical Process Systems sponsored by KOSEF, through a Korea Science and Engineering Foundation(KOSEF) grant funded by the Korean government (MOST) (R01-2007-000-21017-0), and was also supported by the BK 21 project.

6. References

[1] J.F. Scott, Ferroelectric memories, Vol. 3 of the Springer series on *Advanced Microelectronics*, Springer, Heidelberg, April 2000.

[2] T. Hidaka, T. Mayurama, M. Saitoh, N. Mikoshiba, M. Shimizu, T. Shiosaki, L.A. Wills, R. Hiskes, S.A. Dicarolis, and J. Amano, *Appl. Phys. Lett.* 68, 1996, 2358.

[3] M.E. Lines and A.M. Glass, *Principles and Applications of Ferroelectrics and Related Materials*, Oxford University Press, Oxford, England, 1977, p. 525.

[4] W.S. Lee, K.C. Ahn, C.S. Kim, and S.G. Yoon, *J. Vac. Sci. Technol.* B 23 (2005) 1901.

[5] W.S. Lee, K.C. Ahn, H.J. Shin, Y.S. Kim, K.S. No, and S.G. Yoon, *Integr. Ferroelectr.* 73 (2005) 125.

[6] C.B. Eom, R.J. Cava, R.M. Fleming, J.M. Phillips, R.B. VanDover, J.H. Marshall, J.W.P. Hsu, J.J. Krajewski, and W.F. Peck Jr., *Science*, 258 (1992) 1766.

[7] P. Legagneux, G. Garry, D. Dieumegard, C. Schwebel, C. Pellet, G. Gautherin, and J. Siejka, *Appl. Phys. Lett.* 53 (1988) 1506.

[8] D.K. Fork, D.B. Fenner, G.A.N. Connell, J.M. Phillips, and T.H. Geballe, *Appl. Phys. Lett.* 57 (1990) 1137.

[9] D.K. Fork, F.A. Ponce, J.C. Tramontana, and T.H. Geballe, *Appl. Phys. Lett.* 58 (1991) 2294.

[10] Y. Kado and Y. Arita, *J. Appl. Phys.* 61 (1987) 2398.

[11] S.K. Singh and S.B. Palmer, *Ferroelectrics*, 328 (2005) 85

[12] T. Higuchi, Y. Chen, J. Koike, S. Iwashita, M. Ishida, and T. Shimoda, *Jpn. J. Appl. Phys.* 41 (2002) 6867.

[13] W.S. Lee, K.C. Ahn, and S.G. Yoon, *J. Vac. Sci. Technol.* B 23 (2005) 1901.

[14] Q.X. Jia, F. Chu, C.D. Adams, X.D. Wu, M. Hawley, J.H. Cho, A.T. Findikoglu, S.R. Foltyn, J.L. Smith, and T.E. Mitchell, *J. Mater. Res.* 11 (1996) 2263.

[15] W.S. Lee, G.H. Jung, D.H. Kim, S.W. Kim, H.J. Kim, J.R. Park, Y.P. Song, H.K. Yoon, S.M. Lee, I.H. Choi, and S.G. Yoon, *J. of KIEEME*, 18 (2005) 810.

6

Electrocaloric Effect (ECE) in Ferroelectric Polymer Films

S. G. Lu[1], B. Rožič[2], Z. Kutnjak[2] and Q. M. Zhang[1]
[1]*Materials Research Institute and Department of Electric Engineering, The Pennsylvania State University, University Park, PA 16802*
[2]*Jozef Stefan Institute, 1000 Ljubljana*
[1]*USA*
[2]*Slovenia*

1. Introduction

The electrocaloric effect (ECE) is the change in temperature and/or entropy of a dielectric material due to the electric field induced change of dipolar states. Electrocaloric effect in dielectrics is directly related to the polarization changes under electric field.[1-3,6] Hence a large polarization change is highly desirable in order to achieve a large ECE which renders the ferroelectric materials the primary candidates for developing materials with large ECE. Figure 1 illustrates schematically the ECE in a dipolar material. Application of an electric field to the material causes partial alignment of dipoles and consequently a reduction of entropy of the dipolar system. In an isothermal condition, the dipolar material rejects heat $Q=T\Delta S$ to the surrounding, where T is the temperature and ΔS is the isothermal entropy change. Or in an adiabatic process, to keep the total entropy of the material constant, the temperature of the dielectric is increased by ΔT, the adiabatic temperature change which is related to the $Q = C\Delta T$ where C is specific heat capacity of the dielectric. In a reverse process, as the applied electric field is reduced to zero and the dipoles return to the less ordered state (or disordered state), an increase in the entropy of dipolar system occurs and under an isothermal condition, the dielectric will absorb heat Q from the surrounding.

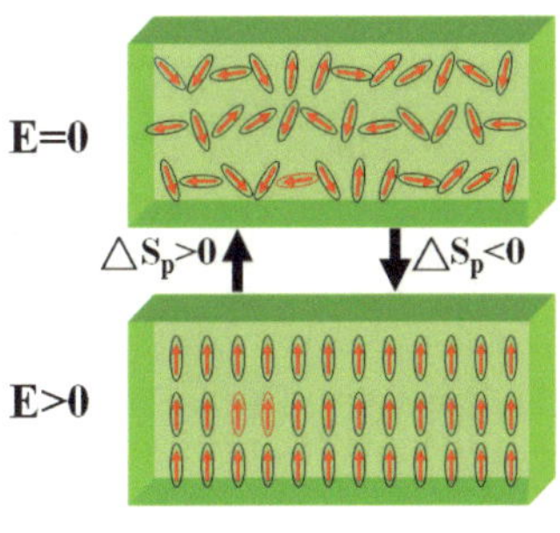

Fig. 1. Schematic drawing of the ECE process in a dipolar material. When E=0, the dipoles orient randomly. When E>0, especially larger than the coercive electric field, the dipoles orient along the electric field direction.

The ECE may provide an effective means of realizing solid-state cooling devices for a broad range of applications such as on-chip cooling and temperature regulation for sensors or other electronic devices. Refrigerations based on ECE have the potential of reaching high efficiency relative to vapor-compression cycle systems, and no green house emission.

Solid-state electric-cooling devices based on the thermoelectric effect (Peltier effect) have been used for many decades (Spanner, 1951; Nolas, Sharp, and Goldsmid, 2001). However, these cooling devices require a large DC current which results in large amount of waste heat through Joule heating. For example, using the typical coefficient of performance (COP) for these devices, e.g. 0.4 to 0.7, 2.4 to 3.5 watts of heat will be generated at the hot end of the system when pumping 1 watt heat from the cold end. Hence, the thermoelectric effect based cooling devices will not meet the requirement of high energy efficiency.

A counterpart of ECE is the MCE, which has been extensively studied for many years due to the findings of giant magnetocaloric effect in several magnetic materials near room temperature (Gschneidner Jr., Pecharsky and Tsokol, 2005; Pecharsky, Holm, Gschneidner Jr. and Rink, 2003). Both ECE and MCE devices exploit the change of order parameter brought about by an external electric or magnetic field. However, the difficulty of generating high magnetic field for MCE devices to reach giant MCE, severely limits their wide applications. This makes the MCE devices difficult to be used widely, especially for miniaturized microelectronic devices, and to achieve high efficiency. In contrast, high electric field can be easily generated and manipulated, which makes ECE based cooling devices attractive and more practical for a broad range of applications.

This chapter will introduce the basic concept of ECE, the thermodynamic considerations on materials with large ECE, and review previous investigations on the ECE in polar crystals, ceramics, and thin films. A newly discovered large ECE in ferroelectric polymers will be presented. Besides, we will also discuss different characterization techniques of ECE such as the direct measurements and that deduced from Maxwell relations, as well as phenomenological theory on ECEs.

2. Thermodynamic considerations on materials with large ECE

2.1 Maxwell relations

In general the Gibbs free energy G for a dielectric material could be expressed as a function of temperature T, entropy S, stress X, strain x, electric field E and electric displacement D in the form

$$G = U - TS - X_i x_i - E_i D_i , \tag{1}$$

where U is the internal energy of the system, the stress and field terms are written using Einstein notation. The differential form of Eq. (1) could be written as

$$dG = -SdT - x_i dX_i - D_i dE_i . \tag{2}$$

Entropy S, strain x_i and electric displacement D_i can be easily expressed when the other two variables are assumed to be constant,

$$S = -\left(\frac{\partial G_1}{\partial T}\right)_{X,D}, x_i = -\left(\frac{\partial G_1}{\partial X_i}\right)_{T,D}, D_i = -\left(\frac{\partial G_1}{\partial E_i}\right)_{T,X}. \tag{3}$$

The Maxwell relation can be derived for (S, T) and (D, E) two pairs of parameters (Line and Glass, 1977),

$$\left(\frac{\partial S}{\partial E_i}\right)_{T,X} = \left(\frac{\partial D_i}{\partial T}\right)_{E,X}. \tag{4}$$

Or

$$\left(\frac{\partial T}{\partial E}\right)_S = \frac{T}{c_E}\left(\frac{\partial D}{\partial T}\right)_E = \frac{Tp_E}{c_E}, \tag{5}$$

where c_E is the heat capacity, p_E the pyroelectric coefficient. Eqs. (4) and (5) indicate the mutually inverse relationships of ECE and the pyroelectric coefficient. Hence, for the ECE materials with a constant stress X imparted, the isothermal entropy change ΔS and adiabatic temperature change ΔT can be expressed as (Line and Glass, 1977)

$$\Delta S = -\int_{E_1}^{E_2}\left(\frac{\partial D}{\partial T}\right)_E dE, \tag{6}$$

$$\Delta T = -\frac{T}{\rho}\int_{E_1}^{E_2}\frac{1}{c_E}\left(\frac{\partial D}{\partial T}\right)_E dE. \tag{7}$$

Equations (4) through (7) indicate that in order to achieve large ΔS and ΔT, the dielectric materials should posses a large pyroelectric coefficient over a relatively broad electric field and temperature range. For ferroelectric materials, a large pyroelectric effect exists near the ferroelectric (F) – paraelectric (P) phase transition temperature and this large effect may be shifted to temperatures above the transition temperature when an external electric field is applied. It is also noted that a large ΔT may be achieved even if ΔS is small when the c_E of a dielectric material is small. However, as will be pointed out in the following paragraph, this is not desirable for practical refrigeration applications where a large ΔS is required.

It is noted that in the temperature region including a first-order FE-PE transition, Eq. (6) should be modified to take into account of the discontinuous change of the polarization ΔD at the transition, i.e.,

$$\Delta S = -\int_0^E\left(\frac{dD}{dT}\right)_E dE + \Delta D\left(\frac{\partial E}{\partial T}\right). \tag{8}$$

Although a few studies on the ECE were conducted in which direct measurement of ΔT was made (Sinyavsky, Pashkov, Gorovoy, Lugansky, and Shebanov, 1989; Xiao, Wang, Zhang, Peng, Zhu and Yang, 1998), most experimental studies were based on the Maxwell relations where the electric displacement D versus temperature T under different electric fields was characterized. ΔS and ΔT were deduced from Eqs. (6) and (7) (see below for details). For dielectric materials with low hysteresis loss and the measurement is in an ideal situation, results obtained from the two methods should be consistent with each other. However, as will be shown later that for the relaxor ferroelectric polymers, the ECE deduced from the

Maxwell relations can be very different from that measured directly and hence the Maxwell relations cannot be used for these materials in deducing ECE. In general, the Maxwell relations are valid only for thermodynamically equilibrium and ergodic systems.
In an ideal refrigeration cycle the working material (refrigerant) must absorb entropy (or heat) from the cooling load while in thermal contact with the load (isothermal entropy change ΔS). The material is then isolated from the load while the temperature is increased due to the application of external field (adiabatic temperature change ΔT). The material is then in thermal contact with the heat sink and entropy that was absorbed from the cooling load is rejected to the heat sink. The working material is then isolated from the heat sink and the temperature is reduced back as the field is reduced. The temperature of the refrigerant will be the same as the temperature of the cooling load when they are contacted. The whole process is repeated to further reduce the temperature of the load. Therefore, both the isothermal entropy change ΔS and the adiabatic temperature change ΔT are the key parameters for the ECE of a dielectric material for refrigeration (Wood and Potter, 1985; Kar-Narayan and Mathur, 2009).

2.2 Phenomenological theory of ECE

Phenomenological theory has been widely utilized to illustrate the macroscopic phenomena that occur in the polar materials, e.g. ferroelectric or ferromagnetic materials near their phase transition temperatures. The general form of the Gibbs free energy associated with the polarization can be expressed as a series expansion in terms of the electric displacement (Line and Glass, 1977)

$$G = \frac{1}{2}\alpha D^2 + \frac{1}{4}\xi D^4 + \frac{1}{6}\zeta D^6, \tag{8}$$

where $\alpha=\beta(T-T_0)$, and β, ξ and ζ are assumed to be temperature-independent phenomenological coefficients. From $\left(\frac{\partial G}{\partial T}\right)_D = -\Delta S$, one can obtain,

$$\Delta S = -\frac{1}{2}\beta D^2 \tag{9}$$

Then the adiabatic temperature change ΔT ($=T\Delta S/c_E$) can be obtained, i.e.

$$\Delta T = -\frac{1}{2c_E}\beta T D^2 \tag{10}$$

Based on Eqs. (9) and (10), the entropy will be reduced when the material changes to a polar state from a non-polar state when an external action, e.g. temperature, electric field or stress, is applied. The entropy change and temperature change are associated with the phenomenological coefficient β and electric displacement D, viz. proportional to β and D^2. Both parameters will affect the ECE values of the materials. A material with large β and large D will generate large ECE entropy change and temperature change near the ferroelectric (F) - paraelectric (P) phase transition temperature.

2.3 ECE in several ferroelectric materials

Based on the literature reported values of β and D, the ECE values of various ferroelectric materials could be estimated. For instance, for $BaTiO_3$, $\beta=6.7\times10^5 (JmC^{-2}K^{-1})$ and D=0.26

C/m^2 (Jona and Shirane, 1993; Furukawa, 1984), ΔS will be approximately 3 J/(kgK). Using the specific heat $c_E = 4.07\times10^2$ J/(kgK) and T_c = 107 °C (Jona and Shirane, 1993; Akay, Alpay Mantese, and Rossetti Jr, 2007), results in a ΔT = 2.8 °C. Similarly, for $Pb(Zr_xTi_{1-x})O_3$ ($0.0<x\leq0.6$), β = 1.88×10^5 and D = 0.39 C/m^2 (Amin, Cross, and Newnham, 1981; Amin, Newnham, and Cross, 1981), one will obtain ΔS = 1.8 J/(kgK). Taking T_c = 250 °C, and c_E = 3.4×10^2 J/(kgK) (PI Ceramic website, 2009), will result in ΔT = 2.7 °C.

For ferroelectric polymers, e.g. P(VDF-TrFE), phenomenological theory predicts large ECE values. For example, P(VDF-TrFE) 65/35 mol% copolymer, with β= 3.5×10^7 $JmC^{-2}K^{-1}$ and D = 0.08 C/m^2 (Furukawa, 1984), will exhibit a ΔS = 62 J/(kgK). Making use of its specific heat capacity c_E = 1.4×10^3 J/(kgK) (Furukawa, Nakajima, and Takahashi, 2006) and Curie temperature T_c=102 °C (Furukawa T, 1984), yields ΔT = 16.6°C. The large ΔS and ΔT values suggest that a large ECE may be achieved in ferroelectric P(VDF-TrFE) copolymers. Furthermore, relaxor ferroelectric polymers based on P(VDF-TrFE), such as P(VDF-TrFE-CFE) 59.2/33.6/7.2 mol% (CFE-chlorofluoroethylene) relaxor ferroelectric terpolymers also have potential to reach a large ECE because the β and D are still large.

It was found that β of ferroelectric ceramics (~ 10^5) is about two orders of magnitude smaller than that of P(VDF-TrFE) (~ 10^7). D however is only several times higher for ceramics, since ΔS ~ βD^2, ΔS is still about one order of magnitude smaller than that of the P(VDF-TrFE) based polymers.

In addition, the heat of F-P phase transition can also be used to assess the ECE (Q=TΔS) in a ferroelectric material at temperatures above the F-P transition. For a very strong order-disorder ceramic system (an example of which is the ferroelectric ceramic triglycine sulphate, TGS), the heat of F-P phase transition is 2.0×10^3 J/kg (corresponding to an entropy change of ΔS ~ 6.1 J /(kgK)). For $BaTiO_3$, F-P heat is smaller 9.3×10^2 J/kg (ΔS ~ 2.3 J/(kgK)) (Jona and Shirane, 1993). In other words, although ceramic materials may exhibit a higher adiabatic temperature change, their isothermal entropy change may not be very high. In contrast, ferroelectric polymers offer higher heat of transition in a F-P transition. For instance, P(VDF-TrFE) 68/32 mol% copolymer shows a heat of F-P transition of more than 2.1×10^4 J/kg (or ΔS ~ 56.0 J/(kgK)) (Neese, Chu, Lu, Wang, Furman and Zhang, 2008). This is approximately 10 times larger than its inorganic counterparts.

3. Investigations on ECE in polar materials

3.1 ECE studies in ferroelectric ceramics and single crystals

The history of ECE study may be traced back to as early as 1930s. In 1930, Kobeko and Kurtschatov did a first investigation on ECE in Rochelle salt (Kobeko and Kurtschatov, 1930), however they did not report any numerical values. In 1963, Wiseman and Kuebler redid their measurements (Wiseman and Kuebler, 1963), obtaining ΔT=0.0036 °C in an electric field of 1.4 kV/cm at 22.2°C. In their study, the Maxwell relation was used to derive $\Delta T(\Delta T = -\frac{T}{c_E}\frac{\partial\alpha}{\partial T}D\Delta D)$, where α=1/ε as defined in Eq. (8) and ε is the permittivity). The isothermal entropy change was 28.0 J/m^3K (1.56×10^{-2} J/(kgK)).

Other studies on inorganic materials used KH_2PO_4 crystal, and $SrTiO_3$, $Pb(Sc_{0.5}Ta_{0.5})O_3$, and $Pb_{0.98}Nb_{0.02}(Zr_{0.75}Sn_{0.20}Ti_{0.05})_{0.98}O_3$ ceramics. For KH_2PO_4 crystal, Maxwell relation was used in the form of $\Delta T = -(T / c_E)\left(\frac{\partial D/\partial T}{\partial D/\partial E}\right)dD$ to obtain ΔT =1 °C for a 11 kV/cm electric field and an entropy change of 2.31×10^3 J/m^3K (or 0.99 J/(kgK)) (Baumgartner, 1950). For $SrTiO_3$, ΔT

=1 °C and ΔS=34.63 J/m^{3}K (6.75 × 10^{-3} J/(kgK)) under 5.42 kV/cm electric field at 4 K from Eq. (7) (Lawless and Morrow, 1977). For $Pb(Sc_{0.5}Ta_{0.5})O_3$, a ΔT =1.5 °C and ΔS of 1.55 × 10^4 J/m^{3}K (1.76 J/(kgK)) were measured directly for a sample under 25 kV/cm field at 25 °C (Sinyavsky and Brodyansky, 1992). For $Pb_{0.98}Nb_{0.02}(Zr_{0.75}Sn_{0.20}Ti_{0.05})_{0.98}O_3$, ΔT =2.5 °C and ΔS=1.73 × 10^4 J/m^{3}K (2.88 J/(kgK)) at 30 kV/cm and 161 °C deduced from Eq. (7) (Tuttle and Payne, 1981).

A direct ECE measurement was carried out for (1-x)$Pb(Mg_{1/3}Nb_{2/3})O_3$-x$PbTiO_3$ (x=0.08, 0.10, 0.25) ceramics near room temperature using a thermocouple when a dc electric field was applied (Xiao, Wang, Zhang, Peng, Zhu and Yang, 1998). A temperature change of 1.4 °C was observed for x=0.08 although at high temperatures (as x increased), this change was reduced. This high ECE can be accounted for by considering the electric field-induced first-order phase transition from the mean cubic phase to 3m phase.

These results indicate that the ECE in ceramic and single crystal materials are relatively small, viz. ΔT<2.5 °C, and ΔS<2.9 J/(kgK), mainly because the breakdown field is low, using applied electric fields that are less than 3 MV/m. Defects existing in bulk materials cause early breakdown and empirically the breakdown electric field was inversely proportional to the material's thickness. For piezoelectric ceramics, the breakdown field (in kV/cm) is related to the thickness (in cm) via the relationship, $E_b=27.2t^{-0.39}$, indicating that thin films are more appropriate for an ECE study. Additionally, the breakdown field of dielectric polymers can be several orders of magnitude higher than ceramics, suggesting polar-polymers are good candidates for ECE investigations.

3.2 ECE in ferroelectric and antiferroelectric thin films

In 2006, Mischenko et al. investigated ECE in sol-gel derived antiferroelectric $PbZr_{0.95}Ti_{0.05}O_3$ thin films near the F – P transition temperature. In their study, films with 350 nm thickness were used to allow for electric fields as high as 48 MV/m. An adiabatic temperature change of 12 °C was obtained (as deduced from Eq. (7)) at 226 °C, slightly above the phase transition temperature (222°C) (Mischenko, Zhang, Scott, Whatmore and Mathur, 2006). Both the high electric field and the high operation temperature near phase transition contribute to the large ΔT (=$T\Delta S/c_E$). On the other hand, its isothermal entropy change is estimated to be 8 J/(kgK), which is not high compared with magnetic alloys exhibiting giant magnetocaloric effect (MCE) near room temperature, where $\Delta S \geq 30$ J/(kgK) was observed (Provenzano, Shapiro and Shull, 2004). As stated previously, for high performance refrigerants, a large ΔS is necessary (Wood and Potter, 1985).

To reduce the operational temperature for large ECE in ceramic thin films, Correia et al. successfully fabricated $PbMg_{1/3}Nb_{2/3}O_3$-$PbTiO_3$ thin films with perovskite structure using $PbZr_{0.8}Ti_{0.2}O_3$ seed layer on Pt/Ti/TiO_2/SiO_2/Si substrates (Correia, Young, Whatmore, Scott, Mathur and Zhang, 2009). A temperature change of ΔT=9 K was achieved at 25 °C. An entropy change of 9.7 J/(kgK) can be deduced. A significant difference for ferroelectric thin films is that the largest ΔT occurs at 25 °C, near the depolarization temperature (18 °C), not above the permittivity peak temperature. The large ECE only happens at field heating. Transitions for stable and metastable polar nanoregions (PNR) to nonpolar regions are accounted for by observed phenomena. Interactions of PNRs are similar to that between the dipoles in a glass. The field-induced phase transition has been observed in PMN-PT single crystals (Lu, Xu and Chen, 2005; Ye and Schmid, 1993). Thermal history has a critical impact on the field- induced phase transition. Relaxor ferroelectrics are of great interest due to their phase transition temperatures being near or at room temperature. The field induced phase transition may produce larger polarization, e.g. induced polarization, <P_d>, which can lead to larger dP/dT as well as large ΔS and ΔT.

For thin film, the substrate must be taken into account as it may exert stresses on the thin film due to the misfit of the lattices and electromechanical coupling from the strain changes under electric field. The free energy of thin film is subject to lateral clamping and may be expressed as (Akay, Alpay Mantese, and Rossetti Jr, 2007)

$$G_{film} = G_0 + \tilde{\alpha}D^2 + \tilde{\xi}D^4 + \tilde{\zeta}P^6 - EP + G_{strain}, \tag{11}$$

where

$$\tilde{\alpha} - \alpha - 2u_m Q_{12}\tilde{C} \tag{12}$$

and

$$\tilde{\beta} = \beta + Q_{12}^2\tilde{C} \tag{13}$$

are the modified phenomenological coefficients, $G_{strain} = u_m^2\tilde{C}$ is the polarization-free strain energy, $\tilde{C} = C_{11} + C_{12} - 2C_{12}^2 / C_{11}$, C_{ij} are the elastic constants at constant polarization, Q_{ij} are the cubic electrostrictive coefficients, and u_m is the in-plane misfit strain. The phase transition temperature varies linearly with the lattice misfit strain via Eq. (12) while the two-dimensional clamping is illustrated by Eq. (13). The excess entropy S_E^{XS} and specific heat ΔC_E of the ferroelectric phase transition follow the form (Akay, Alpay Mantese, and Rossetti Jr, 2007),

$$S_E^{XS}(T,E) = -T\left(\frac{\partial G(D)}{\partial T}\right)_E, \tag{14}$$

$$\Delta C_E(T,E) = -T\left(\frac{\partial^2 G(D)}{\partial T^2}\right)_E. \tag{15}$$

It was found that for $BaTiO_3$ (BTO) thin film deposited on substrate, perfect lateral clamping of BTO will transform the discontinuous phase transition (1st order phase transition) into a continuous one. Accordingly the polarization and the specific heat capacity will be reduced near the phase transition temperature. On the other hand, based on Eqs. (12) and (13), adjustment of misfit strain in epitaxial ferroelectric thin films may vary the magnitude and temperature dependencies of their ECE properties.

4. Large ECE in ferroelectric polymer films

4.1 Direct method to measure ECE

Although most of the ECE studies rely on Maxwell relation (indirect method) to deduce the ECE of a material, it is always desirable to directly measure ECE in a dielectrics as refrigerants in cooling devices, i.e., to directly measure the isothermal entropy change ΔS and adiabatic temperature change ΔT induced by a change in the applied field (direct method). In our study, both the indirect method and direct method were employed to characterize the ECE in polymer films. The direct comparison of the results from two methods can also shed light on how reliable the indirect method is in deducing the ECE from a ferroelectric material.

There are several methods that have been used in measuring the magnetocaloric effect (MCE) in terms of measuring the isothermal entropy change and adiabatic temperature change, such as thermocouple (Dinesen, Linderoth and Morup, 2002; Lin, Xu and Zhang, 2004; Spichkin, Derkachb, Tishin, Kuz'min, Chernyshov, Gschneidner Jr, and Pecharsky, 2007), thermometer (Gopal, Chahine and Bose, 1997), and calorimeter (Tocado, Palacios and Burriel, 2005; Pecharsky, Moorman and Gschneidner, Jr, 1997).

Here, a high resolution calorimeter was used to measure the sample temperature variation due to ECE when an external electric field was applied (Yao, Ema and Garland, 1998). The temperature signal was measured by a small bead thermistor. A step-like pulse was generated by a functional generator to change the applied electric field in the film, and the width of the pulse was chosen so that the sample can reach thermal equilibrium with surrounding bath. Due to the fast electric as well as thermal response (ECE) of the polymer films (in the order of tens of milliseconds (Furukawa, 1989), a simple zero-dimensional model to describe the thermal process can be applied with sufficient accuracy. In a relaxation mode, the temperature T(t) of the whole sample system can be measured, which has an exponential relationship with time, i.e.

$$T(t) = T_{bath} + \Delta T e^{-t/\tau}, \tag{16}$$

where T_{bath} is the initial temperature of the film, ΔT the temperature change of the polymer film. The total temperature change ΔT_{EC} of the whole sample system was measured, which can be expressed as $\Delta T_{EC} = \Delta T \sum C_p^i / C_p^{EC}$. Here, C_p^i represents the heat capacity of each subsystem, C_p^{EC} is the heat capacity of the polymer film covered with electrode. ΔT_{EC} was measured as a function of temperature at constant electric field and as a function of electric field at constant temperature. ΔS can be determined from $T\Delta S = C_p^i \Delta T$.

4.2 ECE in the normal ferroelectric P(VDF-TrFE) 55/45 mol% copolymer

4.2.1 Experimental results of ECE

As indicated in Section 2, the ferroelectric copolymer may produce large ECE near its phase transition temperature. P(VDF-TrFE) 55/45 mol% was chosen because its F-P phase transition is of second-order (continuous), thus avoiding the thermal hysteresis effect associated with the first-order phase transition. In addition, among all available P(VDF-TrFE) copolymers, this composition exhibits the lowest F-P phase transition temperature (~ 70 °C), which is favorable for refrigeration near room temperature.

Polymer films used for the indirect ECE measurement were prepared using a spin-casting method on metalized glass substrates. The film thickness for this study was in the range of 0.4 μm to 1 μm. The free-standing films for the direct ECE measurement were fabricated using a solution cast method and the film thickness is in the range of 4 μm to 6 μm. Figure 2 shows the permittivity as a function of temperature for P(VDF-TrFE) 55/45 mol% copolymers measured at 1 kHz. It can be seen that the thermal hysteresis between the heating and cooling runs is pretty small (~ 1 °C). The remanent polarization as a function of temperature shown in Fig. 3 further indicates a second-order phase transition occurred in the material. The phase transition temperature is about 70 °C, and the glass transition temperature is about -20 °C. At temperature higher than 100 °C, the loss tangent rises sharply, which is associated with the thermally activated conduction.

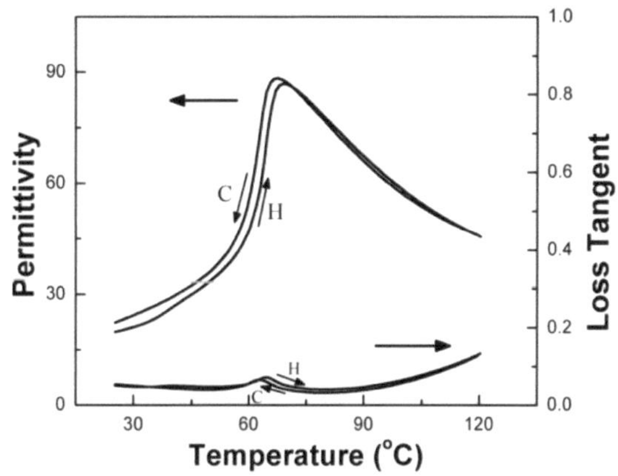

Fig. 2. Permittivity as a function of temperature for P(VDF-TrFE) 55/45 mol% copolymers.

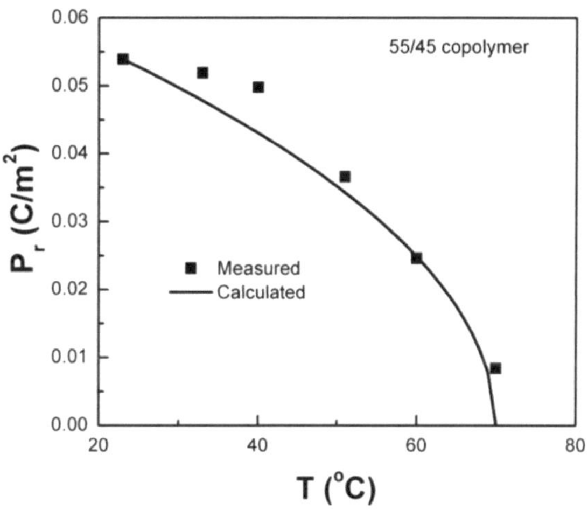

Fig. 3. Remanent polarization as a function of temperature for P(VDF-TrFE) 55/45 mol% copolymers.

Figure 4 shows the electric displacement as a function of electric field measured at various temperatures. At temperatures below the transition temperature, the polymer film is in a ferroelectric state, the normal hysteresis loop is observed while at higher temperatures, the loop becomes slimed, remanent polarization diminishes, and saturation polarization still exists. Hence the electric displacement as a function of electric field at different temperatures can be procured, which is presented in Fig. 5 (Neese, 2009). One can see that the electric displacement monotonically decreases with temperature above the phase transition. The Maxwell relations were used to calculate the isothermal entropy change and adiabatic temperature change as a function of ambient temperature. The results deduced are presented in Figs. 6 and 7.

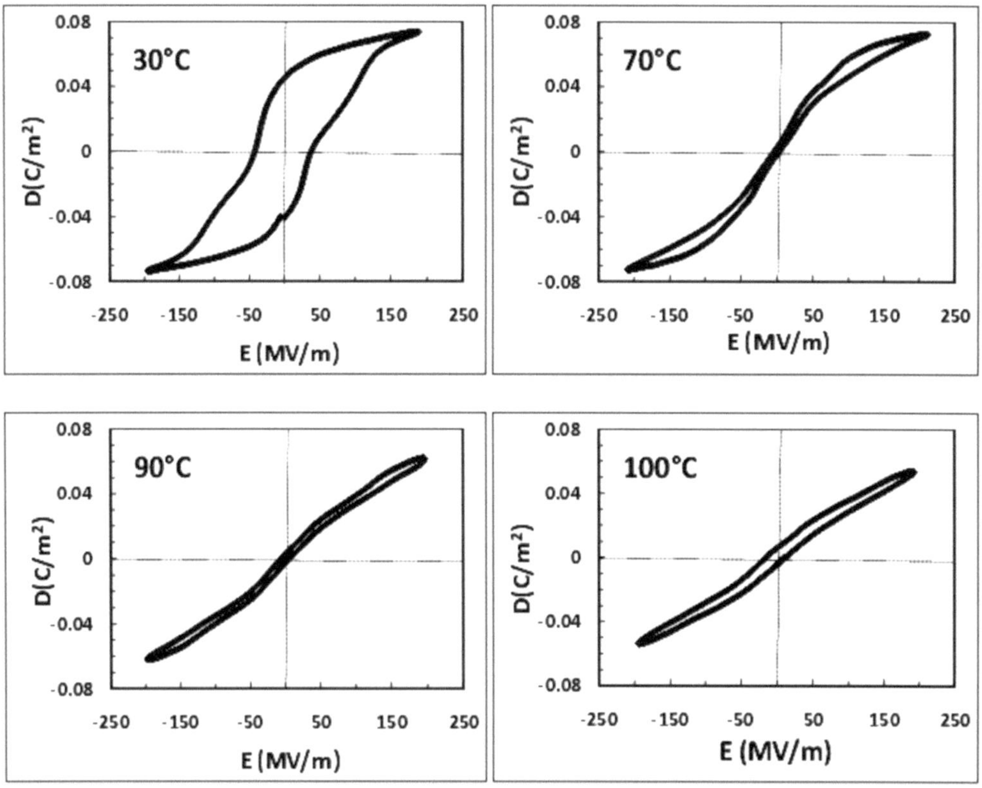

Fig. 4. Electric displacement – electric field hysteresis loops at temperature below and above the phase transition.

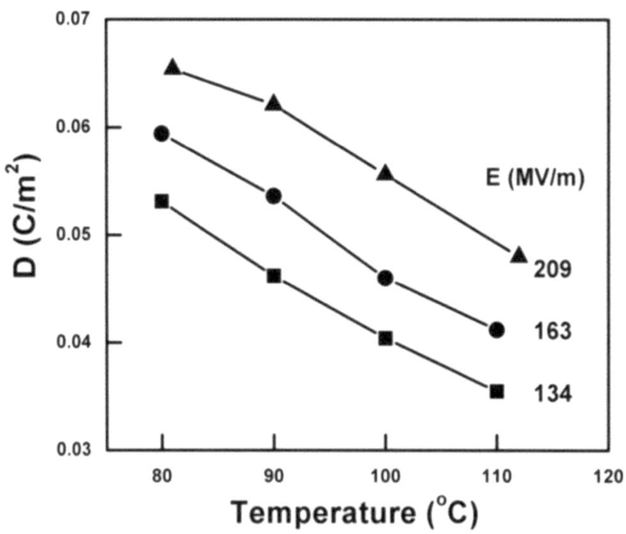

Fig. 5. Electric displacement as a function of temperature at different electric fields for P(VDF-TrFE) 55/45 mol% copolymers.

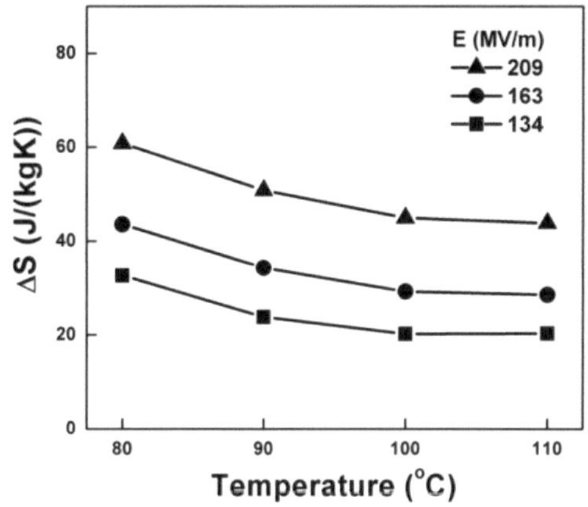

Fig. 6. Isothermal entropy changes as a function of ambient temperature at different electric fields.

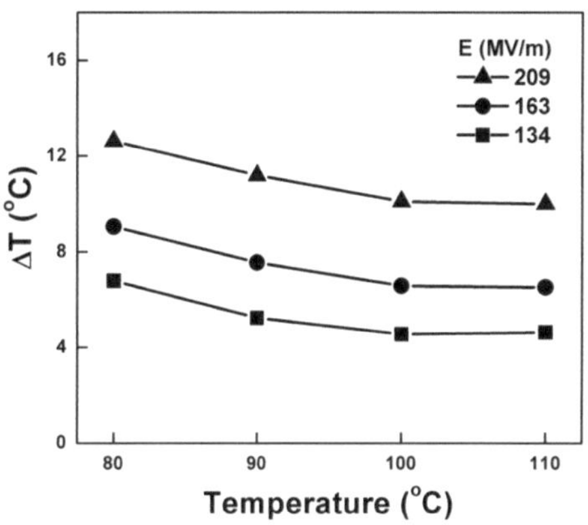

Fig. 7. Adiabatic temperature changes as a function of ambient temperature at different electric fields.

Present in Fig. 8 is the directly measured ΔS and ΔT as a function of temperature measured under several electric fields for the unstretched P(VDF-TrFE) 55/45 mol% copolymer. As can be seen, the ECE effect reaches maximum at the temperature of FE-PE transition. A comparison between the directly measured and deduced ECE indicates that within the experimental error, the ECE deduced from the Maxwell relation is consistent with that directly measured. Therefore, for a ferroelectric material at temperatures above F-P transition, Maxwell relation can be used to deduce ECE.

4.2.2 Phenomenological calculations on ECE

It is well established by many studies (see Fig. 2) that the F-P phase transition of P(VDF-TrFE) 55/45 copolymer is of second-order. For the copolymer with 2nd order phase transition, free energy associated with polarization can be written as

$$G = G_0 + \frac{1}{2}\beta(T - T_c)P^2 + \frac{1}{4}\xi P^4 - EP \tag{17}$$

where G_0 is the free energy of the material not associated with polarization, β and ξ are phenomenological coefficients, that are assumed temperature independent. T_c is the Curie temperature, E the electric field, and P the polarization.
Differentiating G with respect to P yields the relationship between E and P,

$$E = \beta(T - T_c)P + \xi P^3. \tag{18}$$

When E=0 and at $T<T_c$,

$$P^2 = \beta(T - T_c) / \xi. \tag{19}$$

Further differentiating the Eq. (18) yields the reciprocal permittivity,

$$\frac{1}{\varepsilon} = \beta(T - T_c) + 3\xi P^2 \quad (T<Tc). \tag{20}$$

$$\frac{1}{\varepsilon} = \beta(T - T_c) \quad (T\geq Tc). \tag{21}$$

Using Eqs. (19) and (21), the permittivity versus temperature (Fig. 2), and the polarization versus temperature relationships (Fig. 3), β and ξ can be obtained. Their values are, $\beta=2.4\times10^7$ ($JmC^{-2}K^{-1}$), and $\xi=3.9\times10^{11}$ (Jm^5C^{-4}).
Now Eq. (18) can be used to derive the polarization as a function of temperature under different DC bias fields. Before doing the calculation, it should be noted that the F-P transition temperature is a function of DC bias field. This relationship was obtained by directly measuring the permittivity as a function of temperature in different DC bias fields. The results are shown in Fig. 9.
However, the dielectric measurement becomes extremely difficult when $E_{DC}>100$ MV/m. Hence, the relationship of ΔT_c – $E^{2/3}$ (Lines and Glass, 1977) was fitted and extrapolated to obtain T_c at E > 100 MV/m.
The calculated polarization versus temperature relationships under different DC biases are shown in Fig. 10. Based on the D-T data, the ΔS and ΔT can be calculated via Eqs. (6) and (7). Results are shown in Figs. 11 and 12.

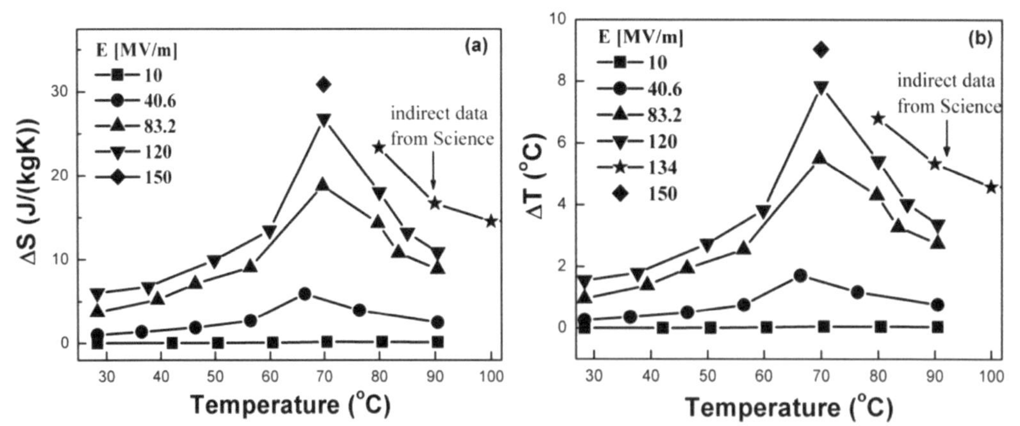

Fig. 8. Directly measured ΔS (a) and ΔT (b) as a function of temperature under several electric fields for unstretched P(VDF-TrFE) 55/45 mol% copolymers.

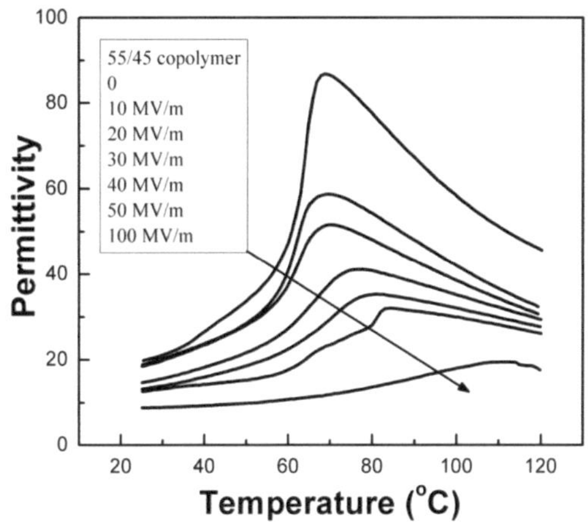

Fig. 9. Permittivity as a function of temperature at 1 kHz in various DC bias fields for 55/45 copolymer.

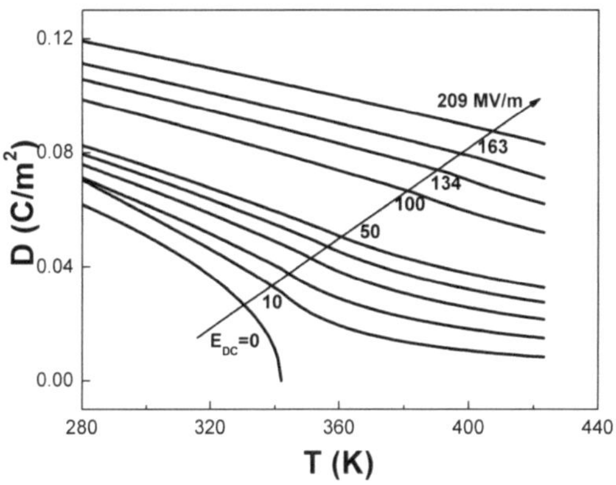

Fig. 10. Polarization versus temperature relationships with various DC biases for 55/45 copolymer.

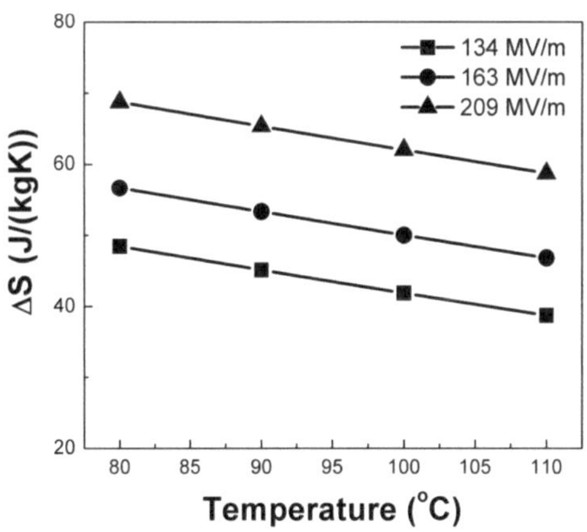

Fig. 11. ECE entropy changes versus temperature for 55/45 copolymer.

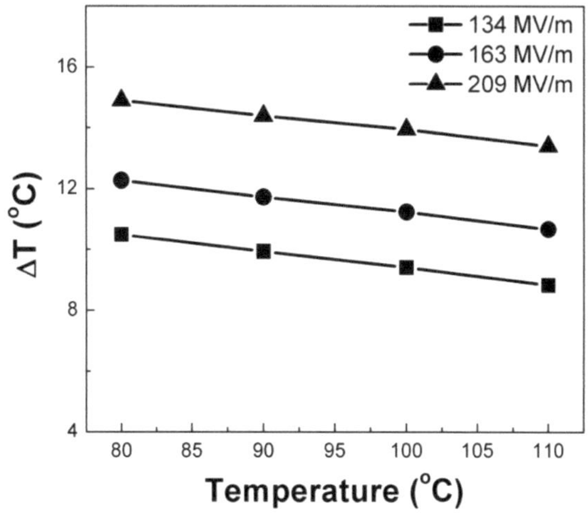

Fig. 12. ECE temperature changes versus temperature for 55/45 copolymer.

Phenomenological calculation indicates that, there is a giant ECE exhibited by P(VDF-TrFE) 55/45 copolymers. The ΔS and ΔT can reach 70 J/(kgK) and 15 °C respectively near the phase transition temperature ~ 70 °C. It can also be seen that ΔS has a linear relationship with D^2 (or P^2), the slope is $1/2\beta$.

4.3 ECE in the relaxor ferroelectric P(VDF-TrFE-CFE) terpolymers

Both the pure relaxor ferroelectric P(VDF-TrFE-CFE) 59.2/33.6/7.2 mol% terpolymer and blends with 5% and 10% of P(VDF-CTFE) copolymer were studied. For the P(VDF-TrFE-CFE) relaxor terpolymer, it was observed that blending it with a small amount of P(VDF-CTFE) 91/9 mol% copolymer [CTFE: chlorotrifluoroethylene] can result in a large increase in the elastic modulus, especially at temperatures above the room temperature, which does not affect the polarization level very much. Such an increase in the elastic modulus

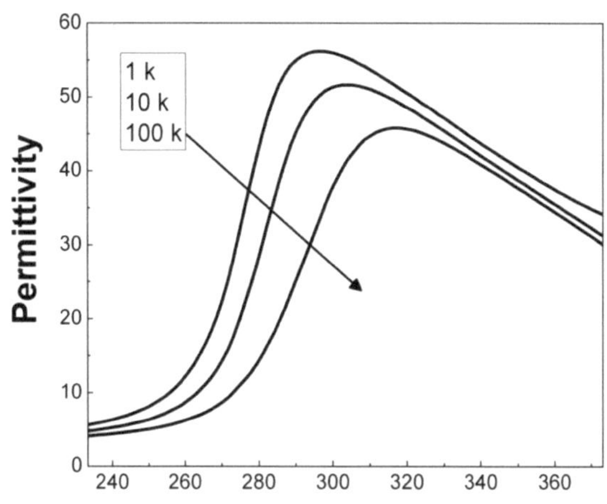

Fig. 13. Permittivity as a function of temperature at 1, 10 and 100 kHz for 59.2/33.6/7.2 mol% terpolymer.

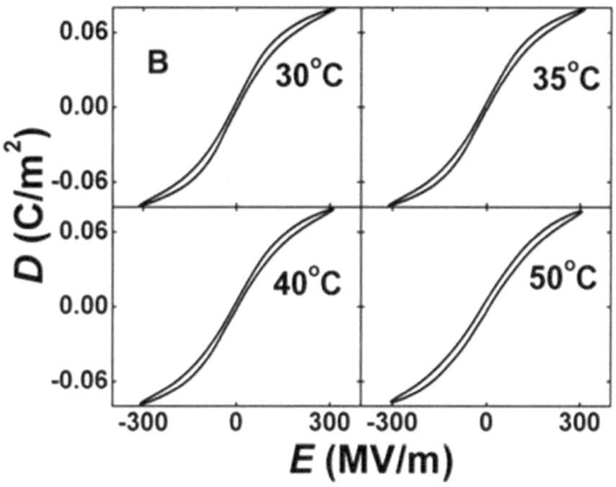

Fig. 14. Electric displacement - electric field hysteresis loops at temperature above the phase transition for 59.2/33.6/7.2 mol% terpolymers.

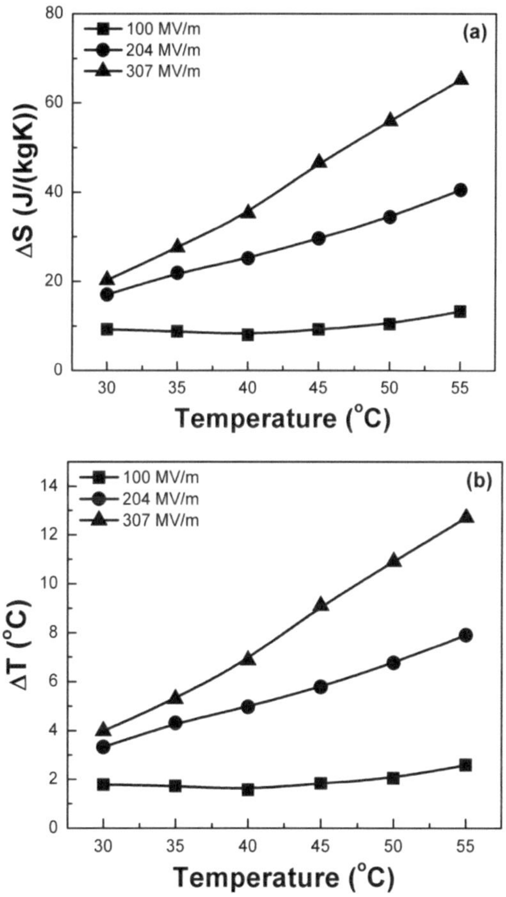

Fig. 15. ECE entropy changes (a) and temperature change (b) versus temperature deduced from Maxwell relation for 59.2/33.6/7.2 mol% terpolymer.

improves the dielectric strength of the blend polymer films and allows the direct measurement of ECE to be carried out to higher fields (>100 MV/m).
Present in Fig. 13 is the dielectric constant data for the terpolymer. The D-E loops for the P(VDF-TrFE-CFE) 59.2/33.6/7.2 mol% terpolymer are presented in Fig. 14, from which ΔS and ΔT are deduced from the Maxwell relatoin as shown in Fig. 15 (a) and 15 (b), respectively. The results show that the terpolymer has a weak ECE at room temperature and increases with temperature. At 55 °C which is the highest temperature measured, a ΔS=55 J/(kgK) and ΔT=12 °C under the field of 307 MV/m are deduced from the Maxwell relation.
The ECE from the direct measurement is presented in Fig. 16, which is for the 59.2/33.6/7.2 mol% terpolymer. The data show quite different behavior compared with Fig. 15. First of all, the directly measured ECE from the relaxor terpolymer is much larger than that deduced from the Maxwell relation. Moreover, the directly measured ECE shows much weak temperature dependence at E < 70 MV/m.

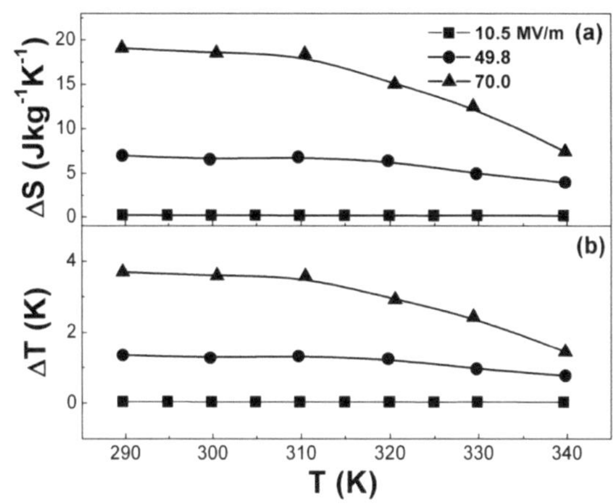

Fig. 16. Directly measured entropy changes (a) and temperature change (b) versus temperature for 59.2/33.6/7.2 mol% terpolymer .

The results indicate that the Maxwell relation is not suitable for ECE characterization for the relaxor ferroelectric polymers even at temperatures above the broad dielectric constant maximum. This is likely caused by the non-ergodic behaviour of relaxor ferroelectric polymers even at temperatures above the dielectric constant maximum while the Maxwell relations are valid only for thermodynamically equilibrium systems (ergodic systems)
We also note that a recent report of ECE deduced from the Maxwell relation on a P(VDF-TrFE-CFE) relaxor ferroelectric terpolymer by Liu et al. (Liu et al. 2010) shows very irregular field dependence of ECE measured at temperatures below 320 K where ECE in fact decreases with field, which is apparently not correct. These results all indicate that the Maxwell relation cannot be used to deduce ECE for the relaxor ferroelectric polymers, or even the relaxor ferroelectric materials in general, even at temperatures far above the freezing transition and the broad dielectric constant peak temperature.

5. Conclusions

General considerations for polar materials to achieve larger ECE were presented based on the phenomenological theory analysis. It is shown that in order to realize large ECE, a dielectric material with a large polarization P as well as large phenomenological coefficient β is required. It is further shown that both the phenomenological consideration and experimental data on heat of ferroelectric-paraelectric transition suggest that ferroelectric P(VDF-TrFE) based polymers have potential to achieve giant ECE. Indeed, experimental results show that the normal ferroelectric P(VDF-TrFE) 55/45 mol% copolymers exhibit a large ECE, i.e., an adiabatic temperature change over 12 °C and an isothermal entropy change over 50 J/(kgK) were obtained. The experimental results also indicate that for the normal ferroelectric materials, the ECE deduced from the Maxwell relation is consistent with that directly measured.

The experimental results on ECE in the relaxor ferroelectric P(VDF-TrFE-CFE) terpolymer were also presented which reveal a very large ECE at ambient condition in the relaxor terpolymers. In contrast to the normal ferroelectric polymers, the ECE deduced from the Maxwell relation for the relaxor terpolymers significantly deviates from that directly measured. The results indicate that the Maxwell relation is not suitable for ECE characterization for the relaxor ferroelectric polymers even at temperatures above the broad dielectric constant maximum. This is likely caused by the non-ergodic behaviour of relaxor ferroelectric polymers even at temperatures above the dielectric constant maximum while the Maxwell relations are valid only for thermodynamically equilibrium systems (ergodic systems) .

As a final point, one interesting question to ask when searching for electrocaloric materials to achieve giant ECE at ambient temperature is how to design dielectric materials to significantly enhance the entropy in the polar-disordered state since ECE is directly related to the entropy difference between the polar-disordered and ordered states in a dielectric material, in other words, how to design a ferroelectric material to increase β while maintaining large D in Eqs. (9) and (10). This is certainly an interesting area of research. The successful outcome will have significant impact on the society, in terms of efficient energy use for refrigeration, new and compact cooling devices which are more environmentally friendly.

6. Acknowledgements

The works at Penn State University was supported by the US Department of Energy, Division of Materials Sciences, under Grant No. DE-FG02-07ER46410. The work at Jozef Stefan Institute was supported by the Slovenian Research Agency. The authors thank B. Neese, B. Chu, Y. Wang, E. Furman, Xinyu Li, and Lee J. Gorny for their contributions to the works presented in this chapter.

7. References

Akcay, G.; Alpay, S. P.; Mantese, J. V. & Rossetti Jr., G. A. (2007). Magnitude of the intrinsic electrocaloric effect in ferroelectric perovskite thin films at high electric fields. *Appl Phys Lett*, 90 (25) (JUN, 2007), 252909/1-3. ISSN: 0003-6951.

Amin, A.; Cross, L. E. & Newnham, R. E. (1981). Calorimetric and phenomenological studies of the $PbZrO_3$-$PbTiO_3$ system. *Ferroelectrics*, 37 (1-4) (1981), 647-650. ISSN: 0015-0193.

Amin, A; Newnham, R. E.; Cross, L. E. & Cox, D. E. (1981). Phenomenological and structural study of a low-temperature phase-transition in the $PbZrO_3$-$PbTiO_3$ system. *J. Solid State Chem*, 37 (2) (1981), 248-255. ISSN: 0022-4596.

Baumgartner, H. (1950). Elektrische sättigungserscheinungen und elektrokalorischer effect von kaliumphosphat KH_2PO_4. *Helv Phys Acta*, 23 (6-7) (1950), 651-696. ISSN: 0018-0238.
Correia, T. M.; Young, J. S.; Whatmore, R. W.; Scott, J. F.; Mathur, N. D. & Zhang, Q. (2009). Investigation of the electrocaloric effect in a $PbMg_{1/3}Nb_{2/3}O_3$-$PbTiO_3$ relaxor thin film. *Appl Phys Lett*, 95 (18) (NOV, 2009), 182904/1-3. ISSN: 0003-6951.
Dinesen, A. R.; Linderoth, S. & Mørop, S. (2002). Direct and indirect measurement of the magnetocaloric effect in a $La_{0.6}Ca_{0.4}MnO_3$ ceramic perovskite. *J. Magn Magn Mater*, 253 (1-2) (DEC, 2002), 28-34. ISSN: 0304-8853.
Epstein, R. & Malloy, K. J. (2009). Electrocaloric devices based on thin-film heat switches. *J Appl Phys*, 106 (6) (SEP, 2009), 064509/1-7. ISSN: 0021-8979.
Furukawa, T. (1984). Phenomenological aspect of a ferroelectric vinylidene fluoride / trifluoroethylene copolymer. *Ferroelectrics*, 57 (1-4) (1984), 63-72. ISSN: 0015-0193.
Furukawa, T. (1989). Piezoelectricity and pyroelectricity in polymers. *IEEE Trans Electr Ins*, 24 (3) (JUN, 1989), 375-394. ISSN: 0018-9367.
Furukawa, T.; Nakajima, T. & Takahashi, Y. (2006). Factors governing ferroelectric switching characteristics of thin VDF/TrFE copolymer films. *IEEE Trans Diel Electr Ins*, 13 (5) (OCT, 2006), 1120-1131. ISSN: 1070-9878.
Gopal, B. R.; Chahine, R. & Bose, T. K. (1997). A sample translatory type insert for automated magnetocaloric effect measurements. *Rev Sci Instrum*, 68 (4) (APR, 1997), 1818-1822. ISSN: 0034-6748.
Gschneidner Jr., K. A.; Pecharsky, V. K. & Tsokol, A. O. (2005). Recent developments in magnetocaloric materials. *Rep Prog Phys*, 68, 1479-1539. ISSN: 0034-4885.
Jona, F. & Shirane, G. (1993). *Ferroelectric Crystals*. Dover Publications, Inc. ISBN: 0486673863, 9780486673868. New York.
Kar-Narayan, S. & Mathur, N. (2009). Predicted cooling powers for multilayer capacitors based on various electrocaloric and electrode materials. *Appl. Phys. Lett.* 95, 242903/1-3.
Kobeko, Von P. & Kurtschatov, J. (1930). Dielektrische eigenschaften der seignettesalzkristalle', *Z Phys*, 66, 192-205. ISSN: 0340-2347.
Kucherov, Y. R. (1997). 'Piezo-pyroelectric energy converter and method', US Patent, US 5644184.
Lang, S. B. (1976). Cryogenic refrigeration utilizing electrocaloric effect in pyroelectric lithium sulfate monohydrate. *Ferroelectrics*, 11 (3-4) (1976), 519-523. ISSN: 0015-0193.
Lawless, W. N. & Morrow, A. J. (1977). Specific heat and electrocaloric properties of a $SrTiO_3$ ceramic at low temperatures. *Ferroelectrics*, 15 (3-4) (1977), 159-165. ISSN: 0015-0193.
Lin, G. C.; Xu, C. D. & Zhang, J. X. (2004). Magnetocaloric effect in $La_{0.80}$-$xCa_{0.20}Sr_xMnO_3$ (x=0.05, 0.08, 0.10). *J Magn Magn Mater*, 283 (2-3) (DEC, 2004), 375-379. ISSN: 0304-8853.
Lines, M. & Glass, A. (1977). *Principles and Applications of Ferroelectrics and Related Materials*, Clarendon Press, ISBN: 0-19-851286-4. Oxford.
Liu, P. F.; Wang, J. L.; Meng, X. J.; Yang, J. Dkhil B. & Chu, J. H. (2010). Huge electrocaloric effect in Langmuir–Blodgett ferroelectric polymer thin films. *New J. Phys.* 12, 023035/1-8. ISSN: 1367-2630.

Lu, S. G., Xu, Z. K. & Chen, H. (2005). Field-induced dielectric singularity, critical exponents, and high-dielectric tunability in [111]-oriented (1−*x*)Pb(Mg1/3Nb2/3)O3-*x*PbTiO3 (*x*=0.24) *Phys Rev* B 72 (5) (AUG, 2005), 054120/1-4. ISSN: 1098-0121.

Mathur, N. & Mischenko, A. (2006). Solid state electrocaloric cooling devices and methods. World Patent, WO 2006/056809.

Mischenko, A. S.; Zhang, Q.; Scott, J. F.; Whatmore, R. W. & Mathur, N. D. (2006). Giant electrocaloric effect in thin-film $PbZr_{0.95}Ti_{0.05}O_3$. *Science,* 311 (5765) (MAR, 2006), 1270-1271. ISSN: 0036-8075.

Neese, B.; Chu, B. J.; Lu, S. G.; Wang, Y.; Furman, E. & Zhang, Q. M. (2008). Large electrocaloric effect in ferroelectric polymers near room temperature. *Science*, 321 (5890) (AUG, 2008), 821-823. ISSN: 0036-8075.

Neese, B. Ph.D. Dissertation, The Pennsylvania State University, 2009.

Newnham, R. E. (2005). *Properties of Materials: anisotropy, symmetry, structure*. Oxford University Press, ISBN: 019852076X/ISBN-13: 9780198520764. Oxford.

Nolas, G.; Sharp, J. & Goldsmid, H. (2001). *Thermoelectrics,* Springer-Verlag, ISBN: 978-3-540-41245-8. Berlin.

Pecharsky, A. O.; Gschneidner Jr., K. A. & Percharsky, V. K. (2003). The giant magnetocaloric effect of optimally prepared $Gd_5Si_2Ge_2$. *J Appl Phys,* 93 (8) (APR, 2003), 4722-4728. ISSN: 0021-8979.

Pecharsky, V. K.; Holm, A. P.; Gschneidner Jr., K. A. & Rink, R. (2003). Massive magnetic-field-induced structural transformation in Gd_5Ge_4 and the nature of the giant magnetocaloric effect. *Phys Rev Lett,* 91 (19) (NOV, 2003), 197204/1-4. ISSN: 0031-9007.

Pecharsky, V. K.; Moorman, J. O. & Gschneidner Jr., K. A. (1997). A 3–350 K fast automatic small sample calorimeter. *Rev Sci Instrum,* 68 (11) (NOV, 1997), 4196-4207. ISSN: 0034-6748.

PI Ceramic website: www.piceramic.de/site/piezo_002.html (accessed: January, 2009).

Provenzano, V.; Shapiro, A. J. & Shull, R. D. (2004). Reduction of hysteresis losses in the magnetic refrigerant $Gd_5Ge_2Si_2$ by the addition of iron. *Nature,* 429 (6994) (JUN, 2004), 853-857. ISSN: 0028-0836.

Sinyavsky, Y. V. & Brodyansky, V. (1992). Experimental testing of electroca;loric cooling with transparent ferroelectric ceramic as a working body. *Ferroelectrics,* 131 (1-4) (1992), 321-325. ISSN: 0015-0193.

Sinyavsky, Y. V.; Pashkov, N. D.; Gorovoy, Y. M. & Lugansky, G. E. & Shebanov, L. (1989). The optical ferroelectric ceramic as working body for electrocaloric refrigeration. *Ferroelectrics,* 90, 213-217. ISSN: 0015-0193.

Spanner, D. C. (1951). The Peltier effect and its use in the measurement of suction pressure. *J. Experm. Botany,* 2 (5), 145-168. ISSN: 0022-0957.

Spichkin, Y. I.; Derkach, A. V.; Tishin, A. M.; Kuz'min, M. D.; Chernyshov, A. S.; Gschneidner Jr., K. A. & Pecharsky, V. K. (2007). Thermodynamic features of magnetization and magnetocaloric effect near the magnetic ordering temperature of Gd. *J. Magn Magn Mater,* 316 (2) (SEP, 2007), e555-e557. ISSN: 0304-8853.

Tocado, L.; Palacios, E. & Burriel, R. (2005). Direct measurement of the magnetocaloric effect in $Tb_5Si_2Ge_2$. *J Magn Magn Mater,* 290-291 (Part 1, SP. Iss SI) (APR, 2005), 719-722. ISSN: 0304-8853.

Tuttle, B. A. & Payne, D. A. (1981). The effect of microstructure on the electrocaloric properties of $Pb(Zr,Sn,Ti)O_3$ ceramics. *Ferroelectrics*, 37 (1-4) (1981), 603-606. ISSN: 0015-0193.

Wiseman, G. G. & Kuebler, J. K. (1963). Electrocaloric effect in ferroelectric Rochelle salt. *Phys Rev*, 131 (5) (1963), 2023-2027. ISSN: 0013-899X.

Wood, M. E. & Potter, W. H. (1985). General analysis of magnetic refrigeration and its optimization using a new concept: maximization of refrigerant capacity. *Cryogenics*, 25 (12) (DEC, 1985), 667-683. ISSN: 0011-2275.

Xiao, D. Q.; Wang, Y. C.; Zhang, R. L.; Peng, S. Q.; Zhu, J. G. & Yang, B. (1998). Electrocaloric properties of $(1-x)Pb(Mg_{1/3}Nb_{2/3})O_3-xPbTiO_3$ ferroelectric ceramics near room temperature. *Mater Chem Phys*, 57 (2) (DEC, 1998), 182-185. ISSN: 0254-0584.

Yao, H.; Ema, K. & Garland, C. W. (1998). Nonadiabatic scanning calorimeter. *Rev Sci Instrum*, 69 (1) (JAN, 1998), 172-178. ISSN: 0034-6748.

Ye, Z. G. & Schmid, H. (1993). Optical dielectric and polarization studies of the electric field-induced phase transition in $Pb(Mg_{1/3}Ng_{2/3})O_3$ [PMN]. *Ferroelectrics*, 145 (1) (AUG, 1993), 83-108. ISSN: 0015-0193.

7

Study on Substitution Effect of $Bi_4Ti_3O_{12}$ Ferroelectric Thin Films

Jianjun Li[1,2], Ping Li[1] and Jun Yu[2]
[1]*University of Electronic Science and Technology of China*
[2]*Huazhong University of Science and Technology*
China

1. Introduction

Nowadays, ferroelectric thin films have attracted considerable attention because of their potential uses in device applications, such as sensors, micro electro-mechanical system (MEMS) and nonvolatile ferroelectric random access memory (NvFRAM) especially (Scott & Paz De Araujo, 1989; Paz De Araujo et al., 1995; Park et al. 1999). Lead zirconate titanate [$PbZr_xTi_{1-x}O_3$ (PZT)] ferroelectric thin film is an early material for NvFRAM. PZT and related ferroelectric thin films, which are most widely investigated, usually have high remanent polarization (P_r). However, they are generally suffered from a serious degradation of ferroelectric properties with polarity switching, when they are deposited on platinum electrodes.

Bismuth-layered perovskite ferroelectric thin films, with the characteristics of fast switching speed, high fatigue resistance with metal electrodes, and good retention, have attracted much attention. Bismuth titanate [$Bi_4Ti_3O_{12}$ (BIT)] is known to be a typical kind of layer-structured ferroelectrics with a general formula $(Bi_2O_2)^{2+}(A_{m-1}B_mO_{3m+1})^{2-}$. Its crystal structure is characterized by three layers of TiO_6 octahedrons regulary interleaved by $(Bi_2O_2)^{2+}$ layers. At room temperature the symmetry of BIT is monoclinic structure with the space group B1a1, while it can be considered as orthorhombic structure with the lattice constant of the *c* axis (c = 3.2843 nm), which is considerably larger than that of the other two axis (a = 0.5445 nm, b = 0.5411 nm). The BIT has a spontaneous polarization in the *a*-*c* plane and exhibits two independently reversible components along the *c* and *a* axis (Takenaka & Sanaka, 1980; Ramesh et al., 1990). It shows spontaneous polarization values of 4 and 50 μC/cm² along the *c* and *a* axis respectively. The ferroelectric properties of these bismuth layer-structured thin films are mostly influenced by the orientation of the films (Simoes et al., 2006). The BIT thin film is highly *c*-axis oriented, thus its spontaneous polarization is much lower than that for *a*-axis oriented (Fuierer & Li, 2002). For applications in NvFRAM devices, ferroelectric materials should have high remanent polarization, low coercive field (E_c), low fatigue rate and low leakage current density. However, BIT thin film has much lower values of switching polarization and suffers from poor fatigue endurance and high leakage current as a result of the internal defects (Uchida et al., 2002). Numerous works have been made to substitute BIT thin film with proper ions to optimize the ferroelectric properties.

In recent years, it was reported that some *A*-site or *B*-site substituted BIT showed large remanent polarizations. In the case of *A*-site substitution in BIT, La-substituted BIT

[$Bi_{3.25}La_{0.75}Ti_3O_{12}$ (BLT)] films exhibited enhanced P_r of 12 μC/cm^2 with high fatigue resistance, which make them applicable to direct commercialization (Chon et al., 2002). Other lanthanides ions, such as Nd^{3+}, Pr^{3+}, Sm^{3+}, etc. result in similar results (Watanabe et al., 2005; Chon et al., 2003; Chen et al., 2004). In the case of *B*-site substitution in BIT, some donor ions such as V^{5+}, Nd^{5+}, W^{6+}, could effectively decrease the space charge density resulting in the improvement of the ferroelectric properties (Kim et al., 2002; Wang & Ishiwara, 2003). For further improvement of the ferroelectric properties, *A* and *B*-site cosubstitution by various ions should be considered because the properties of BIT based materials strongly depend on species of the substituent ions.

In this chapter, we first summarized the researches on the effect of *A*-site or/and *B*-site substitution on microstructures and properties of $Bi_4Ti_3O_{12}$ ferroelectric thin films. Then La/V substituted BIT thin films were deposited by sol-gel method, and the effect of substitution of La^{3+} and V^{5+} on structural and electrical properties of the BIT thin film was investigated.

2. *A*-site or *B*-site substitution

2.1 *A*-site substitution

The properties of different *A*-site substituted BIT thin films were summarized in Table 1. La-substituted BIT (BLT) films exhibited large P_r and low E_c with high fatigue resistance, and BLT thin film has been already applied in commercial NvFRAM product in all *A*-site substituted BIT thin films.

Substitution Ion	Content	Orientation	$2P_r$ (μC/cm^2)	$2E_c$ (kV/cm)	Fatigue Endurance	References
La	0.75	random	24	100	3×10^{10}	Park et al. 1999
Nd	0.85	random	63.6	260	3×10^{10}	Hou et al. 2005
		c	103	190	6.5×10^{10}	Chon et al. 2002
		(104)	40	100	10^9	Garg et al. 2003
Pr	0.3	*a*	92	100	-	Matsuda et al. 2003
	0.85	*c*	40	-	4.5×10^{10}	Chom et al. 2003
	0.9	random	60	104	-	Chen et al. 2004
Sm	0.85	c	49	113	4.5×10^{10}	Chon et al. 2001
Gd	0.6	random	49.6	249	1.45×10^{10}	Kim et al. 2005

Table 1. Summary of the properties of different *A*-site substituted BIT thin films

Yau et al. (Yau et al. 2005) reported the mechanism of polarization enhancement in La-substituted BIT thin films. Independently controlling processing temperature or La substitution could adjust the orientation and enhance the P_r. Increasing La substitution decreased *c* orientation but increased P_r. The lattice paraments *a*, *b* and *c* of $Bi_{4-x}La_xTi_3O_{12}$ films increase monotonously with La content *x* from *x*=0-0.6. The fitted lines (assumed for *x*>0.6), with different positive slopes, indicated different enlargements in *a*, *b* and *c* and an increase in cell volume. The increase in the *a*, *b* and *c* lattice paramenters approached single crystal values with increasing *x*, which also lowered the orthorhombic distortion, i.e., $2(a-b)/(a+b)$. This strongly suggested that the relaxations of structural distortion and strain arise from the La substitution, which also enhanced P_r.

Lee et al. (Lee et al. 2002) reported correlation between internal stress and ferroelectric fatigeue in La-substituted BIT films. When the La content exceeded *x*=0.25, there was little change in the chemical stability of elements. Chemical stability of oxygen ions alone could not fully explain the effect of La substitution on fatigue. The decrease of the strain was saturated at a composition of *x*=0.75, and films showed fatigue-free characteristics with this composition. Thus, internal strain as well as chemical stability of ions play a significant role in the fatigue behavior of BLT films.

2.2 *B*-site substitution

The properties of different *B*-site substituted BIT thin films were summarized in Table 2 (Choi et al., 2004).

Substitution Ion	c-axis orientation fraction (%)	$2P_r$ (calc.) (μC/cm²)	$2P_r$ (meas.) (μC/cm²)
-	97.6	8.6	11.1
V	83.6	11.3	11.9
W	65.2	14.7	14.3
Nb	29.1	21.5	22.9

Table 2. Summary of the properties of different *B*-site substituted BIT thin films

Either *A*-site or *B*-site substitution can improve remanent polarization of BIT thin film, but the mechanism is different with each other, and the effect of *A*-site substitution is prominent. For *A*-site substitution, there is a highly asymmetric double-well potential at TiO_6 octahedro unit adjacent to the interleaving Bi_2O_2 layer along the *c* axis, and it results in the development of remanent polarization along the *c* axis. For *B*-site substitution, the decreasing *c*-axis orientation results in the development of remanent polarization. Simultaneously, the role of *A*-site substitution is to suppress the *A*-site vacancies accompanied with oxygen vacancies which act as space charge. And the role of *B*-site substitution is the compensation for the defects, which cause a fatigue phenomenon and strong domain pinning. For further improvement of the ferroelectric properties, *A* and *B*-site cosubstitution by various ions should be considered because the properties of BIT based materials strongly depend on species of the substituent ions.

3. La/V substitution

3.1 Experimental

The BIT, $Bi_{3.25}La_{0.75}Ti_3O_{12}$ (BLT), $Bi_{4-x/3}Ti_{3-x}V_xO_{12}$ (BTV) and La^{3+}/V^{5+} cosubstituted BIT [$Bi_{3.25-x/3}La_{0.75}Ti_{3-x}V_xO_{12}$ (BLTV)] thin films were prepared on the $Pt/TiO_2/SiO_2$/p-Si(100) substrates by sol-gel processes. The sol-gel method is one of the chemical solution deposition (CSD) methods, which is commonly used as a fabrication method for thin films. The important advantages of sol-gel method are high purity, good homogeneity, lower processing temperatures, precise composition control of the deposition of multicomponent compoudns, versatile shaping, and deposition with simple and cheap apparatus.

The precursor solution for these films were prepared from Bismuth nitrate [$Bi(NO_3)_3 \bullet 5H_2O$], lanthanum nitrate [$La(NO_3)_3 \bullet xH_2O$], titanium butoxide [$Ti(OC_4H_9)_4$] and vanadium oxytripropoxide [$VO(C_3H_7O)_3$]. The bismuth nitrate and lanthanum nitrate were dissolved in 2-methoxyethanol at room temperature to reach clear solution. Titanium butoxide and vanadium oxytripropoxide were stabilized by acetylacetone at room temperature, which was also act as the stabilizer. These related solutions were mixed, then sonicated and refluxed to dissolve the solutes sufficiently and improved the stability of the solutions. The final concentration in these mixed solutions was adjusted to 0.1 mol/L, and the solutions were aged in vessels for 24 h to get the BIT, BLT, BTV and BLTV precursors. A 10 mol% excess amount of bisumth nitrate was used to compensate Bi evaporation during the heat treatment. The precursor solutions were spin-coated on the $Pt/TiO_2/SiO_2$/p-Si(100) substrates at 3700rpm for 30 s. Then the sol films were dried at 300°C for 5 min and pyrolyzed at 400°C for 10 min to remove residual organic compounds. These processes were repeated six times to achieve the desired film thickness. Then the films were annealed at 750°C for 30 min under air ambient in a horizontal quartz-tube furnace to produce the layered-perovskite phase.

The phase identification, crystalline orientation, and degree of crystallinity of the prepared films were studied by a χ' Pert PRO X-ray diffractometer (PANalytical, B V Co., Holland) with Cu-Ka radiation at 40 kV. The surface and cross-section morphologies were investigated using a Sirion 200 field-emission scanning electron microscope (FEI Co., Holland). The local microstructure and local symmetry of the films were also characterized by Raman spectroscopy (LabRam HR800, Horiba Jobin Yvon Co., France). The Raman measurements were performed at room temperature using the 514.5 nm line of an argon ion laser as the excitation source. Pt top electrodes with an area of about 7×10^{-4} cm^2 were deposited by means of sputtering using a shadow mask for electrical measurements. The polarization-voltage (*P*-*V*) hysteresis loops and fatigue of the films were measured by using a RT66A ferroelectric test system (Radiant Technology Inc., USA), and the leakage current behaviors of the films were analyzed using a Keithley 2400 sourceMeter/High Resistance Meter (Keithley Instruments Inc., USA) with a staircase dc-bias mode and appropriate delay time at each voltage step.

3.2 The effect of La/V substitution

The XRD patterns of the BIT, BLT, BTV (x=0.03) BLTV (x=0.03) thin films deposited on the $Pt/TiO_2/SiO_2$/p-Si(100) substrates are shown in Fig. 1. All the diffraction peaks of these film samples can be indexed according to the reference pattern of $Bi_4Ti_3O_{12}$ powder [BIT, JCPDS (Joint Committee on Powder Diffraction Standards) 35-795]. The good agreement in XRD peaks of the BLT, BTV and BLTV films with those of BIT indicates that the lattice structure of these films are similar to that of BIT. It can be seen that all the films are the single phase of layer-structured perovskite and no pyrochlore phase. The BIT film exhibits a high *c*-axis orientation with the *(00l)* peak being of highest intensities, while the BLTV film exhibits a highly random orientation with the (117) peak being of highest intensity. In order to determine the degree of preferred orientation, the volume fraction of *c*-axis-oriented grains in a film sample is defined as

$$\alpha_{(00l)} = \Sigma(I_{00n} / I^{*}_{00n}) / \Sigma(I_{hkl} / I^{*}_{hkl}) \quad (1)$$

Where I_{hkl} is the measured intensity of (*hkl*) for the films, I^{*}_{hkl} is the intensity for powders, and n is the number of reflections (Lu et al., 2005; Bae et al., 2005). The α values obtained for the BIT, BLT, BTV BLTV films corresponding to Fig. 1 are 76.4, 46.1, 52.5 and 24.4% respectively. The BIT film shows *c*-axis preferred orientation, other films show random orientation.

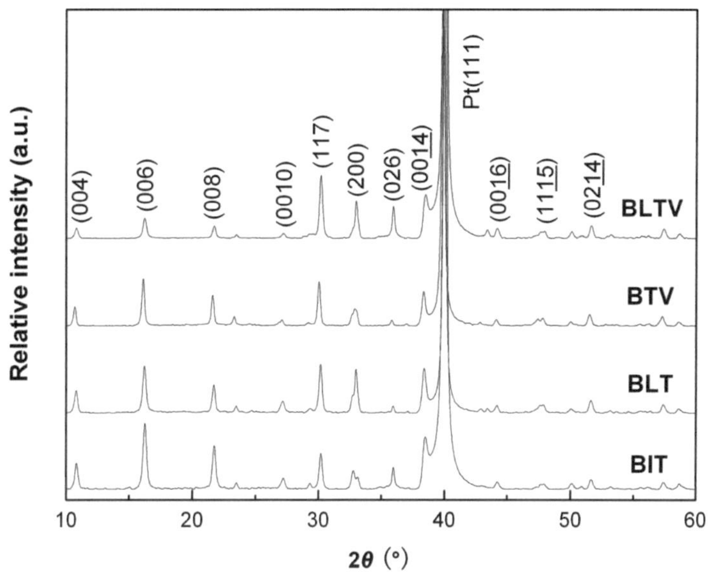

Fig. 1. XRD patterns of the BIT, BLT, BTV and BLTV thin films deposited on the $Pt/TiO_2/SiO_2/p\text{-}Si(100)$ substrates

The BIT film has a strong tabular habit with the growth of *c*-axis orientation, for the energy of the *(00l)* surfaces is lower and consequently the *(h00)* and *(0k0)* faces grow more rapidly. Yau et al. (Yau et al. 2005) reported that, for La-substituted BIT thin films ($Bi_{4-x}La_xTi_3O_{12}$), the degree of *c*-axis orientation decreases with the increase of La content *x* (Yau et al., 2005). Choi et al. reported that, for *B*-site(V^{5+}, W^{6+}, Nb^{5+}) substituted BIT thin films, the degree of *c*-axis orientation decreases greatly with different donor ions substitution (Choi et al., 2004). The substitution may break down the Ti-O chains due to the differences in the electron affinites, which results in the clusters in the precursor favoring the growth of non-*c*-axis orientation. Thus, the BLT, BTV and BLTV thin films with substitition exhibit less highly *c*-axis oriented than the typical unsubstituted BIT thin film, and the result suggests that substitution has an important effect on the orientation of the BIT films.

The lattice constants of the four different film samples are also influenced by the substitution. The lattice constans have been calculated from the XRD patterns and listed in

Table 3. The lattice constant, *a*, decreases and *b* increases as La^{3+} is substituted, while a increases and *b* decreases as V^{5+} is substituted. The orthorhombicity could reflect the variation in the lattice constants, which is defined as $2(a-b)/(a+b)$. The variation of orthorhombicity agrees well with those reported in the literature for La-substituted $Bi_4Ti_3O_{12}$. The decrease of orthorhombicity for the BLT film suggests relaxation of the structural distortion, while the increase for the BTV and BLTV films suggests a increase of the structural distortion and a decrease of the symmetry.

	a (nm)	*b* (nm)	*c* (nm)	$2(a-b)/(a+b)$
BIT	0.5421	0.5391	3.2763	5.55×10^{-3}
BLT	0.5419	0.5392	3.2776	4.96×10^{-3}
BTV	0.5432	0.5386	3.2731	8.50×10^{-3}
BLTV	0.5426	0.5388	3.2756	7.03×10^{-3}

Table 3. Lattice constants and orthorhombicity for the BIT, BLT, BTV and BLTV thin films

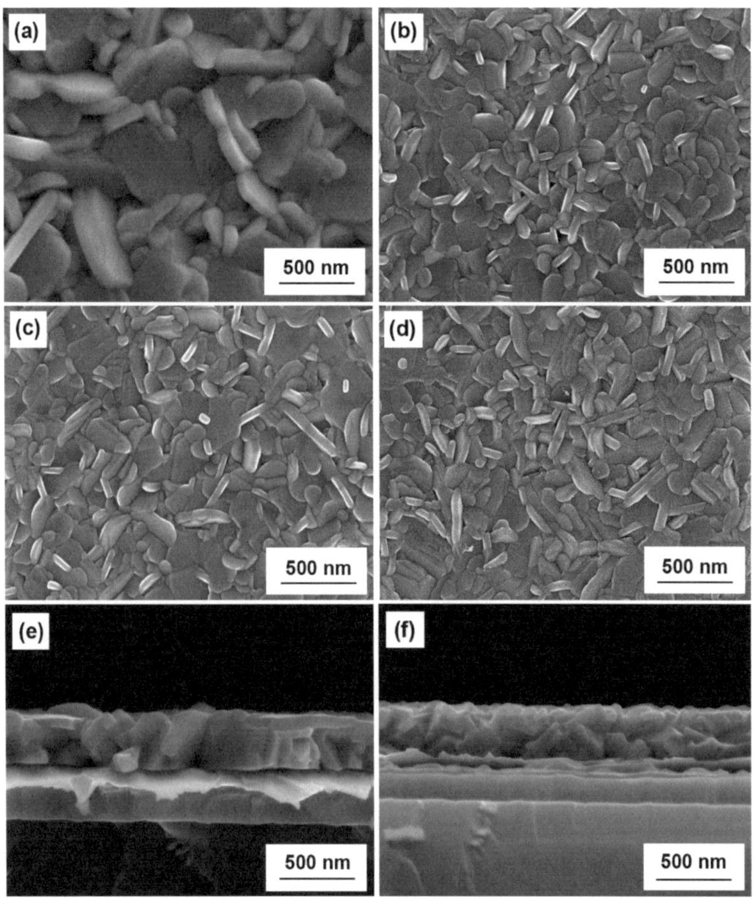

Fig. 2. FE-SEM surface morphologies of the (a) BIT, (b) BLT, (c) BTV and (d) BLTV thin films, and the cross-section micrographs of the (e) BIT and (f) BLTV thin films

Figure 2 shows the field-emission scanning electron microscopy (FE-SEM) surface and cross-section morphologies of these thin films. All the films show dense microstructure without any crack. The thickness of the films is about 360 nm by the cross-section view. From the surface morphologies, it can be seen that the BLTV film is mainly composed of fine rod-like grains with small sizes about 80-160 nm in Fig. 2(d), while the BIT film is mainly composed of large or small plate-like grains with sizes up to 200-500 nm in Fig. 2(a). And it also can been found from the cross-section micrographs, the rod-like grains for the BLTV film slant toward to the film surface in Fig. 2(f), while the large columnar grains for the BIT film are vertical to the film surface in Fig. 2(e). The rod-like grains increase and plate-like grains decrease in the BIT films after substitution of La^{3+} or V^{5+}, and also the size of plate-like grains decreases. For BIT based films, the rod-like grains are related to the growth of random orientation and the plate-like grains are related to the growth of *c*-axis orientation (Lee et al., 2002; Li et al.,2007). In Fig. 1, we can see that the full-width at falf-maximum (FWHM) of the (*00l*) peak decreases after substitution of La^{3+} or V^{5+}, which indicates the restricted growth of (*00l*)-oriented grains. So the results of FE-SEM surface and cross-section morphologies agree with those of the XRD patterns discussed above.

Raman scattering is a powerful probe in studying complex-structured materials because it is highly sensitive to local microstructure and symmetry. The Raman spectra for these films were investigated in the Raman frequency shift range of 100-1000 cm^{-1} as presented in Fig. 3. For bismuth layer-structured ferroelectrics (BLSFs), their phonon modes can generally be classified into two categories: low frequency modes below 200 cm^{-1} and high frequency modes above 200 cm^{-1} (Shulman et al., 2000). The low frequency modes below 200 cm^{-1} are related to large atomic masses, which reflect the vibration of Bi^{3+} ions in $(Bi_2O_2)^{2+}$ layer and *A*-Site Bi^{3+} ions. The high frequency modes above 200 cm-1 reflect the vibration of Ti^{4+} and TiO_6 octahedron. The phonon modes at 118 and 147 cm^{-1} reflect the vibration of *A*-site Bi^{3+} ions in layer-structured perovskite. These modes shift to higher frequencies in the BLT and BLTV films after *A*-site La^{3+} substitution, but almost remain unchanged in the BTV film after *B*-site V^{5+} substitution, which suggests the *B*-site V^{5+} substitution has not affected the *A*-site Bi^{3+} ion. The average masses of La/Bi decrease for La atomic mass is lower than Bi atomic mass (M_{La}/M_{Bi}=139/208), so the related modes show high-frequency shift, which also indicates *A*-site Bi^{3+} ions in BIT film are partly substituted for La^{3+}.

The 231 and 270 cm^{-1} modes are considered to reflect the distortion modes of TiO_6 octahedron. The 231 cm^{-1} mode is Raman inactive when the symmetry of the TiO_6 octahedron is O_h, but it becomes Raman active when distortion occurs in TiO_6 octahedron. The disappearance of the 231 cm^{-1} mode and low-frequency shift of 270 cm^{-1} mode for the BLT film could be explained by a decrease in distortion of TiO_6 due to the influence of La^{3+} ions substitution which may lower the corresponding binding strength and decrease the Raman shift (Zhu et al., 2005). And it is consistent with the decrease of orthorhombicity for the BLT film. While in B-site V^{5+} substituted BTV film, the 231 cm^{-1} mode almost remains unchanged and 270 cm^{-1} mode shifts to a slightly higher frequency. These results suggest that the *A*-site La^{3+} substitution has influenced the *B*-site Ti^{4+} ions in TiO_6 octahedron.

The 537 and 565 cm^{-1} modes are attributed to a combination of stretching and bending of TiO_6 octahedron in BIT. According to references, *A*-site La^{3+} substitution leads to structural disorder and in turn broadens the line of these two modes, thus resulting in the mode at 557 cm^{-1} in the BLT and BLTV films (Wu et al., 2001). The only *B*-site V^{5+} substitution in the BTV film does not affect these two original modes greatly with both shifting to slightly higher frequencies. Thus the variation of the mode frequencies at 537, 557 and 565 cm^{-1} also implies that the A-site La^{3+} substitution exerts influence on Ti^{4+} ions in B sites of the BIT thin film.

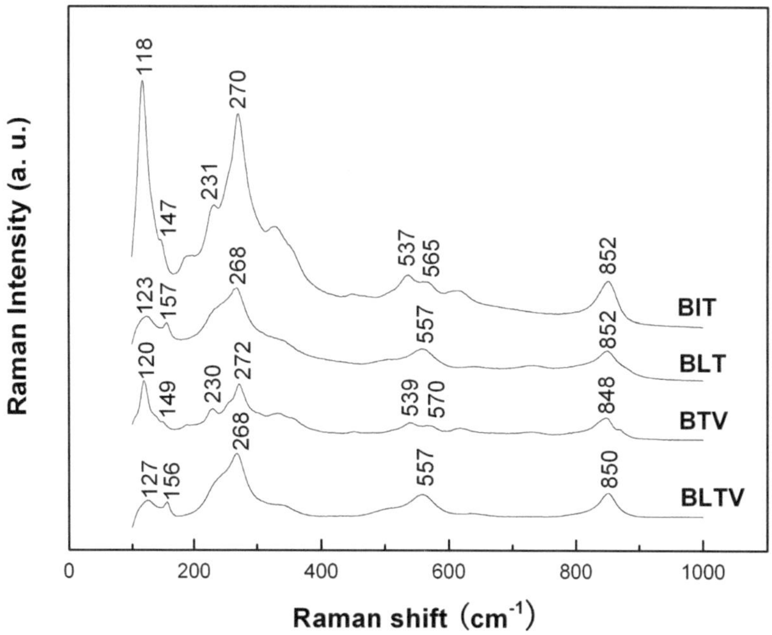

Fig. 3. Raman spectra of the BIT, BLT, BTV and BLTV thin films

The 852 cm^{-1} mode is a pure stretching of TiO_6 octahedron (Yau et al., 2005; Mao et al., 2006). The *A*-site La^{3+} substitution in the BLT film hardly affects the mode, while the *B*-site V^{5+} substitution in the BTV film results in its significant low-frequency shift, indicating that the V^{5+} is entering into the lattice replacing the Ti^{4+} in *B* site and a decrease in O_h symmetry of TiO_6 octahedron, which is consistent with the increase of orthorhombicity for the BTV film. The wave number of the mode of the BLTV film is between the corresponding one of the BLT and BTV films, which should be attributed to the effects of cosubstitution of La^{3+} and V^{5+} in *A* and *B* sites, respectively. On the other hand, for the BLTV film, since V^{5+} is electronically more active than Ti^{4+}, and there is a little asymmetry of TiO_6, a increase in Ti-O hybridization is implied.

Figure 4 shows the *P-V* curves of the BIT, BLT, BTV and BLTV thin film capacitors at a voltage of 12 V. The only *B*-site V^{5+} substitution improves the ferroelectric properties of BIT film slightly. The BLT and BLTV films show well-saturated hysteresis loops. The measured values of $2P_r$ and $2E_c$ are 15.6 μC/cm² and 248 kV/cm for the BIT film, 37.6 μC/cm² and 226 kV/cm for the BLT film, 21.6 μC/cm² and 188 kV/cm for the BTV film, 50.8 μC/cm² and 194 kV/cm for the BLTV film, respectively. For *A*-site substituted BIT film, the E_c usually varys slightly, while for *B*-site substituted BIT film, the E_c becomes smaller (Wang & Ishiwara, 2002). All these substitutions can improve the P_r value for the BIT film, and the BLTV film has the largest P_r with smaller E_c among these films. On one hand, the spontaneous polarization of BIT along the *c*-axis is know to be much smaller than that along the *a*-axis. From this viewpoint, randomly oriented films are considered to be more favorable than *c*-axis oriented films. Therefore the randomly oriented BLTV and BLT films have a more suitable orientation in comparison with other film samples for achieving large P_r. And the BLTV film gets the largest P_r for its least degree of *c*-axis orientation. On the other hand, as hybridization is essential to ferroelectricity, the increase of hybridization may increase the polarization (Cohen, 1992). Thus the increase of Ti-O hybridization inside TiO_6 of the BLTV film results in its excellent ferroelectric properties.

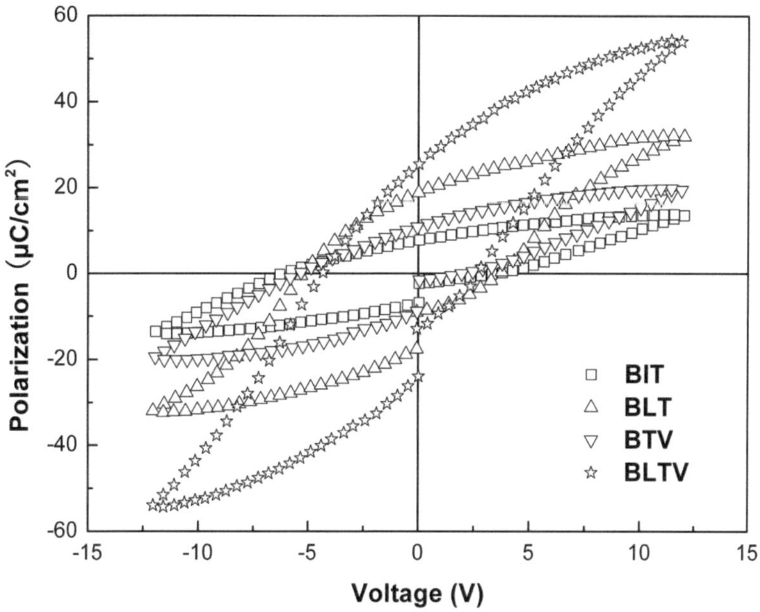

Fig. 4. Polarization-voltage (*P*-*V*) hysteresis loops of the BIT, BLT, BTV and BLTV thin films at the voltage of 12 V

Figure 5 illustrates the fatigue characteristics of these thin films. The test was performed at room temperature using 12 V, 100 kHz bipolar square pulses. From the figure we can see that, as the switching cycles increase, the normalized P_r (the P_r to the initial polarization) of the BLT, BTV and BLTV films decreases silghtly, while that of the BIT film decreases greatly. The P_r of the BIT, BLT, BTV and BLTV films decreases by 24%, 14%, 16% and 10% respectively after 10^{10} cycles.

The strongest fatigue endurance for the BLTV film and good fatigue property for the BLT and BTV films after La^{3+} and V^{5+} substitution are due to the decrease of the defects, such as oxygen vacancies. For BLSF materials, some oxygen vacancies and Bi vacancies are generated unavoidably because of the volatilization of Bi during the annealing processes under high temperatures (Yuan & Or, 2006). The reaction mechanism is as follow:

$$2Bi^{3+} + 3O^{2-} \leftrightarrow Bi_2O_3 + 2v_{Bi}^{'''} + 3v_o^{\cdot\cdot} \quad (2)$$

where $v_o^{\cdot\cdot}$ is the oxygen vacancy with valence of +2, and $v_{Bi}^{'''}$ is the Bi vacancy with valence of -3. The oxygen vacancies are the hole trappers and act as the space charges. And they cause strong domain pinning with repetitive switching cycles, which deteriorate the properties of the ferroelectric films (Park & Chadi, 1998). Better chemical stability of the perovskite layers against oxygen vacancies after substitution of some *A*-site Bi atoms for La atoms is helpful to the fatigue behavior (Kang et al., 1999). For the BTV and BLTV films, V^{5+} substitution for Ti^{4+} and a decrease in Bi content are conducted simultaneously to maintain the charge neutrality, as indicated by the equation of composition ($Bi_{4-x/3}Ti_{3-x}V_xO_{12}$ and $Bi_{3.25-x/3}La_{0.75}Ti_{3-x}V_xO_{12}$). Furthermore, the substitution of the *B*-site Ti^{4+} for the donor-type V^{5+} can suppress the generation of oxygen vacancies due to the charge neutrality restriction (Sun et al., 2006). The reaction mechanisms are as follows:

$$V_2O_5 \leftrightarrow 2V_{Ti}^{\cdot} + 5O_O + 2e' \tag{3}$$

$$\frac{1}{2}O_2 + v_o^{\cdot\cdot} + 2e' \leftrightarrow O_O \tag{4}$$

$$Bi + v_{Bi}''' + 3V_{Ti}^{\cdot} \leftrightarrow Bi_{Bi} + 3V_{Ti} \tag{5}$$

where $V_{Ti}^{\cdot}$ is the V ion with +1 effective charge at the Ti site, V_{Ti} is the V at the Ti, O_O is oxide in the lattice, Bi_{Bi} is Bi in the lattice, and e is the compensatory charge. Thus, v_{Bi}''', the tervalent negative electric center, is neutralized by $V_{Ti}^{\cdot}$, and the oxygen vacancies are suppressed by the higher-valentcation substitution. So the substitution of *A*-site La^{3+} and *B*-site V^{5+} in BIT thin film can improve the properties of the ferroelectric thin film effectively. The BLT and BTV thin films show good fatigue properties.

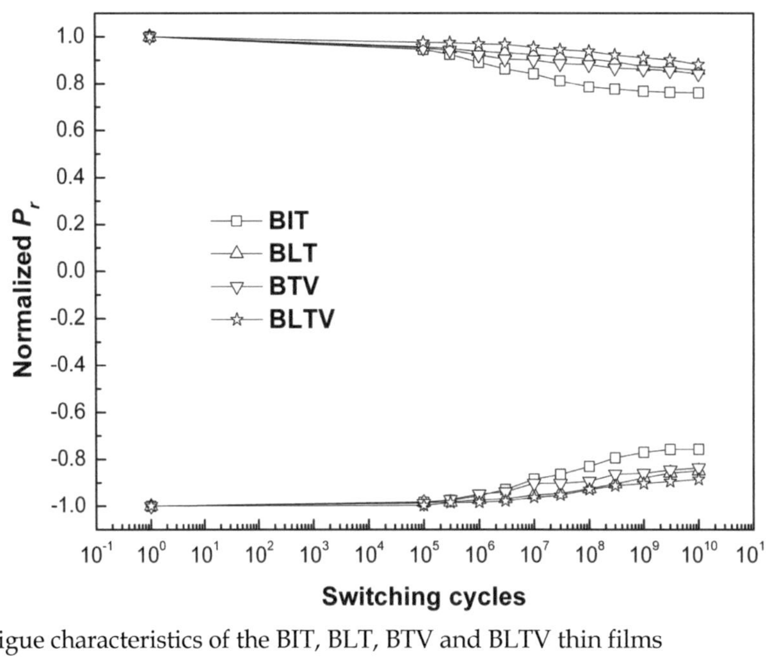

Fig. 5. Fatigue characteristics of the BIT, BLT, BTV and BLTV thin films

The volatile Bi is partly substituted for La, and there are still Bi and oxygen vacancies. There are small amounts of V^{5+} substitution in the present BLTV film, which could avoid more charge defects caused by excessive V^{5+} substitution according to equation (3). The decrease of oxygen vacancies still follows the proposed equation (4) in the BLTV film except for the effect of La^{3+} substitution. The decrease and suppression of oxygen vacancies are enhanced by simultaneous substitutions of La^{3+} and V^{5+} in the BLTV film. Thus, cosubstitution of La^{3+} and V^{5+} in the BLTV film results in its excellent fatigue property.

The dc leakage current is usually one of the most concerned factors for NvFRAM application of ferroelectric thin films, because of its direct relation to power consumption and function failure of devices (Araujo et al., 1990). Therefore leakage current measuremnts are a crucial part of any electrical characterization. During the measurement, the dc voltage over a range of 0-5 V was applied with a step of 0.05 V, and a delay time of 10 s between each step to ensure the collected data of steady state. The leakage current characteristics of current density (J) versus electric field (E) for these thin films at room temperature are shown in Fig. 6. The leakage current density increases gradually at low electric field, but generally in the order of 10^{-9}-10^{-8} A/cm^2 for the BLTV film, 10^{-8}-10^{-7} A/cm^2 for the BLT and BTV films, and 10^{-6}-10^{-5} A/cm^2 for the BIT film below 100 kV/cm. The leakage current density of the BLT and BTV films is smaller than that of the BIT film, and the BLTV film acquires the smallest leakage current density.

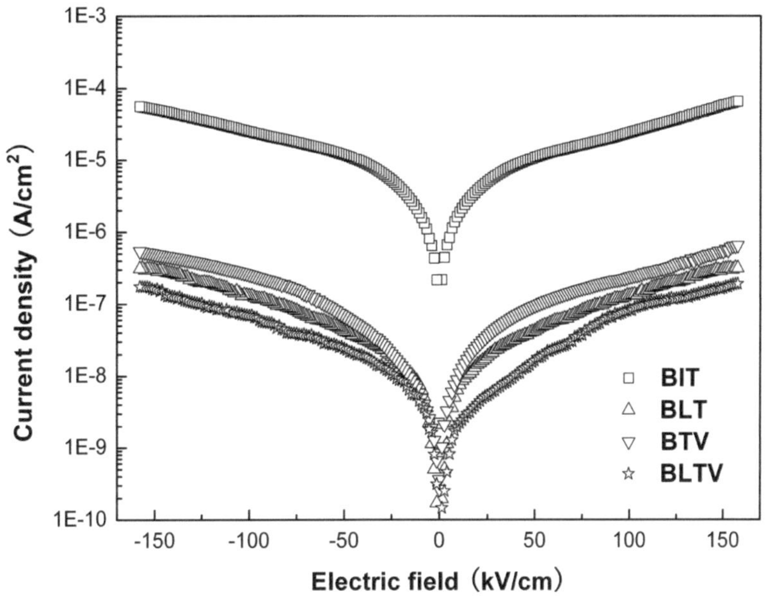

Fig. 6. Electric field dependence of leakage current density for the BIT, BLT, BTV and BLTV thin films

For further study of the effect of La^{3+} and V^{5+} cosubstitution, the leakage current behaviors for the BIT and BLTV films were measured at various temperatures from 20 to 150°C in Fig. 7. The mechanisms of leakage currents are closely related to the temperature for the thermally assisted conduction process. There is a systematic increase in leakage current density with the increase in temperature for both of the films, but the leakage current density for the BIT film increases faster with the increase of the voltage and temperature. Figs 7 (c) and (d) show the plots of log J versus log E of the films in the temperature range from 20 to 150°C. In the low electric field region, the leakage current shows an ohmic behavior with the slope of log J/log E being about 1. The linear region extends up to a onset voltage, at which the space charges start dominating the conduction process. The onset voltage decreases with the increase of the temperature, which is seen from the figures. In the high electric field region, the slope is varying from 2.16 to 2.30 for the BLTV film with the increase of the temperature, while 2.28 to 2.72 for the BIT film, which agrees with the space charge limited current (SCLC) leakage mechanism for both of the films. The larger slop for the BIT film could be caused by more thermally excited electrons from trap levels in the film (Chaudhun & Krupanidhi, 2005). Both electrons and oxygen vacancies domain the space charge mechanism. At low temperatures, oxygen vacancies somehow do not respond to the high electric field, and the highly mobile electrons are the major charge carriers of the current. However, oxygen vacancies play a significant role in the space charge mechanism at high temperatures (Bhattacharyya et al., 2002). From Figs 7 (c) and (d), we can see that the leakage current density increases in three order of magnitude for the BIT film in the high electric field at 150°C(when the temperature was increased from 20 to 150°C), while increases in less than one order of magnitude for the BLTV film. For there are more oxygen vacancies in the BIT film, the leakage current density for the film increases drastically in the high electric field at the high temperature. Therefore, the improved leakage current property is also attributed to the decrease and suppression of oxygen vacancies after La^{3+} and V^{5+} substitution in the thin films.

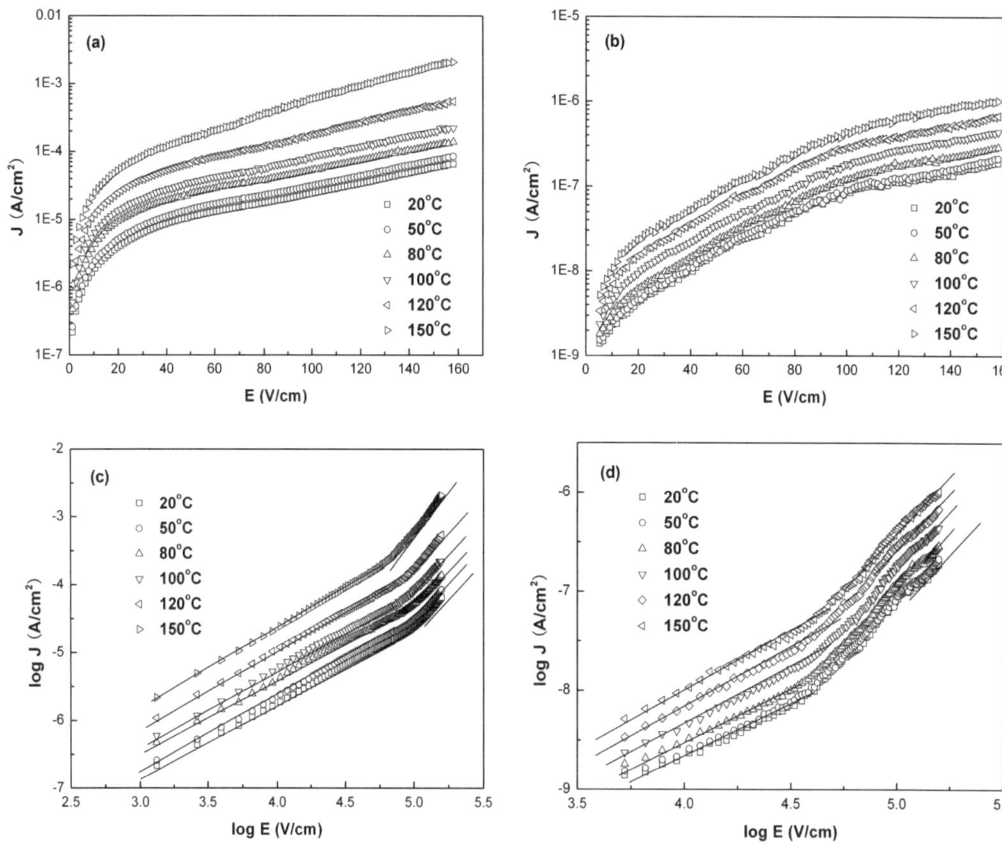

Fig. 7. Leakage current characteristics at the temperature range from 20 to 150°C: J-E plots for the (a) BIT and (b) BLTV thin films; the log J vs log E plots for the (c) BIT and (d) BLTV thin films

3.3 The Effect of V content in BLTV

To acquire the optimal content of V substitutition, $Bi_{3.25-x/3}La_{0.75}Ti_{3-x}V_xO_{12}$ were prepared with different content of V substitutition (x: 0-8%) by sol-gel processes.

The XRD patterns of the BLT, BLTV(x=1%, 3%, 5%, 8%) thin films deposited on the $Pt/TiO_2/SiO_2$/p-Si(100) substrates are shown in Fig. 8. It can be seen that increasing V content do not destroy their crystal structure and all the films are the layered perovskite structure. The BLT film without V substitution exhibits higher intensity of (006) peak. The intensity of this peak decreases with V content increasing from 0 to 0.05, but increases again when V content increasing to 0.08. The volume fraction of c-axis-oriented grains in $BLTV_x$ (0-8%) films calculated according to equation (1) is 31.7, 21.5, 16.9, 20.8 and 39.3 respectively. Thus, the BLTV(3%) film shows the least degree of c-axis orientation.

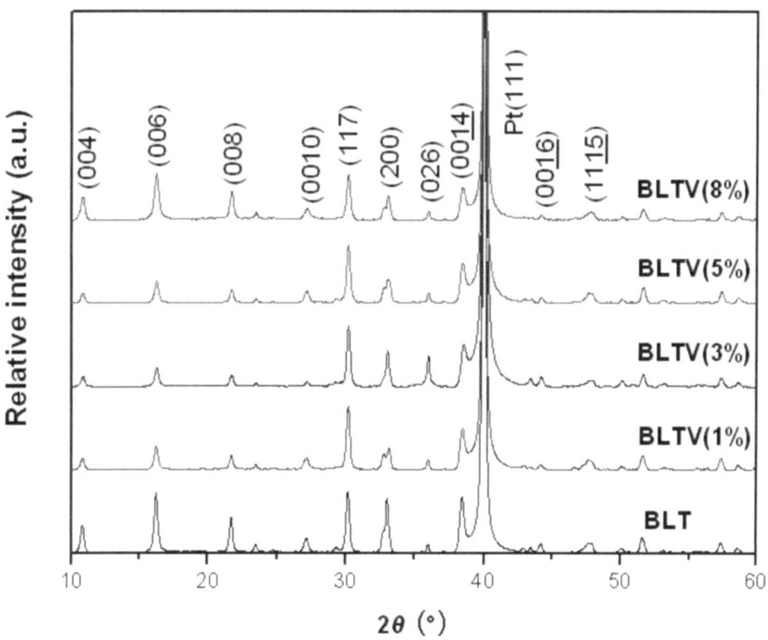

Fig. 8. XRD patterns of the BLTV(x: 0-8%) thin films deposited on the $Pt/TiO_2/SiO_2$/p-Si(100) substrates

Figure 9 shows the FE-SEM surface and cross-section morphologies of the BLTV(x: 0-8%) thin films. All the films show dense microstructure without any crack. From the surface morphologies, it can be seen that the BLT film is mainly composed of fine rod-like and plate-like grains. The rod-like grains increase but the plate-like grains decrease with V content increasing, and there are most rod-like grains when V content is 3% and 5%. Whereas, plate-like grains increase again when V content increasing to 8%. The result of FE-SEM surface morphologies agrees with that of the XRD patterns discussed above.

The Raman spectra for these BLTV(x: 0-8%) thin films were investigated in the Raman frequency shift range of 100-1000 cm^{-1} as presented in Fig. 10. The 852 cm^{-1} mode is a pure stretching of TiO_6 octahedron, Raman shift for BLTV(1%), BLTV(3%), BLTV(5%) and BLTV(8%) thin films is 851, 850, 850 and 848 cm^{-1} respectively at this mode. There are two obvious peaks between 600 and 800 cm^{-1} mode for BLTV(8%) thin film, which indicates a great increase of the structural distortion and a great decrease in O_h symmetry of TiO_6 octahedron.

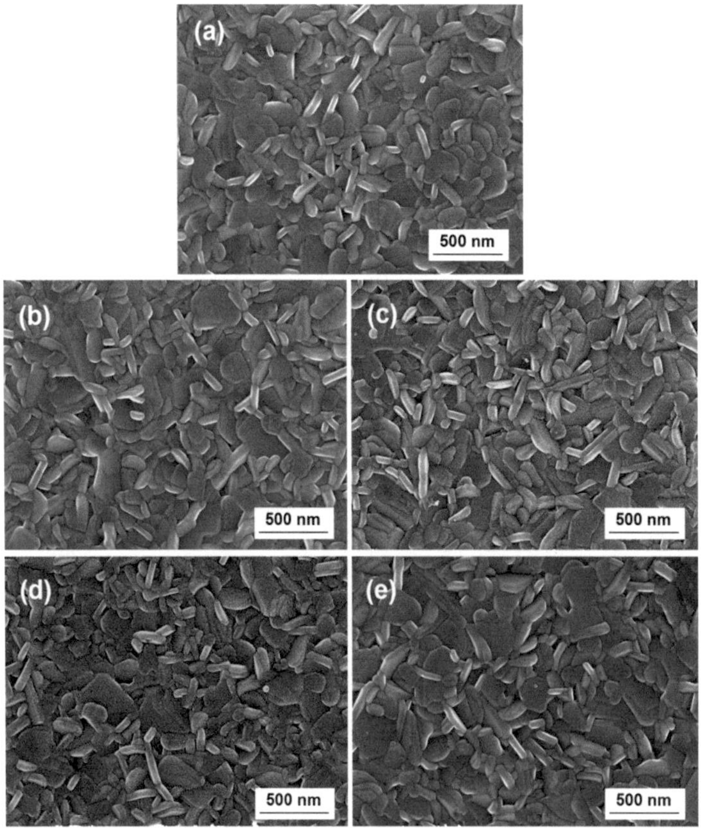

Fig. 9. FE-SEM surface morphologies of the BLTV(x: 0-8%) thin films: (a) BLT, (b) BLTV(1%), (c) BLTV(3%), (d) BLTV(5%) and (e) BLTV(8%)

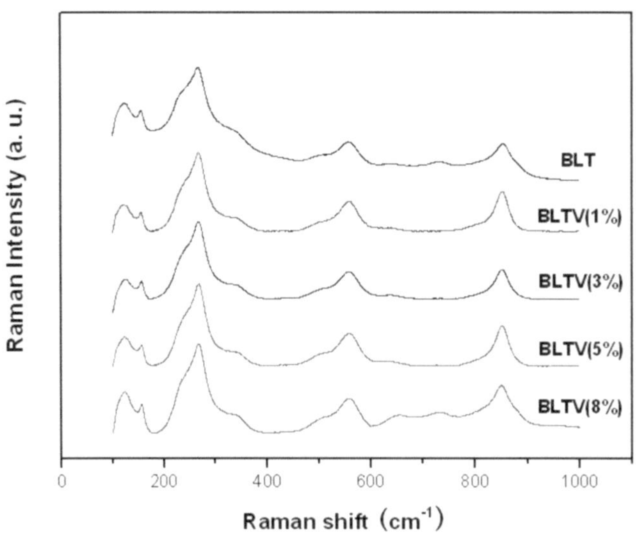

Fig. 10. Raman spectra of the BLTV(x: 0-8%) thin films

Figure 11 shows the *P-V* curves of these BLTV(*x*: 0-8%) thin film capacitors at a voltage of 12 V. All these thin films show well-saturated hystersis loops. With increasing of substitution content, P_r increases. In the range x>3%, P_r decreases with increasing V content. When x=8%, P_r becomes lower than that of BLT and the squareness of the hystersis loop is degraded as well.

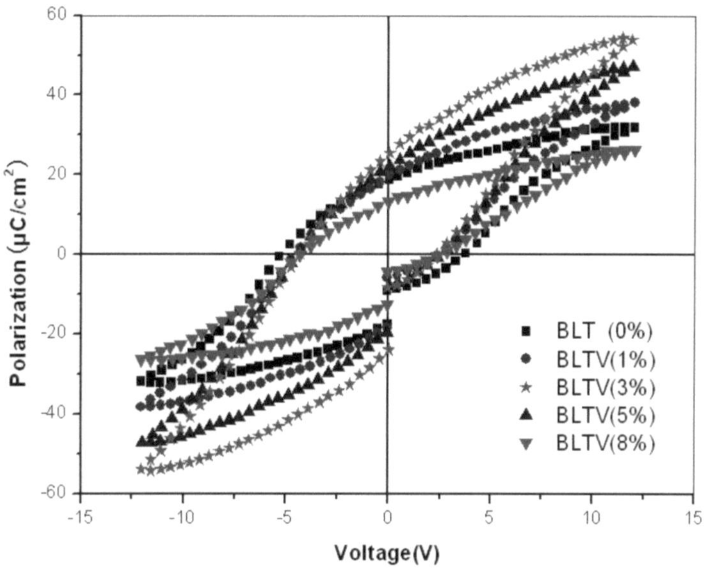

Fig. 11. Polarization-voltage (*P-V*) hysteresis loops of the BLTV(*x*: 0-8%) thin films at the voltage of 12 V

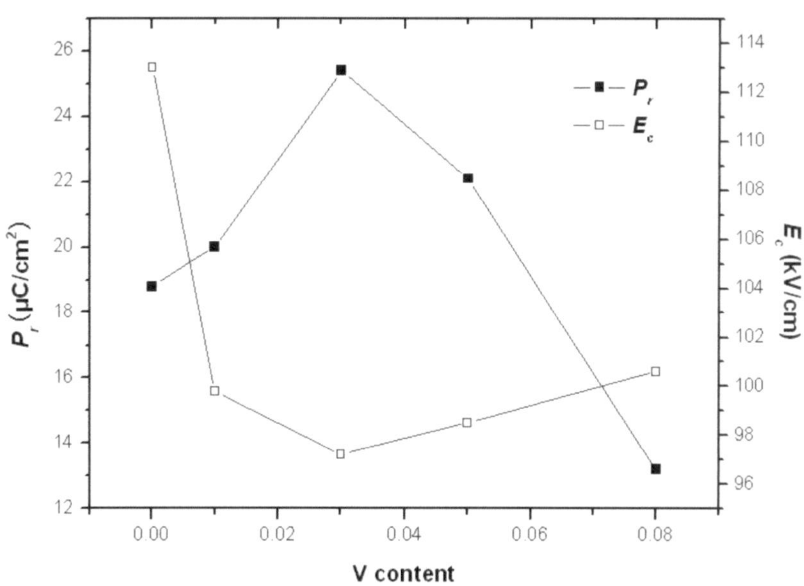

Fig. 12. P_r and E_c of the BLTV(*x*: 0-8%) thin films as functions of the V content

Figure 12 summarizes the variation in the P_r and E_c against the V content in the BLTV(x: 0-8%) thin films. The P_r values under the voltage of 12 V were 18.8, 20.1, 25.4, 22.1 and 13.2 μC/cm^2, respectively, for x = 0, 1%, 3%, 5%, 8%. The BLTV(3%) thin film gets the largest P_r for its least degree of c-axis orientation. On the other hand, as hybridization is essential to ferroelectricity, the increase of hybridization may increase the polarization. Thus the increase of Ti-O hybridization inside TiO_6 of the BLTV(3%) thin film results in its excellent ferroelectric properties. After the substitution content is above 5%, P_r begins to decrease, this is because of a limited solution of V ion in BLTV thin films. The great increase of the structural distortion results in a great decrease of ferroelectric properties.
Figure 13 illustrates the fatigue characteristics of these BLTV(x: 0-8%) thin films. With increasing of substitution content, the fatigue property is improved. In the range x>3%, the fatigue property is worsened with increasing V content. Small amounts of V^{5+} substitution can suppress the generation of oxygen vacancies due to the charge neutrality restriction, but excessive V^{5+} substitution will cause more charge defects which worsen the fatigue property.

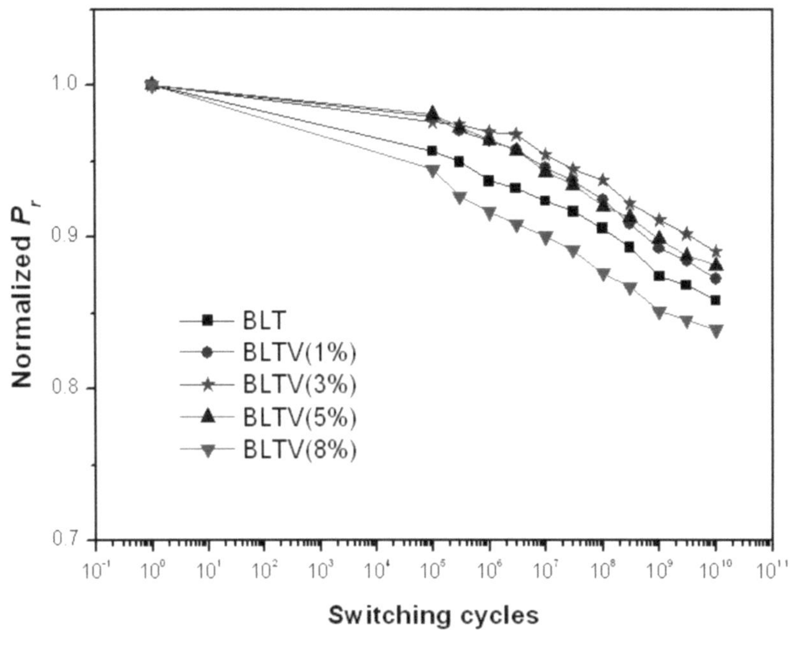

Fig. 13. Fatigue characteristics of the BLTV(x: 0-8%) thin films

4. Conclusion

In conclusion, BIT, BLT, BTV and BLTV thin films were fabricated on the Pt/TiO_2/SiO_2/p-Si(100) substrates by sol-gel processes. The microstructures and electrical properties of these films after substitution of La^{3+} and V^{5+} were investigated. The X-ray diffraction patterns indicate the BLTV thin film shows randomly oriented with fine rod-like grains revealed from the FE-SEM surface and cross-section morphologies. Raman spectra show the *A*-site La^{3+} substitution exerts influence on Ti^{4+} ions in *B* sites of the BIT thin film, TiO_6 (or VO_6) symmetry decreases and Ti-O (or V-O) hybridization increase for V^{5+} substitution. Either *A*-site La^{3+} or *B*-site V^{5+} substitution can improve the P_r value for the BIT thin film, but only the BLT and BLTV film capacitors are characterized by well-saturated *P-V* curves at an applied voltage of 12 V. The BLTV thin film shows the largest $2P_r$ of 50.8 μC/cm^2 with small $2E_c$ of 194 kV/cm among these films. For the BLTV thin film, the fatigue test exhibits the strongest fatigue endurance up to 10^{10} cycles and the leakage current density is generally in the order of 10^{-9}-10^{-8} A/cm^2 below 100 kV/cm at room temperature while increasing in less than one order of magnitude at the temperature range from 20 to 150°C. To acquire the optimal concentration of V substitutition, BLTV were prepared with different concentration of V substitutition (x: 0-8%). The remanent polarization and fatigue properties first increase then decrease with increasing of V content. BLTV(3%) thin film exhibits excellent ferroelectric property and the strongest fatigue endurance up to 10^{10} cycles. When V content increases to 8%, the properties decrease due to slightly large distortion of crystal lattice and charge defects caused by excessive V substitution. The excellent properties of the BLTV thin film are attributed to the effective decrease or suppression of oxygen vacancies after La^{3+} and V^{5+} substitution in the thin film.

5. References

Araujo, C. A.; Mcnillan, L. D.; Melnick, B. M.; Cuchiaro, J. D. & Scott, J. F. (1990). Ferroelectric memories. *Ferroelectrics,* Vol. 104, No. 1, pp. 241-256, ISSN 0015-0193

Bae, J. C.; Kim, S. S.; Choi, E. K.; Song, T. K.; Kim, W. J. & Lee, Y. I. (2005). Ferroelectric properties of lanthanum-doped bismuth titanate thin films grown by a sol-gel method. *Thin Solid Films,* Vol. 472, No. 1-2, pp. 90-95, ISSN 0040-6090

Bhattacharyya, S.; Laha, A. & Krupanidhi, S. B. (2002). Analysis of leakage current conduction phenomenon in thin $SrBi_2Ta_2O_9$ films grown by excimer laser ablation. *J. Appl. Phys.,* Vol. 91, No. 7, pp. 4543-4548, ISSN 0021-8979

Chaudhuri, A. R. & Krupanidhi, S. B. (2005). dc leakage behavior in vanadium-doped bismuth titanate thin films. *J. Appl. Phys.,* Vol. 98, No. 9, 094112, ISSN 0021-8979

Chen, M.; Liu, Z. L.; Wang, Y.; Wang, C. C.; Yang, X. S. & Yao, K. L. (2004). Ferroelectric properties of Pr_6O_{11}-doped $Bi_4Ti_3O_{12}$. *Solid State Communications,* Vol. 130, No. 11, pp. 735-739, ISSN 0038-1098

Chen, M.; Liu, Z. L.; Wang, Y.; Wang, C. C.; Yang, X. S. & Yao, K. L. (2004). Ferroelectric properties and microstructures of Sm-doped $Bi_4Ti_3O_{12}$ ceramics. *Physica B,* Vol. 352, No. 1-4, pp. 61-65, ISSN 0921-4526

Choi, E. K.; Kim, S. S.; Kim, J. K.; Bae, J. C.; Kim, W. J.; Lee, Y. I. & Song, T. K. (2004). Effects of donor ion doping on the orientation and ferroelectric properties of bismuth titanate thin films. *Jpn. J. Appl. Phys.,* Vol. 43, pp. 237-241, ISSN 0021-4922

Chon, U.; Kim, K. B.; Jang, H. M. & Yi, G. C. (2001). Fatigue-free samarium-modified bismuth titanate ($Bi_{4-x}Sm_xTi_3O_{12}$) film capacitors having large spontaneous polarizations. *Appl. Phys. Lett.,* Vol. 79, No. 19, pp. 3137-3139, ISSN 0003-6951

Chon, U.; Jang, H. M.; Kim, M. G. & Chang C. H. (2002). Layered perovskites with giant spontaneous polarizations for nonvolatile memories. *Phys. Rev. Lett.*, Vol. 89, No. 8, 087601, ISSN 0031-9007

Chon, U.; Shim, J. S. & Jang, H. M. (2003). Ferroelectric properties and crystal structure of praseodymium-modified bismuth titanate. *J. Appl. Phys.*, Vol. 93, No. 8, pp. 4769-4775, ISSN 0021-8979

Cohen, R. E. (1992). Origin of ferroelectricity in perovskite oxides. *Nature*, Vol. 358, pp. 136-138, ISSN 0028-0836

Fuierer, P. & Li, B. (2002). Nonepitaxial orientation in sol-gel bismuth titanate films. *J. Am. Ceram. Soc.*, Vol. 85, No. 2, pp. 299-304, ISSN 1551-2916

Garg, A.; Barber, Z. H.; Dawber, M.; Scott, J. F.; Snedden, A. & Lightfoot, P. (2003). Orientation dependence of ferroelectric properties of pulsed-laser-ablated $Bi_{4-x}Nd_xTi_3O_{12}$ films. *Appl. Phys. Lett.*, Vol. 83, No. 12, pp. 2414-2416, ISSN 0003-6951

Hou, F.; Shen, M. R. & Cao, W. W. (2005). Ferroelectric properties of neodymium-doped $Bi_4Ti_3O_{12}$ thin films crystallized in different environments. *Thin Solid Films*, Vol. 471, No. 1-2, pp. 35-39, ISSN 0040-6090

Kim, J. K.; Kim, J.; Song, T. K. & Kim, S. S. (2002). Effects of niobium doping on microstructures and ferroelectric properties of bismuth titanate ferroelectric thin films. *Thin Solid Films*, Vol. 419, No. 1-2, pp. 225-229, ISSN 0040-6090

Kim, S. S.; Song, T. K., Kim, J. K. & Kim, J. (2002). Ferroelectric properties of vanadium-doped $Bi_4Ti_3O_{12}$ thin films deposited by a sol–gel method. *J. Appl. Phys.*, Vol. 92, No. 4, pp. 2213-2215, ISSN 0021-8979

Kim, S. S.; Bae, J. C. & Kim, W. J. (2005). Fabrication and ferroelectric studies of (Bi, $Gd)_4Ti_3O_{12}$ thin films grown on $Pt/Ti/SiO_2/Si$ and p-type Si substrates. *Journal of Crystal Growth*, Vol. 274, No. 3-4, pp. 394-401, ISSN 0022-0248

Lee, J. K.; Kim, C. H.; Suh, H. S. & Hong, K. S. (2002). Correlation between internal stress and ferroelectric fatigeue in $Bi_{4-x}La_xT_3O_{12}$ thin films. *Appl. Phys. Lett.*, Vol. 80, No. 19, pp. 3593-3595, ISSN 0003-6951

Lee, H. N.; Hesse, D.; Zakharov, N. & Gösele, U. (2002). Ferroelectric $Bi_{3.25}La_{0.75}Ti_3O_{12}$ Films of Uniform a-Axis Orientation on Silicon Substrates. *Science*, Vol. 296, No. 5575, pp. 2006-2009, ISSN 0036-8075

Li, J.; Yu, J.; Peng, G.; Wang, Y. B. & Zhou, W. L. (2007). The influence of the thickness of TiO_2 seeding layer on structural and electrical properties of $Bi_{3.15}Nd_{0.85}Ti_3O_{12}$ thin films. *J. Phys. D: Appl. Phys.*, Vol. 40, pp. 3788-3792, ISSN 0022-3727

Lu, C. J.; Qiao, Y.; Qi, Y. J.; Chen, X. Q. & Zhu, J. S. (2005). Large anisotropy of ferroelectric and dielectric properties for $Bi_{3.15}Nd_{0.85}Ti_3O_{12}$ thin films deposited on $Pt/Ti/SiO_2/Si$. *Appl. Phys. Lett.*, Vol. 87, No. 22, 222901, ISSN 0003-6951

Matsuda, H.; Ito, S. & Lijima, T. (2003). Design and ferroelectric properties of polar-axis-oriented polycrystalline $Bi_{4-x}Pr_xTi_3O_{12}$ thick films on Ir/Si substrates. *Appl. Phys. Lett.*, Vol. 83, No. 24, pp. 5023-5025, ISSN 0003-6951

Mao, X. Y.; He, J. H., Zhu, J. & Chen, X. B. (2006). Structural, ferroelectric, and dielectric properties of vanadium-doped $Bi_{4-x/3}Ti_{3-x}V_xO_{12}$. *J. Appl. Phys.*, Vol. 100, No. 4, 044104, ISSN 0021-8979

Park, B. H., Kang, B. S., Bu, S. D., Noh, T. W.; Lee, J. & Jo, W. (1999). Lanthanum-substituted bismuth titanate for use in non-volatile memories. *Nature,* Vol. 401, pp. 682-684, ISSN 0028-0836

Park, C. H. & Chadi D. J. (1998). Microscopic study of oxygen-vacancy defects in ferroelectric perovskites. *Phys. Rev. B,* Vol. 57, R13961, ISSN 1098-0121

Paz De Araujo, C. A., Cuchiaro, J. D., McMillan, L. D., Scott, M. C. & Scott, J. F. (1995). Fatigue-free ferroelectric capacitors with platinum electrodes. *Nature,* Vol. 374, pp. 627-629, ISSN 0028-0836

Ramesh, R.; Luther, K.; Wilkens, B.; Hart, D. L.; Wang, E. & Trascon, J. M. (1990). Epitaxial growth of ferroelectric bismuth titanate thin films by pulsed laser deposition. *Appl. Phys. Lett.,* Vol.57, No. 15, pp.1505-1507, ISSN 0003-6951

Scott, J. F. & Paz De Araujo, C. A. (1989). Ferroelectric memories. *Science,* Vol. 246, No. 4936, pp. 1400-1405, ISSN 0036-8075

Shulman, H. S.; Damjanovic, D. & Setter, N. (2000). Niobium doping and dielectric anomalies in bismuth titanate. *J. Am. Ceram. Soc.,* Vol. 83, No. 3, pp. 528-532, ISSN 1551-2916

Simoes, A. Z.; Pianno, R. F. C.; Ries, A.; Varela, J. A. & Longo, E. (2006). a-b axis-oriented lanthanum doped $Bi_4Ti_3O_{12}$ thin films grown on a TiO_2 buffer layer. *J. Appl. Phys.,* Vol. 100, No. 8, 084106, ISSN 0021-8979

Sun, H.; Zhu, J., Fang, H. & Chen, X. B. (2006). Large remnant polarization and excellent fatigue property of vanadium-doped $SrBi_4Ti_4O_{15}$ thin films. *J. Appl. Phys.,* Vol. 100, No. 7, 074102, ISSN 0021-8979

Takenaka, T. & Sanaka, K. (1980). Grain orientation and electrical properties of hot-forged $Bi_4Ti_3O_{12}$ ceramics. *Jpn. J. Appl. Phys.,* Vol. 19, pp. 31-39, ISSN 0021-4922

Uchida, H.; Yoshikawa, H.; Okada, I.; Matsuda, H.; Lijima, T.; Watanabe, T.; Kojima, T. & Funakubo, H. (2002). Approach for enhanced polarization of polycrystalline bismuth titanate films by Nd^{3+}/V^{5+} cosubstitution. *Appl. Phys. Lett.,* Vol. 81, No.12, pp. 2229-2231, ISSN 0003-6951

Wang, X. S. & Ishiwara, H. (2003). Polarization enhancement and coercive field reduction in W- and Mo-doped $Bi_{3.35}La_{0.75}Ti_3O_{12}$ thin films. *Appl. Phys. Lett.,* Vol. 82, No. 15, pp. 2479-2481, ISSN 0003-6951

Watanabe, T.; Funakubo, H.; Osada, M.; Uchida, H. & Okada, I. (2005). The effects of neodymium content and site occupancy on spontaneous polarization of epitaxial $(Bi_{4-x}Nd_x)Ti_3O_{12}$ films. *J. Appl. Phys.,* Vol. 98, No. 2, 024110, ISSN 0021-8979

Wu, D.; Li, A. D.; Zhu, T.; Li, Z. F.; Liu, Z. G. & Ming, N. B. (2001). Processing- and composition-dependent characteristics of chemical solution deposited $Bi_{4-x}La_xTi_3O_{12}$ thin films. *J. Mater. Res.,* Vol. 16, No. 5, pp. 1325-1332, ISSN 0884-2914

Yau, C. Y.; Palan, R.; Tran, K. & Buchanan, R. C. (2005). Mechanism of polarization enhancement in La-doped $Bi_4T_3O_{12}$ films. *Appl. Phys. Lett.,* Vol. 86, No. 3, 032907, ISSN 0003-6951

Yuan, G. L. & Or, S. W. (2006). Enhanced piezoelectric and pyroelectric effects in single-phase multiferroic $Bi_{1-x}Nd_xFeO_3$ ($x = 0–0.15$) ceramics. *Appl. Phys. Lett.,* Vol. 88, No. 6, 062905, ISSN 0003-6951

Zhu, J.; Chen, X. B.; Zhang, Z. P. & Shen, J. C. (2005). Raman and X-ray photoelectron scattering study of lanthanum-doped strontium bismuth titanate. *Acta. Mater.*, Vol. 53, No. 11, pp. 3155-3162, ISSN 1359-6454

8

Uniaxially Aligned Poly(*p*-phenylene vinylene) and Carbon Nanofiber Yarns through Electrospinning of a Precursor

Hidenori Okuzaki and Hu Yan
University of Yamanashi
Japan

1. Introduction

Nanofibers made of conducting or semiconducting polymers have been extensively studied from both fundamental and technological aspects to understand their intrinsic electrical and mechanical properties and practical use for elecctromagnetic interference shielding, conducting textiles, and applications to high-sensitive sensors and fast-responsive actuators utilizing their high speccific surface area (Okuzaki, 2006).

Poly(*p*-phenylene vinylene) (PPV) has been paid considerable attention due to its properties of electrical conductivity, electro- or photoluminescence, and non-linear optical response, which have potential applications in electrical and optical devices, such as light-emitting diodes (Burrourghes et al., 1991), solar cells (Sariciftci et al., 1993), and field-effect transistors (Geens et al., 2001). Most of the work with regard to PPV exploits thin coatings or cast films (Okuzaki et al., 1999), while a few reports have been investigated on PPV nanofibers by chemical vapor deposition polymerization with nanoporous templates (Kim & Jin, 2001). The electrospinning, a simple, rapid, inexpensive, and template-free method, capable of producing submicron to nanometer scale fibers is applied to fabricate nanofibers of conducting polymers, such as sulphuric acid-doped polyaniline (Reneker & Chun, 1996) or a blend of camphorsulfonic acid-doped polyaniline and poly(ethylene oxide) (MacDiarmid et al., 2001). However, the electrospinning is not applicable to PPV because of its insoluble and infusible nature. Although poly[2-methoxy-5-(2′-ethylhexyloxy)-1,4-phenylenevinylene], an electro-luminescent derivative of PPV, could be electrospun from 1,2-dichloroethane solution as a randomly oriented mesh, the resulting fibers were not uniform exhibiting leaf-like or ribbon-like morphology due to the low viscosity limited by the polymer solubility (Madhugiri et al., 2003).

On the other hand, carbon nanofibers have superior mechanical properties, electrical conductivity, and large specific surface area, which are promising for various potential applications in nanocomposites such as electromagnetic interference shielding (Yang et al., 2005), rechargeable batteries (Kim et al., 2006), and supercapacitors (Kim et al., 2004). Currently, the carbon nanofibers are fabricated by a traditional vapor growth (Endo, 1988), arc discharge (Iijima, 1991), laser ablation, and chemical vapor deposition (Ren et al., 1998), but they involve complicated processes and high equipment costs for the fabrication. Reneker et al. fabricated carbon nanofibers by carbonization of electrospun polyacrylonitrile

(PAN) precursor fibers (Chun et al., 1999). Carbon nanofibers can also be derived from various precursors, such as mesophase-pitch (Park et al., 2003), polybenzimidazol (Kim et al., 2004), and polyimide (Chung et al., 2005). The electrospun fibers are, however, directly deposited on the grounded target as a randomly oriented mesh.
This chapeter deals with a successful fabrication of uniaxially aligned PPV and carbon nanofibers by electrospinning of a soluble precursor and subsequent thermal conversion or carbonization (Fig. 1). The effects of spinning and carbonization conditions on morphology and structure of the resulting nanofiber yarns have been investigated by means of scanning electron microscopy (SEM), Fourier transform infrared (FT-IR), wide-angle X-ray diffraction (WAXD), thermogravimetry (TG), and Raman analyses.

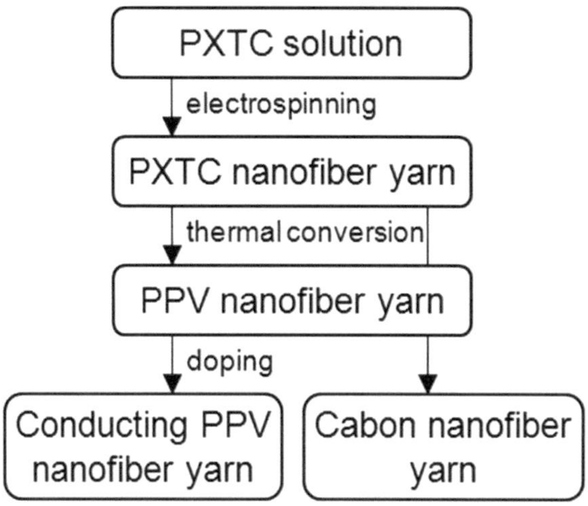

Fig. 1. Flow chart for fabrication of conducting PPV and carbon nanofiber yarns.

2. Experimental

Poly(*p*-xylenetetrahydrothiophenium chloride) (PXTC), a precursor of PPV, was commercially available in the form of 0.25% aqueous solution from Aldrich Inc. The M_w determined by membrane osmometry was 1.3 x 10^5, while low-angle laser-light scattering combined with centrifugation yielded M_w = 9.9 x 10^5 (M_n = 5.0 x 10^5, M_w/M_n = 2.0) (Gagnon et al., 2006). About 1 ml of PXTC solution containing different amount of methanol was poured into a glass syringe (12 mm in diameter) and an electric field was applied to a single-hole spinneret (340 μm in diameter) from a variable high-voltage power supply (Towa Keisoku) capable of applying positive voltages (E) up to 30 kV. The electrospun fibers were collected on a grounded flat plate (10 x 10 cm) covered with an aluminum foil used as a target where the distance between the needle and the target electrode was 20 cm. The viscosity of the PXTC solution was measured at 30 °C with a B-type viscometer (VM-10A, CBC Materials). The diameter of yarn was measured with an optical microscope equipped with a video system (InfiniTube, Edmund Optics). The thermogravimetric (TG) analysis was performed with a TG-DTA (2000S, MAC Science) in a temperature range from room temperature to 1000 °C at a heating rate of 10 °C min^{-1} under an argon atmosphere. The PXTC yarns were subsequently converted to PPV yarns by heat treatment at 250 °C for 12 h in vacuum. FT-IR spectra was measured with a FTIR-8100 (Shimadzu) and wide-angle X-ray diffraction patterns were measured with a RINT (Rigaku). Carbonization of the PXTC yarns was carried out in a quartz tube under vacuum by using an electric furnace (FR100, Yamato). The diameter of thus-carbonized fiber was measured with an FE-SEM (S-4500, Hitachi) and the distribution was evaluated by measuring at least 100 points. The Raman spectra were measured with a laser Raman spectrometer (NRS-2100, JASCO) equipped with an optical microscope. Excitation light from an argon ion laser (514.5 nm) was focused on the fiber surfaces through the optical microscope.

3. Results and discussion

3.1 Spontaneous formation of PXTC yarns by electrospinning

Fig. 2 shows optical images of PXTC yarns spontaneously formed by electrospinning at C = 0.1% and E = 20 kV). When the applied voltage reaches a critical value, the electrostatic force overcomes the surface tension of the PXTC solution, thereby ejecting a jet from the spinneret.

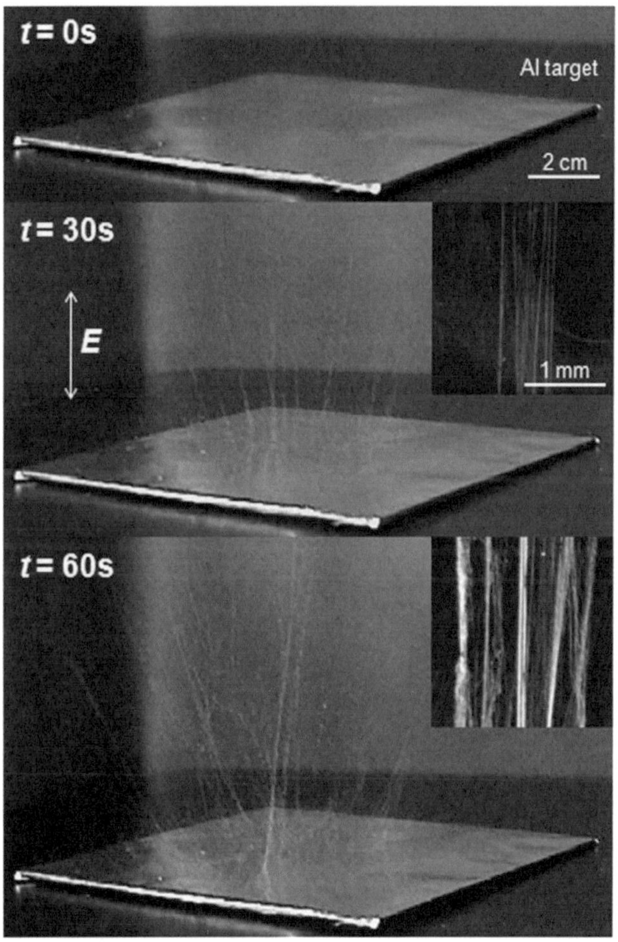

Fig. 2. Spontaneous formation of PXTC yarns by electrospinning at C = 0.1% and E = 20 kV for 0, 30, and 60 s.

Since the PXTC chains and solvent molecules bear the same (positive) charge, they repel each other and the droplets become smaller due to the separation while traveling in air during a few milliseconds from the spinneret to the grounded target. Meanwhile, evaporation of solvent molecules takes place rapidly because the separation of droplets produces high surface area to volume. Conventionally electrospun fibers are directly deposited on the grounded target as a randomly oriented mesh due to the bending instability of the highly charged jet. In contrast, the PXTC is electrospun into centimeters-long yarns vertically on the surface of the aluminum target but parallelly to the electric field where an electrostatic attractive force stretches the yarn vertically on the target electrode, leading to a successive upward growth of the yarn. It should be noted that a few yarns get twisted while swinging due to the bending instability of the jet and grow into a thicker yarn. The unusual formation of the yarns can be explained by the ionic conduction of the PXTC (Okuzaki et al., 2006), i. e., the deposited PXTC fibers discharge through polyelectrolyte chains to the ground target, which prefers the following deposition on the electrospun fibers so as to decrease the gap between the fibers and the spinneret. Once a yarn forms by touching the adjacent fibers, the following deposition may preferably occur on the yarn rather than on the individual fibers because of the low electric resistance.

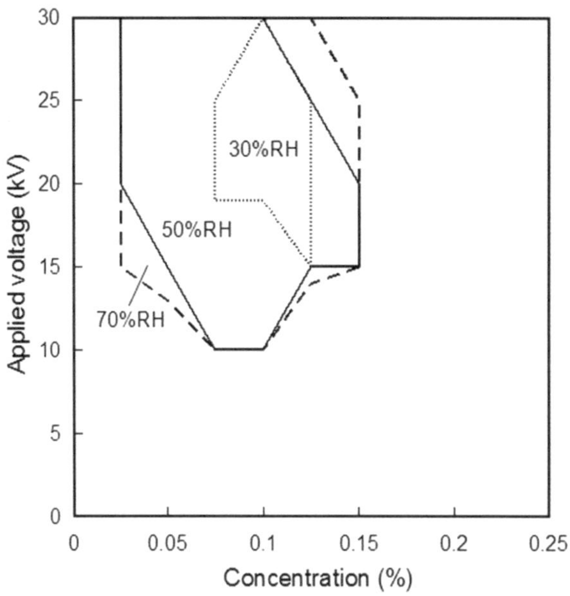

Fig. 3. Effects of applied voltage and concentration on formation of PXTC yarns at various relative humidities.

In order to clarify the mechanism of the yarn formation in more detail, electrospinning was performed at various relative humidities (RHs) and the results were shown in Fig. 3. At 30 %RH, yarn formed only in a area of C = 0.075~0.125% and E = 15~30 kV, while at 70 %RH the area expanded to C = 0.025~0.15% and E = 10~30 kV (Okuzaki et al., 2008). Here, a rise the ambient humidity from 50 to 70 %RH increased in the electric current at E = 20 kV from 1.7 to 2.6 μA and brought about rapid growth of the yarn. On the other hand, the current was less dependent on the humidity in the absence of the PXTC in the solution (0.39-0.46 μA at E = 20 kV). This clearly indicates that the ionic conduction of the PXTC crucially influences the yarn formation. Indeed, the similar phenomenon was observed for other ionic conducting polyelectrolyte such as poly(acrylic cid). Fig. 4 shows influence of PXTC concentration on average diameter of the yarn, spinning rate, and viscosity of the solution. At E = 20 kV, the yarn was formed at C = 0.025%-0.15%, while at C > 0.15%, no yarn was formed at E up to 30 kV since the viscosity of the solution (> 8.8 cP) was too high to maintain a continuous jet from the spinneret. On the other hand, at C < 0.025%, small droplets of the solution were collected on the surface of the aluminum target due to the low viscosity (< 1.2 cP) and surface tension not enough to maintain a stable drop at the spinneret. Thus, methanol crucially increases the volatility of solvent as well as decreases the surface tension, which arises from the fact that dilution with pure water never yields yarn.

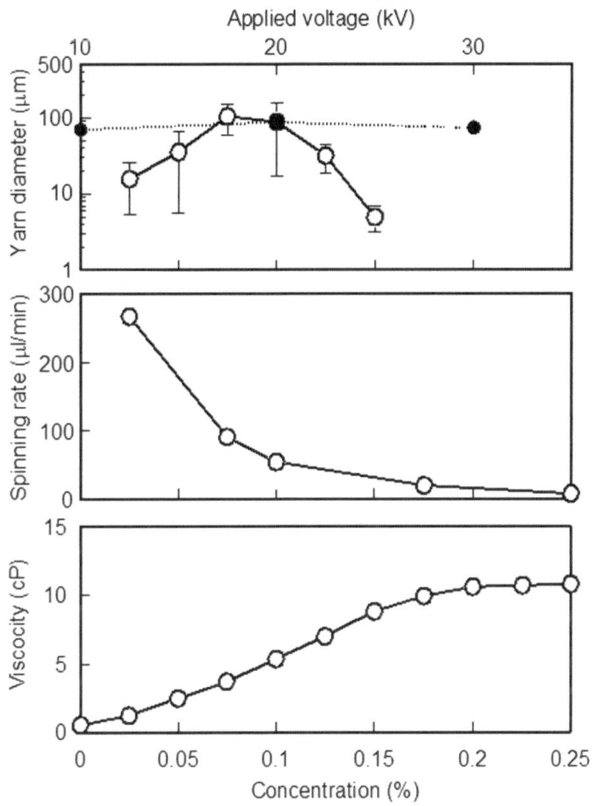

Fig. 4. Dependence of PXTC concentration on average diameter of the yarn and spinning rate at E = 20 kV in 10 min and viscosity of the solution measured at 30 °C with a viscometer. Dependence of applied voltage on average diameter of the yarn at C = 0.1 % in 10 min was also plotted (broken line).

The average diameter of the yarn electrospun at E = 20 kV in 10 min increased from 5 to 100 μm with decreasing of C from 0.15% to 0.025%. This can be explained by the increase of spinning rate, defined as the average volume of the solution electrospun in 1 min, due to the decrease of viscosity. A further decrease of viscosity, however, yields finer yarns due to the decrease of C. The effect of applied voltage on average diameter of the yarn was also demonstrated in Fig. 4 (broken line), where the electrospinning was carried out at a constant

C of 0.1% for 10 min. It is found that the average diameter of the yarn is less dependent on the applied voltage and the value is about 100 μm, where the yarns were as long as the gap between the spinneret and the target electrode (20 cm). A further electrospinning resulted in a deposition of the droplets on the yarns before the solvent evaporated completely, which allowed the yarns to fall down on the target electrode due to their own weight.

3.2 Uniaxially aligned PPV nanofiber yarns by thermal conversion of PXTC

TG curve of the PXTC yarns (solid line) is divided into three major steps in terms of weight loss, as shown in Fig. 5: The first small weight loss at 25-100 °C is associated with the desorption of moisture, and the second one at 100-250 °C is due to the elimination of tetrahydrothiophene and hydrochloric acid, corresponding to the conversion to PPV (as shown in the inset) (Okuzaki et al., 1999). The third weight loss at 500-600 °C continuing gradually up to 1000 °C is due to the carbonation of the PPV that will be discussed later (Ohnishi et al., 1986). The PXTC yarns are subsequently converted to PPV yarns by heat treatment at 250 °C for 12 h in vacuum (Okuzaki et al., 1999). About 42% weight loss of the as-electrospun PXTC yarns in a temperature range of 25-300 °C, decreases to less than 0.01% after the thermal conversion, demonstrating that the precursor unit is completely converted to PPV.

Fig. 6 shows FT-IR spectra of PXTC yarns electrospun at C = 0.1% and E = 20 kV and PPV yarns converted at 250 °C for 12 h in vacuum. The FT-IR spectrum of the resulting PPV yarns shows a clear peak at 965 cm^{-1} assigned to the *trans*-vinylene C–H out-of-plane bend, while an absorption peak at 632 cm^{-1} to the C–S stretch of the PXTC disappeared completely. A clear indication of paracrystalline structure is seen in a wide-angle X-ray diffraction pattern as shown in Fig. 7. The PPV yarns show clear peaks at 2θ = 20.5° (d = 4.33 Å) and

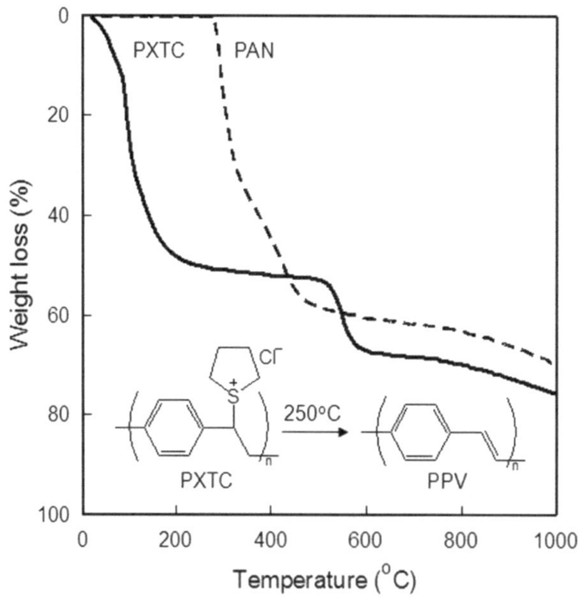

Fig. 5. TG thermograms of the PXTC yarns (solid line) and polyacrylonitrile (PAN: broken line) measured at a heating rate of 10 °C min^{-1} under an argon atmosphere. Inset: thermal conversion of PXTC to PPV.

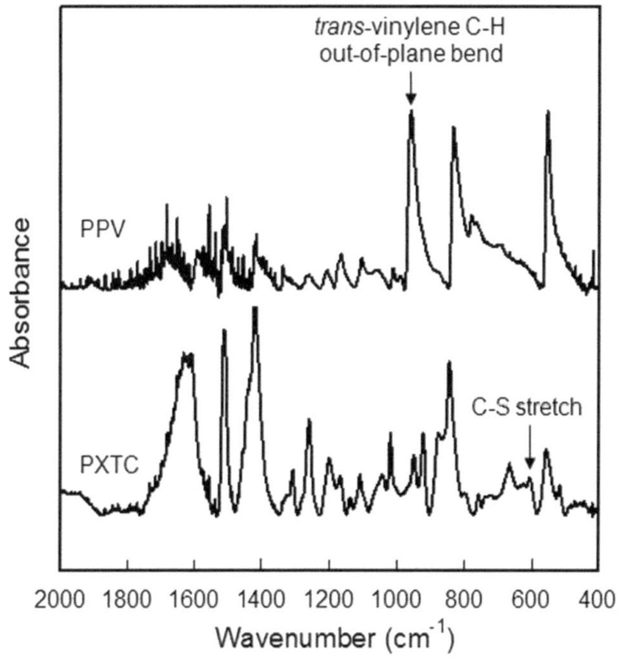

Fig. 6. FT-IR spectra of PXTC yarns electrospun at C = 0.1% and E = 20 kV and PPV yarns converted at 250 °C for 12 h in vacuum.

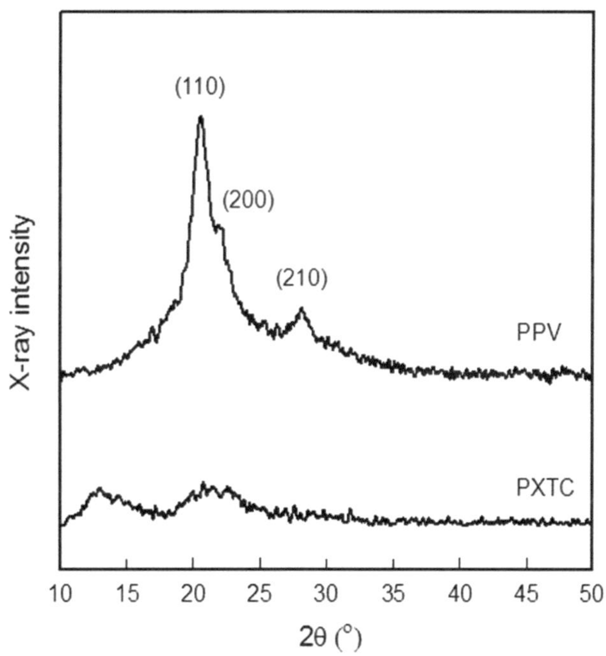

Fig. 7. Wide-angle X-ray diffraction patterns of PXTC yarns electrospun at C = 0.1% and E = 20 kV and PPV yarns converted at 250 °C for 12 h in vacuum.

28.2° (d = 3.16 Å), and a shoulder at 2θ = 22.0° (d = 4.03 Å), corresponding to the diffractions from (110), (210), and (200) planes of the monoclinic unit cell of the PPV crystal, respectively (Masse et al., 1989). Since the as-spun PXTC fibers are amorphous, the crystallization will take place during the thermal conversion to PPV. The degree of crystallinity estimated from the WAXD pattern is *ca.* 45 % and the apparent crystallite size normal to the (210) plane calculated from Scherrer's equation is 75 Å. The diffraction patterns of the side and end views of the PPV yarns are substantially the same, indicating no notable orientation of PPV chains on a molecular level.

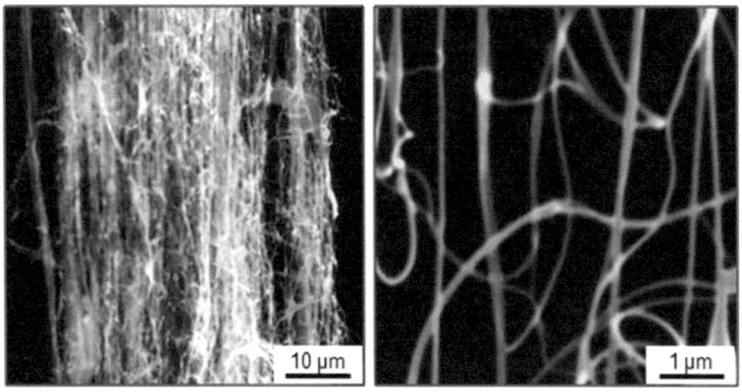

Fig. 8. SEM micrographs of PPV yarn fabricated by thermal conversion of the PXTC yarn elecctrospun at C = 0.1% and E = 20 kV for 10 min.

It is seen from Fig. 8, a μm-thick yarn is composed of numerous PPV nanofibers uniaxially aligned along the axis of the yarn. It is noted that the PPV fibers preserve their morphology even after the elimination reaction. An influence of PXTC concentration on distribution of diameters for the PPV nanofibers is demonstrated in Fig. 9. At C = 0.05~0.15%, a distribution peak of the fiber diameter locates between 100-200 nm, where more than 50% of fibers have a diamter less than 200 nm. Particularly, at C = 0.1%, more than 25% of fibers are finer than 100 nm. In contrast, more than 30% of fibers are thicker than 500 nm with a wide distribution at C = 0.025%, which is ascribed to the association or fusion of fibers since the solvents may not evaporate completely due to the low concentration. Furthermore, effects of electric field on distribution of fiber diameter and orientation of the PPV nanofibers were demonstrated in Fig. 10. Although the distribution of fiber diameter is less dependent on the electric field, orientation of the PPV nanofibers along the axis of the yarn is improved as the voltage becomes higher, in which the ratio of fibers having a tilt angle smaller than 10° increases from 44% (E = 10 kV) to 60% (E = 30 kV). This can be explained by the electrostatic repulsion between positively charged fibers deposited on the yarn (Li et al., 2004) and/or explained by stretching the fibers due to the electrostatic force between the spinneret and the yarn in favor of discharging through the yarn, leading to a configuration in parallel alignment (Okuzaki et al., 2008).

3.3 Conducting PPV nanofiber yarns by acid doping

The doping is performed by dipping the PPV nanofiber yarns in conc. H_2SO_4 at room temperature. After doping, the nanofiber yarns were soaked in a large amount of acetonitrile and dried under vacuum. Fig. 11 shows current-voltage relationship for PPV

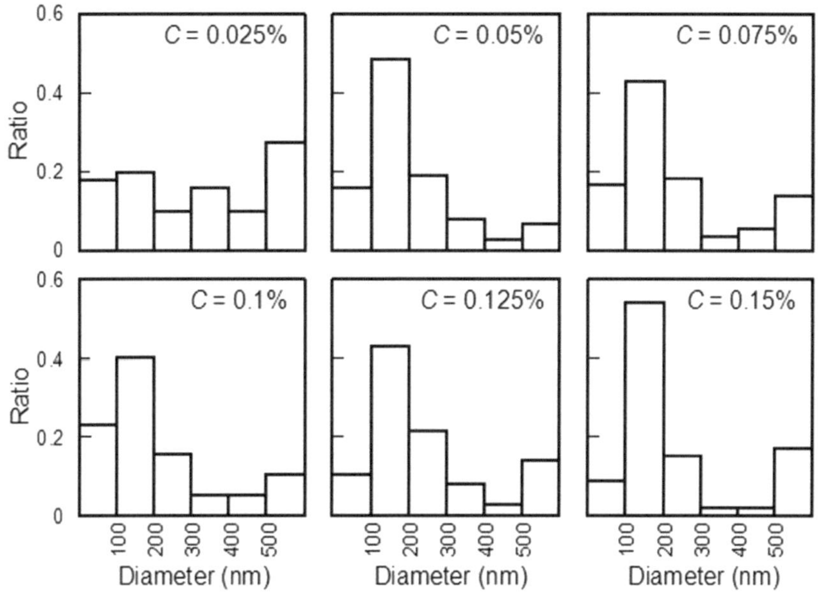

Fig. 9. Dependence of PXTC concentration on distribution of fiber diameter for PPV nanofibers (E = 20 kV).

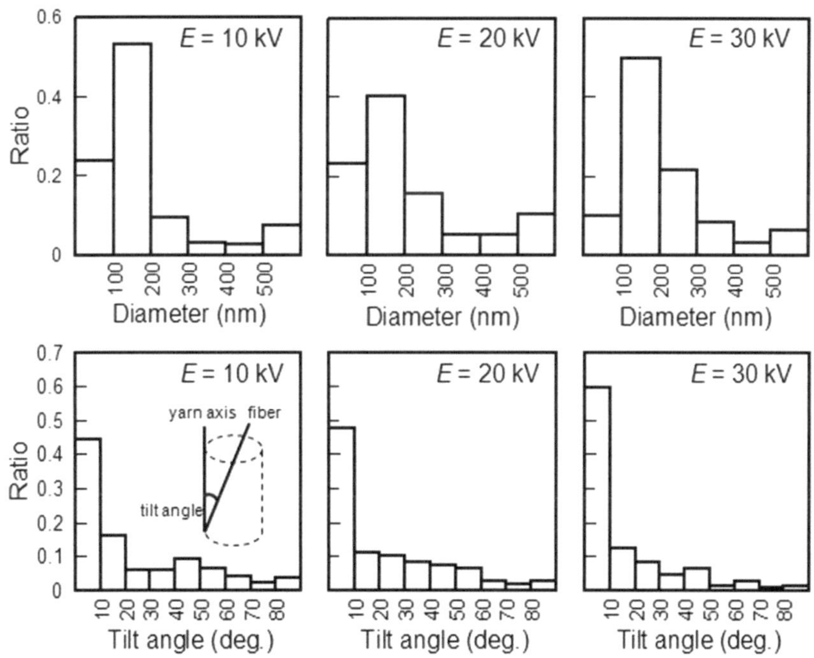

Fig. 10. Dependence of applied voltage on distribution of fiber diameter and orientation of PPV nanofibers along the axis of the yarn (C = 0.1%).

and H_2SO_4-doped PPV nanofiber yarns measured with a digital multimeter (model 2000, Keithley) by a two-probe method where the nanofiber yarns were placed on two gold electrodes kept at an interval of 2 mm. No detectable current was observed in a voltage range of -5~+5 V for the PPV nanofiber yarns, demonstrating that the PPV has extremely low conductivity in the neutral state due to the small carrier density. Upon doping with H_2SO_4, the color changed from yellowish brown to black and apparent current flowed through the nanofiber yarns, where the current almost linearly increased in this voltage range and the value attained 1~2 μA at 5 V (Okuzaki et al., 2008). Although the conductivity of a single nanofiber can not be estimated because mean value of the cross-sectional area in the nanofiber yarns is unknown, the results clearly indicate that the electrical conductivity of the PPV nanofiber yarns increases by doping with H_2SO_4.

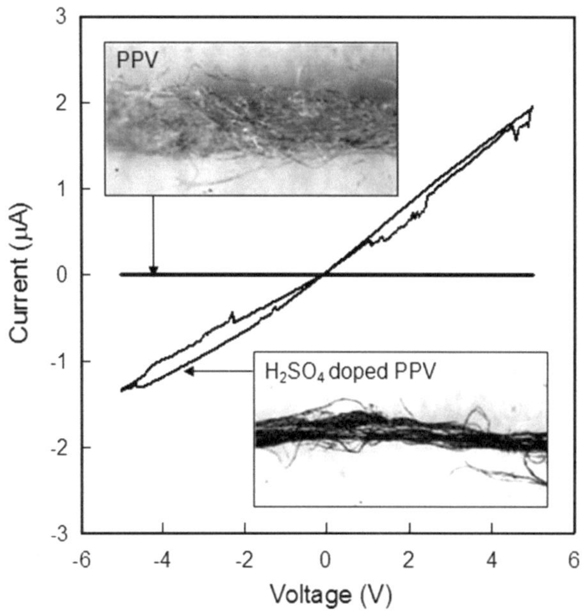

Fig. 11. Current-voltage relationship for PPV and H_2SO_4-doped PPV nanofiber yarns.

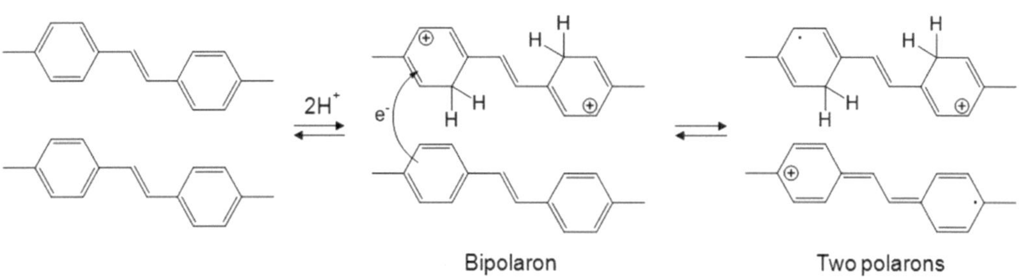

Fig. 12. Schematic illustration of proposed doping mechanism of PPV.

This can be explained by oxidation of PPV to produce charge carriers where anions are intercalated between polymer chains as counter ions (Okuzaki et al., 2005). However, the doping of PPV with protonic acids has not been fully understood yet (Ahlskog et al., 1997). Han et al. have found that non-oxidizing protonic acids can be effective dopants for a wide range of conjugated polymers (Han & Elsenbaumer, 1989). The mechanism of protonic acid doping appears to involve direct protonation of the polymer backbone followed by an internal redox process that gives polarons as the predominant charge defects, similarly to polyaniline (Chiang & MacDiarmid, 1986).

3.4 Uniaxially aligned carbon nanofibers by thermal conversion of PXTC

It is seen from the TG curve as shown in Fig. 5, the third weight loss at 500-600 °C continuing gradually up to 1000 °C is due to the carbonization of the PPV (Ohnshi et al., 1986). As a comparison, the TG curve of PAN, a typical precursor of carbon fiber (broken line), shows a significant weight loss of about 40% between 300 and 500 °C. This corresponds to the carbonation, involving dehydrogenation, cyclization, and polymerization of nitrile groups, and then losing noncarbon elements, such as HCN, NH, and H, in the form of volatile gases. As a result, the carbon yield for the PXTC was 25% at 1000 °C, which was somewhat smaller than that for the PAN (30%).

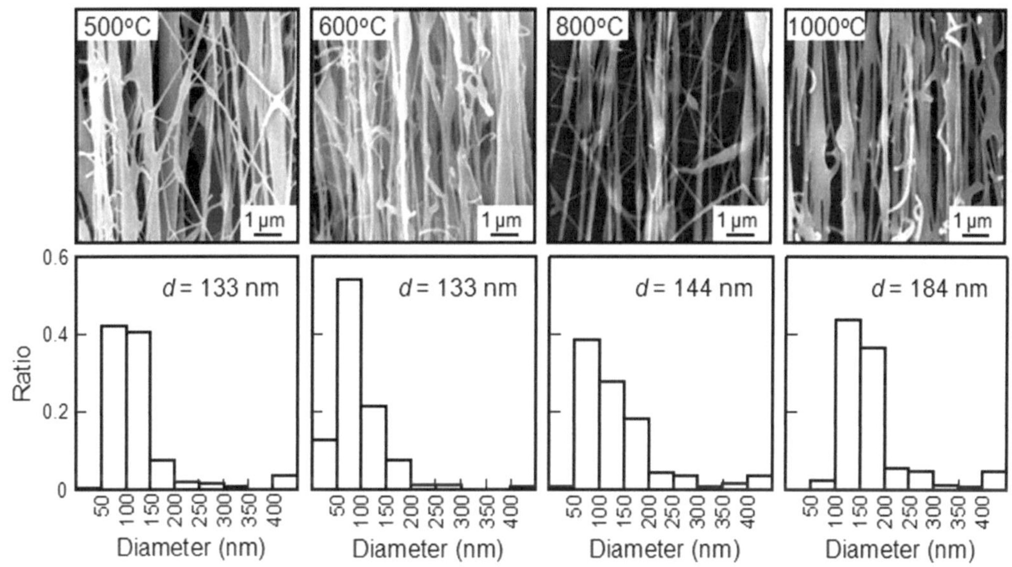

Fig. 13. SEM micrographs and distribution of fiber diameter for various nanofiber yarns thermally treated at various temperatures.

Fig. 13 shows SEM micrographs and distribution of fiber diameter for the PXTC yarns thermally treated at various temperatures. Although we failed to measure SEM micrographs of the as-electrospun PXTC yarns owing to the elimination of tetrahydrothiophene and hydrochloric acid under the radiation of an electron beam, it can be considered that the fiber diameter decreases during the thermal conversion to PPV based on the fact that nearly 50% of weight is lost up to 250 °C (Fig. 5). Agend et al. (2007) fabricated carbon nanofibers by electrospinning of PAN from DMF solution and subsequent carbonization and found that average diameter of PAN nanofibers decreased from 149 to 109 nm by carbonization at 1100 °C. It should be emphasized in Fig. 13 that the fibers preserve their morphology with average diameters d = 133-184 nm. Interestingly, the nanofibers were uniaxially aligned along the axis of the yarn, similarly as observed in PPV nanofibers where about 70% of nanofibers in the yarn had a tilt angle less than 30° (Okuzaki et al., 2008). Furthermore, there was no notable change in the morphological structure and the distribution of fiber diameter even at temperatures higher than 500-600 °C of the third weight loss where the carbonization of the PPV occurs.

Raman spectroscopy is one of the most sensitive tools for providing information on the crystalline structure of carbon fibers. The highly ordered carbon such as a single hexagonal crystal graphite shows a strong peak at 1582 cm^{-1} (G band) from in-plane modes with E_{2g} symmetry, while the disordered carbon due to imperfection or lack of hexagonal symmetry shows a broad peak at 1360 cm^{-1} (D band) (Kim et al., 2004). A clear evidence of the structural change caused by the carbonization is demonstrated in Fig. 14. Two peaks at around 1580 and 1360 cm^{-1}, assigned to G and D bands, respectively, evidently appear in the Raman spectra for the nanofiber yarns thermally treated at temperatures higher than 600 °C, indicating practical transformation of disordered carbon into graphitic structure in the carbonization process. It notes that the carbonization took place without melting process accompanying the third weight loss at 500-600 °C, suggesting that the PPV is intermolecularly condensed to form more compact structures by removal of hydrogen and aromatization (Ohnishi et al., 1986) as shown in Fig. 15, similarly to carbonization of poly(*p*-phenylene) (Fitzer et al., 1971).

Lespade et al. (1984) identified four graphitization indices: (i) the wavenumber of the G band, (ii) the band width of the G band, (iii) the intensity ratio of the D and G bands, and (iv) the band width of the second order phonon. Especially, the intensity ratio of the D and G bands is one of the most important parameters characterizing not only the degree of graphitization but also defects in the carbonaceous materials. Using a mixed Gaussian-Lorentzian curve-fitting procedure, the ratio of the integrated intensity (R) of the D (I_D) and G bands (I_G) is calculated as

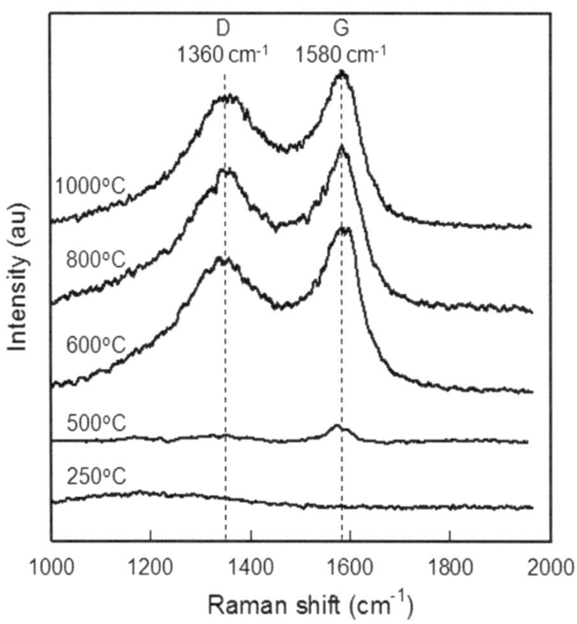

Fig. 14. Raman spectra of various nanofiber yarns thermally treated at various temperatures.

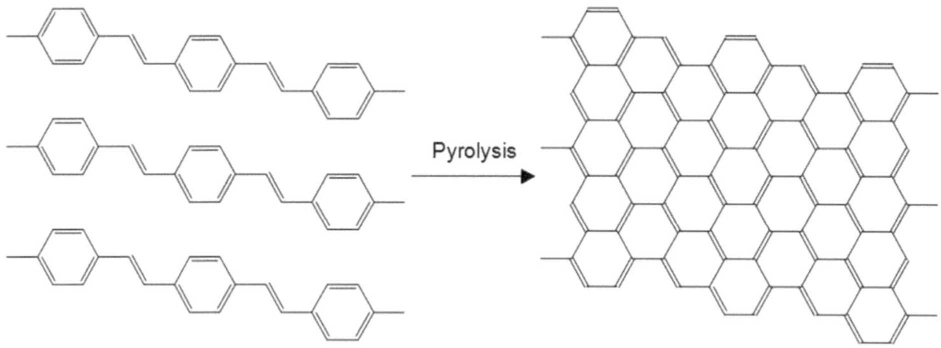

Fig. 15. Schematic illustration of pyrolysis of PPV.

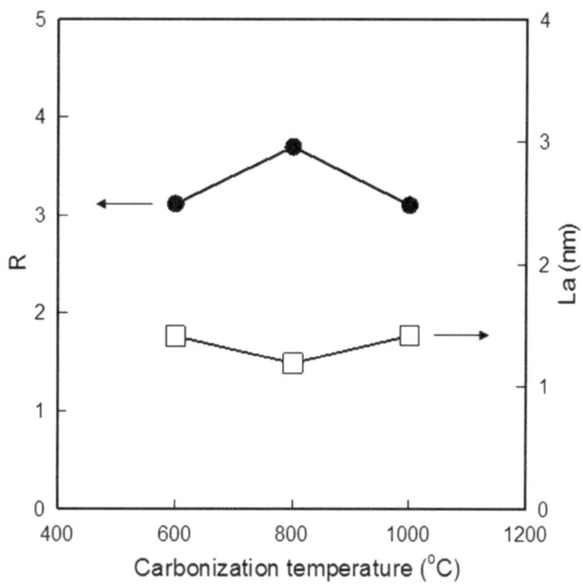

Fig. 16. Dependence of carbonization temperature on R and L_a values for various carbon nanofiber yarns.

$$R = I_D / I_G \tag{1}$$

The R is also sensitive to the concentration of graphite edge planes and/or crystal boundaries relative to standard graphite planes. Knight and White (1989) investigated the relation between R and the graphitic crystallite size (L_a) for Tuinstra and Koenig's original data plus such additional data from the literature (Tuinstra & Koenig, 1970). They found the inverse relation between R and L_a and developed an empirical formula as

$$L_a = 4.4 / R \tag{2}$$

In Fig. 16, the values of R and L_a for the nanofibers carbonized at 600-1000 °C are 3.1-3.7 and 1.2-1.4 nm, respectively, which are similar to those of the PAN-derived carbon nanofibers (Wang et al., 2003). Since the Raman spectra do not change at the carbonization temperatures between 600 and 1000 °C, the variation of the R value is considered to be within the experimental error. Assuming that the I_G and I_D are proportional to the number of the scattering ordered and disordered sp^2 bonding carbon atoms, their molar fractions X_G and X_D can be defined as follows.

$$X_G = I_G / (I_G + I_D) = 1 / (1 + R) \quad (3)$$

$$X_D = I_D / (I_G + I_D) = R / (1 + R) \quad (4)$$

From equations (2)-(4), the higher the degree of graphitization, the higher the X_G and L_a, but the lower the X_D and R. The values of X_G and X_D were calculated as 0.21-0.24 and 0.76-0.79 for the resulting carbon nanofiber yarns, respectively.

4. Conclusion

Unlike conventional electrospun polymer fibers deposited on a target electrode as a randomly oriented mesh, PXTC was electrospun into centimeters-long yarns vertically on the surface of the electrode but parallel to the electric field. The formation of the yarn was strongly influenced by the concentration, applied voltage, and relative humidity. The subsequent thermal conversion of thus-electrospun PXTC yarns was carried out at 250 °C for 12 h in vacuum and resulted in the uniaxially aligned PPV nanofibers with an average diameter of 150 nm. The applied voltage enhanced orientation of PPV nanofibers along the axis of the yarn, while the distribution of fiber diameter was less dependent on the applied voltage. WAXD analysis indicated that the PPV nanofibers exhibited paracrystalline structure with crystallinity and crystallite size of 45% and 75 Å, respectively. Furthermore, doping with sulfuric acid changed the color from yellowish brown to black and brought about a significant increase in the electrical conductivity of the PPV nanofiber yarn. The uniaxially aligned carbon nanofibers with average diameters of 133-184 nm were successfully fabricated by carbonization of the PXTC nanofiber yarns. The graphitic crystallite size and the molar fraction of the ordered sp^2 bonding carbon atoms in the resulting carbon nanofiber yarns carbonized at 60-1000 °C were 1.2-1.4 nm and 0.21-0.24, respectively. Thus, we succeeded in fabricating uniaxially aligned semiconducting, conducting, and carbon nanofiber yarns by electrospinning and subsequent thermal conversion or carbonization of PXTC, which would open up a new field of applications in organic nanoelectronics such as elecctromagnetic interference shielding, conducting textiles, and high-sensitive sensors and fast-responsive actuators utilizing their high speccific surface area.

5. References

Agend, F.; Naderi, N. & Alamdari, R. F. (2007). Fabrication and electrical characterization of electrospun polyacrylonitrile-derived carbon nanofibers, *J. Appl. Polym. Sci.*, 106, 255-259.

Ahlskog, M.; Reghu, M.; Noguchi, T. & Ohnishi, T. (1997). Doping and conductivity studies on poly(*p*-phenylene vinylene), *Synth. Met.*, 89, 11-15.

Burrourghes, J. H.; Bradley, D. D. C.; Brown, A. R.; Marks, R. N.; Mackay, K.; Friend, R. H.; Burns, P. L. & Holmes, B. (1991). Light-emitting diodes based on conjugated polymers, *Nature*, 347, 539-541.

Chiang, J. C. & MacDiarmid, A. G. (1986). 'Polyaniline': Protonic acid doping of the emeraldine form to the metallic regime, *Synth. Met.*, 13, 193-205.
Chun, I.; Reneker, D. H.; Fong, H.; Fang, X.; Deitzel, J.; Beck-Tan, N. & Kearns, K. (1999). Carbon nanofibers from polyacrylonitrile and mesophase pitch, *J. Adv. Mater.*, 31, 36-41.
Chung, G. S.; Jo, S. M. & Kim, B. C. (2005). Properties of carbon nanofibers prepared from electrospun polyimide, *J. Appl. Polym. Sci.*, 97, 165-170.
Endo, M. (1988). Grow carbon fibers in the vapor phase, *CHEMTECH*, 18, 568-576.
Fitzer, E.; Mueller, K. & Schaefer, W. (1971). The chemistry of the pyrolytic conversion of organica compounds to carbon, *Chem. Phys. Carbon*, 7, 237-383.
Gagnon, D. R.; Capistran, J. D.; Karasz, F. E.; Lenz, R. W. & Antoun, S. (1987). Synthesis, doping, and electrical conductivity of high molecular weight poly(*p*-phenylene vinylene), *Polymer*, 28, 567-573.
Geens, W.; Tsamouras, D.; Poortmans, J. & Hadziioannou, G. (2001). Field-effect mobilities in spin-cast and vacuum-deposited PPV-type pentamers, *Synth. Met.*, 122, 191-194.
Han, C. C. & Elsenbaumer, R. L. (1989). Protonic acids: Generally applicable dopants for conducting polymers, *Synth. Met.*, 30, 123-131.
Iijima, S. (1991), Helical microtubules of graphitic carbon, *Nature* (London), 354, 56-58.
Kim, K. & Jin, J.-I. (2001). Preparation of PPV nanotubes and nanorods and carbonized products derived therefrom, *Nano Lett.*, 1, 631-636.
Kim, C.; Kim, Y. J. & Kim, Y. A. (2004). Fabrication and structural characterization of electro-spun polybenzimidazol-derived carbon nanofiber by graphitization, *Solid State Commun.*, 132, 567-571.
Kim, C. K.; Park, S. H.; Lee, W. J.; Yang, K. S. (2004). Characteristics of supercapaitor electrodes of PBI-based carbon nanofiber web prepared by electrospinning, *Electrochim. Acta*, 50, 877-881.
Kim, C.; Park, S. H.; Cho, J. I.; Lee, D. Y.; Park, T. J.; Lee, W. J. & Yang, K. S. (2004). Raman spectroscopic evaluation of polyacrylonitrile-based carbon nanofibers prepared by electrospinning, *J. Raman Spectrosc.*, 35, 928-933.
Kim, C.; Yang, K. S.; Kojima, M.; Yoshida, K.; Kim, Y. J.; Kim, Y. A. & Endo, M. (2006). Fabrication of electrospinning-derived carbon nanofiber webs for the anode material of lithium-ion secondary batteries, *Adv. Funct. Mater.*, 16, 2393-2397.
Knight, D. S. & White, W. B. (1989). Characterization of diamond films by Raman spectroscopy, *J. Mater. Res.*, 4, 385-393.
Lespade, P.; Marchand, A.; Couzi, M. & Cruege, F. (1984). Caracterisation de materiaux carbones par microspectrometrie Raman, *Carbon*, 22, 375-385.
Li, D.; Wang, Y. & Xia, Y. (2004). Electrospinning nanofibers as uniaxially aligned arrays and layer-by-layer stached films, *Adv. Mater.*, 16, 361-366.
MacDiarmid, A. G.; E. Jones, Jr., W.; Norris, I. D.; Gao, J.; Johnson, Jr., A. T.; Pinto, N. J.; Hone, J.; Han, B.; Ko, F. K.; Okuzaki, H. & Llaguno, M. (2001). Electrostatically-generated nanofibers of electronic polymers, *Synth. Met.*, 119, 27-30.
Madhugiri, S.; Dalton, A.; Gutierrez, J.; Ferraris, J. P.; Balkus K. J. Jr. (2003). Electrospun MEH-PPV/SBA-15 composite nanofibers using a dual syringe method, *J. Am. Chem. Soc.*, 125, 14531-14538.

Masse, M. A.; Schlenoff, J. B.; Karasz, F. E. & Thomas, E. L. (1989). Crystalline phases of electrically conductive poly(*p*-phenylene vinylene), *J. Polym. Sci., Polym. Phys.*, 27, 2045-2059.

Ohnishi, T.; Murase, I.; Noguchi, T. & Hirooka, M. (1986). Highly conductive graphite film prepared from pyrolysis of poly(*p*-phenylene vinylene), *Synth. Met.*, 14, 207-213.

Okuzaki, H.; Ikeda, N.; Kubota, I. & Kunugi, T. (1999). Mechanical properties and structure of poly(p-phenylenevinylene) films prepared by the zone-reaction method, *Macromolecules*, 32, 5606-5612.

Okuzaki, H.; Hirata, Y. & Kunugi, T. (1999). Mechanical properties and structure of zone-drawn poly(*p*-phenylene vinylene) films, *Polymer*, 40, 2625-2629.

Okuzaki, H.; Ikeda, N.; Kubota, I. & Takahashi, T. (2005). Zone-reacted poly(*p*-phenylene vinylene) films. Changes in mechsnical properties and microstructure by acid doping, *Synth. Met.*, 153, 165-168.

Okuzaki, H.; Takahashi. T.; Miyajima, N.; Suzuki, Y. & Kuwabara, T. (2006). Spontaneous formation of poly(p-phenylene vinylene) nanofiber yarns through electrospinning of a precursor, *Macromolecules*, 39, 4276-4278.

Okuzaki, H. (2006). Nanofibers of functional polymers fabricated by electrospinning, *Kobunshi*, 55, 126-129.

Okuzaki, H.; Takahashi, T.; Hara, Y. & Yan, H. (2008). Uniaxially aligned carbon nanofibers derived from electrospun precursor yarns, *J. Polym. Sci., Polym. Phys.*, 46, 305-310.

Okuzaki, H.; Takahashi, T.; Hara, Y. & Yan, H. (2008). Conducting nanofiber yarns fabricated by electrospinning, *Sen'i Gakkaishi*, 64, 1, 5-10.

Park, S. H.; Kim, C.; Choi, Y. O. & Yang, K. S. (2003). Preparations of pitch-based CF/ACF webs by electrospinning, *Carbon*, 41, 2655-2657.

Ren, Z. F.; Huang, Z. P.; Xu, J. H.; Bush, P.; Siegal, M. P. & Provencio, P. N. (1998). Synthesis of large arrays of well-alighned carbon nanotubes on glass, *Science*, 282, 1105.

Reneker, D. H. & Chun, I. (1996). Nanometre diameter fibres of polymer, produced by electrospinning, *Nanotechnology*, 7, 216-223.

Sariciftci, N. S.; Braun, D.; Zhang, C.; Srdanov, V. I.; Heeger, A. J.; Stucky, G. & Wudl, F. (1993). Semiconducting polymer-buckminsterfullerene heterojunctions: Diodes, photodiodes, and photovoltaic cells, *Appl. Phys. Lett.*, 62, 585.

Tuinstra, F. & Koenig, J. L. (1970). Raman spectrum of graphite, *J. Chem. Phys.*, 53, 1126.

Wang, Y.; Serrano, S. & Santiago-Aviles, J. J. (2003). Raman characterization of carbon nanofibers prepared using electrospinning, *Synth. Met.*, 138, 423-427.

Yang, S.; Lozano, K.; Lomeli, A.; Foltz, H. D. & Jones, R. (2005). Electromagnetic interference shielding effectiveness of carbon nanofiber/LCP composites, *Composites: Part A*, 36, 691-697.

9

Applications of Carbon Materials for Ferroelectric and Related Materials

Young-Seak Lee, Euigyung Jeong and Ji Sun Im
Chungnam National University
South Korea

1. Introduction

Ferroelectricity is spontaneous electric polarization of a material without an external electric field and the polarization can be reversed by applying an external electric field as shown in Fig.1 (Känzig, 1957; Lines & Glass, 1979). Because the spontaneous polarization of the material is changed by an external stimulus, ferroelectric materials are also piezoelectric and pyroelectric, when the stimuli are force and heat, respectively. Ferroelectric materials have non-linear polarization; thus, they can be used as capacitors with tunable capacitance. Moreover, the hysteresis effect of the spontaneous polarization of ferroelectric materials enables the application of ferroelectric random-access memories (RAMS) for computers and radio-frequency identification (RFID) cards (Buck, 1952). In addition, piezoelectric materials are used for high voltage power source, sensors, and actuators. In these systems, it is evident that ferroelectric and related materials are not used alone. For complete systems, other materials are required to compensate for the poor properties, such as the low conductivity of piezoelectric concrete, or to enhance their performance (Shifeng, 2009; Ishiwara 2009).

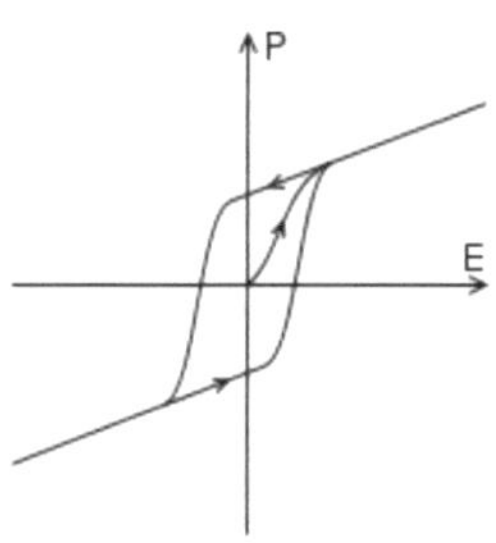

Fig. 1. The unique spontaneous polarization of ferroelectric materials

Carbon materials are used extensively in various forms in a variety of systems, because of their high thermal and chemical stability, and excellent mechanical, electrical, and electrochemical properties, which can be maximized by using an appropriate process. The various useful properties of carbon materials are attributed to their unique structures as shown in Fig.2 (Wei et al., 1993; Cousins, 2003; Frondel & Marvin, 1967; Kroto et al., 1985; Wang et al., 2009). Diamonds are famous for their impressive mechanical properties, which result from the strong covalent bonds based on sp^3 hybridization only between carbon atoms. On the other hand, carbon materials, such as fullerenes and carbon nanotubes

(CNTs), have excellent electrical or electrochemical properties, such as low resistance, due to the π-π conjugations based on sp^2 hybridization. Of the abovementioned carbon materials sp^2 carbon materials are attractive for use with ferroelectric and related materials, due to their excellent electrical properties, such as good conductivity and low resistance.
Therefore, this chapter introduces carbon materials, with their unique and excellent properties, for applications in ferroelectric and related materials.

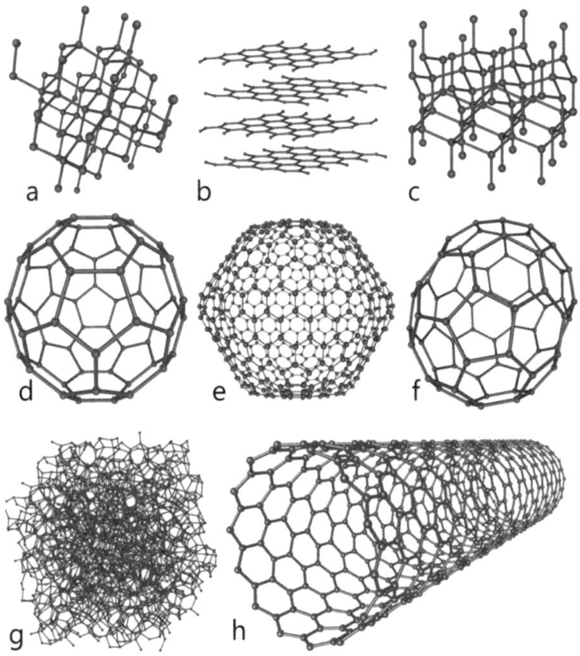

Fig. 2. Carbon materials in various multi-atomic structures with different chemical configurations: (a) diamond, (b) graphite, (c) lonsdaleite, (d) C60, (e) C540, (f) C70, (g) amorphous carbon, and (h) single-walled carbon nanotube

2. Ferroelectric materials assisted by carbon materials

Ferroelectric random access memory (FeRAM) has been considered for non-volatile memories, because it has the lowest power consumption among various semiconductor memories and its operation speed is similar to that of dynamic RAM (DRAM) (Arimoto & Ishiwara, 2004). A single ferroelectric-gate FET (field effect transistor) is the main component of FeRAM. However, fabricating ferroelectric-gate FETs with excellent electrical properties is difficult, due to the diffusion problem. When a ferroelectric material, such as lead zirconate titanate (PZT), is deposited directly on a Si substrate, forming a film, the constituent element of the substrate and film are diffused, or mixed, during the crystallization process. One way to avoid this diffusion problem is to insert an insulating buffer layer between the Si substrate and a ferroelectric film. Fig.3. shows the resulting gate structure, either an MFIS (Fig.3(a)) or MFMIS (Fig.3(b)) structure (M: metal, F: ferroelectrics, I: insulator, and S: semiconductor) (Ishiwara, 2009). However, inserting the buffer layer raises new problems, such as short data retention time and high operation voltage. Using CNT is one of the efficient methods for solving these problems.

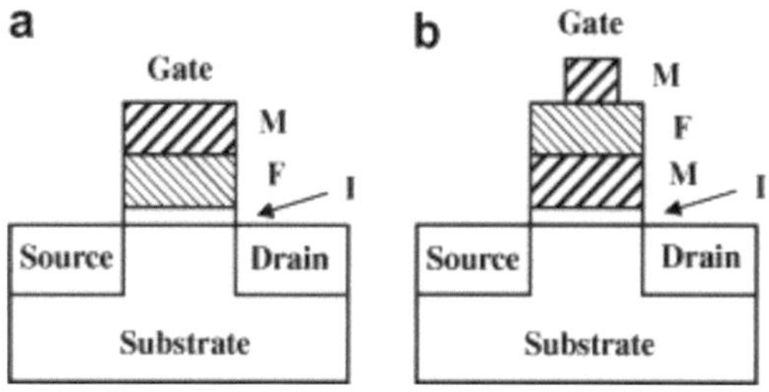

Fig. 3. Schematic drawings of (a) MFIS and (b) MFMIS gate structures (Ishiwara, 2009)

Whereas an insulating buffer layer is usually inserted between a Si substrate and the ferroelectric film in a ferroelectric-gate Si transistor, direct contact between the substrate and the ferroelectric film (MFS structure) can be achieved in a ferroelectric-gate CNT transistor (Fig.4.), due to the absence of dangling bonds on the surface of ideal CNTs. The direct contact in a ferroelectric-gate FET is expected to achieve longer data retention time with a lower operation voltage than that in a FET with MFIS gate structure, because no depolarization field is generated in the ferroelectric film. In addition, the high current drivability of CNTs also enables ferroelectric-gate CNT transistors to be one of the most promising candidates for future high-density non-volatile memories.

The current through a CNT in CNT transistors comes from thermally assisted tunneling through the source Schottky barrier J. Appenzeller, M. Radosavljevic, J. Knoch and Ph. Avouris, *Phys. Rev. Lett.* 92 (2004), p. 048301. (Appenzeller et al., 2004). Thus, the gate voltage lowers the Schottky barrier height and produces a high electric field at the semiconductor surface, rather than modulating the channel conductance. In addition, in the MFS structure, no depolarization field is expected to be generated in the ferroelectric film with zero external voltage. Thus, zero electric field exists for both directions of the remnant polarization. As a result, the polarization direction in the ferroelectric film is hard to be read out by drain current in a FET with drain electrodes and a Schottky barrier source. Therefore, it is easier to discuss the operation characteristics of ferroelectric-gate CNT transistors, semi-quantitatively.

A simplified current modulation model at the source Schottky barrier in a Si transistor can be applied to a ferroelectric-gate CNT after changing the dielectric gate insulator to a ferroelectric (Tsutsui & Asada, 2002). Fig.5 is a one-dimensional energy band diagram at the source edge in a ferroelectric-gate FET and it assumes that the work functions of the gate and source metals are the same. Because a CNT transistor has a very thin semiconductor thickness as shown in Fig.5, the semiconductor can be assumed to be an insulator, when the doping density of the CNT is not very high. This assumption enables the source region in the ferroelectric-gate FET to be expressed as an MFIM structure, which is indicated in the dotted square of Fig.4. In an MFIM structure, the first M is the gate metal, F is a ferroelectric, I is a CNT, and the last M is the source metal. A graphical method can be used in this one-dimensional structure to calculate electric fields in both F and I films easily (Ishiwara, 2001). Interestingly, the depolarization field appears even in a CNT transistor with MFS gate structure.

The average electric field in the semiconductor can be calculated under the assumption that relative dielectric constant of the semiconductor is the same as that of Si (11.8) and the ferroelectric film has the remnant polarization of 3 $\mu C/cm^2$ and a square-shaped P-V hysteresis. Moreover, the calculated average electric field in the semiconductor is 3 MV/cm, which is independent of the semiconductor thickness. Because the calculated value is much higher than the dielectric breakdown field of Si, which is less than 1MV/cm, sufficient charge can be injected into the semiconductor when the electric field lowers the Schottky barrier height. However, the depolarization field in the ferroelectric film is not high, which can be explained by the diagram as shown in Fig.5. The product of film thickness and electric field, which is the voltage across the film, was the same between the ferroelectric and semiconductor films under the zero bias condition. Moreover, holes were injected after the application of negative gate voltage, whereas no holes were injected after the application of a positive gate voltage. Combining this charge injection model with the voltage drop model at the drain Schottky barrier (Tsutsui, 2002) produced I_D–V_D characteristics of a ferroelectric-gate CNT FET.

Ishiwara reported two structures, CNTs deposited with SiO_2/Si and SBT/Pt/Ti/SiO_2/Si structures using a spin-coated method. In the CNTs deposited with SiO_2/Si structure, Si substrate is used as the gate electrode in the MOS-CNT transistors, whereas a Pt film is used in the CNTs deposited with SBT/Pt/Ti/SiO_2/Si ferroelectric-gate FETs. However, both transistors had charge injection-type hysteresis loops. Therefore, the first step for realizing ferroelectric-gate CNT transistors is elimination of the spurious hysteresis loops.

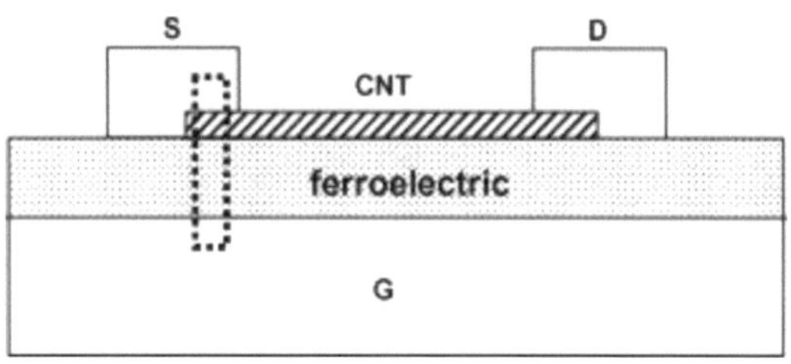

Fig. 4. Schematic drawing of a ferroelectric-gate CNT transistor (Ishiwara, 2009)

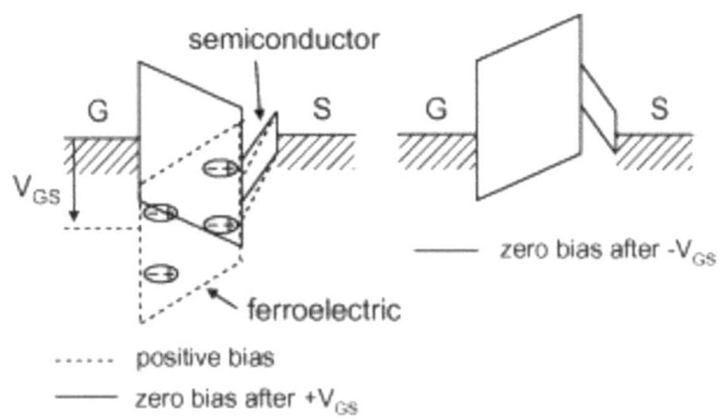

Fig. 5. One-dimensional energy band diagram at the source edge in a ferroelectric-gate Schottky FET (Ishiwara, 2009)

3. Piezoelectric materials assisted by carbon composites

Cement-based piezoelectric composites have been studied for applications, such as sensors and actuators in civil engineering (Li et al., 2002; Cheng et al., 2004; Chaipanich et al., 2007). These sensors and actuators have a great potential to be used for non-destructive performance monitoring of bridges and dams, for example (Sun et al., 2000). The piezoelectric composites have been prepared with a variety of different connectivity patterns, and the 0-3 connectivity is the simplest (Newnham et al., 1978). In a 0-3 cement-based piezoelectric composite, a three-dimensionally connected cement was loaded with a zero-dimensionally connected active piezoelectric ceramic particles. PZT is mainly used for piezoelectric ceramic and has high dielectric constant and density, which results good piezoelectric properties. The 0-3 cement-based piezoelectric composite is also compatible with concrete, the most popular host material in civil engineering. In addition, cement-based composites are easy to manufacture and amenable to mass production (Li et al., 2002; Huang et al., 2004; Cheng et al., 2005). However, the 0-3 cement-based piezoelectric composites are complicated by poling because the ceramic structure does not form a continuously connected structure across the inter-electrode-dimension. The difficulty of poling is mainly due to the high electric impedance and piezoelectric activities of the composites are lower than those of pure ceramics. Thus, the poling field and the poling temperature should be increased to facilitate poling of the ceramic particles in the composites. However, if the poling voltage is too high, the samples can be broken down or if a poling temperature is too high, the mechanical properties of the cement can be weaken. Therefore, improving the electrical conductivity of the cement matrix is one of the efficient methods to increase the polarization of the composite.

Carbon materials have excellent electrical properties and are often used to enhance the conductivity of other composites. Shifeng et al. investigated the effect of carbon black on the properties of 0-3 piezoelectric ceramic/cement composites (Shifeng et al., 2009). The composites were manufactured using sulphoaluminate cement and piezoelectric ceramic $[0.08Pb(Li_{1/4}Nb_{3/4})O_3 \cdot 0.47PbTiO_3 \cdot 0.45PbZrO_3]$[P(LN)ZT] as raw materials with a compression technique. The piezoelectric strain constant d_{33} of the composites with carbon black content was as shown in Fig.9. The piezoelectric strain constant at its maximum was 42% larger than that of the composite without carbon black when 0.3 wt% content of the carbon black was used, and decreased when the content of carbon black was beyond 0.3 wt%. The Maxwell-Wagner model can explain this trend. The following equation (1) gives the ratio of the electric field acting on the ceramic particles and matrix phases in a 0-3 piezoelectric composite with the conductivity (Blythe, 1979).

$$E_1/E_2 = \sigma_2/\sigma_1 \tag{1}$$

where σ_1 and σ_2 are the electric conductivity of the ceramic and the matrix, respectively. It is noteworthy that the electric field working on the ceramic particles is controlled by the ratio (σ_2/σ_1). Because the conductivity of ceramic particles is much higher than that of the cement matrix, σ_2/σ_1 is small. Thus, the addition of a small amount of a conductive material, such as carbon materials, decreases the impedance of the composite and increases the electric conductivity of the composite. As a results, the electric conductivity of the cement matrix increases, resulting in easier poling. Therefore, up to a suitable amount of carbon black addition, the piezoelectric constant d_{33} increases gradually, whereas with more than 0.3 wt% of the amount of carbon black added, the piezoelectric properties is decreased, because the higher voltage could not be established during the poling process. Gong et al. also reported the piezoelectric and dielectric behavior of 0-3 cement-based composites mixed with carbon black (Gong et al., 2009). They fabricated 0-3 cement-based composites from white cement, PZT powder and a small amount of carbon black and found similar results to those of a poling process of the composite at room temperature that was facilitated by the addition of carbon black; the piezoelectric properties of the composite were improved. It was also found that when too much carbon black was added, the piezoelectric properties of the composite decreased, due to the conductive properties of the carbon black. Sun et al. also found that carbon fiber could be used for piezoelectric concrete (Sun et al., 2000).

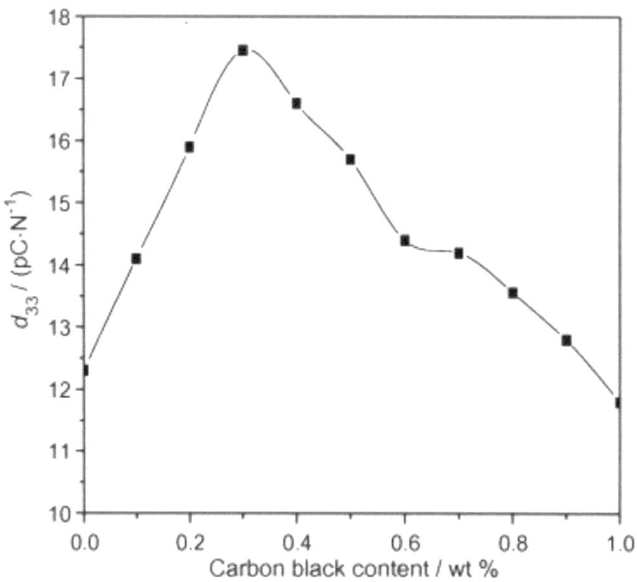

Fig. 6. Variation of piezoelectric strain constant d_{33} of the composites with carbon black content (Shifeng et al., 2009)

As for the piezoelectric cement composites, the piezoelectric properties of the polymer composites can be facilitated by addition of the conductive carbon materials. Sakamoto et al. reported acoustic emission (AE) detection of PZT/castor oil-based polyurethane (PU) with and without graphite doping (Sakamoto et al., 2002). The piezoelectric and pyroelectric properties of the composite increased with graphite doping, due to the enhanced conductivity/reduced resistance caused by graphite doping. Moreover, the piezoelectric coefficient d_{33} varied with the carbon content with similar behavior to that of the piezoelectric cement composites, as shown in Fig.7. As a result, two simulated sources of AE, ball bearing drop and pencil lead break, were detected better with the graphite doped PZT/PU composite. This can be also explained by the aforementioned Maxwell-Wagner interfacial mechanism.

Piezoelectric polymer composites can be also useful as mechanical damping composites (Hori et al., 2001). PZT/carbon black/epoxy resin composites were manufactured and their mechanical and damping properties were investigated. A measure of mechanical of damping intensity, the mechanical loss factor (η), reached its maximum at a certain level of carbon black added and decreased above that as shown in Fig.8. In this composite, carbon black had an additional function: when the mechanical energy from vibrations and noises was transformed into electrical energy (current) by PZT, the electric current was conducted to an external circuit through CB powders and then dissipated as thermal energy through a resistor.

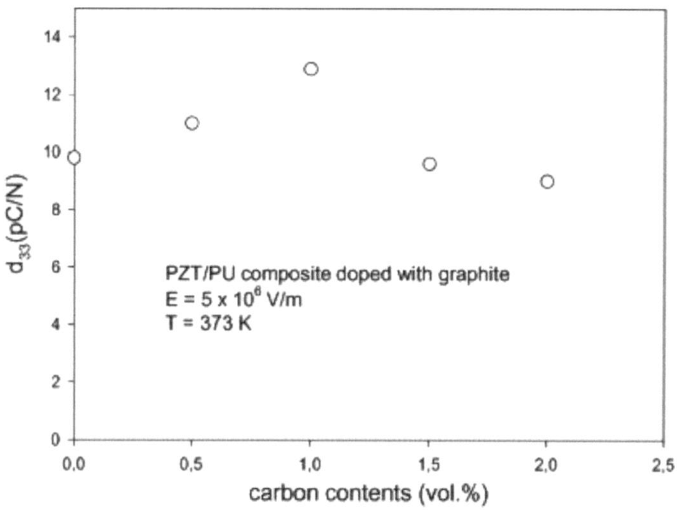

Fig. 7. Variation of the piezoelectric coefficient d_{33} with the carbon contents in the composite film (Sakamoto et al., 2002)

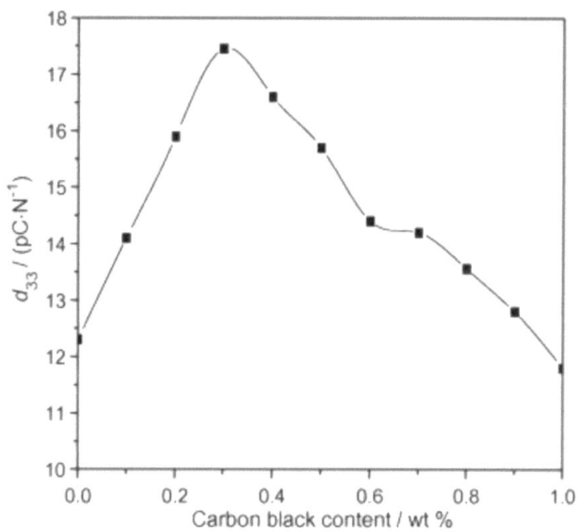

Fig. 8. Variation of piezoelectric strain constant d_{33} of the composites with carbon black content (Hori et al., 2001)

Unlike the studies described so far, Li et al. reported that piezoelectric material improve the properties of carbon materials (Li et al., 2010). The single crystal PMN-PT (lead magnesium niobate-lead titanate, which has a chemical formular of (1-x)[$Pb(Mg_{1/3}Nb_{2/3})O_3$] and x[$PbTiO_3$], was embedded in the two different activated carbons (NAC and HAC), and hydrogen adsorption of the composite was investigated. Hydrogen adsorption of the composite was enhanced due to an electric field generated from the piezoelectric material, and the amount of the enhancement was proportional to the charges generated by the piezoelectric materials. Therefore, higher hydrogen adsorption was achieved because higher pressure creasted more charges (Fig.9 and 10). Hydrogen adsorption at lower temperatures was much greater than that at higher temperatures, because the electric field has a great effect on the bonding of hydrogen molecules more when the kinetic energy of the molecules is lower at lower temperatures (Fig.10).

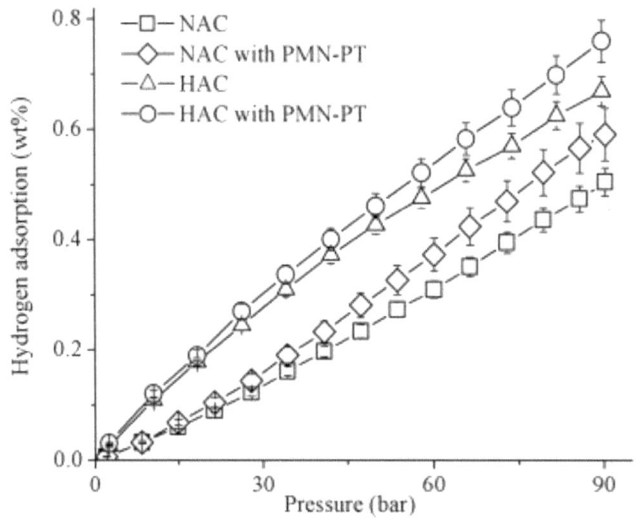

Fig. 9. PMN–PT effect on H_2 adsorption of NAC and HAC at 293 K under various pressures (Li et al., 2010)

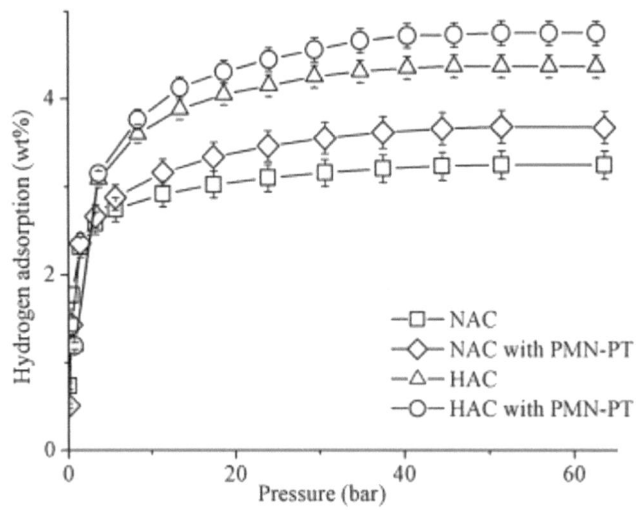

Fig. 10. PMN–PT effect on H_2 adsorption of NAC and HAC at 77 K under various pressures (Li et al., 2010)

5. Conclusion

In this chapter, carbon materials were introduced to assist the application of ferroelectric related materials. Ferroelectric-gate CNT transistors use the unique interfacial and electrical properties of CNTs to a longer data retention time of the transistor. In all of the piezoelectric composites for different applications, up to a certain level of carbon materials added, the piezoelectric properties of the composites increased, as the content of carbon materials increased. But the piezoelectric properties decreased, as the content of carbon materials increased above that level. These behaviors can be explained by Maxwell-Wagner mechanism. Therefore, even if the carbon materials themselves do not have any ferroelectricity, the excellent electrical properties of carbon materials enable them to assist the applications of ferroelectric and related materials.

6. References

Appenzeller, J.; Radosavljevic; Knoch, J. & Avouris, Ph. (2004), Tunneling Versus Thermionic Emission in One-Dimensional Semiconductors, *Physical Review Letters*, 92, 048301-0483304.

Arimoto, Y & Ishiwara H. (2004). Current status of ferroelectric random-access memory, *MRS Bulletin*, 29, 9, 823-828.

Blythe, R. L. (1979). *Electrical Properties of Polymer*, Cambridge University Press, London.

Buck D.A. (1952). Ferroelectrics for Digital Information Storage and Switching, *Report R-212*, M.I.T.

Chaipanich, A.; Jaitanong, N. & Tunkasiri, T. (2007), Fabrication and properties of PZT-ordinary Portland cement composites, *Materials Letters*, 30, 5206-5208.

Cheng, X.; Huang, Sh. & Chang, J. (2004). Piezoelectric and dielectric properties of piezoelectric ceramic-sulphoaluminate cement composites, *Journal of the European Ceramic Society*, 25, 3223-3228.

Cousins, C. S. G. (2003). Elasticity of carbon allotropes. I. Optimization, and subsequent modification, of an anharmonic Keating model for cubic diamond, *Physical Review B*, 67, 024107-024119.

Frondel, C. & Marvin U.B. (1967). Lonsdaleite, a new hexagonal polymorph of diamond, *Nature*, 214, 587–589.

Gong, H.; Li, Z.; Zhang, Y.; Fan, R. (2009). Piezoelectric and dielectric behavior of 0-3 cement-based composites mixed with carbon black, *Journal of the European Ceramic Society*, 29, 2013-2019.

Hori, M; Aoki, T.; Ohira, Y. & Yano, S. (2001). New type of mechanical damping composites composed of piezoelectric ceramics, carbon black and epoxy resin, *Composites A*, 32, 287-290.

Huang, S.; Chang, J. & Cheng, X. (2004), Poling process and piezoelectric properties of lead zirconate titanate/sulphoaluminate cement composites, *Journal of Materials Science*, 39, 23, 6975-6979.

Ishiwara, H. (2001), Current status and prospects of FET-type ferroelectric memories, *Journal of Semiconductor Technology and Science*, 1, 1, 1-14.

Ishiwara, H. (2009). Current status of ferroelectric-gate Si transistors and challenge to ferroelectric-gate CNT transistors, *Current Applied Physics*, 9, S2-S6.

Känzig, W. (1957), Ferroelectrics and Antiferroelectrics, *Solid State Physics*. 4. Academic Press, ISBN 0126077045, New York.

Kroto, H. W.; Heath, J. R.; O'Brien, S. C.; Curl R. F. & Smalley, R. E. (1985). C_{60}: Buckminsterfullerene, *Nature*, 318, 162–163.

Li, X.; Hwang, J.-Y.; Shi, S.; Sun, X. & Zhang, Z. (2010). Effect of piezoelectric material on hydrogen adsorption, *Fuel Processing Technology*, 91, 1087-1089.

Li, Z. J.; Zhang, D. & Wu, K. R. (2002). Cement-based 0-3 piezoelectric composites, *Journal of American Ceramic Society*, 85, 2, 305-313.

Lines M. & Glass A. (1979). *Principles and applications of ferroelectrics and related materials*. Clarendon Press, Oxford.

Newnham, R. E.; Skinner, D. P. & Cross L. E. (1978), Connective and piezoelectric-pyroelectric composites, *Materials Research Bulletin*, 13, 525-536.

Sakamoto, W. K.; Marin-Franch, P. & Das-Gupta, D. K. (2002), Characterization and application of PZT/PU and graphite doped PZT/PU composite, *Sensors and Actuators* A, 100, 2-3, 165, 174.

Shifeng, H.; Xue, L.; Futian, L.; Jun, C.; Dongyu, X. & Xin C. (2009). Effect of carbon black on properties of 0-3 piezoelectric ceramic/cement composites, *Current Applied Physics*, 9, 1191-1194.

Sun, M.; Liu, Q.; Li, Z. & Hu, Y. (2000), A study of piezoelectric properties of carbon fiber reinforced concrete and plain cement paste during dynamic loading, *Cement and Concrete Research*, 30, 1593-1595.

Tsutsui, M. & Asada, M. (2002), Dependence of drain current on gate oxide thickness of p-type vertical PtSi Schottky source/drain metal oxide semiconductor field-effect transistors,*Japanese Journal of Applied Physics*, 41, 1, 54-58.

Tsutsui, M.; Nagai, T & Asada, M. (2002), Analysis and fabrication of P-type vertical PtSi Schottky source/drain MOSFET, *IEICE Transactions on Electronics*, E85-C, 5, 1191-1199.

Wang, X.; Li, Q.; Xie, J.; Jin, Z.; Wang, J.; Li, Y.; Jiang, K. & Fan, S. (2009), Fabrication of Ultralong and Electrically Uniform Single-Walled Carbon Nanotubes on Clean Substrates, *Nano Letters*, 9, 9, 3137–3141.

Wei, L.; Kuo, P. K.; Thomas, R. L.; Anthony, T. & Banholzer, W. (1993). Thermal conductivity of isotopically modified single crystal diamond, *Physical Review Letters*, 70, 3764-3767.

10

Dielectric Relaxation Phenomenon in Ferroelectric Perovskite-related Structures

A. Peláiz-Barranco[1] and J. D. S. Guerra[2]
[1]*Facultad de Física-Instituto de Ciencia y Tecnología de Materiales, Universidad de La Habana. San Lázaro y L, Vedado. La Habana 10400,*
[2]*Grupo de Ferroelétricos e Materiais Multifuncionais, Instituto de Física, Universidade Federal de Uberlândia. 38400-902, Uberlândia – MG,*
[1]*Cuba,*
[2]*Brazil*

1. Introduction

The dielectric relaxation phenomenon in ferroelectric materials reflects the delay (time dependence) in the frequency response of a group of dipoles when submitted to an external applied field. When an alternating voltage is applied to a sample, the dipoles responsible for the polarization are no longer able to follow the oscillations of the electric field at certain frequencies. The field reversal and the dipole reorientation become out-of-phase giving rise to a dissipation of energy. Over a wide frequency range, different types of polarization cause several dispersion regions (Figure 1) and the critical frequency, characteristic of each contributing mechanism, depends on the nature of the dipoles. The dissipation of energy, which is directly related to the dielectric losses, can be characterized by several factors: *i-* the losses associated to resonant processes, characteristics of the elastic displacing of ions and electrons, and *ii-* the dipolar losses, due to the reorientation of the dipolar moment or the displacing of the ions between two equilibrium positions.
For ferroelectric materials the dielectric relaxation mechanisms are very sensitive to factors such as temperature, electric field, ionic substitution, structural defects, etc. The defects depend on either intrinsic or extrinsic heterogeneities due to special heat treatments, ionic substitutions, grain size additives, and grain boundary nature. On the other hand, structural defects may cause modifications of the short and/or long-range interactions in ferroelectric materials. From this point of view, apart from the localized dipolar species, free charge carriers can exist in the material. Several physical processes cause the decay of the electrical polarization: dipolar reorientation, motion of the real charges stored in the material and its ohmic conductivity. The former is induced by thermal excitations, which lead to decay of the resultant dipolar polarization. The second process is related to the drift of the charges stored in the internal field of the sample and their thermal diffusion. With the increase of the temperature, the dipoles tend to gradually disorder owing to the increasing thermal motion and the space charges trapped at different depths are gradually set free. Thus, the electrical conductivity in ferroelectric materials affects the physical properties because of there will be a competition between the ferroelectric phase and free charge carriers.

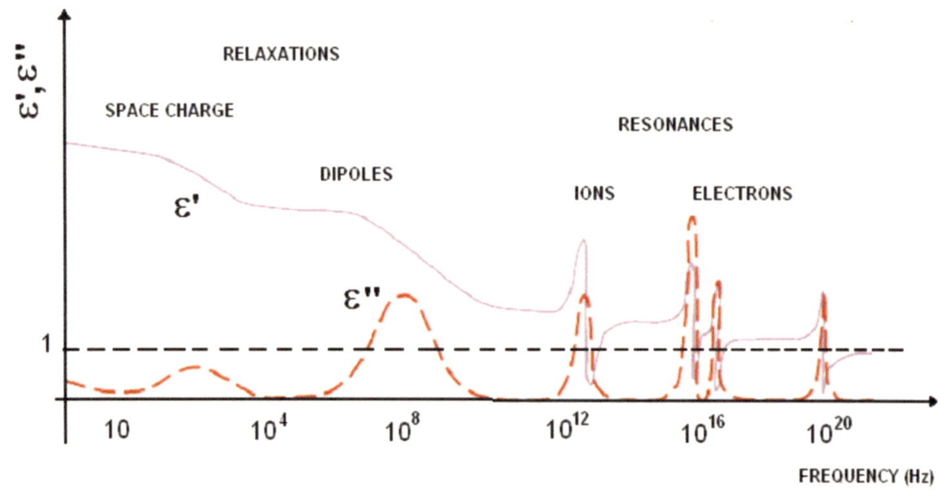

Fig. 1. General representation of relaxation and resonance types

The oxygen vacancies are always related to dielectric relaxation phenomenon, as well as to the electrical conductivity for ferroelectric perovskite-related structures, considering that these are the most common mobiles species in such structures (Jiménez & Vicente, 1998; Bharadwaja & Krupanidhi, 1999; Chen et al., 2000; Islam, 2000; Yoo et al., 2002; Smyth, 2003; Verdier et al., 2005). For ABO_3-type perovskite structures, the BO_6 octahedra play a critical role in the demonstration of the ferroelectric properties. It has been reported that the ferroelectricity could be originated from the coupling of the BO_6 octahedra with the long-range translational invariance via the A-site cations (Xu, 1991). In this way, the breaking of the long-range translational invariance results in the dielectric relaxation phenomenon. The degree of the coupling between neighboring BO_6 octahedra will be significantly weakened by introducing defects, such as vacancies. On the other hand, for the layered ferroelectric perovskites (Aurivillius family, $[Bi_2O_2]^{2+}[A_{n-1}B_nO_{3n+1}]^{2-}$), the oxygen vacancies prefer to stay in the Bi_2O_2 layers, where their effect upon the polarization is considered to be small, instead of the octahedral site, which controls the polarization. The origin of the dielectric behavior for these materials have been associated to a positional disorder of cations on A- or B-sites of the perovskite layers that delay the evolution of long-range polar ordering (Blake et al., 1997; Ismunandar & Kennedy, 1999; Kholkin et al., 2001; Haluska & Misture, 2004; Huang et al., 2006).

For at least several decades, the dielectric response of ferroelectric materials (polycrystals, single crystals, liquids, polymers and composites) has been of much interest to both experimentalists and theorists. One of the most attractive aspects in the dielectric response of ferroelectric materials is the dielectric relaxation phenomenon, which can show the direct connection that often exists between the dipolar species and the charge carriers in the materials. Researchers typically fit the complex dielectric permittivity data according to a relaxation theoretical model, which is representative of the physical processes taking place in the system under investigation.

The complex dielectric permittivity (ε) can be expressed as:

$$\varepsilon(\omega) = \varepsilon'(\omega) - i\varepsilon''(\omega) \tag{1}$$

where ω is the angular frequency; ε' is the real component, which is in phase with the applied field; ε'' is the imaginary component, which is in quadrature with the applied field. Both components of the complex dielectric permittivity are related each other by the Kramers-Kronig relations.

The Debye's model (Debye, 1929), which considers not-interacting dipoles, proposes the following expression for the complex dielectric permittivity:

$$\varepsilon(\omega)=\varepsilon_\infty+\frac{\varepsilon_s-\varepsilon_\infty}{1+i\omega\tau} \tag{2}$$

where τ is the relaxation time, ε_s is the statical dielectric permittivity (at very low frequencies) and ε_∞, the dielectric permittivity at high frequencies.

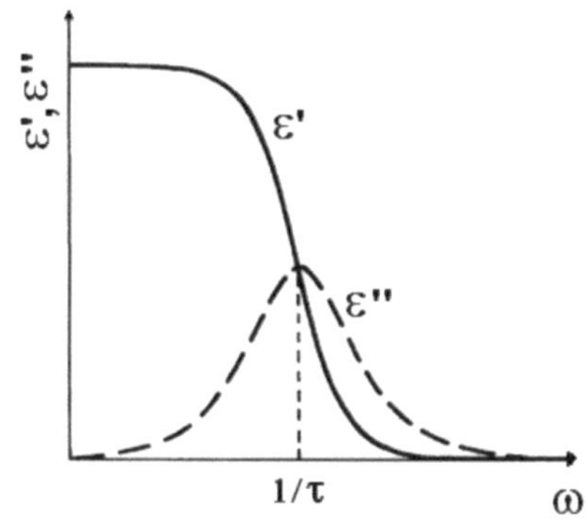

Fig. 2. Frequency dependence for the real and imaginary components of the dielectric permittivity from the Debye's model

This was the first model proposed to evaluate the dielectric relaxation. However, the experimental results in polar dielectric materials have showed that corrections to that simple model are necessary. The Cole-Cole's model (Cole & Cole, 1941) introduces a correction considering a distribution function for the relaxation time (expression 3).

$$\varepsilon(\omega)=\varepsilon_\infty+\frac{\varepsilon_s-\varepsilon_\infty}{1+(i\omega\tau)^\alpha} \tag{3}$$

From that model, the real (ε') and imaginary (ε'') parts of the dielectric permittivity (Cole & Cole, 1941) can be written as:

$$\varepsilon'(\omega)=\varepsilon_\infty+\left(\frac{\Delta\varepsilon'}{2}\right)\left\{1-\frac{\sinh(\beta z)}{\cosh(\beta z)+\cos\left(\beta\frac{\pi}{2}\right)}\right\} \tag{4}$$

$$\varepsilon''(\omega)=\left(\frac{\Delta\varepsilon'}{2}\right)\frac{\sin\left(\beta\frac{\pi}{2}\right)}{\cosh(\beta z)+\cos\left(\beta\frac{\pi}{2}\right)} \tag{5}$$

where $z=\ln(\tau)$, $\Delta\varepsilon'=\varepsilon_S-\varepsilon_\infty$, and $\beta=(1-\alpha)$, where α shows the deformation of the semicircle arc in the Cole-Cole plot, i.e. it is the angle from the ε' axis to the center of the semicircle arc.
The temperature dependence for the relaxation time follows the Arrhenius dependence given as:

$$\tau = \tau_o \exp\left(\frac{Ea}{k_B T}\right) \tag{6}$$

where E_a is the activation energy of the relaxation process, k_B is the Boltzmann's constant, T is the temperature and τ_o is the pre-exponential factor. By using this dependence, the activation energy value of the dielectric relaxation process can be calculated and the mechanism, or mechanisms, associated to it be evaluated.
On the other hand, it has been previously commented that for ferroelectric materials (for dielectrics in general) the charge carriers have to be taking into account in the dielectric relaxation mechanisms. When an electric field is applied to the material, there is the known reorientation of the dipoles but also the displacement of the charge carriers. Therefore, the electrical conductivity behavior should be considered.
In order to evaluate the frequency dependent conductivity, the Universal Relaxation Law (Jonscher, 1996) can be used, where the electrical conductivity can be expressed as:

$$\sigma(\omega) = \sigma_{dc}\left[1 + \left(\frac{\omega}{\omega_H}\right)^n\right] + A\omega \tag{7}$$

where σ_{dc}, ω_H, n and A are the dc conductivity, the onset frequency of the ac conductivity (mean frequency of the hopping process), the exponent and the weakly temperature dependent term, respectively. The first and the second terms of the equation (7) refer to the universal dielectric response (UDR) and to the nearly constant loss (NCL), respectively.
The power-law frequency dependent term (UDR) originates from the hopping of the carriers with interactions of the inherent defects in the materials, while the origin of the NCL term (the linear frequency dependent term) has been associated to rocking motions in an asymmetric double well potential (Nowick & Lim, 2001). For NCL term, the electrical losses occur during the time regime while the ions are confined to the potential energy minimum (León et al., 2001).
Both the dc conductivity and the hopping frequency are found to be thermally activated following an Arrhenius dependence (equations 8 and 9), being U_{dc} and U_H the activation energies of the dc conductivity and the hopping frequencies of the carriers, respectively, and σ_o and ω_o the pre-exponential factors. It has been interpreted by using different theoretical models (Funke, 1993; Ngai, 1993; Jonscher, 1996) as indicating that the ac conductivity originates from a migration of ions by hopping between neighboring potential wells, which eventually gives rise to dc conductivity at the lowest frequencies.

$$\sigma_{dc} T = \sigma_o \exp\left(-\frac{U_{dc}}{k_B T}\right) \tag{8}$$

$$\omega_H = \omega_o \exp\left(-\frac{U_H}{k_B T}\right) \tag{9}$$

The objective of the present chapter is conducted to discuss the dielectric relaxation phenomenon in ferroelectric perovskite-related structures considering the relaxation mechanisms and the influence of the vacancies on them, so that the results can be more easily understood, from the physical point of view. Examples are given from the previously reported work of the present authors, as well as from the literature, with preference to the formers.

2. Ferroelectric ABO_3 perovskites.

Perovskite type oxides of general formula ABO_3 are very important in material, physics and earth sciences because they exhibit excellent physical properties, which make them especial candidates for a wide application range in the electro-electronic industry (Lines, 1977). They are known for their phase transitions, which may strongly affect their physical and chemical properties. Perovskite structure-type oxides exist with all combination of cation oxidation states, with the peculiar characteristics that for ABO_3 compounds one of the cations, traditionally the A-site ion, is substantially larger than the other one (B-site). This structure seem to be a particularly favorable configuration, because it is found for an extremely wide range of materials, where its basic pattern is frequently found in compounds that differ significantly from the ideal composition. From a direct review of the current literature, it has been established as well that important defects in the perovskite structure are directly related to vacancies of all three sub-lattices, electrons, holes, and substitutional impurities (Lines, 1977). Such a chemical defects strongly depend on the crystal structure as well as on the chemical properties of the constituent chemical species. The structure influences the types of lattice defects that may be formed in significant concentrations, and also influences the mobilities of the defects and hence of the chemical species. These mobilities determine whether or not defect equilibrium can be achieved within pertinent times at several temperatures, and at which temperature, during the cooling process of the material, these defects become effectively quenched. The charges and size of the ions affect the selection of the most favored defects, and their ability to be either oxidized or reduced determines the direction and amount of non-stoichiometry and the resulting enhanced electronic carrier concentrations. The volatility of a component, for instance, can also affect the equilibration and defect choice. This aspect is of particular importance for Pb-based compounds, due to the volatility of PbO during the synthesis of a specified material, which commonly achieves very high treatment temperatures (Xu, 1991). Thus, for perovskites structure-type systems, the partial substitution of A- or B-site ions promotes the activation of several conduction mechanisms (carrier doping).

In the last few years, extensive studies have been carried out by doping at A- and/or B-site by various researchers. It has been suggested that the introduction of different elements, which exhibit the dissimilar electronic configuration each other, should lead to dramatic effects associated with the electronic configuration mismatch between the ions located at the same A- or B-site. In this way, dielectric dispersions related to a conductivity phenomenon, which obeys the Arrhenius´s dependence, have been reported for $Ba_{1-x}Pb_xTiO_3$ ceramics (Bidault et al., 1994). Dielectric anomalies at the high-temperature region have been studied for $(Pb,La)TiO_3$, $BaTiO_3$ and $(Pb,La)(Zr,Ti)O_3$ systems (Keng et al., 2003). The authors have showed that the dielectric anomalies are related to the competition phenomenon of the dielectric relaxation and the electrical conduction by oxygen vacancies.

On the other hand, the dielectric spectra of $(Ba_{0.85}Sr_{0.15})TiO_3$ ceramics have been study in the paraelectric phase showing the contribution of the dc conduction to the dielectric relaxation (Li & Fan, 2009). For $KNbO_3$ ceramics, the temperature and frequency dependence of the dielectric and the conductivity properties have been studied, showing that the dielectric relaxation can be attributed to the hopping of oxygen vacancies in the six equivalent sites in the perovskite structure (Shing et al., 2010).

This section will show some experimental and theoretical results, for ABO_3-type perovskite structure systems, where the BO_6 octahedra play a critical role in the demonstration of the ferroelectric properties. First, it will be presented the $(Pb_{1-x}La_x)(Zr_{0.90}Ti_{0.10})_{1-x/4}O_3$ ceramic system (hereafter labeled as PLZT x/90/10), with x = 2, 4 and 6 at%, whose results has been discussed in the framework of the analysis of the effect of oxygen vacancies on the electrical response of soft doping $Pb(Zr,Ti)O_3$ ceramics in the paraelectric state (Peláiz-Barranco et al., 2008a; Peláiz-Barranco & Guerra, 2010). On the other hand, the frequency and temperature dielectric response and the electrical conductivity behavior around the ferroelectric–paraelectric phase transition temperature will be presented in the $(Pb_{0.88}Sm_{0.08})(Ti_{1-x}Mn_x)O_3$, with x = 0, 1 and 3 at.% (named as PSTM-x) ferroelectric ceramic system (Peláiz-Barranco et al., 2008b). In this case, the contribution of the conductive processes to the dielectric relaxation for the studied frequency range has been discussed considering also the oxygen vacancies as the most mobile ionic defects in perovskites, whose concentration seems to increase with the manganese content.

2.1 $(Pb_{1-x}La_x)(Zr_{0.90}Ti_{0.10})_{1-x/4}O_3$ ceramic system.

Figure 3 shows the temperature dependence of the real (ε′) and imaginary (ε′′) parts of the dielectric permittivity (A and B, respectively), for the studied PLZT x/90/10 compositions, at 20 kHz. The dielectric properties results revealed no frequency dependent dispersion of the dielectric parameters and a non-diffusive phase transition for all the studied compositions. However, it can be observed that the temperature corresponding to the maximum in real and imaginary dielectric permittivities (T_m) is strongly dependent on the lanthanum content. Additional anomalies in ε′(T) were observed in the paraelectric phase range for low frequencies, characterize by an increase of the dielectric parameter with the increase of the temperature (Peláiz-Barranco et al., 2008a; Peláiz-Barranco & Guerra, 2010). Moreover, this anomaly is more pronounced when the lanthanum content increases. As shown in the temperature dependence of the imaginary part of the dielectric permittivity (Figure 3(B)), very high values of this parameter are observed at the corresponding temperature range, even at not low frequencies. Such results lead to infer that the anomalous behavior in the real dielectric permittivity with the increase of the temperature, and the high values of ε′′, are attributed to the same mechanism, that is to say, to the conductivity losses (Peláiz-Barranco et al., 2008a). It can be also noted that, for the PLZT 4/90/10 and 6/90/10 compositions, the maximum imaginary dielectric permittivity, which is associated to the paraelectric-ferroelectric phase transition (Figure 3(B) inset), can not be easily observed as in the case of the PLZT 2/90/10 composition. The observed behavior for the PLZT 4/90/10 and 6/90/10 compositions has been associated to the quickly increase of the losses factor above the phase transition temperature (T_m), which is more pronounced for such compositions than for the PLZT 2/90/10 composition.

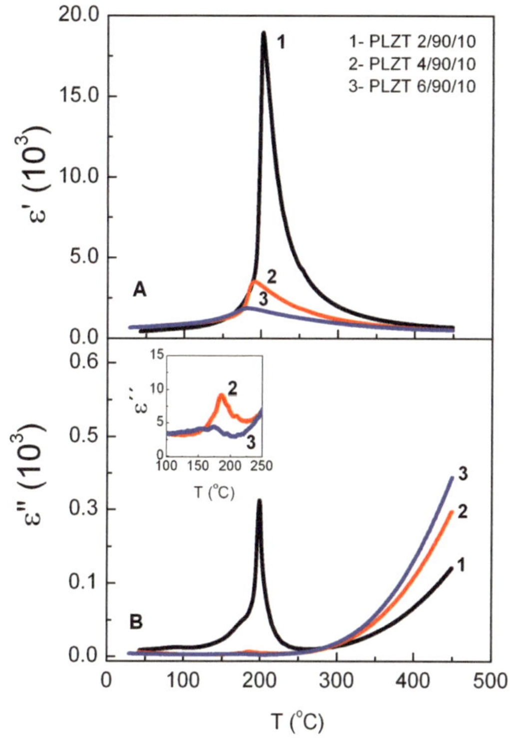

Fig. 3. Temperature dependence of the real (ε′) and imaginary (ε″) parts of the dielectric permittivity for the PLZT $x/90/10$ (x = 2, 4 and 6 at%) compositions, at 20 kHz

With the increase of measurement frequency, the observed behavior in the paraelectric phase gradually decreases, and almost disappears in the studied temperature range. Figure 4 shows the temperature dependences of ε′ and ε″ (A and B, respectively) for the PLZT 4/90/10 composition, at several frequencies. As can be seen, the high temperature anomaly almost disappears for the higher frequencies, showing that it could be related to a low-frequency relaxation process. It is well known that in perovskite ferroelectrics the oxygen vacancies could be formed in the process of sintering due to the escape of oxygen from the lattice (Kang et al., 2003). Thus, this anomaly could be correlated with a low-frequency relaxation process due to oxygen vacancies. Conduction electrons could be created by the ionization of oxygen vacancies, according to the expressions (10) and (11), where V_O, V'_O and V''_O represent the neutral, single- and doubly-ionized oxygen vacancies, respectively.

$$V_o \leftrightarrow V_o^{\cdot} + e' \tag{10}$$

$$V_o^{\cdot} \leftrightarrow V_o^{\cdot\cdot} + e' \tag{11}$$

The positions of the electrons depend on the structure characteristics, temperature range, and some other factors. It was, however, shown that the oxygen vacancies lead to shallow the electrons level. These electrons are easy to be thermally activated becoming conducting electrons. To elucidate the physical mechanism of this behavior in the paraelectric state, the electrical conductivity has been analyzed in the paraelectric phase (Peláiz-Barranco et al., 2008a).

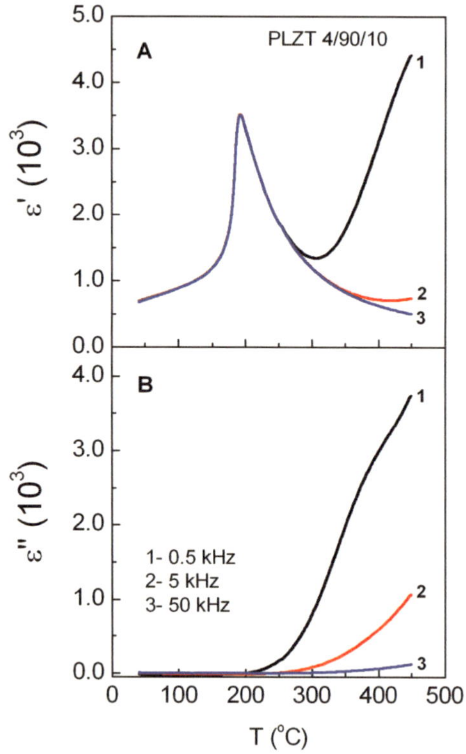

Fig. 4. Temperature dependence of the real (ε′) and imaginary (ε′′) parts of the dielectric permittivity, at several frequencies for the PLZT 4/90/10 composition

By using the Jonscher's relation expressed on the equation (7), the frequency dependence of the electrical conductivity for the studied compositions was analyzed in the paraelectric state, at several temperatures (Peláiz-Barranco et al., 2008a). The temperature dependence of the dc conductivity and the hopping frequency also followed an Arrhenius' behavior, according to the equations (8) and (9), respectively. The obtained activation energy values for both cases are shown in the Table 1.

Composition	U_{dc} (eV)	U_H (eV)
PLZT 2/90/10	0.56	1.21
PLZT 4/90/10	0.49	1.15
PLZT 6/90/10	0.46	0.99

Table 1. Activation energy values for dc conductivity contribution, U_{dc}, and the hopping process, U_H, for the studied compositions (Peláiz-Barranco et al., 2008a)

According to the obtained results, some considerations can be taken into account in order to explain the observed behaviors. It is well known that during the synthesis and process of the ceramic system, i.e. the powders calcinations and the sintering stages, especially for the $Pb(Zr,Ti)O_3$ system (PZT), there exists a high volatility of lead oxide because of the high temperatures operation. Such lead volatilization provides both fully-ionized cationic lead (V''_{Pb}) vacancies and anionic oxygen vacancies (V''_O). On the other hand, following the Eyraud's model (Eyraud et al., 1984; Eyraud et al., 2002), it is assumed that the lanthanum valence, as doping in the PZT system, has a strong influence in the ionization state of extrinsic lead and oxygen vacancies. In order to compensate the charge imbalance, one La^{3+} ion occupying a Pb^{2+} site (on the A-site of the perovskite structure) generates one half of singly ionized lead vacancies (V'_{Pb}) rather than one doubly ionized vacancy (V''_{Pb}). Thus, in the studied PLZT *x*/90/10 compositions it should be considered different types of defects, related to V'_{Pb}, V''_{Pb}, V'_{O} and V''_{O} vacancies, whose contribution depends on the analyzed temperature range. A very low concentration of neutral lead and oxygen vacancies are considered around and above the room temperature.

At low temperatures, the lead vacancies are quenched defects, which are difficult to be activated. They could become mobile at high temperatures with activation energy values around and above 2 eV (Guiffard et al., 2005). However, to the best of our knowledge the ionization state of lead vacancies remains unknown. On the other hand, it has been reported (Verdier et al., 2005; Yoo et al., 2002) that the oxygen vacancies exist in single ionized state with activation energy values in the range of 0.3–0.4 eV. For Pb-based perovskite ferroelectrics, activation energies values in the range of 0.6–1.2 eV are commonly associated to doubly-ionized oxygen vacancies (Smyth, 2003; Moos et al., 1995; Moos & Härdtl, 1996).

For PLZT 4/90/10 and PLZT 6/90/10 compositions, the results have suggested that the dc conductivity could be related to single ionized states (V'_{O}). For the PLZT 2/90/10 composition, a higher activation energy value suggests a lower oxygen vacancies concentration (Steinsvik et al., 1997), whose values are closer to those values associated to doubly-ionized oxygen vacancies (V''_{O}) (Smyth, 2003; Moos et al., 1995; Moos & Härdtl, 1996). Thus, the dc conductivity is assumed to be produced according to the reaction given by the expression (11), decreasing the number of V'_{O}, which could contribute for the lower anomaly observed for the PLZT 2/90/10 composition. The released electrons may be captured by Ti^{4+} and generates a reduction of the valence, following the relation (12).

$$Ti^{4+} + e' \leftrightarrow Ti^{3+} \quad (12)$$

Thus, the conduction process can occur due to the hopping of electrons between Ti^{4+} and Ti^{3+}, leading to the contribution of both single and doubly-ionized oxygen vacancies and the hopping energy between these localized sites for the activation energy in the paraelectric phase region for the studied PLZT compositions.

Considering the oxygen vacancies as the most mobile defects in the studied PLZT compositions, it has been analyzed their influence on the dielectric relaxation processes (Peláiz-Barranco et al., 2008a; Peláiz-Barranco & Guerra, 2010). It is known that the spontaneous polarization originating from the ionic or dipoles displacement contributions is known to be the off-center displacement of Ti^{4+} ions, from the anionic charge center of the oxygen octahedron for the PLZT system (Xu, 1991). The presence of oxygen vacancies would distort the actual ionic dipoles due to the Ti^{4+} ions. The decay of polarizations due to the distorted ionic dipoles could be the cause for the dielectric relaxation processes. However, usually the activation energy values associated to the relaxations involving thermal motions of Ti^{4+} (Peláiz-Barranco & Guerra, 2010) are higher than those observed in the studied PLZT *x*/90/10 compositions , showing that it should not be a probable process.

The obtained U_H values have suggested that the hopping process could be related to the doubly-ionized oxygen vacancies motion (Smyth, 2003; Moos et al., 1995; Moos & Härdtl, 1996). The short-range hopping of oxygen vacancies, similar to the reorientation of the dipoles, could lead to the dielectric relaxation.

On the other hand, the frequency and temperature behavior of the complex dielectric permittivity was analyzed (Peláiz-Barranco & Guerra, 2010) considering the Cole-Cole model, given by equations (4) and (5). The main relaxation time, obtained by the fitting by using both equations, followed the Arrhenius dependence (equation (6)) in the studied temperature range. The activation energy values for the relaxation processes are shown in Table 2 (Peláiz-Barranco & Guerra, 2010). The results have been associated with ionized oxygen vacancies. Thus, following the previously discussion, the dielectric relaxation phenomenon could be related to the short-range hopping of oxygen vacancies.

Composition	Ea (eV)
PLZT 2/90/10	0.48
PLZT 4/90/10	0.40
PLZT 6/90/10	0.37

Table 2. Activation energy values of the relaxation process for the studied compositions (Peláiz-Barranco & Guerra, 2010)

2.2 $(Pb_{0.88}Sm_{0.08})(Ti_{1-x}Mn_x)O_3$ ceramic system.

Figure 5 depicts the temperature dependence of ε′ and ε″ for the PSTM-x samples, for three selected frequencies, as example of the investigated behavior for the whole frequency range. Two peaks or inflections were observed for the real part of the dielectric permittivity at the studied frequency range. The first one, which was observed around 340°C for all the compositions, was associated to the temperature of the paraelectric-ferroelectric phase transition (T_m), according to previous reports for these materials (Takeuchi et al., 1983; Takeuchi et al., 1985). For all the cases, the paraelectric-ferroelectric phase transition temperature (T_m) did not show any frequency dependence, which is typical of 'normal' paraelectric-ferroelectric phase transitions (Xu, 1991). On the other hand, the imaginary dielectric permittivity showed a peak at the same temperature, to that observed for the real part of the dielectric permittivity (T_m), confirming the 'normal' characteristic of the phase transition. In fact, not frequency dispersion of the temperature of the maximum dielectric permittivity was observed. However, a sudden increase of ε″ was obtained for temperatures around 400°C for all the studied compositions, which could be associated to high conductivity values that promote the increase of the dielectric losses.

The presence of two peaks in the temperature dependence of ε′ has been observed in $(Pb_{0.88}Ln_{0.08})(Ti_{0.98}Mn_{0.02})O_3$ (being Ln = La, Nd, Sm, Gd, Dy, Ho, Er) ferroelectric ceramics (Pérez-Martínez et al., 1997; Peláiz-Barranco et al., 2009a). In this way, for small-radius-size ions (Dy, Ho and Er) a strong increase of the tetragonality has been observed (Pérez-Martínez et al., 1997), even for values higher than those observed for pure lead titanate. Both peaks were associated to paraelectric-ferroelectric phase transitions concerning two different contributions to the total dielectric behavior of the samples; one, on which the rare earths ions occupy the A-sites and the other one where the ions occupy the B-sites of the perovskite structure (Pérez-Martínez et al., 1997). For high-radius-size ions (La, Nd, Sm and Gd) the analysis has shown that, even when it could be possible an eventual incorporation of the rare earth into the A- and/or B-sites of the perovskite structure, both peaks could not be associated to the paraelectric-ferroelectric phase transitions. In this case, the observed peak at lower temperatures has been associated to the paraelectric-ferroelectric phase transition, whereas the hopping of oxygen vacancies has been considered as the cause for the dielectric anomaly at higher temperatures (Peláiz-Barranco et al., 2009a).

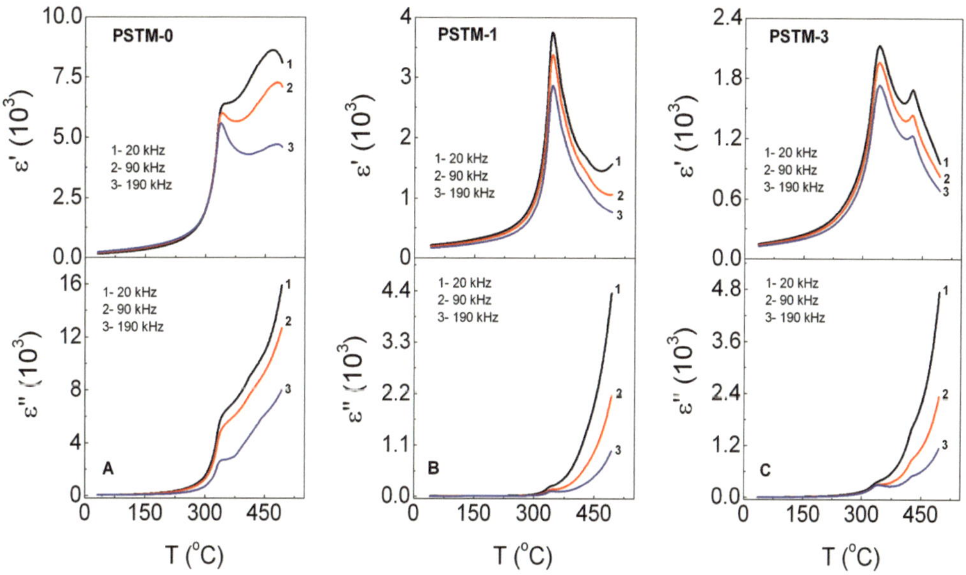

Fig. 5. Temperature dependence of the real (ε′) and imaginary (ε′′) parts of the dielectric permittivity for the PSTM-0 (A), PSTM-1 (B) and PSTM-3 (C) compositions

Considering these results, two studies were carried out on these samples; the first one (Peláiz-Barranco et al., 2008b) concerning the dielectric relaxation phenomenon and the electrical conductivity behavior for temperatures around the paraelectric-ferroelectric phase transition (around 340°C); the second one concerning the electrical conductivity behavior around the second peak (Peláiz-Barranco & González-Abreu, 2009b).

The frequency dependence of the ε′ and ε′′ for the studied PSTM-x compositions below T_m (Peláiz-Barranco et al., 2008b) was analyzed by using the Cole-Cole model (equations (4) and (5)). The temperature dependence for the mean relaxation times, obtained from the fitting of the experimental data by using equations (4) and (5) revealed an Arrhenius' behavior, according to the equation (6). The obtained values for the activation energy were 0.89 eV, 0.81 eV and 0.62 eV, for the PSTM-0, PSTM-1 and PSTM-3 compositions, respectively, which clearly show that the dielectric relaxation processes in the studied samples are closely related to the oxygen vacancies, which have been reported as the most mobile ionic defects in perovskites (Poykko & Chadi, 2000). According to previously reported results (Steinsvik et al., 1997), the activation energy for ABO_3 perovskites decreases with the increase of the oxygen vacancies content. As observed, the activation energy values for the studied samples decreases with the increase of the manganese content, which suggests an increase of oxygen vacancies concentration (Steinsvik et al., 1997). On the other hand, previous studies on electronic paramagnetic resonance (EPR) in earth rare and manganese modified lead titanate ceramics, have shown that a Mn^{4+}/Mn^{2+} reduction takes place during the sintering process (Ramírez-Rosales et al., 2001). As a consequence, oxygen vacancies have to be created to compensate for the charge imbalance. The increase of manganese content should promote a higher concentration of oxygen vacancies, which agrees with the decrease of activation energy values. Similarly to the results obtained for the PLZT $x/90/10$ ceramics, the obtained activation energy values for the PSTM-x compositions were found to be lower that those for relaxations involving thermal motions of Ti^{4+} (Maglione & Belkaoumi, 1992). So that, the observed relaxation process for the studied ceramics could be attributed to the decay of polarization in the oxygen defect-related dipoles due to their hopping conduction.

In order to better understand the conduction mechanism related to the observed relaxation process, the experimental data for the PSTM-*x* compositions were fitted by considering the equations (7), (8) and (9). Results of the dc conductivity (σ_{dc}) and the hopping frequency (ω_H) are shown in the Figure 6, for temperatures below T_m. The solid lines on Figures 6(A) and 6(B), represent the fitting using the equations (8) and (9), respectively. The obtained activation energy values (for dc conductivity contribution, U_{dc}, and the hopping process, U_H) are shown in the Table 3.

Composition	**U_{dc} (eV)**	**U_H (eV)**
PSTM-0	1.19	0.98
PSTM-1	0.94	0.90
PSTM-3	0.81	0.71

Table 3. Activation energy values (for dc conductivity contribution, U_{dc}, and the hopping process, U_H) for the studied compositions, obtained from the fitting of the experimental data by using the equations (8) and (9) (Peláiz-Barranco et al., 2008b)

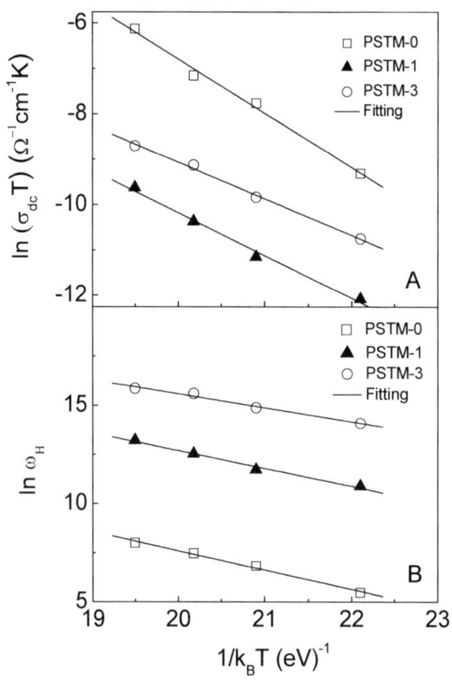

Fig. 6. Arrhenius' dependence for the dc conductivity (σ_{dc}) and the hopping frequency (ω_H) (A and B, respectively), below T_m, for the studied PSTM-*x* compositions

As can be seen, the activation energy values for the dc conductivity are very close to the activation energy values of the ionic conductivity by oxygen vacancies in perovskite type ferroelectric oxides (Jiménez & Vicente, 1998; Bharadwaja & Krupanidhi, 1999; Chen et al., 2000; Smyth, 2003). Thus, it can be concluded that oxygen vacancies are the most likely charge carriers operating in these ceramics below T_m. At room temperature, the oxygen vacancies exhibit a low mobility, whereby the ceramic samples exhibit an enhanced resistance. However, with rising temperature, they are activated and contribute to the observed electrical behavior. In addition, the dielectric relaxations occurring at low frequency are related to the space charges in association with these oxygen vacancies, which can be trapped at the grain boundaries or electrode-sample interface.

On the other hand, the activation energy values for the hopping processes (U_H) are not very far from the U_{dc} values, showing that the hopping processes could be related to the movement of the oxygen vacancies. The short-range hopping of oxygen vacancies, similar to the reorientation of the dipole, could lead to the relaxation processes.

For temperatures above T_m, i.e. in the paraelectric phase, it was carried out the same analysis in the studied samples (Peláiz-Barranco et al., 2008b). The electrical conduction was associated with the doubly ionized oxygen vacancies and the relaxation processes were related to the distorted ionic dipoles by the oxygen vacancies.

Concerning the second peak (Peláiz-Barranco & González-Abreu, 2009b), a detailed study of the electrical conductivity behavior was carried out by using the Jonscher´s formalism. The electrical conduction was associated to the doubly-ionized oxygen vacancies and its influence on the relaxation processes was analyzed.

3. Bi-layered ferroelectric perovskites.

The materials of the Bi–layered structure family were first described by Aurivillius in 1949. The corresponding general formula is $[Bi_2O_2]^{2+}[A_{n-1}B_nO_{3n+1}]^{2-}$, where A can be K^+, Sr^{2+}, Ca^{2+}, Ba^{2+}, Pb^{2+}, etc.; B can be Ti^{4+}, Nb^{5+}, Ta^{5+}, W^{6+}, etc., and n is the number of corner sharing octahedral forming the perovskite like slabs, which is usually in the range 1-5. The oxygen octahedra blocks, responsible for ferroelectric behavior, are interleaved with $(Bi_2O_2)^{2+}$ layers resulting in a highly anisotropic crystallographic structure where the *c* parameter, normal to $(Bi_2O_2)^{2+}$ layers, is much greater than a and *b* parameters of the orthorhombic cell. For this family, ferroelectricity is strongly depending on the crystallographic orientation of the materials, being the aim of continuing research (Lee et al, 2002; Watanabe et al., 2006). It is well-known that these have the majority polarization vector along the *a*-axis in a unit cell and that the oxygen vacancies prefer to stay in the Bi_2O_2 layers, where their effect upon the polarization is thought to be small, and not in the octahedral site that controls polarization.

The layered perovskites are considered for nonvolatile memory applications because of the characteristics of resistance to fatigue (Chen et al., 1997). The frequency response of $SrBi_2Ta_2O_9$ and $SrBi_2Nb_2O_9$ ceramics has been studied considering the bulk ionic conductivity to evaluate the electrical fatigue resistance. The results have showed much higher conductivity values than those of the PZT perovskite ferroelectrics, which has been considered as the cause of the good fatigue resistance. There is an easy recovery of the oxygen vacancies from traps, which limits the space charge generated during the polarization reversal process.

An interesting feature of these materials is that some of them allow cation site mixing among atoms positions (Mahesh Kumar & Ye, 2001), especially between the bismuth and the A-site of the perovskite block (Blake et al., 1997). For example, it has been reported (Ismunandar & Kennedy, 1999) that both Sr^{2+} and Ba^{2+} can occupy the Bi sites in $ABi_2Nb_2O_9$ (A=Sr,Ba). For the studied materials, the degree of disorder is related to the change in the cell volume, which is much lower than that reported in $PbBi_2Nb_2O_9$. For $SrBi_2Ta_2O_9$ (Mahesh

Kumar & Ye, 2001), the substitution of Fe^{3+} for Sr^{2+} increases the ferroelectric-paraelectric transition temperature and provides a higher resistivity value. However, the substitution of Ca^{2+} for Bi^{3+} in the bismuth layers increases the electrical conductivity.
The Aurivillius family exhibit a dielectric behavior characteristic for ferroelectric relaxors (Miranda et al., 2001; Kholkin et al., 2001), e.g. i) marked frequency dispersion in the vicinity of temperature (T_m) where the real part of the dielectric permittivity (ε') shows its maximum value, ii) the temperature of the corresponding maximum for ε' and the imaginary part of the dielectric permittivity (ε'') appears at different values, showing a frequency dependent behavior, iii- the Curie-Weiss law is not fulfilled for temperatures around T_m. The materials have attracted considerable attention due to their large remanent polarization, lead-free nature, relatively low processing temperatures and other characteristics (Miranda et al., 2001; Kholkin et al., 2001; Lee et al, 2002; Nelis et al, 2005; Huang et al., 2006; Watanabe et al., 2006). The origin of the relaxor behavior for these materials has been associated to a positional disorder of cations on A or B sites of the perovskite blocks that delay the evolution of long-rage polar ordering (Miranda et al., 2001).
For $BaBi_2Nb_2O_9$ ceramics (Kholkin et al., 2001), it has been studied the frequency relaxation of the complex dielectric permittivity in wide temperature and frequency ranges. The reported relaxation time spectrum is qualitative difference from that of conventional relaxor ferroelectrics; this shifts rapidly to low frequencies on cooling without significant broadening. It has been discussed considering a reduced size of polarization clusters and their weak interaction in the layered structure.
For bismuth doped $Ba_{1-x}Sr_xTiO_3$ ceramics, it has been showed that the Bi^{3+} doping decreases the maximum of the real part of the dielectric permittivity, whose temperature also shifts to lower temperatures for the lower Sr^{2+} concentrations (Zhou et al, 2001). A relaxor behavior has been reported for these ceramics, suggesting a random electric field as the responsible of the observed behavior.
The complex dielectric permittivity spectrum in the THz region was studied in $SrBi_2Ta_2O_9$ films (Kadlec et al., 2004). The lowest-frequency optical phonon revealed a slow monotonic decrease in frequency on heating with no significant anomaly near the phase transitions. The dielectric anomaly near the ferroelectric phase transition was discussed considering the slowing down of a relaxation mode. It was also discussed the loss of a centre of symmetry in the ferroelectric phase and the presence of polar clusters in the intermediate ferroelastic phase.
For $BaBi_4Ti_4O_{15}$ ceramics (Bobić et al., 2010), the dielectric relaxation processes have been described by using the empirical Vogel–Fulcher law. The electrical conductivity behavior has suggested that conduction in the high-temperature range could be associated to oxygen vacancies.
The previous results correspond to part of the studies carried out in the last years concerning the dielectric and electrical conductivity behaviors of Bi–layered structure materials. However, a lot of aspects of these remain unexplored, especially concerning the dielectric relaxation phenomenon and the conductivity mechanisms. It has been the motivation of the present authors, and other colleagues, to study the dielectric relaxation in one of these systems.

3.1 $Sr_{1-x}Ba_xBi_2Nb_2O_9$ ceramic system.

It has been previously commented that $SrBi_2Nb_2O_9$ is a member of the Aurivillius family (Blake et al., 1997; Ismunandar & Kennedy, 1999; Haluska & Misture, 2004; Nelis et al., 2005;

Huang et al, 2006), consisting of octahedra in the perovskite blocks sandwiched by two neighboring $(Bi_2O_2)^{2+}$ layers along the *c*-axis in a unit cell. The divalent Sr^{2+} cation located between the corner-sharing octahedra can be totally or partially replaced by other cations, most commonly barium (Haluska & Misture, 2004; Huang et al, 2006). The barium modified $SrBi_2Nb_2O_9$ ceramics have showed that the incorporation of Ba^{2+} in the Sr^{2+} sites (A sites of the perovskite) provides a complex dielectric response showing a transition from a normal to a relaxor ferroelectric (Huang et al, 2006). On the other hand, it has been analyzed the barium preference for the bismuth site, which occur to equilibrate the lattice dimensions between the $(Bi_2O_2)^{2+}$ layers and the perovskite blocks (Haluska & Misture, 2004).

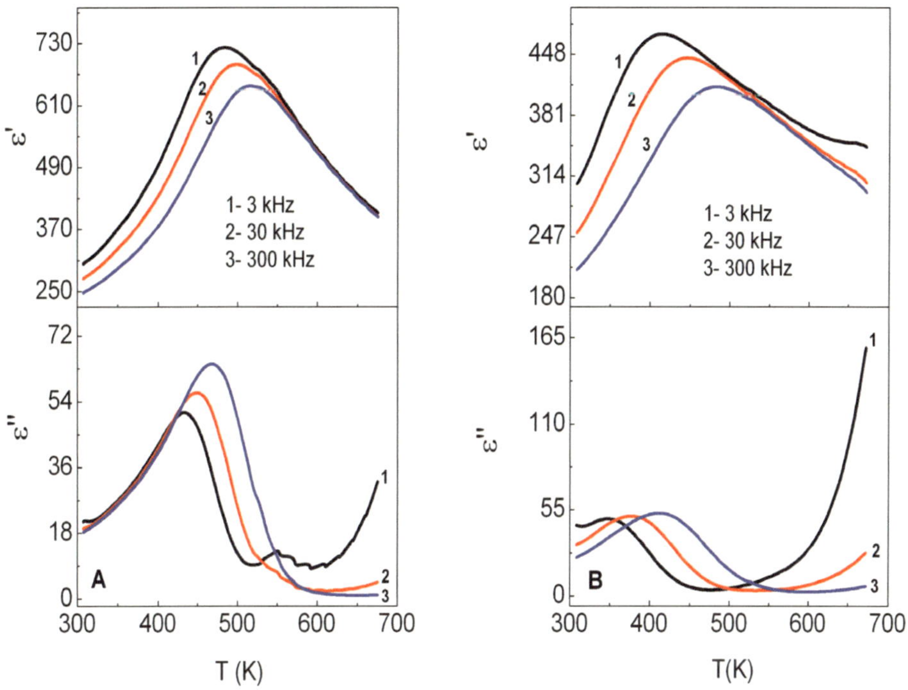

Fig. 7. Temperature dependence of the real (ε') and imaginary (ε'') parts of the dielectric permittivity for A)- $Sr_{0.5}Ba_{0.5}Bi_2Nb_2O_9$ and B)- $Sr_{0.1}Ba_{0.9}Bi_2Nb_2O_9$, at various frequencies

$Sr_{1-x}Ba_xBi_2Nb_2O_9$ ferroelectric ceramics were analyzed in a wide frequency range (100 Hz to 1 MHz) for temperatures below, around and above the temperature where the real part of the dielectric permittivity showed a maximum value. Three compositions were considered, i.e. x=50, 70 and 90 at%. Typical characteristics or relaxor ferroelectrics were observed in the studied samples (González-Abreu et al., 2009; González Abreu, 2010). Figure 7 shows the temperature dependence of the real (ε') and imaginary (ε'') parts of the dielectric permittivity for $Sr_{0.5}Ba_{0.5}Bi_2Nb_2O_9$ and $Sr_{0.1}Ba_{0.9}Bi_2Nb_2O_9$, at various frequencies, as example of the observed behavior in the studied frequency range for the three compositions. On the other hand, an important influence of the electrical conductivity mechanisms was considered from the temperature dependence of the imaginary part of the dielectric permittivity, which increases with temperature, especially for the lower frequencies.

The origin of the relaxor behavior for these ceramics has been explained by considering a positional disorder of cations on A or B sites of the perovskite blocks that delay the evolution of long-range polar ordering (Miranda et al., 2001). For Ba^{2+} doped $SrBi_2Nb_2O_9$ ceramics, a higher frequency dependence of the dielectric parameters has been observed than that of the undoped $SrBi_2Nb_2O_9$ system (González Abreu, 2010). The results have been discussed by considering the incorporation of a bigger ion into the A site of the perovskite block. The Ba^{2+} ions not only substitute the Sr^{2+} ions in the A-site of the perovskite block but enter the $(Bi_2O_2)^{2+}$ layers leading to an inhomogeneous distribution of barium and local charge imbalance in the layered structure.

Following the Cole-Cole model (Cole & Cole, 1941), the real and imaginary part of the dielectric permittivity were fitted by using equations (4) and (5) in a wide temperature range (González-Abreu et al., 2009; González Abreu, 2010). Figures 8 and 9 show the experimental data (solid points) and the corresponding theoretical results (solid lines) at a few representative temperatures for $Sr_{0.5}Ba_{0.5}Bi_2Nb_2O_9$ and $Sr_{0.1}Ba_{0.9}Bi_2Nb_2O_9$, respectively, as example of the observed behavior in the studied temperature range for the three compositions.

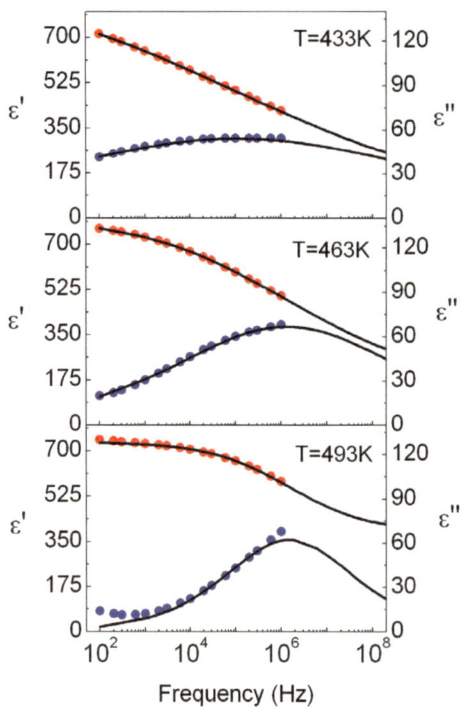

Fig. 8. Frequency dependence of the real ε′ (•) and imaginary ε′′ (•) parts of the dielectric permittivity, at a few representatives temperatures, for $Sr_{0.5}Ba_{0.5}Bi_2Nb_2O_9$. Solid lines represent the fitting by using equations (4) and (5)

Two relaxation processes were evaluated for the studied compositions from the temperature dependence of the relaxation time (González-Abreu et al., 2009; González Abreu, 2010), which was obtained from the fitting by using the Cole-Cole model. The first one takes place around the temperature range where the high dispersion of the real part of the dielectric permittivity was observed (see Figure 7), i.e. below and around the region where the maximums of ε´ are obtained in the studied frequency range. The temperature dependence of the relaxation time was found that follows the Vogel-Fulcher law (González-Abreu et al., 2009; González Abreu, 2010), which describes the typical temperature dependence of the relaxation time for relaxor ferroelectrics. From this point of view, it could be considered that the main origin of the first dielectric relaxation process could be associated to the relaxor-like ferroelectric behavior of the studied ceramics.

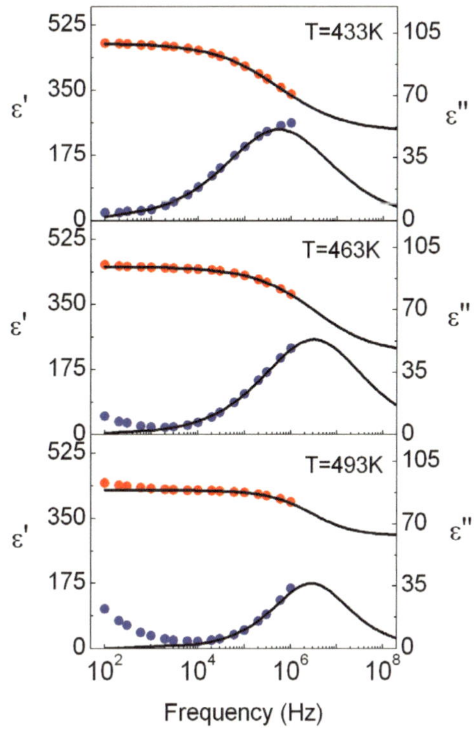

Fig. 9. Frequency dependence of the real ε′ (•) and imaginary ε′′ (•) parts of the dielectric permittivity, at a few representatives temperatures, for $Sr_{0.1}Ba_{0.9}Bi_2Nb_2O_9$. Solid lines represent the fitting by using equations (4) and (5)

The temperature dependence of the relaxation time for second relaxation process was fitted by using the known Arrhenius law given by equation 6 (González-Abreu et al., 2009; González Abreu, 2010). This second process was observed for temperatures where an important contribution of the electrical conductivity mechanisms was considered from the temperature dependence of the imaginary part of the dielectric permittivity. It is known that Ba^{2+} ions could occupy the A-sites at the octahedra in the perovskite blocks and the bismuth site in the layered structure for $Sr_{1-x}Ba_xBi_2Nb_2O_9$ ceramics (Haluska & Misture, 2004). The electrical charge unbalance caused by the trivalent Bi^{3+} ion substitution for the divalent Ba^{2+} ions is compensated by the creation of oxygen vacancies. Then, it could be suggested that the hopping of the electrons, which appears due to the ionization of the oxygen vacancies, could contribute to the dielectric relaxation and its long-distance movement contributes to the electrical conduction.

However, other important contribution should be considered. For relaxor ferroelectrics, microdomains can be observed even at temperature regions far from the region where the maximums of ε´ are obtained, i.e. far from the temperature of the relaxor behavior peak (Burns, 1985). Thus, it has been suggested that the dielectric relaxation process could result from the contribution of the interaction between the dipoles, which forms the microdomains existing at higher-temperature side of the relaxor peak, and the electrons that are due to the ionization of the oxygen vacancies.

4. Summary

In this chapter, the dielectric relaxation phenomenon was discussed in ferroelectric perovskite-related structures considering the relaxation mechanisms and the influence of the vacancies on them. A schematic diagram illustrating the various relaxation processes in the frequency spectrum was showed and some important models used to analysis the dielectric relaxation were presented. Examples, which evidence relations between the relaxation phenomena and defects in the materials, were given.

5. Acknowledgements

The authors wish to thank to the Third World Academy of Sciences for financial support (RG/PHYS/LA No. 99-050, 02-225 and 05-043), to ICTP for financial support of Latin-American Network of Ferroelectric Materials (NET-43), and to FAPESP and CNPq Brazilian agencies. Special thanks to the Ferroelectric Materials Group from Havana University, Cuba.

6. References

Bharadwaja, S. S. N. & Krupanidhi, S. B. (1999). Growth and study of antiferroelectric lead zirconate thin films by pulsed laser ablation. *Journal of Applied Physics*, 86, No. 10, (November 1999) 5862-5869, ISSN 0003-695.

Bidault, O.; Goux, P.; Kehikech, M.; Belkaoumi, M. & Maglione, M. (1994). Space-charge relaxation in perovskites. *Physical Review B*, 49 (March 1994) 7868-7873, ISSN 1098-0121.

Blake, S. M.; Falconer, M. J.; McCreedy, M. & Lightfoot, P. Cation disorder in ferroelectric Aurivillius phases of the type $Bi_2ANb_2O_9$ (A=Ba, Sr, Ca). *Journal of Materials Chemistry*, 7, No. 8, (August 1997) 1609-1613, ISSN 0959-9428.

Bobić, J. D.; Vijatović, M. M.; Greičius, S.; Banys, J. & Stojanović, B. D. (2010). Dielectric and relaxor behavior of $BaBi_4Ti_4O_{15}$ ceramics. *Journal of Alloys and Compounds*, 499 (March 2010) 221–226, ISSN 0925-8388.

Burns, G. (1985). Crystalline ferroelectrics with a glassy polarization phase. *Phase Transitions*, 5, No. 4 (September 1985) 261-275, ISSN 0141-1594.

Chen A.; Yu, Z. & Cross, L. E. (2000). Oxygen-vacancy-related low-frequency dielectric relaxation and electrical conduction in Bi:$SrTiO_3$. *Physical Review B*, 62, No. 1, (July 2000) 228-236, ISSN 1098-0121.

Chen, Tze-Chiun; Thio, Chai-Liang & Desu, S. B. (1997). Impedance spectroscopy of $SrBi_2Ta_2O_9$ and $SrBi_2Nb_2O_9$ ceramics correlation with fatigue behavior. *Journal of Materials Researchs*, 12, No. 10 (October 1997) 2628-2637.

Cole, K. S. & Cole, R. H. (1941). Dipersion and absorption in dielectrics. *Journal of Chemical Physics*, 9 (April 1941) 341-351, ISSN 0021-9606.

Debye, P. *Polar Molecules* (1929), Chemical Catalogue Company, New York.

Eyraud, L.; Eyraud, P. & Claudel, B. (1984). Influence of simultaneous heterovalent substitutions in both cationic sites on the ferroelectric properties of PZT type ceramics. *Journal of Solid State Chemistry*, 53, No. 2, (July 1984) 266-272, ISSN 0022-4596.

Eyraud, L.; Eyraud, P.; Lebrun, L.; Guiffard, B.; Boucher, E.; Audigier, D. & Guyomar, D. (2002). Effect of (Mn, F) co-doping on PZT characteristics under the influence of external disturbances. *Ferroelectrics*, 265, No. 1, (2002) 303-316, ISSN 0015-0193.

Funke, K. (1993). Jump relaxation in solid electrolytes. *Progress in Solid State Chemistry*, 22, No. 2 (May 1993) 111-195, ISSN: 0079-6786.

González-Abreu, Y.; Peláiz-Barranco, A.; Araújo, E. B. & Franco Júnior, A. (2009). Dielectric relaxation and relaxor behavior in bi-layered perovskites. *Applied Physics Letters*, 94, No. 26 (June 2009) 262903, ISSN 0003-6951.

González-Abreu, Y. *Relajación dieléctrica y conductividad eléctrica en cerámicas ferroeléctricas tipo perovskita* (2010), Master in Physics Science Thesis, Physics Faculty, Havana University, Cuba.

Guiffard, B., Boucher, E., Eyraud, L., Lebrun, L. & Guyomar, D. (2005). Influence of donor co-doping by niobium or fluorine on the conductivity of Mn doped and Mg doped PZT ceramics. *Journal of the European Ceramic Society*, 25, No. 12, (March 2005) 2487-2490, ISSN 0955-2219.

Haluska, M. S. & Misture, S. T. (2004). Crystal structure refinements of the three-layer Aurivillius ceramics $Bi_2Sr_{2-x}A_xNb_2TiO_{12}$ (A=Ca,Ba; x= 0, 0.5, 1) using combined X-ray and neutron powder diffraction. *Journal of Solid State Chemistry*, 177, No. 6, (July 2004) 1965-1975, ISSN 0022-4596.

Huang, S.; Feng, Ch.; Chen, L. & Wang, Q. (2006). Relaxor Behavior of $Sr_{1-x}Ba_xBi_2Nb_2O_9$ Ceramics. *Journal of the American Ceramic Society*, 89, No. 1, (January, 2006) 328-331, ISSN 0002-7820.

Ismunandar & Kennedy, B. J. Effect of temperature on cation disorder in $ABi_2Nb_2O_9$ (A=Sr, Ba)", *Journal of Materials Chemistry*, 9, No. 2, (February 1999) 541-544, ISSN 0959-9428.

Jiménez, B. & Vicente, J. M. (1998). Oxygen defects and low-frequency mechanical relaxation in Pb-Ca and Pb-Sm titanates. *Journal of Physics D: Applied Physics*, 31, No. 4, (February 1998) 446-452, ISSN 0022-3727.

Jonscher, A. K. *Universal Relaxation Law* (1996), Chelsea Dielectrics Press Ltd, ISBN 0950871125, London.

Kadlec, F.; Kamba, S.; Kužel, P.; Kadlec, C.; Kroupa, J. & Petzelt, J. (2004). High-temperature phase transitions in $SrBi_2Ta_2O_9$ film: a study by THz spectroscopy. *Journal of Physics: Condensed Matter*, 16 (September 2004) 6763–6769, ISSN 0953-8984.

Kang, B. S.; Choi, S. K. & Park, C. H. (2003). Diffuse dielectric anomaly in perovskite-type ferroelectric oxides in the temperature range of 400-700°C. *Journal of Applied Physics*, 94, No. 3, (August 2003) 1904-1911, ISSN 0021-8979.

Kholkin, A. L.; Avdeev, M; Costa, M. E. V.; Baptista, J. L. & Dorogotsev, S. N. (2001). Dielectric relaxation in Ba-based layered perovskites. *Applied Physics Letters*, 79, No. 5, (July 2001) 662-664, ISSN 0003-6951.

Lee, H. N.; Hesse, D.; Zakharov, N. & Gösele, U. (2002). Ferroelectric $Bi_{3.25}La_{0.75}Ti_3O_{12}$ films of uniform a-axis orientation on silicon substrates. *Science*, 296 (June 2002) 2006-2009, ISSN 0036-8075.

León, C.; Rivera, A.; Várez, A. ; Sanz, J. ; Santamaría, J. & Ngai, K. L. (2001). Origin of Constant Loss in Ionic Conductors. *Physical Review Letters*, 86, No. 7, (February 2001) 1279-1282, ISSN 0031-9007.

Li, Z. & Fan, H. (2009). Polaron relaxation associated with the localized oxygen vacancies in $Ba_{0.85}Sr_{0.15}TiO_3$ ceramics at high temperatures. *Journal of Applied Physics*, 106 (2009) 054102, ISSN 0021-8979.

Lines, M. E. & Glass, A. M. *Applications of Ferroelectric and Related Materials* (1977), Claderon Press, ISBN 0198512864, Oxford.

Maglione, M. & Belkaoumi, M. (1992). Electron relaxation-mode interaction in $BaTiO_3$-Nb. *Physical Review B*, 45, No. 5, (February 1992) 2029-2034, ISSN 0163-1829.

Mahesh Kumar, M. & Ye, Z. -G. (2001). Dielectric and electric properties of donor- and acceptor-doped ferroelectric $SrBi_2Ta_2O_9$. *Journal of Applied Physics*, 90 (July 2001) 934-941, ISSN 0003-695.

Miranda, C.; Costa, M. E. V.; Avdeev, M.; Kholkin, A. L. & Baptista, J. L. (2001). Relaxor properties of Ba-based layered perovskites. *Journal of European Ceramic Society*, 21 (October 2001) 1303-1306, ISSN 0955-2219.

Moos, R.; Menesklou, W. & Härdtl, K. H. (1995). Hall-mobility of undoped n-type conducting strontium-titanate single-crystals between 19-K and 1373-K. *Applied Physics A - Materials Science & Processing*, 61, No. 4, (October 1995) 389-395, ISSN 0721-7250.

Moos, R. & Härdtl, K. H. (1996). Electronic transport properties of $Sr_{1-x}La_xTiO_3$ ceramics. *Journal of Applied Physics*, 80, No. 1, (July 1996) 393-400, ISSN 0021-8979.

Nelis, D.; Mondelaers, D.; Vanhoyland, G.; Hardy, A.; Van Werde, K.; Van den Rul, H.; Van Bael, M. K.; Mullens, J.; Van Poucke, L. C. & D'Haen, J. (2005). Synthesis of strontium bismuth niobate ($SrBi_2Nb_2O_9$) using an aqueous acetate–citrate precursor gel: thermal decomposition and phase formation. *Thermochimica Acta*, 426 (February 2005) 39-48, ISSN 0040-6031.

Ngai, K. L. (1993). Analysis of NMR and conductivity-relaxation measurements in glassy Li_2S-SiS_2 fast-ion conductors. *Physical Review B*, 48, No. 18, (November 1993) 13481-13485, ISSN 1098-0121.

Nowick, A. S. & Lim, B. S. (2001). Electrical relaxations: Simple versus complex ionic systems. *Physical Review B*, 63, No. 18, (April 2001) 184115, ISSN 1098-0121.

Peláiz-Barranco, A. ; Guerra, J. D. S. ; López-Noda, R. & Araújo, E. B. (2008). Ionized oxygen vacancy-related electrical conductivity in $(Pb_{1-x}La_x)(Zr_{0.90}Ti_{0.10})_{1-x/4}O_3$ ceramics. *Journal of Physics D: Applied Physics*, 41, (October 2008) 215503, ISSN 0022-3727.

Peláiz-Barranco, A. ; González-Abreu, Y. & López-Noda, R. (2008). Dielectric relaxation and conductivity behavior in modified lead titanate ferroelectric ceramics. *Journal of Physics: Condensed Matter*, 20, (November 2008) 505208, ISSN 0953-8984.

Peláiz-Barranco, A.; Guerra, J. D. S.; Calderón-Piñar, F.; Aragó, C.; García-Zaldívar, O.; López-Noda, R; Gonzalo, J. A. & Eiras, J. A. (2009). Dielectric response features and oxygen migration on rare earth modified lead titanate ferroelectric ceramics. *Journal of Materials Science*, 44, No. 1, (January 2009) 204-211, ISSN 0022-2461.

Peláiz-Barranco, A. & González-Abreu, Y. (2009). Oxygen vacancies related electrical response in modified lead titanate ceramics. *Solid State Communications*, 149 (October 2009) 2082-2084, ISSN 0038-1098.

Peláiz-Barranco, A. & Guerra, J. D. S. (2010). Dielectric relaxation related to single-ionized oxygen vacancies in $(Pb_{1-x}La_x)(Zr_{0.90}Ti_{0.10})_{1-x/4}O_3$ ceramics. *Materials Research Bulletin*, 45, (September 2010) 1311–1313, ISSN 0025-5408.

Pérez-Martínez, O.; Saniger, J. M.; Torres-García, E.; Flores, J. O.; Calderón-Piñar, F.; Llópiz, J. C. & Peláiz-Barranco, A. (1997). Inclusion of Dy, Ho and Er in B sites of modified lead titanate. *Journal of Matterials Science Letters*, 16, No. 14, (July 1997) 1161-1163, ISSN 0261-8028.

Poykko, S. & Chadi, D. J. (2000). First principles study of Pb vacancies in $PbTiO_3$. *Applied Physics Letters*, 76, No. 4 (January 2000) 499-501, ISSN 0003-6951.

Ramírez-Rosales, D.; Zamorano-Ulloa, R. & Pérez-Martínez, O. (2001). Electron spin resonance study of the conversion of Mn^{4+} to Mn^{2+} in the $Pb_{1-x}Eu_xTi_{1-y}Mn_yO_3$ ceramic system. *Solid State Communications*, 118, No. 7, (May 2001) 371-376, ISSN 0038-1098.

Saiful Islam, M. (2000). Ionic transport in ABO_3 perovskite oxides: a computer modelling tour. *Journal of Materials Chemistry*, 10, No. 4 (April 2000) 1027-1038, ISSN 0959-9428.

Singh, G.; Tiwari, V. S. & Gupta, P. K. (2010). Role of oxygen vacancies on relaxation and conduction behavior of $KNbO_3$ ceramic. *Journal of Applied Physics*, 107 (2010) 064103, ISSN 0021-8979.

Smyth, D. M. (2003). Comments on the defect chemistry of undoped and acceptor-doped $BaTiO_3$. *Journal of Electroceramics*, 11, No. 1-2, (September 2003) 89-100, ISSN 1385-3449.

Steinsvik, S.; Bugge, R.; Gjonnes, J.; Tafto, J. & Norby, T. (1997). The defect structure of $SrTi_{1-x}Fe_xO_{3-y}$ (x=0-0.8) investigated by electrical conductivity measurements and electron energy loss spectroscopy (EELS). *Journal of Physics and Chemistry of Solids*, 58, No. 6, (July 1997) 969-976, ISSN 0022-3697.

Takeuchi, H.; Jyomura, S.; Nakaya, C. & Ishikawa, Y. (1983). Samarium-substituted lead titanate ceramics for high frequency ultrasonic probes. *Japanese Journal of Applied Physics*, 22, No. 2, (1983) 166-169, ISSN 0021-4922.

Takeuchi, H.; Jyomura, S. & Nakaya, C. (1985). New piezoelectric materials for ultrasonic transducers. *Japanese Journal of Applied Physics*, 24, No. 2, (1985) 36-40, ISSN 0021-4922.

Verdier, C.; Morrison, F. D.; Lupascu, D. C. & Scott, J. F. (2005). Fatigue studies in compensated bulk lead zirconate titanate. *Journal of Applied Physics*, 97, No. 2, (December 2005) 024107, ISSN 0003-695.

Wachsmuth, B.; Zschech, E.; Thomas, N. W.; Brodie, S. G.; Gurman, S. J.; Baker, S. & Bayliss, S. C. (1993). Structure Model of Aurivillius Compounds. *Physica Status Solidi (a)*, 135 (January 1993) 59-71, ISSN 0031-8965.

Watanabe, T. & Fuankubo, H. (2006). Controlled crystal growth of layered-perovskite thin films as an approach to study their basic properties. *Journal of Applied Physics*, 100 (May 2006) 051602, ISSN 0003-695.

Xu, Y. *Ferroelectric Materials and Their Applications* (1991), Elsevier Science Publishers B.V., ISBN 0444883541, Netherlands.

Yoo, H. I.; Song, C. R. & Lee, D. K. (2002). $BaTiO_{3-\delta}$: Defect structure, electrical conductivity, chemical diffusivity, thermoelectric power, and oxygen nonstoichiometry. *Journal of Electroceramics*, 8, No. 1, (July 2002) 5-36, ISSN 1385-3449.

Zhou, L.; Vilarinho, P. M. & Baptista, J. L. (2001). Dielectric properties of bismuth doped $Ba_{1-x}Sr_xTiO_3$ ceramics. *Journal of European Ceramic Society*, 21 (April 2001) 531-534, ISSN 0955-2219.

11

The Ferroelectric-Ferromagnetic Composite Ceramics with High Permittivity and High Permeability in Hyper-Frequency

Yang Bai
The University of Science and Technology Beijing
China

1. Introduction

With the rapid development of portable electronic products and wireless technology, many electronic devices have evolved into collections of highly integrated systems for multiple functionality, faster operating speed, higher reliability, and reduced sizes. This demands the multifunctional integrated components, serving as both inductor and capacitor. As a result, low temperature co-fired ceramics (LTCC) with integrated capacitive ferroelectrics and inductive ferrites has been regarded as a feasible solution through complex circuit designs. However, in the multilayer LTCC structure consisting of ferroelectrics and ferrites layers, there are always many undesirable defects, such as cracks, pores and cambers, owing to the co-firing mismatch between different material layers, which will damage the property and reliability of end products (Hsu & Jean, 2005). A single material with both inductance and capacitance are desired for true integration in one element. For example, if the materials with both high permeability and permittivity are used in the anti electromagnetic interference (EMI) filters, the size of components can be dramatically minimized compared to that of conventional filters composed of discrete inductors and capacitors. Because little single-phase material in nature can meet such needs (Hill, 1999), the development of ferroelectric-ferromagnetic composite ceramics are greatly motivated.

Many material systems, such as $BaTiO_3$ / NiCuZn ferrite, $BaTiO_3$ / MgCuZn ferrite, $Pb(Zr_{0.52}Ti_{0.48})O_3$ / NiCuZn ferrite, $Pb(Mg_{1/3}Nb_{2/3})O_3$-$Pb(Zn_{1/3}Nb_{2/3})O_3$-$PbTiO_3$ / NiCuZn ferrite and $Bi_2(Zn_{1/3}Nb_{2/3})_2O_7$ / NiCuZn ferrite, were investigated and found exhibit fine dielectric and magnetic properties. In these reports, spinel ferrites, such as NiCuZn ferrite, were always used as the magnetic phase of composite ceramics, because they are mature materials for LTCC inductive components. However, the cut-off frequency of spinel ferrites is limited below 100MHz by the cubic crystal structure, so the resulting composite ceramics can not be used in hyper-frequency or higher frequency range. To keep up with the trend towards higher frequency for electronic technology, hexagonal ferrites, including Y-type hexagonal ferrite $Ba_2Me_2Fe_{12}O_{22}$ and Z-type hexagonal ferrite $Ba_3Me_2Fe_{24}O_{41}$ (Me=divalent transition metal), should be used in the composite ceramics.

Co_2Z hexagonal ferrite has high permeability and low loss in hyper-frequency, but the very high sintering temperature (>1300°C) works against its application in LTCC. Y-type hexagonal ferrite has a bit lower permeability, but the excellent sintering behavior makes it a good candidate of magnetic material in LTCC. To achieve high dielectric permittivity, lead-based relaxor ferroelectric ceramics is a good choice as the ferroelectric phase in the composite ceramics owing to its high dielectric permittivity and low sintering temperature.

In this chapter, we summarize the co-firing behavior, microstructure and electromagnetic properties of the composite ceramics for hyper-frequency. The material system is mainly focused on a composite ceramics composed of $0.8Pb(Ni_{1/3}Nb_{2/3})O_3$-$0.2PbTiO_3$ (PNNT) and $Ba_2Zn_{1.2}Cu_{0.8}Fe_{12}O_{22}$ (BZCF), which has excellent co-firing behavior and good electromagnetic properties in hyper-frequency, and some other composite ceramics are also involved in some sections.

2. The co-firing behavior, phase composition and microstructure

2.1 The co-firing behavior and densification

Due to the different sintering temperatures and shrinkage rates of ferroelectric phase and ferromagnetic phase, remarkable co-firing mismatch often occurs and results in undesirable defects, such as cracks and cambers. As a result, the property of composite ceramics and the reliability of end products are damaged. Thanks to the existence of large amount of grain boundaries to dissipate stress, the composite ceramics with powder mixture have much better co-firing behavior than the multilayered composite ceramics. Although the mismatch of densification rate is alleviated to a larger extent, a good sintering compatibility between ferroelectric and ferromagnetic grains is still required for better co-firing match. The starting temperature of shrinkage and the point of maximum shrinkage rate are both important for the co-firing behavior of composite ceramics. Some research indicates that the composite ceramics exhibits an average sintering behavior between two phases and the shrinkage rate curve of composite ceramics is between those of two component phases (Qi et al., 2008).

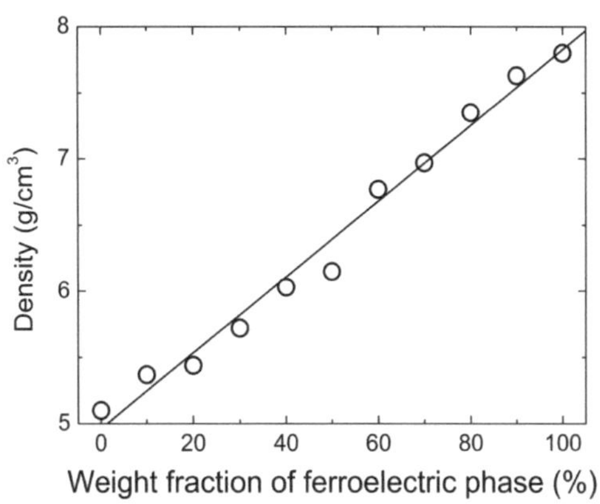

Fig. 1. The density of the sintered PNNT-BZCF composite ceramics as a function of the weight fraction of ferroelectric phase

Y-type hexagonal ferrite has a lower sintering temperature of 1000~1100°C, which is similar to that of lead-based ferroelectric ceramics. For example, in PNNT-BZCF composite material, BZCF has a sintering temperature of 1050°C, same as that of PNNT. Hence, the composite system has good co-firing behavior for each composition (Bai et al., 2007). After sintered at 1050°C, all the samples exhibit a high density, above 95% of theoretical density. Fig. 1 shows the composition dependence of the density of sintered PNNT-BZCF composite ceramics. Since the density of a composite material is the weighted average of those of constituent phases, it increases linearly with the rise of weight fraction of ferroelectric phase. Similar relationship is obtained in other composite materials (Qi et al., 2004 & Shen et al., 2005).

2.2 The element diffusion

The element diffusion always occurs between two phases during the sintering process at high temperature. The thickness of diffusion layer and element distribution influence microstructure and properties of composite ceramics. The diffusion coefficient is determined by the ion's radius and charge. Table 1 shows the radius of some ions commonly used in ferroelectric ceramics and ferrite.

Ba^{2+}	Pb^{2+}	Ti^{4+}	Nb^{5+}	Fe^{3+}	Fe^{2+}	Co^{2+}	Zn^{2+}	Cu^{2+}	Mn^{3+}	Ni^{2+}
0.135	0.120	0.068	0.07	0.076	0.064	0.074	0.074	0.072	0.066	0.072

Table 1. The radius of some ions

It is always thought that Ba^{2+} ion does not diffuse due to the large radius, which has been confirmed by experiments. Although Pb^{2+} also has large radius, the low vapor pressure make it to easily escape from lattice at high temperature. The deficiency of Pb^{2+}in the lattice can result in the formation of pyrochlore phase. For the metallic ion in ferrite, the diffusion coefficient can be ranked as $D_{Co}>D_{Fe}>D_{Zn}>D_{Ni},D_{Cu}$ based on the experiments of atomic emission spectrometry (AES) and electron probe micro-analyzer (EPMA). Fig. 2 (a) and (b) show the backscattered electron image and element distribution around the interface in the composite ceramics consisting of $Pb(Mg_{1/3}Nb_{2/3})O_3$ and NiCuZn ferrite. The diffusion of different ions between ferroelectric grain and ferromagnetic grain is clear.

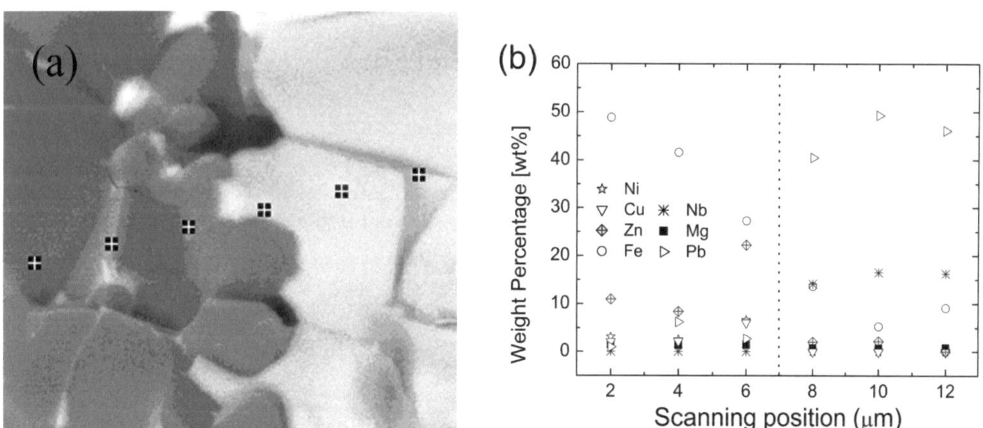

Fig. 2. (a) the backscattered electron image and (b) element distribution around the interface of the composite ceramics of $Pb(Mg_{1/3}Nb_{2/3})O_3$-NiCuZn ferrite

The element diffusion can influence the microstructure and electromagnetic properties of composite ceramics. For example, the grains at the interface of two phases may grow abnormally large. To alleviate the element diffusion, lowering the sintering temperature is a feasible solution.

2.3 Phase composition crystal structure

During the co-firing process, chemical reactions may take place at the interface of two phases and produce some new phases, which affect the properties of co-fired composite ceramics. For example, pyrochlore phase is often formed in the composite ceramics with lead-based ferroelectric ceramics due to Pb volatilization. Fig. 3 shows a X-ray diffraction (XRD) spectrum of the composite ceramics of 40wt%$PbMg_{1/3}Nb_{2/3}O_3$-60wt%NiCuZn ferrite, which clearly shows the existence of pyrochlore phase. If the sintering temperature is lowered below 1000°C, the volatilization of Pb can be largely reduced, so does the formation of pyrochlore phase.

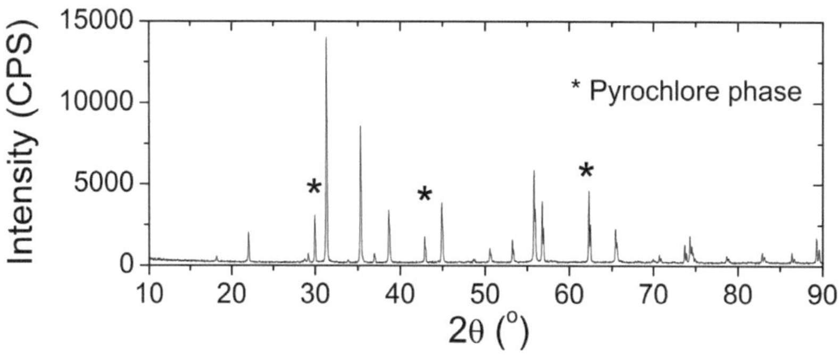

Fig. 3. XRD spectrum of the composite ceramics of 40wt%$PbMg_{1/3}Nb_{2/3}O_3$-60wt%NiCuZn ferrite

Fig. 4 compares the XRD spectra of PNNT-BZCF composite ceramics before and after sintering process. According to the XRD spectra, no other phase is found after co-firing process, i.e. no obvious chemical reaction takes place between PNNT and BZCF during the sintering process of 1050°C.

It is clear that only perovskite phase can be detected in the XRD spectra of samples either before or after co-firing process, if the weight fraction of PNNT is higher than or equals to 0.8. It is because that the crystal structure of Y-type hexagonal ferrite is more complex than the perovskite structure of ferroelectric phase, which results in a much lower electron density. In addition, for the sample with same volume fraction of PNNT and BZCF, the XRD intensity of perovskite phase is much stronger than that of BZCF due to the same reason. With the rise of BZCF amount, the intensities of its diffraction peaks gradually enhance. Up to the weight fraction of ferroelectric phase is as low as x=0.1, the XRD peaks of perovskite phase are still obvious in the XRD spectrum.

The crystal morphology and orientation may be changed due to the different densification characters of two phases in the co-firing process (Bai et al., 2009). It is noticed from Fig. 4 that the relative intensity of the diffraction peaks of Y-type hexagonal ferrite changes after the co-firing process. For the green samples, the primary diffraction peak of BZCF is at 30.4° corresponding to (110) plane, which is same as the pure Y-type hexagonal ferrite; while the primary peak is at 32° corresponding to (10$\underline{13}$) plane for the sintered samples, which is the secondary peak of pure Y-type hexagonal ferrite. This variation of XRD intensities of Y-type hexagonal ferrite after co-firing process is well indexed by comparing the sintered samples of x=0 and x=0.1 in Fig. 5. The change reflects a lattice distortion induced by the internal stress.

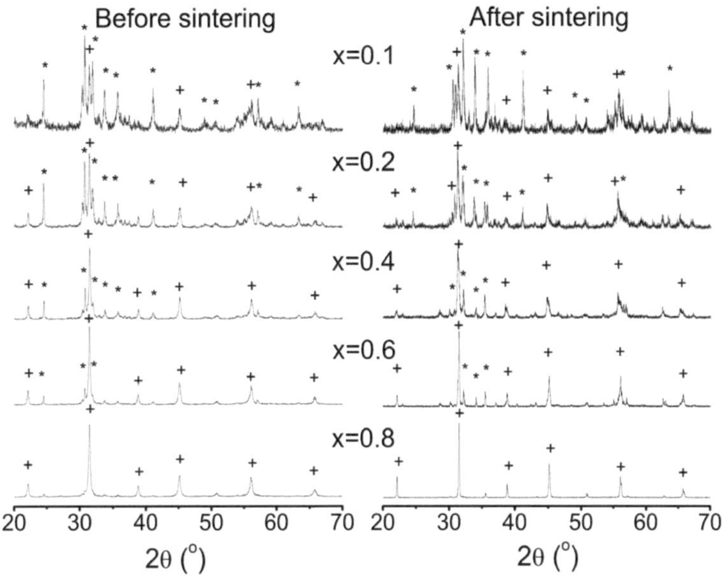

Fig. 4. XRD spectra of PNNT-BZCF composite ceramics before and after sintering at 1050°C (+ : perovskite phase, PNNT; * : Y-type hexagonal ferrite, BZCF).

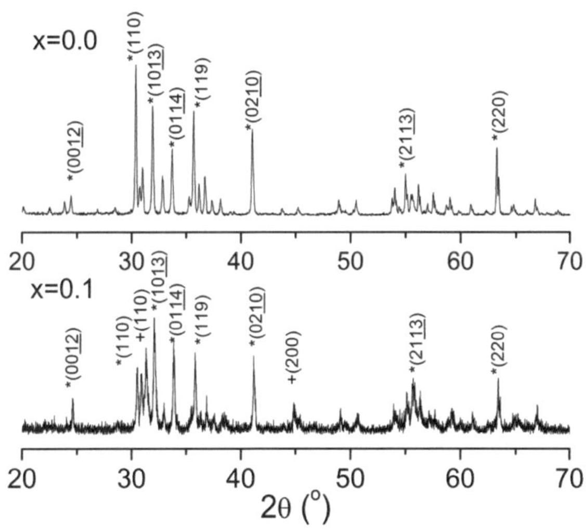

Fig. 5. XRD spectra comparison of 10wt%PNNT-90wt%BZCF composite ceramics and pure Y-type hexagonal ferrite (+ : perovskite phase, PNNT; * : Y-type hexagonal ferrite, BZCF)

2.4 Microstructure

When the sintering temperatures of two constituent phases are greatly different, the co-firing mismatch will result in various defects in the microstructure of co-fired composite ceramics. If two phases have similar sintering temperature, the co-firing mismatch will be slight.

Fig. 6. The microstructure of PNNT-BZCF composite ceramics (a) x=0 (b) x=0.2 (c) x=0.3 (d) x=0.5 (e) x=0.7 (f) x=0.9 [(a) is secondary electron image, while (b)-(f) are backscattered electron images.]

Fig. 6 (a)–(f) show the scanning electron microscope (SEM) images of the microstructure of PNNT-BZCF composite ceramics. In backscattered electron image, the grains of PNNT and BZCF respectively appear white and gray due to the difference of molecular weights of the elements in them. The sintered samples exhibit dense microstructures for each composition and the grains of PNNT and BZCF distribute homogeneously. It indicates that this composite system has a fine co-firing behavior over a wide composition range, which thanks to the same sintering temperature of PNNT and BZCF.

The average size of ferroelectric or ferrite grains decreases with the rise of corresponding phase amount. For example, as BZCF content is low, few ferrite grains are besieged by large amount of PNNT grains. It becomes difficult for small ferrite grains to merge with the neighboring likes. With the increase of ferrite's content, the chance of amalgamation of small grains rises, and then grains grow larger. The thing is same for PNNT grains.

From the SEM images, it is noticed that the grain morphology of ferrite changes obviously with composition. In the sample of pure Y-type hexagonal ferrite ($x=0$), the grains are platelike and many of them are of hexagonal shape [Fig. 6 (a)]. In the co-fired ceramics [Fig. 6(b)-(f)], the planar grains of hexagonal ferrite become equiaxed crystals just as those ferroelectric grains. During the co-firing process, the grain growth of two constituent phases is affected each other. Because equiaxed crystal is more favorable for a compact-stack microstructure than planar crystal, the surrounding equiaxed grains of PNNT modulate the grain growth of BZCF particles and assimilate their grain shape into equiaxed crystal during the co-firing process. It is well known that the internal stress is unavoidable in the co-fired ceramics. In BZCF-PNNT composite ceramics, the compact-stacked grains and the change of BZCF's grain morphology suggest the existence of internal stress and lattice distortion, which are also reflected in XRD spectra as discussed in prior section.

3. The static electromagnetic properties

3.1 The ferroelectric hysteresis loop

For the ferroelectric-ferromagnetic composite ceramics, the ferroelectric or ferromagnetic character is determined by the corresponding phase, while the magnetoelectric effect is always weak. To examine the ferroelectricity of composite ceramics, the ferroelectric polarization–electric field (P-E) hysteresis loop is the most important character.

For PNNT-BZCF composite ceramics, the P-E hysteresis loops are observed over the whole composition range (Fig. 7), which implies the ferroelectric nature of composite ceramics. The maximum polarization P_{max} decreases with the reduction of ferroelectric phase due to dilution effect, which indicates that the ferroelectricity of composite ceramics originates from the nature of ferroelectric phase.

It is also noted that the shape of P-E loop varies with composition. The sample with high PNNT amount ($x>0.8$) has fine and slim hysteresis loop, while the sample with relative less ferroelectric phase has an open-mouth-shaped P-E loop. It is because that the ferrite has much lower electric resistivity of about 10^6 Ω cm than that of ferroelectric ceramics (above 10^{11} Ω cm). In the ferroelectric-ferromagnetic composite ceramics, the ferrite grains serve as a conductive phase in the electric measurement, especially under a high electric field. If the ferrite content is low, the small ferrite grains are besieged by the ferroelectric grains with high resistivity and there is no conductive route in the microstructure. As a result, the composite ceramics has high resistivity and low leak current. With the rise of ferrite amount, the percolation occurs in the composite system and the resistivity drops remarkably (Qi et al., 2004 & Bai et al., 2007). The large leak current results in an open-mouth-shaped P-E loop.

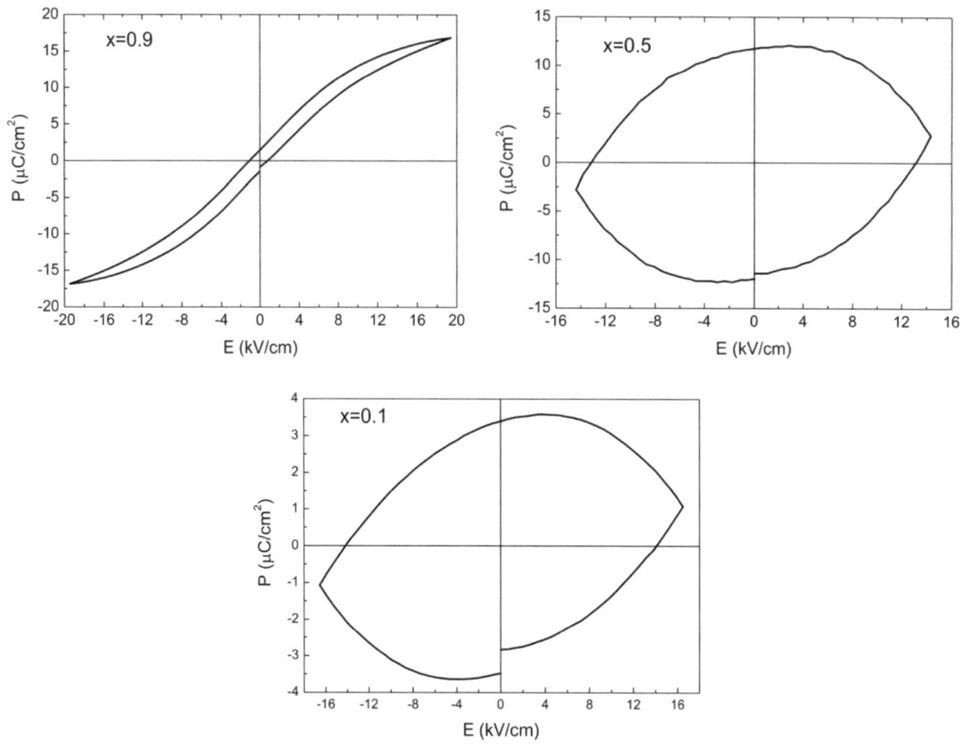

Fig. 7. P-E hysteresis loops of PNNT-BZCF composite ceramics (a) x=0.9; (b) x=0.5; (c) x=0.1

3.2 The ferromagnetic hysteresis loop

The magnetic hysteresis loop is the best experimental proof for the ferromagnetic nature of materials. Up to now, none of ferroelectric ceramics exhibits ferromagnetic character at room temperature, so the ferromagnetic behaviors of composite ceramics are dominated by the ferrite phase. For the application in high frequency, soft magnetic material is needed for the composite materials.

The magnetic hysteresis loops of PNNT-BZCF composite ceramics is plotted in Fig. 8. The ferromagnetic characters of composite ceramics are only inherited from those of the magnetic phase of Y-type hexagonal ferrite, so all the samples exhibit soft magnetic character with low coercive force H_c and low remnant magnetization M_r. The coexistence of magnetic hysteresis loop and P-E loop implies that the PNNT-BZCF composite ceramics have both ferromagnetic and ferroelectric properties at room temperature, which also confirms the possibility to achieve both high permittivity and permeability.

Fig. 9 shows the composition dependence of M_s, M_r and H_c for PNNT-BZCF composite ceramics before and after sintering process. For the green sample before sintering, M_s and M_r both decrease monotonously with the reduction of BZCF amount, while H_c keeps a constant. The magnetic properties of green samples are dominated by the nature of individual magnetic particles and there is little interaction between constituent phases due to loose microstructure. The linear decrease of M_s and M_r is only a result of dilution effect. The small ferrite particles and lots of defects in microstructure endow the green samples a relatively high H_c, which is insensitive to the variation of composition.

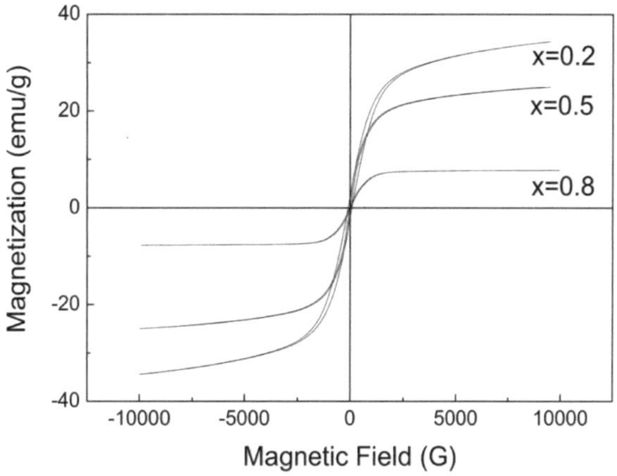

Fig. 8. The magnetic hysteresis loops of PNNT-BZCF composite ceramics

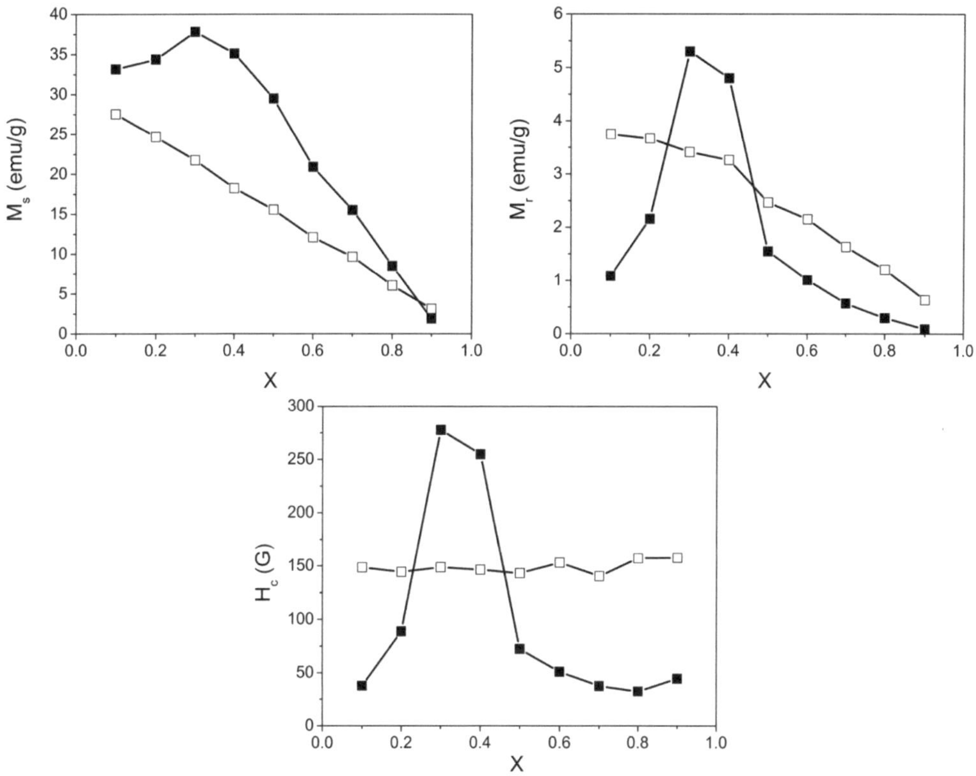

Fig. 9. The composition dependence of (a) M_s, (b) M_r and (c) H_c for the composite ceramics of PNNT-BZCF before (□) and after (■) sintering process

After the sintering process, M_s decreases monotonously with the reduction of ferrite amount if the mechanical interaction between two constituent phases is weak. For example, M_s varies near linearly with the ferrite content in $BaTiO_3$-NiCuZn ferrite or PMNZT-NiCuZn ferrite composite ceramics, where ferroelectric phase and ferrite phases are both of equiaxial grains and little internal stress is produced after co-firing process (Qi et al., 2004).

If the microstructure varies notable after co-firing process, the magnetic properties of composite ceramics will be affected (Bai et al., 2009). With the reduction of BZCF amount in PNNT-BZCF composite materials, M_s, M_r and H_c of the sintered samples increase first, reach a maximum at x=0.3, and then decrease (Fig. 9). The enhancement of M_s in the range of $0.1<x<0.3$ originates from the internal stress produced in the co-firing process. It is reported that the structural distortion can generate spontaneous magnetization in several composite materials (Kanai et al., 2001 & Kumar et al., 1998). In PNNT-BZCF composite ceramics, the enhancement of M_s in the range of $0.1<x<0.3$ is also thought as a result of the internal stress induced structural distortion, which has been detected by XRD spectra and SEM images. A more notable enhancement of M_r and H_c is observed in the range of $0.1<x<0.3$, because M_r and H_c are more sensitive to the microstructure. The stress on ferrite grains increases the resistance of domain wall's motion and spin rotation, so the magnetization reversal under external magnetic field becomes more difficult, which is reflected as the increase of M_r and H_c. When ferrite's amount decreases further, the dilution effect dominates the magnetic properties of composite ceramics, and then M_s, M_r and H_c decline monotonically.

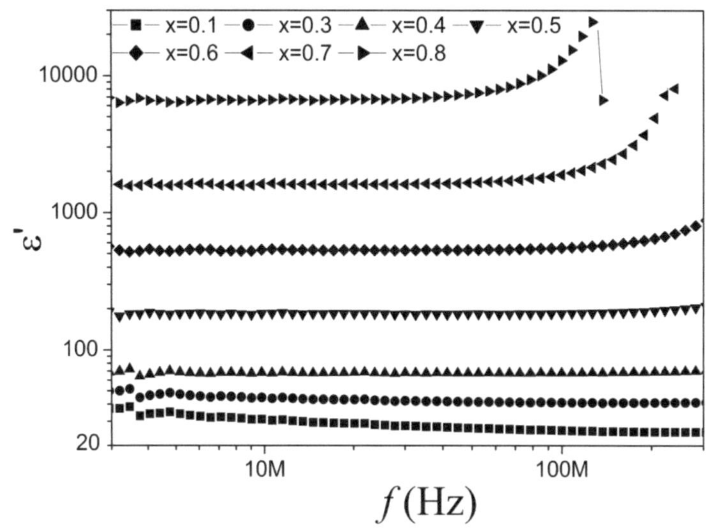

Fig. 10. Frequency dependence of permittivity of PNNT-BZCF composite ceramics

4. The permittivity and permeability in hyper-frequency

4.1 Permittivity

Owing to the polarization of dipolar, ferroelectric ceramics always have significant permittivity higher than several thousands, while the permittivity of ferrite may be as low as ~20. The dielectric mechanism of ferrite is associated with the conduction mechanism, which is attributed to the easy electron transfer between Fe^{2+} and Fe^{3+}. Although some ferrite containing large amount of Fe^{2+} ions has high permittivity of several thousands, such as MnZn ferrite, the low electric resistivity limits its application in high frequency. When the sintering temperature is lower than 1100°C or the sample is sintered under high partial pressure of oxygen, the sample has high resistivity and low permittivity of ~20, which is suitable for the application in high frequency.

The permittivity of two-phased composite material is always between those of two constituent phases and controlled by their relative volume fractions. Since the permittivity of ferroelectric ceramics is much higher than that of ferrite, the composite ceramics with more ferroelectric phase have higher permittivity. Fig. 10 shows the frequency dispersion of permittivity of PNNT-BZCF composite ceramics. The permittivity increases monotonically with the rise of PNNT amount. For example, the permittivity increases from 30 to 6600 (@ 10MHz) when the weight fraction of PNNT rises from 0.1 to 0.8.

4.2 Permeability

The permeability of nonmagnetic ferroelectric ceramics is always one, while the soft magnetic ferrites have high permeability. Due to the inverse proportion of permeability and cut-off frequency, the permeability turns lower in higher frequency range, which is associated with the magnetic structure of material. For example, the permeability of NiZn and MnZn spinel ferrites is higher than several thousands below MHz frequency range, but it turns very low above MHz, that is attributed to their cubic structure. The hexagonal ferrites, especially Y-type hexagonal ferrite, have planar magnetocrystalline anisotropy, which endows them high permeability above 100MHz. Because the ferrite has much higher permeability than ferroelectric ceramics, the permeability of composite ceramics increases with the rise of ferrite's amount (Fig. 11).

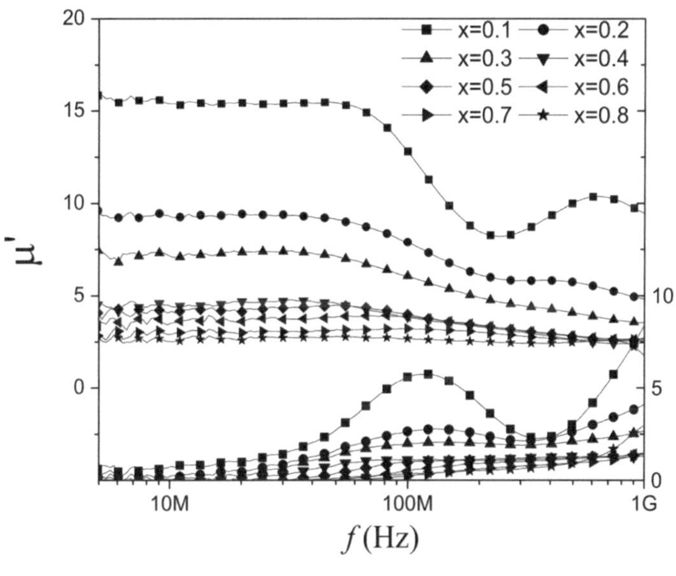

Fig. 11. Frequency dependence of real part and imaginary part of complex permeability of the composite ceramics of PNNT-BZCF (x=0.1~0.8)

4.3 The theoretical prediction

The effective macroscopic electromagnetic properties of a composite material are determined by the intrinsic characters of constituent phases and their relative volume fractions, so some mixture theories and equations have been established based on an equivalent dipole representation to predict the electromagnetic properties of a composite material. In this section, three most popular mixture theories are introduced, including mixture law, Maxwell-Garnett equations (containing MGa, MGb) and Bruggeman effective medium theory (EMT).

The mixture law is the simplest mixture theory to predict the properties of a composite material. According to the structure type of a composite material, the mixture law has different forms, such as parallel connection model and series connection model. For the composite material with powder mixture, the mixture law has a form as

$$\ln \Psi^* = f_a \ln \Psi_a + f_b \ln \Psi_b \tag{1}$$

where Ψ^*, Ψ_a and Ψ_b are the effective dielectric permittivity or magnetic permeability of composite material and two constituent phases. The f_a and f_b refer to the volume fraction of two phases and $f_a + f_b = 1$.

Further, some mixture theories are developed based on an equivalent dipole representation of the mixture, where the effective macroscopic electromagnetic properties of composite material are modeled as the intrinsic dipole moments per unit volume of each constituent phase and the relative volume fraction. It is assume that the isolated particles of constituent phases are embedded in a matrix host. The electric and magnetic intrinsic dipole moments of component phases, as well those of matrix host, are used to calculate the effective macroscopic properties of composite material. In static (or quasistatic) regime, a general form of the mixture equation was established based on the assumption that the components of isolated particles are embedded in a contiguous host medium (Aspnes, 1982). It can be expressed as

$$\frac{\Psi^* - \Psi_h}{\Psi^* + 2\Psi_h} = f_a \frac{\Psi_a - \Psi_h}{\Psi_a + 2\Psi_h} + f_b \frac{\Psi_b - \Psi_h}{\Psi_b + 2\Psi_h} \tag{2}$$

where Ψ_h is the effective permittivity or permeability of the host medium. If the host material is chosen as either phase ***a*** or ***b***, Maxwell-Garnett mixture equations, including MGa and MGb equations, are obtained (Maxwell Garnett, 1904 & 1906). If phase ***a*** is chosen as the host material, Equation (2) can be simplified to MGa equation as

$$\frac{\Psi^* - \Psi_a}{\Psi^* + 2\Psi_a} = f_b \frac{\Psi_b - \Psi_a}{\Psi_b + 2\Psi_a} \tag{3}$$

Similarly, MGb equation is derived as

$$\frac{\Psi^* - \Psi_b}{\Psi^* + 2\Psi_b} = f_a \frac{\Psi_a - \Psi_b}{\Psi_a + 2\Psi_b} \tag{4}$$

Different values of Ψ^* can be calculated using MGa and MGb equations for a composite material with a given volume fraction of particles. It is because the properties of matrix host are dominated until the volume fraction of isolated particle closely approaches unity. This expression works fairly well provided the inclusions make up a small fraction of the total volume. However, the Maxwell-Garnett model omits the variation of microstructure, so the imbedded phase never percolates even when the matrix has obviously inverted.

To characterize the microstructural inversion, Bruggeman effective medium theory is formed (Bruggeman, 1935), where the host material is chosen as the mixture itself ($\Psi^* = \Psi_h$), and then Equation (2) is reduced as

$$0 = f_a \frac{\Psi_a - \Psi_h}{\Psi_a + 2\Psi_h} + f_b \frac{\Psi_b - \Psi_h}{\Psi_b + 2\Psi_h} \tag{5}$$

Bruggeman effective medium theory assumes that component ***a*** and ***b*** are both embedded in the effective medium itself and are not treated as contiguous constituents, so it can predict percolation of either phase when its volume fraction is over 1/3. Its predictions exhibit a significant improvement compared with MG equations. This formalism has more applicability for composites formed by the constituents with similar mechanical properties.

In this section, component ***a*** and ***b*** are chosen as ferroelectric ceramics and ferrite, respectively. To exclude the influence of frequency dispersion, the experimental data of permittivity or permeability are accessed in region where the value is steady within a wide frequency range.

Fig. 12 compared the measured permittivity and calculated values by different equations for PNNT-BZCF composite ceramics. Mixture law and MGa equation give good predictions of permittivity for the composite ceramics with less ferroelectric phase, while the calculated results greatly deviates from the experimental data if PNNT's amount is large. On the contrary, MGb equation works well only if PNNT's amount is very high. EMT result matches the experimental data well if one phase has much higher volume fraction than the other, but it does not work well when two phases have comparable volume fraction. In addition, MGa and MGb equations offer upper and lower limits for the permittivity of PNNT-BZCF composite ceramics.

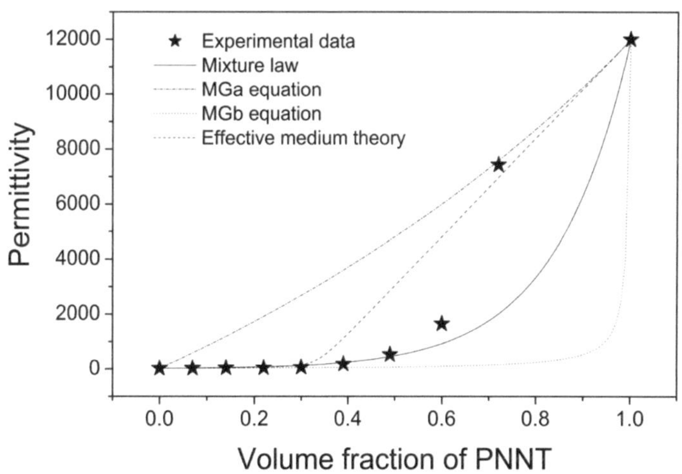

Fig. 12. The composition dependence of measured and calculated permittivity of PNNT-BZCF composite ceramics

The theoretical predicted permeability and experimental data of PNNT-BZCF composite ceramics are shown in Fig. 13. The prediction by MGa equation fits the experimental data well over the whole range of compositions, while those of other equations are higher than the measured data.

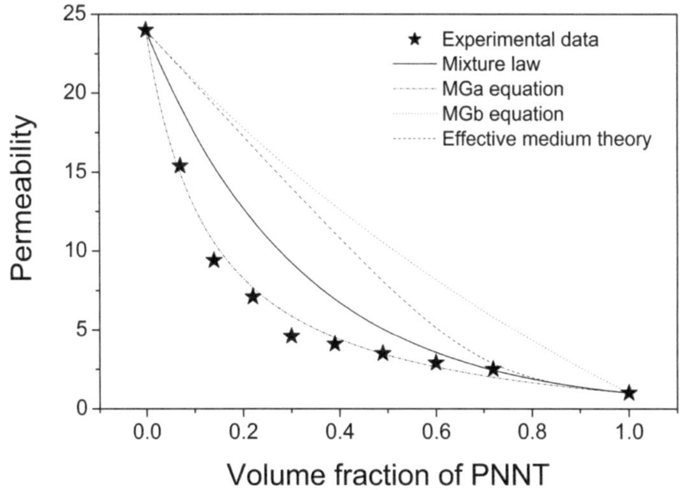

Fig. 13. The composition dependence of measured and calculated permeability of the composite ceramics of PNNT-BZCF

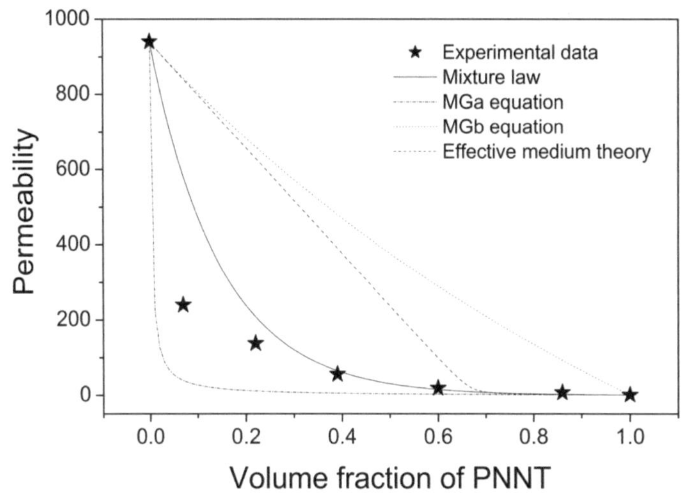

Fig. 14. The composition dependence of measured and calculated permeability of the composite ceramics of PNNT-NiCuZn ferrite

To further check the applicability of these mixture theories for permeability, the composite ceramics of PNNT-NiCuZn ferrite is discussed, where NiCuZn ferrite has a high permeability of ~950 (Shen et al., 2005). Fig. 14 compared the measured permeability and the values calculated by different equations. The permeability predicted by mixture law, MGa and EMT equations matches the experimental data well when the ferrite amount is relatively low. For the composite material with high volume fraction of magnetic phase, all the equations can not give precise predictions. Although an exact prediction is not presented, the MGa and MGb predictions give the upper and lower limits to the permeability of composition material.

The mixture equation based on a single simple model of microstructure may be inadequate to predict the effective macroscopic dielectric or magnetic properties over the whole composition range, so more complex equations with two or more models will be used to achieve wider applicability and more precise prediction.

5. The electromagnetic resonance character in hyper-frequency

5.1 The dielectric resonance

The frequency dispersion character is as important as the values of permittivity and permeability for the application in high frequency range. Different dispersion characters are needed by different applications. For example, capacitor or inductor requires a stable permittivity or permeability and low loss in a certain frequency range, so the resonance limits its working frequency range; while filter or EMI component needs high loss around the dielectric or magnetic resonance frequency. In a composite material, the frequency dispersion is determined by the intrinsic properties of constituent phases and affected by the interaction between them.

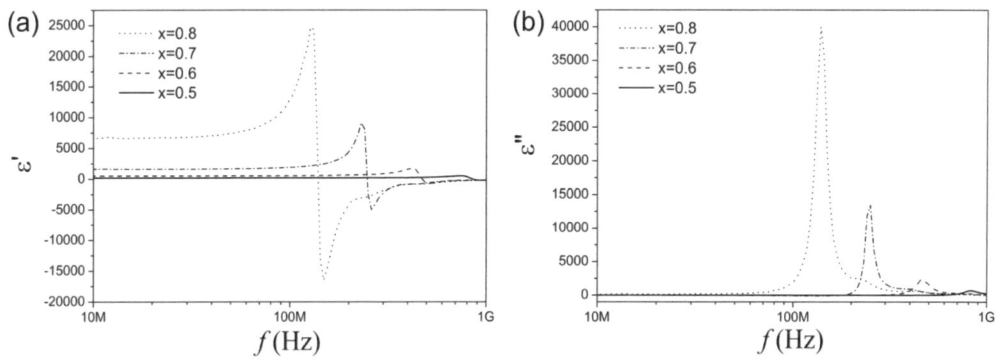

Fig. 15. Frequency dispersion of (a) real part and (b) imaginary part of permittivity of PNNT-BZCF composite ceramics

The frequency dispersion of real part and imaginary part of permittivity of PNNT-BZCF composite ceramics is shown in Fig. 15 (a) and (b). A strong dielectric resonance peak is observed above 100MHz, which originates from the dipole's vibration (Bai et al., 2006) or the followed piezoelectric vibration (Ciomaga et al., 2010). The resonance frequency increases with the reduction of PNNT amount and shifts out of the upper limit of measurement when the weight fraction of PNNT is lower than 0.4 (Fig. 16). In addition, the resonance peak turns flatter with the reduction of PNNT amount, which is characterized as the variation of half peak breadth in Fig. 17. The change of the shape of resonance peak implies that the dielectric response tends to transform from resonance to relaxation.

In addition to the intrinsic properties of constituent phases, the electromagnetic interaction between ferroelectric and ferromagnetic phases influences the frequency and shape of resonance peak. The charged particles in ferroelectric phase vibrate under the force of external electric field. When the frequency of alternating electric field matches the nature frequency of the charged particles' vibration, dielectric resonance occurs. In the ferroelectric-ferromagnetic composite material, the spatial inhomogeneous electromagnetic field around ferrite grains will disturb the charged particles' motion in ferroelectric phase and change their nature frequency. The equivalent damping for the charged particles' motion increases with the enhancement of magnetic phase, so the dielectric response changes from resonance to relaxation gradually. Since the macroscopic frequency spectra of permittivity reflects the statistical average effect of microscale charged particles, the resonance peak turns flatter and shifts to higher frequency with the rise of ferrite amount.

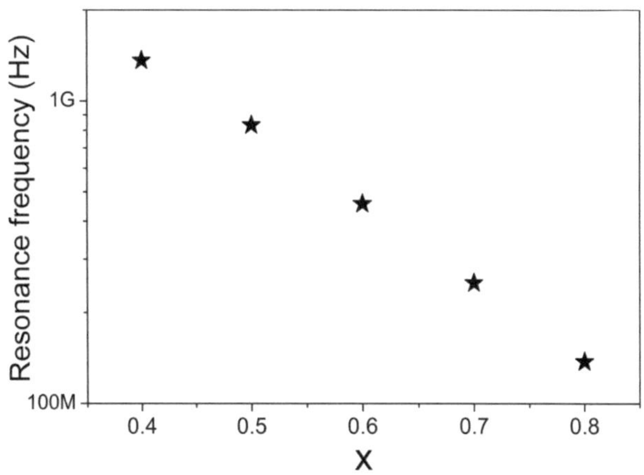

Fig. 16. The composition dependence of dielectric resonance frequency of PNNT-BZCF composite ceramics

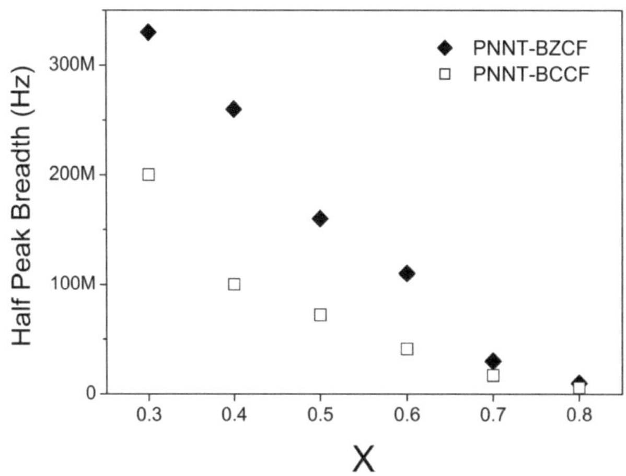

Fig. 17. The composition dependence of the half peak breadth of dielectric resonance peak for PNNT-BZCF and PNNT-BCCF composite ceramics

The electromagnetic interaction between two phases is affected by the permeability of magnetic phase. Fig. 17 (a) compares the dielectric dispersions of PNNT-BZCF and PNNT-$Ba_2Co_{1.2}Cu_{0.8}Fe_{12}O_{22}$ (BCCF) composite ceramics with same composition ratio, where BZCF and BCCF have identical properties in sintering character, microstructure, permittivity, and electric resistivity, except for in permeability. The permeability of BZCF (>20) is much higher than that of BCCF (~3.5). From Fig. 18, two composite materials have same dielectric behavior of permittivity except for the distinctly different resonance characters. The dielectric resonance peak of PNNT-BCCF composite ceramics is narrow and sharp, while that of PNNT-BZCF composite ceramics is much wider and smoother. The comparison of half peak breadth shows the contrast in Fig. 17. The induced magnetic field around ferrite particles is enhanced with the permeability of ferrite, but the variation of electromagnetic environment is not strong enough to change the value of permittivity within low frequency range and can only vary the resonance character, which is sensitive to surrounding condition.

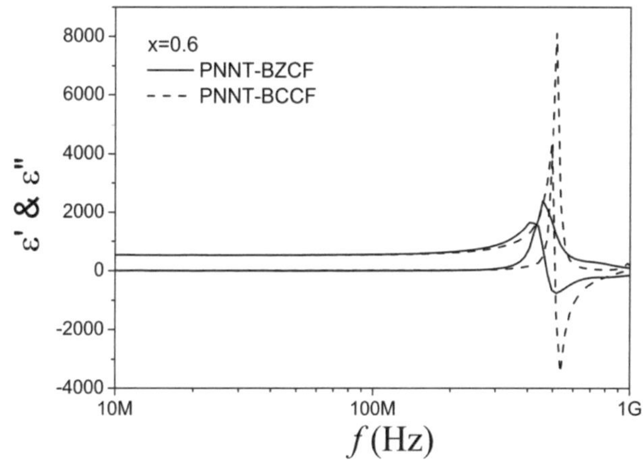

Fig. 18. The comparison of the dielectric frequency spectra of PNNT-BZCF and PNNT-BCCF composite ceramics

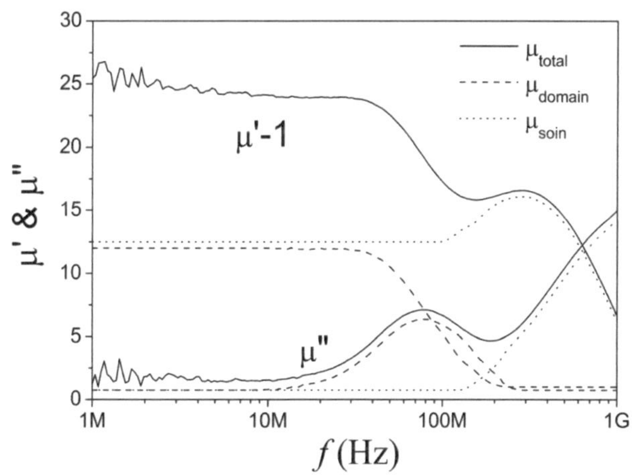

Fig. 19. Frequency spectrum of permeability of BZCF and the divided contributions of domain wall motion and spin rotation

5.2 The magnetic resonance

The frequency dispersion of permeability of ferroelectric-ferromagnetic composite ceramics is determined by the nature of magnetic phase. In the frequency spectra of a soft magnetic ferrite, there will be some kinds of resonances, such as magnetic domain wall resonance and spin resonance. As shown in Fig. 19, there are two resonance peaks in the frequency spectra of permeability, where the low-frequency peak originates from the domain wall resonance and the high-frequency peak results from the spin resonance (Bai et al., 2004). The permeability can be divided into two parts according to the contributions of domain wall motion and spin rotation, $\mu_{total} = \mu_{domain} + \mu_{spin}$, as shown in Fig. 19.

For the composite ceramics with high ferrite fraction, there are still two resonance peaks in the frequency spectra (Fig. 19). With the reduction of BZCF amount, the permeability decreases, and the resonance peaks turn flatter and weaker gradually on account of dilution effect.

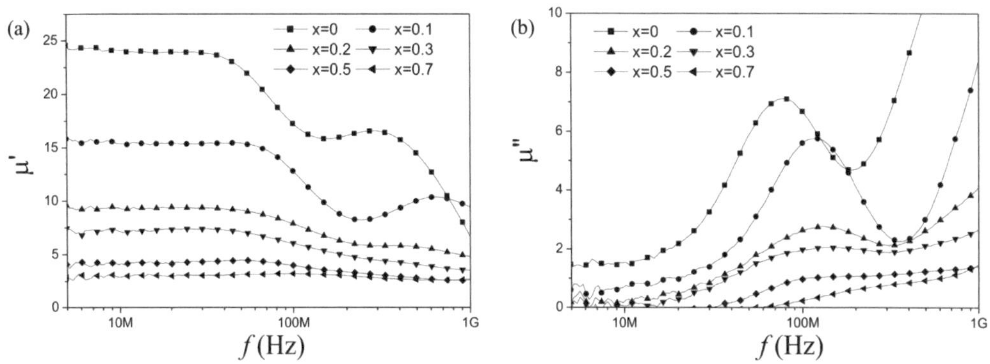

Fig. 20. Frequency dispersion of (a) real part and (b) imaginary part of complex permeability of PNNT-BZCF composite ceramics

The frequency dispersion of composite ceramics' permeability is influenced by the microstructure, such as internal stress. For the samples with large BZCF amount (x<0.3), two resonance peaks appear in the frequency spectra of permeability. Spin rotation is much sensitive to the internal stress on magnetic particles, so the spin resonance peak disappears when x>0.2. In contrast, the domain wall resonance peak exists up to x=0.7.

6. Summary

With the rapid development of electronic products, the multi-functional ferroelectric–ferromagnetic composite materials are great desired by various novel electronic components and devices. Then various composite systems and preparation methods were widely investigated and encouraging progresses have been made. To avoid co-firing mismatch and achieve a fine microstructure, the materials with similar densification behavior are desired as the constituent phases in the composite ceramics. And low sintering temperature is needed not only by the technical requirement of LTCC but also to avoid the element diffusion, volatilization and formation of other phase.

To keep up with the trend towards higher frequency for electronic technology, the composite materials with both high permittivity and permeability in hyper frequency is developed in recent years. The co-fired composite ceramics of $0.8Pb(Ni_{1/3}Nb_{2/3})O_3$-$0.2PbTiO_3/Ba_2Zn_{1.2}Cu_{0.8}Fe_{12}O_{22}$ are mainly introduced in this chapter, which has excellent co-firing behavior, dense microstructure and good electromagnetic properties. Owing the intrinsic characters of constituent phases, the composite ceramics exhibit both ferromagnetic

and ferroelectric properties over a wide composition range, and it has both high permittivity and high permeability in hyper-frequency, which can be tuned by the relative fraction of phases.
In the development of novel composite materials, the prediction of effective electromagnetic properties is greatly needed for material design. In this chapter three popular mixture theories, mixture law, Maxwell-Garnett equations and Bruggeman effective medium theory, have been introduced. In most cases, these theories can give a rough prediction for permittivity and permeability, or provides upper and lower limits, while the predictions departs from the experimental data great in some conditions. Hence, more complex equations with two or more models are being investigated.

7. References

Aspnes, D. E. (1982). Local-field effects and effective-medium theory: A microscopic perspective, *Am. J. Phys.*, vol. 50, pp.704-709

Bai, Y.; Zhou, J.; Gui, Z. & Li, L. (2004). Frequency dispersion of complex permeability of Y-type hexagonal ferrites, *Mater. Lett.*, vol. 58, pp. 1602-1606

Bai, Y.; Zhou, J.; Sun, Y.; Li, B.; Yue, Z.; Gui, Z. & Li, L. (2006). Effect of electromagnetic environment on the dielectric resonance in the ferroelectric-ferromagnetic composite, *Appl. Phys. Lett.*, vol. 89, pp. 112907

Bai, Y.; Zhou, J.; Gui, Z.; Li, L & Qiao, L. (2007). A ferromagnetic ferroelectric cofired ceramic for hyper-frequency, *J. Appl. Phys.*, vol. 101, pp. 083907

Bai, Y.; Xu, F.; Qiao, L. & Zhou, J. (2009). The modulated grain morphology in co-fired composite ceramics and its influence on magnetic properties, *J. Magn. Magn. Mater.*, vol. 321, pp. 148-151

Ciomaga, C. E.; Dumitru, I.; Mitoseriu, L.; Galassi, C.; Iordan, A. R.; Airimioaeic M. & Palamaruc M. N. (2010). Magnetoelectric ceramic composites with double-resonant permittivity and permeability in GHz range: A route towards isotropic metamaterials, *Scripta Mater.*, vol. 62, pp. 610–612

Bruggeman, D. A. G. (1935). Berechnung verschiedener physikalischer Konstantenvon heterogenen Substanzen, *Ann. Phys. (Leipzig)*, vol. 24, pp. 636-679

Hill, N. A. (2000). Why are there so few magnetic ferroelectrics? *J. Phys. Chem. B.*, vol. 104, pp. 6694-6709

Hsu, R. T. & Jean, J. H. (2005). Key factors controlling camber behavior during the cofiring of bi-layer ceramic dielectric laminates, *J. Am. Ceram. Soc.*, vol. 88, pp. 2429-2434

Kanai, T.; Ohkoshi, S.; Nakajima, A.; Nakajima, A.; Watanabe, T. & Hashimoto, K. (2001). A ferroelectric ferromagnet composed of $(PLZT)_x(BiFeO_3)_{1-x}$ solid solution, *Adv. Mater.*, vol. 13, pp. 487-490

Kumar, M. M.; Srinath, S.; Kumar, G. S. & Suryanarayana, S. V. (1998). Spontaneous magnetic moment in BiFeO3-BaTiO3 solid solutions at low temperatures, *J. Magn. Magn. Mater.*, vol. 188, pp. 203-212

Maxwell Garnett, J. C., (1904). Colours in metal glasses and in metallic films, *Philos. Trans. R. Soc. London*, vol. 203, pp. 385-420

Maxwell Garnett, J. C., (1906). Colours in metal glasses, in metallic films, and in metallic solutions. II, *Philos. Trans. R. Soc. London*, vol. 205, pp. 237-288

Qi, X.; Zhou, J.; Yue, Z.; Li, M. & Han, X. (2008). Cofiring behavior of ferroelectric ferromagnetic composites, *Key Eng. Mater.*, vol. 368-372, pp. 573-575

Qi, X.; Zhou, J.; Li, B.; Zhang, Y.; Yue, Z.; Gui, Z. & Li L. (2004). Preparation and spontaneous polarization-magnetization of a new ceramic ferroelectric-ferromagnetic composite, *J. Am. Ceram. Soc.*, vol. 87, pp. 1848-1852

Qi, X.; Zhou, J.; Yue, Z.; Gui, Z.; Li L. & Buddhudu S. (2004). A ferroelectric ferromagnetic composite material with significant permeability and permittivity, *Adv. Func. Mater.*, vol. 14, pp. 290-296

Shen, J.; Bai, Y.; Zhou, J. & Li, L. (2005). Magnetic properties of a novel ceramic ferroelectric-ferromagnetic composite, *J. Am. Ceram. Soc.*, vol. 88, pp. 3440-3443

12

Aging-Induced, Defect-Mediated Double Ferroelectric Hysteresis Loops and Large Recoverable Electrostrain Curves in Mn-Doped Orthorhombic $KNbO_3$-Based Lead-Free Ceramics

Siu Wing Or
Department of Electrical Engineering, The Hong Kong Polytechnic University, Hung Hom, Kolwoon, Hong Kong

1. Introduction

Aging is a physical phenomenon in many ferroelectric materials characterized by the spontaneous changes of dielectric, ferroelectric, and piezoelectric properties with time (Jaffe et al., 1971; Lambeek & Jonker, 1986; Schulze & Ogino, 1988; Uchino, 2000). Aging is generally considered to be detrimental because it tends to limit the application viability of ferroelectric materials in terms of reliability and stability. Recently, a series of studies show that aging is useful and valuable to intensionally induce anomalous double (or constricted) ferroelectric hysteresis (*P*–*E*) loops, and hence large recoverable electrostrain (*S*–*E*) curves, in impurity- or acceptor-doped tetragonal ferroelectric titanates, such as barium titanate ($BaTiO_3$) (Lambeek & Jonker, 1986; Ren, 2004; Zhang & Ren, 2005; Zhang & Ren, 2006). Specifically, these aging effects can provide an alternative way of both physical interest and technological importance to modify or enhance the electromechanical properties of tetragonal ferroelectrics. From the phenomenological respects, the aging effects can be described by a gradual stabilization of ferroelectric domain structure by defects (i.e., dopant, vacancy or impurity) (Arlt & Rebels, 1993; Damjanovic, 1998; Hall & Ben-Omran, 1998). In fact, various stabilization theories, including the grain-boundary theory, surface-layer model, domain-wall theory, and volume theory, have been proposed over the past decades (Okasaki & Sakata, 1962; Takahashi, 1970; Carl & Hardtl, 1978; Lambeck & Jonker, 1978; Lambeek & Jonker, 1986; Robels & Arlt, 1993). Among them, the domain-wall-pinning effect has been accepted as a general mechanism of aging (Lambeek & Jonker, 1986; Ren, 2004). It is only quite recently that the volume effect based on the symmetry-conforming principle of point defects was proposed and recognized as the intrinsic governing mechanism of ferroelectric aging (Lambeek & Jonker, 1986; Zhang & Ren, 2006; Yuen et al. 2007).

Compared to tetragonal ferroelectrics, orthorhombic ferroelectrics are as interesting and as important, since orthorhombic ferroelectric phase lies widely in ferroelectrics similar to tetragonal ferroelectric phase (Yamanouchi et al., 1997; Saito et al., 2004; Wang et al., 2006). In particuar, orthorhombic $A^{+}B^{5+}O_3$ alkaline niobates ($KNbO_3$) are promising candidates for

lead-free piezoelectric applications due to their good piezoelectric properties and high Curie temperatures (Yamanouchi et al., 1997; Saito et al., 2004). However, literature report on the aging effects in this class of (lead-free) ferroelectrics remains essentially insufficient till today (Feng & Or, 2009).
More recently, we have investigated the aging effects in an Mn-doped orthorhombic $KNbO_3$-based $[K(Nb_{0.90}Ta_{0.10})O_3]$ lead-free ceramic: $K[(Nb_{0.90}Ta_{0.10})_{0.99}Mn_{0.01}]O_3$ so as to provide a relatively complete picture about the aging on both orthorhombic and tetragonal ferroelectrics for the related communities (Feng & Or, 2009). In this work, we present the aging-induced double P–E loops and recoverable S-E curves in the ceramic, and show that aging in the orthorhombic ferroelectric state is capable of inducing an obvious double P–E loop accompanying a recoverable electrostrain as large as 0.15% at 5 kV/mm at room temperature. Such aging effects are interpreted by a point defect-mediated reversible domain switching mechanism of aging driven by a symmetry-conforming short-range ordering (SC-SRO) of point defects. Large nonlinear electrostrains in excess of 0.13% over a broad temerpature range of 25–140 °C are also demonstrated, suggesting potential application of the aging effects to modify or enhance the electromechanical properties of environmentally-friendly (lead-free) ceramics.

2. Ceramic preparation and property measurements

2.1 Ceramic preparation

The Mn-doped orthorhombic $A^+B^{5+}O_3$ alkaline niobate ($KNbO_3$)-based $[K(Nb_{0.90}Ta_{0.10})O_3]$ lead-free ceramic: $K[(Nb_{0.90}Ta_{0.10})_{0.99}Mn_{0.01}]O_3$ was synthesized using a conventional solid-state reaction technique (Yamanouchi et al., 1997; Saito et al., 2004). The parental compound $K(Nb_{0.90}Ta_{0.10})O_3$ was essentially based on $KNbO_3$ but was modified by adding 10% Ta to the Nb site. As shown in Fig. 1, such modification served to shift the cubic (paraelectric)-tetragonal (ferroelectric) phase transition temperature T_C and the tetragonal (ferroelectric)-orthorhombic (ferroelectric) phase transition temperature T_{O-T} to the lower temperature side, besides making the "hard" material to be relatively "soft" (Triebwasser, 1959). To formulate the ceramic, 1.0 mol.% Mn was added to the B-site of $K(Nb_{0.90}Ta_{0.10})O_3$ as the acceptor dopant.

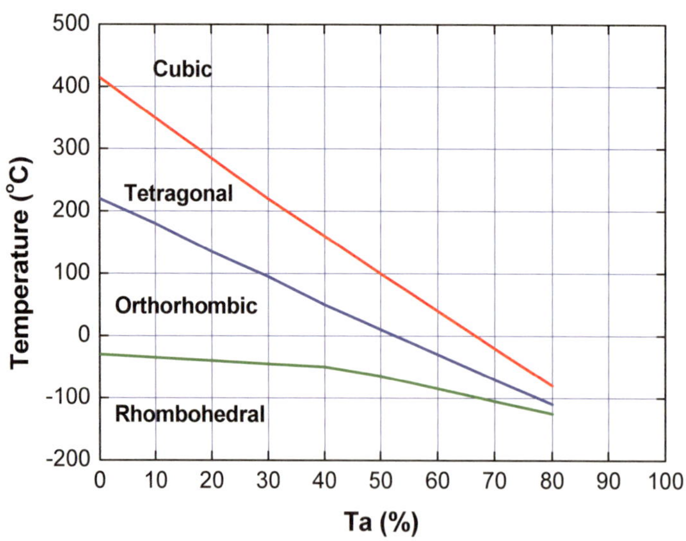

Fig. 1. Phase diagram of $K(Nb_{1-x}Ta_x)O_3$ compound

The starting chemicals were K_2CO_3 (99.5%), Nb_2O_5 (99.9%), Ta_2O_5 (99.9%), and MnO_2 (99%). Calcination was done at 850 °C for 4 h in a K_2O-rich atmosphere, while sintering was carried out at 1050 °C for 0.5 h in air. In order to remove the historical effect, all the as-prepared samples were deaged by holding them at 500 °C for 1 h followed by an air-quench to room temperature. The quenched and deaged samples are designated as "fresh samples". Some fresh samples were aged at 130 °C for 5 days, and the resulting samples are denoted as "aged samples".

2.2 Property measurements

The temperature dependence of dielectric constant of the fresh samples was evaluated at different frequencies using a LCR meter (HIOKI 3532) with a temperature chamber. The bipolar and unipolar ferroelectric hysteresis (*P*–*E*) loops and electrostrain (*S*–*E*) curves for the aged and fresh samples were measured at a frequency of 5 Hz using a precision ferroelectric test system (Radiant Workstation) and a photonic displacement sensor (MTI 2000) under various temperatures in a temperature-controlled silicon oil bath (Fig. 2).

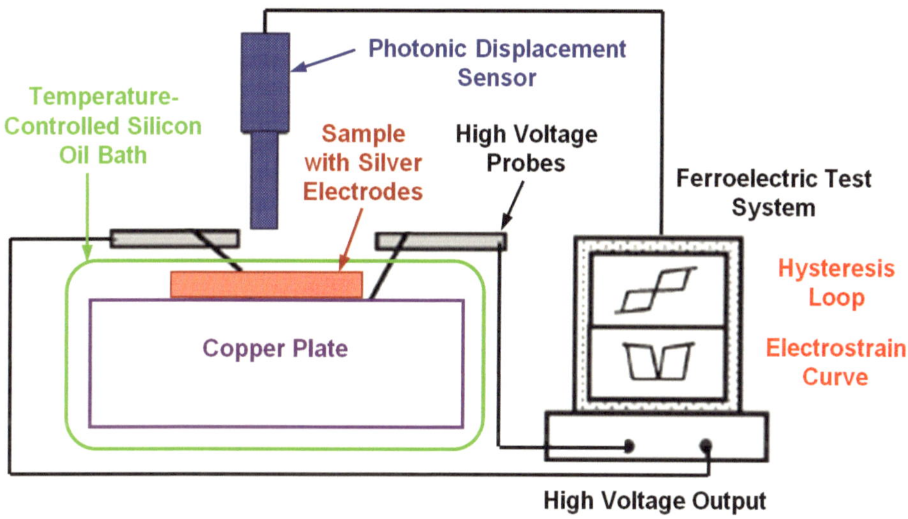

Fig. 2. Experimental setup for measuring bipolar and unipolar ferroelectric hysteresis (*P*–*E*) loops and electrostrain (*S*–*E*) curves

3. Results and discussion

3.1 Temperature dependence of dielectric constant

Fig. 3 shows the temperature dependence of dielectric constant for the fresh samples at three different frequencies of 0.1, 1, and 10 kHz. Three distinct dielectric peaks are observed at about 326, 148, and -15 °C, respectively. X-ray diffraction (XRD) characterization indicates that they correspond to the cubic (paraelectric)-tetragonal (ferroelectric) phase transition temperature T_C, the tetragonal (ferroelectric)-orthorhombic (ferroelectric) phase transition temperature $T_{O\text{-}T}$, and the orthorhombic (ferroelectric)-rhombohedral (ferroelectric) phase transition temperature $T_{R\text{-}O}$, respectively (Triebwasser, 1959). Therefore, our samples have a rhombohedral (*R*) structure for temperatures below -15 °C, an orthorhombic (*O*) structure for temperatures ranging from -15 to 148 °C, a tetragonal (*T*) structure for temperatures varying from 148 to 326 °C, and a cubic (*C*) structure for temperatures above 326 °C.

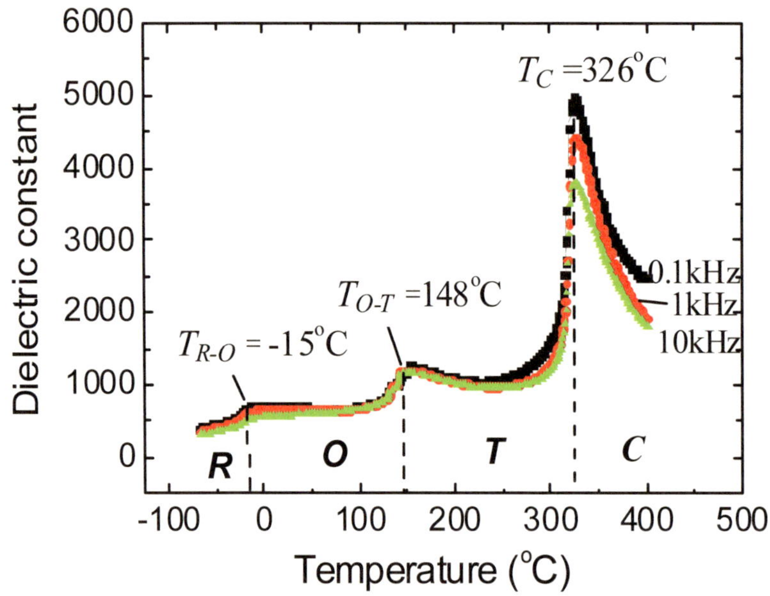

Fig. 3. Temperature dependence of dielectric constant for the fresh samples at three different frequencies of 0.1, 1, and 10 kHz. *C*=cubic, *T*=tetragonal, *O*=orthorhombic, and *R*=rhombohedral

3.2 Room-temperature bipolar and unipolar ferroelectric hysteresis loops and electrostrain curves

Fig. 4 illustrates the bipolar and unipolar ferroelectric hysteresis (*P*–*E*) loops and electrostrain (*S*-*E*) curves for the aged and fresh samples at room temperature. In contrast with the normal bipolar *P*–*E* loop for the fresh samples, the aged samples in Fig. 4(a) possess an interesting bipolar double *P*–*E* loop, very similar to that of the aged acceptor-doped tetragonal ferroelectrics such as the $A^{2+}B^{4+}O_3$ system (Ren, 2004; Zhang & Ren, 2005; Zhang & Ren, 2006). Moreover, a large recoverable electrostrain of 0.15% at 5 kV/mm, accompanying the double *P*–*E* loop, is achieved in our aged samples. This recoverable *S*–*E* curve is indeed different from the butterfly irrecoverable *S*–*E* curve as obtained in the fresh samples due to the existence of a recoverable domain switching in the aged samples but an irrecoverable domain switching in the fresh samples. Fig. 4(b) shows the unipolar *P*–*E* loops and *S*–*E* curves for the aged and fresh samples. It is clear that a large polarization *P* of about 22 μC/cm² is obtained at 5 kV/mm for the aged samples compared to a much smaller *P* of about 6 μC/cm² in the fresh samples at the same field level. With the large *P*, a large nonlinear electrostrain of 0.15% at 5 kV/mm is available for the aged samples owing to the reversible domain switching. It is noted that this electrostrain not only is 2.5 times larger than the fresh samples, but also exceeds the "hard" lead zirconate titanate (PZT) value of 0.125% at 5 kV/mm (Park & Shrout, 1997). It is also noted that the electrostrain in our fresh samples (having a small quantity of Mn acceptor dopant) is not obviously different from that in the undoped $K(Nb_{0.90}Ta_{0.10})O_3$ ceramic, and similarly large electrostrain has been reported recently on the aged tetragonal $K(Nb_{0.65}Ta_{0.35})O_3$-based ceramics (Feng & Ren, 2007).

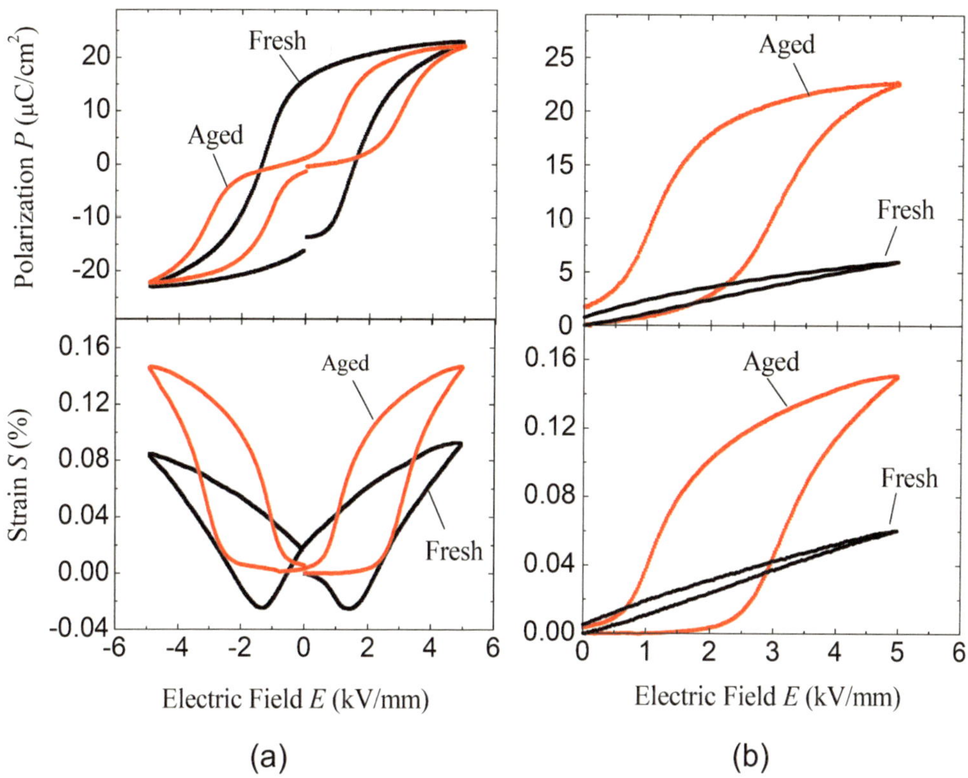

Fig. 4. (a) Bipolar and (b) unipolar ferroelectric hysteresis (*P–E*) loops and electrostrain (*S-E*) curves for the aged and fresh samples at room temperature

3.3 Physical interpretation by a point defect-mediated reversible domain switching mechanism of aging

Although our orthorhombic $K[(Nb_{0.90}Ta_{0.10})_{0.99}Mn_{0.01}]O_3$ ceramic has different crystal symmetry from tetragonal ferroelectric $BaTiO_3$, they all belong to perovskite ABO_3 structure. This lets us to believe that the observed aging effects in our aged orthorhombic samples can be explained according to a point defect-mediated reversible domain switching mechanism of aging driven by a symmetry-conforming short-range ordering (SC-SRO) of point defects (i.e., acceptor ions and vacancies) adopted successfully in $BaTiO_3$ (Ren, 2004; Zhang & Ren, 2005; Zhang & Ren, 2006). In fact, when acceptor dopant Mn^{4+}/Mn^{3+} ions displace the central Nb^{5+}/Ta^{5+} ions of the *B*-site in the aged $K[(Nb_{0.90}Ta_{0.10})_{0.99}Mn_{0.01}]O_3$ samples, oxygen vacancies V_O form at the O^{2-} sites to maintain the charge neutrality, resulting in point defects (i.e., defect dipoles) with the central acceptor dopants. Fig. 5 depicts how such aging effects are produced in a single-crystal grain of our aged samples. Some associated remarks are included as follows.

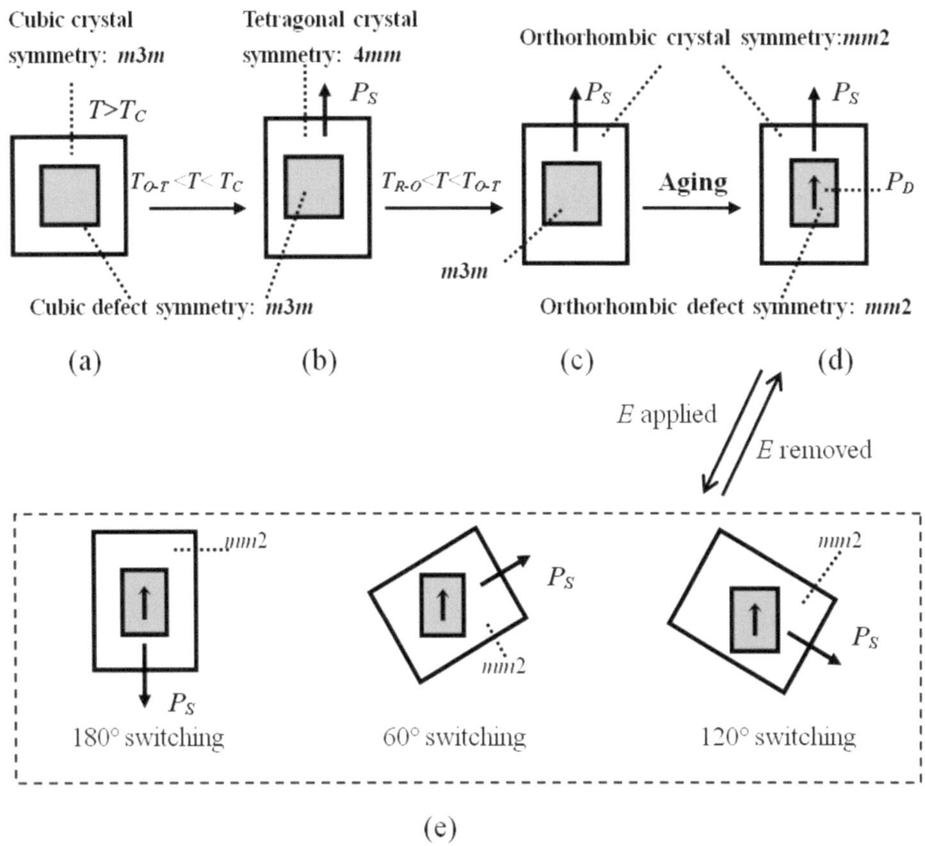

Fig. 5. Crystal and defect symmetries of a single-crystal grain in (a) a fresh $K[(Nb_{0.90}Ta_{0.10})_{0.99}Mn_{0.01}]O_3$ sample at $T>T_C$, (b) the fresh sample at $T_{O\text{-}T}<T<T_C$, (c) the fresh sample at $T_{R\text{-}O}<T<T_{O\text{-}T}$, and (d) an aged sample at room temperature. (e) Electric field E-induced switching of the 180°, 60°, and 120° ferroelectric domains in the aged sample at room temperature

1. When the fresh samples are just sintered and their temperature T is still above the Curie point T_C (i.e., $T>T_C$), its single-crystal grain exhibits a cubic crystal symmetry $m3m$, and point defects naturally show a conforming cubic defect symmetry $m3m$, as shown in Fig. 5(a).
2. At $T_{O\text{-}T}<T<T_C$, the single-crystal grain of the fresh samples shows a tetragonal crystal symmetry $4mm$, due to the displacement of positive and negative ions along the [001] crystallographic axis, producing a nonzero spontaneous polarization P_S as shown in Fig. 5(b). However, the short-range ordering (SRO) distribution of point defects keeps the same cubic defect symmetry $m3m$ as that in the cubic paraelectric phase because the diffusionless paraferroelectric transition cannot alter the original cubic SRO symmetry of point defects (Ren, 2004).
3. At $T_{R\text{-}O}<T<T_{O\text{-}T}$, the single-crystal grain of the fresh samples exhibits an orthorhombic crystal symmetry $mm2$, owing to the ferroelectric-ferroelectric phase transition from the tetragonal to orthorhombic structure, producing a nonzero P_S along the [110]

crystallographic axis as shown in Fig. 5(c). Again, the SRO distribution of point defects still keeps the same cubic defect symmetry *m*3*m* because of fast cooling. As a result, two unmatched symmetries (i.e., the orthorhombic crystal symmetry and the cubic defect symmetry) exist simultaneously in the fresh ferroelectric state [Fig. 5(c)]. According to the SC-SRO principle (Ren & Otsuka, 2000), such a state is energetically unstable and the samples tend to a symmetry-conforming state.

4. After aging at 130 °C for 5 days in the ferroelectric state, the cubic defect symmetry *m*3*m* changes gradually into a polar orthorhombic defect symmetry *mm*2, while the single-crystal grain of the aged samples has a polar orthorhombic crystal symmetry *mm*2, as shown in Fig. 5(d). Such a change is realized by the migration of V_O during aging, and the polar orthorhombic defect symmetry creates a defect polarization P_D, aligning along the spontaneous polarization P_S direction [Fig. 5(d)].
5. When an electric field E is initially applied in opposition to P_S of the aged orthorhombic samples [Fig. 5(e)], an effective switching of the available 180° ferroelectric domains is induced, contributing to a small polarization at low E (<1.5 kV/mm), as shown in Fig. 4(b). Continuing a larger applied E (>1.5 kV/mm), non-180° domain switching (mainly 60° and 120° domain switching according to the polar orthorhombic crystal symmetry) is induced, but the polar orthorhombic defect symmetry and the associated P_D cannot have a sudden change [Fig. 5(e)]. Hence, the unchanged defect symmetry and the associated P_D cause a reversible domain switching after removing E. Consequently, an interesting macroscopic double P–E loop and a large recoverable S–E curve are produced as in Fig. 4. For the fresh samples, since the defect symmetry is a cubic symmetry and cannot provide such an intrinsic restoring force, we can only observe a normal macroscopic P–E loop and a butterfly S–E curve due to the irreversible domain switching [Fig. 4(a)].

It should be noted that the microscopic description for the orthorhombic $KNbO_3$-based ferroelectrics is very similar to that for acceptor-doped tetragonal ferroelectric titanates (Ren, 2004; Zhang & Ren, 2005; Zhang & Ren, 2006). The observed aging effects originate essentially from the inconformity of the crystal symmetry with the defect symmetry after a structural transition. This may be the intrinsic reason why macroscopic double P-E loops and recoverable S–E curves are achieved in different ferroelectric phases and different ferroelectrics. Such aging mechanism, based on the SC-SRO principle of point defects, is insensitive to crystal symmetry and constituent ionic species, indicating a common physical origin of aging.

3.4 Effect of temperature on ferroelectric hysteresis loops and electrostrain curves

Fig. 6 plots the unipolar ferroelectric hysteresis (P–E) loops and electrostrain (S–E) curves for the aged samples at five different temperatures of 25, 80, 120, 140, and 160 °C in order to investigate their temperature stabilities for applications. The insets show the temperature dependence of maximum polarization P_{max} and maximum strain S_{max} of the aged samples at 5 kV/mm. It can be seen that the aging-induced high P_{max} in excess of 19 μC/cm^2 and large S_{max} in excess of 0.13% can be persisted up to 140 °C, reflecting a good temperature stability for the effects. Above 140 °C, both the unipolar P–E loop and S–E curve become normal, while P_{max} and S_{max} decrease significantly. This can be ascribed to the destruction of defect symmetry and migration of V_O as a result of the exposure to high temperature and the approach of the tetragonal phase ($T_{O\text{-}T}$ =148 °C). Thus, point defects cannot provide a restoring force for a reversible domain switching so that the obvious P–E loop becomes a normal loop and the recoverable S–E curve vanishes.

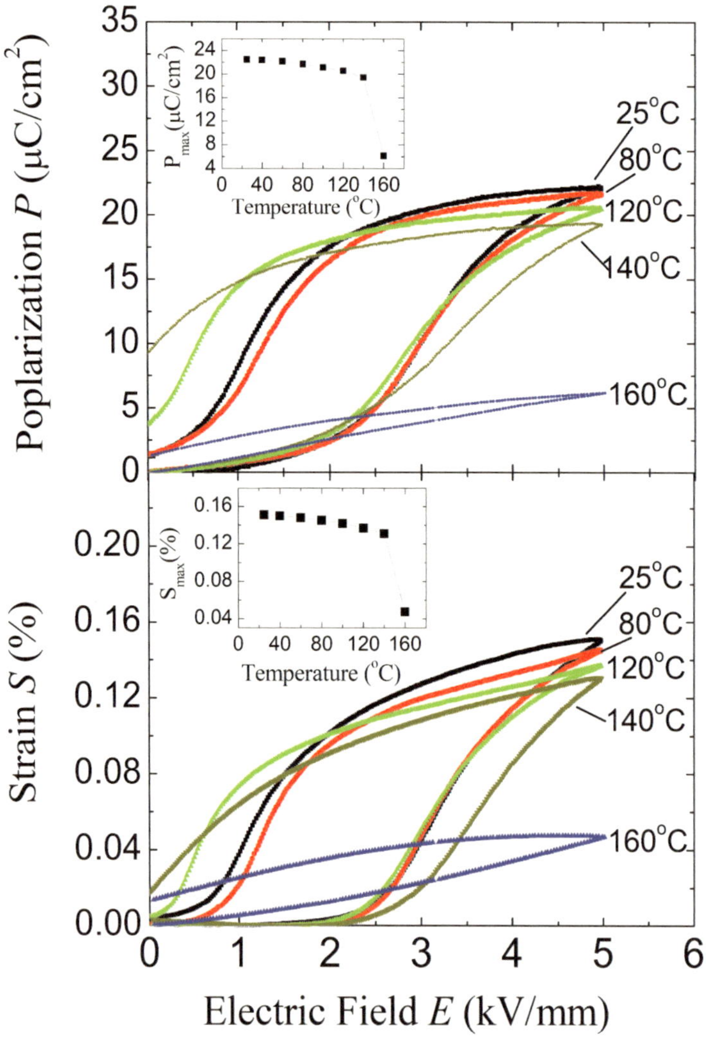

Fig. 6. Unipolar ferroelectric hysteresis (P–E) loops and electrostrain (S–E) curves for the aged samples at five different temperatures of 25, 80, 120, 140, and 160 °C. The insets show the temperature dependence of maximum polarization P_{max} and maximum strain S_{max} of the aged samples at 5 kV/mm

4. Conclusion

In summary, we have investigated the aging-induced double ferroelectric hysteresis (P–E) loops and recoverable electrostrain (S–E) curves in an Mn-doped orthorhombic $KNbO_3$-based $[K(Nb_{0.90}Ta_{0.10})O_3]$ lead-free ceramic: $K[(Nb_{0.90}Ta_{0.10})_{0.99}Mn_{0.01}]O_3$. Obvious double P–E

loops and large recoverable *S–E* curves with amplitudes in excess of 0.13% at 5 kV/mm have been observed in the aged samples over a wide temperature range of 25–140 °C. The observations have been found to have striking similarities to tetragonal ferroelectrics, besides following a point defect-mediated reversible domain switching mechanism of aging driven by a symmetry-conforming short-range ordering (SC-SRO) of point defects. Such aging effects, being insensitive to crystal structure and constituent ionic species, provide a useful way to modify or enhance the electromechanical properties of lead-free ferroelectric material systems.

5. Acknowledgements

This work was supported by the Research Grants Council and the Innovation and Technology Fund of the Hong Kong Special Administration Region (HKSAR) Government under Grant Nos PolyU 5266/08E and GHP/003/06, respectively.

6. References

Arlt, G. & Rebels U. (1993). Aging and fatigue in bulk ferroelectric perovskite ceramics. *Integrated Ferroelectrics*, 3, 343–349

Carl, K. & Hardtl, K. H. (1977). Electrical after-effects in Pb(Ti, Zr)O_3 ceramics. *Ferroelectrics*, 17, 473-486

Damjanovic, D. (1998). Ferroelectric, dielectric and piezoelectric properties of ferroelectric thin films and ceramics. *Reports on Progress in Physics*, 61, 1267-1324

Feng, Z. Y. & Ren, X. (2007). Aging effect and large recoverable electrostrain in Mn-doped $KNbO_3$-based ferroelectrics. *Applied Physics Letters*, 91, 032904

Feng, Z. Y. & Or, S. W. (2009). Aging-induced, defect-mediated double ferroelectric hysteresis loops and large recoverable electrostrains in Mn-doped orthorhombic $KNbO_3$-based ceramics. *Journal of Alloys and Compounds*, 480, L29-L32

Hall, D. A. & Ben-Omran, M. M. (1998). Ageing of high field dielectric properties in $BaTiO_3$-based piezoceramics. *Journal of Physics: Condensed Matter*, 10, 9129–9140

Jaffe, B.; Cook, W. R. & Jaffe, H. (1971). *Piezoelectric Ceramics*, Academic Press, 0123795508, New York

Lambeck, P. V. & Jonker, G. H. (1978). Ferroelectric domain stabilization in $BaTiO_3$ by bulk ordering of defects. *Ferroelectrics*, 22, 729-731

Lambeek, P. V. & Jonker, G. H. (1986). The nature of domain stabilization in ferroelectric perovskites. *Journal of Physics and Chemetry of Solids*, 47, 453–461

Okasaki, K. & Sakata, K. (1962). Space charge polarization and aging of barium titanate ceramics. *Electrotechnical Journal of Japan*, 7, 13–18

Park, S. -E. & Shrout, T. R. (1997). Ultrahigh strain and piezoelectric behavior in relaxor based ferroelectric single crystals. *Journal of Applied Physics*, 82, 1804-1811

Ren, X. & Otsuka, K. (2000), Universal symmetry property of point defects in crystals. *Physical Review Letters*, 85, 1016–1019

Ren, X. (2004). Large electric-field-induced strain in ferroelectric crystals by point-defect-mediated reversible domain switching. *Nature Materials*, 3, 91–94

Robels, U. & Arlt, G. (1993). Domain wall clamping in ferroelectrics by orientation of defects. *Journal of Applied Physics*, 73, 3454-3460

Schulze, W. A. & Ogino, K. (1988). Review of literature on aging of dielectrics. *Ferroelectrics*, 87, 361–377

Takahashi, M. (1970). Space charge effect in lead zirconate titanate ceramics caused by the addition of impurities. *Japnese Journal of Applied Physics*, 9, 1236–1246

Triebwasser, S. (1959). Study of ferroelectric transitions of solid-solution single crystals of $KNbO_3$-$KTaO_3$. *Physical Review*, 114, 63–70

Uchino, K. (2000). *Ferroelectric Device*, Mercel Dekker, Inc., 0-8247-8133-3, New York

Wang, X. X.; Or, S. W.; Lam, K. H.; Chan, H. L. W.; Choy, P. K. & Liu, P. C. K. (2006). Cymbal actuator fabricated using $(Na_{0.46}K_{0.46}Li_{0.08})NbO_3$ lead-free piezoceramic. *Journal of Electroceramics*, 16, 385–388

Yamanouchi, K.; Odagawa, H.; Kojima, T. & Matsumura, T. (1997). *Electronic Letters*, 33, 193 Saito, Y.; Takao, H.; Tani, T.; Nonoyama, T.; Takatori, K.; Homma, T.; Nagaya, T. & Nakamura, M. (2004). Lead free Piezoceramics. *Nature*, 432, 84–87

Yuen, G. L.; Yang, Y. & Or, S. W. (2007). Aging-induced double ferroelectric hysteresis loops in $BiFeO_3$ multiferroic ceramic. *Applied Physics Letters*, 91, 122907

Zhang, L. X. & Ren, X. (2005). *In situ* observation of reversible domain switching in aged Mn-doped $BaTiO_3$ single crystals. *Physical Revuew B*, 71, 174108

Zhang, L. X. & Ren, X. (2006). Aging behavior in single-domain Mn-doped $BaTiO_3$ crystals: Implication for a unified microscopic explanation of ferroelectric aging. *Physical Revuew B*, 73, 094121

13

Effects of B-site Donor and Acceptor Doping in Pb-free $(Bi_{0.5}Na_{0.5})TiO_3$ Ceramics

Yeon Soo Sung and Myong Ho Kim
Changwon National University
Korea

1. Introduction

$Pb(Zr,Ti)O_3$ (PZT)-based ceramics have been used extensively in various dielectric, piezoelectric, and ferroelectric applications because of their superior properties and versatility (Jaffe et al., 1971). The lead (Pb) content of PZT materials, however, poses serious problems for human health and the environment that can no longer be overlooked. It is inevitable that PZT will be replaced with Pb-free materials. PZT-based ceramics are still in use only because there are no materials that exhibit properties comparable to those of PZT.

The scientific community has intensively searched for and studied Pb-free piezoelectric materials. Among the few Pb-free materials available, solid solutions based on $(Bi_{0.5}Na_{0.5})TiO_3$ (BNT) (Isupov, 2005; Zhou et al., 2004; Park et al., 1996; Tu et al., 1994; Zvirgzds et al., 1982; Pronin et al., 1980; Jaffe et al., 1971) and $(Na_{0.5}K_{0.5})NbO_3$ (NKN) (Dai et al., 2009; Wang et al., 2007; Saito et al., 2004; Tennery & Hang, 1968; Haertling, 1967; Jaeger & Egerton, 1962; Egerton & Dillon, 1959) with a morphotropic phase boundary (MPB) have been considered as potential candidates for replacing PZT.

Each material has its merits and demerits. In terms of the piezoelectric coefficient (d_{33}), which is a key property, NKN-based ceramics appear to be the better choice considering their d_{33} of ~416 pC/N (Saito et al., 2004), which is high compared to the ~231 pC/N of BNT-based ceramics (Lin et al., 2006). However, NKN-based ceramics exhibit poor reproducibility because of hygroscopic problems. Furthermore, the range of temperature stability is narrow because those properties are mostly due to polymorphic phase transition. Regarding durability, BNT-based ceramics appear to be the better choice because, once sintered, they do not degrade as seriously as NKN-based ceramics do; this durability is an important factor in practical applications.

BNT has an ABO_3 perovskite structure with an A-site complex having rhombohedral symmetry at room temperature (Jaffe et al., 1971). It is known that BNT exhibits strong ferroelectricity with a high remnant polarization, coercive field, and conductivity. Because of its large coercive field and relatively high conductivity, BNT itself is difficult to pole, and consequently, its piezoelectric properties are not as appropriate for practical applications as those of PZT. As in the case of PZT, BNT-based solid solutions made with tetragonal $(Bi_{0.5}K_{0.5})TiO_3$ (BKT) or $BaTiO_3$ (BT) (Yu & Ye, 2008; Li et al., 2005; Kreisel et al., 2000; Takenaka et al., 1991) have been studied to improve the general properties by finding the MPB. MPB compositions of BNT-based binary, ternary, and complex solid solutions have been found, but the properties obtained by the formation of the MPB are still inferior to those of PZT.

In improving the properties of solid solutions, the characteristics of end members should be thoroughly evaluated to understand the composition-dependent responses of solid solutions. However, there have been only a few reports on end member compounds; most of the studies have attempted to improve important material properties. The basic properties of BNT are not completely understood; unclear aspects remain. Regarding the understanding of BNT, there are reports on A-site modifications (Sung et al., 2010; Zuo et al., 2008; Kimura et al., 2005; Nagata et al., 2005), but there are only a few on B-site doping.

In this work, the effects of B-site doping in BNT ceramics were studied using a donor Nb^{5+} and an acceptor Mn^{3+} for B-site Ti^{4+}. Two key properties, dielectric depolarization temperature (T_d) and piezoelectric coefficient (d_{33}), were evaluated in relation to microstructure and phase purity. Then, they were compared with donor and acceptor effects in PZT (Jaffe et al., 1971; Zhang et al., 2008; Erhart et al., 2007; Zhang & Whatmore, 2003; Randall et al., 1998; Park & Chadi, 1998; Gerson, 1960) to identify similarities and differences between BNT and PZT ceramics. In addition, the electrical conductivity of BNT was measured and discussed as a function of the oxygen partial pressure and temperature.

2. Experimental

Powders of at least 3 N purity Bi_2O_3 (99.9%), Na_2CO_3 (99.95%), TiO_2 (99.9+%), Nb_2O_5 (99.9%), and Mn_2O_3 (99.999%) were used to prepare $(Bi_{0.5}Na_{0.5})(Ti_{1-x}D_x)O_3$ (D = Nb, Mn) ceramics through solid-state processing. In handling the raw powders, hygroscopic Na_2CO_3 was thoroughly dried in a dry-oven until no change in weight occurred. Then it was quickly weighed in air; if not, it absorbs moisture from the air and increases in weight, thus comprising an incorrect composition from the beginning.

The compositions of the samples were controlled to be x = 0–1 mol % for Nb doping and 0–2 mol % for Mn doping. The powders of each composition were ball-milled using yttria-stabilized zirconia balls and anhydrous ethanol to keep the powders from water. After milling, powders were dried and calcined twice at 780 °C and 800 °C for 2 h in air to prevent the premature loss of any component during calcination and to have homogeneous powders after calcination. Next, they were mixed with 5 wt% polyvinyl alcohol (PVA) aqueous solution to 0.5 wt% and screened using a 150 μm sieve for pelletizing. Pellets of 10 or 18 mm in diameter and ~1 mm in thickness were made by uniaxial pressing at 150 MPa. Then, all the samples were sintered at 1150 °C for 2 h in air. During sintering, the heating rate was controlled to burn out the PVA at around 500 °C.

The Archimedes principle was applied to estimate the apparent densities of the sintered pellets, and these values were compared with the theoretical densities. X-ray diffraction (XRD) patterns generated by a diffractometer with Cu Kα radiation (λ = 1.541838 Å) at 40 kV and 30 mA were taken from the polished surfaces of sintered pellets and analyzed for identifying phases. Scanning electron microscopy (SEM) with energy dispersive spectroscopy was used to detect any secondary phase formation as well as to examine grain morphology that occurred during sintering.

To measure their properties, the samples were polished on both sides down to 0.5 mm in thickness using #400, 800, and 1200 emery papers. They were painted with Ag paste and then cured at 650 °C for 0.5 h in air. Poling of samples was carried out at a dc field of 40

kV/cm for 0.5 h in silicone oil at room temperature using a high voltage supply (Keithley 248), and leakage current was monitored using an electrometer (Keithley 6514).
After aging for 24 h, the room temperature d_{33} was measured using a piezo d_{33} meter (ZJ-6B, IACAS) at 0.25 N and 110 Hz. The dielectric constant (ε) and the loss tangent (tan δ) at various frequencies on heating and cooling were measured using an impedance analyzer (HP4192A). The planar electromechanical coupling factors (k_p) and mechanical quality factors (Q_m) of the samples were calculated from the resonance frequencies (f_r), the antiresonance frequencies (f_a), the resonant impedances (Z_r), and the capacitances (C_p) at 1 kHz, measured using an impedance gain phase analyzer (HP4194A). Electric field dependent polarization (*P-E)* hysteresis loops were measured using a Sawyer-Tower circuit to measure remnant polarization (P_r) and coercive field (E_c).

3. Results

The weight losses of the samples that occurred during sintering were consistently less than 1 wt%. This was thought to primarily result from the evaporation of moisture, the PVA, and possibly some of the Bi and Na oxides. The relative apparent densities of the pellets after sintering were all above ~95% of the theoretical density, indicating that the samples were consistently prepared under the processing conditions used in this work.

3.1 Phase and microstructure

Figure 1 shows the XRD patterns of $(Bi_{0.5}Na_{0.5})(Ti_{1-x}D_x)O_3$ ceramics with D = Nb or Mn substituting for Ti in BNT. In the case of a donor Nb doping, no secondary peak was observed up to 1 mol %, and all the peaks were indexed to be rhombohedral. In the case of an acceptor Mn doping, XRD patterns were also phase-pure up to 2 mol %.

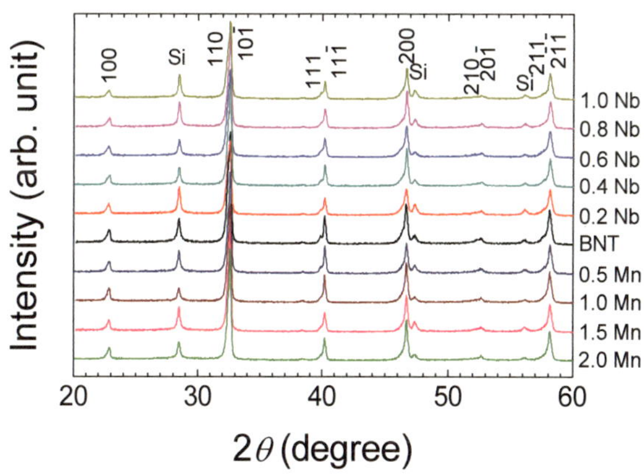

Fig. 1. XRD patterns of $(Bi_{0.5}Na_{0.5})(Ti_{1-x}D_x)O_3$ ceramics (D = Nb, Mn) sintered at 1150 °C for 2 h in air. Si peaks are of 5 N silicone powder used as an internal standard for calibration

The phase purity of the sintered $(Bi_{0.5}Na_{0.5})(Ti_{1-x}D_x)O_3$ with D = Nb or Mn pellets was confirmed by SEM images. No secondary phase was observed, as shown in Fig. 2. Regarding the grain morphology, the grain shapes in general were well developed in both

Nb and Mn doping, indicating that the sintering condition was consistent for both donor- and acceptor-doped samples.

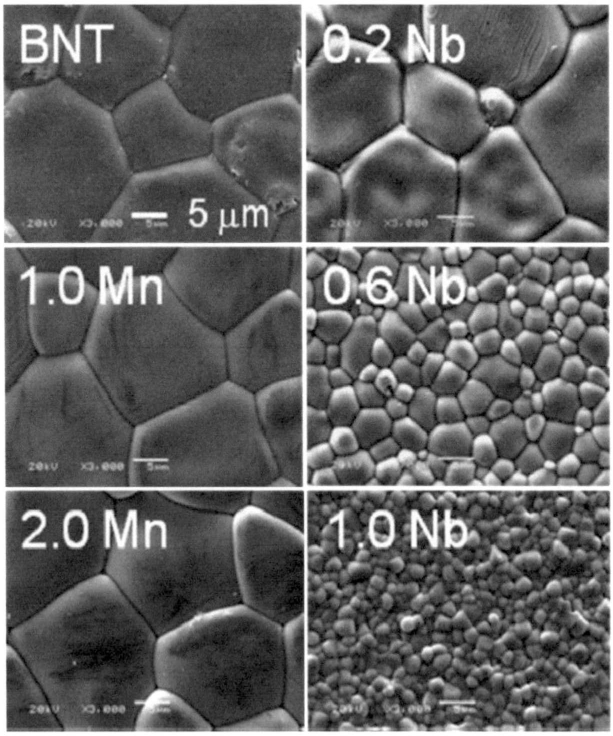

Fig. 2. SEM secondary electron images of the surfaces of $(Bi_{0.5}Na_{0.5})(Ti_{1-x}D_x)O_3$ pellets (D = Nb, Mn) sintered at 1150 °C for 2 h in air

An obvious difference observed among the images is the grain size relative to doping; the grain size decreased with Nb doping and increased with Mn doping, depending on the amount of Nb and Mn doping. The decrease in grain size in the case of Nb donor doping can be explained by A-site cation vacancies created by B-site donor doping required to maintain charge neutrality in the lattice. These cation vacancies exist along grain boundaries rather than inside grains, which is thermodynamically more stable. Grain boundaries would be pinned by these defects, inhibiting grain growth and resulting in relatively small grains in the case of donor doping (Yi et al., 2002; Lewis et al., 1985; Lucuta et al., 1985). Mn acceptor doping, on the other hand, induces oxygen vacancies instead of A-site vacancies required to maintain charge neutrality in the lattice. It appears that grain growth was not inhibited by Mn doping inducing oxygen vacancies, as shown in Fig. 2. These different roles of donor and acceptor that were observed in the microstructure affected the properties of BNT in different ways.

3.2 Piezoelectric properties

For various piezoelectric applications, d_{33}, k_p, and Q_m are key properties that need to be evaluated. As shown in Fig. 3, d_{33} gradually increased with Nb donor doping and gradually decreased with Mn acceptor doping. As mentioned above, B-site donor doping induces A-site vacancy to maintain charge neutrality in the lattice. This A-site vacancy facilitates domain walls motion. Consequently, domains alignment during poling improves, resulting in a higher d_{33}.

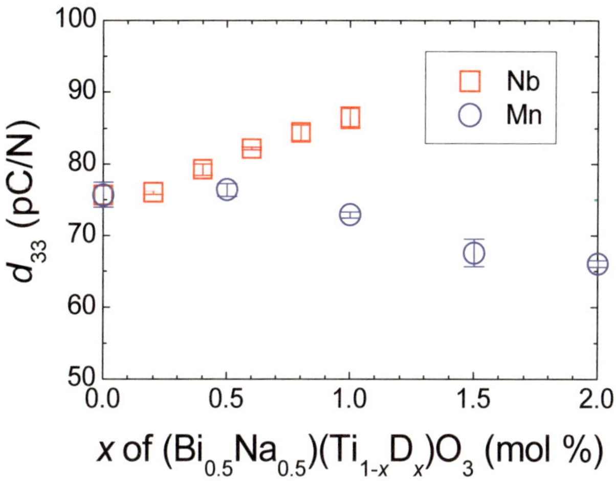

Fig. 3. Piezoelectric coefficient (d_{33}) of $(Bi_{0.5}Na_{0.5})(Ti_{1-x}D_x)O_3$ ceramics (D = Nb, Mn)

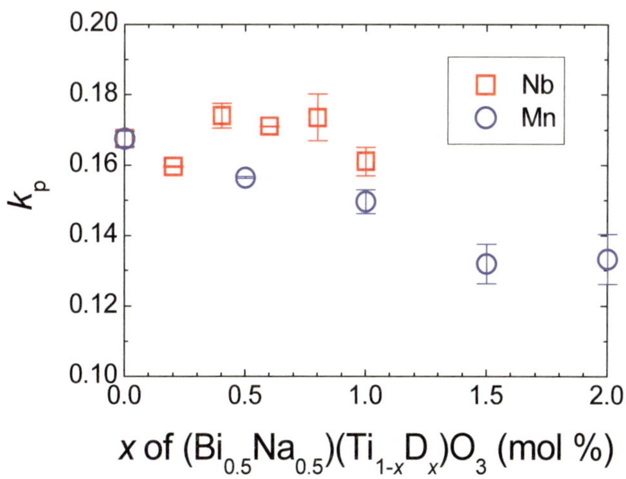

Fig. 4. Electromechanical coupling factor (k_p) of $(Bi_{0.5}Na_{0.5})(Ti_{1-x}D_x)O_3$ ceramics (D = Nb, Mn)

B-site acceptor doping, on the other hand, reduces A-site vacancy or induces oxygen vacancy to maintain charge neutrality in the lattice. Acceptor doping produces the results that are opposite to those produced by donor doping, as shown in Fig. 3. Oxygen vacancy is relatively mobile; this pins domain walls (Zhang & Whatmore, 2003; Park & Chadi, 1998), resulting in a lower d_{33} unless the vacancy becomes immobile by forming defect complexes with other defects (Hu et al., 2008; Zhang et al., 2008; Erhart et al. 2007). The increase from donor doping and the decrease from acceptor doping in the d_{33} of BNT ceramics are generally consistent with the variations from doping in the d_{33} of PZT-based ceramics (Jaffe et al., 1971). Regarding the microstructure, d_{33} appears to be inversely related to grain size.

The trend observed in k_p, as shown in Fig. 4, was similar to that observed in d_{33}. A slight increase from Nb donor doping and a gradual decrease from Mn acceptor doping occurred; this is consistent with the typical outcomes of B-site donor and acceptor doping in PZT.

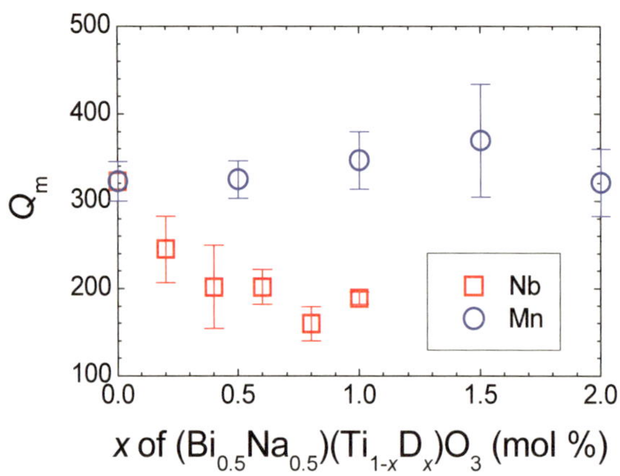

Fig. 5. Mechanical quality factor (Q_m) of $(Bi_{0.5}Na_{0.5})(Ti_{1-x}D_x)O_3$ ceramics (D = Nb, Mn)

Q_m, on the contrary, decreased with Nb donor doping and increased with Mn acceptor doping, as shown in Fig. 5. Regarding mechanical loss, it is expected that the grain boundary would cause more loss than the grain itself. As grain size increases, grain boundary area decreases, yielding a higher Q_m. This result can be coupled with the variation in grain size according to doping, as shown in Fig. 2. Q_m became smaller as grain size became smaller with Nb donor doping, but it became slightly larger as grain size became slightly larger with Mn acceptor doping.

3.3 Dielectric properties

For measuring dielectric properties according to changes in doping, the samples were pre-poled to expose T_d, as indicated by the arrow marks in Fig. 6. With Nb donor doping both room temperature ε and tan δ increased while with Mn acceptor doping, ε did not change significantly and tan δ decreased, as shown in Fig. 6. These dielectric results of BNT with B-site donor and acceptor doping are similar to those of PZT with doping, although the two are dielectrically different materials.

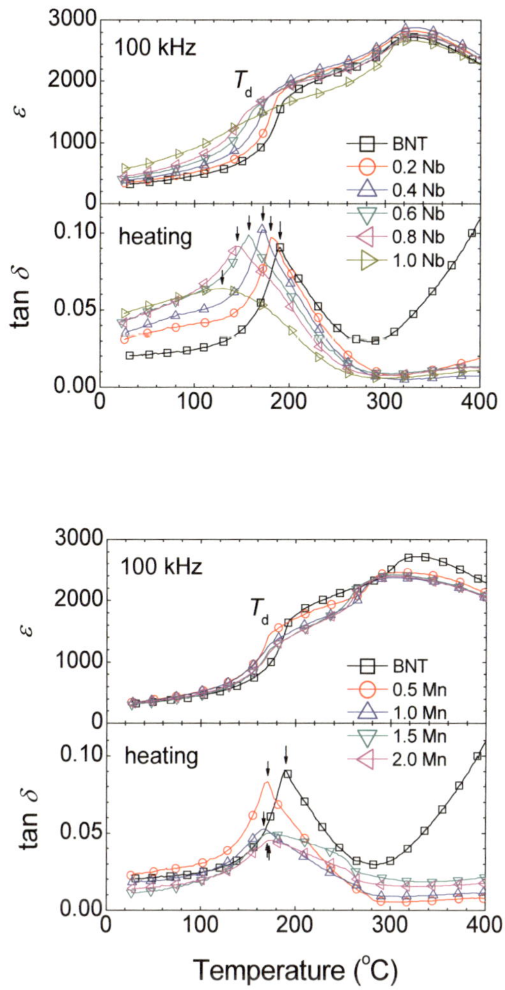

Fig. 6. Dielectric constant (ε) and loss tangent (tan δ) of $(Bi_{0.5}Na_{0.5})(Ti_{1-x}D_x)O_3$ ceramics (D = Nb, Mn). Arrow marks indicate the depolarization temperature (T_d)

Two transitions appeared; the temperature maximum (T_m) at ε maximum (ε_m) and T_d. In the case of Nb donor doping, T_m did not change significantly and T_d decreased gradually, as shown in Fig. 6. In the case of Mn acceptor doping, both T_{max} and T_d decreased to 0.5 mol % then did not change significantly with further Mn doping, as shown in Fig. 6. The variations in T_d of BNT ceramics with donor and acceptor doping are shown in Fig. 7.

BNT is dielectrically different from PZT. As a relaxor exhibiting frequency dependence and a broad transition, BNT is dielectrically lossy because of the to space-charge contribution (Hong & Park, 1996). A polymorphic phase transition occurs as lattice symmetry changes from rhombohedral to tetragonal, then to cubic symmetry. In addition, BNT has a unique transition, T_d, unlike PZT. For practical applications, this T_d of ~200 °C is considered to be a critical drawback of BNT because its temperature stability is lower than that of PZT.

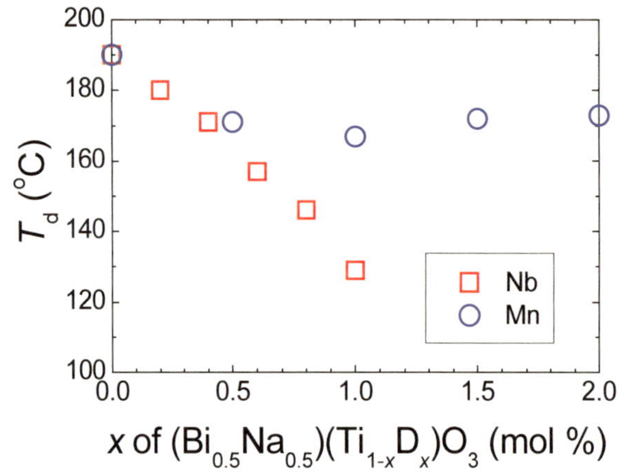

Fig. 7. Depolarization temperature (T_d) of $(Bi_{0.5}Na_{0.5})(Ti_{1-x}D_x)O_3$ ceramics (D = Nb, Mn)

3.4 Electrical properties

In oder to determine the electrical conductivity of BNT ceramics, the resistivity of the sample was measured as a function of temperature (T) and oxygen partial pressure (pO_2), then converted into electrical conductivity (σ).

The sample was attached to Pt wires (ϕ = 0.2 mm) according to the four probe method to eliminate the non-ohmic contact. The low oxygen pressure was established using an O_2/Ar mixture. Figure 8 shows σ of BNT ceramics in the range of T = 700–900 °C and $pO_2 = 10^0$–10^{-5} atm.

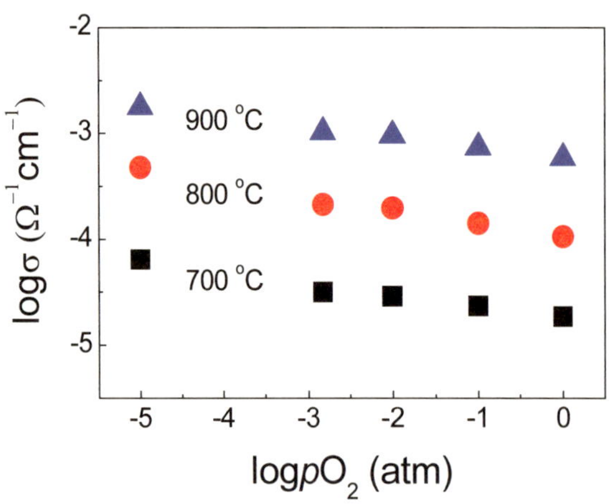

Fig. 8. Electrical conductivity (σ) of BNT ceramics as a function of oxygen partial pressure (pO_2) at 700, 800, and 900 °C

In the experimental results, the $d\log\sigma/d\log pO_2$ slopes are clearly negative in the temperature range tested. Therefore, the conduction behavior of BNT can be considered as *n*-type in the elevated temperature range.

3.5 Other properties

In addition to the above-mentioned piezoelectric and dielectric properties, other properties, such as ferroelectric and electrical properties, must be considered in relation to practical applications. For instance, E_c, P_r, and leakage current density (J_{leak}) are important properties that need to be characterized. Table 1 summarizes certain properties measured with various B-site donors and acceptors.

As shown in Table 1, with donor doping, E_c decreased and P_r increased. Similar results were obtained regardless of the donor material (Nb^{5+}, Ta^{5+}, or W^{6+}). With acceptor doping, on the other hand, E_c increased and P_r decreased. Overall, there were consistent trends in the case of donors, but the results with acceptors were not as consistent as those with donors.

Regarding J_{leak} at room temperature, there were some difficulties in taking the measurement; J_{leak} was very low at zero field and was only ~10^{-8} A/cm² at a dc poling field of 40 kV/cm. There were no consistent and reliable data values in the literature to compare because of differences among samples that were caused by different sample preparation methods.

	BNT	donor			acceptor	
	Ti^{4+}	Nb^{5+}	*Ta^{5+}	†W^{6+}	Mn^{3+}	Sc^{3+}
ionic radius (Å)	0.605	0.64	0.64	0.60	0.645	0.745
phase purity (mol %)	yes	1.0	0.8	0.4	2.0	0.8
grain size (μm)	~20	~2	~5	~5	>20	~15
d_{33} (pC/N)	74	87	84	84	66	77
k_p	0.17	0.17	-	0.16	0.13	0.16
Q_m	320	160	202	180	369	269
ε	324	↑	-	↑	-	-
tan δ	0.02	↑	-	-	↓	↑
T_d (°C)	190	129	-	-	167	-
E_c @ 60 Hz (kV/cm)	*~41	*~24	~18	~20	-	-
P_r @ 60 Hz (μC/cm²)	*~35	-	-	~40	-	-
J_{leak} @ 40 kV/cm (A/cm²)	~10^{-8}	-	-	-	-	-

Table 1. Effects of various B-site donors and acceptors in $(Bi_{0.5}Na_{0.5})(Ti_{1-x}D_x)O_3$ ceramics. The ionic radius of Ti^{4+} is given for BNT in comparison with donors and acceptors at coordination number 6 (Shannon, 1976). The values for phase purity are not the solubility limits of doping but merely indicate the phase purity confirmed experimentally up to that point. The variation in the values for BNT is thought to be mostly due to differences in sample preparation. *(Yeo et al., 2009) and †(Cho et al., 2008)

4. Discussion

PZT has been studied for decades and its properties have been categorized in detail, while only a few reports and studies are available for BNT. The effects of donor and acceptor doping on both BNT and PZT ceramics are summarized in Table 2. The trends related to donor and acceptor doping observed in BNT ceramics are similar to those observed in PZT, but there are also some differences.

For ABO_3 perovskite, as mentioned above, B-site donor doping induces A-site vacancies while acceptor doping reduces A-site vacancies and induces oxygen vacancies to maintain charge neutrality in the lattice. A-site vacancies facilitate domain walls motion (Gerson, 1960), which explains a high d_{33} and dielectric loss (Jaffe et al, 1971). Oxygen vacancies, on the other hand, are mobile and pin domain walls; this restricts domain switching, resulting in a low d_{33} and dielectric loss. Oxygen vacancies, however, can be tied up with cation vacancies forming defect complexes that make them immobile (Hu et al., 2008; Zhang et al., 2008; Erhart et al. 2007; Zhang & Whatmore, 2003; Park & Chadi, 1998). If this occurs, domain walls motion is not hindered but grain growth is hindered by grain boundary pinning caused by defect complexes (Sung et al., 2010).

Coupling the piezoelectric values with the microstructural variations in Fig. 2, a higher d_{33} was obtained from smaller grains with Nb donor doping, and a lower d_{33} was obtained from larger grains with Mn acceptor doping. This implies that, in the case of Nb donor doping, the domain walls must have been unpinned, leading to a high d_{33}, while grain boundaries must have been pinned, causing small grains. The situation was reversed in the case of Mn acceptor doping. Domain walls must have been pinned, leading to a low d_{33}, while grain boundaries must have been unpinned, causing large grains. This indicates that oxygen vacancies, if they were formed by Mn acceptor doping must have been tied up by the formation of defect complexes.

	donor		acceptor	
	‡PZT	BNT	‡PZT	BNT
d_{33}	↑	↑	↓	↓
k_p	↑	-	↓	↓
Q_m	↓	↓	↑	↑
ε	↑	↑	↓	-
$\tan\delta$	↑	↑	↓	↓
T_c or T_d	-	↓	-	↓
E_c	↓	↓	↑	-
P_r	-	↑	-	-
J_{leak}	↓	-	↑	-

Table 2. Comparison of the effects of donor and acceptor doping in BNT ceramics (given in Table 1) with those in PZT ceramics ‡(Jaffe et al., 1971). For acceptor doping effects, Mn was counted, but Sc was not

Two key properties involved in the practical application of BNT-based ceramics are d_{33} and T_d. T_d especially is a limiting factor because the piezoelectric properties of BNT-based ceramics diminish above T_d. For improved temperature stability, therefore, it is important to have a higher T_d. However, d_{33} has exhibited an inverse relationship with T_d. In other words, as T_d increases, d_{33} decreases, and vice versa (Sung et al., 2010). This is similar to the relationship between d_{33} and Curie temperature (T_c) in $PbTiO_3$, as d_{33} increases, T_c decreases, and vice versa.

With acceptor doping, both d_{33} and T_d decreased then remained constant. With donor doping, on the other hand, these two properties exhibited an inverse relationship. This inverse relationship was observed in Bi-based titanate end members as well as their solid solutions, but it has not been clearly explained, partly because T_d is a unique transition in BNT-based ceramics that cannot be explained in terms of the PZT database. Sung et al. suggested lattice distortion as a common factor that affects the two properties inversely. If lattice distortion decreases, d_{33} increases through better poling by electric field, but T_d decreases through easier depoling by temperature.

For doping effects in general, all the explanations have been given in terms of A-site and oxygen-site vacancies. In terms of the ionic sizes, Mn^{3+}, Ti^{4+}, Nb^{5+}, Ta^{5+}, and W^{6+} (but not Sc^{3+}) are similar, as shown in the comparison with Ti^{4+} in Table 1. In cases of similar ionic size, there will be minimal change in the shape and size of the lattice. With a small amount of doping, therefore, it would be difficult to detect any variation in the XRD patterns. Indeed, as shown in Fig. 9, there was no apparent change in peak positions caused by Nb donor or Mn acceptor doping. It would be interesting to know if any variation in lattice parameters occurs in the case of a substantial amount of doping with a relatively large difference in ionic radii.

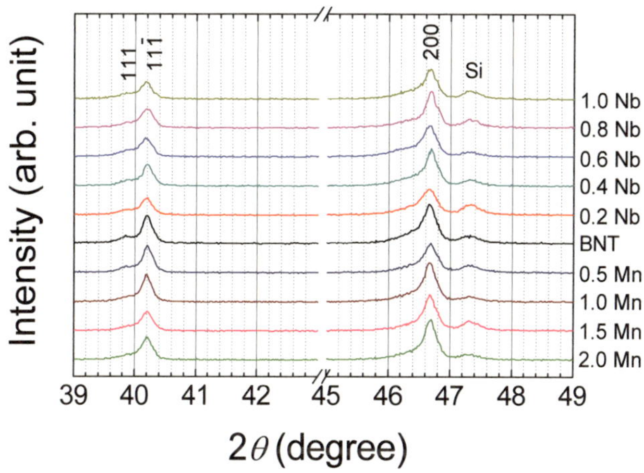

Fig. 9. Detailed XRD patterns of $(Bi_{0.5}Na_{0.5})(Ti_{1-x}D_x)O_3$ ceramics (D = Nb, Mn), from the patterns shown in Fig. 1

There are some data regarding donor doping effects in BNT-based ceramics, but little information is available about acceptor doping effects. This situation could exist because acceptors with 3+ valence can be on either the B-site Ti or the A-site vacancy, considering the volatile A-site Bi^{3+} and Na^{1+}. For dornors with 5+ or 6+, on the other hand, it is unlikely that they would occupy the A-site vacancy instead of the B-site Ti. Moreover, even if they did, the resulting outcomes would still be considered as donor doping effects. As shown in Table 2, among the data on donor effects, consistent trends can be identified. Grain size, Q_m, T_d, and E_c decreased, while d_{33} and P_r increased. These trends become reversed with acceptor doping.

Finally, it cannot be emphasized enough that hygroscopic Na_2CO_3 must be completely dried before and during weighing. Otherwise, the nominal compositions of the samples will be incorrect. If the ionic radius of an acceptor with 3+ valence is larger than that of Ti^{4+}, doping of an acceptor with 3+ valence on Na^{1+} vacancy, instead of the Ti^{4+} site, might occur. In addition, there is the possibility of an acceptor with 3+ valence substituting for A-site Bi^{3+} instead of for B-site Ti^{4+}. This appears to explain the inconsistent results and donor-like effects (see Table 1) that were observed in Sc^{3+}, which has a larger ionic radius than the others do as compared with that of Ti^{4+}. To confirm the acceptor effects in BNT, the next experiment should include the doping of acceptors with 2+ valence and small differences in ionic radii.

5. Summary

The effects of a B-site donor, Nb^{5+}, and an acceptor, Mn^{3+}, for Ti^{4+} in Pb-free $(Bi_{0.5}Na_{0.5})(Ti_{1-x}D_x)O_3$ (D = Nb or Mn) ceramics were studied in terms of microstructure and important properties. Regarding the microstructure, all the samples retained a perovskite structure with rhombohedral symmetry with both Nb donor and Mn acceptor doping. No secondary phase was seen up to x = 1.0 mol % of Nb-doped BNT ceramics and x = 2.0 mol % of Mn-doped BNT ceramics. The grain size, however, decreased with Nb doping and slightly increased with Mn doping. In addition to producing variation in microstructure, Nb and Mn yielded opposite piezoelectric and dielectric outcomes. Higher d_{33}, lower Q_m, and smaller grain size were obtained from Nb doping, while lower d_{33}, higher Q_m, and larger grain size were obtained from Mn doping. These results indicate a correlation between piezoelectric properties and grain size. Regarding the dielectric transition, T_d decreased with Nb doping, but it decreased slightly then held steady with Mn doping. Overall, the donor effects observed in BNT ceramics were similar to those observed in PZT ceramics, but acceptor effects need to be studied further in order to produce consistent data. At elevated temperatures, BNT exhibited n-type conduction.

6. Acknowledgments

This work was supported by the National Research Foundation of Korea Grant funded by the Korean Government (2009-0088570) and the Basic Science Research Program through the National Research Foundation of Korea (NRF) funded by the Ministry of Education, Science and Technology (2010-0025055). This research was also financially supported by the Ministry of Education, Science and Technology (MEST) and Korea Institute for Advancement of Technology (KIAT) through the Human Resource Training Project for Regional Innovation.

7. References

Cho, J. H.; Yeo, H. G.; Sung, Y. S.; Song, T. K.; & Kim, M. H. (2008). Dielectric and piezoelectric characteristics of W-doped lead-free $(Bi_{0.5}Na_{0.5})(Ti_{1-x}W_x)O_3$ ceramics. *New Physics: Sae Mulli* (The Korean Physical Society) 57, 409-416.

Dai, Y. J.; Zhang, X. W.; Chen, K. P.; (2009). Morphotropic phase boundary and electrical properties of $K_{1-x}Na_xNbO_3$ lead-free ceramics. *Appl. Phys. Lett.*, 94, 042905.

Egerton, L. & Dillon, D. M. (1959). Piezoelectric and dielectric properties of ceramics in the system potassium-sodium niobate. *J. Am. Ceram. Soc.*, 42, 438-442.

Erhart, P.; Eichel, R.; Traskelin, P. & Albe, K. (2007). Association of oxygen vacancies with impurity metal ions in lead titanate. *Phys. Rev. B*, 76, 174116.

Gerson, R. (1960). Variation in ferroelectric characteristics of lead zirconate titanate ceramics due to minor chemical modifications. *J. Appl. Phys.*, 31, 188-194.

Haertling, G. H. (1967). Properties of hot-pressed ferroelectric alkali niobate ceramics. *J. Am. Ceram. Soc.*, 50, 329-330.

Hong, K. S. & Park, S. E. 1996. Phase relations in the system of $(Na_{1/2}Bi_{1/2})TiO_3$-$PbTiO_3$. II. Dielectric property. *J. Appl. Phys.*, 79, 388-392.

Hu, G. D.; Fan, S. H.; Yang, C. H. & Wu, W. B. (2008). Low leakage current and enhanced ferroelectric properties of Ti and Zn codoped $BiFeO_3$ thin film. *Appl. Phys. Lett.*, 92, 192905.

Isupov, V. A. (2005). Ferroelectric $Na_{0.5}Bi_{0.5}TiO_3$ and $K_{0.5}Bi_{0.5}TiO_3$ perovskites and their solid solutions. *Ferroelectrics*, 315, 123-147.

Jaeger, R. E. & Egerton, L. (1962). Hot pressing of potassium-sodium niobates. *J. Am. Ceram. Soc.*, 45, 209-213.

Jaffe, B.; Cook, Jr. W. R.; & Jaffe, H. (1971). *Piezoelectric Ceramics*, Academic Press, New York. ISBN 0-12-379550-8.

Kimura, T.; Fukuchi, E. & Tani, T. (2005). Fabrication of textured bismuth sodium titanate using excess bismuth oxide. *Jpn. J. Appl. Phys.*, 44, 8055-8061.

Kreisel, J.; Glazer, A. M.; Jones, G.; Thomas, P.; Abello, L. & Lucazeau, G. (2000). An x-ray diffraction and Raman spectroscopy investigation of A-site substituted perovskite compounds: the $(Na_{1-x}K_x)_{0.5}Bi_{0.5}TiO_3$ $(0 \le x \le 1)$ solid solution. *J. Phys. Condens. Mat.*, 12, 3267-3280.

Lewis, G. V.; Catlow, R. A. & Casselton, R. E. W. (1985). PTCT effect in $BaTiO_3$. *J. Am. Ceram. Soc.*, 68, 555-558.

Li, Y.; Chen, W.; Zhou, J.; Xu, Q.; Sun, H. & Liao, M. (2005). Dielectric and ferroelectric properties of lead-free $Na_{0.5}Bi_{0.5}TiO_3$-$K_{0.5}Bi_{0.5}TiO_3$ ferroelectric ceramics. *Ceram. Int.*, 31, 139-142.

Lin, D.; Xiao, D.; Zhu, J. & Yu, P. (2006). Piezoelectric and ferroelectric properties of $[Bi_{0.5}(Na_{1-y}K_xLi_y)_{0.5}]TiO_3$ lead-free piezoelectric ceramics. *Appl. Phys. Lett.*, 88, 062901.

Lucuta, P. GR.; Constantinescu, FL. & Barb, D. (1985). Structural dependence on sintering temperature of lead zirconate-titanate solid solutions. *J. Am. Ceram. Soc.*, 68, 533-537.

Nagata, H.; Shinya, T.; Hiruma, Y.; Takenaka, T.; Sakaguchi, I. & Haneda, H. (2005). Piezoelectric properties of bismuth sodium titanate ceramics. *Ceram. Trans.* 167, 213-221. ISBN 978-1-57498-188-9.

Park, C. H. & Chadi, D. J. (1998). Microscopic stud of oxygen-vacancy defects in ferroelectric perovskites. *Phys. Rev. B*, 57, R13961-R13964.

Park, S. E.; Chung, S. J.; Kim, I. T. (1996). Ferroic phase transitions in $(Na_{1/2}Bi_{1/2})TiO_3$ crystals. *J. Am. Ceram. Soc.*, 79, 1290-1296.

Pronin, I. P. ; Syrnikov, P. P.; Isupov, V. A.; Egorov, V. M. & Zaitseva, N. V. (1980). Peculiarities of phase transitions in sodium-bismuth titanate. *Ferroelectrics*, 25, 395-397.

Randall, C. A.; Kim, N.; Kucera, J.; Cao, W. & Shrout, T. R. (1998). Intrinsic and extrinsic size effects in fine-grained morphotropic-phase-boundary lead zirconate titanate ceramics. *J. Am. Ceram. Soc.*, 81, 677-688.

Saito, Y.; Taka, H.; Tani, T.; Nonoyama, T.; Takatori, K.; Homma, T.; Nagaya, T. & Nakamura M. (2004). Lead-free piezoceramics. *Nature*, 432, 84-87.

Shannon, R. D. 1976. Revised effective ionic radii and systematic studies of interatomic distances in halides and chalcogenides. *Acta Cryst.* A32, 751-767.

Sung, Y. S.; Kim, J. M.; Cho, J. H.; Song, T. K.; Kim, M. H.; Chong, H. H.; Park, T. G.; Do, D. & Kim, S. S. (2010). Effects of Na nonstoichiometry in $(Bi_{0.5}Na_{0.5+x})TiO_3$ ceramics. *Appl. Phys. Lett.*, 96, 022901.

Sung, Y. S.; Kim, J. M.; Cho, J. H.; Song, T. K.; Kim, M. H.; Park, T. G. (2010). Roles of lattice distortion in $(1\text{-}x)(Bi_{0.5}Na_{0.5})TiO_3\text{-}xBaTiO_3$ ceramics. *Appl. Phys. Lett.*, 96, 202901.

Takenaka, T.; Maruyama, K. & Sakata, K. (1991). $(Bi_{1/2}Na_{1/2})TiO_3\text{-}BaTiO_3$ system for lead-free piezoelectric ceramics. *Jpn. J. Appl. Phys.*, 30, 2236-2239.

Tennery, V. J. & Hang, K. W. (1968). Thermal and X-ray diffraction studies of the $NaNbO_3\text{-}KNbO_3$ system. *J. Appl. Phys.*, 39, 4749-4753.

Tu, C. S.; Siny, I. G. & Schmidt, V. H. (1994). Sequence of dielectric anomalies and high-temperature relaxation behavior in $Na_{1/2}Bi_{1/2}TiO_3$. *Phys. Rev. B*, 49, 11550-11559.

Wang, Y.; Damjanovic, D.; Klein, N.; Hollenstein, E. & Setter, N. (2007). Compositional inhomogeneity in Li- and Ta-modified $(K,Na)NbO_3$ ceramics. *J. Am. Ceram. Soc.*, 90, 3485-3489.

Yeo, H. G.; Sung, Y. S.; Song, T. K.; Cho, J. H.; Kim, M. H. & Park, T. G. (2009). Donor doping effects on the ferroelectric and the piezoelectric properties of Pb-free $(Bi_{0.5}Na_{0.5})TiO_3$ ceramics. *J. Korean. Phys. Soc.*, 54, 896-900.

Yi, J. Y.; Lee, J. K. & Hong, K. S. (2002). Dependence of the microstructure and the electrical properties of lanthanum-substituted $(Na_{1/2}Bi_{1/2})TiO_3$ on cation vacancies. *J. Am. Ceram. Soc.*, 85, 3004-3010.

Yu, H. & Ye, Z. G. (2008). Dielectric, ferroelectric, and piezoelectric properties of the lead-free $(1\text{-}x)(Na_{0.5}Bi_{0.5})TiO_3\text{-}xBiAlO_3$ solid solution. *Appl. Phys. Lett.*, 93, 112902.

Zhou, Z. H.; Xue, J. M.; Li, W. Z.; Wang, J.; Zhu, H. & Miao, J. M. (2004). Ferroelectric and electrical behavior of $(Na_{0.5}Bi_{0.5})TiO_3$ thin films. *Appl. Phys. Lett.*, 85, 804-806.

Zvirgzds, J. A. ; Kapostins, P. P.; Zvirgzde, J. V. & Kruzina, T. V. (1982). X-ray study of phase transitions in ferroelectric $Na_{0.5}Bi_{0.5}TiO_3$. *Ferroelectrics*, 40, 75-77.

Zuo, R.; Su, S.; Wu, Y.; Fu, J.; Wang, M. & Li, L. (2008). Influence of A-site nonstoichiometry on sintering, microstructure and electrical properties of $(Bi_{0.5}Na_{0.5})TiO_3$ ceramics. *Mater. Chem. Phys.*, 110, 311-315.

Zhang, Z.; Wu, P.; Lu, L. & Shu, C. (2008). Defect and electronic structures of acceptor substituted lead titanate. *Appl. Phys. Lett.*, 92, 112909.

Zhang Q. & Whatmore, R. W. (2003). Improved ferroelectric and pyroelectric properties in Mn-doped lead zirconate titanate thin films. *J. Appl. Phys.*, 94, 5228-5233.

14

Enhanced Dielectric and Ferroelectric Properties of Donor (W^{6+}, Eu^{3+}) Substituted SBT Ferroelectric Ceramics

Indrani Coondoo[1] and Neeraj Panwar[2]
[1]National Physical Laboratory, Dr. K. S. Krishnan Marg,New Delhi-110012,
[2]Department of Physics, University of Puerto Rico, San Juan, PR-00931
[1]India
[2]USA

1. Introduction

In recent years, there has been a tremendous interest in ferroelectric materials from perspective of their potential applications in electronic devices such as non-volatile random access memories (NvRAMs), pyroelectric infrared detectors and optical switches [1-4]. On account of their read-write speed, non volatility, low operating power and radiation hardness, NvFRAMs are promising candidates for substituting silicon based electrically erasable programmable read-only memories (EEPROMs) and flash EEPROMs. The materials for memory applications are required to possess the following properties: a large remanent polarization (P_r), a low coercive field (E_c) and sufficient fatigue endurance against repetitive polarization switching etc. [5]. Among ferroelectrics, lead zirconate titanate $PbZr_{1-x}Ti_xO_3$ (PZT) has been investigated extensively and has been found to be the most promising material for NvFRAM applications. However, apart from Pb toxicity, PZT suffers from serious degrading problems such as fatigue, ageing and leakage current that hinder the usability of this material in devices [6-10]. Among them fatigue is an important reliability issue for NvRAM devices which is defined as a decrease in switchable polarization with increasing number of polarization reversals [11].

During the search for an alternate ferroelectric material for these applications, it was found that bismuth oxide layered structures (BLSFs) originally synthesized by Aurivillius [12] are the most suitable candidates due to their relatively high Curie temperature (T_c), low dielectric dissipation and anisotropic nature originating from their layered structure [13-15]. The Aurivillius family of layered bismuth-oxides encompasses many ferroelectric materials, a fact which was known since the pioneering work of Smolenskii [16] and Subbarao [17], more than forty years ago. $SrBi_2Ta_2O_9$ (SBT), which is an n=2 member of the Aurivillius family of layered compounds, proved to be a versatile material for multifarious applications. Since Araujo et. al. [18] reported the fatigue-free behavior of $SrBi_2Ta_2O_9$ (SBT), it has gained importance among Pb-free ferroelectric memory materials [19-28]. Araujo recognized the inherent advantage of BLSFs over other ferroelectric materials on account of the fact that the former have intermediate bismuth oxide layers between the ferroelectric units [29].

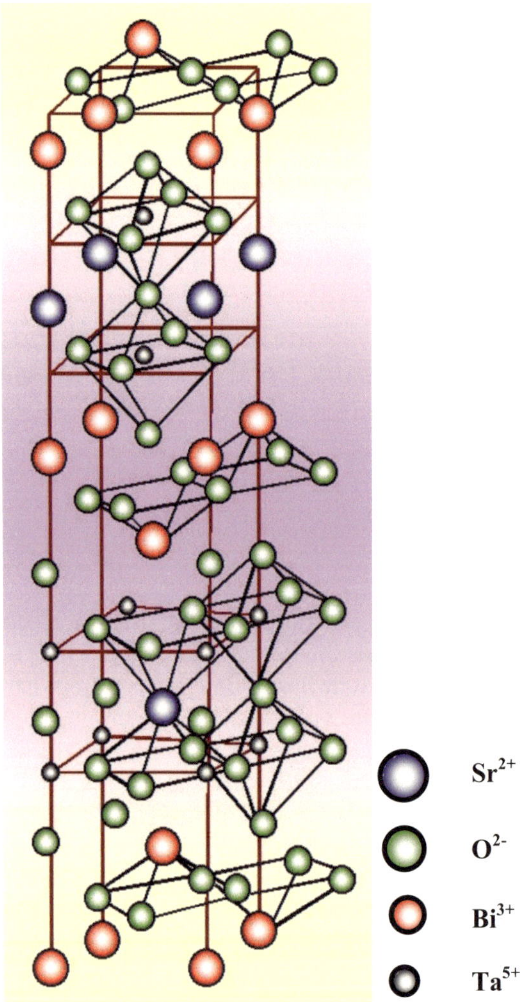

Fig. 1. The crystal structure of $SrBi_2Ta_2O_9$

Fig. 1 depicts the crystal structure of $SrBi_2Ta_2O_9$ consisting of $(Bi_2O_2)^{2+}$ layers and perovskite-type $(SrTa_2O_7)^{2-}$ units with double TaO_6 octahedral layers [30]. Besides improvement in the fatigue characteristics, a longer polarization retention time, lesser tendency to take an imprint [30,31], high dielectric constant and low switching fields have also been observed in SBT [32], which are better than those of PZT [18, 33-34]. The high fatigue endurance to repetitive switching of polarization ($\approx 10^{12}$ switching cycles) is believed to originate from the anisotropic nature of SBT. It has been argued that the $(Bi_2O_2)^{2+}$ layers control the electronic response [35-36] and perform the primary function in preventing degradation of remanent charge [35,37]. The ferroelectricity arises mainly in the perovskite blocks [15, 35, 38-40]. The dielectric constant peak corresponding to a ferro-paraelectric transition has been reported in the range of 300–320 °C [35, 38].Though SBT shows improved ferroelectric properties [41-42], its piezoelectric properties have not been investigated in detail. Kholkin et. al. [43] reported on the electromechanical properties of SBT but the maximum piezoelectric coefficient that was measured was quite small.

In this chapter effect of tungsten and europium substitution on dielectric, conductivity, piezoelectric and ferroelectric properties of SBT ceramics have been carried out and the results are presented.

2. Experimental techniques

The polycrystalline $SrBi_2Ta_2O_9$ (SBT) ferroelectric ceramics doped with W (tungsten) and Eu (europium) having the following compositions were prepared by the solid state reaction technique:

$$SrBi_2(Ta_{1-x}W_x)_2O_9;\ x = 0.0, 0.025, 0.050, 0.075, 0.10, 0.20$$

$$Sr_{1-x}Eu_xBi_2Ta_2O_9;\ \ x = 0.0, 0.025, 0.050, 0.10, 0.20$$

The starting chemicals used were strontium carbonate ($SrCO_3$), bismuth oxide (Bi_2O_3), tantalum oxide (Ta_2O_5), tungsten oxide (WO_3) and europium oxide (Eu_2O_3). The chemicals were weighed in stoichiometric proportions as mentioned above. The weighed powders were mixed and thoroughly ground and passed through sieve of appropriate mesh size. The ground powder was calcined at 900 °C in air for two hours. Thereafter, the calcined powder was pressed into disk shaped pellets. The pellets were sintered in air at 1200 °C. The sintered pellets were polished to a thickness of nearly 1mm. High temperature conductive silver paste was used for electroding the parallel surfaces. After applying the silver paste, the pellets were cured at 550°C for half an hour before electrical characterization. X-ray diffractograms of all the samples were recorded for the structural analysis using Bruker X-ray diffractometer with CuK_α radiation of wavelength 1.54439 Å in the range from $10° \leq 2\theta \leq 70°$ at a scanning rate of 0.05° / second. The SEM micrographs of the fractured surfaces of the samples were obtained using the Cambridge Stereo Scan 360 scanning electron microscope. A Solartron 1260 Gain-phase impedance analyzer was used for measuring the dielectric constant and dielectric loss in the present work. The d.c conductivity was measured using Keithley's 6517A electrometer. Hysteresis measurements were done at room temperature using an automatic PE loop tracer based on Sawyer-Tower circuit at switching frequency of 50 Hz. The piezoelectric coefecient, d_{33} was measured using a Berlincourt d_{33} meter.

3. Results and discussion

3.1 XRD analysis

The observed XRD patterns of the studied samples having different concentrations of tungsten and europium are shown in Fig. 2(a) and (b), respectively. The XRD patterns of the $SrBi_2(Ta_{1-x}W_x)_2O_9$ samples show the characteristic peaks of SBT. The peaks have been indexed with the help of a computer program - POWDIN [44] using the observed interplanar spacing d. It is observed that the single phase layered perovskite structure is maintained in the range $0.0 \leq x \leq 0.05$. An unidentified peak of very low intensity is observed in the composition with $x > 0.05$. In all the diffraction patterns, peaks shift slightly towards higher diffraction angle with increasing W concentration implying a decrease in lattice parameters. This can be understood from the fact that the ionic radius of W^{6+} (0.60Å) is smaller in comparison to Ta^{5+} (0.64Å). On the other hand no extra peaks are observed in the XRD patterns of $Sr_{1-x}Eu_xBi_2Ta_2O_9$ samples. It can therefore be concluded that the single phase layered perovskite structure is maintained in all the SEBT samples. The calculated lattice parameters for both the series are tabulated (Table 1).

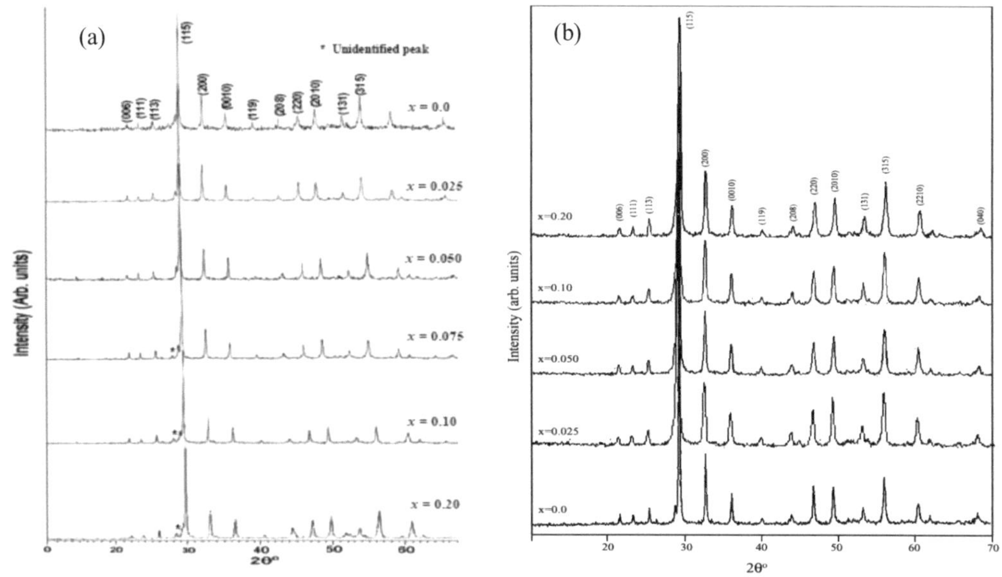

Fig. 2. XRD patterns of the (a) $SrBi_2(Ta_{1-x}W_x)_2O_9$ and (b) $Sr_{1-x}Eu_xBi_2Ta_2O_9$ samples

W Conc	*a* (Å)	*b* (Å)	*c* (Å)	Volume (Å³)	Eu Conc.	*a* (Å)	*b* (Å)	*c* (Å)	Volume (Å)³
0	5.5212	5.5139	24.9223	758.7182	**0**	5.5212	5.5139	25.0122	761.4550
0.025	5.5314	5.5202	25.1079	766.6555	**0.025**	5.5262	5.5187	25.0282	763.2960
0.05	5.5270	5.5199	25.0585	764.4969	**0.05**	5.5256	5.5158	24.9765	761.2364
0.075	5.5251	5.5045	25.0567	762.0472	**0.1**	5.5209	5.5101	24.9555	759.1641
0.1	5.5242	5.5060	25.085	762.9915	**0.2**	5.5167	5.5042	24.9349	757.1487
0.2	5.5233	5.4939	25.0861	761.2241					

Table 1. Lattice parameters and unit cell volume for $SrBi_2(Ta_{1-x}W_x)_2O_9$ and $Sr_{1-x}Eu_xBi_2Ta_2O_9$ samples

3.2 Morphological studies

Figs. 3 and 4 show the surface morphology of the SBTW and SEBT samples respectively. A systematic study of the micrographs reveals porous and loosely packed grains for the pure sample. From the figures one can notice an increase in the average grain size, homogeneity and relatively packed microstructure with W and Eu substitution. Randomly oriented and anisotropic plate-like grains are observed in all the W substituted samples. The average grain size in the SEBT sample with x = 0.025 is ~ 5-6 μm while that in the sample with x = 0.20 the size increases to ~ 7-8 μm.

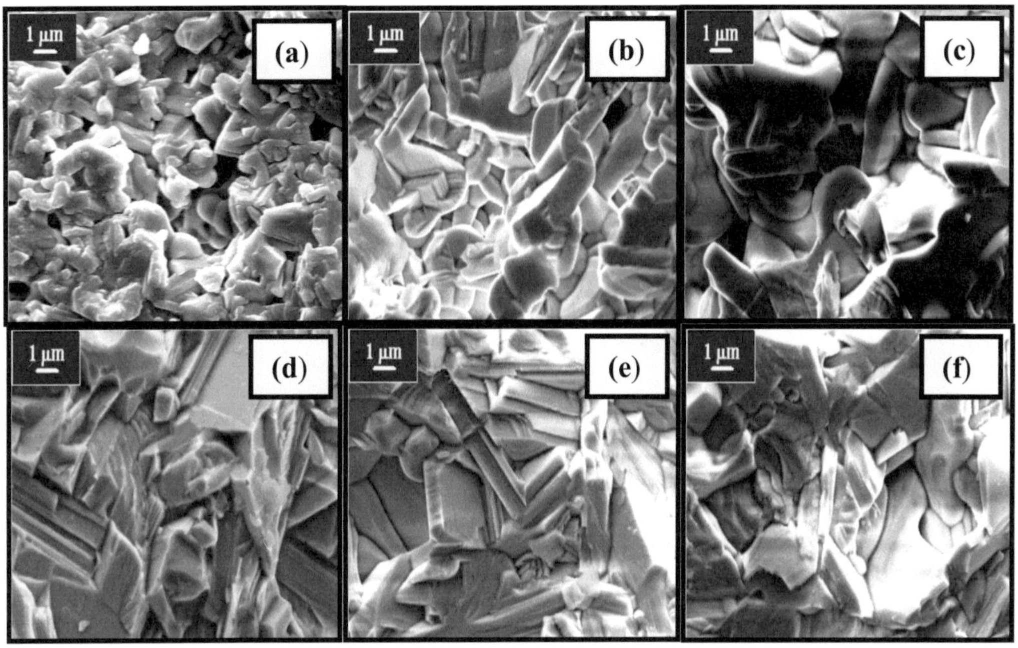

Fig. 3. SEM images with (a) x = 0.0, (b) x = 0.025, (c) x = 0.05, (d) x = 0.075, (e) x = 0.10 and (f) x = 0.20 of $SrBi_2(Ta_{1-x}W_x)_2O_9$ samples

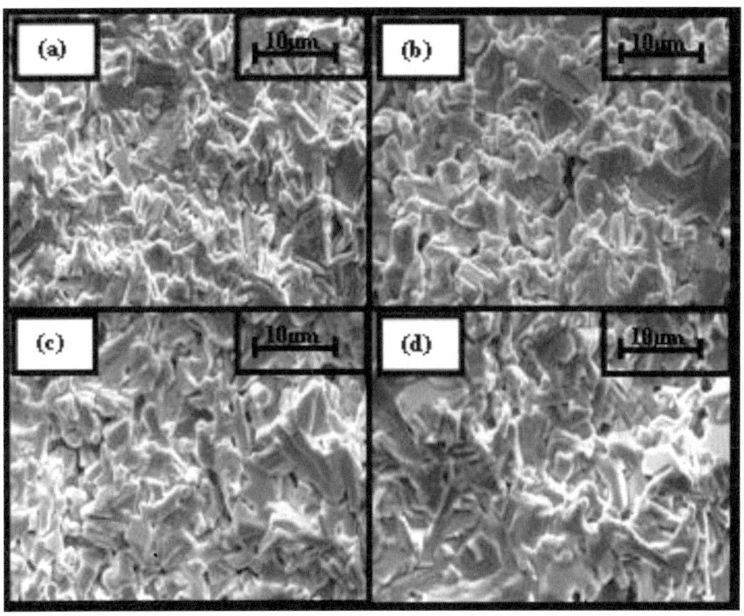

Fig. 4. SEM images with (a) x = 0.0, (b) x = 0.025, (c) x = 0.05, (d) x = 0.10 of $Sr_{1-x}Eu_xBi_2Ta_2O_9$ samples

3.3 Dielectric studies

It is well known that the dielectric permittivity and loss of ferroelectric materials in most cases depend upon the composition, grain size, secondary phases, *etc* [45]. Since bismuth layered perovskites are anisotropic in nature, its dielectric behavior is often influenced by the crystal structure [45]. All ferroelectric materials are characterized by a transition temperature known as Curie temperature (T_c) at which the dielectric constant is maximum. At temperatures $T>T_c$, the crystal does not exhibit ferroelectricity while for $T<T_c$ it is ferroelectric [46].

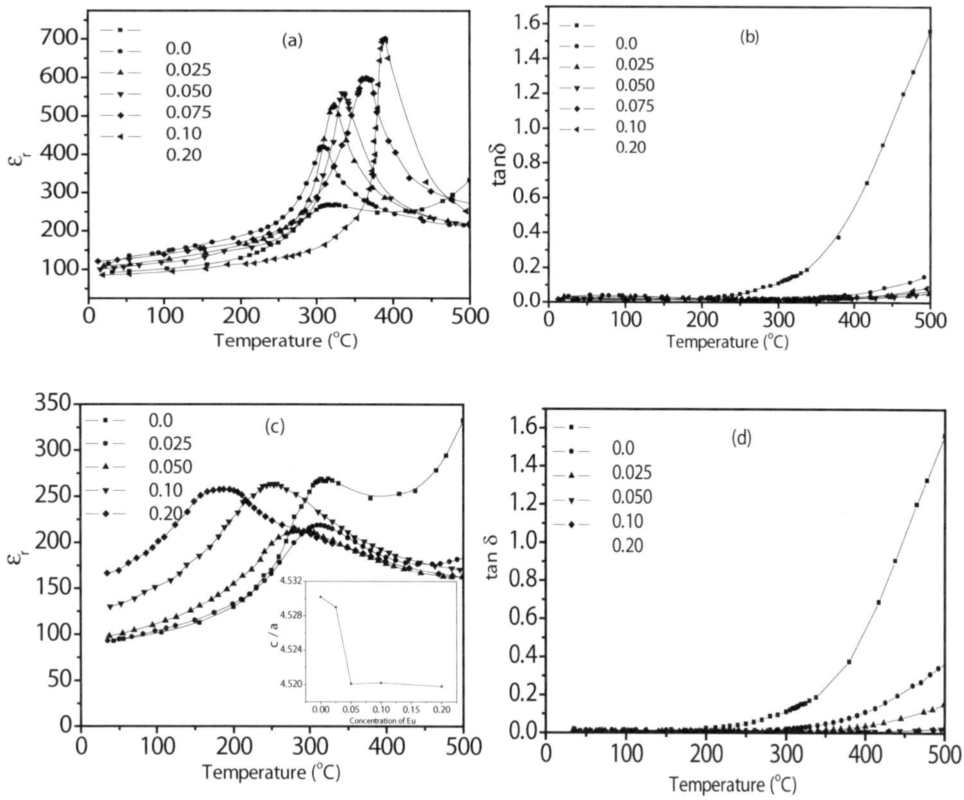

Fig. 5. (a) Temperature dependence of dielectric permittivity and (b) loss tangent for SBTW samples. (c) Temperature dependence of dielectric permittivity and (d) loss tangent for SEBT samples. The inset in (c) shows variation of tetragonal strain (c/a) with Eu concentration

The phase transition temperature of a sample is deduced from its dielectric permittivity (ε_r) versus temperature curve. Fig. 5(a) shows the temperature dependence of ε_r at selected frequencies of 100 kHz in SBTW samples. All the doped samples exhibit sharp transition at their respective Curie temperatures (T_c). A shift in T_c to higher temperatures and a corresponding increase in the peak dielectric constant values with increasing concentration of tungsten are observed.

The observed variation in T_c & peak-ε_r with concentration is explained in the ensuing paragraph. Generally in isotropic perovskite ferroelectrics, doping at *B*-site (located inside an oxygen octahedron) with smaller ions results in the shift of the Curie point to a higher temperature, leading to a larger polarization due to the enlarged "rattling space" available for smaller *B*-site ions [37]. However, in the anisotropic layered-perovskites, the crystal structure may not change as freely as that in the isotropic perovskites with doping due to the structural constraint imposed by the $(Bi_2O_2)^{2+}$ interlayer [47-48]. On comparing the variation of in-plane lattice parameters *a* and *b* (Table 1) with tungsten concentration, we observe that with increasing tungsten concentration a decrease in the lattice parameters *a* and *b* is observed. It is this enhancement of ferroelectric structural distortion along with the introduction of cation vacancies at the *A*-site that lead to an eventual increase in T_c value [21, 25, 47, 49-50]. In Fig. 5 (a) it is observed that peak-ε_r increases with increasing tungsten concentration. The high T_c which is indicative of enhanced polarizability [21, 36, 51-52], explains the increase in peak ε_r. Moreover, since the valency of the substitutional cation (W^{6+}) is higher than the Ta^{5+}, the substitution creates cationic vacancies at Sr-site ($V_{Sr}^{''}$) to maintain electrical neutrality of the lattice structure [22, 47, 53]. In the tungsten doped samples, because of the constraint of maintaining overall charge neutrality of the structure, substitution of W^{6+} ions for Ta^{5+} in the structure result in the formation of cation vacancies at *A*-sites. For substitution of two W^{6+} ions, one *A*-site (Sr site) remains vacant. The process can be represented as:

$$Null = W_{Ta}^{\bullet} + \frac{1}{2} V_{Sr}^{''} \quad (1)$$

where $W_{Ta}^{\bullet}$ represents tungsten replacing tantalum site and $V_{Sr}^{''}$ denotes the Sr-vacant site. It has been reported that cation vacancies make the domain motion easier and increase the dielectric permittivity [53-54] and thus an increase in ε_r with increasing W content is observed. There is also a possibility that the microstructural development due to compositional deviation from the stoichiometry affect the dielectric properties. It is expected that the domain walls are quite free in their movement in larger grains than smaller sized grains, since grain boundaries contribute additional pinning points for the moving walls [55-56]. Increase in the grain size makes the domain wall motion easier which results in an increase in the dielectric permittivity [41]. Since the grain size increases with sintering W concentration (Fig. 3), an increase in peak - ε_r is observed.

Fig. 5 (b) shows the tangent loss (tanδ) as a function of temperature in W-doped samples measured at 100 kHz. It is observed that tungsten doping in SBT reduces dielectric loss significantly. The dissipation in ferroelectric materials occurs due to various causes such as domain wall relaxation, space charge accumulation at grain boundaries, dipolar losses, dc conductivity, *etc.* [45]. The presence of oxygen vacancies $V_o^{\bullet\bullet}$, which act as space charge and contribute to the electrical polarization can be related to the dielectric loss [50, 57-58]. The substitution of W^{6+} for Ta^{5+} in SBT results in the formation of cation vacancies which effectively reduces the concentration of oxygen vacancies which in turn significantly reduce the dielectric loss. Reduction in loss have been reported in other donor doped BLSFs also [25, 49].

Fig. 5(c) shows the temperature dependence of dielectric permittivity at 100 kHz for SEBT samples. The substitution of Eu in SBT lowers and broadens the ferro-paraelectric phase transition temperature. The broadened peaks, as observed in Fig. 5c, indicate that transition in all the samples is of diffuse type, an important characteristic of a disordered perovskite structure [59]. The broad peak implies that the ferroelectric - paraelectric phase transition does not occur at discrete temperature but over a temperature range [11]. The broadening or diffuseness of peak occurs mainly due to two mechanisms. One is due to the substitution disordering in the arrangement of cations at one or more crystallographic sites in the lattice structure leading to heterogenous domains [60]. Another possible explanation of broadened peak is due to the defect induced relaxation at high temperature [60-61].

Comprehensive structural analysis of SBT with the aid of neutron diffraction and Raman scattering have shown disorder in the distribution of the Sr ions and Bi ions [62-63]. Recent independent studies by Ismunandar *et. al.* [64] and Blake *et. al.* [62] have reported that there is a significant degree of Sr/Bi cation disorder in the $SrBi_2Nb_2O_9$ compounds. Thus, it is possible that cation disorder between Sr and Bi sites also occurs in the tantalum analogues [51, 66]. Macquart *et. al.* [67], provided conclusive evidence of cation disorder in $ABi_2Ta_2O_9$, where A = Ca, Ba, Sr. As already mentioned pristine $SrBi_2Ta_2O_9$ is not perfectly stoichiometric and contains a certain amount of inherent defects (*e.g.* oxygen vacancies) resulting from the volatilization of Bi_2O_3 at high temperatures. When Bi_2O_3 is lost, bismuth and oxygen vacancy complexes are formed in the $(Bi_2O_2)^{2+}$ layers [58, 68]. Hence, such disorder in the arrangements of cations at A-sites (in perovskite- like unit) and Bi sites (in Bi_2O_2 layers) is likely to be present. It is because of the cation disordering, that ions at A and B site are not homogenously distributed on a microscopic scale. Such compositional fluctuations lead to a microscopic heterogeneity in the structure that consists of microdomains having slightly different chemical compositions [69-70]. These microdomains with different chemical compositions will have different ferro-paraelectric transition temperatures. This distribution of T_c induces a gradual ferroelectric transition leading to broadening of the peak [48, 69-71]. In the context of above discussions, it is highly probable that some Eu^{3+} ions, in addition to occupying Sr^{2+} sites, also occupy the available bismuth vacant sites in the Bi_2O_2 layer. In other words, there is a possibility of inhomogenous distribution of Eu ions in perovskite blocks and $(Bi_2O_2)^{2+}$ layers. This can be expressed as:

$$Eu^{\bullet\bullet\bullet} + V_{Bi}^{'''} = Eu_{Bi} = Null \tag{2}$$

The above expression denotes the occupancy of Eu ion ($Eu^{\bullet\bullet\bullet}$) at the vacant Bi site ($V_{Bi}^{'''}$); since Eu has +3 charge and Bi vacancy has effective -3 charge, they neutralize each other. Thus, it is reasonable to believe that the observed broadening of dielectric peak in Eu substituted SBT, is due to the oxygen vacancy-induced-dielectric relaxation [60-61, 72] and not a result of diffused phase transition as in a relaxor ferroelectric since we did not observe frequency dependence of T_c for SEBT.

The fall in Curie temperature for SEBT samples can be understood in terms of the observed tetragonal strain variation (inset of Fig. 5c). Tetragonal strain is the internal strain in the lattice, which is reported to affect the phase transition temperature [52, 73-75]. Smaller value of strain indicates that lesser amount of thermal energy is required for the phase transition and therefore a decrease in T_c is expected with a decrease in the strain, as indeed observed. Dielectric measurements reported in other works have also shown that the introduction of rare-earth ions at the *A* site in various perovskites and layered oxides, decreases ferroelectric phase transition temperature [76-78]. Curie temperature of Pr substituted SBT has been reported to be lower than that of SBT [77]. Dielectric measurements of La-substituted $PbTiO_3$ have revealed a decrease in T_c [79]. It is also reported that with increasing La content in SBT, the dielectric peak broadens and the Curie temperature decreases [42]. Nd has been substituted at *A*-site in $Bi_4Ti_3O_{12}$ (BIT) which resulted in decrease of T_c [80]. Vaibhav *et. al.* [74] have reported that La substitution in $SrBi_2Nb_2O_9$ result in a decrease of T_c with broadened peak. Watanabe *et. al.* [80] reported that lanthanoids like Nd and Pr substitutions effectively decrease T_c in BIT. Therefore, since Eu is a rare-earth ion, it is plausible that with increase in the content of europium, T_c decreases with broadened peak around the transition temperature. In addition, it is also known that the incorporation of cations into $(Bi_2O_2)^{2+}$ layers reduces their electrostatic influence on the perovskite blocks, which might further contribute to the lowering of phase-transition [48, 81]. As discussed above, some of the Eu^{3+} also occupies the available vacant bismuth sites in the bismuth-oxide layer and therefore a decrease in the phase transition temperature is observed. Possibly all the above discussed mechanisms contribute partially to the observed dielectric behavior of the studied compositions.

Fig. 5(d) shows the temperature dependence of tangent loss at 100 kHz for SEBT samples. It is observed that europium doping reduces the dielectric loss. The loss of Bi_2O_3 during sintering process results in the formation of $V_o^{\bullet\bullet}$ in SBT [82]. The additional charge in case of donor substituted compounds is compensated through the formation of cation vacancies to maintain the overall charge neutrality of the unit structure. Similar observations have been made in rare-earth (La^{3+},Nd^{3+}, Ce^{3+}, Sm^{3+}) substituted SBT [39, 42,76-77]. Based on the above reports and the present observations it is reasonable to believe that the substitution of trivalent Eu ions for divalent Sr ions results in the formation of cation vacancies at the *A* sites. For maintaining the charge neutrality of the unit structure, on the substitution of two Eu^{3+} ions, one *A*-site (Sr site) remains vacant. The corresponding formation of vacancy can be represented as:

$$Null = Eu_{Sr}^{\bullet} + \frac{1}{2}V_{sr}^{''} \tag{3}$$

where $V_{Sr}^{''}$ denotes the strontium vacant site; $Eu_{Sr}^{\bullet}$ denotes Eu occupying the Sr site. Thus, it can be inferred that in Eu added SBT, the charge difference between Sr^{2+} and Eu^{3+} is compensated by the formation of Sr vacancies $V_{Sr}^{''}$. Since $V_{Sr}^{''}$ are effectively negatively charged and $V_o^{\bullet\bullet}$ are effectively positively charged, these $V_{Sr}^{''}$ reduce the number of $V_o^{\bullet\bullet}$ and thereby reduce the dielectric loss.

3.4 dc conductivity

The electrical conductivity of ceramic materials encompasses a wide range of values. The conductivity is usually strongly dependent upon temperature and composition [83]. In insulators charge carriers are regarded as defects in the perfect crystalline order, and the consideration of charge transport leads necessarily to consideration of point defects and their migration [84].

Pure SBT has inherent oxygen vacancies that are effectively doubly positively charged, and thus behave as acceptor-excess material [85]. The apparent net excess of acceptor content in pure SBT can be suppressed by addition of donors that reduces the oxygen vacancy concentration [85]. It has been reported that conductivity in $Bi_4Ti_3O_{12}$ (BIT) can be significantly decreased with the addition of donors, such as Nb, Sb and Ta [86-88]. Makovec et. al. [39] reported that the minimum conductivity in air-sintered $BaBi_4Ti_4O_{15}$ ceramics was obtained by the substitution of ~ 5 mol% of the Ti^{4+} ions with Nb^{5+} donors, and this resulted in a conductivity decrease of two orders of magnitude. Therefore, it is reasonable to believe that the conductivity in SBT can be suppressed by donor addition.

Fig. 6(a) and (b) shows the temperature dependence of dc conductivity (σ_{dc}) for SBTW and SEBT samples, respectively. The curves show that the conductivity increases with temperature. This suggests the presence of negative temperature coefficient of resistance (NTCR), which is a characteristic of insulators [84]. It is observed that d.c. conductivity reduces with both W and Eu substitution in SBT. Two predominant conduction mechanisms indicated by slope changes in two different temperature regions are observed. Such change in the slope in the vicinity of the Curie temperature is attributed to the differences in the activation energy values in the ferroelectric and paraelectric regions. The temperature region of ~300 °C to ~700 °C in these ceramics corresponds to the intrinsic ionic conduction range [86, 89]. In Table 2 the activation energies in the intrinsic conduction region, calculated using the Arrhenius equation for all the studied samples are given. The E_a value of the W and Eu substituted samples is much higher than that of the pure sample.

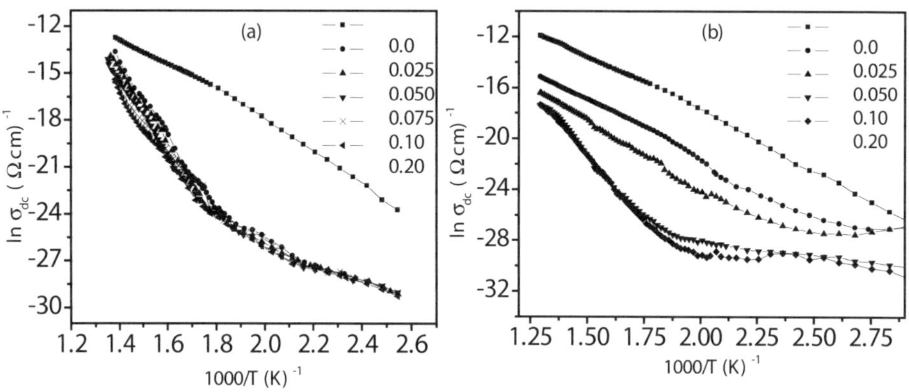

Fig. 6. Variation of dc conductivity with temperature for (a) SBTW and (b) SEBT samples

W in SBT	E_a (eV)	Eu in SBT	E_a (eV)
0.0	0.76	**0.0**	0.69
0.025	1.92	**0.025**	0.77
0.05	1.96	**0.05**	0.96
0.075	1.98	**0.1**	1.59
0.10	1.86	**0.2**	1.73
0.20	1.74		

Table 2. Activation energy (E_a) values obtained in high temperature region for $SrBi_2(Ta_{1-x}W_x)_2O_9$ and $Sr_{1-x}Eu_xBi_2Ta_2O_9$ samples

It is known that the major disadvantage of the layer-structured perovskite materials for certain applications is their relatively high conductivity [90]. The high conductivity observed in layer-structured perovskite materials is related to the presence of oxygen vacancies which are effectively positively charged defects or acceptor impurities. The concentration of these defects and, consequently, the material's conductivity, thus, increases by acceptor doping and decreases by donor doping [83-84, 89, 91-93]. Similarly in our case we observed a decrease in conductivity with donor (W and Eu) substitution in SBT. The constraint of maintaining the overall charge neutrality of the structure generates cation vacancies at Sr sites. These cation vacancies thus formed (represented by Eqs. 1 and 3), effectively reduces the concentration of oxygen vacancies. The effect of both W and Eu substitution on the concentration of oxygen vacancies is verified from the dc conductivity measurements since a decrease in dc conductivity is observed in both cases.
It is observed in Table 2 that the E_a values increases with increasing Eu content. As the concentration of Eu increases, more and more oxygen vacancies are compensated for and thereby the energy needed for the charge carriers to migrate (by hopping) also increases, leading to higher activation energy. As a result, the conductivity decreases because there are fewer oxygen vacancies available with sufficient energy to move around and the energy barrier between the scarcer vacancies increases [86]. Whereas in W-substituted samples E_a is observed to increase with W concentration up to x = 0.05; however, beyond that it decreases (Table 2). The decrease in the activation energy for samples with $x > 0.05$ suggests an increase in the concentration of mobile charge carriers [94]. This observation can be ascribed to the existence of multiple valence states of tungsten. Since tungsten is a transition element, the valence state of W ions in a solid solution most likely varies from W^{6+} to W^{4+} depending on the surrounding chemical environment [95-96]. When W^{4+} are substituted for Ta^{5+} sites, oxygen vacancies would be created, i.e. one oxygen vacancy would be created for every two tetravalent W ions entering the crystal structure increasing the concentration of mobile charge carriers and consequently a decrease in the E_a beyond $x > 0.05$.
The observed variation in conductivity with tungsten and europium content in SBT is consistent with dielectric loss, which also reduces with increasing tungsten concentration and has been explained in light of contribution from oxygen vacancies.

3.5 Ferroelectric properties

The P-E loops measured at 50 Hz and room temperature for SBTW and SEBT samples are shown in Fig. 7 (a) and (b), respectively. It is observed that W-substitution results in the formation of well-defined hysteresis loops. The optimum tungsten content for maximum $2P_r$ (~ 25 $\mu C/cm^2$) is observed to be x = 0.075.
It is known that ferroelectric properties are affected by compositional modification, microstructure and lattice defects like oxygen vacancies within the structure of the materials [15-26, 97]. In hard ferroelectrics, with lower-valent substituents the associated oxide vacancies are likely to assemble in the vicinity of domain walls [43, 98-99]. These domains are locked by the defects and their polarization switching is difficult, leading to a decrease in P_r and an increase in E_c. Moreover, theoretical calculations [100] have also shown that the increase in space charge density brings about a decrease in P_r. On the other hand, in soft ferroelectrics, with higher-valent substituents, the defects are cation vacancies whose

mobility is extremely low below T_c. Thus, the interaction between cation vacancies and domain walls is much weaker than that in hard ferroelectrics [38, 43] leading to an increase in P_r values. Watanabe et. al. [101] reported a remarkable improvement in ferroelectric properties in the $Bi_4Ti_3O_{12}$ ceramic by adding higher valent cation, V^{5+} at the Ti^{4+} site. Recently, significantly large P_r value has been reported for W-substituted BIT sintered sample [102]. It has also been reported that cation vacancies generated by donor doping make domain motion easier and enhance the ferroelectric properties [103]. Also it is known that domain walls are relatively free in large grains and are inhibited in their movement as the grain size decreases [45]. In the larger grains, domain motion is easier which results in larger P_r [104-105].

Based on the obtained results and above discussion, it can be understood that in pure SBT, the oxygen vacancies assemble at sites like domain boundaries leading to a strong domain pinning. Hence well-saturated P-E loops for pure SBT are not obtained. Whereas, in the W-substituted samples, the associated cation vacancy formation suppresses the concentration of oxygen vacancies. A reduction in the number of oxygen vacancies reduces the pinning effect on the domain walls, leading to enhanced remnant polarization and lower coercive field. Another possible reason could be attributed to the increase in grain size of SBTW, as observed in SEM micrographs (Fig. 3). In the present study, the grain size is observed to increase with increasing W concentration; however, the remanent polarization does not monotonously increase with increasing W concentration (Fig. 7a). It is observed that beyond $x > 0.075$ P_r values decreases. Therefore, besides the effect of grain size and reduction in oxygen vacancies, other factors also influence variation in remnant polarization.

The decrease in the value of $2P_r$ for $x > 0.075$, seems possibly due to the presence of secondary phases (observed in XRD diffractograms) which hampers the switching process of polarization [106-110]. Also this could be explained on the basis of the inference drawn w.r.t. dc conductivity study in SBTW. We concluded that beyond $x > 0.05$, the number of charge carriers increases in the form of oxygen vacancies. This increase in oxygen vacancies beyond $x > 0.05$ leads to pinning of domain walls and thus a reduction in the P_r values is observed [111].

Fig. 7 (b) shows the P-E loops of the SEBT samples. It is observed that the remanent polarization increase with increase in europium content. The maximum $2P_r$ ~14 $\mu C/cm^2$ is observed in the sample with $x = 0.20$. A lot of studies on the influence of rare-earth ion substitution in simple perovskite ferroelectrics have shown improved ferroelectric properties [88, 112-115]. The substitution of smaller lanthanoid ions like Samarium (Sm^{3+}), Neodium (Nd^{3+}) and lanthanum (La^{3+}) etc. into BIT is reported to enhance remnant polarization [25, 116-118]. $SrBi_4Ti_4O_{15}$, which has a crystalline structure similar to BIT, is another typical BLSF ($m = 4$) that shows enhanced ferroelectric properties as a result of La doping [119]. Noguchi et. al. [76] have reported that La, Nd, and Sm substitution in SBT show a large P_r compared to pure SBT. These results suggest that the smaller rare earth ion substitution enhances ferroelectric properties. Also, the substitution of Sr^{2+} by Eu^{3+} in the SBT lattice, results in the formation of vacancies at the A site ($V_{Sr}^{''}$) that suppresses the concentration of oxygen vacancies leading to the observed enhancement of remnant polarization (Fig. 7b).

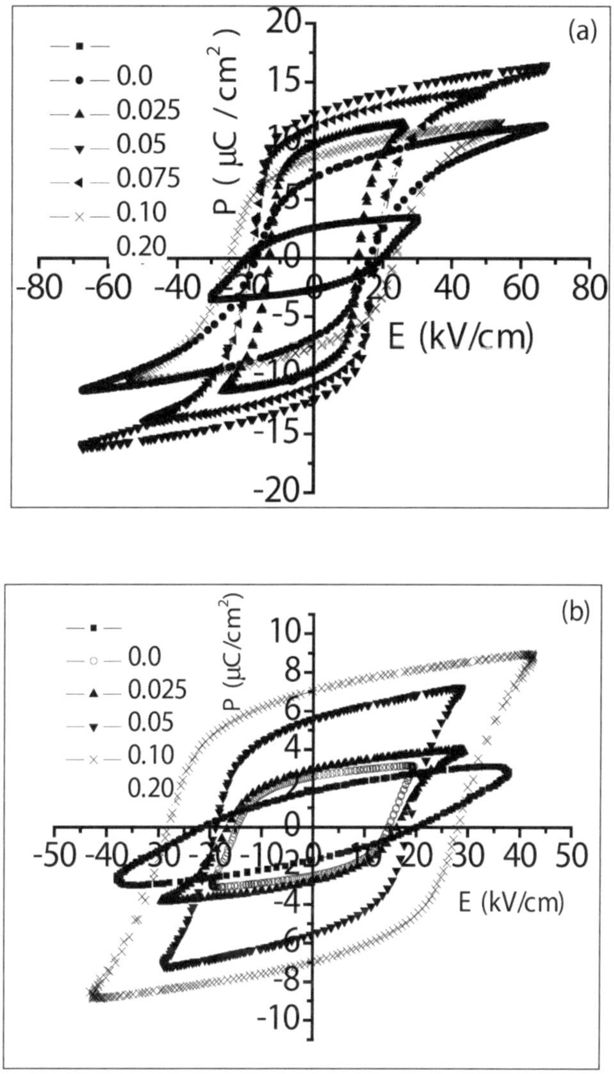

Fig. 7. P-E hysteresis loops for (a) SBTW and (b) SEBT samples

3.6 Piezoelectric studies

Piezoelectric ceramics are widely used for electromechanical transducers and hydrostatic sensing applications. Among the available materials, lead zirconate titanate (PZT) exhibits a large piezoelectric coefficient, d_{33} (400–600 pC/N) [120]. In recent years, lead-free materials such as bismuth-layered structured ferroelectrics (BLSFs) have been attracting attention for piezoelectric device applications, and are found suitable for fine tolerance resonators with excellent frequency stability [121]. Ando *et. al.* [122] have reported the effects of compositional modifications in strontium bismuth niobate (SBN)-based BLSF materials in improving the piezoelectric properties. Kholkin *et. al.* [43] has reported on the electromechanical properties of SBT but the maximum polarization and piezoelectric coefficient that were measured were quite small. Since BLSFs generally have high Curie temperature and high coercive field at room temperature, it is necessary to perform poling at high temperature [123]. However, relatively high conductivity often prevents the application of high electrical field during the high-temperature poling treatment [123-124]. The conductivity, therefore, should be reduced in BLSFs to increase the poling efficiency.

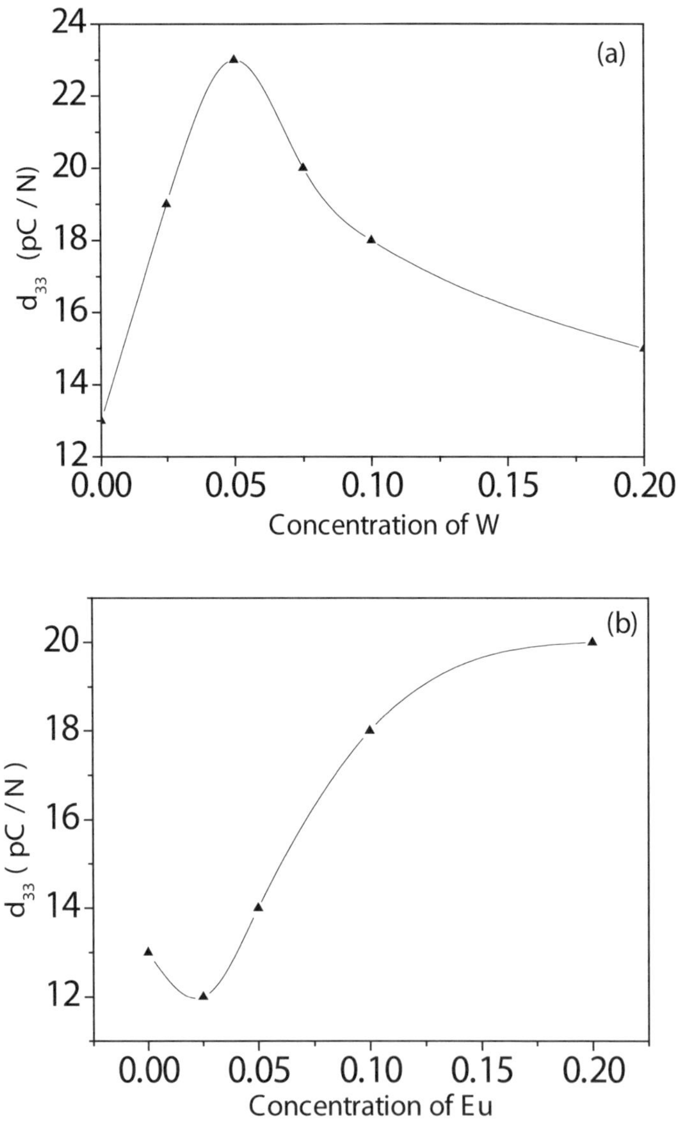

Fig. 8. Variation of d_{33} in (a) SBTW and (b) SEBT samples

Fig. 8(a) and (b) shows the variation of piezoelectric charge coefficient d_{33} with x in SBTW and SEBT samples, respectively. The d_{33} increases with x up to $x = 0.05$, however, d_{33} did not improve significantly. A decrease in d_{33} values is observed for the samples with x > 0.05. The piezoelectric coefficient, d_{33}, increases from 13 pC/N in the sample with x = 0.0 to 23 pC/N in the sample with x = 0.05 in SBTW. Whereas in SEBT samples, the coefficient is observed to increase from 13 (x =0.0) to 20 pC/N (x = 0.20). The above observation can be explained on the basis of conductivity behavior observed in W- and Eu-substituted SBT samples. As observed, electrical conductivity reduces with both donor substituents (Fig. 6a and b) in SBT. This decrease in conductivity upon donor substitution improves the poling efficiency and thus higher d_{33} values are obtained. It has also been reported that the incorporation of higher valent ions accompanied with cation vacancies at the *A* site in BLSFs, improves not only the ferroelectric property but also the piezoelectric one [49, 125]. The piezoelectric constants have also been found to increase with increase in grain size [49]. The report that the vacancies at *A* sites improve the piezoelectric properties due to increased wall mobility, supports the present observation in both cases. Moreover, since the grain size increases with both W and Eu content (Figs. 3 and 4), it is reasonable to believe that the increase in grain size will also contribute to the increase in d_{33} values. The decrease in the value of d_{33} for samples with x > 0.05 in SBTW samples is possibly due to the presence of secondary phases [1, 126-127] in the samples discussed earlier with respect to XRD observations (Fig. 2a).

4. Conclusions

The addition of tungsten in SBT is observed to be effective in improving dielectric, electrical, ferroelectric and piezoelectric properties. The X-ray diffractograms show the formation of the single phase layered structure up to W concentration x ≤ 0.05, beyond which an unidentified peak is observed though its intensity is very small. Scanning Electron Microscopy (SEM) photographs reveal that W addition in SBT is effective in improving the microstructure, as, well developed dense microstructure with large grains are seen. The average grain size increases with increase in W content. The substitution of the smaller W^{6+} ions for Ta^{5+} ions in SBT is found to be effective in improving the dielectric properties. Dielectric constant (ε_r) and the Curie temperature (T_c) increases with increasing W content. The dielectric loss reduces significantly with increase in doping level. The maximum T_c of ~ 390 °C is observed in the sample with x = 0.20 as compared to ~ 320 °C for the pure sample. The peak - ε_r increases from ~ 270 in the sample with x = 0.0 to ~ 700 for the composition with x = 0.20. The temperature dependence of the electrical conductivity shows that tungsten doping results in the decrease of conductivity by as much as 2-3 orders of magnitude compared to pure SBT. All the tungsten-doped ceramics have higher $2P_r$ than that in the pure sample. The maximum $2P_r$ (~25 $\mu C/cm^2$) is obtained in the composition with x=0.05. The d_{33} values increase with increasing W content up to x ≤ 0.05. The value of d_{33} in the composition with x = 0.05 is ~ 23 pC/N as compared to ~ 13 pC/N in the pure sample.

For the europium-substituted samples, the single phase structure is maintained for the entire concentration range. The lattice parameters decrease with increase in europium concentration. The average grain size increases with increasing Eu concentration. It is

observed that the studied samples undergo a diffused transition near T_c. The dielectric loss and conductivity reduces with increasing Eu content. An increase in remanent polarization has been observed with increasing Eu content. The maximum $2P_r$ value of ~ 14 $\mu C/cm^2$ is observed in the sample with x = 0.20 as compared to 4 $\mu C/cm^2$ for the pure sample. A maximum d_{33} of ~ 20 pC/N is obtained in the sample with x = 0.20 as compared to 13 pC/N for the pure sample.
Thus, addition of both the donor cations, W^{6+} and Eu^{3+} in SBT, is found to be effective in improving microstructural, electrical, ferroelectric and piezoelectric properties. Addition of both the donors, result in increase in grain size which makes domain motion easier and thereby enhance the dielectric permittivity, remanent polarization and d_{33} values. Also, the associated cation vacancies that are formed to maintain the charge neutrality in the structure is known to make domain motion easier. This contributes further in enhancing the various properties mentioned above. Also, W and Eu doping in SBT results in reduced dielectric loss and conductivity. Such compositions with low conductivity and high P_r values should be excellent materials for highly stable ferroelectric memory devices.

5. References

[1] J. F. Scott and C. A. P. de Araujo, *Science*, 246, 1400 (1989).
[2] G. H. Haertling, *J. Vac. Sci. Tech.*, 9, 414 (1990).
[3] J. J. Lee, C. L. Thio and S. B. Desu, *J. Appl. Phys.*, 78, 5073 (1995).
[4] S. Dey and R. Zuleeg, *Ferroelectrics*, 108, 37 (1990).
[5] A. Kitamura, Y. Noguchi and M. Miyayama, *Mater. Lett.*, 58, 1815 (2004).
[6] H. N. Al-Shareef, A. I. Kingon, X. Chen, K. R. Bellur and O. Auciello, *J. Mater. Res.*, 9, 2986 (1994).
[7] W. L. Warren, D. Dimos, B. A. Tuttle, R. D. Nasby and G. E. Pike, *Appl. Phys. Lett.*, 65, 1018 (1994).
[8] J. F. Chang and S. B. Desu, *J. Mater. Res.*, 9, 955 (1994).
[9] K. Amanuma, T. Hase and Y. Miyasaka, *Jpn. J. Appl. Phys.*, 33, 5211 (1994).
[10] P. C. Joshi and S. B. Krupanidhi, *J. Appl. Phys.*, 72, 5827 (1992).
[11] Y. Zhu, X. Zhang, P. Gu, P. C. Joshi and S. B. Desu, *J. Phys.: Condens. Matter*, 9, 10225 (1997).
[12] B. Aurivillius, *Ark. Kemi*, 1, 463 (1949).
[13] B. Aurivillius, *Ark. Kemi*, 1, 499 (1949).
[14] R. L. Withers, J. G. Thompson, L. R. Wallenberg, J. D. FitzGerald, J. S. Anderson and B. G. Hyde, *J. Phys. C: Solid State Phys.*, 21, 6067 (1988).
[15] R. L. Withers, J. G. Thompson and A. D. Rae, *J. Solid State Chem.*, 94, 404 (1991).
[16] G. A. Smolenskii, V. A. Isupov and A. I. Agranovskaya, *Sov. Phys. Solid State*, 3, 651 (1959).
[17] E. C. Subbarao, *J. Phys. Chem. Solids*, 23, 665 (1962).
[18] C. A. P. de Araujo, J. D. Cuchiaro, L. D. Macmillan, M. C. Scott & J. F. Scott, *Nature* (London) 374, 627 (1995).
[19] M. Hirose, T. Suzuki, H. Oka, K. Itakura, Y. Miyauchi and T. Tsukada, *Jpn. J. Appl. Phys.*, 38, 5561 (1999).
[20] M. Kimura, T. Sawada, A. Ando and Y. Sakabe, *Jpn. J. Appl. Phys.*, 38, 5557 (1999).

[21] T. Takeuchi, T. Tani and Y. Saito, *Jpn. J. Appl. Phys.*, 39, 5577 (2000).
[22] T. Mihara, H. Yoshimori, H. Watanabe and C. A. P. de Araujo, *Jpn. J. Appl. Phys.*, 34, 5233 (1995).
[23] K. Kato, C. Zheng, J. M. Finder, S. K. Dey and Y. Torii, *J. Am. Ceram. Soc.*, 81, 1869 (1998).
[24] E. Tokumitsu, G. Fujii and H. Ishiwara, *Mater. Res. Symp. Proc.*, 493, 459 (1998).
[25] B. H. Park, B. S. Kang, S. D. Bu, T. W. Noh, J. Lee and W. Joe, *Nature*, 401, 682 (1999).
[26] K. Ishikawa and H. Funakubo, *Appl. Phys. Lett.*, 75, 1970 (1999).
[27] A. Gruverman, A. Pignolet, K. M. Satyalakshmi, M. Alexe, N. D. Zakharov and D. Hesse, *Appl. Phys. Lett.*, 76, 106 (2000).
[28] J. Lettieri, M. A. Zurbuchen, Y. Jia, D. G. Schlom, S. K. Streiffer and M.E. Hawley, *Appl. Phys. Lett.*, 77, 3090 (2000).
[29] A. R. James, S. Balaji and S. B. Krupanidhi, *Mater. Sci. Engi.* B64, 149 (1999).
[30] R. Jain, V. Gupta and K. Sreenivas, *Mater. Sci. Engi.* B78, 63 (2000).
[31] J. Zhang, Z. Yin and M. S. Zhang, *Appl. Phys. Lett.*, 81(25) 4778 (2002).
[32] K. Amanuma, T. Hase and Y. Miyasaka, *Appl. Phys. Lett.*, 66, 2 (1995).
[33] J. F. Scott, *Phys. World*, 46, 22 (1995).
[34] O. Auciello, J. F. Scott and R. Ramesh, *Phys. Today*, 51, 22 (1998).
[35] S. K. Kim, M. Miyayama and H. Yanagida, *Mater. Res. Bull.*, 31, 121(1996).
[36] M. G. Stachiotti, C. O. Rodriguez, C. A. Draxl & N. E. Christensen, *Phys. Rev.*, B61, 14434 (2000).
[37] Y. Noguchi, M. Miyayama and T. Kudo, *Phys. Rev.*, B63, 214102 (2001).
[38] A. D. Rae, J. G. Thompson and R. L. Withers, *Acta. Crystallogr. Sect.*, B48, 418 (1992).
[39] Y. Shimakawa, Y. Kubo, Y. Nakagawa, T. Kamiyama, H. Asano and F. Izumi, *Appl. Phys. Lett.*, 74, 1904 (1998).
[40] S. M. Blake, M. J. Falconer, M. McCreedy and P. Lightfoot, *J. Mater. Chem.*, 7, 1609 (1997).
[41] A. Li, H. Ling, D. Wu, T. Yu, M. Wang, X. Yin, Z. Liu and N. Ming, *Appl. Surf. Sci.*, 173, 307 (2001).
[42] T. Noguchi, T. Hase and Y. Miyasaka, *Jpn. J. Appl. Phys.*, 35, 4900 (1996).
[43] A. L. Kholkin, K. G. Brooks and N. Setter, *Appl. Phys. Lett.*, 71, 2044 (1997).
[44] Wu, E., POWD, *An interactive powder diffraction data interpretation and indexing program* Ver2.1, School of Physical Science, Flinders University of South Australia, Bedford Park S.A. JO42AU.
[45] R. R. Das, P. Bhattacharya, W. Perez & R. S. Katiyar, *Ceram. Int.*, 30, 1175 (2004).
[46] M. Deri, *Ferroelectric Ceramics* (Akademiai Kiado, Budapest, 1966).
[47] Y. Wu, C. Nguyen, S. Seraji, M. J. Forbess, S. J. Limmer, T. Chou and G. Z. Cao, *J. Amer. Ceram. Soc.*, 84, 2882 (2001).
[48] P. D. Martin, A. Castro, P. Millan and B. Jimenez, *J. Mater. Res.*, 13, 2565 (1998).
[49] H. T. Martirena and J. C. Burfoot, *J. Phys. C: Solid State Phys.*, 7, 3182 (1974).
[50] Y. Wu, C. Nguyen, S. Seraji, M. J. Forbess, S. J. Limmer, T. Chou and G. Z. Cao, *J. Amer. Ceram. Soc.*, 84, 2882 (2001).
[51] Y. Shimakawa, Y. Kubo, Y. Nakagawa, T. Kamiyama, H. Asano and F. Izumi, *Phys. Rev.*, B61, 6559 (2000).
[52] C. Prakash and A. S. Bhalla, *Ferroelectrics*, 262, 321 (2001).
[53] Y. Noguchi, M. Miyayama and T. Kudo, *J. Appl. Phys.*, 88, 2146 (2000).

[54] K. Singh, D. K. Bopardik and D. V. Atkare, *Ferroelectrics*, 82, 55 (1988).
[55] E. C. Subbarao, *Integrat. Ferro.*, 12, 33 (1996).
[56] Y. Wu, S. J. Limmer, T. P. Chou and C. Nguyen, *J. Mater. Sci. Lett.*, 21, 947 (2002).
[57] I. Coondoo, A. K. Jha and S. K. Agarwal, *Ceram. Int.*, 33, 41 (2007).
[58] I. Coondoo, A. K. Jha and S.K. Agarwal, *J. Eur. Ceram. Soc.*, 27, 253 (2007).
[59] R. Rai, S. Sharma and R. N. P. Choudhary, *Solid State Commu.*, 133, 635 (2005).
[60] R. Rai, S. Sharma and R. N. P. Choudhary, *Mater. Lett.*, 59 3921 (2005).
[61] A. Chen, Y. Zhi and L.E. Cross, *Phys. Rev.*, B 62, 228 (2000).
[62] S. M. Blake, M. J. Falconer, M. McCreedy and P. Lightfoot, *J. Mater. Chem.*, 7, 1609 (1997).
[63] S. Kojima, *J. Phys. : Condens. Matter*, 10, L327 (1998).
[64] Ismunandar and B. J. Kennedy, *J. Mater. Chem.*, 9, 541 (1999).
[65] Y. Shimakawa, Y. Kudo, Y. Nakagawa, T. Kamiyama, H. Asano, and F. Izumi, *Phys. Rev.* B 61, 6559 (2000).
[66] Y. Noguchi, M. Miyayama, K. Oikawa and T. Kamiyama: Submitted *to Phys. Rev.*B.
[67] R. Macquart and B. J. Kennedy and Y. Shimakawa, *J. Solid State Chem.*, 160, 174 (2001).
[68] I. Coondoo, A. K. Jha and S.K. Agarwal, (*Proceedings of the NSFD-XIII*) Nov 23 - 25 'o4, pp293.
[69] A. Chen, Y. Zhi, and J. Zhi, *Phys. Rev.*, B 61, 957 (2000).
[70] A. Chen, J. F. Scott, Y. Zhi, H. Ledbetter, and J. L. Baptista, *Phys. Rev.* B59, 6661 (1999).
[71] L. E. Cross, *Ferroelectrics*, 76, 241 (1987).
[72] I. Coondoo and A. K. Jha, *Solid State Commu.*, 142, 561 (2007).
[73] V. Shrivastava, A. K. Jha and R. G. Mendiratta, *Solid State Commu.*, 133, 125 (2005).
[74] V. Shrivastava, A.K. Jha and R.G. Mendiratta, *Physica* B 371, 337 (2006).
[75] I. Coondoo, A. K. Jha, S.K. Agarwal and N. C. Soni, *J. Electroceram.*, 16, 393(2006).
[76] Y. Noguchi, M. Miyayama, K. Oikawa, T. Kamiyama, M. Osada, and M. Kakihana, *Jpn. J. Appl. Phys.*, 41, 7062 (2002).
[77] Y. Noguchi, A. Kitamura, L. Woo, M. Miyayama, K. Oikawa and T. Kamiyama, *J. Appl. Phys.* 94, 6749 (2003).
[78] T. Volk, L. Ivleva, P. Lykov, D. Isakov, V. Osiko and M. Wohlecke, *Appl. Phys. Lett.*, 79, 854 (2001).
[79] T.-Y. Kim and H. M. Jang, *Appl. Phys. Lett.*, 77, 3824 (2000).
[80] T. Watanabe, T. Kojima, T. Sakai, H. Funakubo, M. Osada, Y. Noguchi and M. Miyayama, *J. Appl. Phys.*, 92, 1518 (2002).
[81] B. Frit and J. P. Mercurio, *J. Alloys and Compounds*, 188, 2694 (1992).
[82] B. H. Park, S. J. Hyun, S. D. Bu, T. W. Noh, J. Lee, H. D. Kim, T. H. Kim and W. Jo, *Appl. Phys. Lett.* 74, 1907 (1999).
[83] M. Villegas, A. C. Caballero, C. Moure, P. Duran and J. F. Fernandez, *J. Eur. Ceram. Soc.*, 19, 1183 (1999).
[84] I. Pribosic, D. Makovec and M. Drofenik, *J. Eur. Ceram. Soc.*, 21, 1327 (2001).
[85] T. Baiatu, R. Waser and K. H. Hardtl, *J. Am. Ceram. Soc.*, 73,1663 (1999).
[86] J. J. Dih and R. M. Fulrath, *J. Am. Ceram. Soc.*, 61, 448 (1978).
[87] W. L. Warren, K. Vanheusdan, D. Dimos, G. E. Pike and B. A. Tuttle, *J. Am. Ceram. Soc.*, 78, 536 (1995).
[88] Q. Tan and D. Viehland: *Philos. Mag.* B 80,1585 (2000).
[89] C. Voisard, D. Damnajovic and N. Setter, *J. Eur. Ceram. Soc.*, 19, 1251 (1999).

[90] D. Makovec, I. Pribosic, Z. Samardzija and M. Drofenik, *J. Am. Ceram. Soc.*, 84, 2702 (2001).
[91] H. S. Shulman, M. Testorf, D. Damjanovic and N. Setter, *J. Am. Ceram. Soc.*, 79, 3124 (1996).
[92] R. Gerson and H. Jaffe, *J. Phys. Chem. Solids,* 24, 979 (1963).
[93] M. Takahashi, *Jpn. J. Appl. Phys.*, 10, 643 (1971).
[94] B. H. Venkataraman and K. B. R. Varma, *J. Phys. Chem. Solids* 66 (10), 1640 (2005).
[95] C. D. Wagner, W. M. Riggs, L. E. Davis and F. J. Moulder, *Handbook of X-ray Photoelectron Spectroscopy*, Perkin Elmer Corp., Chapman & Hall, 1990.
[96] S. Shannigrahi and K. Yao, *Appl. Phys. Lett.* 86(9) 092901 (2005).
[97] M. Miyayama, T. Nagamoto and O. Omoto, *Thin Solid Films*, 300, 299 (1997).
[98] T. Friessnegg, S. Aggarwal, R. Ramesh, B. Nielsen, E. H. Poindexter and D. J. Keeble, *Appl. Phys. Lett.*, 77, 127 (2000).
[99] Q. Tan, J. Li and D. Viehland, *Appl. Phys. Lett.*, 75, 418 (1999).
[100] I. K. Yoo and S. B. Desu, *Mater. Sci. Engg.*, B13, 319 (1992).
[101] T. Watanabe, H. Funakubo, M. Osada, Y. Noguchi and M. Miyayama, *Appl. Phys. Lett.*, 80, 100 (2002).
[102] Y. Noguchi, I. Miwa, Y. Goshima and M. Miyayama, *Jpn. J. Appl. Phys.*, 39, L1259 (2000).
[103] S. Takahashi and M. Takahashi, *Jpn. J. Appl. Phys.*, 11, 31 (1972).
[104] S. B. Desu, P. C. Joshi, X. Zhang and S. O. Ryu, *Appl. Phys. Lett.*, 71, 1041 (1997).
[105] M. Nagata, D. P. Vijay, X. Zhang and S. B. Desu, *Phys. Stat. Sol.*, (a)157, 75 (1996).
[106] J. J. Shyu and C. C. Lee, *J. Eur. Ceram. Soc.*, 23, 1167 (2003).
[107] I. Coondoo, A. K. Jha and S.K. Agarwal, *Ferroelectrics*, 356, 31 (2007).
[108] T. Sakai, T. Watanabe, M. Osada, M. Kakihana, Y. Noguchi, M. Miyayama and H. Funakubo, *Jpn. J. Appl. Phys.*, 42, 2850 (2003).
[109] C. H. Lu and C. Y. Wen, *Mater. Lett.*, 38, 278 (1999).
[110] A. Li, D. Wu, H. Ling, T. Yu, M. Wang, X. Yin, Z. Liu and N. Ming, *Thin Sol. Films*, 375, 215 (2000).
[111] I. Coondoo, S.K. Agarwal and A. K. Jha, *Mater. Res. Bull.*, 44, 1288 (2009).
[112] H. D. Sharma, A. Govindan, T. C. Goel and P. K. C. Pillai, *J. Mater. Sci. Lett.*, 15, 1424 (1996).
[113] H. B. Park, C.Y. Park, Y. S. Hong, K. Kim and S. J. Kim, *J. Am. Ceram. Soc.*, 82, 94 (1999).
[114] A. Garg and T. C. Goel, *J. Mater. Sci.*, 11, 225 (2000).
[115] T. J. Boyle, P. G. Clem, B. A. Tuttle, G. L. Brennecka, J. T. Dawley, M. A. Rodriguez, T. D. Dunbar andW. F. Hammetter, *J. Mater. Res.*, 17, 871 (2002).
[116] T. Takenaka and K. Sakata, *Ferroelectrics*, 38, 769 (1981).
[117] D. Wu, A. Li, T. Zhu, Z. Liu and N. Ming, *J. Appl. Phys.*, 88, 5941 (2000).
[118] U. Chon, K. B. Kim and H. M. Jang, *Appl. Phys. Lett.*, 79, 2450 (2001).
[119] J. Zhu, W. P. Lu, X. Y. Mao, R. Hui and X. B. Chen, *Jpn. J. Appl. Phys.*, 42, 5165 (2003).
[120] S. L. Swartz, *IEEE Trans. Electr. Insul.*, 25, 935 (1990).
[121] T. Ogawa and Y. Numamoto, *Jpn. J. Appl. Phys.*, 41, 7108 (2002).
[122] A. Ando, T. Sawada, H. Ogawa, M. Kimura, and Y. Sakabe, *Jpn. J. Appl. Phys.*, 41, 7057 (2002).
[123] I. S. Yi and M. Miyayama, *Jpn. J. Appl. Phys.*, 36, L1321 (1997).

[124] C. Moure, L. Lascano, J. Tartaj and P. Duran, *Ceram. Int.*, 29, 91 (2003).
[125] C. Fujioka, R. Aoyagi, H. Takeda, S. O. Kamura and T. Shiosaki, *J. Eur. Ceram. Soc.*, 25, 2723 (2005).
[126] R. Jain, V. Gupta, A. Mansingh and K. Sreenivas, *Mater. Sci. Engg.*, B112, 54 (2004).
[127] R. Jain, A. K. S. Chauhan, V. Gupta and K. Sreenivas, *J. Appl. Phys.*, 97, 124101 (2005).

Part 2

Ferroelectrics and Its Applications: A Theoretical Approach

15

Non-Equilibrium Thermodynamics of Ferroelectric Phase Transitions

Shu-Tao Ai
Linyi Normal University
People's Republic of China

1. Introduction

It is well known that the Landau theory of continuous phase transitions is a milestone in the process of the development of phase transition theories. Though it does not tally with the nature of phase transitions in the critical regions, the Landau theory as a phenomenological one has been very successful in many kinds of phase transitions such as the ferroelectric phase transitions, i.e. the vast studies centering on it have been carried out. In the ferroelectric case, we should pay attention to the Landau theory extended by A.F. Devonshire to the first-order phase transitions (Devonshire, 1949; Devonshire, 1951; Devonshire, 1954). This daring act was said to be successful.

However, the Landau theory is based on the equilibrium (reversible) thermodynamics in essence. Can it deal with the outstanding irreversible phenomenon of first-order ferroelectric phase transitions, which is the "thermal hysteresis"? The Landau-Devonshire theory attributes the phenomenon to a series of metastable states existing around the Curie temperature T_C. In principle, the metastable states are not the equilibrium ones and can not be processed by using the equilibrium thermodynamics. Therefore, we believe that the extension of Devonshire is problematic though it is successful in mathematics. The real processes of phase transition were distorted. In Section 2, we will show the unpleasant consequence caused by the metastable states hypothesis, and the evidence for the non-existence of metastable states, i.e. the logical conflict. Then in Section 3, we will show the evidence (experimental and theoretical) for the existence of stationary states to a ferroelectric phase transition. In Section 4 and 5, we will give the non-equilibrium (irreversible) thermodynamic description of phase transitions, which eliminates the unpleasant consequence caused by the metastable states hypothesis. At last, in Section 6 we will give the non-equilibrium thermodynamic explanation of the irreversibility of ferroelectric phase transitions, i.e. the thermal hysteresis and the domain occurrences in ferroelectrics.

2. Limitations of Landau-Devonshire theory

The most outstanding merit of Landau-Devonshire theory is that the Curie temperature and the spontaneous polarization at Curie temperature can be determined simply. However, in the Landau-Devonshire theory, the path of a first-order ferroelectric phase transition is believed to consist of a series of metastable states existing around the Curie temperature T_C. This is too difficult to believe because of the difficulties encountered (just see the following).

2.1 Unpleasant consequence caused by metastable states hypothesis

Basing on the Landau-Devonshire theory, we make the following inference. Because of the thermal hysteresis, a first-order ferroelectric phase transition must occur at another temperature, which is different from the Curie temperature. The state corresponding to the mentioned temperature (actual phase transition temperature) is a metastable one. Since the unified temperature and spontaneous polarization can be said about the metastable state, we neglect the heterogeneity of system actually. In other words, every part of the system, i.e. either the surface or the inner part, is of equal value physically. When the phase transition occurs at the certain temperature, every part of the system absorbs or releases the latent heat simultaneously by a kind of action at a distance. (The concept arose in the electromagnetism first. Here, it maybe a kind of heat transfer). Otherwise, the heat transfer in system, with a finite rate, must destroy the homogeneity of system and lead to a non-equilibrium thermodynamic approach. The unpleasant consequence, i.e. the action at a distance should be eliminated and the life-force should be bestowed on the non-equilibrium thermodynamic approach.

In fact, a first-order phase transition process is always accompanying the fundamental characteristics, called the co-existence of phases and the moving interface (phase boundary). This fact reveals that the phase transition at various sites can not occur at the same time. Yet, the phase transition is induced by the external actions (the absorption or release of latent heat). It conflicts sharply with the action at a distance.

2.2 Evidence for non-existence of metastable states: logical conflict

In the Landau-Devonshire theory, if we neglect the influence of stress, the elastic Gibbs energy G_1 can be expressed with a binary function of variables, namely the temperature T and the electric displacement D (As G_1 is independent of the orientation of $\boldsymbol{D}$, here we are interested in the magnitude of $\boldsymbol{D}$ only)

$$G_1 = g_1(T,D) \tag{1}$$

The long-standing, close correlation between the analytical dynamics and the thermodynamics implies that Equation (1) can be taken as a scleronomic constraint equation

$$f_1(G_1,T,D) = G_1 - g_1(T,D) = 0 \tag{2}$$

where G_1,T,D are the generalized displacements. The possible displacements $\mathrm{d}G_1,\mathrm{d}T$ and $\mathrm{d}D$ satisfy the following equation

$$\mathrm{d}G_1 - \frac{\partial g_1}{\partial T}\mathrm{d}T - \frac{\partial g_1}{\partial D}\mathrm{d}D = 0 \tag{3}$$

In the Landau-Devonshire theory, the scleronomic constraint equation, i.e. Equation (1) is expressed in the form of the power series of D (For simplicity, only the powers whose orders are not more than six are considered)

$$G_1 = G_{10} + \frac{1}{2}\alpha D^2 + \frac{1}{4}\beta D^4 + \frac{1}{6}\gamma D^6 \tag{4}$$

where α,β,γ are the functions of T, and G_{10} is the elastic Gibbs energy of paraelectric phase. The relation between G_1 and D at various temperatures, which belongs to first-order phase transition ferroelectrics is represented graphically in Figure 1. The electric displacements which correspond with the bilateral minima of G_1 are identified as $\pm D^*$, and the electric displacement which corresponds with the middle minimum of G_1 equals zero. The possible electric displacements should be the above ones which correspond with all the minima of G_1.

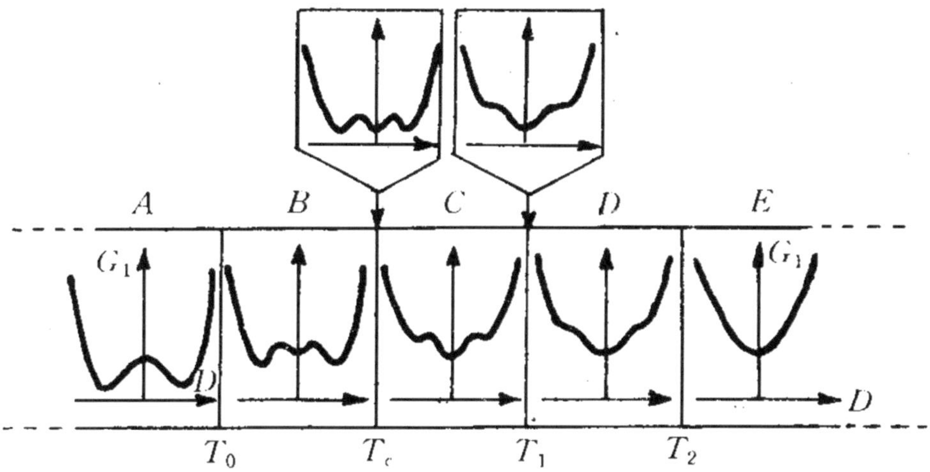

Fig. 1. The relation between elastic Gibbs energy G_1 and zero field electric displacement D belongs to the ferroelectrics at various temperatures, which undergoes a first-order phase transition (Zhong, 1996).

Equivalently, imposed on the generalized displacements G_1, T, D is a constraint, which is

$$\frac{\partial G_1}{\partial D} = 0 \tag{5}$$

So, the possible displacement $\mathrm{d}D$ should be the following:

$$\mathrm{d}D^*, -\mathrm{d}D^*, 0-\left(\pm D^*\right) = \mp D^*, \pm D^* - 0 = \pm D^*, 0 .$$

After all, if our discussions are limited in the equilibrium thermodynamics strictly, there must be the third constraint, i.e. the equilibrium D and T should satisfy

$$h\left(D,T\right) = 0 \tag{6}$$

where h is a binary function of the variables D and T. It can be determined by the principle of minimum energy

$$G_1 = \min \tag{7}$$

for certain T. Then, the metastable states are excluded. Thus, the thermal hysteresis does not come into being. The corollary conflicts with the fact sharply. This reveals that the first-order ferroelectric phase transition processes must not be reversible at all so as not to be dealt with by using the equilibrium thermodynamics.

How can this difficulty be overcome? An expedient measure adopted by Devonshire is that the metastable states are considered. However, do they really exist?
Because the metastable states are not the equilibrium ones, the relevant thermodynamic variables or functions should be dependent on the time t. In addition, the metastable states are close to equilibrium, so the heterogeneity of system can be neglected. Here, the elastic Gibbs energy G_1' should be

$$G_1' = g_2(T,D,t) \tag{8}$$

For the same reason as was mentioned above, Equation (8) can be regarded as a rheonomic constraint on the generalized displacements G_1',T,D

$$f_2(G_1',T,D,t) = G_1' - g_2(T,D,t) = 0 \tag{9}$$

In this case, the possible displacements $\mathrm{d}G_1',\mathrm{d}T$ and $\mathrm{d}D$ satisfy the following equation

$$\mathrm{d}G_1' - \frac{\partial g_2}{\partial T}\mathrm{d}T - \frac{\partial g_2}{\partial D}\mathrm{d}D - \frac{\partial g_2}{\partial t}\mathrm{d}t = 0 \tag{10}$$

Comparing Equation (3) with Equation (10), we may find that the possible displacements here are not the same as those in the former case which characterize the metastable states for they satisfy the different constraint equations, respectively. (In the latter case, the possible displacements are time-dependent, whereas in the former case they are not.). Yet, the integral of possible displacement $\mathrm{d}D$ is the possible electric displacement in every case. The possible electric displacements which characterize one certain metastable state vary with the cases. A self-contradiction arises. So the metastable states can not come into reality.
What are the real states among a phase transition process? In fact, both the evolution with time and the spatial heterogeneity need to be considered when the system is out of equilibrium (Gordon, 2001; Gordon et al., 2002; Ai, 2006; Ai, 2007). Just as what will be shown in Section 3, the real states should be the stationary ones, which do not vary with the time but may be not metastable.

3. Real path: existence of stationary states

The real path of a first-order ferroelectric phase transition is believed by us to consist of a series of stationary states. At first, this was conjectured according to the experimental results, then was demonstrated reliable with the aid of non-equilibrium variational principles.

3.1 Conjecture of stationary states based on experiments

Because in the experiments the ferroelectric phase transitions are often achieved by the quasi-static heating or cooling, we conjectured that they are stationary states processes (Ai, 2006). The results on the motion of interface in ferroelectrics and antiferroelectrics support our opinion (Yufatova et al., 1980; Dec, 1988; Dec & Yurkevich, 1990). From Figure 2, we may find that the motion of interface is jerky especially when the average velocity v_a is small. A sequence of segments of time corresponding to the states of rest may be found. This reveals that in these segments of time (characteristic time of phase transition) the stationary

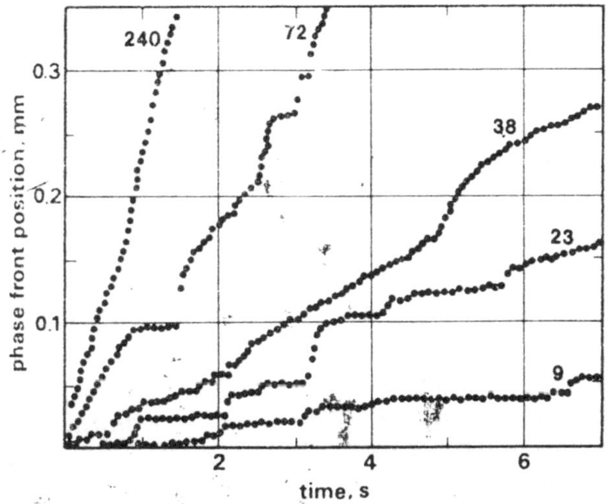

Fig. 2. The position of phase boundary as a function of time for $NaNbO_3$ single crystal for the various values of average velocity v_a (the values in μ m/s given against the curves). (Dec & Yurkevich, 1990).

distributions of temperature, heat flux, stress, etc. may be established. Otherwise, if the motion of interface is continuous and smooth, with the unceasing moving of interface (where the temperature is T_C) to the inner part, the local temperature of outer part must change to keep the temperature gradient ∇T of this region unchanged for it is determined by $\pm l\rho \boldsymbol{v}' = \boldsymbol{J}_q^{diff} = -\kappa \nabla T$, where l is the latent heat (per unit mass), ρ is the mass density, $\boldsymbol{v}'$ is the velocity of interface (where the phase transition is occurring), $\boldsymbol{J}_q^{diff}$ is the diffusion of heat, i.e. heat conduction, κ is the thermal conductivity (and maybe a tensor). Then, the states are not stationary.

3.2 Theoretical evidence based on non-equilibrium variational principles

The non-equilibrium variational principles are just the analogue and generalization of the variational principles in the analytical dynamics. The principle of least dissipation of energy, the Gauss's principle of least constraint and the Hamiltonian prinple etc. in the non-equilibrium thermodynamics play the fundamental roles as those in the analytical dynamics. They describe the characteristics of stationary states or determine the real path of non-equilibrium processes.

For the basic characteristic of non-equilibrium processes is the dissipation of energy, we define the dissipation function ϕ as

$$\phi = \sigma_s - \pi \tag{11}$$

where σ_s is the rate of local entropy production and π is the external power supply (per unit volume and temperature). Let $\lambda = \{\lambda_i\}$ represents the set of extensive, pseudo-thermodynamic variables and its derivative with the time t, i.e. $\dot{\lambda}$ represents the thermodynamic fluxes. And we define a thermodynamic force X_{Ti} and a dissipative force X_{Di} as

$$X_{Ti}(\lambda)=\frac{\partial\sigma_s}{\partial\dot{\lambda}_i} \tag{12}$$

$$X_{Di}(\dot{\lambda})=\frac{\partial\phi}{\partial\dot{\lambda}_i} \tag{13}$$

Let $\xi=\{\xi_i\}$ represent the deviation from a given non-equilibrium stationary state and provided it is small

$$\xi=\lambda-\lambda^{(0)} \tag{14}$$

$$|\xi/\lambda|<<1 \tag{15}$$

where $\lambda^{(0)}$ corresponds with the stationary state. The thermodynamic force can be expanded in terms of the deviation

$$X_{Ti}(\lambda)=\chi_{Ti}^{(0)}+\chi_{Ti}^{(1)}(\xi)+\chi_{Ti}^{(2)}(\xi^2)+\cdots \tag{16}$$

The linear variation in the thermodynamic force can be written as

$$\chi_{Ti}^{(1)}(\xi)=S_{ij}\xi_j \tag{17a}$$

(sum up with respect to j). In certain diagonal representation, Equation (17a) can be re-written as

$$\chi_{Ti}^{(1)}(\xi)=S_i\xi_i \tag{17b}$$

(not sum up with respect to i). S is a negatively-definite matrix.
The dissipation function ϕ can be expanded in terms of the deviation

$$\phi=\phi^{(0)}+\delta\phi+\delta^2\phi+\cdots \tag{18}$$

Because of the least dissipation of energy at stationary states, the linear variation in dissipation function should be zero

$$\delta\phi=0 \tag{19}$$

The quadratically non-linear variation in dissipation function can be written as

$$\delta^2\phi=\frac{1}{2}R_{ij}\dot{\xi}_i\dot{\xi}_j \tag{20}$$

(sum up with respect to i,j). For the same reason, it must be non-negative. So, R is a positively-definite matrix. According to Equation (13), the dissipation force adopts the following form

$$X_{Di}=\chi_{Di}^{(1)}(\dot{\xi})+\cdots \tag{21}$$

with

$$\chi_{Di}^{(1)}\left(\dot{\xi}\right)=R_{ij}\dot{\xi}_j \tag{22a}$$

(sum up with respect to *j*). In the certain diagonal representation, Equation (22a) can be re-written as

$$\chi_{Di}^{(1)}\left(\dot{\xi}\right)=R_i\dot{\xi}_i \tag{22b}$$

(not sum up with respect to *i*).
In view of Equation (14) and (16), the entropy production σ_s is given by the expansion

$$\begin{aligned}\sigma_s &= \sigma_s^{(0)}+\delta\sigma_s+\delta^2\sigma_s+\cdots\\ &= X_{Ti}\cdot\dot{\lambda}_i\\ &= \left(\chi_{Ti}^{(0)}+\chi_{Ti}^{(1)}+\chi_{Ti}^{(2)}+\cdots\right)\cdot\left(\dot{\lambda}_i^{(0)}+\dot{\xi}_i\right)\\ &= \chi_{Ti}^{(0)}\cdot\dot{\lambda}_i^{(0)}+\chi_{Ti}^{(0)}\cdot\dot{\xi}_i+\chi_{Ti}^{(1)}\cdot\dot{\lambda}_i^{(0)}\\ &\quad+\chi_{Ti}^{(1)}\cdot\dot{\xi}_i+\chi_{Ti}^{(2)}\cdot\dot{\lambda}_i^{(0)}+\cdots\end{aligned} \tag{23}$$

Correspondingly

$$\sigma_s^{(0)}=\chi_{Ti}^{(0)}\cdot\dot{\lambda}_i^{(0)} \tag{24}$$

$$\delta\sigma_s=\chi_{Ti}^{(0)}\cdot\dot{\xi}_i+\chi_{Ti}^{(1)}\cdot\dot{\lambda}_i^{(0)} \tag{25}$$

$$\delta^2\sigma_s=\chi_{Ti}^{(1)}\cdot\dot{\xi}_i+\chi_{Ti}^{(2)}\cdot\dot{\lambda}_i^{(0)} \tag{26}$$

(sum up with respect to *i*). Due to the minimum entropy production at stationary states, the linear variation in the entropy production should be zero

$$\delta\sigma_s=0 \tag{27}$$

Similarly, the power function π can be expanded about the stationary state, i.e.

$$\pi=\pi^{(0)}+\delta\pi+\delta^2\pi+\cdots \tag{28}$$

And correspondingly, we can define

$$X_{Ei}\left(\lambda\right)=\frac{\partial\pi}{\partial\dot{\lambda}_i}=\chi_{Ei}^{(0)}+\chi_{Ei}^{(1)}+\chi_{Ei}^{(2)}+\cdots \tag{29}$$

Thus, repeating the steps between Equations (23) and (26), we have

$$\pi^{(0)}=\chi_{Ei}^{(0)}\cdot\dot{\lambda}_i^{(0)} \tag{30}$$

$$\delta\pi=\chi_{Ei}^{(0)}\cdot\dot{\xi}_i+\chi_{Ei}^{(1)}\cdot\dot{\lambda}_i^{(0)} \tag{31}$$

$$\delta^2\pi = \chi_{Ei}^{(1)} \cdot \dot{\xi}_i + \chi_{Ei}^{(2)} \cdot \dot{\lambda}_i^{(0)} \tag{32}$$

(sum up with respect to i).
Introducing the expansions, i.e. Equations (18), (23) and (28) into Equation (11) and setting equal terms of the same order, we obtain at first order

$$\delta\pi = \chi_{Ei}^{(0)} \cdot \dot{\xi}_i + \chi_{Ei}^{(1)} \cdot \dot{\lambda}_i^{(0)} = 0 \tag{33}$$

(sum up with respect to i). Further, we may find that

$$\frac{\chi_{Ti}^{(0)}}{\chi_{Ti}^{(1)}} = \frac{\chi_{Ei}^{(0)}}{\chi_{Ei}^{(1)}} = -\frac{\dot{\lambda}_i^{(0)}}{\dot{\xi}_i} \tag{34}$$

At second order we find

$$\frac{1}{2}\chi_{Di}^{(1)} = \chi_{Ti}^{(1)} - \chi_{Ei}^{(1)} \tag{35}$$

(in Equations (34), (35), not sum up with respect to i).
By analogy, here we can give the thermodynamic Gauss's principle of least constraint which makes the system choose a real path. Let X_T be the active force and $-X_E$ be the constraining force. The thermal inertia may be R_i. Then the constraint Z can be defined as

$$Z = \frac{1}{2R_i}\left(X_{Di} - X_{Ti}\right)^2 = \frac{1}{2R_i}X_{Ei}^2 \tag{36}$$

(sum up with respect to i). The real path makes Z be a minimum, i.e.

$$\delta Z = 0 \tag{37}$$

Assume that

$$\chi_{Ei}^{(1)} = V_{ij}\xi_j \tag{38a}$$

(sum up with respect to j) and in the diagonal representation

$$\chi_{Ei}^{(1)} = V_i\xi_i \tag{38b}$$

(not sum up with respect to i). Henceforth, we will make some discussions in two cases separately.
No external power supply. Then

$$X_{Ei} = 0 \tag{39}$$

According to Equation (35), at second order level, we have

$$\frac{1}{2}\chi_{Di}^{(1)} = \chi_{Ti}^{(1)} \tag{40}$$

In view of Equations (17b) and (22b), we obtain the evolution of deviation ξ_i

$$\xi_i = \xi_i(0)e^{\frac{2S_i}{R_i}t} \quad (41)$$

where $\xi_i(0)$ is the initial value of ξ_i. Equation (41) defines the real path with the addition that R_i should be a suitable value R_i^*. It can be determined by Equations (34) and (41)

$$R_i^* = -\frac{2\chi_{Ti}^{(0)}}{\dot{\lambda}_i^{(0)}} \quad (42)$$

The external power supply exists. Similarly, we can obtain the evolution of deviation ξ_i

$$\xi_i = \xi_i(0)e^{\frac{2(S_i - V_i)}{R_i}t} \quad (43)$$

If V_i, R_i assume the suitable values V_i^{**}, R_i^{**}, the system choose a real path. They can be determined by Equations (34) and (43)

$$V_i^{**} = \frac{\chi_{Ei}^{(0)}}{\chi_{Ti}^{(0)}} S_i \quad (44)$$

$$R_i^{**} = \frac{2\left(\chi_{Ei}^{(0)} - \chi_{Ti}^{(0)}\right)}{\dot{\lambda}_i^{(0)}} \quad (45)$$

(in every equation from Equation (39) to (45), not sum up with respect to i).

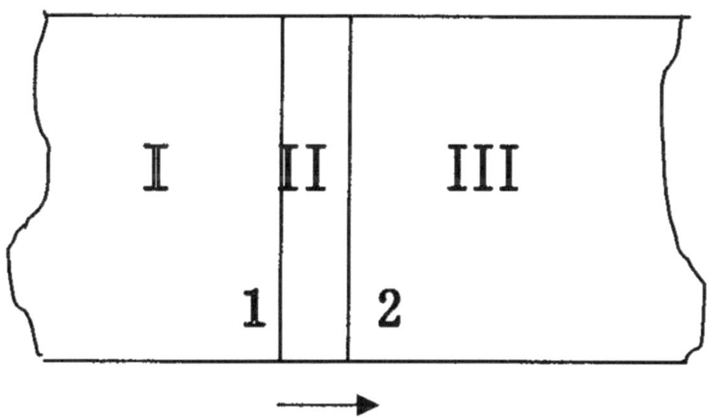

Fig. 3. Three types of regions and their interfaces in the ferroelectric-paraelectric system in which a first-order phase transition is occurring. (Ai et al., 2008)

Both the real paths in the two cases reveal that the deviations decrease exponentially when the system regresses to the stationary states. Stationary states are a kind of attractors to non-equilibrium states. The decreases are steep. So the regressions are quick. It should be noted that we are interested in calculating the change in the generalized displacements during a macroscopically small time interval. In other words, we are concerned with the determination of the path of an irreversible process which is described in terms of a finite difference equation. In the limit as the time interval is allowed to approach zero, we obtain the variational equation of thermodynamic path.

So, if the irreversible process is not quick enough, it can be regarded as the one that consists of a series of stationary states. The ferroelectric phase transitions are usually achieved by the quasi-static heating or cooling in the experiments. So, the processes are not quick enough to make the states deviate from the corresponding stationary states in all the time. In Figure 3, three types of regions and their interfaces are marked I, II, III, 1, 2 respectively. The region III where the phase transition will occur is in equilibrium and has no dissipation. In the region I where the phase transition has occurred, there is no external power supply, and in the region II (i.e. the paraelectric-ferroelectric interface as a region with finite thickness instead of a geometrical plane) where the phase transition is occurring, there exists the external supply, i.e. the latent heat (per unit volume and temperature). According to the former analysis in the two cases, we may conclude that they are in stationary states except for the very narrow intervals of time after the sudden lose of phase stability.

4. Thermo-electric coupling

In the paraelectric-ferroelectric interface dynamics induced by the latent heat transfer (Gordon, 2001; Gordon et al., 2002), the normal velocity of interface v_n was obtained

$$v_n = \frac{1}{l\rho}\left[k_{fer}\left(\nabla T\right)_{fer} - k_{par}\left(\nabla T\right)_{par}\right]\cdot \boldsymbol{n} \tag{46}$$

where l is the latent heat (per unit mass), ρ is the density of metastable phase (paraelectric phase), k_{fer} is the thermal conductivity coefficient of ferroelectric phase, k_{par} is the thermal conductivity coefficient of paraelectric phase, $\left(\nabla T\right)_{fer}$ is the temperature gradient in ferroelectric phase part, $\left(\nabla T\right)_{par}$ is the temperature gradient in paraelectric phase part, $\boldsymbol{n}$ is the unit vector in normal direction. The temperature gradients can be studied from the point of view that a ferroelectric phase transition is a stationary, thermo-electric coupled transport process (Ai, 2006).

4.1 Local entropy production

In the thermo-electric coupling case, the Gibbs equation was given as the following

$$T\mathrm{d}s = \mathrm{d}u - \boldsymbol{E}\cdot \mathrm{d}\boldsymbol{D} - \sum_i \mu_i \mathrm{d}n_i \tag{47}$$

where $T, \boldsymbol{E}, \boldsymbol{D}$ is the temperature, the electric field intensity and the electric displacement within a random small volume, respectively; s, u, μ_i, n_i is the entropy density, the internal energy density, the chemical potential and the molar quantity density in the small volume, respectively. And there, it was assumed that the crystal system is mechanically-free (No force is exerted on it). Differentiating Equation (47) and using the following relations

$$\frac{\partial u}{\partial t} + \nabla\cdot \boldsymbol{J}_u = 0 \tag{48}$$

$$\frac{\partial \boldsymbol{D}}{\partial t} = \frac{\partial(\varepsilon_0 \boldsymbol{E} + \boldsymbol{P})}{\partial t} = \boldsymbol{J}_P \tag{49}$$

$$\frac{\partial n_i}{\partial t} + \nabla \cdot \boldsymbol{J}_{ni} = 0 \tag{50}$$

we have

$$\frac{\partial s}{\partial t} = -\frac{1}{T}\nabla \cdot \boldsymbol{J}_u - \frac{\boldsymbol{E}}{T} \cdot \boldsymbol{J}_P + \sum_i \frac{\mu_i}{T} \nabla \cdot \boldsymbol{J}_{ni} \tag{51}$$

where $\boldsymbol{J}_u, \boldsymbol{J}_P, \boldsymbol{J}_{ni}$ is the energy flux, the polarization current and the matter flux, respectively; $\boldsymbol{P}$ is the polarization. $\boldsymbol{J}_u$ should consist of three parts: the energy flux caused by the heat conduction, the energy flux caused by the charge transport, the energy flux caused by the matter transport

$$\boldsymbol{J}_u = \boldsymbol{J}_q + \phi \boldsymbol{J}_P + \sum_i \mu_i \boldsymbol{J}_{ni} \tag{52}$$

where $\boldsymbol{J}_q$ is the heat flux, ϕ is the electrical potential and satisfies

$$\boldsymbol{E} = -\nabla \phi \tag{53}$$

Then we deduce the following

$$\frac{\partial s}{\partial t} = -\nabla \cdot \left(\frac{\boldsymbol{J}_q + \phi \boldsymbol{J}_P}{T} \right) + \boldsymbol{J}_q \cdot \nabla \left(\frac{1}{T} \right) + \boldsymbol{J}_P \cdot \nabla \left(\frac{\phi}{T} \right) - \sum_i \frac{\boldsymbol{J}_{ni}}{T} \cdot \nabla \mu_i \tag{54}$$

If we define a entropy flux $\boldsymbol{J}_s$ and a rate of local entropy production σ_s as

$$\boldsymbol{J}_s = \frac{\boldsymbol{J}_q + \phi \boldsymbol{J}_P}{T} \tag{55}$$

$$\sigma_s = \boldsymbol{J}_q \cdot \nabla \left(\frac{1}{T} \right) + \boldsymbol{J}_P \cdot \nabla \left(\frac{\phi}{T} \right) - \sum_i \frac{\boldsymbol{J}_{ni}}{T} \cdot \nabla \mu_i \tag{56}$$

Equation (54) can be written as

$$\frac{\partial s}{\partial t} + \nabla \cdot \boldsymbol{J}_s = \sigma_s \tag{57}$$

This is the local entropy balance equation. We know, the system is in the crystalline states before and after a phase transition so that there is no diffusion of any kind of particles in the system. So, $\boldsymbol{J}_{ni} = 0$. The local entropy production can be reduced as

$$\sigma_s = \boldsymbol{J}_q \cdot \nabla \left(\frac{1}{T} \right) + \boldsymbol{J}_P \cdot \nabla \left(\frac{\phi}{T} \right) \tag{58}$$

We know the existence of ferroics is due to the molecular field. It is an inner field. So we must take it into account. Here, the electric field should be the sum of the outer electric field $\boldsymbol{E}_o$ and the inner electric field $\boldsymbol{E}_i$

$$\boldsymbol{E} = \boldsymbol{E}_o + \boldsymbol{E}_i \tag{59}$$

Correspondingly, there are the outer electrical potential ϕ_o and the inner electrical potential ϕ_i and they satisfy

$$\boldsymbol{E}_o = -\nabla \phi_o \tag{60}$$

$$\boldsymbol{E}_i = -\nabla \phi_i \tag{61}$$

If the outer electric field is not applied, ϕ_o can be a random constant. There is no harm in letting the constant equal to zero. Then the entropy production equals

$$\sigma_s = \boldsymbol{J}_q \cdot \nabla\left(\frac{1}{T}\right) + \boldsymbol{J}_P \cdot \nabla\left(\frac{\phi_i}{T}\right) \tag{62}$$

According to the crystal structures of ferroelectrics (Lines & Glass, 1977), we know the polarization current $\boldsymbol{J}_P$ originates from the displacement or ordering of ions in ferroelectrics. We may consider it as the transport of charges influenced by the inner electric field.

4.2 Description of phase transitions

Assume the outer electric field is not applied. Here are the thermodynamic fluxes $\boldsymbol{J}_q, \boldsymbol{J}_P$ and the corresponding thermodynamic forces $\boldsymbol{X}_q, \boldsymbol{X}_P$

$$\boldsymbol{X}_q = \nabla\left(\frac{1}{T}\right) \tag{63}$$

$$\boldsymbol{X}_P = \nabla\left(\frac{\phi_i}{T}\right) \tag{64}$$

$\boldsymbol{J}_i$ can be expanded linearly with $\boldsymbol{X}_j$ $(i, j = q, P)$

$$\boldsymbol{J}_q = \boldsymbol{L}_{qq} \cdot \boldsymbol{X}_q + \boldsymbol{L}_{qP} \cdot \boldsymbol{X}_P \tag{65}$$

$$\boldsymbol{J}_P = \boldsymbol{L}_{Pq} \cdot \boldsymbol{X}_q + \boldsymbol{L}_{PP} \cdot \boldsymbol{X}_P \tag{66}$$

where $\boldsymbol{L}_{qq}, \boldsymbol{L}_{qP}, \boldsymbol{L}_{Pq}$ and $\boldsymbol{L}_{PP}$ are the transport coefficients, which are four second-order tensors. The Onsager relations tell us (Lavenda, 1978)

$$\boldsymbol{L}_{qP} = \boldsymbol{L}_{Pq} \tag{67}$$

According to the reasons expounded in (Ai, 2006), we may regard the force $\boldsymbol{X}_P$ of the region where a ferroelectric phase transition is occurring as a large constant roughly in the characteristic times of phase transition. For the ferroelectric phase transition may be regarded as a stationary state process, the principle of minimum entropy production must be satisfied (Lavenda, 1978)

According to Equations (62), (65), (66), we have

$$\sigma_s = \boldsymbol{L}_{qq} : \boldsymbol{X}_q \boldsymbol{X}_q + 2\boldsymbol{L}_{qP} : \boldsymbol{X}_q \boldsymbol{X}_P + \boldsymbol{L}_{PP} : \boldsymbol{X}_P \boldsymbol{X}_P \tag{68}$$

If there is no restriction on X_q and X_P, according to the conditions on which the entropy production is a minimum

$$\left(\frac{\partial \sigma_s}{\partial X_q}\right)_{X_P} = 2L_{qq} \cdot X_q + 2L_{qP} \cdot X_P = 2J_q = 0 \tag{69}$$

$$\left(\frac{\partial \sigma_s}{\partial X_P}\right)_{X_q} = 2L_{Pq} \cdot X_q + 2L_{PP} \cdot X_P = 2J_P = 0 \tag{70}$$

we know the stationary states are equilibrium ones actually. If we let X_q (or X_P) be a constant, according to Equation (70) (or (69)) we know J_P (or J_q), which is corresponded with another force X_P (or X_q), should be zero.
Then, a first-order ferroelectric phase transition can be described by the second paradigm. Since the force X_P of the region where the phase transition is occurring is a large constant, the flux J_q of the region should be zero (but $J_P \neq 0$). This states clearly that the pure heat conduction and the heat conduction induced by the thermo-electric coupling cancel out each other so as to release or absorb the latent heat. It is certain that the latent heat passes through the region where the phase transition has occurred (at the outside of the region where the phase transition is occurring) and exchanges itself with the thermal bath. Accompanied with the change of the surface's temperature and the unceasing jerky moving of the region where the phase transition is occurring, a constant temperature gradient is kept in the region where the phase transition has occurred, i.e. the force X_q is a constant. So, the flux $J_P = 0$ (but $J_q \neq 0$). This states clearly that the electric displacement of the region where the phase transition has occurred will not change but keep the value at Curie temperature or zero until the phase transition finishes. Differently, the region where the phase transition will occur should be described by the first paradigm for there is no restriction on the two forces X_P, X_q. The states of this region are equilibrium ones. So the temperature gradient ∇T should be zero.

4.3 Verification of interface dynamics

Considering that $(\nabla T)_{par} \approx 0$ for the region where the phase transition will occur (i.e. the paraelectric phase part) can be regarded as an equilibrium system, we modify equation (46) as

$$v_n = \frac{k_{fer} (\nabla T)_{fer} \cdot \boldsymbol{n}}{l\rho} \tag{71}$$

In order to compare it with experiments, we make use of the following values which are about $PbTiO_3$ crystal: ρ =7.1g/cm^3 (Chewasatn & Milne, 1994), l =900cal/mol (Nomura & Sawada, 1955), k_{fer} =8.8$\times$10^{5}erg/cm . s . K (Mante & Volger, 1967). The value of the velocity of the interface's fast motion, which has been measured by the experiments, is 0.5mm/s (Dec, 1988). According to equation (71), we calculate the corresponding temperature gradient to be 57.35K/cm. However, in (Dec, 1989) it is reported that the experimental temperature gradient varies from 1.5 to 3.5K/mm while the experimental velocity of interface's motion varies from 732 to 843μm/s. Considering the model is rather rough, we may conclude that the theory coincides with the experiments.

5. Thermo-electro-mechanical coupling

The comprehensive thermo-electro-mechanical coupling may be found in the ferroelectric phase transition processes. Because there exists not only the change of polarization but also the changes of system's volume and shape when a ferroelectric phase transition occurs in it, the mechanics can not be ignored even if it is mechanically-free, i.e. no outer force is exerted on it. To a first-order ferroelectric phase transition, it occurs at the surface layer of system firstly, then in the inner part. So, the stress may be found in the system.
Since one aspect of the nature of ferroelectric phase transitions is the thermo-electro-mechanical coupling, we take the mechanics into account on the basis of Section 4, where only the thermo-electric coupling has been considered. This may lead to a complete description in the sense of continuum physics. In this section, the mathematical deducing is reduced for it is complicated. The details may be referred in (Ai, 2007).

5.1 Deformation mechanics

For a continuum, the momentum equation in differential form can be written as

$$\nabla \cdot \boldsymbol{\sigma} + \rho \boldsymbol{f} = \rho \boldsymbol{a} \tag{72}$$

where $\boldsymbol{\sigma}, \boldsymbol{f}, \boldsymbol{a}, \rho$ is the stress, the volume force exerted on unit mass, the acceleration and the mass density, respectively. Let $k = \frac{1}{2}\boldsymbol{v}^2$ be the local kinetic energy density (per unit mass), with $\boldsymbol{v}$ is the velocity. Then

$$\frac{\mathrm{d}k}{\mathrm{d}t} = \boldsymbol{v} \cdot \boldsymbol{a} \tag{73}$$

where t is the time. In terms of

$$\nabla \cdot (\boldsymbol{v} \cdot \boldsymbol{\sigma}) = \boldsymbol{v} \cdot (\nabla \cdot \boldsymbol{\sigma}) + (\nabla \boldsymbol{v}) : \boldsymbol{\sigma} \tag{74}$$

$$\boldsymbol{\sigma} : (\nabla \boldsymbol{v}) = \boldsymbol{\sigma} : \boldsymbol{d} \tag{75}$$

we can deduce the following balance equation of mechanical energy basing on Equations (72) (73)

$$\rho \frac{\mathrm{d}k}{\mathrm{d}t} - \nabla \cdot (\boldsymbol{v} \cdot \boldsymbol{\sigma}) = \rho \boldsymbol{v} \cdot \boldsymbol{f} - \boldsymbol{\sigma} : \boldsymbol{d} \tag{76}$$

which is in differential form and in Lagrangian form. Or in Eulerian form

$$\frac{\partial(\rho k)}{\partial t} + \nabla \cdot (\rho k \boldsymbol{v} - \boldsymbol{v} \cdot \boldsymbol{\sigma}) = \rho \boldsymbol{v} \cdot \boldsymbol{f} - \boldsymbol{\sigma} : \boldsymbol{d} \tag{77}$$

where $\boldsymbol{d} = \frac{1}{2}(\nabla \boldsymbol{v} + \boldsymbol{v} \nabla) = \frac{1}{2}\left[\nabla \boldsymbol{v} + (\nabla \boldsymbol{v})^T\right]$ is the rate of deformation or strain rate (the superscript "T" means transposition).
To a ferroelectric phase transition, $\boldsymbol{f}$ and $\boldsymbol{\sigma}$ may be the nominal volume force and stress, which are the embodiments of the actions of thermo-electro-mechanical coupling and are two inner fields. Generally, they are the sums of real and nominal volume force or stress

$$\boldsymbol{f} = \boldsymbol{f}^{real} + \boldsymbol{f}^{nom} \tag{78}$$

$$\boldsymbol{\sigma} = \boldsymbol{\sigma}^{real} + \boldsymbol{\sigma}^{nom} \tag{79}$$

The nominal volume force and stress are not zero until the eigen (free) deformation of system finishes in phase transitions. If they are zero, the eigen (free) deformation finishes.

5.2 Local entropy production and description of phase transitions

The Gibbs equation was given as the following

$$T\mathrm{d}s + \sum_i \mu_i \mathrm{d}n_i + \frac{1}{\rho}\boldsymbol{E}\cdot\mathrm{d}\boldsymbol{D} + \boldsymbol{f}\cdot\mathrm{d}\boldsymbol{r} + \frac{1}{\rho}\nabla\cdot(\boldsymbol{v}\cdot\boldsymbol{\sigma})\mathrm{d}t = \mathrm{d}u + \mathrm{d}k \tag{80}$$

where $T, \boldsymbol{E}, \boldsymbol{D}$ is the local temperature, the local electric field intensity and the local electric displacement, respectively; s, u, n_i, μ_i is the local entropy density (per unit mass), the local internal energy density (per unit mass), the local molar quantity density (per unit mass) and the chemical potential, respectively; $\boldsymbol{r}$ is the displacement vector. If the outer electric field is not applied, the quantity $\boldsymbol{E}$ is the inner electric field $\boldsymbol{E}^{in}$ only (Ai, 2006).
Make the material derivative of Equation (80) with t, then obtain

$$\rho\frac{\mathrm{d}s}{\mathrm{d}t} + \frac{\rho}{T}\sum_i \mu_i \frac{\mathrm{d}n_i}{\mathrm{d}t} + \frac{1}{T}\boldsymbol{E}^{in}\cdot\frac{\mathrm{d}\boldsymbol{D}}{\mathrm{d}t} + \frac{\rho}{T}\boldsymbol{f}\cdot\frac{\mathrm{d}\boldsymbol{r}}{\mathrm{d}t} + \frac{1}{T}\nabla\cdot(\boldsymbol{v}\cdot\boldsymbol{\sigma}) = \frac{\rho}{T}\frac{\mathrm{d}e}{\mathrm{d}t} \tag{81}$$

where e is the total energy with $e = u + k$. We know, the system is in the crystalline states before and after a phase transition so that there is no diffusion of any kind of particles in the system. So, n_i as the local molar quantity (per unit mass) does not change with t, i.e. $\frac{dn_i}{dt} = 0$. $\frac{d\boldsymbol{D}}{dt}$ stands for the polarization current $\boldsymbol{J}_P$, while $\frac{d\boldsymbol{r}}{dt}$ stands for the velocity $\boldsymbol{v}$, $\boldsymbol{E}^{in} = -\nabla\phi^{in}$ (ϕ^{in} stands for the inner electric potential).
After the lengthy and troublesome deduction (Ai, 2007), we obtain the local entropy balance equation in Lagrangian form

$$\rho\frac{\mathrm{d}s}{\mathrm{d}t} = -\nabla\cdot\boldsymbol{J}_s + \sigma_s \tag{82}$$

with the entropy flux $\boldsymbol{J}_s$

$$\boldsymbol{J}_s = \frac{\boldsymbol{J}_q^{diff} + \phi^{in}\boldsymbol{J}_P - \boldsymbol{v}\cdot\boldsymbol{\sigma}}{T} \tag{83}$$

and the rate of local entropy production σ_s

$$\sigma_s = \boldsymbol{J}_q^{diff}\cdot\nabla\left(\frac{1}{T}\right) + \boldsymbol{J}_P\cdot\nabla\left(\frac{\phi^{in}}{T}\right) - \boldsymbol{\sigma}:\left[\boldsymbol{v}\nabla\left(\frac{1}{T}\right)\right] - \rho\boldsymbol{f}\cdot\frac{\boldsymbol{v}}{T} \tag{84}$$

where $\boldsymbol{J}_q^{diff}$ is the diffusion of heat, i.e. heat conduction.

Here are the thermodynamic fluxes $\boldsymbol{J}_q^{diff}, \boldsymbol{J}_P, \boldsymbol{\sigma}(=\boldsymbol{J}_\sigma), \rho\boldsymbol{f}(=\boldsymbol{J}_f)$ and the corresponding thermodynamic forces $\boldsymbol{X}_q, \boldsymbol{X}_P, \boldsymbol{X}_\sigma, \boldsymbol{X}_f$

$$\boldsymbol{X}_q = \nabla\left(\frac{1}{T}\right) \tag{85}$$

$$\boldsymbol{X}_P = \nabla\left(\frac{\phi^{in}}{T}\right) \tag{86}$$

$$\boldsymbol{X}_\sigma = -\boldsymbol{v}\nabla\left(\frac{1}{T}\right) \tag{87}$$

$$\boldsymbol{X}_f = -\frac{\boldsymbol{v}}{T} \tag{88}$$

$J_i(i=q,P,\sigma,f)$ can be expanded linearly with $X_j(j=q,P,\sigma,f)$

$$\boldsymbol{J}_q = \boldsymbol{L}_{qq}\cdot\boldsymbol{X}_q + \boldsymbol{L}_{qP}\cdot\boldsymbol{X}_P + \boldsymbol{L}_{q\sigma}:\boldsymbol{X}_\sigma + \boldsymbol{L}_{qf}\cdot\boldsymbol{X}_f \tag{89}$$

$$\boldsymbol{J}_P = \boldsymbol{L}_{Pq}\cdot\boldsymbol{X}_q + \boldsymbol{L}_{PP}\cdot\boldsymbol{X}_P + \boldsymbol{L}_{P\sigma}:\boldsymbol{X}_\sigma + \boldsymbol{L}_{Pf}\cdot\boldsymbol{X}_f \tag{90}$$

$$\boldsymbol{J}_\sigma = \boldsymbol{L}_{\sigma q}\cdot\boldsymbol{X}_q + \boldsymbol{L}_{\sigma P}\cdot\boldsymbol{X}_P + \boldsymbol{L}_{\sigma\sigma}:\boldsymbol{X}_\sigma + \boldsymbol{L}_{\sigma f}\cdot\boldsymbol{X}_f \tag{91}$$

$$\boldsymbol{J}_f = \boldsymbol{L}_{fq}\cdot\boldsymbol{X}_q + \boldsymbol{L}_{fP}\cdot\boldsymbol{X}_P + \boldsymbol{L}_{f\sigma}:\boldsymbol{X}_\sigma + \boldsymbol{L}_{ff}\cdot\boldsymbol{X}_f \tag{92}$$

where $\boldsymbol{L}_{qq}, \boldsymbol{L}_{PP}, \boldsymbol{L}_{ff}, \boldsymbol{L}_{qP}, \boldsymbol{L}_{Pq}, \boldsymbol{L}_{qf}, \boldsymbol{L}_{fq}, \boldsymbol{L}_{Pf}, \boldsymbol{L}_{fP}$ are nine second-order tensors, $\boldsymbol{L}_{q\sigma}, \boldsymbol{L}_{\sigma q}, \boldsymbol{L}_{P\sigma}, \boldsymbol{L}_{\sigma P}, \boldsymbol{L}_{\sigma f}, \boldsymbol{L}_{f\sigma}$ are six third-order tensors, $\boldsymbol{L}_{\sigma\sigma}$ is a fourth-order tensor. The Onsager relations read

$$L_{ij} = L_{ji} \tag{93}$$

So the rate of local entropy production can be re-written as

$$\sigma_s = \sum_{i,j=1}^{4} L_{ij}X_iX_j \tag{94}$$

For convenience, we have modified the superscripts q, P, σ, f to be 1,2,3,4.

In view of Section 3, the states of a ferroelectric phase transition are stationary. So the principle of minimum entropy production is satisfied. If we keep the former k forces being constants, i.e. $X_i = const(i=1,2,\cdots,k,k<4)$, the conditions on which the local entropy production is a minimum are

$$\left(\frac{\partial\sigma_s}{\partial X_i}\right)_{X_j(\neq X_i)} = \sum_{j=1}^{4}\left(L_{ij}+L_{ji}\right)X_j = 2J_i = 0,\left(i=k+1,\cdots,4\right) \tag{95}$$

This reveals that the fluxes corresponding to the left $4-k$ forces are zero. Of course, if there are no restrictions on $X_i\left(i=1,2,3,4\right)$, all the flues are zero.

We may describe a ferroelectric phase transition by using the two paradigms above similarly as do in Section 4. To a first-order ferroelectric phase transition, the forces $\boldsymbol{X}_P, \boldsymbol{X}_\sigma, \boldsymbol{X}_f$ of the region where the phase transition is occurring can be regarded as three large constants roughly in the characteristic times of phase transition because the electric displacement, the volume and the shape change suddenly. So, the flux $\boldsymbol{J}_q^{diff}$ of the region should be zero (but $\boldsymbol{J}_P \neq 0, \boldsymbol{\sigma} \neq 0, \rho \boldsymbol{f} \neq 0$). This states clearly that the pure heat conduction and the heat conduction induced by the thermo-electric coupling and the thermo-mechanical coupling cancel out each other so as to release or absorb the latent heat. The phase transition occurs at the surface layer firstly, which is mechanically-free. So, when the phase transition occurs in this region, the flux $\boldsymbol{\sigma}$ maybe the nominal stress $\boldsymbol{\sigma}^{nom}$ only, which does work to realize the transformation from the internal energy to the kinetic energy. When the phase transition occurs in the inner part, the flux $\boldsymbol{\sigma}$ should be the sum of $\boldsymbol{\sigma}^{real}$ and $\boldsymbol{\sigma}^{nom}$ because the sudden changes of inner part's volume and shape have to overcome the bound of outer part then $\boldsymbol{\sigma}^{real}$ arises. The region where the phase transition is occurring, i.e. the phase boundary is accompanied with the real stress $\boldsymbol{\sigma}^{real}$ usually, which does work to realize the transformation from the kinetic energy to the internal energy. This has been predicted and described with a propagating stress wave. (Gordon, 1991).

It is certain that the latent heat passes through the region where the phase transition has occurred (at the outside of the region where the phase transition is occurring) and exchange itself with the thermal bath. For $\pm l\rho \boldsymbol{v}_a = \boldsymbol{J}_q^{diff} = -\kappa \nabla T$, a constant temperature gradient ∇T is kept in the region where the phase transition has occurrd, i.e. the force $\boldsymbol{X}_q$ at every site is a constant (which does not change with the time but may vary with the position). So, the fluxes $\boldsymbol{J}_P = \boldsymbol{\sigma} = \rho \boldsymbol{f} = 0$ (but the flux $\boldsymbol{J}_q^{diff} \neq 0$). This states clearly that the electric displacement $\boldsymbol{D}$ will not change but keep the value at Curie temperature or zero until the phase transition finishes and $\boldsymbol{\sigma}^{real} = -\boldsymbol{\sigma}^{nom}$ in this region. Because the electric displacement $\boldsymbol{D}$ and the strain (or deformation) are all determined by the crystal structure of system, $\boldsymbol{J}_P = 0$ reveals that $\boldsymbol{D}$ of this region does not change so does not the crystal structure then does not the strain (or deformation). According to (Gordon, 1991), we know the region where the phase transition has occurred is unstressed, i.e. $\boldsymbol{\sigma}^{real} = 0$, then $\boldsymbol{\sigma}^{nom} = 0$. This reveals that the eigen (free) strain (or deformation) of system induced by the thermo-electro-mechanical coupling of phase transition is complete and the change of it terminates before the phase transition finishes. The two deductions coincide with each other. $\boldsymbol{\sigma}^{real}$ may relaxes via the free surface.

The region where the phase transition will occur should be in equilibrium because there are no restrictions on the forces $\boldsymbol{X}_q, \boldsymbol{X}_P, \boldsymbol{X}_\sigma, \boldsymbol{X}_f$. Whereas, according to (Gordon, 1991), the region is stressed, i.e. $\boldsymbol{\sigma}^{real} \neq 0$. To the heating process of phase transition, this may lead to a change of the spontaneous polarization of this region because of the electro-mechanical coupling (piezoelectric effect).

An immediate result of the above irreversible thermodynamic description is that the action at a distance, which is the kind of heat transfer at phase transitions, is removed absolutely. The latent heat is transferred within a finite time so the occurrence of phase transition in the inner part is delayed. (Of course, another cause is the stress, just see Section 6) In other words, the various parts absorb or release the latent heat at the various times. The action at a distance does not affect the phase transition necessarily.

5.3 Relation between latent heat and spontaneous polarization

By now, among the latent heat, the spontaneous polarization, the stress and the strain, only the latent heat and the spontaneous polarization have been measured often for first-order ferroelectric phase transitions. So, we will consider a simplified case of thermo-electric coupling only so as to establish the relation between the latent heat and the spontaneous polarization in the realm of non-equilibrium thermodynamics.

All the quantities of the region where the phase transition has occurred are marked with the superscript "I"; all the quantities of the region where the phase transition is occurring are marked with the superscript "II"; and all the quantities of the region where the phase transition will occur are marked with the superscript "III". Let's consider the heating processes of phase transition firstly. In the region where the phase transition has occurred,

$$\boldsymbol{J}_q^{I-para} = \boldsymbol{L}_{qq}^{I-para} \cdot \boldsymbol{X}_q^{I-para} = l\rho \boldsymbol{v}_a \tag{96}$$

where we have ignored the difference between the mass density of ferroelectric phase and that of paraelectric phase (almost the same) and denote them as ρ, $\boldsymbol{v}_a$ is the average velocity of interface. In the region where the phase transition is occurring,

$$\boldsymbol{J}_q^{II} = \boldsymbol{L}_{qq}^{II} \cdot \boldsymbol{X}_q^{II} + \boldsymbol{L}_{qP}^{II} \cdot \boldsymbol{X}_P^{II} = 0 \tag{97}$$

$$\boldsymbol{J}_P^{II} = \boldsymbol{L}_{Pq}^{II} \cdot \boldsymbol{X}_q^{II} + \boldsymbol{L}_{PP}^{II} \cdot \boldsymbol{X}_P^{II} \tag{98}$$

The heat which is transferred to the region where the phase transition is occurring is absorbed as latent heat because the pure heat conduction and the heat conduction induced by the thermo-electric coupling cancel out each other. So,

$$l\rho \boldsymbol{v}_a = \boldsymbol{L}_{qq}^{II} \cdot \boldsymbol{X}_q^{II} \tag{99}$$

According to Eq. (97)-(99), we work out

$$\boldsymbol{J}_P^{II} = l\rho \left[\boldsymbol{L}_{Pq}^{II} \cdot \left(\boldsymbol{L}_{qq}^{II} \right)^{-1} - \boldsymbol{L}_{PP}^{II} \cdot \left(\boldsymbol{L}_{qP}^{II} \right)^{-1} \right] \cdot \boldsymbol{v}_a \tag{100}$$

where the superscript "-1" means reverse. While

$$\int \boldsymbol{J}_P^{II} \mathrm{d}t = -\boldsymbol{D}^{spon} = \varepsilon_0 \nabla \phi^{in-III} - \boldsymbol{P}^{spon-III} = -\boldsymbol{P}^{spon-III} \tag{101}$$

where we utilized the boundary condition of $\boldsymbol{D}$ and considered the region where the phase transition will occur is in equilibrium, the superscript "spon" means spontaneous. So,

$$\boldsymbol{P}^{spon} = \boldsymbol{P}^{spon-III} = -\int \boldsymbol{J}_P^{II} \mathrm{d}t \tag{102}$$

The relation between the latent heat and the spontaneous polarization are obtained. In the cooling processes of phase transition,

$$-l\rho \boldsymbol{v}_a = \boldsymbol{L}_{qq}^{I-ferr} \cdot \boldsymbol{X}_q^{I-ferr} + \boldsymbol{L}_{qP}^{I-ferr} \cdot \boldsymbol{X}_P^{I-ferr} = \boldsymbol{L}_{qq}^{II} \cdot \boldsymbol{X}_q^{II} \tag{103}$$

$$\int \boldsymbol{J}_P^{II'} \mathrm{d}t = \boldsymbol{D}^{spon-I} = -\varepsilon_0 \nabla \phi^{in-I} + \boldsymbol{P}^{spon-I} \tag{104}$$

Repeating the above steps, we obtain

$$\boldsymbol{P}^{spon-I} = \int \boldsymbol{J}_P^{II'} \mathrm{d}t + \varepsilon_0 \nabla \phi^{in-I} \tag{105}$$

with

$$\boldsymbol{J}_P^{II'} = -l\rho \left[\boldsymbol{l}_{Pq}^{II} \cdot \left(\boldsymbol{L}_{qq}^{II}\right)^{-1} - \boldsymbol{L}_{PP}^{II} \cdot \left(\boldsymbol{L}_{qP}^{II}\right)^{-1} \right] \cdot \boldsymbol{v}_a = -\boldsymbol{J}_P^{II} \tag{106}$$

Then we find that $\boldsymbol{P}^{spon-III}$ (serves as the equilibrium polarization) is not equal to $\boldsymbol{P}^{spon-I}$ (serves as the non-equilibrium polarization). For the region where the phase transition will occur is stressed, there is some difference between $\boldsymbol{P}^{spon}\left(=\boldsymbol{P}^{spon-III}\right)$ and the equilibrium spontaneous polarization without the affects of stress $\boldsymbol{P}^{spon'}$ because of the piezoelectric effect.

6. Irreversibility: thermal hysteresis and occurrences of domain structure

6.1 Thermal hysteresis

The "thermal hysteresis" of first-order ferroelectric phase transitions is an irreversible phenomenon obviously. But it was treated by using the equilibrium thermodynamics for ferroelectric phase transitions, the well-known Landau-Devonshire theory (Lines & Glass,1977). So, there is an inherent contradiction in this case. The system in which a first-order ferroelectric phase transition occurs is heterogeneous. The occurrences of phase transition in different parts are not at the same time. The phase transition occurs at the surface layer then in the inner part of system. According to the description above, we know a constant temperature gradient is kept in the region where the phase transition has occurred. The temperature of surface layer, which is usually regarded as the temperature of the whole system in experiments, must be higher (or lower) than the Curie temperature. This may lead to the thermal hysteresis.

No doubt that the shape and the area of surface can greatly affect the above processes. We may conclude that the thermal hysteresis can be reduced if the system has a larger specific surface and, the thermal hysteresis can be neglected if a finite system has an extremely-large specific surface. So, the thermal hysteresis is not an intrinsic property of the system.

The region where the phase transition will occur can be regarded as an equilibrium system for there are no restrictions on the forces $\boldsymbol{X}_q, \boldsymbol{X}_P, \boldsymbol{X}_\sigma, \boldsymbol{X}_f$. In other words, the forces and the corresponding fluxes are zero in this region. To a system where a second-order ferroelectric phase transition occurs, the case is somewhat like that of the region where a first-order ferroelectric phase transition will occur. The spontaneous polarization , the volume and the shape of system are continuous at the Curie temperature and change with the infinitesimal magnitudes. This means $\boldsymbol{X}_q, \boldsymbol{X}_P, \boldsymbol{X}_\sigma, \boldsymbol{X}_f$ and $\boldsymbol{J}_q^{diff}, \boldsymbol{J}_P, \boldsymbol{\sigma}, \rho \boldsymbol{f}$ can be arbitrary infinitesimal magnitudes. The second-order phase transition occurs in every part of the system

simultaneously, i.e. there is no the co-existence of two phases (ferroelectric and paraelectric). So, there is no the latent heat and stress. The thermal hysteresis disappears.
The region where a first-order ferroelectric phase transition will occur is stressed. This reveals that the occurrences of phase transition in the inner part have to overcome the bound of outer part, where the phase transition occurs earlier. This may lead to the delay of phase transition in the inner part.

6.2 Occurrences of domain structure

Though the rationalization of the existence of domain structures can be explained by the equilibrium thermodynamics, the evolving characteristics of domain occurrences in ferroelectrics can not be explained by it, but can be explained by the non-equilibrium thermodynamics.
In the region where the phase transition is occurring, the thermodynamic forces $\boldsymbol{X_P}\left(=\nabla\left(\frac{\phi^{in}}{T}\right)\right)$, $\boldsymbol{X_\sigma}\left(=-\boldsymbol{v}\nabla\left(\frac{1}{T}\right)\right)$, $\boldsymbol{X_f}\left(=-\frac{\boldsymbol{v}}{T}\right)$ can be regarded as three large constants in the characteristic time of transition and the thermodynamic flux $\boldsymbol{J_q^{diff}}=0$ (but $\boldsymbol{J_P}\neq 0$, $\sigma\neq 0$, $\rho\boldsymbol{f}\neq 0$). The local entropy production (cf. Equation (84)) reduces to

$$\sigma_s=\boldsymbol{J_P}\cdot\nabla\left(\frac{\phi^{in}}{T}\right)-\sigma:\left[\boldsymbol{v}\nabla\left(\frac{1}{T}\right)\right]-\rho\boldsymbol{f}\cdot\frac{\boldsymbol{v}}{T} \tag{107}$$

Now, we are facing a set of complicated fields of T, $\boldsymbol{v}$ and ϕ^{in}, respectively. Assume that the phase transition front is denoted by S. The points included in S stand for the locations where the transition is occurring. Because the transition occurs along all directions from the outer part to the inner part, we may infer that the orientations of $\nabla\left(\frac{\phi^{in}}{T}\right)$, or of $-\boldsymbol{v}\nabla\left(\frac{1}{T}\right)$ and or of $-\frac{\boldsymbol{v}}{T}$ vary continuously such that they are differently oriented at defferent locations.
There are always several (at least two) symmetry equivalent orientations in the prototype phase (in most cases it is the high temperature phase), which are the possible orientations for spontaneous polarization (or spontaneous deformation or spontaneous displacement). Therefore, the spontaneous polarization, the spontaneous deformation and the spontaneous displacement must take an appropriate orientation respectively to ensure σ_s is a positive minimum when the system transforms from the prototype (paraelectric) phase to the ferroelectric (low temperature) phase. The underlying reasons are that

$$\boldsymbol{J}_P=\frac{\mathrm{d}\boldsymbol{D}}{\mathrm{d}t}=\frac{\mathrm{d}\left(\varepsilon_0\boldsymbol{E}^{in}+\boldsymbol{P}\right)}{\mathrm{d}t} \tag{108}$$

$$\sigma=\boldsymbol{L}:\varepsilon \tag{109}$$

$$\rho\boldsymbol{f}=\rho\frac{\mathrm{d}\boldsymbol{v}}{\mathrm{d}t}-\nabla\cdot\sigma \tag{110}$$

where L, ε, and ε_0 is the modulus of rigidity, the strain and the permittivity of vacuum, respectively. Therefore, P, ε at different locations will be differently oriented. The domain structures in ferroelectrics thus occur.

It seems that the picture of domain occurrences for first-order ferroelectric phase transition systems should disappear when we face second-order ferroelectric phase transition systems. This is true if the transition processes proceed infinitely slowly as expounded by the equilibrium thermodynamics. But any actual process proceeds with finite rate, so it is irreversible. Then the above picture revives.

In (Ai et al., 2010), the domain occurrences in ferromagnetics can be described parallelly by analogy. And the case of ferroelastic domain occurrences is a reduced, simpler one compared with that of ferroelectrics or ferromagnetics.

It is well known that the Landau theory or the Curie principle tells us how to determine the symmetry change at a phase transition. A concise statement is as follows (Janovec, 1976): for a crystal undergoing a phase transition with a space-group symmetry reduction from G_0 to G, whereas G determines the symmetry of transition parameter (or vice versa), it is the symmetry operations lost in going from G_0 to G that determine the domain structure in the low-symmetry phase. The ferroic phase transitions are the ones accompanied by a change of point group symmetry (Wadhawan, 1982). Therefore, the substitution of "point group" for the "space group" in the above statement will be adequate for ferrioc phase transitions. From the above statement, the domain structure is a manifestation of the symmetry operations lost at the phase transition. In our treatment of the domain occurrences in ferroics, we took into account the finiteness of system (i.e. existence of surface) and the irreversibility of process (asymmetry of time). The finiteness of system makes the thermodynamic forces such as $\nabla\left(\frac{\phi^{in}}{T}\right)$, $-\boldsymbol{v}\nabla\left(\frac{1}{T}\right)$, $-\frac{\boldsymbol{v}}{T}$ have infinite space symmetry. The infinite space symmetry, combined with the asymmetry of time, reproduces the symmetry operations lost at the phase transition in the ferroic phase. It can be viewed as an embodiment of time-space symmetry.

After all, for the domain structures can exist in equilibrium systems, they are the equilibrium structures but not the dissipative ones, for the latter can only exist in systems far from equilibrium (Glansdorff & Prigogine, 1971).

7. Acknowledgments

I acknowledge the supports of the Natural Science Foundation Programs of Shandong Province (Grant No. Y2008A36).

8. References

Ai, S.T. (2006). Paraelectric-Ferroelectric Interface Dynamics Induced by Latent Heat Transfer and Irreversible Thermodynamics of Ferroelectric Phase Transitions. *Ferroelectrics*, Vol. 345, No.1, (June, 2006) 59-66, ISSN 1563-5112 (online), 0015-0193 (print).

Ai, S.T. (2007). Mechanical-Thermal-Electric Coupling and Irreversibility of Ferroelectric Phase Transitions. *Ferroelectrics,* Vol.350, No.1 (May, 2007) 81-92, ISSN 1563-5112 (online), 0015-0193 (print).

Ai, S.T.; Xu, C.T.; Wang, Y.L.; Zhang, S.Y.; Ning, X.F. & Noll, E. (2008). Comparison of and Comments on Two Thermodynamic Approaches (Reversible and Irreversible) to

Ferroelectric Phase Transitions. *Phase Transit.*, Vol.81, No.5, (May, 2008) 479-490, ISSN 1029-0338(online), 0141-1594 (print).

Ai, S.T.; Zhang, S.Y.; Ning, X.F.; Wang, Y.L. & Xu C.T. (2010). Non-Equilibrium Thermodynamic Explanation of Domain Occurrences in Ferroics. *Ferroelectrics Lett.*, Vol.37, No.2, (March, 2010) 30-34, ISSN 1563-5228 (online), 0731-5171 (print).

Chewasatn, S. & Milne, S.T. (1994). Synthesis and Characterization of $PbTiO_3$ and Ca and Mn Modified $PbTiO_3$ Fibres Produced by Extrusion of Diol Based Gels. *J. Mat. Sci.*, Vol.29, No.14, (January, 1994) 3621-3629, ISSN 1573-4803 (online), 0022-2461 (print).

Dec, J. (1988). Jerky Phase-Front Motion in $PbTiO_3$ Crystal. *J. Phys. C*, Vol.21, No.7, (March, 1988) 1257-1263, ISSN 0022-3719 (print).

Dec, J. (1989). The Phase Boundary as a Kink. *Ferroelectrics*, Vol.89, No.1, (January, 1989) 193-200, ISSN 1563-5112 (online), 0015-0193 (print).

Dec, J. & Yurkevich, J. (1990). The Antiferroelectric Phase Boundary as a Kink. *Ferroelectrics*, Vol. 110, No.1, (October, 1990) 77-83, ISSN 1563-5112 (online), 0015-0193 (print).

Devonshire, A.F. (1949). Theory of Barium Titanate (Part I). *Phil. Mag.*, Vol.40, No.10, (October, 1949) 1040-1063, ISSN 1478-6443 (online), 1478-6435(print).

Devonshire, A.F. (1951). Theory of Barium Titanate (Part II). *Phil. Mag.*, Vol.42, No.10, (October, 1951) 1065-1079, ISSN 1478-6443 (online), 1478-6435(print).

Devonshire, A.F. (1954). Theory of Ferroelectrics. *Adv. Phys.*, Vol.3, No.2, (May, 1954) 85-130, ISSN 1460-6976 (online), 0001-8732 (print).

Glansdorff, P. & Prigogine, I. (1971). *Thermodynamic Theory of Structure, Stability and Fluctuations*, Wiley-Interscience, ISBN 978-0471302803, New York.

Gordon, A. (1991). Propagation of Solitary Stress Wave at First-Order Ferroelectric Phase Transitions. *Phys. Lett. A*, Vol.154, No.1-2, (March, 1991) 79-80, ISSN 0375-9601.

Gordon, A. (2001). Finite-Size Effects in Dynamics of Paraelectric-Ferroelectric Interfaces Induced by Latent Heat Transfer. *Phys. Lett. A*, Vol.281, No.5-6, (April, 2001) 357-362, ISSN 0375-9601.

Gordon, A.; Dorfman, S. & Fuks, D. (2002). Temperature-Induced Motion of Interface Boundaries in Confined Ferroelectrics. *Phil. Mag. B*, Vol.82, No.1, (January, 2002) 63-71, ISSN 1463-6417 (online), 1364-2812 (print).

Janovec, V. (1976). A Symmetry Approach to Domain Structures. *Ferroelectrics*, Vol.12, No.1, (June, 1976) 43-53, ISSN 1563-5112 (online), 0015-0193 (print).

Lavenda, B.H. (1978). *Thermodynamics of Irreversible Processes*, Macmillan, ISBN 0-333-21616-4, London and Basingstoke.

Lines, M.E. & Glass, A.M. (1977). *Principles and Applications of Ferroelectrics and Related Materials*, Clarendon, ISBN 978-0198507789, Oxford.

Mante, A.J.H. & Volger, J. (1967). The Thermal Conductivity of $PbTiO_3$ in the Neighbourhood of its Ferroelectric Transition Temperature. *Phys. Lett. A*, Vol.24, No.3, (January, 1967) 139-140, ISSN 0375-9601.

Nomura, S. & Sawada, S. (1955). Dielectric Properties of Lead-Strontium Titanate. *J. Phys. Soc. Japan*, Vol.10, No.2, (February, 1955) 108-111, ISSN 1347-4073 (online), 0031-9015 (print).

Wadhawan, V.K. (1982). Ferroelasticity and Related Properties of Crystals. *Phase Transit.*, Vol.3, No.1, (March, 1982) 3-103, ISSN 1029-0338 (online), 0141-1594 (print).

Yufatova, S.M.; Sindeyev, Y.G.; Garilyatchenko, V.G. & Fesenko, E.G. (1980). Different Kinetics Types of Phase Transformations in Lead Titanate. *Ferroelectrics*, Vol.26, No.1 (January, 1980) 809-812, ISSN 1563-5112 (online), 0015-0193 (print).

Zhong, W.L. (1996). *Physics of Ferroelectrics (in Chinese)*, Science Press, ISBN 7-03-005033-9/O. 834, Beijing.

16

Theories and Methods of First Order Ferroelectric Phase Transitions

C L Wang
Shandong University
P R China

1. Introduction

Most ferroelectrics in practical applications are of first order phase transition, such as barium titanate, lead zirconium titanate, etc. However, theoretical investigation of first order phase transition is much more complicated than that of second order phase transitions. Therefore in the chapter, theoretical treatments of ferroelectric phase transition of first order are summarized. In the next section, results from thermodynamic theory are discussed. This kind of theory is often called Landau theory, Landau-Devonshire theory or Landau-Devoshire-Ginzburg theory in literatures. Basic concepts and definitions such as characteristic temperatures, thermal hysteresis and field induced phase transition are presented. In the third section, effective field approach for first order phase transition is formulised. Even though this approach is a very simple statistic physics method, it supplies a very helpful approach to understand many physical phenomena of first order phase transitions. In the fourth section, results from Ising model in a transverse field under mean field approximation are discussed. Then the equivalence of the model with effective field approach is demonstrated. Also application of Monte Carlo simulation on Ising model with four-spin couplings is included. In the last part, understanding of physical property of relaxor ferroelectrics from point view of first order phase transition is addressed.

2. Thermodynamic theory of ferroelectrics

In this section, thermodynamic theory of first order ferroelectric phase transition is formalized. Free energy is analyzed at first. Then definition of characteristic temperatures is discussed. Afterwards temperature dependence of polarization and hysteresis loops are shown for static situation. At last dynamic behavior of polarization under external electric field is presented.

Formalism of thermodynamic description, or Landau theory for ferroelectrics can be found in many classical books (Blinc & Zeks ,1974; Lines & Glass, 1977), or from website like wikipedia (http://en.wikipedia.org/wiki/Ferroelectricity). Based on Landau theory, the free energy of a ferroelectric material, in the absence of an electric field and applied stress may be written as a Taylor expansion in terms of the order parameter, polarization (P). If a sixth order expansion is used (i.e. eighth order and higher terms truncated), the free energy is given by:

$$\Delta G=\frac{1}{2}\alpha_0\left(T-T_0\right)\left(P_x^2+P_y^2+P_z^2\right)+\frac{1}{4}\alpha_{11}\left(P_x^4+P_y^4+P_z^4\right)+\frac{1}{2}\alpha_{12}\left(P_x^2P_y^2+P_y^2P_z^2+P_z^2P_x^4\right)$$
$$+\frac{1}{6}\alpha_{111}\left(P_x^6+P_y^6+P_z^6\right)+\frac{1}{6}\alpha_{112}\left[P_x^4\left(P_y^2+P_z^2\right)+P_y^4\left(P_z^2+P_x^2\right)+P_z^4\left(P_x^2+P_y^2\right)\right] \quad (1)$$
$$+\frac{1}{6}\alpha_{123}P_x^2P_y^2P_z^2$$

where ΔG stands for the free energy difference of ferroelectric phase and that of paraelectric phase, P_x, P_y, and P_z are the components of the polarization vector in the x, y, and z directions respectively. The coefficients, α_i,α_{ij},α_{ijk} must be consistent with the crystal symmetry. To investigate domain formation and other phenomena in ferroelectrics, these equations are often used in the context of a phase field model. Typically, this involves adding a gradient term, an electrostatic term and an elastic term to the free energy. In all known ferroelectrics, α_0 and α_{111} are of positive values. These coefficients may be obtained experimentally or from *ab-initio* simulations. For ferroelectrics with a first order phase transition, α_{11} is negative, and α_{11} is positive for a second order phase transition.
The ferroelectric properties for a cubic to tetragonal phase transition may be obtained by considering the one dimension expression of the free energy which is:

$$\Delta G=\frac{1}{2}\alpha_0\left(T-T_0\right)P_x^2+\frac{1}{4}\alpha_{11}P_x^4+\frac{1}{6}\alpha_{111}P_x^6 \quad (2)$$

The shape of above free energy is schematically shown in Fig. 1 at different temperatures for first order phase transitions. In the following numerical calculations, parameter values are set as α_0=1, T_0=1, α_{11}=-1and α_{111}=1 in arbitrary unit for simplicity.

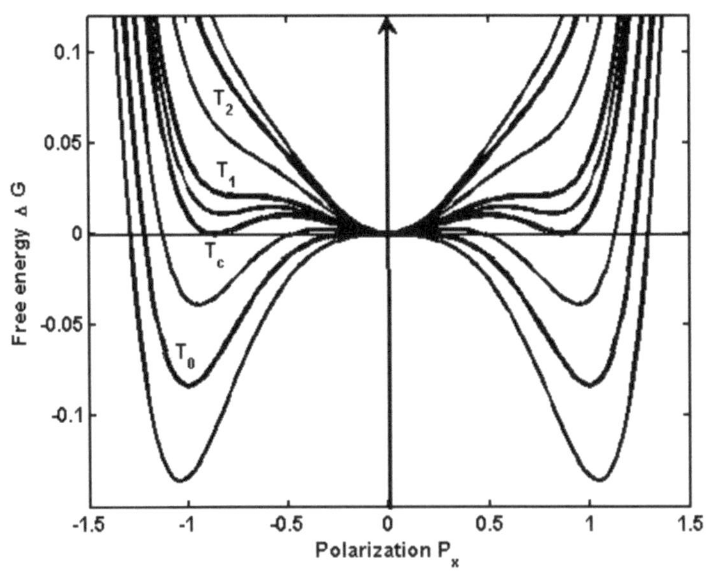

Fig. 1. Free energy for a first order ferroelectric phase transition at different temperatures

There are four characteristic temperatures in the phase transition process, i.e., Curie-Weiss temperature T_0, Curie temperature T_c, ferroelectric limit temperature T_1 and limit temperature of field induced phase transition T_2. The Curie-Weiss temperature can be easily accessed experimentally from the Curie-Weiss law of dielectric constant ε at paraelectric phase, i.e.,

$$\varepsilon_p = \frac{C}{T - T_0} \tag{3}$$

In above expression, C is the Curie-Weiss constant, subscript p of ε stands for paraelectric phase. However the Curie temperature is less accessible experimentally. This temperature measures the balance of the ferroelectric phase and the paraelectric phase. At this temperature, the free energy of ferroelectric phase is the same as that of paraelectric phase. When temperature is between T_0 and T_c, ferroelectric phase is stable and paraelectric phase is meta-stable, this can be easily seen in Fig.1. When the temperature is between T_c and T_1, ferroelectric phase is in meta-stable state while paraelectric phase is stable. When the temperature is higher than T_1, ferroelectric phase disappears. Normally, this temperature is corresponding to the peak temperature of dielectric constant when measured in heating cycle. In other words, peak temperature of dielectric constant measured in heating cycle is the ferroelectric limit temperature T_1, not the Curie temperature T_c in a more precise sense. Between temperature T_1 and T_2, ferroelectric state still can be induced by applying an external electric field. The polarization versus the electric field strength is a double hysteresis loop, which is very similar with that observed in anti-ferroelectric materials. When the temperature is higher than T_2, only paraelectric phase can exist.
The characteristic temperatures T_c, T_1 and T_2 can be easily determined from Eq.(2) of free energy as following. The Curie temperature T_c can be obtained from the following two equations;

$$\Delta G = \frac{1}{2}\alpha_0 \left(T_c - T_0\right) P_x^2 + \frac{1}{4}\alpha_{11} P_x^4 + \frac{1}{6}\alpha_{111} P_x^6 = 0 \tag{4}$$

$$\frac{\partial \Delta G}{\partial P_x} = \alpha_0 \left(T_c - T_0\right) P_x + \alpha_{11} P_x^3 + \alpha_{111} P_x^5 = 0 \tag{5}$$

The first equation means that the free energy of ferroelectric phase is same as that of paraelectric phase, and the second equation implies that the free energy of ferroelectric phase is in minimum. From above two equations, we can have the expression of the Curie temperature T_c as

$$T_c = T_0 + \frac{3}{16}\frac{\alpha_{11}^2}{\alpha_0 \alpha_{111}} \tag{6}$$

At the ferroelectric limit temperature T_1, free energy has an inflexion point at Ps, the spontaneous polarization. As can be seen from Fig. 1, when temperature is below T_1, free energy has are three minima, i.e., at P=±Ps, and P=0. Above temperature T_1, there is only one minimum at P=0. The spontaneous polarization can be obtained from the minimum of the free energy as,

$$\begin{gathered} \frac{\partial \Delta G}{\partial P_x} = \alpha_0 \left(T - T_0\right) P_x + \alpha_{11} P_x^3 + \alpha_{111} P_x^5 = 0 \\ P_x \left[\alpha_0 \left(T - T_0\right) + \alpha_{11} P_x^2 + \alpha_{111} P_x^4\right] = 0 \end{gathered} \tag{7}$$

Since $P_x = 0$ corresponds to a free energy maxima in the ferroelectric phase, the spontaneous polarization Ps is obtained from the solution of the equation:

$$\alpha_0 (T - T_0) + \alpha_{11} P_x^2 + \alpha_{111} P_x^4 = 0 \tag{8}$$

which is

$$P_x^2 = \frac{1}{2\alpha_{111}} \left[-\alpha_{11} \pm \sqrt{\alpha_{11}^2 - 4\alpha_0 \alpha_{111} (T - T_0)} \right] \tag{9}$$

Hence at temperature T_1, we have

$$\alpha_{11}^2 - 4\alpha_0 \alpha_{111} (T_1 - T_0) = 0 \tag{10}$$

In a more explicitly form,

$$T_1 = T_0 + \frac{1}{4} \frac{\alpha_{11}^2}{\alpha_0 \alpha_{111}} \tag{11}$$

Between temperature T_1 and T_2, there are still inflexion points, which means ferroelectric phase can be induced by an external electric field. When temperature is above T_2, the inflexion points disappear. By taking the second derivative of the free energy Eq.(2), we can have the solution of the inflexion points,

$$\frac{\partial^2 \Delta G}{\partial P_x^2} = \alpha_0 (T_2 - T_0) + 3\alpha_{11} P_x^2 + 5\alpha_{111} P_x^4 = 0 \tag{12}$$

$$P_x^2 = \frac{1}{10\alpha_{111}} \left[-3\alpha_{11} \pm \sqrt{(3\alpha_{11})^2 - 20\alpha_0 \alpha_{111} (T - T_0)} \right] \tag{13}$$

That is the polarization at the inflexion points. The limit temperature of the induced phase transition T_2 can be determined by

$$(3\alpha_{11})^2 - 20\alpha_0 \alpha_{111} (T_2 - T_0) = 0 \tag{14}$$

or

$$T_2 = T_0 + \frac{9}{20} \frac{\alpha_{11}^2}{\alpha_0 \alpha_{111}} \tag{15}$$

The temperature dependence of the spontaneous can be calculated from Eq.(9) by elimination of solutions yielding a negative square root (for either the first or second order phase transitions) gives:

$$P_x = \sqrt{\frac{1}{2\alpha_{111}} \left[-\alpha_{11} \pm \sqrt{\alpha_{11}^2 - 4\alpha_0 \alpha_{111} (T - T_0)} \right]} \tag{16}$$

The temperature dependence of the spontaneous polarization from above equation is shown in Fig. 2. The four characteristic temperatures are denoted by light arrows. The dark arrows indicate the temperature cycles. Theoretical temperature hysteresis is $\Delta T=T_1-T_0$. Experimentally, the temperature hysteresis could be larger than this value because of the polarization relaxation process.

The hysteresis loop (P_x versus E_x) may be obtained from the free energy expansion by adding an additional electrostatic term, $E_x P_x$, as follows:

$$\Delta G = \frac{1}{2}\alpha_0 (T - T_0) P_x^2 + \frac{1}{4}\alpha_{11} P_x^4 + \frac{1}{6}\alpha_{111} P_x^6 - E_x P_x \quad (17)$$

$$\frac{\partial \Delta G}{\partial P_x} = \alpha_0 (T - T_0) P_x + \alpha_{11} P_x^3 + \alpha_{111} P_x^5 - E_x = 0 \quad (18)$$

$$E_x = \alpha_0 (T - T_0) P_x + \alpha_{11} P_x^3 + \alpha_{111} P_x^5 \quad (19)$$

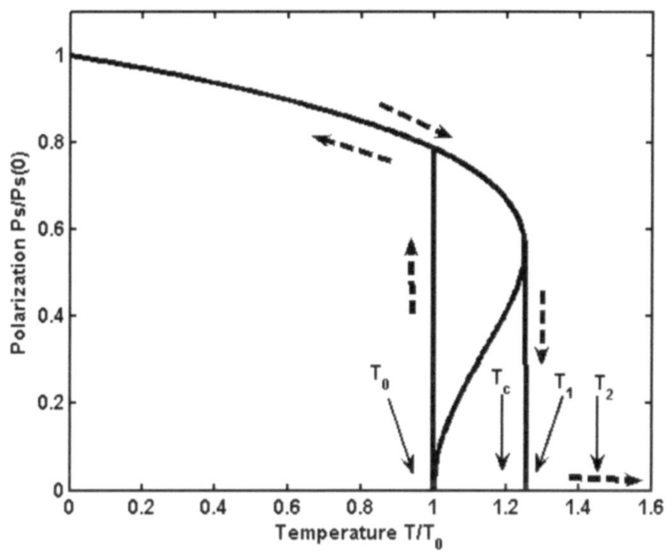

Fig. 2. Temperature dependence of the spontaneous polarization

Plotting E_x as a function of P_x and reflecting the graph about the 45 degree line gives an 'S' shaped curve when temperature is much lower than the transition temperatures, as can be seen from curves in Fig. 3. The central part of the 'S' corresponds to a free energy local maximum, since the second derivative of the free energy ΔG respect polarization P_x is negative. Elimination of this region and connection of the top and bottom portions of the 'S' curve by vertical lines at the discontinuities gives the hysteresis loop. Temperatures are labeled by each curve, label of $0.7T_0$ in Fig. 3 is for temperature $T=0.7T_0$, and $1.2T_2$ is for $T=1.2T_2$, T_{12} stands for $T=(T_1+T_2)/2$. When the temperature is below Curie temperature T_c, normal ferroelectric hysteresis loops can be obtained. When the temperature is between T_1 and T_2, double hysteresis loop or pinched loop could be observed. That means ferroelectric state is induced by the applied electric field. When the temperature is higher than T_2, the polarization versus electric field becomes a non-linear relation, see $1.2T_2$ curve in Fig. 3. It should point out that curves in Fig. 3 are obtained under static electric field. Experimental measurements usually are performed using time dependent electric field, mostly in sine form. Therefore the hysteresis loops obtained experimentally might be different from the shapes shown in Fig. 3.

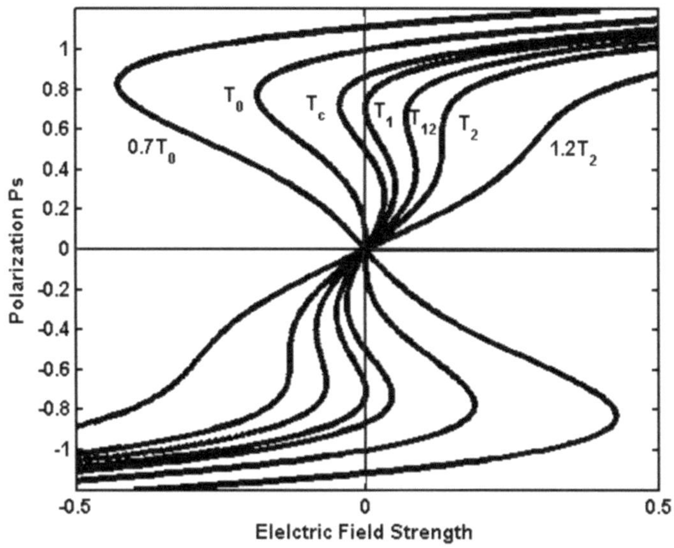

Fig. 3. Static hysteresis loops at different temperatures

Dynamic behavior of ferroelectrics from theoretical simulation could be more helpful for understanding the experimental observing. The dynamic property of ferroelectrics can be studied using Landau-Khalantikov equation (Blinc & Zeks, 1974)

$$\frac{dP}{dt} = -\Gamma \frac{\delta \Delta G}{\delta P} \tag{20}$$

where Γ is the coefficient of thermodynamic restoring force. This equation has been employed to investigate the switching characters of asymmetric ferroelectric films (Wang et al., 1999) and the effects of external stresses on the ferroelectric properties of Pb(Zr,Ti)TiO_3 thin films (Song et al., 2003). By inserting free energy (17) into Eq.(20), we have

$$\frac{dP}{dt} = -\Gamma\left[\alpha_0 (T - T_0) P_x + \alpha_{11} P_x^3 + \alpha_{111} P_x^5 - E_x\right] \tag{21}$$

From above equation, hysteresis loops at different temperature and frequency can be obtained with the electric field is in sine form,

$$E_x = E_0 \sin(\omega t) \tag{22}$$

The hysteresis loops showing in Fig. 4 are at Curie temperature T_c, T_1, $(T_1+T_2)/2$, T_2 and $1.2T_2$ with frequency ω=0.001, E_0=0.5 and Γ=2.0. At Curie temperature T_c, a normal ferroelectric hysteresis loop is obtained. At temperature T_1, the hysteresis loop is pinched. Between T_1 and T_2, double hysteresis loop is obtained as expected, see the blue curve in Fig. 4 for temperature T= $(T_1+T_2)/2$. However, at temperature T_2, double hysteresis loop can be still observed because of the finite value of relaxation time. Higher than T_2, no hysteresis can be observed, but a non-linear P-E relation curve. Similar shapes of hysteresis loops have been observed in $Pb_xSr_{1-x}TiO_3$ ceramics recently(Chen et al., 2009).

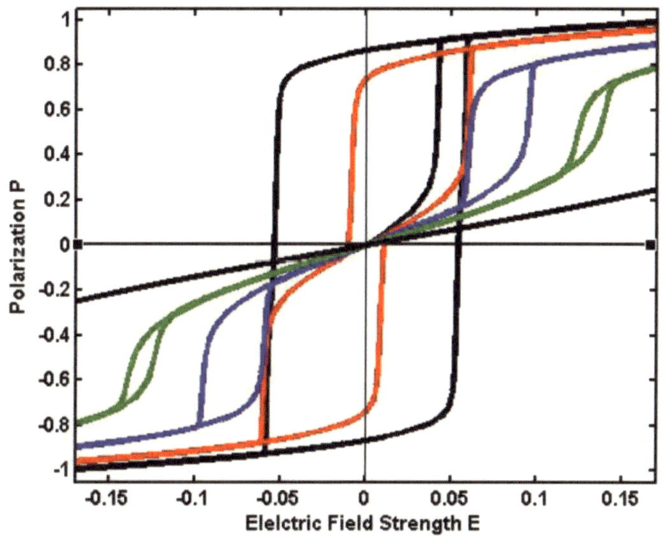

Fig. 4. Dynamic hysteresis loops at different temperatures with frequency ω=0.001. Dark loop is at Curie temperature T_c, red curve is for T_1, blue curve for $(T_1+T_2)/2$, green curve for T_2, and dark line for $T=1.2T_2$.

To get insight understanding of the influence of the frequency on the shape of the hysteresis, more hysteresis loops are presented in Fig. 5. The frequency is set as ω=0.001 for green curves, ω=0.01 for red curves and ω=0.03 for blue curves at Curie temperature T_c, T_1, $(T_1+T_2)/2$ and at T_2 respectively. At Curie temperature T_c, see Fig. 5(a), the coercive field increases with increasing of measure frequency, but the spontaneous polarization is less influenced by the frequency. At temperature T_1, this temperature corresponding to the peak temperature of dielectric constant when measured in heating cycle, pinched hysteresis loop can be observed at low frequency, see the green curve in Fig. 5(b). At higher frequency, the loop behaves like a normal ferroelectric loop, see blue curve in Fig.5(b). When temperature is between T_1 and T_2, as shown in Fig. 5(c) for temperature at $T=(T_1+T_2)/2$, typical double hysteresis loop can be observed at low frequency. When the frequency is higher, it becomes a pinched loop. This trend of loop shape can be kept even when temperature is up to T_2, see curves in Fig. 5(d).

Temperature dependence of polarization at different temperature cycles can be also obtained from Eq.(21) with a constant electric field E=0.01. The results are shown in shown in Fig. 6 with different Γ, the coefficient of thermodynamic restore force. The dark line is for the static polarization, red curves are field heating, blue curves are field cooling. Solid lines are for Γ=100, and dash-dotted lines are for Γ=10. The temperature hysteresis from the static theory, as indicated by the two dark dash lines, is smaller than that from Landau-Khalanitkov theory. In other words, temperature hysteresis ΔT measured experimentally would be usual larger than that from static Landau theory. Also the temperature hysteresis ΔT depends on Γ, since Γ is related with the relaxation time. A larger Γ represents small relaxation time, or quick response of polarization with electric field. Hence ferroelectrics with long relaxation time, i.e., small Γ, would expect a very larger temperature hysteresis.

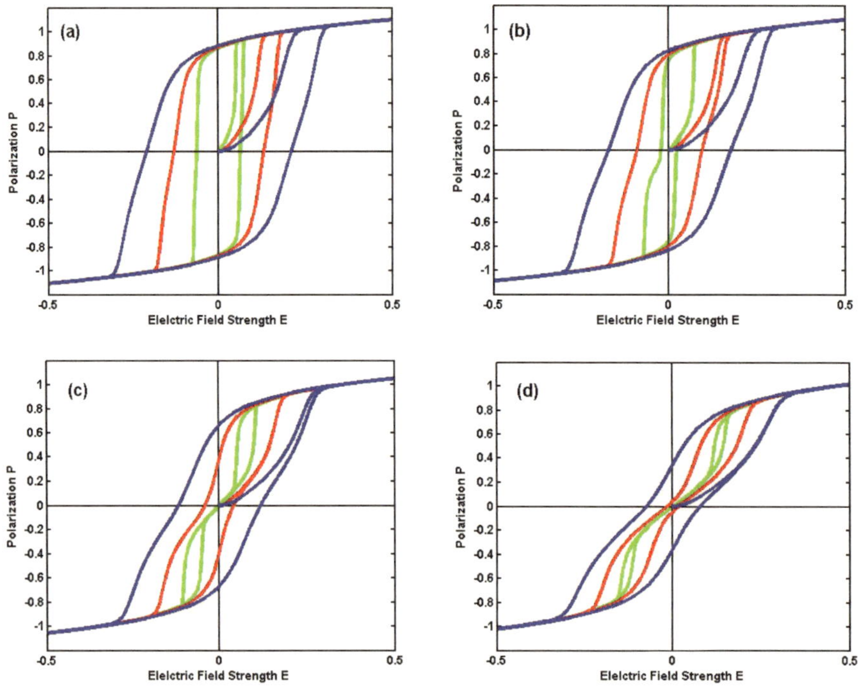

Fig. 5. Hysteresis loops with frequency ω=0.001 for green curves, ω=0.01 for red curves and ω=0.03 for blue curves at temperature (a) T_c; (b) T_1; (c) $(T_1+T_2)/2$, (d) T_2

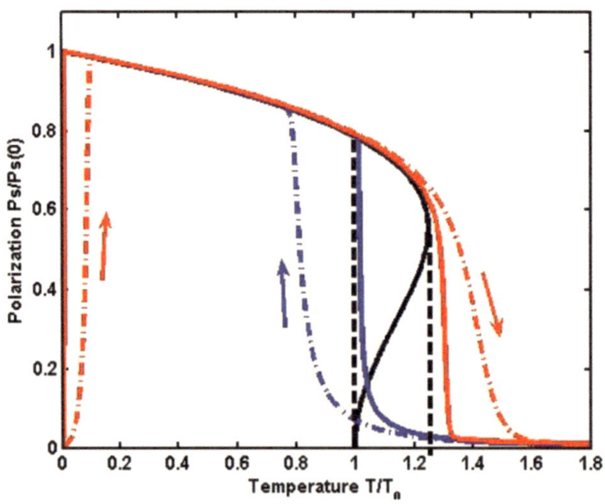

Fig. 6. Temperature dependence of polarization at different temperature cycles. Dark line is for the static polarization. Red curves are field heating, blue curves are field cooling. Solid lines are for short relaxation time(larger Γ), and dash-dot lines are for lone relaxation time (small Γ).

3. Effective field approach

The effective field approach has proved to be a simple statistic physics method but valuable way to describe phase transitions (Gonzalo, 2006). The main supposition of this model is that each individual dipole is influenced, not only by the applied electric field, but by every dipole of the system. In its simplest form, which takes into account only dipole interactions, describes fairly well the main features of continuous ferroelectric phase transitions, i.e. second order phase transition. The inclusion of quadrupolar and higher order terms into the effective field expression is necessary for describing the properties of discontinuous or first order phase transitions (Gonzalo et al., 1993; Noheda et al., 1993, 1994). The effective field approach has turned out to be successful explaining the composition dependence of the Curie temperature in mixed ferroelectrics systems (Ali et al., 2004; Arago et al., 2006). A quantum effective field approach has also been developed for phase transitions at very low temperature (Gonzalo, 1989; Yuan et al., 2003; Arago et al., 2004) in ferro-quantum paraelectric mixed systems. In this section, quantum effective field approach is adopted to reveal the influence of the zero point energy on first order phase transitions (Wang et al., 2008). We can see that when the zero point energy of the system is large enough and the ferroelectric phase is suppressed, a phase transition-like temperature dependence of the polarization can be observed by applying an electric field.
The effective field, as described in detail in Gonzalo's book (Gonzalo, 2006), can be expressed as

$$E_{eff} = E + \beta P + \gamma P^3 + \delta P^5 + \cdots \tag{23}$$

where E is the external electric field, and the following terms correspond to the dipolar, quadrupolar, octupolar, etc., interaction. By keeping the first two terms, i.e. dipolar and quadrupolar interaction, gives account of the first order transition.
From statistical considerations, the equation of state is,

$$P = N\mu \tanh\left(\frac{E_{eff}\mu}{k_B T}\right) = N\mu \tanh\left(\frac{(E + \beta P + \gamma P^3)\mu}{k_B T}\right) \tag{24}$$

where N is the number of elementary dipoles per unit volume, μ is the electric dipole moment, k_B is the Boltzmann constant, and T is the absolute temperature. The Curie temperature is given by

$$T_C = \frac{\beta N \mu^2}{k_B}$$

The explicit form of the equation of state can be rewritten from Eq.(24),

$$E = \frac{k_B T}{\mu} \tanh^{-1}\left(\frac{P}{N\mu}\right) - \beta P - \gamma P^3 \tag{25}$$

In order to handle easier this expression, following normalization quantities are introduced,

$$e \equiv \frac{E}{\beta N \mu}; \quad p \equiv \frac{P}{N\mu}; \quad t \equiv \frac{T}{T_C} = \frac{k_B T}{\beta N \mu^2}; \quad g = \frac{\gamma N^2 \mu^2}{\beta}$$

so Eq. (25) is rewritten as

$$p = \tanh\left(\frac{e + p + g \cdot p^3}{t}\right) \tag{26}$$

or

$$e = t \cdot \tanh^{-1} p - p - gp^3 \tag{27}$$

and as in absence of external field $e = 0,\ p = p_S$

$$t = \frac{p_S + gp_S^3}{\tanh^{-1} p_S} \tag{28}$$

Fig.7 shows the plot of the normalized spontaneous polarization p_S versus normalized temperature t obtained from Eq. (28) for several values of the parameter g. As it is shown in the discussion of the role of the quadrupolar interaction in the order of the phase transition (see Gonzalo et al., 1993; Noheda et al., 1993; Gonzalo, 2006), values of g smaller than 1/3 correspond to a second order, or continuous phase transition, and values larger than 1/3 indicate that the transition is discontinuous, that is, first order. In this case, a spontaneous polarization p_θ exists at temperature $T_\theta > T_C$, and then $\tau_\theta > 1$, being $\Delta t = t_\theta - 1$ the corresponding reduced thermal hysteresis, which is the signature of the first order transitions.

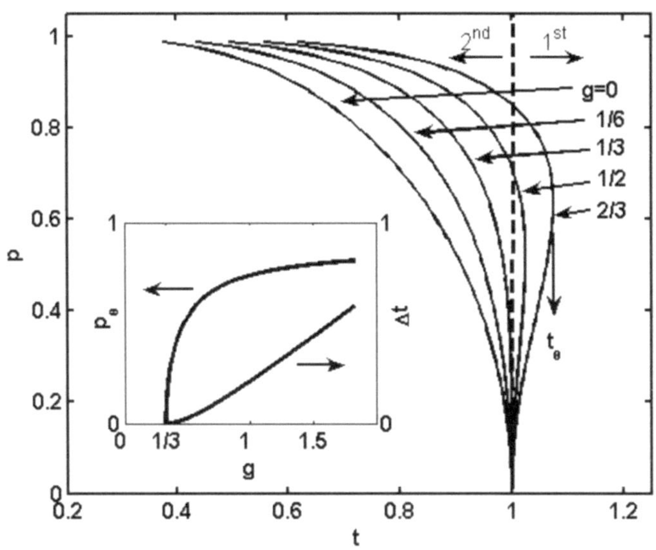

Fig. 7. Temperature dependence of the spontaneous polarization for different quadrupolar interaction coefficient g. The inset shows the polarization discontinuity $p_\theta(g)$ and the thermal hysteresis temperature $\Delta t(g)$ in a first order transition.(Wang et al., 2008)

In order to determine $p_\theta(g)$, deriving in Eq. (28),

$$\left.\frac{\partial t}{\partial p_S}\right|_{p_\Theta} = 0$$

and $p_\theta(g)$ can be obtained from

$$\left(1-p_\theta^2\right)\left(1+3gp_\theta^2\right)\tanh^{-1}p_\theta-\left(p_\theta+gp_\theta^3\right)=0 \tag{29}$$

Substituting $p_\theta(g)$ in Eq. (28) again we obtain $t_\theta(g)$. The inset of Fig.7 shows $p_\theta(g)$ and $\Delta t(g)$ respectively. It can be seen that Δt grows almost linearly with g and that p_θ approaches to a saturation value that, resolving Eq.(29), turns out to be $p_\theta = 0.8894$ when $g \to \infty$.
When a phase transition takes place at very low temperature it is necessary to consider quantum effect. The energy of the system is no longer the classical thermal energy k_BT, but the corresponding energy of the quantum oscillator,

$$E=\hbar\omega_0\left(\frac{1}{2}+\langle n\rangle\right)$$

being $E_0 = \hbar\omega_0/2$ the zero point energy, and $<n>$ the average number of states for a given temperature T. From this quantum energy expression we can obtain a new temperature scale T^Q (see Arago et al., 2004) defined as,

$$\hbar\omega_0\left(\frac{1}{2}+\frac{1}{e^{\hbar\omega_0/k_BT}-1}\right)\equiv k_BT^Q \Rightarrow T^Q\equiv\frac{\hbar\omega_0}{2k_B\tanh\left(\hbar\omega_0/2k_BT\right)} \tag{30}$$

and then, the corresponding quantum normalized temperature, $t^Q \equiv T^Q/T_C$. If we introduce a new normalization for the zero point energy

$$\Omega\equiv\frac{E_0}{k_BT_C}=\frac{\hbar\omega_0/2}{\beta N\mu^2}$$

so we can rewrite,

$$t^Q\equiv\frac{\Omega}{\tanh(\Omega/t)} \tag{31}$$

and Eq. (27) and (28) become respectively,

$$e=t^Q\tanh^{-1}p-p-gp^3 \tag{32}$$

$$t^Q\equiv\frac{\Omega}{\tanh(\Omega/t)}=\frac{p_S+gp_S^3}{\tanh^{-1}p_S},\quad e=0 \tag{33}$$

The temperature dependence of the spontaneous polarization can be found from above equation for a given value of g and different values of the parameter Ω. Fig. 8 plots $p_S(t)$ with $g = 0.8$ to ensure it is a first order transition. The influence of the zero-point energy is quite obvious: when it is small, the phase transition is still of normal first order one. As the zero point energy increases, both the transition temperatures and the spontaneous polarization decrease. The Curie temperature goes to zero for $\Omega = 1$, but no yet t_θ, neither the saturation spontaneous polarization does. From the definition of the normalized zero point energy Ω, we can see then that the Curie temperature goes to zero when the zero point energy is the same as the classical thermal energy k_BT_C. Imposing again the condition of the zero slope,

$(\partial t/\partial p_S)p_\theta = 0$, we can obtain $p_\theta(\Omega)$, and then $t_\theta(\Omega)$, which must be zero when the ferroelectric behavior will be completely depressed. In this way we work out the zero point energy critical value (Ω_{cf} = 1.1236 for g=0.8) that would not allow any ordered state. Furthermore, from the condition $t_\theta(\Omega_{cf}, g) = 0$, we will find the relationship between the critical zero point energy Ω_{cf} and the strength of the quadrupolar interaction given by the coefficient g. Fig. 9 plots Ω_{cf} (g) that indicates that Ω_{cf} grows almost linearly with g, specially for larger values of g. This means that ferroelectrics with strong first order phase transition feature needs a relative large critical value of zero point energy to depress the ferroelectricity.

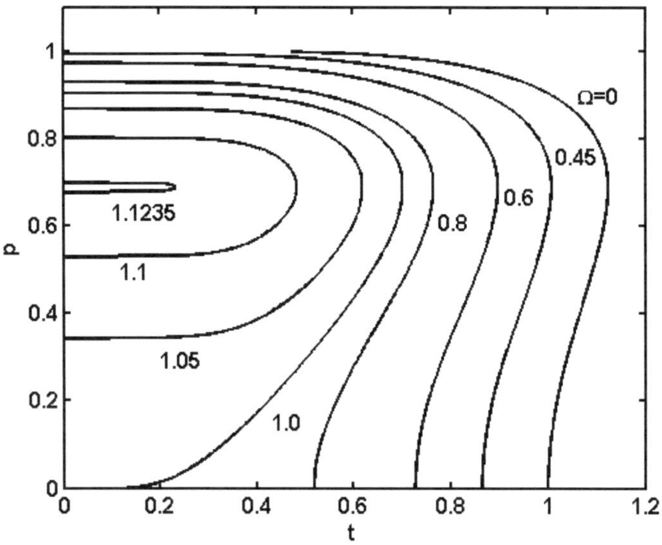

Fig. 8. Temperature dependence of the spontaneous polarization at different zero point energy. All curves correspond to a quadrupolar interaction coefficient g=0.8. (Wang et al., 2008)

Above results prove that a ferroelectric material, with strong quadrupolar interaction, undergoes a first order transition (g>1/3) unless its zero point energy reaches a critical value, Ω_{cf}, because in such case the phase transition is inhibited. However, an induced phase transition must be reached by applying an external electric field. Let be a system with g= 0.8, and Ω=1.6, that is a first order ferroelectric with a zero point energy above the critical value and, hence, no phase transition observed. And let us apply a normalized electric field e that will produce a polarization after Eq. (32). Fig. 10 plots $p(t)$ for different values of the electric field. It can be seen that when it is weak (see curves corresponding to 0.01 and 0.02), the polarization attains quickly a saturation value, similar to what is found in quantum paraelectrics. The curve of e=0.03 (see the dashed line) is split into two parts. The lower part would represent a quantum paraelectric state, but the upper part stands for a kind of ferroelectric state. Therefore there exists a critical value between e=0.02 and 0.03, which is the minimum electric field for inducting a phase transition. For 0.04<e<0.06 the induced polarization curve shows a discontinuous step, but as the electric field increases, 0.07, 0.08 and so on, the polarization changes continuously from a large value at low temperature to a relative small value at high temperature, showing a continuous step. So there is another critical electric field somewhere in between 0.05<e<0.07, separates the discontinuous step and the continuous step of the induced polarization. In fact, this tri-critical point would be around e=0.06. It is also important to remark that the above-mentioned features of the field induced phase transitions have been observed in lead magnesium niobate (Kutnjak et al., 2006), which is a well-known ferroelectric relaxor.

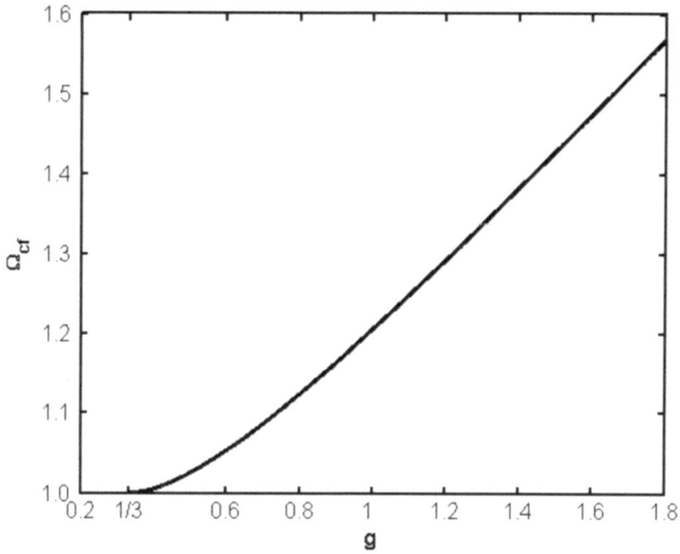

Fig. 9. Quadrupolar interaction dependence of the critical zero point energy Ω_{cf} that depress ferroelectricity. (Wang et al., 2008)

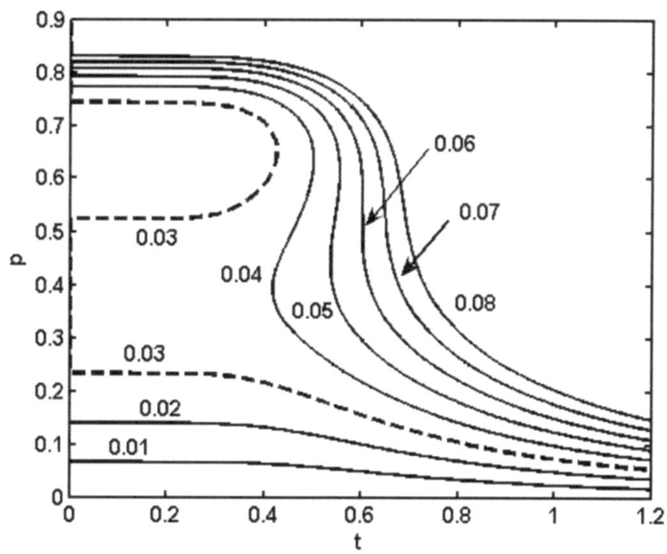

Fig. 10. Temperature dependence of the field induced polarization for Ω=1.15 and g=0.8. The parameter is the strength of the electric field. (Wang et al., 2008)

For further understanding the field induced phase transitions, Fig. 11 shows the hysteresis loops obtained numerically from Eq. (32) and corresponding to the curves displayed in Fig. 10. Double hysteresis loops can be observed when the temperature is lower than a critical point, around t=0.6, suggesting that ferroelectricity can always be induced at very low temperature. When the temperature is higher than this critical value, there is no hysteresis loop and we can just observe a non-linear p–e behaviour (see for instance the case for t=0.7). However, when the electric field is lower than $e \approx 0.025$, as indicated by the dashed arrow in Fig. 11, no hysteresis loop can be observed. That is the case corresponding to the curves e =0.01 and 0.02 in Fig. 10. The critical electric field able to induce a phase transition logically increases with the increasing of temperature, so at lower temperature region in Fig. 10, we can always have field induced ferroelectricity when the applied electric field is strong enough.

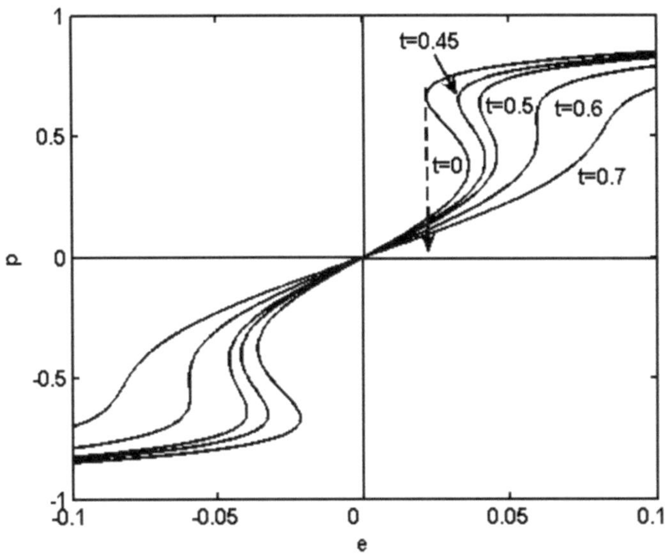

Fig. 11. Hysteresis loops at different temperatures for Ω=1.15 and g=0.8. The dashed arrow indicates the minimum electric field needed to induce a ferroelectric state at t=0. (Wang et al., 2008)

To understand the influence of the zero point energy on the shape of hysteresis loop, here we take the case of g=0.8 at t=0 as example. The corresponding quantum temperature, after Eq. (31), is

$$t^{Q}(t=0)=\Omega$$

and then, Eq. (32) becomes

$$e=\Omega\tanh^{-1}p-p-gp^{3} \tag{34}$$

Fig. 12 plots the hysteresis loops calculated after Eq. (34). When the zero point energy is smaller than the critical value Ω_{cf} (Ω_{cf}=1.1236, in this case) a normal hysteresis loop is obtained, where the coercive field decreases as zero point energy increases. For the critical value, a double hysteresis loop with zero coercive field is found. But if we continue increasing the zero point energy, we arrive to a point where no hysteresis loop is found at all. That suggests that above this other critical value, be Ω_{cp}, there is no way to induce a phase transition, even applying a strong field, and the system remains always in a paraelectric state.

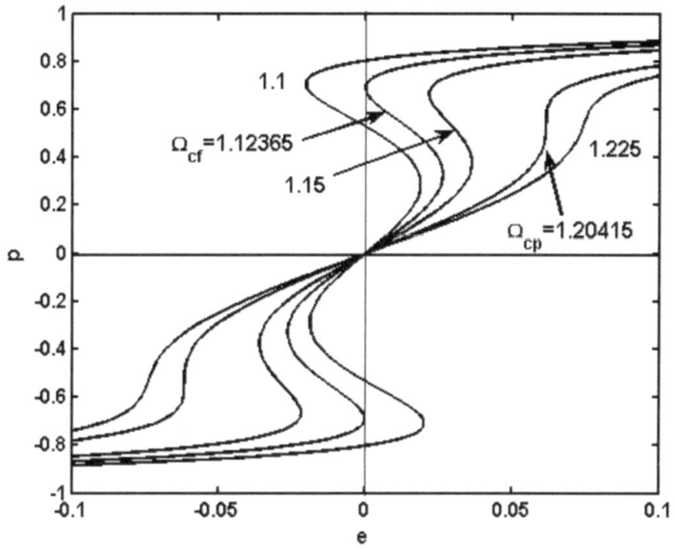

Fig. 12. Hysteresis loops at zero temperature for different zero point energy values. The critical value of Ω_{cf} is the minimum zero point energy for the system to have ferroelectricity, while the critical value of Ω_{cp} is the maximum zero point energy for the system to get field induced ferroelectricity. (Wang et al., 2008)

From the analysis of the hysteresis loops in Fig.12 we can determine the critical electric field needed to induce phase transition imposing the conditions,

$$\left(\frac{\partial e}{\partial p}\right)_{p_c} = 0, \qquad \left(\frac{\partial^2 e}{\partial p^2}\right) > 0$$

and deriving in Eq. (34) we get,

$$\frac{\Omega}{1-p_c^2} - 1 - 3gp_c^2 = 0, \qquad \left(\frac{2\Omega}{\left(1-p_c^2\right)^2} - 6g\right)p_c > 0 \tag{35}$$

hence we obtain,

$$p_c^2 = \frac{(3g-1)+\sqrt{(3g-1)^2 6g(\Omega-2)}}{6g} \tag{36}$$

Substituting p_c (g, Ω) obtained after Eq. (36) into Eq. (34) we can get the coercive field e_c (p_c) at zero temperature. Besides, as it is showed in Fig. 12, when the zero point energy attains its critical value Ω_{cf}, the coercive field turns out to be zero, so this is another way to check

the quadrupolar interaction dependence of the critical zero point energy $\Omega_{cf}(g)$ as displayed in Fig.9. It must be noted the full accordance between the two calculations.
At the second critical value of the zero point energy, Ω_{cp}, the hysteresis loop becomes an inflexion *e-p* curve as it can also be observed in Fig.12, so in this case both Eq. (35) must be equal to zero, and by solving the system of equations, it results,

$$\Omega_{cp} = \frac{(3g+1)^2}{12g}, \quad p_{cp} = \sqrt{\frac{3g-1}{6g}} \tag{37}$$

The phase diagram at zero temperature is shown in Fig. 13. It displays the role of both the quadrupolar interaction strength and the zero point energy on first order phase transitions. On the top of this diagram we find just paraelectric state (PE), while on the bottom left it appears only second order phase transitions (FE: 2nd order) that correspond to $g < 1/3$ and $\Omega \leq 1$. On the bottom right ($g > 1/3$) there are the first order transitions (FE: 1st order) as the critical values of the zero point energy (Ω_{cf} >1) that depress the ferroelectricity grow monotonously with g. Above the curve $\Omega_{cf}(g)$ there are induced electric field phase transitions (first order also). They are limited by another curve $\Omega_{cp}(g)$ given by Eq. (37) indicating that no phase transition can be observed when the zero point energy of the system is greater than this value.

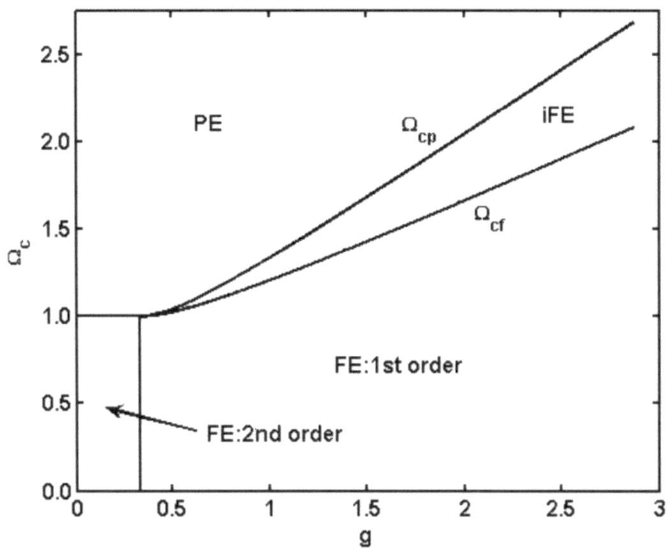

Fig. 13. Phase diagram of zero point energy critical value Ω_c versus quadrupolar interaction coefficient g at zero temperature. The second order phase transition region is denoted as "FE:2nd order", "FE:1st order" indicates a normal first order transition, "iFE" is the region of induced ferroelectric phase and on the top, "PE" corresponds to the paraelectric phase. (Wang et al., 2008)

From above calculations in the framework of effective field approach, it can be seen that a phase transition can be induced by applying an electric field in a first order quantum paraelectric material. There exist two critical values of the zero point energy, one is Ω_{cf} that depress ferroelectricity, and another is Ω_{cp} above which it is impossible to induce any kind of phase transition independently of the value of the electric field applied. Phase diagram is presented to display the role of both the quadrupolar interaction strength and the zero point energy on the phase transitions features.

4. Ising model with a four-spin interaction

Ising model in a transverse field with a four-spin interaction has been used to study the first-order phase transition properties in many ferroelectric systems. Under the mean-field approximation, first-order phase transition in ferroelectric thin films (Wang et al., 1996; Jiang et al., 2005) or ferroelectric superlattices (Qu, Zhong & Zhang, 1997; Wang, Wang & Zhong, 2002) have been systematically studied. Using the Green's function technique, the first-order phase transition properties in order-disorder ferroelectrics (Wang et al., 1989) and ferroelectric thin films (Wesselinowa, 2002) have been investigated. Adopting the higher-order approximation to the Fermi-type Green's function, the first-order phase transition properties have been studied in the parameter space with respect to the ratios of the transverse field and the four-spin interaction to the two-spin interaction for ferroelectric thin films with the uniform surface and bulk parameters (Teng & Sy,2005). These works prove that Ising model with four-spin interaction is a successful model for studying first order phase transition of ferroelectrics. In the following, basic formulism under mean-field approximations and Monte Carlo simulation are presented, and the results are discussed.

4.1 Mean field approximation

The Hamiltonian of the transverse field Ising model with a four-spin interchange interaction term is (Wang et al., 1989; Teng & Sy, 2005):

$$H = -\sum_{i} \Omega_i S_i^x - \sum_{\langle i,j \rangle} J_{ij}^{(2)} S_i^z S_j^z - \sum_{\langle i,j,k,l \rangle} J_{ijkl}^{(4)} S_i^z S_j^z S_k^z S_l^z \tag{38}$$

where Ω is the transverse field, S^x and S^z are the x and z components of pseudospin-1/2 operator, $J^{(2)}$ and $J^{(4)}$ are the two-spin and four-spin exchang interaction, subscript i,j,k,l are site number, summation run over only the nearest-neighbor sites. Under the mean field approximation (Blinc & Zeks ,1974), solutions of Hamiltonian Eq.(38) is,

$$\begin{cases} \langle S_i^x \rangle = \dfrac{\Omega}{2F_i} \tanh\left(\dfrac{F_i}{2k_B T} \right) \\ \langle S_i^y \rangle = 0 \\ \langle S_i^z \rangle = \dfrac{J\langle S_i^z \rangle + J^4 \langle S_i^z \rangle^3 + 2\mu E}{2F_i} \tanh\left(\dfrac{F_i}{2k_B T} \right) \end{cases}$$

where

$$F_i = \sqrt{\Omega^2 + \left(\sum_{<ij>} J_{ij} \langle S_j^z \rangle + \sum_{<ijkl>} J_{ijkl}^4 \langle S_j^z \rangle \langle S_k^z \rangle \langle S_l^z \rangle - 2\mu E \right)^2}$$

For a homogenous system

$$\langle S_i^z \rangle = \langle S_j^z \rangle = \langle S_k^z \rangle = \langle S_l^z \rangle = \langle S^z \rangle$$

$$J = \sum_j J_{ij}, \quad J' = \sum_{jkl} J_{ijkl}^4, \quad F = F_i$$

Therefore we have,

$$R = \langle S_i^z \rangle = \frac{JR + J'R^3 + 2\mu E}{2F} \tanh\left(\frac{F}{2k_B T}\right) \tag{39}$$

$$F = \sqrt{\Omega^2 + \left(JR + J'R^3 + 2\mu E\right)^2} \tag{40}$$

The polarization is related to the z-component spin average as,

$$P = 2N\mu \langle S_i^z \rangle = 2N\mu R \tag{41}$$

Following normalizations are introduced to handle easier this expression,

$$g' = \frac{J'}{J}, \omega = \frac{\Omega}{J}, t = \frac{4k_B T}{J}, \tau = \frac{F}{J}, e = \frac{4\mu E}{J}, p = \frac{P}{N\mu} = 2R$$

Eq.(39) and (40) can be rewritten as,

$$p = \frac{p + g'p^3/4 + e}{2\tau} \tanh\left(\frac{2\tau}{t}\right) \tag{42}$$

$$\tau = \sqrt{\omega^2 + \left(\frac{p}{2} + g'\frac{p^3}{8} + \frac{e}{2}\right)^2} \tag{43}$$

In the absence of transverse field, i.e., $\omega=0$ in Eq.(43), and by defining quadrupolar contribution factor g as,

$$g = \frac{g'}{4} = \frac{J^4}{4 \cdot J}$$

Eq.(42) can be rewritten in the same form in Eq. (26), which is obtained from effective field approach. The equivalence of effective field approach and Ising model with four spin couplings under mean field approximation is completely approved. Therefore the critical value of relative quadrupolar contribution is $g_c=1/3$ under mean field approximation for occurrence of first order phase transition.

4.2 Monte Carlo simulation

A well known and useful method to study phase transitions is by means of the Monte Carlo simulations (Binder, 1984). In particular, phase transitions in Ising systems of relatively low dimensions are sufficient to carry out numerical simulations in the vicinity of the transition temperature, providing a good empirical basis to investigate the asymptotic behavior at the phase transition (Gonzalo & Wang, 2008). Here Monte Carlo simulation is applied to the

Ising model with four-spin coupling for studying the phase transition behavior, especially checking the critical value from the four-spin coupling strength, or quadrupolar contribution (Wang et al., 2010).
Monte Carlo method with metropolis algorithm has been used to simulate 3D Ising cubic lattice. The Hamiltonian is same as in Eq.(38), but without the first term, i.e., without including the tunneling term. In this case, the critical value of the four-spin coupling contribution under mean field approximation is

$$g_c = \frac{12J'}{J} = \frac{2J'}{J} = \frac{1}{3}, \quad (J'/J)_c = 1/6$$

The critical four-spin coupling strength is **J'/J=1/6** from mean field theory, which is a reference value for the Monte Carlo simulations. Periodic boundary condition has been used in the simulations. The lattice size is denoted as N=L×L×L, where L is edge length. In the Monte Carlo simulation, edge length L=20, 30, 100 are used. Monte Carlo steps are chosen different for different lattices size to achieve an adequate accurate of the results.

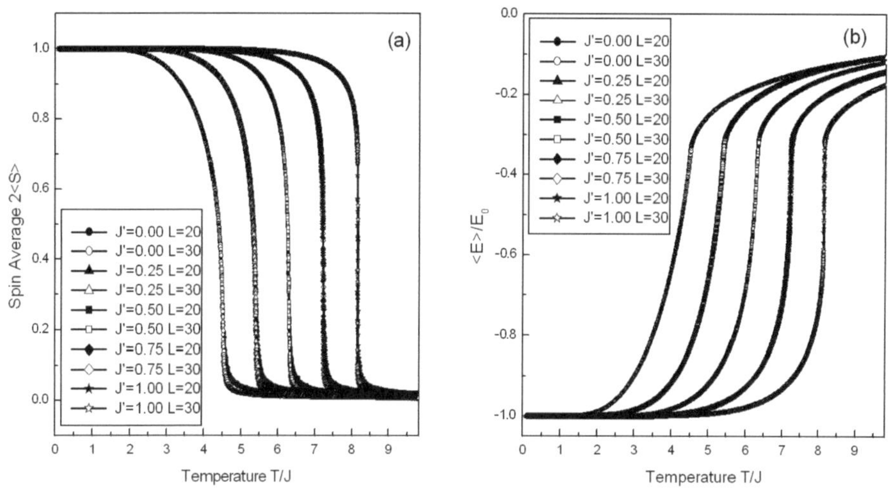

Fig. 14. Temperature dependence of spin average (a) and reduced average energy of cubic lattice with edge length L=20 and 30 for different four-spin coupling strength J'/J. (Wang et al., 2010)

Simulation is started with relative small lattice size, with edge length L=20 and 30. The temperature dependence of the spin average and the reduced average energy with different four-spin coupling strength are presented in Fig.14. The reduced average energy is defined as the energy with respective to the ground state energy E_0 in the temperature of zero Kelvin. The curve for lattice size L=20 is marked by solid symbols, and that for L=30 is marked by open symbols. The temperature is in the scale of two-site coupling J. As the four-spin coupling strength increases from J'/J =0 to J'/J =1, the transition temperature is shifted to higher temperature. The decreasing of the average spin and the increasing of the reduced average energy, with increasing temperature around the transition temperature becomes more rapidly as J'/J increases. The general difference between L=20 and that of L=30 is marginal except around the transition temperatures. Even though the four-spin coupling

strength (J′/J) is much larger than that of the mean field value 1/6, the first order transition characteristic is not such obvious in these two lattice sizes.
The temperature dependence of spin average and the reduced average energy of lattice size 100 is shown in Fig.15 for different four-spin coupling strength J′/J. The value of J′/J increases from 0.0 to 1.1 with increment of 0.1, corresponding the curves from left to right. The general behavior seems similar with that for L=20 and 30 as shown in Fig.14. However, with increasing of the four-spin coupling strength J′/J, as can be seen in the most right side curve of J′/J=1.1 in Fig.15(a), the spin average drops down around the transition temperature very quickly, showing a discontinuous feature. Similar discontinuous feature can be also seen in the reduced average energy curves in Fig.15(b). The reduced average energy goes up very sharply around the transition temperature for J′/J=1.1, as shown in the most right side curve in Fig.15(b). From Fig.15(b), we also notice that there is an inflexion point around the transition temperatures. The reduced average energy at this point is about 1/3 of the ground state energy when J′/J is larger. This condition could supply a criterion for determine the transition temperature in the first order phase transition.

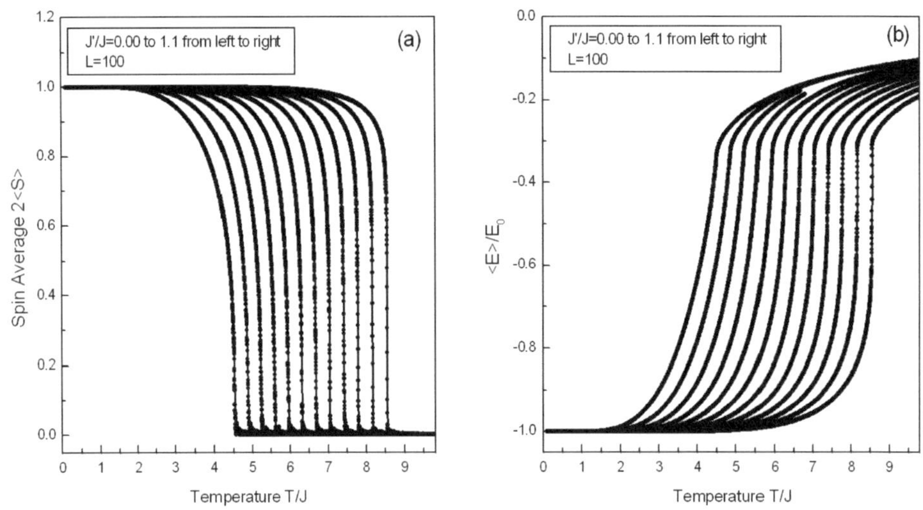

Fig. 15. Temperature dependence of spin average (a) and reduced average energy (b) of lattice size L =100 for with different four-spin coupling strength J′/J. (Wang et al., 2010)

To determine the transition temperature, we appeal the calculation of the Binder cumulant. The temperature dependence of the Binder cumulant for J′/J= 1.0 with different lattice size are shown in Fig.16. For lattice size of L=10, 20, 25, 30 and 40, the Binder cumulant does cross the point at T_c=8.148J. For lattice size L is larger, see L= 50, 70, 100 and 150, the Binder cumulant misses the cross point, and drops down from 2/3 to a very small value at higher temperatures. That means that the transition temperature can not be obtained from the Binder cumulant when the four-spin coupling strength J′/J is larger. It is believed that the Binder cumulants in Fig.16 suggesting the transition is of second order when the lattice size is smaller than L=40, and the transition is of first order when the lattice size is larger than L=40. The lattice size L=40 is around the critical lattice size for four-spin coupling strength J′/J =1. This also implies that the transition temperature of first order transition is lattice size dependent.

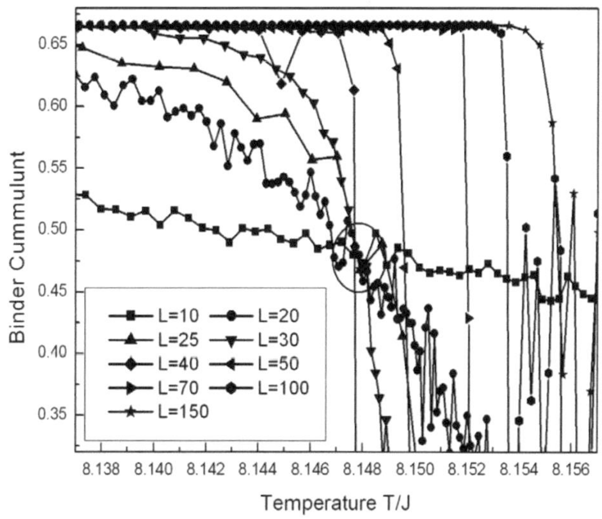

Fig. 16. Binder cumulants for J'/J= 1.0 with different lattice sizes. The circle indicates the cross point of the Binder cumulants for lattice size L<40. (Wang et al/, 2010)

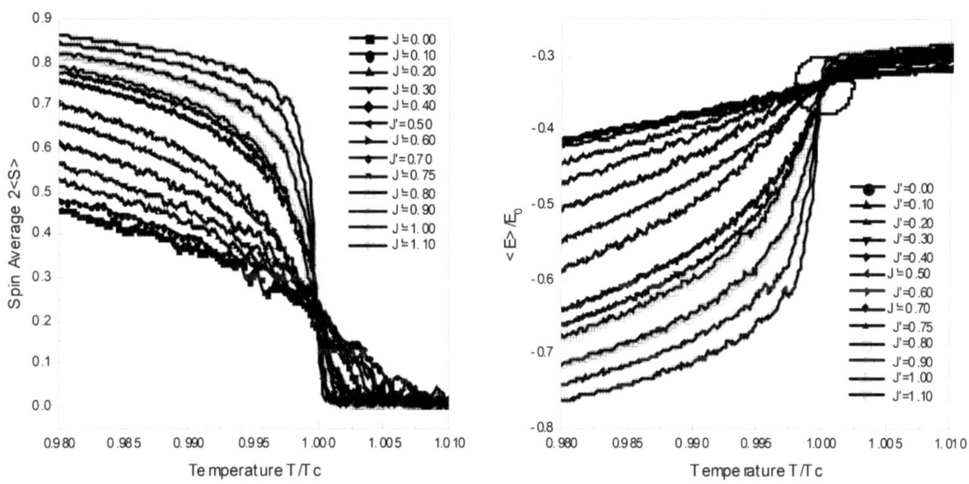

Fig. 17. Temperature dependence of spin average of lattice sizes L=100 for different J'/J around the transition temperature. Temperature is scaled by the transition temperature determined by the criterion $<E>/E_0=1/3$. (Wang et al., 2010)

Fig.17 shows the temperature dependence of spin average and the reduced average energy of lattice size 100 as shown in Fig.15, but with temperature rescaled by the transition temperature T_C. The transition temperature T_C is determined by the criterion of the averaged energy being one third of the ground state energy, i.e., temperature at $<E>/E_0=1/3$ being the transition temperature, as circled in Fig.17(b).

Results from Monte Carlo simulations on Ising cubic lattices with four spin couplings suggest that, (1) critical value of four-spin coupling strength for occurrence of first order phase transition is larger than that of mean field theory; (2) the critical value is lattice dependence. When lattice size is smaller, the phase transition is still of second order; (3) when the phase transition is of first order, the transition temperature can be determined by the average energy being a third of the ground state energy. However, this criterion has not been justified rigorously.

5. Ferroelectric relaxors

Ferroelectric relaxors have been drawn much attention because of their high electro-mechanical performance and unusual ferroelectric properties. Two review articles (Ye, 1998; Bokov & Ye, 2006) have summarized the achievements of recent researches on ferroelectric relaxors, especially for lead magnesium nibate (PMN). Basically, there are two categories of explanations about the fundamental physics of their unusual properties. One is based upon the randomness of their compositions and structures, such as Smolenskii's theory and spherical random bond random field model. Another is presumably based upon the experimental phenomenon, such as macro-micro-domain and super-paraelectric model. In this section, a general explanation of the properties observed in ferroelectric relaxors is proposed after analysis of the later category models (Wang et al., 2009). Field induced phase transition and diffused phase transition are reproduced within the framework of effective field approach.

Interpretation is started with the experimental results of field induced phase transition in PMN. The temperature dependence of the polarization under different electric field strengths has been obtained (Ye & Schmid, 1993; Ye, 1998). From this relation we can understand that (1) there is no ferroelectric phase transition in the whole temperature range, since there is no spontaneous polarization as the temperature goes down to zero Kelvin; (2) ferroelectricity can be induced by an external electric field. These imply that the phase transition in PMN is of first order, but the ferroelectricity is depressed in the whole temperature range. To understand these characteristics of ferroelectric relaxors, we can recall the temperature dependence of the spontaneous polarization in a typical normal first order ferroelectric phase transition is shown in Fig. 2. This implies that PMN is in a paraelectric state, but not far from the ferroelectric state.

Apart from the field induced phase transition, the following features should be also found within this temperature range in a normal first order ferroelectric phase transition: (1) a very long relaxation time, because of the critical slow down as the temperature is near the critical temperature. (2) super-paraelectric behavior. Normally there should be a double hysteresis loop observed in this temperature range. However, the double hysteresis loop could be reduced to a super-paraelectric shape because of the long relaxation time of the critical slow down. (3) macro-micro-domain crossover. As the temperature is much lower than the critical temperature, single domain or macro-domain is expected since it is in ferroelectric state; as the temperature is much higher than the critical temperature, no domain will be observed as it is in paraelectric state. Around the critical temperature, macro-micro-domain crossover is expected, i.e., polar nano-regions are forming in this temperature range. All these features have been observed in the ferroelectric relaxors like PMN.

The depression of the ferroelectricity in PMN reminds us of the case of quantum paraelectrics $SrTiO_3$. Therefore the field induced phase transition in a first order phase transition depressed by quantum fluctuation has been investigated within the framework of effective field approach by inclusive of zero point energy (Wang et al., 2008). The temperature dependence of the induced polarization under different electric field is shown in Fig.10. The polarization p and temperature t are in dimensionless unit, the numbers marked in the Fig.10 are the strength of electric field.

The major difference between the temperature dependence of the polarization in PMN (Ye & Schmid, 1993; Ye, 1998) and Fig.10 appears at lower temperature range. The polarization of PMN is still difficult to recover by an external field. This suggests that the zero point energy at lower temperature could be much larger than that in the higher temperature. In other words, the zero point energy might be increased with decreasing of temperature.

Another evidence of the existence of larger zero point energy can be found from the diffused phase transition in PMN. For the quantum temperature scale with constant zero point energy, the temperature dependence of the dielectric constant has been obtained from Monte Carlo simulation on a Ising model (Wang et al., 2002). As the zero point increases, the transition temperature or the peak temperature of the dielectric constant shifts to a lower temperature. As the zero point energy is larger than the critical value, quantum paraelectric feature is obtained. When the zero point energy increases further, the dielectric constant decreases at low temperature. Therefore if the zero point energy changes with temperature and has a larger value only at lower temperature, the dielectric constant will increase as the temperature increases at lower temperature side, and decreases as the temperature increases at higher temperature side. A round dielectric peak will be formed around the temperature of zero point energy dropping down.

From above analysis, a temperature dependent form of zero point energy is proposed in the following form (Wang et al., 2009)

$$\hbar\omega = \frac{\hbar\omega_0}{1 + e^{\alpha(T - T_X)}} \tag{44}$$

where the zero point energy ω changes around temperature T_x from ω_0 at lower temperature to a relative lower value with crossover rate α. By using the same technique of effective field approach as in Refs.(Wang et al., 2008; Yuan et al., 2003 Gonzalo, 2006), with the quantum temperature scale in Eq.(32) and the zero point energy in Eq.(44), the field induce phase transition and diffused phase transition are obtained and shown in Fig.18. All parameters are in dimensionless scale, e stands for the electric field, and k is for the dielectric constant. From Fig.18 we can see that the basic features of field induce phase transition and diffused phase transition observed in relaxor ferroelectric are reproduced.

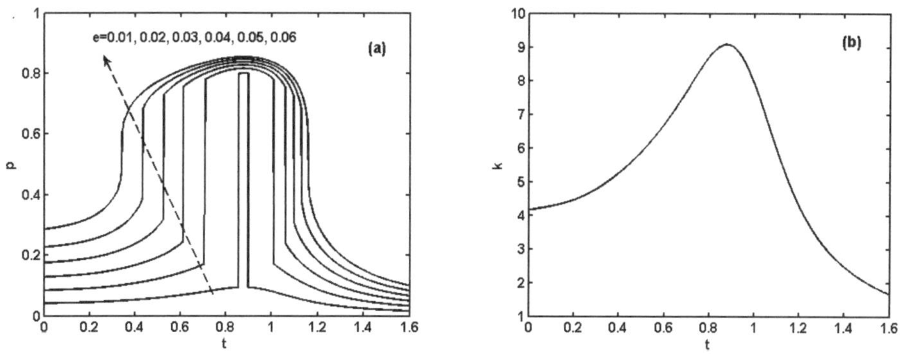

Fig. 18. (a) Field induced phase transition and (b) diffused phase transition. (Wang et al., 2009)

The dimensionless quantum temperature scale used in Fig.18 is shown in Fig.19. The solid line represents the quantum temperature scale, and the dashed line is for the real temperature scale. The phase transition temperature is marked by the arrow. That means that the state evolution of relaxor ferroelectric misses the phase transition temperature as the temperature decreases, and re-entry of the paraelectric state. The kind of non-ergodic behavior is schematically shown in Fig.19(b) for better understanding.

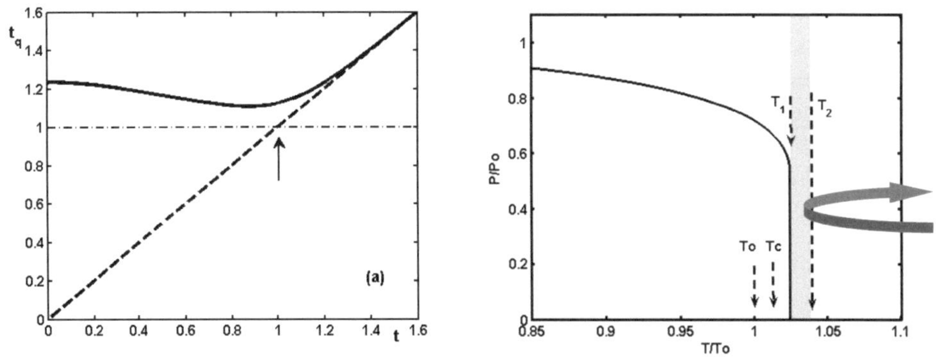

Fig. 19. (a) Quantum temperature scale and (b) non-ergodic behavior. (Wang et al., 2009)

Overall, relaxor ferroelectrics like PMN are in a tri-critical state of long range ferroelectric order, thermal fluctuations and quantum fluctuations. This critical state can be described by a quantum temperature scale. Following features can be understood from this issue: diffuse phase transition, field induced phase transition, long relaxation time, as well as super-paraelectric state, micro-macro-domain crossover and nonergodic behavior. However, a more adequate expression of temperature dependence of zero point energy is still needed for better description of the physical behavior in the relaxor ferroelectrics.

6. Final remarks

Theoretical methods and models for studying ferroelectric with first order phase transition are definitely not limited to the contents in this chapter. Hopefully results from these theoretical techniques can provide useful information for understanding experimental observations. Heavy-computer relying methods, such as first-principle calculations and molecular dynamic simulations etc, have been applied to investigate the physical properties of ferroelectrics. However, fully understanding of the origin of ferroelectrics might need more efforts from both theoretical and experimental side.

7. References

Ali, R.A.; Wang, C.L.; Yuan, M.; Wang, Y. X. & Zhong, W. L. (2004). Compositional dependence of the Curie temperature in mixed ferroelectrics: effective field approach. *Solid State Commun.*, 129, 365 -367

Arago, C.; Garcia, J.; Gonzalo, J. A.; Wang, C. L.; Zhong, W. L. & Xue, X. Y. (2004). Influence of quantum zero point energy on the ferroelectric behavior of isomorphous systems. *Ferroelectrics*, 301, 113 -119

Aragó, C.; Wang, C. L. & Gonzalo, J. A. (2006). Deviations from Vegard's law in the Curie temperature of mixed ferroelectric solid solutions. *Ferroelectrics*, 337, 233-237

Binder, K. (1984). *Applications of Monte Carlo Method in Statistical Physics*, 2nd Edition, Springer-Verlag, ISBN *038712764X*, Berlin

Blinc, R. & Zeks, B. (1974). *Soft modes in ferroelectrics and antiferroeletrics*, North-Holland Publishing Company, ISBN 0 7204 1462 8, Amsterdam

Bokov, A. A. & Ye, Z. G. (2006). Recent progress in relaxor ferroelectrics with perovskite structure. *Journal of Materials Science*, 41, 31-52

Chen, H. W.; Yang, C.R.; Zhang, J.H.; Pei, Y.F. & Zhao, Z. (2009). Microstructure and antiferroelectric-like behavior of $Pb_xSr_{1-x}TiO_3$ ceramics. *J Alloys and Compounds*, 486, 615-620

Gonzalo, J. A. (1989). Quantum effects and competing interactions in crystals of the mixed rubidium and ammonium dihydrogen phosphate systems. *Phys. Rev. B*, 39, 12297-12299

Gonzalo, J. A.; Ramírez, R.; Lifante, G. & Koralewski, M. (1993). Dielectric discontinuities and multipolar interactions in ferroelectric perovskites, *Ferroelectrics Letters Section*, 15, 9-16

Gonzalo, J. A. (2006). *Effective Field Approach to Phase Transitions and Some Applications to Ferroelectrics*, World Scientific Press, ISBN 981-256-875-1, Singapore

Gonzalo, J. A. & Wang, C. L. (2008). *Monte Carlo approach to Phase Transitions*, Asociacion Espanola Ciencia y Cultura, ISBN 978-84-936537-9-8，Madrid

Jiang, Q. & Tao, Y. M. (2005). Critical line between the first-order and the second-order phase transition for ferroelectric thin films described by TIM. *Phys. Lett. A*, 336, 216-222

Kutnjak, Z.; Petzelt, J. & Blinc, R. (2006). The giant electromechanical response in ferroelectric relaxors as a critical phenomenon. *Nature*, 441, 956-959

Lines, M. E. & Glass A. M. (1997). *Principles and Applications of Ferroelectrics and Related Materials*, Clarendon Press, ISBN 0 19 851286 4, Oxford

Noheda, B.; Lifante, G. & Gonzalo, J.A. (1993). Specific heat and quadrupole interactions in uniaxial ferroelectrics, *Ferroelectrics Letters Section*, 15 , 109-114

Noheda, B.; Koralewski, M.; Lifante, G. & Gonzalo J. A. (1994). Dielectric discontinuities and multipolar interactions in ferroelectric perovskites, *Ferroelectrics Letters Section*, 17, 25-31

Qu, B. D.; Zhong, W. L. & Zhang, P. L. (1997). Ferroelectric superlattice with first-order phase transition. *Ferroelectrics*, 197, 23-26

Song, T. K.; Kim, J. S.; Kim, M.H.; Lim, W.; Kim, Y. S. & Lee, J. C. (2003). Landau-Khalatnikov simulations for the effects of external streese on the ferroelectric properties of $Pb(Zr,Ti)O_3$ thin films. *Thin Soild Films*, 424, 84 -87

Teng, B. H. & Sy, H. K. (2005). Transition regions in the parameter space based on the transverse Ising model with a four-spin interaction. *Europhys. Lett.*, 72, 823-829

Wang, C. L.; Qin, Z. K. & Lin, D. L. (1989). First order phase transition in order-disorder ferroelectrics. *Phys. Rev. B*, 40, 680-685

Wang, C. L.; Zhang, L.; Zhong, W. L. & Zhang, P.L. (1999). Switching characters of asymmetric ferroelectric films. *Physics Letters A*, 254, 297-300

Wang, C. L.; Garcia, J.; Arago, C., Gonzalo, J. A. & Marques, M. I. (2002). Monte Carlo simulation of quantum effects in ferroelectric phase transitions with increasing zero-point energy, *Physica A*, 312, 181-186

Wang, C. L.; Li, J.C.; Zhao, M.L.; Zhang, J.L.; Zhong, W.L.; Aragó, C.; Marqués, M.I. & Gonzalo, J.A. (2008). Electric field induced phase transition in first order ferroelectrics with large zero point energy. *Physica A*, 387, 115-122

Wang, C. L.; Li, J. C.; Zhao, M. L.; Liu, J. & Zhang, J. L. (2009). Ferroelectric relaxor as a critical state. *Science in China Series E: Technological Sciences*, 52, 123-126

Wang, C. L.; Li, J.C.; Zhao, M. L.; Marqués, M.I.; Aragó, C. & Gonzalo, J. A. (2010). Monte Carlo simulation of first order phase transitions, *Ferroelectrics*, (in press)

Wang, X. S.; Wang, C. L. & Zhong, W. L. (2002). First-order phase transition in ferroelectric superlattice described by the transverse Ising model. *Solid State Commun.* 122, 311-315

Wang, Y. G.; Zhong, W. L. & Zhang, P. L. (1996). Surface effects and size effects on ferroelectrics with a first-order phase transition. *Phys. Rev. B*, 53, 11439-11443

Wesselinowa, J. M. (2002). Properties of ferroelectric thin films with a first order phase transitions. *Solid State Commun.*, 121, 89-92

Ye, Z. G. & Schmid, H. (1993). Optical, Dielectric and Polarization Studies of the Electric Field-Induced Phase Transition in $Pb(Mg_{1/3}Nb_{2/3})O_3$ [PMN] *Ferroelectrics*, 145, 83-108

Ye, Z. G. (1998). Relaxor Ferroelectric Complex Perovskites: Structure, Properties and Phase Transition, *Key Engineering Materials*, 155-156, 81-122

Yuan, M.; Wang, C. L.; Wang, Y. X.; Ali, R. A. & Zhang, J. L. (2003). Effect of zero-point energy on the dielectric behavior of strontium titanate. *Solid State Commun.*, 127, 419-421

17

Electroacoustic Waves in a Ferroelectric Crystal with of a Moving System of Domain Walls

Vilkov E.A. and Maryshev S.N.
[1]*Kotel'nikov Institute of Radio Engineering and Electronics of Russian Academy of Sciences, Ul'yanovsk Branch,*
[2]*Moscow Institute of Physics and Technology*
Russia

1. Introduction

In addition to conventional applications of ferroelectric crystals that exhibit piezoactivity, such crystals find wide application in the production of domain electronic devices based on the interaction between acoustic waves and ferroelectric domains (acousto-domain interaction) [Esayan et al, 1974]. In the course of the general development of technology of phonon crystals the ferroelectric with periodic domain structures (PDSs) have attracted a considerable interest recently [Golenishchev-Kutuzov et al, 2003]. The interest in such structures with periodically varying elastic, electric, and other physical parameters dates back to the 1960s since, by that time, the possibilities offered by single-domain crystals in the realization of the spectral characteristics of elastic and electromagnetic waves had been exhausted. First of all a ferroelectrics with PDSs differ from usual phonon crystals by ability of its reorganization under external action due to rather fast moving domain walls(DW) [Golenishchev-Kutuzov et al, 2003].

The spectra of elastic oscillations of a ferroelectric with a static PDS may significantly differ from the spectra of a conventional single-domain sample [Li, et al, 1991; Lyubimov & Sannikov, 1979]. This difference is primarily manifested in the formation of forbidden and allowed bands, which are absent in the continuous spectrum of waves in a single-domain crystal. To the full this conclusion extends to oscillation modes of localized at DWs, in particular - to electroacoustic interfacial waves (EIWs) [Maerfield & Tournois, 1971], which were widely studied in ferroelectrics with static PDSs, representing a superlattice [Li, et al, 1991; Nougaoui, 1987]. Considerable modification of spectral parameters of EIWs in superlattices such as boundary localization of oscillation and phase speed was explained in [Lyubimov & Sannikov, 1979; Li, 1987] as result of interaction between electrosound oscillations of neighbo(u)ring DW.

At an early stage of research the possibility of the realization of PDSs with various periods of the lattices of domain walls and the possibility of vary lattice period [Golenishchev-Kutuzov et al, 2003] have shown the prospects of application of the domain superlattices. It was found out, in particular, that crystals with PDSs are effective as broadband piezoconverters of acoustic waves, and with the greater efficiency of transformation, than have monodomain crystals [Batanova & Golenishchev-Kutuzov, 1997]. Other fields of application (the acoustic spectroscopy of polydomain ferroelectrics and processing of the signal information) have demanded of studying of effects of a refraction of acoustic waves from packages of DWs of ferroelectrics. So, in development of results of work [Esayan et al, 1974] it has been shown, that appropriate variations in the period of the domain lattice, crystal parameters, and conditions of wave propagation make it possible to control the height, number, and location of Bragg reflections peaks [Shuvalov & Gorkunova, 1999; Shenderov,1997], thus providing prospects for various applications.

All the results listed above are related to the static case, i.e. they are obtained for domain lattices with motionless DW. It means that the manipulation of PDSs expressed by moving DWs occurs not during the process of acousto-domain interaction. The overlap of the procedure of adjustment PDSs and the process of acousto-domain interaction is interesting due to a number of circumstances.

First, this overlap demands modeling of a supernumerary situation of work of the acousto-domain device at "failure" of device domain structure. Second, it is necessary for research of additional opportunities, for example, scanning of a crystal by means of EIWs on moving DW [Gulyaev et al, 2000] or realization of ideas put forward by Auld [Auld,1973] of the Doppler transformation of a spectrum of frequencies of bulk acoustic waves refracted by moving DW [Shevyakhov, 1990]. Thus, it is expedient to generalize the results of the research of acoustic properties of static superlattices of ferroelectric to the case of dynamic superlattices whose distinctive feature is adjustable moving domain boundaries. Below, on the basis of our works [Vilkov, 2008; Vilkov, 2009; Vilkov et al, 2009] the propagation of guided EIWs and bulk electrosound waves in a dynamic superlattice of equidistant uniformly moving 180-degree DW is considered. Let's notice, that until now the research of acoustic effects in crystals of ferroelectrics with moving DW was limited to cases of non-collective acousto-domain interactions:a single DW [Shevyakhov, 1990; Gulyaev et al, 2000], a strip domain (i.e., a pair of DWs) [El'meshkin & Shevyakhov, 2006] and a structure of a two adjacent strip domains [Bugaev et al, 2005].

2. Basic equations for a shear wave in ferroelectric in the frame of the DW at rest

In acoustoelectronics, the elastic deformation and electric-field intensity E_j normally serve as independent quantities that determine the state of a piezoactive ferroelectric medium. The elastic deformation is determined by the deformation tensor

$$u_{ik} = \frac{1}{2}\left(\frac{\partial u_i}{\partial x_k} + \frac{\partial u_k}{\partial x_i}\right), \quad (1)$$

For the adiabatic deformation, the equations of state are written as follows [Sosnin & Strukov, 1970]

$$\begin{aligned} T_{ik} &= c_{iklm}u_{lm} - e_{j,ik}E_j \\ D_p &= \varepsilon_{pq}E_q + 4\pi e_{p,rm}u_{rm} \end{aligned} . \quad (2)$$

where T_{ik} is the mechanical stress tensor, which is a second- rank tensor (as well as deformation tensor u_{ik} given by expression (1)), and c_{iklm}, $e_{j,ik}$ and ε_{pq} are the tensors of elastic, piezoelectric, and dielectric moduli of a crystal. Note that summation over the repeated subscripts is involved in Eq. (2), where D_p is the electric induction. The initial equations are the equation of motion for a crystal

$$\rho \frac{\partial^2 u_i}{\partial t^2} = \frac{\partial T_{ik}}{\partial x_k} \tag{3}$$

and the Maxwell equations in the quasi-static representation,

$$E_i = -\frac{\partial \varphi}{\partial x_i}, \quad \frac{\partial D_k}{\partial x_k} = 0 . \tag{4}$$

Here, differential operator $\partial / \partial x_i$ is considered a vector, so that tensor summation over repeated subscript k is involved in the second line of expression (4); φ is the electric potential; and ρ is the crystal density.

For the consistent analysis of Eqs. (1)–(4), we specify the symmetry of the crystal and the propagation configuration of acoustic waves. We assume that, 4*mm* ferroelectric has a crystallographic putting such that, in a crystallographic system the fourth-order symmetry axis that is parallel to the *z* axis of laboratory frame *xOyz*. For the given crystal symmetry initial Eqs. (1)–(4) can be presented after some transformations in the form of

$$4\pi e_{1,5}\left(\frac{\partial^2 u_z}{\partial x^2}+\frac{\partial^2 u_z}{\partial y^2}\right)+4\pi e_{3,3}\frac{\partial^2 u_z}{\partial z^2}+4\pi(e_{1,5}+e_{3,1})\frac{\partial}{\partial z}\left(\frac{\partial u_x}{\partial x}+\frac{\partial u_y}{\partial y}\right)=\varepsilon_1\left(\frac{\partial^2 \varphi}{\partial x^2}+\frac{\partial^2 \varphi}{\partial y^2}\right)+\varepsilon_3\frac{\partial^2 \varphi}{\partial z^2} \tag{5}$$

$$\begin{aligned}
\rho\frac{\partial^2 u_x}{\partial t^2} &= c_{11}\frac{\partial^2 u_x}{\partial x^2}+c_{66}\frac{\partial^2 u_x}{\partial y^2}+c_{44}\frac{\partial^2 u_x}{\partial z^2}+(c_{12}+c_{66})\frac{\partial^2 u_y}{\partial x \partial y}+(c_{13}+c_{44})\frac{\partial^2 u_z}{\partial x \partial z}+(e_{1,5}+e_{3,1})\frac{\partial^2 \varphi}{\partial x \partial z},\\
\rho\frac{\partial^2 u_y}{\partial t^2} &= c_{66}\frac{\partial^2 u_y}{\partial x^2}+c_{11}\frac{\partial^2 u_y}{\partial y^2}+c_{44}\frac{\partial^2 u_y}{\partial z^2}+(c_{12}+c_{66})\frac{\partial^2 u_x}{\partial x \partial y}+(c_{13}+c_{44})\frac{\partial^2 u_z}{\partial y \partial z}+(e_{1,5}+e_{3,1})\frac{\partial^2 \varphi}{\partial y \partial z},\\
\rho\frac{\partial^2 u_z}{\partial t^2} &= c_{44}\left(\frac{\partial^2 u_z}{\partial x^2}+\frac{\partial^2 u_z}{\partial y^2}\right)+c_{33}\frac{\partial^2 u_z}{\partial z^2}+(c_{13}+c_{44})\left(\frac{\partial^2 u_x}{\partial x \partial z}+\frac{\partial^2 u_y}{\partial y \partial z}\right)+e_{1,5}\left(\frac{\partial^2 \varphi}{\partial x^2}+\frac{\partial^2 \varphi}{\partial y^2}\right)+e_{3,3}\frac{\partial^2 \varphi}{\partial z^2}.
\end{aligned} \tag{6}$$

The received system of the equations (6) together with the equation (5) describes dynamic behaviour of a piezoelectric material of a class 4mm(6mm,∞m) and often is called as system of the equations of piezoacoustics. We assume also that, the propagation of acoustic waves is in the direction perpendicular to the high-order symmetry axis. In these conditions the solution is not depend on coordinate z and the system of the equations (6) are split into two independent systems of equations. First of them

$$\begin{aligned}
\rho\frac{\partial^2 u_x}{\partial t^2} &= c_{11}\frac{\partial^2 u_x}{\partial x^2}+c_{66}\frac{\partial^2 u_x}{\partial y^2}+(c_{12}+c_{66})\frac{\partial^2 u_y}{\partial x \partial y},\\
\rho\frac{\partial^2 u_y}{\partial t^2} &= c_{66}\frac{\partial^2 u_y}{\partial x^2}+c_{11}\frac{\partial^2 u_y}{\partial y^2}+(c_{12}+c_{66})\frac{\partial^2 u_x}{\partial x \partial y},
\end{aligned} \tag{7}$$

and describe horizontally polarized waves: $\mathbf{u} \perp z$. The second system added by the equation (5)

$$\rho\frac{\partial^2 u_z}{\partial t^2} = c_{44}\left(\frac{\partial^2 u_z}{\partial x^2}+\frac{\partial^2 u_z}{\partial y^2}\right)+e_{1,5}\left(\frac{\partial^2 \varphi}{\partial x^2}+\frac{\partial^2 \varphi}{\partial y^2}\right),$$
$$4\pi e_{1,5}\left(\frac{\partial^2 u_z}{\partial x^2}+\frac{\partial^2 u_z}{\partial y^2}\right)=\varepsilon_1\left(\frac{\partial^2 \varphi}{\partial x^2}+\frac{\partial^2 \varphi}{\partial y^2}\right). \quad (8)$$

describes horizontally polarized waves: $\mathbf{u}\,||\,z$.

Waves of horizontal and vertical polarization do not interact among themselves. At the same time, elastic displacements in vertically polarized waves are not accompanied by the generation of the electric field, so that the crystal behaves toward these waves as a conventional elastic medium in the absence of the piezoelectric effect. In the further analysis, we disregard vertically polarized waves owing to the absence of the acousto-domain interaction. Horizontally polarized waves with the particle displacements along the z axis are piezoelectrically active and accompanied by the electric field in the propagation plane. (see Fig. 1).

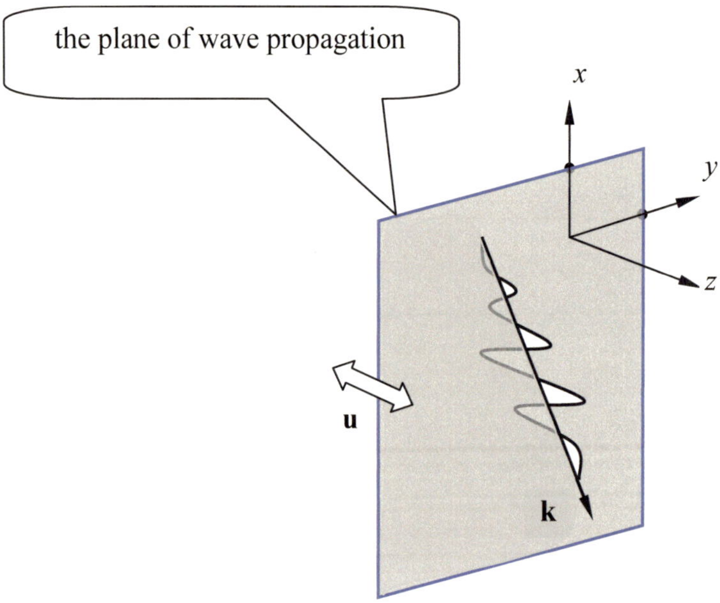

Fig. 1. The geometry of the propagation of a shift wave of horizontal polarization

For final writing of the equations (8) we consider, that the piezoelectric modulus changes the sign in the neighboring domains divided by 180-degree DWs because of the antiparallel orientation of the polar directions, i.e. the plane of wave propagation

$$e^{(j)} = (-1)^{j+1} e_{15}\ , \quad e_{15} > 0\,. \quad (9)$$

where j= 1 and 2 stand for domains of the "+" and "–" types with positive and negative piezoelectric moduli, respectively. As a result we write down of the equation (8) in the form of

$$\rho\frac{\partial^2 u_j}{\partial t^2} = c_{44}\nabla^2 u_j + e^{(j)}\nabla^2\varphi_j\,,\quad 4\pi e^{(j)}\nabla^2 u_j = \varepsilon_1\nabla^2\varphi_j\,, \quad (10)$$

where $\nabla^2=(\partial^2/\partial x^2+\partial^2/\partial y^2)$ is the Laplace operator, and shift displacement u_z have designated as u_j according number of domain j.

Domain walls can move in ferroelectrics under specific external actions. In this case, to construct a solution in domains, it is necessary to consider the equation (10) in the frame of the DW at rest $\tilde{x}0\tilde{y}\tilde{z}$. Because $V_D << c$ (c is the velocity of light, V_D is the velocity of DW), we employ the Galilean transformation for the connection of coordinates:

$$\tilde{x}=x\,,\qquad \tilde{y}=y-V_D t\,,\qquad \tilde{t}=t. \tag{11}$$

Using the relationships between the differential operators according to (11):

$$\frac{\partial}{\partial y}\leftrightarrow\frac{\partial}{\partial \tilde{y}}\,,\quad \frac{\partial}{\partial x}\leftrightarrow\frac{\partial}{\partial \tilde{x}}\,,\qquad \frac{\partial}{\partial t}\leftrightarrow\frac{\partial}{\partial \tilde{t}}-V_D\frac{\partial}{\partial \tilde{y}}\,, \tag{12}$$

and expressions (11) and (12), we can easily represent Eqs. (10) as

$$\left[\frac{1}{c_{44}^{*2}}\left(\frac{\partial}{\partial \tilde{t}}-V_D\frac{\partial}{\partial \tilde{y}}\right)^2-\tilde{\nabla}^2\right]u_j=0\,,\qquad \tilde{\nabla}^2\Phi_j=0 \tag{13}$$

In these equations, $\overset{*}{c}_{44}=c_{44}+4\pi e_{15}^2/\varepsilon_1$, and Φj is the component of total potential φ_j in the jth domain,

$$\varphi_j=-\frac{4\pi e^{(j)}}{\varepsilon_1}u_j+\Phi_j\ \ , \tag{14}$$

which corresponds to the electric field induced from the boundaries $\tilde{y}_n=y_n-V_D t$ by piezopolarization charges [Lyubimov & Sannikov, 1979].

3. The electroacoustic bulk propagating waves in a ferroelectric with a system of moving periodic domain structure

3.1 The statement of the problem

In this section we consider the influence of a uniform motion of domain boundaries forming a dynamic superlattice of the tetragonal ferroelectric, on spectral properties of bulk electrosound waves (EW). The comparison of the phonon spectra of static and dynamic superlattices allows a more detailed description of specific wave processes realized in periodic structures. The results obtained indicate that even slow DW motions cause significant (detectable) variations in the wave spectra.

The schematic of the problem is presented in Fig. 2. We assume that, in a crystallographic system with the fourth-order symmetry axis that is parallel to the z axis of laboratory frame $xOyz$, a $4mm$ ferroelectric represents a superlattice of 180-degree DWs with period $2d$ ($d>>\Delta$ is the distance between neighboring DWs and Δ is the DW thickness). The superlattice uniformly moves at the velocity $V_D || y$ [010]||. Positive velocity V_D corresponds to the superlattice motion codirectional with the y axis. To eliminate the significant structural sensitivity of DWs to the regime of motion related to the generation of both spontaneous polarization and spontaneous deformation in domains [Sosnin. & Strukov, 1970; Vainshtein,

1988], we restrict the consideration to subsonic DW velocities. We also assume that the ferroelectric is far from the phase transition. Then, the DW motion is realized at V_D = const and the current coordinates are given by $y_n = V_D t + nd$, where t is time; n=0, ±1, ±2,... and DWs are geometrically thin and structureless objects ($k\Delta << 1$, where k is the EW wave number).

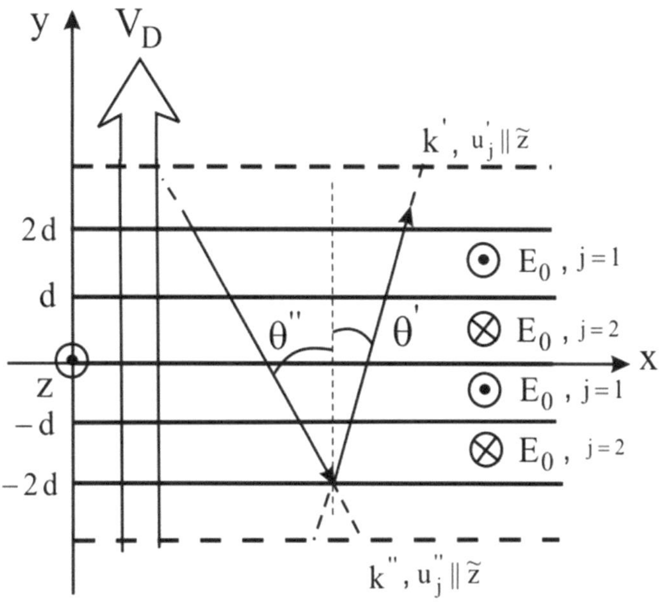

Fig. 2. Schematic representation of a domain superlattice with an electroacoustic eigenwave. The dashed lines indicate that two elementary cells are reproduced in the positive and negative directions of the Y axis and form an infinite superlattice. Vectors k′ and k′′show the direction of the EW propagation. The arrow shows the direction of the lattice motion. Here, E_0 is the electric-field intensity in domains

It is known [Vainshtein, 1988] that the above conditions are valid for DWs in $BaTiO_3$-type ferroelectrics at ultrasonic frequencies. It is commonly accepted [Maerfield & Tournois, 1971; Auld,1973] that, in this case, the ferroelectric can be considered as a piezoelectric crystal that is periodically twinned with respect to the planes $y = y_n$. With regard to the analysis of propagation of only horizontally polarized shear waves in the (001) crystal plane, we assume that antiparallel polarizations in the neighboring domains are related to the sign alternation of the piezoelectric modulus according equation (9). At that, there are in eq.(9) $j = 1$ for $(n\text{-}1)d + V_D t < y < nd + V_D t$ and $j = 2$ for $nd + V_D t < y < (n+1)d + V_D t$ $(n = 0, \pm 2, \pm 4,\ldots)$. The remaining characteristics of domains are identical.

Then, we assume that the electroacoustic eigenwaves with the wave vector = $k' = (k'\sin\theta', k'\cos\theta', 0)$ propagate in the XOY plane at angle θ' relative to the normal to the plane of 180-degree DWs in the positive direction of the y axis (Fig. 2). For the electroacoustic eigenwaves propagating in the XOY plane at angle θ'' to the normal to the DW plane in the negative direction of the y axis, we employ another wave vector, = $k'' = (k''\sin\theta'', k''\cos\theta'', 0)$ in accordance with another expected Doppler shift [Shevyakhov, 1990].

The system of equations for these waves in the neighboring domains with numbers $j = 1$ and 2, which results from the reduction of Eqs. (3) and (4) in accordance with the cyclic Bloch conditions in the case of 180-degree DWs, is represented by equations (9),(13), (14). The fields on the moving DWs of the lattice exhibit the phase conjugation for the wave characteristics of EWs that travel in the direction opposite to that of the DW propagation. Hence, the following relationships are valid [Shevyakhov, 1990]:

$$\cos\theta' = \frac{\cos\theta''(1+V_D^2/v^2)+2V_D/v}{1+2(V_D/v)\cos\theta''+V_D^2/v^2}, \qquad k_y' = -k_y'' + \frac{2k_y''+2(V_D/v)k''}{1-V_D^2/v^2}. \tag{15}$$

Here, v is the EW phase velocity in a single-domain crystal. In the case of a static lattice (V_D =0), the wave characteristics of the oppositely directed eigenwaves are identical: $\theta = \theta'' = \theta'$, $k_y = k_y'' = k_y'$.

To ensure the interaction between an EW with wave vector k′ and a DW moving away, we must provide for faster (in the signal meaning) wave propagation relative to the DW. In accordance with [Shevyakhov, 1990], this can be realized through imposition of a limitation on propagation angle θ′, $\theta' < \theta^{**}$. Owing to the coincidence of the phase and group velocities of an EW in a singledomain crystal, the critical angle $\theta^{**} = \arccos(-V_D/v)$ determines the equality of the projection of the EW group velocity on the direction of the DW motion and the DW velocity.

When $\theta' > \theta^{**}$, , the DW motion along the direction of the DW displacement is faster than the motion of any plane train of EWs with the same angle of incidence θ′. From the physical point of view, this situation corresponds to the vanishing acousto-domain interaction. In addition to the condition $\theta' < \theta^{**}$, we assume that the EW wavelength is significantly smaller than the crystal dimension. Under such conditions, the boundary effects at the outer interfaces of the ferroelectric and its shape weakly affect the behavior of waves and, hence, can be disregarded.

3.2 Dispersion equation

We assume that the domains with numbers 1 and 2 are located between the DWs with the coordinates $\tilde{y}$ = 0, *d*, 2*d* (Fig. 2). We search for the solution to Eqs. (13) in the region of the first domain in the following representation:

$$\begin{aligned} u_1(\tilde{y}) &= (A_1\exp(i\,k_y''\tilde{y}) + A_2\exp(-i\,k_y'\,\tilde{y}))\exp(i(k_x\,\tilde{x}-\Omega\tilde{t})),\\ \Phi_1(\tilde{y}) &= (C_1\exp\,(k_x\,\tilde{y}) + C_2\exp(-k_x\,\tilde{y}))\exp(i\,(k_x\,\tilde{x}-\Omega\tilde{t})) \end{aligned} \tag{16}$$

Similar expressions can be written in the region of the second domain:

$$\begin{aligned} u_2(\tilde{y}) &= (B_1\exp(i\,k_y''\tilde{y}) + B_2\exp(-i\,k_y'\,\tilde{y}))\exp(i(k_x\,\tilde{x}-\Omega\tilde{t})),\\ \Phi_2(\tilde{y}) &= (D_1\exp\,(k_x\,\tilde{y}) + D_2\exp(-k_x\,\tilde{y}))\exp(i\,(k_x\,\tilde{x}-\Omega\tilde{t})). \end{aligned} \tag{17}$$

To join the domain fields in the frame of the DW at rest on the interface $\tilde{y} = d$ in the nonrelativistic quasistatic approximation, we employ the conventional continuity conditions [Auld, 1973] for shear displacements; potentials; the shear components of the stress tensor,

$$T_{yz}^{(j)} = c_{44}\frac{\partial u_j}{\partial \tilde{y}} - e^{(j)}\frac{\partial \varphi_j}{\partial \tilde{y}} \tag{18}$$

and the y components of the electric induction,

$$D_y^{(j)} = \varepsilon_1\frac{\partial \varphi_j}{\partial \tilde{y}} - 4\pi e^{(j)}\frac{\partial u_j}{\partial \tilde{y}}. \tag{19}$$

Thus, with allowance for expressions (14), (18), and (19), we obtain the following boundary conditions:

$$\begin{gathered}\varphi_1(\tilde{y}) = \varphi_2(\tilde{y}), \qquad \frac{\partial \Phi_1}{\partial \tilde{y}} = \frac{\partial \Phi_2}{\partial \tilde{y}}, \qquad u_1 = u_2 \\ \overset{*}{c}_{44}\frac{\partial u_1}{\partial \tilde{y}} + e_{15}\frac{\partial \Phi_1}{\partial \tilde{y}} = \overset{*}{c}_{44}\frac{\partial u_2}{\partial \tilde{y}} - e_{15}\frac{\partial \Phi_2}{\partial \tilde{y}},\end{gathered} \tag{20}$$

To simplify the further analysis, we introduce the following quantities:

$$\sigma_{1,2}(\tilde{y}) = \frac{\partial \Phi_{1,2}}{\partial \tilde{y}}, \qquad \alpha_{1,2}(\tilde{y}) = \overset{*}{c}_{44}\frac{\partial u_{1,2}}{\partial \tilde{y}} \pm e_{15}\frac{\partial \Phi_{1,2}}{\partial \tilde{y}}. \tag{21}$$

In the expressions for $u_{1,2}$, $\alpha_{1,2}$, $\varphi_{1,2}$, and $\sigma_{1,2}$ we represent arbitrary constants $A_{1,2}$, $B_{1,2}$, $C_{1,2}$, $D_{1,2}$ in terms of their values on the boundary $\tilde{y}=0$ (for the domain layer with the subscript j = 1) and on the boundary $\tilde{y}=d$ (for the domain layer with the subscript j = 2) and substitute these representations into expressions (14),(16), (17) and (21). The resulting formulas are as follows:

$$u_{1,2}(\tilde{y}) = m\,u_{1,2} + \frac{m_1}{c_{44}^*(k_y'' + k_y')}\alpha_{1,2} + 0\,\varphi_{1,2} \pm \frac{e_{15}\,m_1}{c_{44}^*(k_y'' + k_y')}\sigma_{1,2},$$

$$\begin{gathered}\alpha_{1,2}(\tilde{y}) = -\left(\frac{\overset{*}{c}_{44}k_y'' k_y' m_1}{(k_y'' + k_y')} + \overset{*}{c}_{44}K^2 k_x \mathrm{sh}(k_x\tilde{y})\right)u_{1,2} + m'\,\alpha_{1,2} \pm e_{15}\,k_x\,\mathrm{sh}(k_x\tilde{y})\varphi_{1,2} \pm e_{15}(\mathrm{ch}(k_x\tilde{y}) - m)\sigma_{1,2}, \\ \varphi_{1,2}(\tilde{y}) = \pm\frac{\overset{*}{c}_{44}K^2(m - \mathrm{ch}(k_x\tilde{y}))}{e_{15}}u_{1,2} \pm \frac{K^2 m_1}{e_{15}(k_y'' + k_y')}\alpha_{1,2} + \mathrm{ch}(k_x\tilde{y})\,\varphi_{1,2} + \left(-\frac{K^2\,m_1}{(k_y'' + k_y')} + \frac{\mathrm{sh}(k_x\tilde{y})}{k_x}\right)\sigma_{1,2},\end{gathered} \tag{22}$$

$$\sigma_{1,2}(\tilde{y}) = \pm\left(-\frac{\overset{*}{c}_{44}K^2 k_x}{e_{15}}\right)\mathrm{sh}(k_x\tilde{y})\,u_{1,2} + 0\,\alpha_{1,2} + k_x\mathrm{sh}(k_x\tilde{y})\varphi_{1,2} + \mathrm{ch}(k_x\tilde{y})\sigma_{1,2}\,,$$

where the values of the fields $u_{1,2}$, $\alpha_{1,2}$, $\varphi_{1,2}$, and $\sigma_{1,2}$ on the right-hand side of the identities for the subscript j = 1 correspond to $\tilde{y}=0$ and the plus sign is selected. For the subscript j = 2, we use the field values at $\tilde{y}=d$ and choose the minus sign. In expressions (22), we introduce quantity $K^2 = 4\pi e_{15}^2/\varepsilon_1 \overset{*}{c}_{44}$, which is the squared coefficient of the electromechanical coupling of the crystal for the shear waves that propagate in the (001) basis plane. For brevity, we use the following notation in expressions (22):

$$m' = 2\frac{(k_y^{''}k_y^{'})^{1/2}}{(k_y^{''}+k_y^{'})}\cos\left[\frac{(k_y^{''}+k_y^{'})\tilde{y}}{2}+i\ln\left(\frac{k_y^{''}}{k_y^{'}}\right)^{\frac{1}{2}}\right]\exp\left[i\frac{(k_y^{'}-k_y^{''})\tilde{y}}{2}\right]$$

$$m = 2\frac{(k_y^{''}k_y^{'})^{1/2}}{(k_y^{''}+k_y^{'})}\cos\left[\frac{(k_y^{''}+k_y^{'})\tilde{y}}{2}+i\ln\left(\frac{k_y^{'}}{k_y^{''}}\right)^{\frac{1}{2}}\right]\exp\left[i\frac{(k_y^{'}-k_y^{''})\tilde{y}}{2}\right],$$

$$m_1 = 2\sin\left[\frac{(k_y^{''}+k_y^{'})\tilde{y}}{2}\right]\exp\left[i\frac{(k_y^{'}-k_y^{''})\tilde{y}}{2}\right].$$

Thus, for $V_D = 0$, , we have $m = \cos(k_y\tilde{y})$ and $m_1 = 2\sin(k_y\tilde{y})$ with allowance for expression (15).

We represent the fields in the domain with the subscript $j = 1$ on the DW $\tilde{y} = 0$ in terms of their values at arbitrary point $\tilde{y}$ of the given layer. For the domain with the subscript $j = 2$, we represent the fields on the DW $\tilde{y} = d$ in terms of their values at an arbitrary point of this layer. For this purpose, we find matrices that are inverses of the 4 ×4 square matrices consisting of the coefficients of $u_{1,2}$, $\alpha_{1,2}$, $\varphi_{1,2}$ and $\sigma_{1,2}$in expression (18). At $\tilde{y} = d$, these inverse matrices are represented as

$$\mathbf{M}_j(d) = \begin{pmatrix} \frac{m}{\delta} & -\frac{m_1}{c_{44}^{*}(k_y^{''}+k_y^{'})\,\delta} & 0 & \frac{e_j\,m_1}{c_{44}^{*}(k_y^{''}+k_y^{'})\,\delta} \\ \frac{c_{44}^{*}k_y^{''}k_y^{'}}{(k_y^{''}+k_y^{'})\delta}m_1 + K^2\mu^{*}k_x sh(k_x d) & \frac{m'}{\delta} & -e_j sh(k_x d) & e_j(-m/\delta + ch(k_x d)) \\ \frac{c_{44}^{*}K^2}{e_j}(m/\det - ch(k_x d)) & \frac{-K^2 m_1}{e_j(k_y^{''}+k_y^{'})\delta} & ch(k_x d) & \frac{-K^2 m_1}{e_j(k_y^{''}+k_y^{'})\delta} - \frac{sh(k_x d)}{k_x} \\ \frac{c_{44}^{*}K^2 k_x sh(k_x d)}{e_j} & 0 & -k_x sh(k_x d) & ch(k_x d) \end{pmatrix} \quad (23)$$

where $\delta = mm' + (k_y^{'}\,k_y^{''}\,m_1^2)/(k_y^{'}+k_y^{''})^2$. The matrices $\mathbf{M}_1 = \mathbf{M}_1(d)$ and $\mathbf{M}_2 = \mathbf{M}_2(d)$, which establish the relationships between the fields at the beginning and end of a domain layer, are called transition matrices [Bass, 1989]. It is known [Bass, 1989] that the product $\mathbf{M} = \mathbf{M}_1{\cdot}\mathbf{M}_2$ determines the transition matrix on the period of the structure. We do not present the elements of this matrix, since the corresponding expressions are cumbersome even for a static superlattice.

Using the transition matrix, we derive a system of algebraic equations to find the dispersion equation. This system is two times smaller than that obtained according to the conventional approach from [Bass, 1989]. To find the dispersion equation for an infinite structure, we need to determine the eigenvalues $\lambda = \exp(2i\kappa d)$ of matrix $\mathbf{M}$. For this purpose, we must find and set to zero the determinant $\mathrm{Det}[\mathbf{M} - \lambda\mathbf{E}]=0$, where $\mathbf{E}$ is the 4 ×4 unit matrix and κ is the Bloch vector, which represents the transverse wave number averaged over the period of the structure. The corresponding transformations yield the EW dispersion equation in the case of a moving superlattice:

$$\lambda^4+\lambda^3\left[\frac{-(m^2+m'^2)+}{\delta^2}+\frac{2}{\delta^2}\frac{k_y^{'}k_y^{''}m_1^2}{(k_y^{''}+k_y^{'})^2}-2\,\mathrm{ch}(2k_xd)+\frac{8K^2k_xm_1\,\mathrm{sh}(k_xd)}{\delta(k_y^{''}+k_y^{'})}\right]+$$

$$+\lambda^2\left[1+\frac{1}{\delta^2}+\frac{16K^4k_x^2m_1^2}{\delta^2(k_y^{''}+k_y^{'})^2}\mathrm{sh}^2(k_xd)+\frac{2\,\mathrm{ch}(2k_xd)}{\delta^2}\left((m^2+m'^2)-\frac{2k_y^{''}k_y^{'}m_1^2}{(k_y^{''}+k_y^{'})^2}\right)+\right.$$

$$\left.+\frac{8k_xK^2m_1\mathrm{sh}(k_xd)}{(k_y^{''}+k_y^{'})}\left(\frac{1+\delta}{\delta^2}\right)-\frac{8K^2k_x(m+m')m_1\,\mathrm{sh}(2k_xd)}{\delta^2(k_y^{''}+k_y^{'})}\right]+$$

$$+\lambda\left[\frac{-(m^2+m'^2)}{\delta^2}+\frac{2}{\delta^2}\frac{k_y^{'}k_y^{''}m_1^2}{(k_y^{''}+k_y^{'})^2}-\frac{2\,\mathrm{ch}(2k_xd)}{\delta^2}+\frac{8K^2k_xm_1\,\mathrm{sh}(k_xd)}{\delta^2(k_y^{''}+k_y^{'})}\right]+\frac{1}{\delta^2}=0. \tag{24}$$

The features of the interaction between an elastic wave and the moving lattice are determined by the DW velocity. However, expression (24) implicitly contains a dependence on velocity V_D owing to the presence of wave vector $k_y^{'}$ and quantities m, m', m_1, and δ, which are represented in terms of this wave vector. Therefore, this implicit dependence can be demonstrated only numerically, because of the complexity of the resulting expressions.

In order to consider the case of a static lattice ($V_D = 0$), we make the following substitutions in expression (24): $m, m' \to \cos(k_y\tilde{y})$, $\delta \to 1$, $k_y^{'} = k_y^{''} \to k_y$, and $m_1 \to 2\sin(k_y\tilde{y})$. Then, adding quantities $Q(k_x, k_y)\cos^2(k_yd)$ and $Q(k_x, k_y)\mathrm{ch}^2(k_yd)$, ($Q(k_x, k_y) = 16K^2(k_x/k_y)\sin(k_yd)\mathrm{sh}(k_xd)$) to the coefficient of λ^2 in Eq. (24) and subtracting the aforementioned quantities from this coefficient, we obtain after some algebra the dispersion equation for a static lattice:

$$\lambda^4-(S_1+S_2)\lambda^3+(2+S_1S_2+U)\lambda^2-(S_1+S_2)\lambda+1=0\ . \tag{25}$$

Here, we use the following notation:

$$S_1=2\cos(2k_yd)-4\frac{k_x}{k_y}K^2\sin(k_yd)\mathrm{sh}(k_xd),$$

$$S_2=2\,\mathrm{ch}(2k_xd)-4\frac{k_x}{k_y}K^2\sin(k_yd)\mathrm{sh}(k_xd),$$

$$U=16K^2\frac{k_x}{k_y}\sin(k_yd)\mathrm{sh}(k_xd)(\cos(k_yd)-\mathrm{ch}(k_xd))^2.$$

A solution to Eq. (25) is a reciprocal fourth-order polynomial [Korn & Korn, 1968] and can be analytically represented as

$$\lambda_{1,2} = \frac{y_1 \pm \sqrt{y_1^2 - 4}}{2}, \quad y_1 = \frac{(S_1 + S_2)}{2} + \sqrt{\frac{(S_1 - S_2)^2 - 4U}{4}},$$
$$\lambda_{3,4} = \frac{y_2 \pm \sqrt{y_2^2 - 4}}{2}, \quad y_2 = \frac{(S_1 + S_2)}{2} - \sqrt{\frac{(S_1 - S_2)^2 - 4U}{4}}. \tag{26}$$

When term U in the coefficient of λ^2 equals zero ($k_x = 0$ or $K = 0$ and, hence, $S_1 = 2\cos(k_y d)$ and $S_2 = 2\text{ch}(2k_x d)$), we obtain the following result from expression (26):

$$\lambda_{1,2} = \frac{S_1 \pm \sqrt{S_1^2 - 4}}{2}, \quad \lambda_{3,4} = \frac{S_2 \pm \sqrt{S_2^2 - 4}}{2}. \tag{27}$$

Note that $\lambda_{1,2}$ and $\lambda_{3,4}$ are the roots of the two corresponding quadratic equations. Hence, Eq. (25) can be written as $(\lambda^2 - S_1\lambda + 1)(\lambda^2 - S_2\lambda + 1) = 0$ and the solutions can be found with the use of the Vieta theorem and the substitution $\lambda_1 + \lambda_1^{-1} = \cos(2i\kappa d)$ ($\lambda = \exp(2i\kappa d)$, $S_2 = 2\text{ch}(2k_x d)$). Thus, we obtain

$$\cos(2\kappa_1 d) = \cos(2k_y d). \quad \cos(2\kappa_2 d) = \text{ch}(2k_x d)., \tag{28}$$

For the case under study, this factorization of Eq. (25) corresponds to the extinction of the interaction between acoustic and electric oscillations that is due to the absence of the piezoelectric effect ($K = 0$) or the absence of near-boundary electric oscillations in the domains $\Phi_j = 0$ [Shuvalov & Gorkunova,1999] under the conditions for the EAW normal propagation ($k_x = 0$) when the acousto-domain interaction vanishes. A similar result is valid for Eq. (20), which can be represented also as a product of the dispersion relations for independent elastic and electric subsystems.

For the case under study, this factorization of Eq. (25) corresponds to the extinction of the interaction between acoustic and electric oscillations that is due to the absence of the piezoelectric effect ($K = 0$) or the absence of near-boundary electric oscillations in the domains $\Phi_j = 0$ [Shevyakhov, 1990] under the conditions for the EW normal propagation ($k_x = 0$) when the acousto-domain interaction vanishes. A similar result is valid for Eq. (24), which can be represented also as a product of the dispersion relations for independent elastic and electric subsystems.

When k_x=0 or $K = 0$, the first equation from (28) becomes identity, a result that means that a purely elastic wave[1] propagates in the crystal along the normal to the domain planes with propagation constant k_y (i.e., a bulk shear wave is excited in an infinite crystal). It follows from the second identity that the Bloch number is either purely imaginary ($K = 0$), i.e., a solution does not exist, owing to the formation of a continuous forbidden band) or (k_x=0) equal to $2\pi n / (2d)$, where n is integer. In the second case (k_x=0), the boundary conditions are satisfied if $A_1 = 0$, $B_1 = 0$, $A_2 = 0$, $B_2 = 0$, and $B_2 = 0$ (i.e., all of the fields are identically zero). Therefore, the solution is degenerate and must be disregarded. Excluding both variants ($k_x = 0$ and $K = 0$) as the variants that do not allow the acousto-domain interaction, we assume that Eq. (24) always describes coupled electroacoustic oscillations for a static superlattice.

---

[1] At $K \neq 0$, this wave is accompanied by the in-phase oscillations of the electric field (see the first term in expression (14)).

3.3 Solution and analysis of the dispersion equation

A numerical solution to Eq. (24) that has been obtained with the use of expression (27) is shown in Fig. 3. In this study, the calculations are performed for a barium titanate crystal (($BaTiO_3$) with the following parameters: the crystal density ρ= 5 g/cm³, $K^2 \approx 0.37$, the velocity of the transverse waves in the absence of the polarizing field $c_{t0} = (c_{44} / \rho)^{1/2} = 2\times10^5$ cm/s, $\varepsilon_1 = 5\times10^3$, and $e_{15} \approx 3\cdot10^6$ g/(cm s). Figures 3a and 3b respectively demonstrate the real and imaginary parts of the roots of the dispersion equation for the static (dashed lines) and moving (solid lines) superlattices. The results are obtained for the propagation angle $\theta'' = \pi/3$ and the positive direction of DW motion ($V_D > 0$). The numbers of the curves correspond to the root numbers (λ_1, λ_2, λ_3, λ_4). The roots $\lambda_{1,2} = \exp(2i\kappa_{1,2}d)$ in expression (27) (Fig. 3a, curves *1, 2*) are purely real. This means that the Bloch wave numbers $\kappa_{1,2}$ are purely imaginary for any wave number *k* and correspond to the modes that are forbidden for the given periodic structure. Note that the shapes of curves *1* and *2* remain almost unchanged when the motion is taken into account. The roots $\lambda_{3,4} = \exp(2i\kappa_{3,4}d)$ (Fig. 3a, curves *3, 4*) describe propagating waves. Below, we consider only the spectral properties of the propagating eigenmodes of the superlattice.

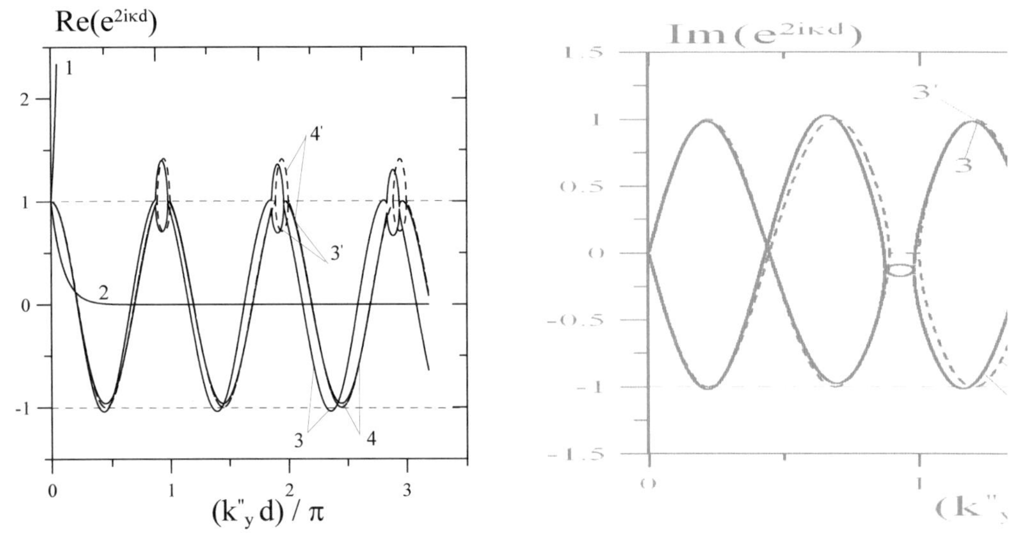

Fig. 3. (a) Real parts of the roots λ_1, λ_2, λ_3 (3, 3′), and λ_4(4, 4′) and plotted as a function of $(k''_y d)/\pi$ for various values of V_D: (1, 2, 3', 4') 0 and (3, 4) 0.01v. Curves 3' and 4' coincide except in the region of the loop-shaped segments, (b) Imaginary parts of the roots λ_3 (3, 3′), and λ_4 (4, 4′) plotted as a function of $(k''_y d)/\pi$ for various values of V_D: (3, 4) 0.01 v, 3′, 4′

It is seen from Fig. 3a that, in the static case, the real parts of roots *3* and *4* are cosines whose arguments contain the Bloch number. The imaginary parts of these roots are the sines of the above argument. It follows from the comparison of Figs. 3a and 3b that curves *3* and *4* describe counterpropagating waves. Indeed, the spectral characteristics of the modes of the static superlattice that propagate in the opposite directions must differ only by the sign of the Bloch wave number: The positive and negative Bloch wave numbers respectively correspond to the waves that propagate in the positive and negative direction of the $\tilde{y}$ axis.

The calculations show that an increase in propagation angle θ'' leads to an increase in the width of the loop fragment (Fig. 3) on the curve of the real part of the root. In accordance with Fig. 3b, the imaginary part of the root is absent for this spectral interval in the case of the dashed curves. Thus, the Bloch wave number, which is the propagation constant averaged over the lattice period, is purely imaginary and the wave propagation is impossible, so that the corresponding fragment is a forbidden band [Balakirev & Gilinskii, 1982].

The appearance of forbidden bands is obvious. In accordance with Fig. 3a, the Bloch vector exhibits variations in the transmission band, so that an integer number of half eigenwaves is realized at the band edges. At each boundary, this circumstance corresponds to the in phase summation of the forward wave and the wave that is reflected from the edge whose distance from the boundary equals the period of the structure. In contrast, the forward and reflected (antiphase) waves are mutually cancelled in the band gaps.

An increase in angle θ'' results in a simultaneous increase in the widths of allowed and forbidden bands. However, their number decreases in the range of wave vectors $0 < k'' < 2 \times 10^5$ cm^{-1} (the frequency range $0 < \omega'' < 5 \times 10^{10}$ s^{-1}), which is common to all of the plots. For the given domain size $d = 10^{-4}$ cm, dimensionless parameter kd ranges from 0 to 20. Thus, the calculations are performed almost in the absence of limitations on this parameter.

In addition, it is seen from Fig. 3 that the DW motion causes the Doppler shift of eigenwaves, which eliminates the degeneration of the roots of the dispersion relation. This fact is manifested in the graphs: The real parts of roots λ_3, and λ_4 for the counterpropagating EWs are not equal and are described by different curves. Curve *4* corresponds to the wave that propagates in the direction opposite to the $\tilde{y}$ axis, and curve *3* corresponds to the wave that propagates along the $\tilde{y}$ axis. In accordance with Fig. 2, at $V_D > 0$, , the forbidden bands exhibit a shift to the long-wavelength region. The comparison with the calculated results obtained for $V_D < 0$ shows that, relative to the forbidden band at $V_D = 0$, the forbidden bands are symmetrically shifted to the shortwavelength region.

For relatively large $(\kappa_{3,4}d)/\pi$, we observe a larger difference between curves *3* and *4*. This means that, as the number of the oscillation mode increases, the effect of the DW motion on the EW spectrum strengthens. This result is in agreement with the results from [Vilkov, 2007], where the spectral properties of magnetostatic waves are analyzed with allowance for the motion of a DW superlattice. In the spectral fragments with0 $V_D \neq 0$ (Fig. 3) that coincide with the spectral forbidden band of the static superlattice, a shifted forbidden band is formed. Thus, the amplitudes of eigenwaves in this band of the moving superlattice contain oscillating factors. In addition, new forbidden bands emerge at $V_D \neq 0$ in the spectral fragments where the real and imaginary parts of the roots of the dispersion equation are greater than unity. This circumstance is due to the fact that the lattice motion results in an additional phase shift between the counterpropagating waves.

The calculated results show that, if the Bloch wave number in the first spectral band in Fig. 3 is approximately equal to k_y'' at small angles θ'', a significant difference between κ and k_y'' can be realized at large angles. For example, at $\theta = 80°$, the propagation constant κ averaged over the superlattice period is approximately two times greater than k_y'', a result that indicates a significant effect of the superlattice on the spectrum of bulk EWs. The lattice motion leads to a difference between the Bloch wave numbers of the counter propagating waves and, hence, to differences between the EW propagation velocities and between the field profiles that characterize EWs. Thus, the mutual nonreciprocity of the EW propagation induced by the lattice motion needs further analysis.

3.4 Calculation of the EW displacement profiles and phase

To calculate the EAW displacement profiles, we employ the periodicity condition from [Bass et al, 1989]: On the boundaries with the coordinates $\tilde{y}=0$ and $\tilde{y}=2d$, the fields can differ only by the phase factor (the Floquet theorem); i.e.,

$$u(0)=u(2\,d)\exp(2\,i\kappa\,d),\qquad \varphi(0)=\varphi(2\,d)\exp(2\,i\kappa\,d),\qquad \Phi(0)=\Phi(2\,d)\exp(2\,i\kappa\,d)\,,\quad (29)$$

Substituting expressions for the fields (16) and (17) with allowance for formulas (25) into boundary conditions (20) on two boundaries of an elementary cell ($\tilde{y}=0$ and $\tilde{y}=d$), we obtain a system of eight homogeneous algebraic equations for amplitudes $A_{1,2}$, $B_{1,2}$, $C_{1,2}$ and $D_{1,2}$.

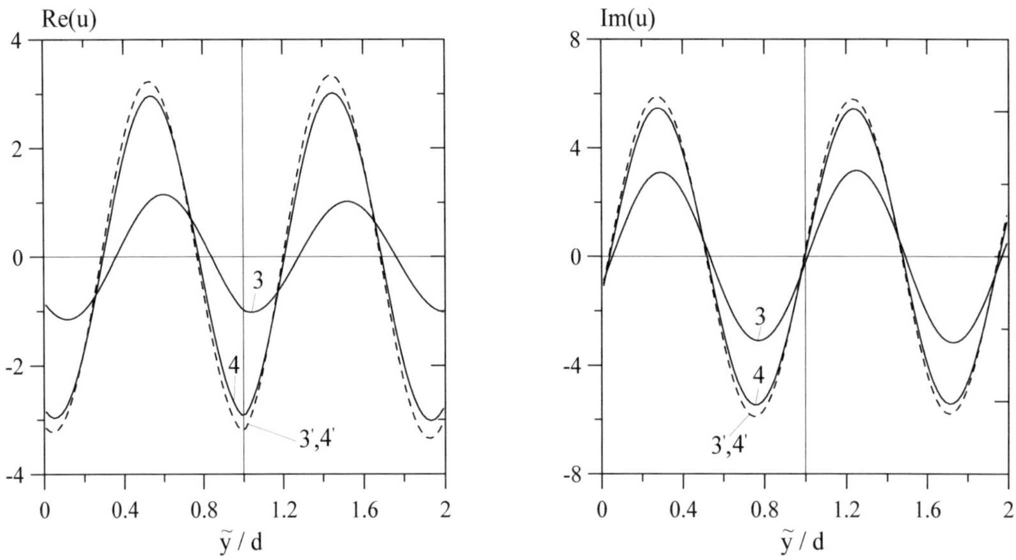

Fig. 4. (a) Real imaginary parts of the amplitude profile of the EW shear displacement for the wave k vector = 131300 cm–1 corresponding to the allowed band, θ′′ = 60°, and V_D = (dashed lines) 0 and (3, 4) 0.01v, (b) Imaginary parts of the amplitude profile of the EW shear displacement for the wave k vector = 131300 cm–1 corresponding to the allowed band, θ′′=60°, and V_D = (dashed lines) 0 and (3, 4) 0.01v

Assuming that one of the amplitudes (e.g., A_1) equals unity, we solve seven of eight equations and represent all of the amplitudes in terms of the selected one. Then, we substitute the resulting expressions into formulas (16) and (17) and, with allowance for the solutions to Eq. (25), find the desired field profiles within the lattice period $0<\tilde{y}<2d$.

Figure 3 demonstrates the real and imaginary parts of the shear wave displacement in the crystal with the static (dashed curves) and moving (solid curves) superlattices for the positive motion of the lattice ($V_D>0$). Figure 4 corresponds to the wave vector falling in the EW spectral band in Fig. 3. It is seen that the real and imaginary parts of the displacement respectively correspond to the symmetric and antisymmetric modes. Note that the presented set of mode profiles is universal for given k'' and the sum of the lattice spatial harmonics,

$$u(\tilde{y}) = \sum_{n=-\infty}^{\infty} u_n \exp\left[i\left(\kappa + \frac{2\pi n}{d} \right)\tilde{y} \right]. \tag{30}$$

This circumstance is obvious because, in accordance with expression (30), physically nonequivalent states in the periodic structure correspond only to the range $-\pi/d < \kappa < \pi/d$. It is seen from Fig. 4 that the amplitudes of counterpropagating waves coincide for the static lattice and significantly differ for the moving lattice. In particular, the amplitude of the wave propagating along the axis (curve *3*) is approximately two times smaller than the amplitude of the wave propagating in the opposite direction (curve *4*). This amplitude imbalance of the counterpropagating waves added to the misphasing can be interpreted via a variation produced in the wave energy owing to an external source that provides for the DW motion. The mechanism that controls the amplitude variations involves the Doppler frequency shifts: The wave with the maximum Doppler shift (Fig. 4, curve *3*) exhibits the largest amplitude difference relative to the wave of the static superlattice (Fig. 4, dashed line). In contrast, note minor variations in the amplitude of the wave (Fig. 4, curve *4*) whose spectral characteristics other than κare initially identical to the spectral characteristics of the wave in the static lattice.

Each of the harmonics in expression (30) is characterized by the same displacement profile, whereas the phase velocities are different [Balakirev & Gilinskii, 1982]:

$$\begin{aligned} |\vec{v}''_{fn}| &= \omega''[(\kappa_4 + 2\pi n/d)^2 + k_x^2]^{-1/2} \\ |\vec{v}'_{fn}| &= \omega'[(\kappa_3 + 2\pi n/d)^2 + k_x^2]^{-1/2} \end{aligned} \tag{31}$$

Here, $|\vec{v}'_{fn}|$ and $|\vec{v}''_{fn}|$ are the magnitudes of the phase velocities of the waves that propagate in the direction of the PDS motion and in the opposite direction. For the static

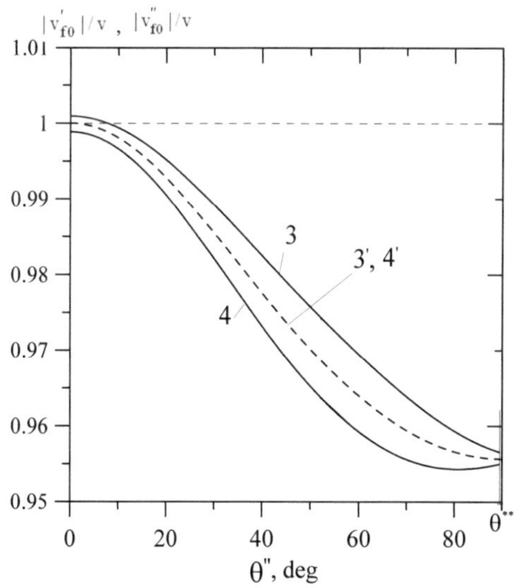

Fig. 5. Plots of the EW phase velocities vs. the propagation angle θ'' for V_D = (dashed line) 0 and (*3, 4*) 0.01*v*. Curves *3* and *4* correspond to the EWs propagating along and opposite to the *y* axis, respectively

lattice, we have $|\vec{v}''_{fn}|=|\vec{v}'_{fn}|$ It follows from expression (31) that, for the nth harmonic, the phase velocity can be infinitesimal. For certainty, we set the phase velocity of the Bloch wave in the following way: $|\vec{v}''_f|=|\vec{v}''_{f0}|$, and $|\vec{v}'_f|=|\vec{v}'_{f0}|$. The dependences of the phase velocities of the Bloch waves on propagation angle θ'' calculated for $V_D > 0$are presented in Fig. 5.

We do not consider the limiting angles θ'' = 0°, and θ'' = 90°and the nearest vicinities for the following reasons. In the first case (θ'' = 0°), the electric-field retardation must be taken into account in the correct analysis of the acousto-domain interaction [Balakirev & Gilinskii, 1982]. For a moving PDS, the transition θ'' → 90° is impossible because of the limitation θ' < θ** related to the termination of the acousto-domain interaction. It is seen from Fig. 5 that the greatest difference (about 2%) between phase velocities $|\vec{v}'_{fn}|$ and $|\vec{v}''_{fn}|$ is realized in the interval 60° < θ < 70° and can be experimentally detected. The phase-velocity nonreciprocity can be caused by the Doppler separation of the frequencies of counterpropagating waves and a simultaneous variation in the Bloch wave numbers that is due to different spatial–temporal periodicities in two opposite directions induced by the uniformly moving DW superlattice.

4. Reflection of electroacoustic waves from a system of moving domain walls in a ferroelectric

4.1 The statement of the problem

In the previous section we have shown, that significant modification of the spectrum of modes of shear waves are possible owing to motion of boundaries of a superlattice of a ferroelectric. It can be assumed, that motion of domain boundaries will exert the strong influence on reflection of electroacoustic waves from a lattice of domain boundaries. The important controlling role of the velocity of domain-wall motion is indirectly confirmed by the results obtained in [Shevyakhov, 1990], which indicate that the interaction of a bulk electroacoustic wave with a single moving domain wall in the case of a noticeable change in the amplitude coefficients (the range where the angles of incidence are not very small) is accompanied by the Doppler frequency transformation. By analogy with the static case [Shuvalov & Gorkunova, 1999], it can be expected that, for a system of moving domain walls, the reflection can be considerably enhanced in the direction of Bragg angles. At the same time, the velocity of domain-wall motion will serves as a new parameter that is convenient for controlling the reflection and transmission of waves in combination with their frequency shifts. In this section the interaction of electroacoustic waves with a periodic domain structure formed in a tetragonal ferroelectric by a finite number of uniformly moving 180-degree domain walls is considered in the quasi-static approximation.

The schematic diagram of the problem is depicted in Fig. 6. We consider the same the ferroelectric and the same the periodic structure, as in the previous section, but at that periodic structure is formed by finite number 180-degree domain boundaries. It is assumed that, in the y direction, the domain-wall lattice, which consists of $2N$ domains for the structure "+–" (Fig. 6a) or $2N+1$ domains for the structure "++" (Fig. 6b), has a period $2d$, so that $d \gg \Delta$ (where d is the distance between neighboring domain walls and Δ is the domain-wall thickness). On both sides, the lattice uniformly moving at the velocity V_D || y|| [010] is surrounded by semi-infinite single-domain crystal regions (the external numbers (with respect to the lattice) of domains are $n=0$ and $2N+1$ for the structure "+–" and $n=0$ and $2N+2$ for the structure "++"). In order to avoid a significant structural sensitivity of domain walls to the motion regime, we will restrict our consideration to the velocities, we will restrict our consideration to the velocity of shear waves in a single-domain sample [Sosnin. & Strukov 1970; Vainshtein,1988]. Under the above conditions, the motion of domain walls can be considered to be specified ($V_D = \text{const}$) with the current coordinates $y_m = V_D t + md$, , where t is the time and m = 0, ±1, ±2,.... Correspondingly, the domain walls are assumed to be geometrically thin and structureless ($k\Delta<<1$, where k is the wave number of the electroacoustic wave).

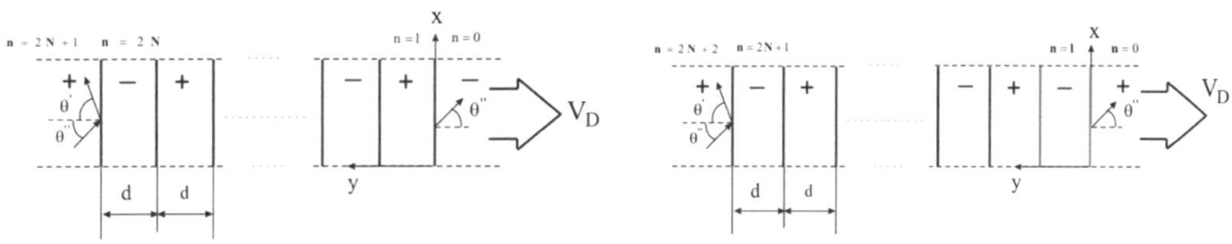

(a) (b)

Fig. 6. (a) Schematic diagram of the problem: a moving domain lattice is surrounded by semi-infinite single-domain crystal regions with (a) opposite (structure "+-")directions of polarization. Tilted vectors indicate the direction of propagation of the electroacoustic wave. The arrow indicates the direction of the lattice motion, (b) Schematic diagram of the problem: a moving domain lattice is surrounded by semi-infinite single-domain crystal regions with identical (structure "++") directions of polarization. Tilted vectors indicate the direction of propagation of the electroacoustic wave. The arrow indicates the direction of the lattice motion

We assume, that j= 1 and 2 stand for domains of the "+" and "–" types with positive and negative piezoelectric moduli, respectively according Eq(9). Other differences between domains are absent. We also assume that the domain with the number n= 0 is a domain of the "–" type for the structure "+-" and a domain of the "+" type for the structure "++" and that the domain with the number n= $2N$+ 1 or n= $2N$+ 2 is a domain of the "+" type (Fig. 6). Moreover, it also is assumed that the electroacoustic wave with the wave $\mathbf{k}'' = (k''\sin\theta'', k''\cos\theta'', 0)$, propagates in the $x0y$ plane and is incident on the domain lattice at the angle θ'' ($\theta'' \neq 0$). In view of the expected Doppler frequency shift [Shevyakhov, 1990], another wave vector $\mathbf{k}' = (k'\sin\theta', k'\cos\theta', 0)$ is assigned to reflected waves (Fig. 6). The initial equations remain without changes and to be defined by the equations (9), (13), (14). As a result of the phase conjugation of the fields at moving domain walls of the lattice, the wave characteristics of the reflected and transmitted electroacoustic waves are related by the expressions (15).

In order to avoid the analysis of the additional refraction scheme when the reflected wave becomes adjusting with respect to the incident wave, it is necessary to introduce the constraint on the angle of incidence $\theta'' < \theta^*$ [Shevyakhov, 1990], where

$$\theta^* = \arccos[-2\,(V_D / v) / (1 + (V_D / v)^2)]. \tag{32}$$

4.2 Technique for calculating the transmittance and reflectance of electroacoustic waves

In the domain with the number n = 2N + 1 for the structure "+-" or with the number n = 2N + 2 for the structure "++," we have the incident and reflected electroacoustic waves and the electric field wave localized at the boundary $\tilde{y}$ = 2Nd or $\tilde{y}$ = 2Nd + d (Fig. 6). The shear displacements and the potentials of these waves as the solutions to Eqs. (13) can be written in the form

$$\begin{aligned} u_{2N+1,2N+2}(\tilde{y}) &= (\exp(i\,k_y''(\tilde{y}-2N\,d)) + R\exp(-i\,k_y'(\tilde{y}-2N\,d)))\exp(i(k_x\tilde{x}-\Omega\tilde{t})) \\ \Phi_{2N+1,2N+2}(\tilde{y}) &= (C\exp(-k_x(\tilde{y}-2N\,d))\exp(i\,(k_x\tilde{x}-\Omega\tilde{t})) \end{aligned}, \qquad (33)$$

where the amplitude of the incident wave is taken to be unity, R is the reflectance, and C is the amplitude of the electric wave potential. Correspondingly, for the domain with the number n = 0 for both structures, we obtain

$$\begin{aligned} u_0(\tilde{y}) &= W\exp(i\,k_y''\tilde{y})\exp(i(k_x\tilde{x}-\Omega\tilde{t})) \\ \Phi_0(\tilde{y}) &= D\exp(k_x\tilde{y})\exp(i\,(k_x\tilde{x}-\Omega\tilde{t})) \end{aligned}. \qquad (34)$$

Here, W is the transmittance of the electroacoustic wave and D is the amplitude of the localized electric field wave. The solutions to Eqs. (13) in the region within the lattice for a domain of the "+" type are sought in the following form:

$$\begin{aligned} u_1(\tilde{y}) &= (A_1\exp(i\,k_y''\tilde{y}) + A_2\exp(-i\,k_y'\tilde{y}))\exp(i(k_x\tilde{x}-\Omega\tilde{t})) \\ \Phi_1(\tilde{y}) &= (C_1\exp(k_x\tilde{y}) + C_2\exp(-k_x\tilde{y}))\exp(i\,(k_x\tilde{x}-\Omega\tilde{t})) \end{aligned}, \qquad (35)$$

In the region of a domain of the "-" type, the solution is represented in a similar way

$$\begin{aligned} u_2(\tilde{y}) &= (B_1\exp(i\,k_y''\tilde{y}) + B_2\exp(-i\,k_y'\tilde{y}))\exp(i(k_x\tilde{x}-\Omega\tilde{t})) \\ \Phi_2(\tilde{y}) &= (D_1\exp(k_x\tilde{y}) + D_2\exp(-k_x\tilde{y}))\exp(i\,(k_x\tilde{x}-\Omega\tilde{t})) \end{aligned}, \qquad (36)$$

In order to match the fields of the domains in the rest system at the domain walls, in the quasi-static nonrelativistic approximation, we use the standard requirements for the continuity of the shear displacements, the potentials, the shear components of the stress tensor, and the electric induction components y (see Eqs(20)). At the same time, The matrices $\mathbf{M}_1 = \mathbf{M}_1(d)$ and $\mathbf{M}_2 = \mathbf{M}_2(d)$ (see Eqs.(23)) relate the fields at the initial and final points of the same domain layer. Their product determines the transition matrix at the period of the structure $\mathbf{M} = \mathbf{M}_1\mathbf{M}_2$.

Let us assume that, in expressions (20) and (22), the quantities determining the entire set of waves in the domain are designated as u_0, α_0, φ_0, and σ_0 in the domain with the number n = 0. Then, their relation to the quantities u_{2N}, α_{2N}, φ_{2N}, and σ_{2N} in the domain with the number $n = 2N$ in the structure "+-" can be found using the 4-by-4 transformation matrix $\mathbf{M}$ at the period of the structure as follows:

$$\begin{pmatrix} u_0 \\ \alpha_0 \\ \varphi_0 \\ \sigma_0 \end{pmatrix} = \mathbf{M}^N \begin{pmatrix} u_{2N} \\ \alpha_{2N} \\ \varphi_{2N} \\ \sigma_{2N} \end{pmatrix} \qquad (37)$$

where N is the exponent of the matrix $\mathbf{M}$ and, simultaneously, the number of "whole" unit cells formed by domains of the "-" and "+" type in the lattice. For the structure "++," the relation of the fields of the domain with the number $n = 0$ to the fields of the last domain in the lattice will be somewhat different as a result of the additional transformation of the waves by the last domain; that is,

$$\begin{pmatrix} u_0 \\ \alpha_0 \\ \varphi_0 \\ \sigma_0 \end{pmatrix} = \mathbf{M}^N \mathbf{M}_2 \begin{pmatrix} u_{2N+1} \\ \alpha_{2N+1} \\ \varphi_{2N+1} \\ \sigma_{2N+1} \end{pmatrix} \tag{38}$$

The reflectance for the structure "+ –" (structure "++") can be determined from the boundary conditions at the domain wall between the domain with the number $n = 2N$ ($n = 2N + 1$) and the domain with the number $n = 2N + 1$ ($n = 2N + 2$), i.e., by equating the corresponding wave parameters at $\tilde{y} = 2Nd$ ($\tilde{y} = 2Nd + d$). Then, we multiply the obtained identities by $\mathbf{M}^N$ ($\mathbf{M}^N\mathbf{M}_2$) from the left and right and, with the use of relationships (20)-(22), (33), (34), (37), and (38), find the systems of four equations for determining the reflectance and transmittance: for the structure "+–,"

$$\begin{pmatrix} W \\ -(ic_{44}^{*} k_y^{''})W - e_{15} k_x D \\ -(4\pi e_{15} / \varepsilon_1)W + D \\ D k_x \end{pmatrix} = \mathbf{M}^N \begin{pmatrix} 1+R \\ -(ic_{44}^{*} k_y^{''}) + (ic_{44}^{*} k_y^{'})R - e_{15} k_x C \\ (4\pi e_{15} / \varepsilon_1)(1+R) + C \\ -C k_x \end{pmatrix} \tag{39}$$

and for the structure "++,"

$$\begin{pmatrix} W \\ -(ic_{44}^{*} k_y^{''})W + e_{15} k_x D \\ (4\pi e_{15} / \varepsilon_1)W + D \\ D k_x \end{pmatrix} = \mathbf{M}^N \mathbf{M}_2 \begin{pmatrix} 1+R \\ -(ic_{44}^{*} k_y^{''}) + (ic_{44}^{*} k_y^{'})R - e_{15} k_x C \\ (4\pi e_{15} / \varepsilon_1)(1+R) + C \\ -C k_x \end{pmatrix} \tag{40}$$

Here, D, C, R, and W are unknown quantities. The expressions obtained for the reflectance and transmittance from relationships (39) and (40) are cumbersome and difficult for the analysis. In this respect, further calculations of the reflectance were performed numerically. As follows from [Shuvalov &. Gorkunova, 1999], the analytical formulas for the reflectance of electroacoustic waves from the static lattice can be obtained in a simple form only within the short-wavelength approximation. The magnitude of the reflectance in this approximation has the form [Shuvalov &. Gorkunova, 1999]

$$|R|^2 = \frac{q_\theta^4}{q_\theta^4 + \sin^2(\kappa d) / \sin^2(2(N+1)\kappa d)} \tag{41}$$

where $\cos(2\kappa d) = 2[\cos(k_y d) - q_\theta^2 \sin(k_y d)]^2 - 1$, $q_\theta^2 = K^2 \mathrm{tg}(\theta^{''})$, and κ is the Bloch wave number.

4.3 Calculation of the reflectance of electroacoustic waves from the moving lattice

Before proceeding to the analysis of the influence of the domain-wall motion on the reflectance spectrum of electroacoustic waves, it should be noted that the results of the numerical calculations of the reflectance from the system of equations (40) in the case of the static lattice are in complete agreement with those obtained in [Shuvalov &. Gorkunova, 1999]. In particular, in the short-wavelength range, the value of $|R|$ determined from the approximate formula (41) agrees well with the result of the exact numerical calculations. Since the number of peaks (the main peak plus secondary peaks) in one band gap is equal to the number of domains in the lattice, in our work, the modification of the spectrum due to the domain-wall motion was demonstrated using small numbers N in order to provide the clearness of the results.

The dependences of the magnitude of the reflectance of electroacoustic waves on the reduced normal component of the wave vector at the fixed angle of incidence ($\theta'' = 30°$) for the structures "++" and "+-" are plotted in Figs. 7 and 8, respectively. The calculated data presented in these and subsequent figures were obtained for the equidistant lattice of domains with $d = 10^{-4}$ cm in the barium titanate crystal $BaTiO_3$ with the following parameters: the density of the crystal is $\rho = 5$ gr/sm^3, $K^2 \approx 0.37$, the velocity of transverse waves in the absence of piezoelectric effect is $v = (c_{44} / \rho)^{1/2} = 2{\cdot}10^5$ cm/s. In Figs. 7 and 8, the dashed lines show the dependences $|R|\ (2k_y^{''}d/\pi)$ for the static lattice according to the calculations from the system of equations (40). The thick lines in Figs. 7 and 8 depict the dependences $|R|\ (2k_y^{''}d/\pi)$ for the lattice moving away (the direction of domain-wall motion is opposite to the direction of the Y axis, $V_D < 0$). It can be seen that the domain-wall motion noticeably modifies the reflectance spectrum of electroacoustic waves: all peaks in the spectrum are broadened, increase in the intensity, and are shifted toward the short-wavelength range. In this case, the larger the ratio $(2k_y^{''}d/\pi)$, the larger the shift, so that the maximum of the magnitude of the reflectance can give way to its minimum.

It can be seen from the behavior of the thin lines in Figs. 7 and 8 that, in the case of the approaching lattice (the direction of domain wall motion coincides with the direction of the Y axis, $V_D > 0$), the changes in the spectrum are as follows: the peaks in the reflectance spectrum are narrowed, decrease in the intensity, and are shifted toward the long-wavelength range. In this case, the larger the ratio $(2k_y^{''}d/\pi)$, the larger the shift. Furthermore, it was revealed that the higher the velocity V_D, the stronger the manifestation of the above changes in the spectrum. This effect of the shift in the spectra for the moving lattice ($V_D < 0$, $V_D > 0$) with respect to the spectrum of the static lattice is explained by the Doppler shift in the frequency of the electroacoustic wave due to its interaction with the moving domain walls and, in actual fact, represents an analog of Mandelstam -Brillouin scattering [Fabelinskii, 1968].It can be seen from Figs. 3,8 and Fig.3a that, when the wave number corresponds to the band gap of the Bloch spectrum, the magnitude of the reflectance reaches a maximum; i.e., there appears a Bragg peak. The condition for the appearance of this peak is a correlated reflection of electroacoustic waves from all domain walls in the lattice.

A comparison of the reflectance spectra of electroacoustic waves for the structures "++" (Fig. 7) and "+-" (Fig. 8) reveals several main differences. The first difference between the two reflectance spectra manifests itself in the range of the wave number $k = 0$, i.e., for an infinite wavelength. At $k \to 0$, the reflectance tends to zero for the structure "++" and to the reflectance for a single domain wall for the structure "+-" [Shevyakhov, 1990]. Physically, this difference in the behavior of the spectra can be explained as follows. The shear wave

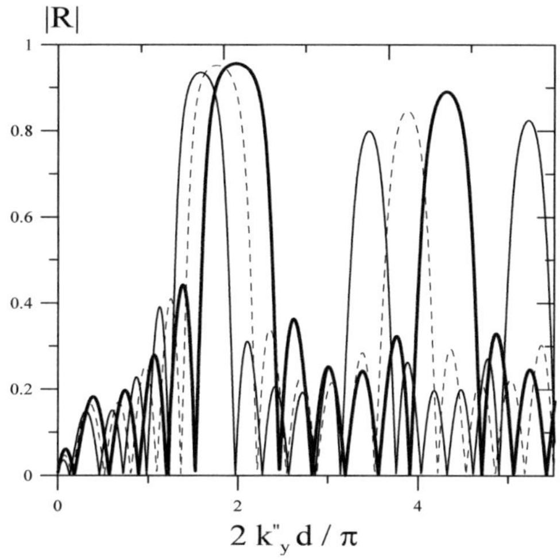

Fig. 7. Dependences of the magnitude of the reflectance |R| on the quantity $(k_y^{''} d)/\pi$ for the structure "++" consisting of five domains (N = 2) at the velocities V_D = (dashed line) 0, (thin line) 0.1v, and (thick line) –0.1v

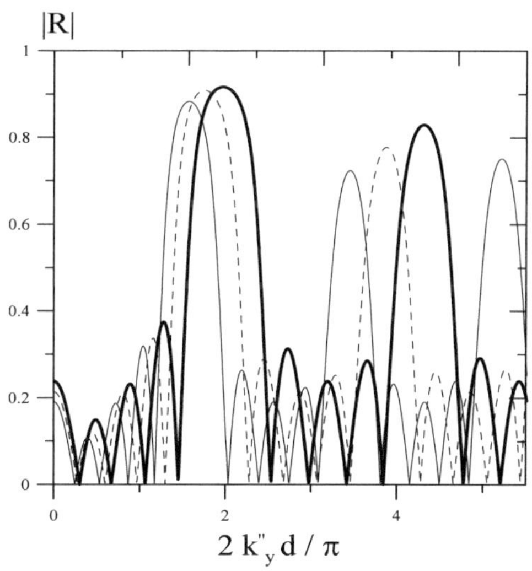

Fig. 8. Dependences of the magnitude of the reflectance |R| on the quantity $(k_y^{''} d)/\pi$ for the structure "+ –" consisting of four domains (N = 2) at the velocities V_D = (dashed line) 0, (thin line) 0.1v, and (thick line) –0.1v

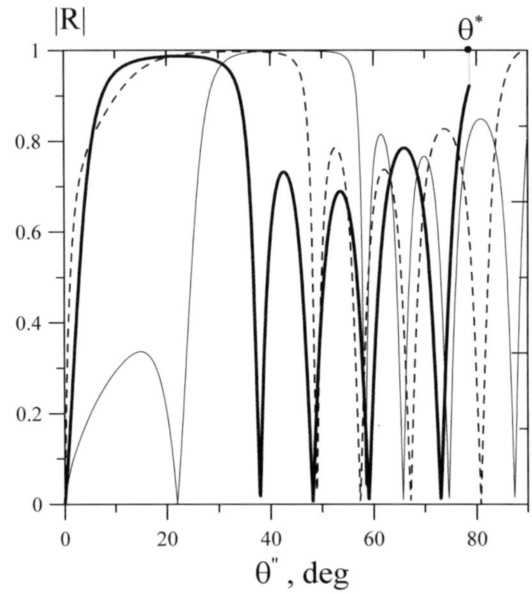

Fig. 9. Dependences of the magnitude of the reflectance | R| on the angle of incidence θ'' for the structure "++" consisting of nine domains (N = 4) in the case of k'' = 31941 sm[-1] at the velocities V_D = (dashed line) 0, (thin line) 0.1v, and (thick line) –0.1v

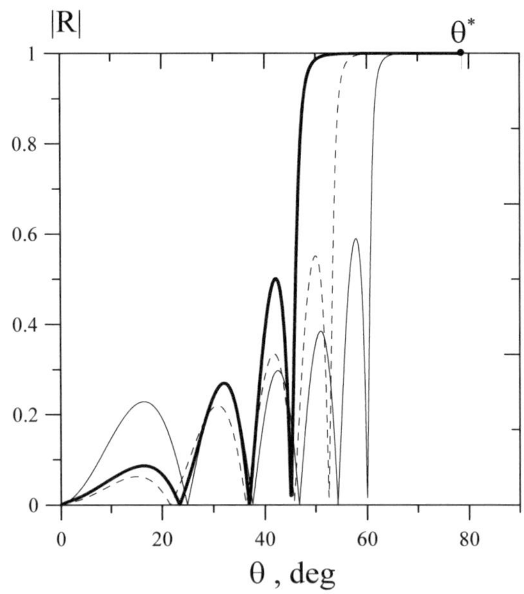

Fig. 10. Dependences of the magnitude of the reflectance | R| on the angle of incidence θ'' for the structure "++" consisting of five domains (N = 2) in the case of k'' = 55000 cm–1 at the velocities V_D = (dashed line) 0, (thin line) 0.1v, and (thick line) –0.1v

with the wavelength considerably larger than $2Nd + d$ is "insensitive" to the structure "++," and the electroacoustic wave propagates in the ferroelectric as in a single-domain sample in which the reflection is absent; i.e. $|R| \to 0$ at $k \to 0$ (Fig. 7). For the electroacoustic wave with the wavelength $\lambda \to \infty$, the structure "+-" is represented as a single domain wall, which is confirmed by the results of numerical calculations (Fig. 8).

The second difference lies in the fact that, at the center of the allowed band, the reflectance spectrum for the structure "++" is characterized by $|R| = 0$, whereas the reflectance spectrum for the structure "+-" always contains the secondary maximum. Finally, the third evident difference manifests itself in the number of peaks (the main peak plus secondary peaks) in one band gap: their number is always odd for the structure "++" and always even for the structure "+-."

The dependences of the magnitude of the reflectance $|R|$ of electroacoustic waves on the angel of incidence θ'' of the electroacoustic wave on the lattice for the structure "++" are plotted in Figs. 9 and 10. It can be seen from Fig. 9 that, for the wave number corresponding to the center of the first band gap, it is possible to choose the condition providing an almost total reflection ($|R| \approx 1$) in the range of small angles of incidence. The domain-wall motion leads to the fact that the reflectance peaks shift toward smaller (larger) angles of incidence for the lattice moving away (approaching) as compared to the reflectance peaks for the static lattice. In this case, the number of Bragg maxima for the approaching lattice increases by unity. It can also be seen that, for the lattice moving away, the dependence represented by the thick curve, according to condition (32) is limited by the angles of incidence corresponding to $\theta'' < \theta^*$. When the condition $k_x d >> 1$ is satisfied, the magnitude of the reflectance reaches unity even in the range of grazing angles of incidence, which can be clearly seen from Fig. 10. It should be noted that the strict fulfillment of the inequality $\exp(-k_x d) << 1$ represents the condition for the existence of the Marfeld–Tournois mode at the outer surface of the domain-wall lattice [Maerfield & Tournois, 1971].

5. Electroacoustic waves confined by a moving domain wall superlattice of a ferroelectric crystal

5.1 The statement of the problem

For completeness of the description of influence of motion of domain boundaries on a spectrum of electroacoustic waves, there is a necessity of consideration of propagation of electroacoustic interfacial waves boundary guided by a DW in a ferroelectric with a lattice of moving domain boundaries. In this section the dispersion properties of electroacoustic wave modes confined by a superlattice of 180° domain walls uniformly moving in a tetragonal ferroelectric crystal are considered. The crystal symmetry, wave polarization type, and the propagation geometry are the same as in [Golenishchev-Kutuzov et al, 2003; Li et al, 1991; Lyubimov & Sannikov, 1979], the only additional assumption being that the consideration is restricted to modes localized on the DWs of the superlattice.

Geometry of the problem is presented in Fig. 11, where the dynamic superlattice of equidistant 180° DWs with the (010) orientation is shown in the accompanying frame of reference $\tilde{x}0\tilde{y}\tilde{z}$. In virtue of the inequality $V_D << c$, where c is the speed of light and V_D is the velocity of DW motion, this frame is related to the laboratory one $x0yz$ via the Galilean transformation (11). As the main structure-forming element of the supperlattice, chose a pair

of adjacent stripe domains of a ferroelectric with the interfaces $\tilde{y}$ =0, d, and 2d shown in Fig. 11 with dark. The rest domains are formed by translation of this pair to whole number n of lattice periods 2d in the positive (n>0) or negative (n<0) direction of the $\tilde{y}$. axis. Ascribe the number j=1 to the lower domain of the pair of a lattice unit cell at $0<\tilde{y}<d$ and the number j=2to the upper domain at $d<\tilde{y}<2d$. Thus, each domain of the lattice will be defined by the translation number n=0, ±1, ±2, ±3, ... and the pair number j=1 or j=2; the value n=0 corresponds to the unit (initial) cell.
In view of the aforesaid, under the conditions of specified DW motion at the velocity V_D || y || [010] the current coordinates in the laboratory frame are $y_n = V_D t + nd$. Considering that antiparallel polarizations in domains of the lattice unit cell are related to the sign alteration of piezoelectric modulus e_{15}, which is the only active one under the given conditions (9). Extend the validity of this condition over the entire lattice, assuming that in Eq. (9) we have

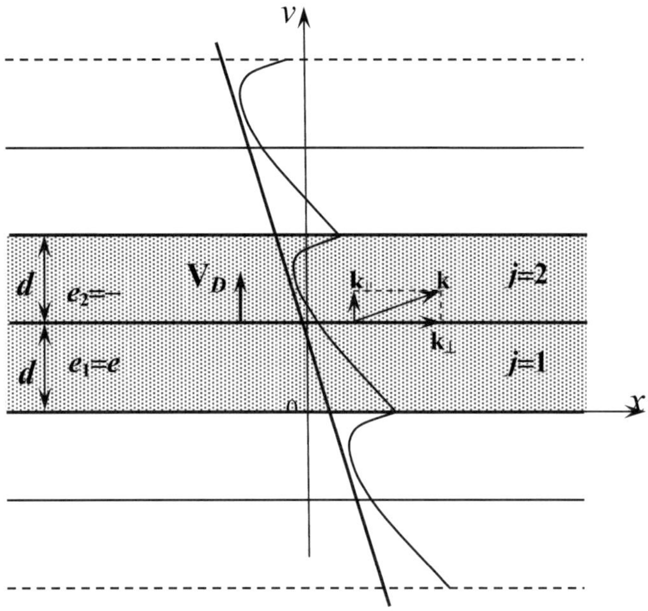

Fig. 11. Schematic diagram showing geometry of the problem and the distribution of shear displacements along the tilted front of a partial EIW with $\kappa = \pi/d$

$$j = \begin{cases} 1, & \text{for} \quad nd + V_D t < y < (n+1)d + V_D t \\ 2, & \text{for} \quad (n+1)d + V_D t < y < (n+2)\,d + V_D t \end{cases}. \tag{42}$$

The lattice cell fields are described by the solutions of the equations (13)-(14). Then, assuming that the waves propagate in the plane $\tilde{x}0\tilde{y}$ in the positive direction of the $\tilde{x}$ axis, with regard of proportional displacements u_j and potentials Φ_j with the exponential coefficient $\exp[i(k_{\parallel}\,\tilde{x} - \Omega\tilde{t}\,)]$, we conclude that the EIWs are noncollinear: $\mathbf{k}=(k_{\parallel}, k_{\perp}, 0)$, $k_{\perp} \neq 0$ (Fig. 11). Deviation of the EIW front from the orthogonal position is determined by the transverse component of wave vector $k_{\perp}$ [Gulyaev et al, 2000]:

$$k_{\perp} = \frac{\Omega}{c_t}\frac{\beta}{1-\beta^2}\,, \tag{43}$$

where $\beta = V_D/c_t$.

Assume also that the EIW length is much less than the characteristic size of a crystal. Under such conditions, the interfacial effects at the external interfaces of a ferroelectric, as well as the shape of the latter, will not have a considerable effect on the behavior of the electroacoustic waves and can be disregarded.

5.2 Derivation of the EIW dispersion equation for the dynamic lattice of a ferroelectric crystal

Solutions of Eqs. (13)-(14) within the pair of domains numbered with the indices j=1 and 2 of the unit cell can be presented as

$$\begin{aligned} u_1(\tilde{y}) &= [A_1\exp(-s\tilde{y}) + A_2\exp(s\tilde{y})]\exp(ik_{\perp}\tilde{y})\exp[i(k_{||}\tilde{x} - \Omega\tilde{t})], \\ \Phi_1 &= [C_1\exp(k_{||}\tilde{y}) + C_2\exp(-k_{||}\tilde{y})]\exp[i(k_{||}\tilde{x} - \Omega\tilde{t})], \end{aligned} \qquad (0 < \tilde{y} < d)\,, \tag{44}$$

$$\begin{aligned} u_2(\tilde{y}) &= [B_1\exp(-s\tilde{y}) + B_2\exp(s\tilde{y})]\exp(ik_{\perp}\tilde{y})\exp[i(k_{||}\tilde{x} - \Omega\tilde{t})], \\ \Phi_2 &= [D_1\exp(k_{||}\tilde{y}) + D_2\exp(-k_{||}\tilde{y})]\exp[i(k_{||}\tilde{x} - \Omega\tilde{t})], \end{aligned} \qquad (d < \tilde{y} < 2d)\,, \tag{45}$$

where $\Omega = \omega(1-\beta^2)$ is the EIW frequency in the accompanying frame of reference and s is the coefficient of interfacial localization

$$s = \frac{1}{1-\beta^2}\sqrt{k_{||}^2(1-\beta^2) - \frac{\Omega^2}{c_t^2}}\,, \tag{46}$$

related to frequency ω in the laboratory frame by the formula $\omega = c_t\sqrt{k_{||}^2(1-\beta^2)^{-1} - s^2}$. To sew the domain fields at the internal interface $\tilde{y} = d$, we will use the standard conditions (20) of continuity of shear displacements, potentials, shear components of the stress tensor $T_{yz}^{(j)}$, and y -components of the electric inductions .

For fields with structure of type (44), (46) procedure of construction of a matrix of transition and the dispersive equation will be same, as well as for waves of bulk propagation (see section 3 and 4), but there are the replacements in (13), (18) - (21) $\tilde{y}\,, \tilde{x}$ – directional derivative according Eqs (44), (46). We do not present the intermediate calculations here and at once write down the dispersion relation of EIWs in the following form:

$$\lambda^4 + \lambda^3 Q(k_{||}, s) + \lambda^2 P(k_{||}, s) + \lambda R(k_{||}, s) + \frac{1}{\delta^2} = 0\,, \tag{47}$$

where the eigenvalues $\lambda = \exp(2i\kappa d)$ of matrix of transition and κ is the Bloch wavenumber characterizing the degree of phase synchronism of the electroacoustic oscillations on DWs over the lattice period. At the specified geometric and material parameters, quantities Q, P, and R serving as coefficients in Eq. (47) represent the following functions of spectral variables $k_{||}$ and s:

$$Q(k_{||},s) = -2\,\mathrm{ch}(2sd)\delta^{-1} - 2\,\mathrm{ch}(2k_{||}d) + \frac{8K^2 k_{||}\mathrm{sh}(sd)\mathrm{sh}(k_{||}d)}{s\delta^{1/2}},$$

$$P(k_{||},s) = 1 + \frac{1}{\delta^2} + \frac{16K^2 k_{||}}{s}\left[\mathrm{sh}(k_{||}d)\mathrm{sh}(sd)\left(\frac{K^2 k_{||}}{\delta s}\mathrm{sh}(k_{||}d)\mathrm{sh}(sd) + \delta^{1/2}(1+\delta)\right)\right] + \tag{48}$$
$$+\frac{4\,\mathrm{ch}(2k_{||}d)\mathrm{ch}(2sd)}{\delta} - \frac{8K^2 k_{||}\mathrm{sh}(2sd)\mathrm{sh}(2k_{||}d)}{\delta s},$$

$$R(k_{||},s) = -2\frac{\mathrm{ch}(2sd)}{\delta} - \frac{2\,\mathrm{ch}(2k_{||}d)}{\delta^2} + \frac{8K^2 k_{||}\mathrm{sh}(sd)\mathrm{sh}(k_{||}d)}{\delta^{3/2}},$$

$$\delta = \exp(-2ik_{\perp}d).$$

The roots of Eq. (47), along with expressions (48), relate spectral EIW parameters $k_{||}$ and s to the Bloch wavenumber. Holding to the interpretation of the dispersion law as the functional coupling $s=s(k_{||})$ accepted in the theory of electroacoustic waves [Maerfield & Tournois, 1971 and Gulyaev, 1969], it would be reasonable to consider Bloch wavenumber κ as an additional parameter. The occurrence of this parameter owes to the conditions of translational symmetry of the solution describing EIWs over the entire set of allowed values $-\pi/d < \kappa < \pi/d$. Thus, at the designated κ value, Eq. (48) along with expressions (49) establish the dispersion properties of a partial EIW.

Features of a spectrum of the partial EIWs confined by a moving lattice are determined by the velocity of DW motion. In Equation (47), the parametric dependence on V_D is implicit and implemented through wave vector component $k_{\perp}$ from (43). In view of inconvenience of the obtained expressions and, above all, transcendence of Eq. (47), the dependences of the EIW dispersion spectra on the velocity of DW motion can be demonstrated only numerically. However, multiplying Eq. (47) by δ and dividing it by λ^2 in accordance to Eq. (48), we obtain the simpler form:

$$a^2 + a\left[4\cos(k_{\perp}d) + 4\cos(2\kappa d - k_{\perp}d) - 8\,\mathrm{ch}(sd)\mathrm{ch}(k_{||}d)\right] + \cos(4\kappa d - 2k_{\perp}d) - 4\cos(2\kappa d)\mathrm{ch}(2sd) -$$
$$-4\cos(2\kappa d - 2k_{\perp}d)\mathrm{ch}(2k_{||}d) + 2\cos(2k_{\perp}d) + 4\,\mathrm{ch}(2sd)\mathrm{ch}(2k_{||}d) = 0, \tag{49}$$

where $a = 4k_{||}K^2\mathrm{sh}(sd)\mathrm{sh}(k_{||}d)/s$.. Then, solving Eq. (49) with respect to a, one can obtain a dispersion relation in the form more convenient for an analysis and numerical calculations. In particular, if the Bloch wavenumber $\kappa = \pi/(2d)$ is taken (in the periodic structure only the κ values lying within the first allowed band $-\pi/d < \kappa < \pi/d$ correspond to the physically nonequivalent states), from Eq. (49) we obtain

$$s = K^2 k_{||}\left\{\mathrm{th}(k_{||}d)\mathrm{th}(sd)\frac{[1 \pm \sin^2(k_{\perp}d)/\mathrm{ch}(sd)]}{[1 - \sin^2(k_{\perp}d)]/\mathrm{ch}^2(sd)}\right\}. \tag{50}$$

Expression (50) shows, in particular, that while for a static lattice ($k_{\perp} = 0$) the only mode exists, in the case of a moving lattice ($k_{\perp} \neq 0$) the spectrum splits into two different modes corresponding to the alternating "+" and "-" signs. This splitting can be considered as the

removal of a degeneracy of the interaction of acoustoelecric oscillations of DWs of a unit cell due to their transverse motion taking place for $\kappa = \pi/(2d)$.
To turn to the static lattice at $\kappa \neq \pi/(2d)$, one should take $k_\perp = 0$ in Eq. (49). After solving the quadratic equations with respect to a and making some transformations, we obtain the explicit representation of the EIW dispersion equation:

$$s = -k_{||}K^2 \left\{ \frac{\mathrm{sh}(k_{||}d)\mathrm{sh}(sd)}{[\cos(\kappa d) \mp \mathrm{ch}(k_{||}d)][\cos(\kappa d) \pm \mathrm{ch}(sd)]} \right\}. \tag{51}$$

According to Eq. (51), the partial EIW spectrum in a ferroelectric with the static lattice initially includes two modes for a fixed value of the Bloch wavenumber different from $\pi/(2d)$. This result could be expected from the results reported in [Lyubimov & Sannikov, 1979] for a stripe domain. Note, however, that at finite values of the lattice period no strict correspondence between the spectra of modes (49) and (51) in the limit $V_D \to 0$ and the spectra from [Lyubimov & Sannikov, 1979] is observed due to the conditions of translational symmetry of solution (11). On the other hand, the transition to the results of [Lyubimov & Sannikov, 1979] under the condition $d \to \infty$ is excluded from Eqs. (49) and (51) at all, as it requires additional consideration of another boundary problem for the structure of alternating domains with different thicknesses. The only available analytical confirmation of correspondence remains the fact that in the limit $d \to \infty$, when the transition to an individual DW occurs, from Eqs. (49) and (51) we immediately have the expected result $s = k_{||}K^2$ [Maerfield & Tournois, 1971].

5.3 Dispersion spectra of modes of noncollinear electroacoustic interfacial waves of a dynamic superlattice of a ferroelectric crystal

It is reasonable to start the study of spectra of the partial EIWs of a dynamic superlattice with the case $\kappa = \pi/d$, when the electroacoustic oscillations of DWs distant from one another by a lattice period ($\lambda_B = 2d$ is length of Bloch wave) are inphase. One can see from Eq. (49) that the results for this case are identical to those for $\kappa = 0$ and $\kappa = -\pi/d$. Another feature is that the transition to the case $\kappa = -\pi/d$ can be implemented by inversion of the velocity of DW motion: $V_D \to -V_D$. Physically, it means that the dispersion spectra of EIW modes under the condition of synchronism of the electroacoustic oscillations of DWs are independent of the chosen direction of DW motion; i. e., the lattice reveals reciprocity of the transverse distribution of fields and EIW propagation.
The values of Bloch wavenumber $\kappa = \pm\pi/d$ determine the boundaries of the first allowed band. A schematic of the dispersion EIW spectra for the case $\kappa = \pm\pi/d$ is presented in Fig. 12. The dispersion branches showed by dashed lines correspond to the static lattice; the presence of DW motion is reflected by bold solid lines. Thin straight lines represent the linear EIW spectra on a single DW, static (β=0) [Maerfield & Tournois, 1971] or moving ($\beta \neq 0$) [Gulyaev et al, 2000]. A dashed straight line shows the linear spectrum of volume SH waves propagating in a single-domain crystal.
The overall picture of the spectrum of modes of the partial EIWs at $\kappa = \pm\pi/d$ outwardly resembles the picture of the spectrum of modes of a stripe domain: there are only two modes and the high-frequency one has a lower cutoff frequency (black point in the figure); the low-frequency mode is present over the entire frequency range. The change in the high-frequency asymptotes of the spectra due to DW motion is the same as that for a single DW

[Gulyaev et al, 2000] or a stripe domain [Bugaev et al, 2005]. However, there are principle distinctions between the spectra of stripe domain modes and the spectra in Fig. 12.

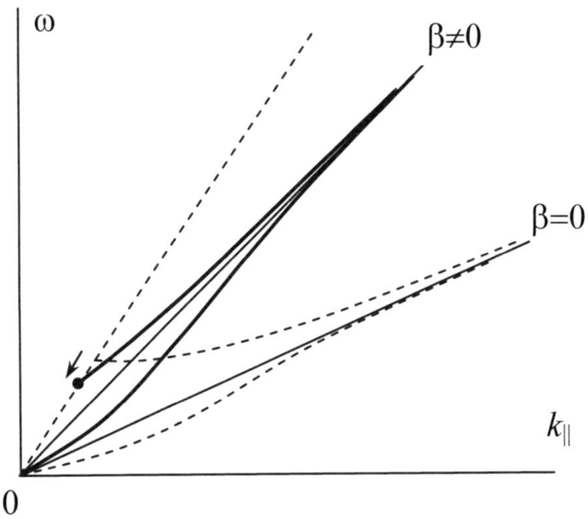

Fig. 12. General view of the mode spectrum of the partial EIW for the lattice with the Bloch wavenumbers $\kappa = \pm\pi/d$

First, as the velocity of DW motion increases, the cutoff point of the high-frequency mode in Fig. 12 shifts, as the arrow indicates, along the spectrum of volume waves towards lower frequencies, i. e., in the opposite direction as compared to the case of a moving stripe domain. A consequence of this behavior is the limiting situation at $\beta \to 1$, when the high- and low-frequency branches merge with the linear spectrum of volume waves (the Marfeld--Turnois asymptote at $\beta \to 1$ approaches the volume wave spectrum) and the region of EIW existence degenerates into the spectral line. This degeneracy of the EIW into an ordinary shear wave propagating in the direction of DW motion corresponds to "tripping" the shear waves and the moving DWs due to the absence of piezoelectric polarization charges [Gulyaev et al, 2000]. This result is quite expectable, since the number of DWs moving at the sound velocity is of no importance, as in the case of the superlattice.

Second, the low-frequency branch exhibits a qualitatively different character of dispersion. Indeed, at low frequencies the dispersion branch for a stripe domain occurs below the spectrum of volume waves but above the Marfeld--Turnois mode asymptote. In other words, the phase velocity of the wave lies between the values of the EIW velocity on a single DW and the velocity of a volume SH wave. Meanwhile, in the lattice spectrum the dispersion branch of the low-frequency mode lies completely below the Marfeld--Turnois asymptote. This type of the EIWs for a static lattice was mentioned previously in [Li, et al, 1991], where the resemblance of their dispersion to the dispersion of flexural modes in thin planes was pointed out.

Third, the mode spectra of a stripe domain correspond to different types of the transverse distribution of shear displacements, including symmetric for the low-frequency branch (symmetric mode) and antisymmetric for the high-frequency branch (antisymmetric mode). In the case of a lattice, the calculations using formulas (51) with the dispersion EIW indices obtained from Eq. (47) show that the distributions of the shear displacements across the unit cell are qualitatively similar for both modes and correspond mainly to the symmetric type. Obviously, the DW motion causes the symmetry break. Figure 11 demonstrates such a quasi-symmetric profile shown, for better illustration, in a strongly emphasized form.

When there are only two modes at $\kappa = \pm\pi/d$, the degenerated case of EIW propagation takes place, i. e., there is no splitting of modes upon induced DW motion. Possibility of this splitting in other cases was analytically demonstrated in the previous section by the example of the lattice with $\kappa = \pi/(2d)$. The shape of the partial EIW mode spectrum for the lattice with this Bloch wavenumber calculated from Eq. (50) is presented in Fig. 13.

As was mentioned above, in the static case Eq. (50) has the only root. The corresponding mode is shown in Fig. 13 by a dashed curve and is purely antisymmetric judging by the character of the transverse distribution of the shear displacements. This naturally follows from the fact that at $\kappa = \pi/(2d)$ the electroacoustic oscillations on DWs distant by a lattice period are antiphase. The absence of a symmetric mode is explained by piezoelectric inconsistency of the symmetrically distributed shear displacements and the resulting stresses in an internal DW of the unit cell due to the alteration of a sign of the piezoelectric modulus [Li, et al, 1991].

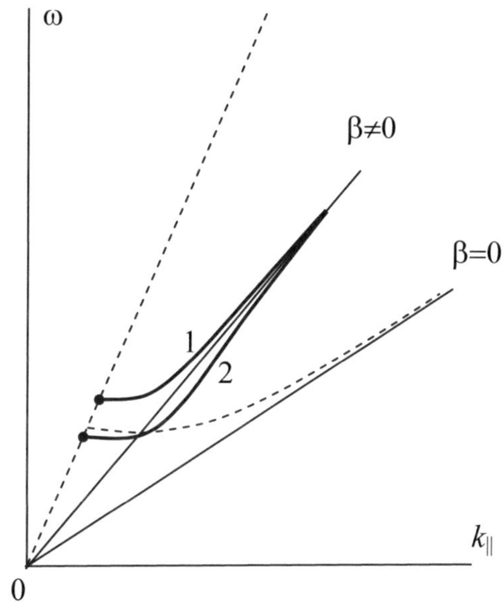

Fig. 13. Spectrum of the partial EIW modes for the Bloch wavenumber $\chi = \pi(2d)$: 1 and 2 are the branches split by DW motion

Motion of DWs splits the dashed branch of the antisymmetric mode into two branches 1 and 2 shown in Fig. 13 by bold solid lines. In the high-frequency limit, the spectra of the split modes are asymptotically drawn to the changed (turned towards the dashed line of the volume wave spectrum) Marfeld--Turnois asymptote. With the change in the direction of DW motion (alteration of signs in Eq. (50) from $\pm$ to $\mp$), branches 1 and 2 in Fig. 13 change places. Clearly, this does not break the overall picture of the spectrum. In this sense, the conclusion on reciprocity of the transverse field distribution and EIW propagation revealed by the lattice of moving DWs retains its validity in the considered variant.

One can see from Eq. (50) that the effect of the inversion of the velocity of DW motion on the spectrum of modes of the partial EIW is equivalent to that of the transformation $\kappa \to -\kappa$. Thus, the picture of the spectra in Fig. 13 includes the case $\kappa = -\pi/(2d)$. This conclusion is general and can be extended to any case $\kappa \neq \pi/(2d)$. The most typical situation of the transformation of the partial EIW mode spectrum by the lattice of moving DWs is that for $\kappa \neq \pi/(2d)$ in the static case there are already two modes (dashed curves in Fig. 14) instead of the only mode split into pairs by motion, as in Fig. 13. The transformations $V_D \to -V_D$ or $\kappa \to -\kappa$ yield intrapair rearrangement of the split modes $1 \leftrightarrow 2$ and $3 \leftrightarrow 4$ (solid curves), which validates general independence of spectra representation on the direction of DW motion.

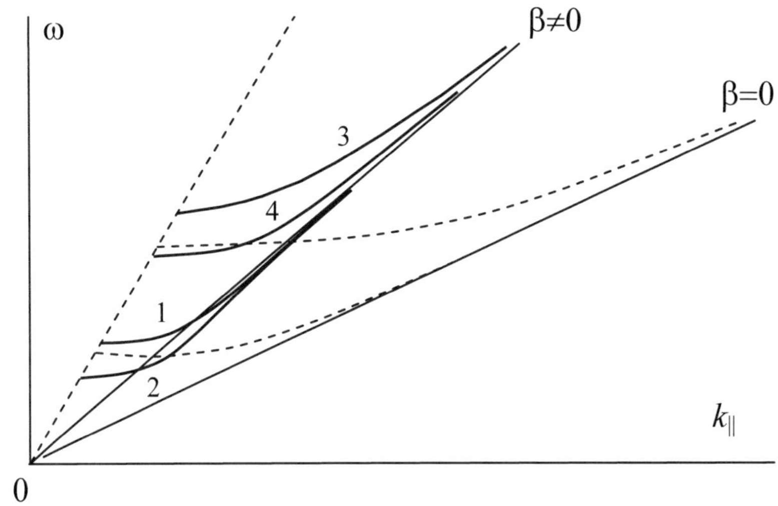

Fig. 14. Spectrum of the partial EIW modes for the arbitrary Bloch wavenumber $\kappa \neq \pi/(2d)$and $\kappa \neq \pi/d$: 1, 2 and 3, 4 are the pair branches split by DW motion

Concerning the structure of the transverse distribution of the shear displacements for the split branches, one should pay attention on its noticeable closeness to the transverse distribution for a generative mode of a partial EIW of the static lattice. In the general case, the generative mode reveals mixed signs of symmetry-antisymmetry occurring depending on closeness of value $2\kappa d$ to values 2π or π. If the situation corresponds to that illustrated in Fig. 13, the transverse distributions of the displacements for both split modes slightly differ from a strictly symmetric form at high velocities of DW motion (β>0.7) and from one another.

Figure 15 depicts the dependence of the frequency of an electroacoustic wave on the Bloch wavenumber for some fixed wavenumber. The dispersion curves shown by dashed lines correspond to the static lattice; in the presence of DW motion, they are shown by solid lines. The low-frequency group of curves 1 corresponds just to the EIWs. The high-frequency group of curves 2 separated from group 1 by the band gap corresponds to the first allowed band for the electroacoustic waves of volume propagation [Vilkov, 2008] (presented for comparison). There is the only allowed band for the EIWs; it lies below all the allowed bands of the volume propagation spectra. In both cases, DW motion splits the spectrum of any electroacoustic wave of the static lattice into pairs (spectral doublet). In the general case

$\kappa \neq \pi/(2d)$ and $\kappa \neq \pi/d$ for the EIWs at the chosen Bloch number and wave vector we have four branches, as not one but two modes corresponding to different symmetries of the distribution of the shear displacements across the cell undergo pair splitting[2]. One can see from Fig. 15 that although the values of the Bloch number $\kappa = \pm\pi/d$ determine the boundaries of the first allowed band, they do not determine the frequency region of location of the dispersion branches of all the rest partial EIWs. There exist intermediate Bloch wavenumbers for which the frequency at a fixed wavenumber lies above or below as compared to the case of the Bloch wavenumber $\kappa = \pm\pi/d$. This feature is more pronounced for the dynamic lattice.

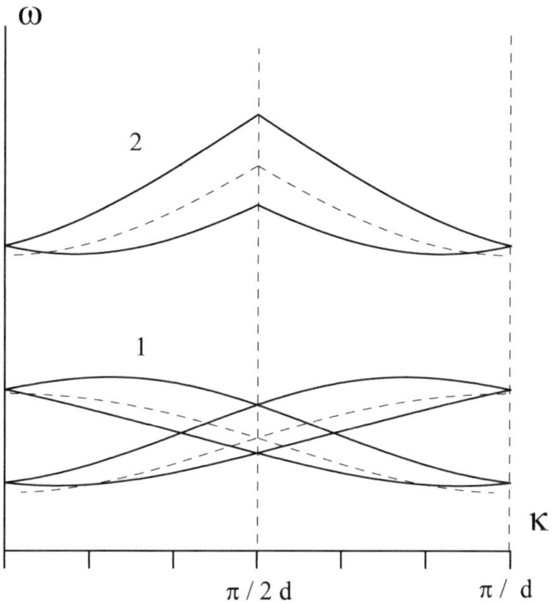

Fig. 15. Dependence of the frequency of an electroacoustic wave on the Bloch wavenumber at a fixed *k* value

6. Conclusions

The effect of the uniform motion of ferroelectric DWs that form a dynamic superlattice on the spectral properties of EWs has been analyzed for the first time. The results obtained indicate that even slow DW motions cause significant (detectable) variations in the wave spectra. Moreover, the spectrum of reflection of electroacoustic waves from a periodic domain structure formed by a finite number of uniformly moving 180-domain walls in the tetragonal ferroelectric was analyzed. It was established that, owing to the interference of reflected waves, the magnitude of the reflectance for the wave number corresponding to the band gap of the Bloch spectrum tends to unity for both static and moving domain structures. The inclusion of the domain-wall motion leads to shift of the Bragg reflectance

---

[2] For the regions not close to $\kappa = \pi/(2d)$, and $\kappa = \pi/d$, two solid curves lying above in frequency correspond to the quasi-symmetric distribution.

peaks toward the long-wavelength (short-wavelength) range in the case of the counter (one-way) motion of domain walls. The higher the velocity of domain-wall motion, the larger the shift of the peaks. The intensity of these peaks decreases (increases) as compared to that in the case of static domain walls. It was demonstrated that, apart from the appropriate choice of the lattice parameters and the wave characteristics of electroacoustic waves, the range of the total reflection at small angles of incidence and those close to grazing angles of incidence can be controlled by specifying the velocity of domain walls.

In the chapter the dispersion properties of the noncollinear EIWs of the dynamic superlattice of the equidistant uniformly moving 180° DWs of ferroelectric crystals have been also described. It has been shown that due to DW motion the partial Bloch spectra of modes of the interfacial electroacoustic waves which are not related to the boundaries of the first allowed band undergo pair splitting into the high- and low-frequency branches changing places during the inversion of the DW velocity. The partial modes setting the limits of the allowed band for the EIWs are not subjected to the mentioned splitting, invariant to the inversion of the velocity of DW motion, and transformed by DW motion in the high-frequency asymptote similar to the waves on individual domain walls.

It was revealed that the velocity of domain-wall motion serves as a new parameter that is convenient for controlling the reflection and transmission of waves in combination with their frequency shifts. This circumstance leads to additional experimental possibilities of the acoustic spectroscopy of polydomain ferroelectrics by means of the application of the acoustic nonreciprocity of the crystals that is induced by DW motion. The induced acoustic nonreciprocity and the Doppler frequency conversion can be used for the development of sensors and acoustoelectronic devices for data conversion with the frequency output.

7. References

Auld B.A. (1973). Acoustic Field and Waves in Solids Vol.2, *Wiley*-Interscience, ISBN 0471037028, New York.

Balakirev, M. K. & Gilinskii, I. A. (1982). Waves in Piezoelectric Crystals, *Nauka*, ISBN 5-BO-B-БA-82, Novosibirsk.

Bass, F.G., Bulgakov, A. A., & Tetervov A. P. (1989). High-Frequency Properties of Semiconductors with Superlattices, *Nauka*, ISBN 5-02-014021-X, Moscow.

Batanova, N.L. & Golenishchev-Kutuzov A.V (1997). Acoustic Waves in Ferro- and Piezoelectrics with a Periodic Electrical Relief. *J. Acoust. Phys.* Vol.43, No 4, pp.468-471, ISSN 1063-7710.

Bugaev, A. S., Maryshev S. N., & Shevyakhov N. S. (2005) Electroacoustic Waves at the Interfaces between Two Strip Domains Moving in a Ferroelectric. // *Journ. Communications Technol. and Electronics*. Vol. 50. No 9. pp. 1031-1037, ISSN 1064-2269.

El'meshkin, O. Yu., & Shevyakhov, N.S. (2006) Mode spectrum of noncollinear electroacoustic boundary-guided waves in a ferroelectric with a moving strip domain *J. Acoust. Phys.* Vol.43, No 1, pp. 56-61, ISSN 1063-7710

Esayan, S. Kh., Lemanov, V. V., & Smolenskiĭ G. A. (1974) Reflection and refraction of elastic waves on the domain boundaries in ferroelectric crystals $Gd_2(MoO_4)_3$. *Sov. Phys. Dokl.*, Vol 19, No 7, pp. 393-401.

Fabelinskii, I. L. (1968). Molecular Scattering of Light. PLenum Press, New York.

Golenishchev-Kutuzov, A. V., V. A. Golenishchev-Kutuzov, V. A., and R. I. Kallimullin, (2003) Induced Domain Structures in Electrically and Magnetically Ordered Materials, *Fizmatlit*, ISBN: 978-5-9221-0449-4 Moscow.

Gulyaev, Yu. V. (1969) Electroacoustic surface waves in solids. *JETP Lett.*, Vol. 9, No 1 pp. 37-38, ISSN 0021-3640.

Gulyaev, Yu. V., El'meshkin, O. Yu., & Shevyakhov N. S. (2000) Electroacoustic surface waves on moving boundaries // *Journ. Communications Technol. and Electronics*. Vol. 45. No 3. pp. 351-356, ISSN 1064-2269

Korn, G.A. & Korn, T.M. (1968), Mathematical Handbook for Scientist and Engineers, Second Edition, McGraw- Hill Book Company, ISBN 10: 0-07-035370-0, ISBN 13: 9780070353701, Inc., New York.

Li, Xingjiao. (1987.) Analysis of the field of the ferroelectric domain layer mode. *Journ. Appl. Phys.*, Vol. 61. No 6. pp. 2327-2334, ISNN 0021-8979.

Li, Xingjiao; Li, Yibing; Lei, Yiwu & Cross, L. E. (1991). The ferroelectric domain layer interface wave in multiple domain layered structure. *Journ. Appl. Phys.*, Vol. 70, No 6, pp.3209-3214, ISSN 0021-8979.

Lyubimov, V. N. & Sannikov D. G. (1979) Surface acoustic waves on 180-degree domain boundaries in ferroelectrics of type barium titanate crystal. *Kristallografiya*, Vol 24 No 1, pp.5-10, ISSN 0023-4761. .

Maerfield C. & Tournois P. (1971) Pure Shear Elastic Surface Wave Guided by the Interface of Two Semi-Infinite Media. *Appl. Phys. Lett.*, Vol. 19, No 4. pp. 117-118, ISSN 0003-6951

Nougaoui A., Rouhani, B.Djafari. (1987) Dynamics of infinite and semi-infinite piezoelectric superlatticies: shear horizontal waves and effective medium approximation. *Surf. Sci.*, Vol. 185. No 1-2. pp. 154-174.

Shevyakhov, N. S. (1990) Reflection of a transverse wave by a moving domain wall in a tetragonal ferroelectric. *Akust. Zh.* Vol. 36, No 1, pp.160-165, ISSN 0320-7919.

Shenderov E.L. (1997) Reflection of a plane sound wave from a semi-infinite periodic transversely isotropic set of layers. *J. Acoust. Soc. Am.*, Vol. 101, No 3, pp.1239-1249, ISSN 0001-4966.

Shuvalov A.L. & A.S. Gorkunova A.S. (1999) Transverse acoustic waves in piezoelectric and ferroelectric antiphase superlattices. Phys. Rev. B, Vol. 59, No 14, pp.9070-9077.

Sosnin, A. S. & B. A. Strukov. (1970) *Introduction to Ferroelectricity*, Vysshaya Shkola, Moscow.

Vainshtein L. A. (1988). Electromagnetic Waves, *Radio i Svyaz*, ISBN 5256000640, Moscow.

Vilkov, E.A., (2007) Spectrum of magnetostatic waves in a ferromagnet with a moving superlattice of domain walls. *Physics of the Solid State*, Vol. 48, No 9, pp. 1754-1759, ISSN 1063-7834.

Vilkov, E.A. (2008) Spectral properties of electroacoustic waves in a ferroelectric with a moving periodic domain structure. *Physics of the Solid State*, Vol. 50, No 8, pp. 1519-1526, ISSN 1063-7834.

Vilkov, E.A. (2009) Reflection of electroacoustic waves from a system of moving domain walls in a ferroelectric. *Physics of the Solid State*, Vol. 51, No 2, pp. 343-350, ISSN 1063-7834.

Vilkov, E.A., Maryshev S.N. & Shevyakhov (2009) N.S. Electroacoustic waves of a moving domain wall superlattice in a ferroelectric crystal. *Technical Physics Letters*, Vol. 35, No 4, pp.326-330, ISSN 1063-7850.

18

Ferroelectric Optics:Optical Bistability in Nonlinear Kerr Ferroelectric Materials

Abdel-Baset M. A. Ibrahim[1], Mohd Kamil Abd Rahman[1], and Junaidah Osman[2]

[1]*School of Physics and Material Sciences, Faculty of Applied Sciences, Universiti Teknologi MARA, 40450 Shah Alam, Selangor,*

[2]*School of Physics, Universiti Sains Malaysia, 118000 USM, Penang, Malaysia*

1. Introduction

Ferroelectric oxides with perovskite structure such as $PbLaTiO_3$ (PLT), $BaTiO_3$, $PbTiO_3$, $SrBi_2Ta_2O_9$ and $LiNbO_3$ are very attractive class of materials which possess numerous useful properties such as high dielectric constant, large spontaneous polarization, and remarkable optical nonlinearity. Potential applications of these materials include real-time holography, correlation filtering and various novelty filter applications (Sutherland 1996 and Eaton 1991). They are also popular materials for the fabrication of nonvolatile memories (Ramesh 2001). Over the past few years, ferroelectric oxides have been widely investigated for various nonlinear optical applications (Shi 2006; Xuan 1998; Zhang 1999; Zhao 1996) especially for optical switches. Optical switches are devices invented to perform multiplexing at very fast speeds and with less delay than the customary switches works with electronic signals. A ferroelectric optical switch is expected to allow the processing of millions of signals at a speed of terahertz.

To obtain optical bistability phenomenon, two ingredients are necessary, a nonlinear process and a feedback mechanism (Gibbs, 1977, 1979, and 1985). In all optical systems, the feedback can be "distributed", "extrinsic" or "intrinsic". In multilayer systems with alternating nonlinear materials, the feedback is "distributed"; it arises from the interaction of the propagating wave with many cross-sections of a nonlinear medium. In a Fabry-Perot (FP) resonator, the feedback is "extrinsic"; it arises as a result of reflection from the mirrors placed at its interface. In a single nonlinear layer, the feedback can be "intrinsic" or "mirrorless"; it arises in each elementary oscillator due to the strong local nonlinear response of an individual atom or molecule.

In the usual or the standard analysis to study the optical bistability (Marburger 1978, Gupta 1987, Biran 1990, Danckaert 1989, Shen 1984, and Haelterman 1989) in nonlinear optics, the governing equation for optical propagation within the nonlinear medium is a nonlinear wave equation in the electric field derived from Maxwell's equations. The usual constitutive relation between the nonlinear polarization and the electric field is then obtained by expanding the nonlinear polarization as a Taylor series in the electric field. The usefulness of this constitutive relation is that the polarization is a natural source term in the Maxwell's equation. Even though such constitutive relation is used to describe majority of the nonlinear optical phenomena, it is not essential or unique. Goldstone and Garmire (Goldstone 1984), in their work on the intrinsic bistability in semi-infinite nonlinear medium, used an inverted form of the usual constitutive relation. They expressed the electric field in terms of the total polarization ussing the nonlinear Duffing anharmonic oscillator equation and solved the nonlinear wave equation in terms of polarization as the independent variable. They stressed that the usual analysis in nonlinear optics is not suitable to describe a potentially important class of bistable interactions which result from intrinsic material bistability.

The advantage of Duffing constitutive relation is that the driving field is treated as dependant on the material response which allows to account for the optical bistability results from the intrinsic feedback mechanisim even in microscopic domain. An additional advantage of using the Duffing Oscillator over the usual constitutive relation is that the exact nature of the nonlinear susceptibility $\chi^{(3)}$ of the material is not required because the nonlinearity of the system is automatically contained in the induced polarization (Ibrahim 2007). Moreover if we are dealing with operating wavelength in the neighborhood of resonance where the nonlinear material usually exhibits a huge third order nonlinearity , the usual constitutive relation becomes really questionable since the undepleted wave approximation is clearly violated.

Recently, Murgan *et al.* (2002) have derived expressions of the tensor elements for various second- and third-order nonlinear optical effects including optical Kerr effect for bulk FE materials having various symmetries. They have shown that many of these elements have large linear and nonlinear optical coefficients even in the visible and near-infrared frequency regions. Particularly near resonance, the FE materials becomes highly nonlinear. They have found that it is the combination of the temperature divergence and the resonant frequency, which is typically in the THz region, dependence that underlies their large values. For these reasons, it is believed that the Maxwell-Duffing analysis is more suitable for investigating optical bistability in Kerr FE materials especially when the operating frequency is in the resonance region.

In this chapter, the Maxwell-Duffing approach will be applied to investigate the optical bistability in ferroelectric materials with Kerr nonlinearity. For ferroelectric materials, the Landau-Khalatnikov dynamical equation is used with anharmonic potential as the constitutive relation. Such nonlinear binding potential is provided from the Landau-Devonshire free energy for bulk ferroelectric exhibiting second-order phase transitions. A nonlinear polarization equation is derived and integrated across the ferroelectric medium. Through the application of the exact boundary conditions, expressions for reflectance R, transmittance **T**, are derived as a function of the polarization P and the driving field $\boldsymbol{E}$. Results for both Fabry-Perot resonators filled with a ferroelectric medium and for a ferroelectric slab without coating mirrors will be presented. The nonlinear response of the polarization, reflectance and transmittance as a function of the electric field incident amplitude will be illustrated. The effect of thickness, operating frequency, and temperature on the bistable characteristics of the FE slab will also demonstrated. In the case of FP resonator with partially reflecting mirrors, the effect of mirror reflectivity on the optical bistability is studied. The possibility of obtaining a reliable optical switch from such system will be explored. The examples shown in this chapter are based on the available experimental data of $BaTiO_3$.

2. Mathematical formulation

Consider a Fabry Pérot resonator filled with bulk ferroelectric crystal and coated with a pair of thin identical partially-reflecting mirrors as illustrated in Fig. 1. A high intensity incident infrared radiation is impinging the material at normal incidence. The nonlinear ferroelectric material ($BaTiO_3$) is assumed in the ferroelectric phase and exhibits a second-order like phase transitions. To derive a nonlinear polarization wave equation for medium 2, we begin by considering Landau-Devonshire free energy F expression written in terms of the polarization $\boldsymbol{P}(z,t)$ as following (Lines and Glass, 1977)

$$F(\boldsymbol{P},T)=\frac{\alpha}{2\varepsilon_0}\boldsymbol{P}^2+\frac{\beta}{4\varepsilon_0^{\ 2}}\boldsymbol{P}^4-\boldsymbol{E}.\boldsymbol{P} \tag{1}$$

The parameter $\alpha=a(T-T_c)$ is temperature-dependent with a being the inverse of the Curie constant, T is the thermodynamic temperature, and T_c is the Curie temperature. The parameter β is the nonlinear coefficient; it is material-dependent with mechanical dimension m^3J^{-1} and ε_0 is the dielectric permittivity of vacuum. The term $\boldsymbol{E}.\boldsymbol{P}$ accounts for the coupling of the far infra-red (FIR) radiation with the driving field $\boldsymbol{E}$. The response of a FE material exposed may be described by the time-dependent Landau-Khalatnikov dynamical equation of motion in terms of polarization, $\boldsymbol{P}$, as

$$M\frac{d^2\boldsymbol{P}}{dt^2}+\Gamma\frac{d\boldsymbol{P}}{dt}=-\frac{\partial F}{\partial \boldsymbol{P}} \tag{2}$$

In the above, M is the inertial coefficient with mechanical dimension $Kg.m^3A^{-2}.s^{-2}$. The term $\Gamma\, d\boldsymbol{P}/dt$ represents the linear loss and Γ is a damping parameter with mechanical dimension $Kg.m^3.A^{-2}.s^{-3}$. The driving field $\boldsymbol{E}$ in the FE medium is considered to have a form of uniform time-harmonic plane wave propagating in the negative z-direction at fundamental frequency ω

$$\boldsymbol{E}_2(z,t)=\frac{1}{2}\left[E_2(z)\exp(-i\omega t)+E_2^*(z)\exp(+i\omega t)\right] \tag{3}$$

In equation (3), $E_2(z)$ and $E_2^*(z)$ are the electric field amplitude in the ferroelectric medium and it complex conjugate respectively. The total polarization $\boldsymbol{P}(z,t)$ is also considered to be time harmonic, in phase, and propagates in same direction as the $\boldsymbol{E}$ field, which is

$$\boldsymbol{P}(z,t)=\frac{1}{2}\left[P(z)\exp(-i\omega t)+P^*(z)\exp(+i\omega t)\right] \tag{4}$$

In equation (4), $P(z)$ and $P^*(z)$ are the polarization amplitude and its complex conjugate respectively. Therefore, substituting (1), (3) and (4) into (2) gives the following time-independent Landau-Khalatnekov equation

$$E_2(z)=\left[-M\omega^2-i\Gamma\omega+a(T-T_c)/\varepsilon_0\right]P(z)+\left(3\beta/4\varepsilon_0^{\ 2}\right)\left|P(z)\right|^2P(z) \tag{5}$$

Equation (5) is the time-independent form of equation (2); it describes the electric field in the ferroelectric medium in terms of polarization and other material parameters. In deriving equation (5), the third-harmonic term is usually ignored. The corresponding magnetic field is derived from equation (5) using the relation $H_{2,x}(z)=(i/\omega\mu_0)\left(dE_{2,y}/dz\right)$, where here for simplicity we have considered E_2 to be purely polarized in the y-direction $\left(0,E_y,0\right)$, and H is purely polarized in the x-direction $\left(H_x,0,0\right)$. Therefore,

$$H_{2,x}(z)=\frac{i}{\omega\mu_0}\frac{3\sqrt{3}}{8}\left\{4\left[-M\omega^2-i\Gamma\omega+a(T-T_c)/\varepsilon_0\right]\frac{dP(z)}{dz}+\frac{\beta}{\varepsilon_0{}^2}\left[3P^2(z)\frac{dP^*(z)}{dz}+6|P(z)|^2\frac{dP(z)}{dz}\right]\right\} \tag{6}$$

In linear régime $(\beta=0)$, equation (5) may be combined with the linear equation $P=\varepsilon_0\chi(\omega)E$ to obtain the linear dielectric function $\varepsilon(\omega)$ for ferroelectric medium

$$\varepsilon(\omega)=\varepsilon_\infty+\left[-M\omega^2-i\Gamma\omega+a(T-T_c)/\varepsilon_0\right]^{-1} \tag{7}$$

From equation (7), the linear refractive index of the FE medium may be evaluated as $n_2=\left[\varepsilon(\omega)\right]^{1/2}$. ε_∞ is the high-frequency limit of the dielectric function $\varepsilon(\omega)$. Equation (7) is essentially similar to that of typical dielectric except that it is temperature-dependent function. For convenience in the numerical work, it is helpful to scale the relevant equations and use dimensionless variables (Lines and Glass 1977). Therefore the dimensionless parameters are being introduced;

$$e_2=E_2/E_c,\quad f=\omega/\omega_0,\quad p=P/P_s,\quad t=T/T_c,\quad u=\omega_0 z/c \tag{8}$$

Equation (8) shows that the coercive field of ferroelectric material at zero temperature E_c is used to scale the dimensional electric field inside the FE medium to give the scaled electric field e_2. In similar fashion, the resonance frequency ω_0 is used to scale the operating frequency ω to give a scaled operating frequency f. The polarization P and the thermodynamic temperature T are scaled in terms of spontaneous polarization at zero temperature P_s and the Curie temperature T_c respectively. Finally, the thickness z is scaled by dividing out c/ω_0 to give a scaled thickness $u=\omega_0 z/c$. In fact, any physical variable can be made dimensionless just by dividing out a constant with similar dimension. For helpful discussion about scaling analysis of physical equations, the reader is referred to Snieder (2004). Therefore, substituting the scaled parameters of equation (8) into equation (5), we obtain the following dimensionless form of Landau-Khalatnikov equation;

$$e_2=\frac{3\sqrt{3}}{8}\left[4\left(t-1-mf^2-ifg_F\right)+3|p(u)|^2\right]p(u) \tag{9}$$

In equation (9), the coefficient $m=\left[M\omega_0^2\varepsilon_0/aT_c\right]$ is the scaled inertial coefficient while $g_F=\left[\Gamma\omega_0\varepsilon_0/aT_c\right]$ is the scaled damping parameter. To describe the propagation in the ferroelectric medium, the time-independent electromagnetic wave equation $d^2E_2/dz^2+\left(\omega^2\varepsilon_\infty/c^2\right)E_2+\omega^2\mu_0P=0$ is employed. However, this equation has to be converted to dimensionless form using the scaled parameters in equation (8) as well. This yields the following scaled form of the electromagnetic wave equation;

$$\frac{d^2 e_2}{du^2} + f^2 \varepsilon_\infty e_2 + f^2 \frac{P_0}{E_0 \varepsilon_0} p(u) = 0 \tag{10}$$

Substituting the electric field expression from equation (9) into the wave equation (10), the following nonlinear polarization equation is obtained;

$$2\left[2\hbar + 3|p|^2\right]\frac{d^2 p}{du^2} + 3p^2 \frac{d^2 p^*}{du^2} + 12p\frac{dp}{du}\frac{dp^*}{du} + 6p^*\left[\frac{dp}{du}\right]^2 + f^2\varepsilon_\infty\left[4(\hbar+\xi)+3|p|^2\right]p = 0 \tag{11}$$

Equation (11) is a nonlinear equation describes the evolution of the polarization in a ferroelectric medium with thickness u. For simplicity, we have introduced the scaled coefficients ξ and $\hbar$ in equation (11), where $\xi = (2\sqrt{3}/9)(P_s/E_c\varepsilon_0\varepsilon_\infty)$ and $\hbar = t - 1 - mf^2 - ifg_F$. For ferroelectric material exhibits a second-order phase transitions, the coercive field at zero temperature is $E_c = +\sqrt{4a^3T_c^3/27\varepsilon_0\beta}$ while the spontaneous polarization at zero temperature is $P_s = \sqrt{aT_c\varepsilon_0/\beta}$. Upon substituting the value of P_s and E_c, the value of ξ reduces to $\xi = (\varepsilon_\infty aT_c)^{-1}$ which is basically a constant value for each specific material. The coefficient $\hbar$ is also important since it contains contributions from thermodynamic temperature t, operating frequency f, and the damping parameter g_F.

To obtain numerical solution, it is helpful to eliminate the term d^2p^*/du^2 from equation (11). This can be done as follows; first, the complex conjugate of equation (11) is obtained. Second, the term d^2p^*/du^2 is eliminated between equation (11) and its complex conjugate. This leads to the following nonlinear propagation equation,

$$\begin{aligned}&\left[16|\hbar|^2 + 24(\hbar+\hbar^*)|p|^2 + 27|p|^4\right]\frac{d^2p}{du^2} + 12\left(2\hbar^* + 3|p|^2\right)p^*\left[\frac{dp}{du}\right]^2 - 18p^3\left[\frac{dp^*}{du}\right]^2 \\ &+12\left(4\hbar^* + 3|p|^2\right)p\frac{dp}{du}\frac{dp^*}{du} + f^2\varepsilon_\infty p\left[16\hbar^*(\hbar+\xi) + 12(2\hbar+\xi)|p|^2 + 9|p|^4\right] = 0\end{aligned} \tag{12}$$

In equation (12), the coefficient $\hbar^* = t - 1 - mf^2 + ifg_F$ is the complex conjugate of $\hbar$. Equation (12) may be integrated numerically across the ferroelectric medium as an initial value problem to evaluate the desired polarization.

3. Analysis of the Fabry-Perot Interferometer

The analysis to find the complex reflection r, and transmission coefficients τ, is basically similar to the standard analysis in linear optics (Born & Wolf 1980); where $\boldsymbol{R} = |r|^2$ and $\boldsymbol{T} = |\tau|^2$ represent the reflected and transmitted intensities respectively. Referring to Fig. 1, the electric fields in medium1 and 3 are assumed to have the form of a plane wave propagating in free space with propagation constants $k_1 = k_3 = k_0 = \omega n_0/c$ and $n_0 = 1$. Therefore, we may write

$$E_1 = E_0\left[\exp(-ik_1 z) + r\exp(ik_1 z)\right] \tag{13}$$

$$H_1 = (E_0 k_1/\omega\mu_0)[\exp(-ik_1 z) - r\exp(ik_1 z)] \tag{14}$$

$$E_3 = \tau E_0 \exp[-ik_3(z+L)] \tag{15}$$

$$H_3 = k_3(E_0/\omega\mu_0)\tau\exp[-ik_3(z+L)] \tag{16}$$

where, E_0 is the amplitude of the incident electric field. At top interface, The tangential components of the electric field E is continuous with $E_1^{||}(z=0) = E_2^{||}(z=0)$ where E_1 and E_2 are substituted from equation (13) equation (5) respectively. The standard scaling procedure then yields the following expression for complex reflection coefficient r;

$$r = \frac{3\sqrt{3}}{8e_0}\left[4\hbar p_t(u) + 3p_t(u)|p_t(u)|^2\right] - 1 \tag{17}$$

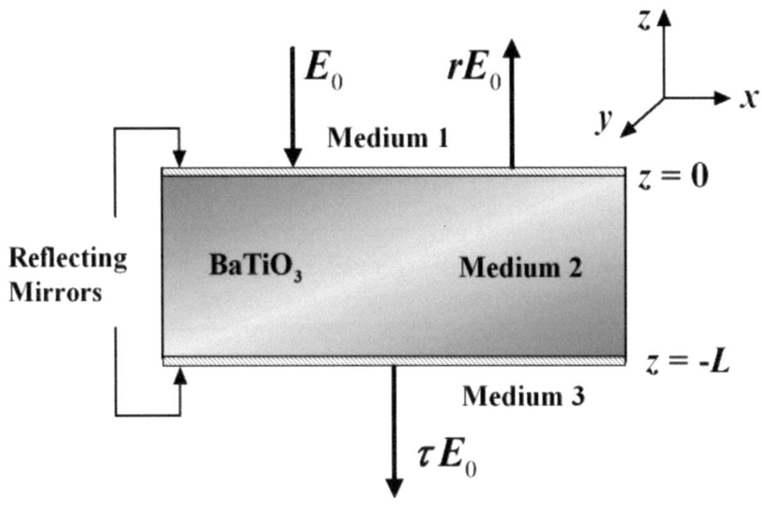

Fig. 5.1. Geometry of the Fabry-Pérot resonator.

The subscript t of p_t in equation (17) refers to the polarization at top interface. Due to the existence of the mirrors at both interfaces , the boundary conditions for the magnetic field at top interface becomes $H_{1x}(z) - \eta E_{1y}(z) = H_{2x}(z)$ (Lim, 1997) where H_{1x}, E_{1y} , and H_{2x} are represented by equations (6), (13), and (14) respectively. The parameter $\eta = \eta_a - i\eta_b$ $= \sigma_M\delta_M - i\omega\varepsilon_0\varepsilon_M\delta_M$ is the mirror coefficient with conductivity σ_M, thickness δ_M, and permittivity of the mirror medium ε_M respectively. For perfect dielectric mirror with conductivity $\sigma_M \to 0$ the term η_a becomes zero. In such a case the wave propagates into the mirror material without attenuation. Experimentally such coating mirror can be designed to meet the required reflectance at optimized wavelength using various metallic or dielectric materials. The standard scaling procedure, then yields following dimensionless equation for the magnetic field at top interface;

$$[(1-r)+(1+r)\eta_s]fe_0 = i\frac{27}{4}\left[\hbar + i\frac{3}{2}|p_t|^2\right]\frac{dp_t}{du} + i\frac{81}{16}p_t^2\frac{dp_t^*}{du} \tag{18}$$

In equation (18), $\eta_s = c\mu_0\eta = \eta_{s,a} - i\eta_{s,b}$ accounts for the scaled mirror parameter and for purly dielectric mirror $\eta_{s,a} = 0$ and η_s reduces to $-i\eta_{s,b}$. If we eliminate the complex reflection coefficient r between equations (17) and (18), the following equation is obtained;

$$e_0 = \frac{3}{16f}\left\{\sqrt{3}\left(1-\eta_s\right)f\left[4\hbar p_t + 3|p_t|^2 p_t\right] + 9i\left[2\left(2\hbar + 3|p_t|^2\right)\frac{dp_t}{du} + 3p_t{}^2\frac{dp_t^*}{du}\right]\right\} \tag{19}$$

Equation (19) will be used later to evaluate the amplitude of incident electric field e_0 numerically as a function the polarization at top interface. In similar fashion, the boundary conditions at the bottom boundary $z = -L$ are applied. Continuation of the tangential components of E at $z = -L$ $\left(E_2^{||} = E_3^{||}\right)$ yields an expression for the complex transmission coefficient;

$$\tau = \frac{3\sqrt{3}}{8e_0}\left[4\hbar + 3|p_b(u)|^2\right]p_b(u) \tag{20}$$

In the above, the subscript b in the polarization p_b refers to the bottom boundary. On the other hand, the boundary conditions for the H-field $\left(H_{3x} + \eta E_{3y} = H_{2x}\right)$ are also applied where H_{2x}, E_{3y}, and H_{3x} are represented by equations (6), (15), and (16) respectively. The standard scaling procedure, then yields the following dimensionless equation

$$\tau f e_0\left(1+\eta_s\right) = i\frac{27}{4}\left[\hbar + \frac{3}{2}|p_b|^2\right]\frac{dp_b}{du} + i\frac{81}{16}p_b^2\frac{dp_b^*}{du} \tag{21}$$

Substituting the complex transmission coefficient τ from equation (20) into equation (21), and then eliminating the derivative dp_b^*/du from the resultant equation, the following equation is obtained;

$$\frac{dp_b}{du} = \frac{-i\frac{\sqrt{3}}{9}fp_b\left[6\left(1-\eta_s^*\right)\left(4\hbar^* + 3|p_b|^2\right)|p_b|^2 + 4\left(1-\eta_s\right)\left(2\hbar^* + 3|p_b|^2\right)\left(4\hbar + 3|p_b|^2\right)\right]}{\left[4\left(2\hbar + 3|p_b|^2\right)\left(2\hbar^* + 3|p_b|^2\right) - 9|p_b|^4\right]} \tag{22}$$

In the former equation, the coefficient $\hbar^*$ and η_s^* are the complex conjugates of $\hbar$ and η_s respectively. Equation (22) is used to evaluate the derivative dp_u/du for arbitrary values of p_b at the bottom interface z = -L. Both p_b and dp_b/du are used as initial conditions to integrate equation (12) across the ferroelectric medium. . It should be noted that the top boundary z = 0 is u = 0 in the scaled unit while the bottom boundary z = -L is u = -l where $u = \omega_0 z/c$ and $l = \omega_0 L/c$.

4. Intrinsic optical bistability in ferroelectrics

Recently, experimental results concerning intrinsic optical bistability in a thin layer of $BaTiO_3$ monocrystal were presented (Ciolek in 2006). The intrinsic optical bistability in the $BaTiO_3$ monocrystal was achieved through the interaction of two lasers without the application of any optical resonator or external feedback. Further, experimental results concerning optical bistability of polarization state of a laser beam, induced by the optical Kerr effect of the B_5NH_4 monocrystal was recently observed (Osuch 2004). The measurements were performed by the means of an ellipsometer of a special construction, which allows for the simultaneous measurement of all four polarization parameters of the laser light beam. Other examples of experimentally demonstrated intrinsic optical bistability with different setups of laser sources and geometries of samples have been reported (Hehlen 1994, Pura 1998, Hehlen 1999 & Przedmojski 1978). Therefore, it is equally important to investigate the intrinsic as well as extrinsic optical bistability in FE material and here comes the advantage of Maxwell-Duffing approach over the standard approach. Mathematically, for FE slab without partially reflecting mirrors, the mirror parameter is set to zero $(\eta_s = 0)$ in the relevant equations. Therefore, we will show graphical results of polarization, reflectance, and transmittance versus the electric field input intensity for FE slab as well as for FP resonator.

5. Material aspects

Generally speaking, the mathematical formulation presented here to investigate the optical bistability is valid for any ferroelectric insulating crystal. Particularly, ferroelectrics with high Kerr nonlinearity and photorefractivity. However, in order to obtain more realistic results, material parameters used in simulation are based on published data of $BaTiO_3$. We should point out that below the Curie temperature T_c, all $BaTiO_3$ phase transitions are of the first-order type except that the transition from the cubic to tetragonal phase is a first-order transition close to second-order transitions. Therefore, close to T_c the 6th order term has to be added to the free energy F in equation (1) apart from the type of the transition since at T_c the coefficient β is zero (Ginzburg 2005). However, well below the transition temperature ($T < T_c$) the form provided in equation (1) may be used as an approximation provided that only tetragonal symmetry is considered.

To integrate equation (12) numerically, it is necessary to evaluate certain material-dependent parameters such as $m = \left[M\omega_0^2 \varepsilon_0 / aT_c \right]$, damping coefficient $g_F = \left[\Gamma \omega_0 \varepsilon_0 / aT_c \right]$, and the coefficient $\xi = \left(\varepsilon_\infty a T_c \right)^{-1}$. To determine these scaled parameters, it is necessary to know the dimensional parameters for $BaTiO_3$ such as Curie temperature T_c, the inverse of the Curie constant a, resonance ω_0, and ε_∞. The value of T_c for $BaTiO_3$ used here is $120\ ^o\boldsymbol{C}$ which gives $T_c = 393.15\ \boldsymbol{K}$. We note that, some ferroelectric literature show different values of T_c which slightly differ from $120\ ^o\boldsymbol{C}$. However, $BaTiO_3$ single crystals obtained are usually not so pure because they are grown by the flux method which makes their Curie point usually about $120\ ^o\boldsymbol{C}$ (Mitsui 1976).

The inverse of the Curie constant a is $1/C$ where $C = 1.7\times10^5\ \boldsymbol{K}$ (Mitsui 1976). It should be noted that, several ferroelectric books uses the free energy density F in CGS units where $a = 4\pi / C$ $^oC^{-1}$. For example, as in Fatuzzo (1967), the a parameter becomes $a = 4\pi / C = 7.4\times10^{-5}\ ^o\boldsymbol{C}^{-1}$. Here, the SI units of measurements are adopted for all dimensional physical variables. It should also be noted that other values of the Curie constant C (Within the range $0.8\times10^4 - 1.7\times10^5$) have been reported which differs considerably. It seems that the method of preparation and the electronic conductivity of the samples have great influence on the Curie constant. For further details, the reader is referred to Seitz (1957). To estimate the resonance ω_0 for $BaTiO_3$, we use the temperature-dependent relation $\omega_0 = \left[-2a\left(T - T_c\right)/\varepsilon_0 M \right]^{1/2}$ for FE material exhibiting a second-order phase

transitions. Knowing the value of M for BaTiO3 to be 6.44×10^{-21} $\mathrm{JmA^{-2}}$ (Murgan 2004), ω_0 is found to be $\omega_0 = 1.437\times10^{13}(T_c - T)^{1/2}$. At room temperature, ω_0 becomes $\approx 1.43\times10^{14}$ Hz. Other fixed material parameters are damping parameter $\Gamma = 3.32\times10^{-5}$ $\mathrm{Kg.m^3.A^{-2}.s^{-3}}$ (Murgan 2004), and the high frequency limit of the dielectric function $\varepsilon_\infty = 3.84$ (Dawber 2005). With these values for the dimensional parameters a, T_c, ω_0, Γ, and ε_∞, the scaled input parameters like m, g and ξ may be calculated.
Since the dimensional polarization amplitude P is scaled in terms of the spontaneous polarization P_s at zero temperature. Therefore, the value of P_s at zero temperature is required. An early measurement of spontaneous polarization P_s by Merz (1949) shows $P_s \approx 0.16$ $\mathrm{C.m^{-2}}$ at room temperature then the value drops to $P_s \approx 0.1$ $\mathrm{C.m^{-2}}$ at zero temperature. However, here we will consider the value of P_s at zero temperature based on a later measurement on a very good $BaTiO_3$ crystal by Kanzig (1949) and confirmed by Merz in (1953). The later experiment shows a value of P_s = 0.26 $\mathrm{C.m^{-2}}$ at room temperature, then it drops to $P_s \approx 0.22$ $\mathrm{C.m^{-2}}$ at zero temperature. The discrepancies between the earlier and the later measurements of P_s were attributed to domains which can not be reversed easily (Seitz 1957). The spontaneous polarization curve P_s as a function of temperature (-140 °C - 120 °C) obtained by Merz (1953) for $BaTiO_3$ may be also found in various FE books such as Cao (2004) and Rabe (2007).
Because both the dimensional electric field amplitude inside the FE medium E_2 and the incident electric field amplitude E_0 are scaled in terms of the coercive field at zero temperature. Therefore, the value of E_c at zero temperature is also required. First, we discuss the estimated value of E_c using thermodynamic theory and its agreement with the experimentally observed value for $BaTiO_3$. It is possible to estimate the value of E_c using the relation $E_c = +\sqrt{4a^3T_c^3/27\varepsilon_0\beta}$ once the value of the nonlinear coefficient β is known. To do so, we may use the relation $P_s^2 = a\varepsilon_0(T_c - T)/\beta$ which yields $\beta = \varepsilon_0 a(T_c - T)/P_s^2$. Substituting the value of $P_s \approx 0.22$ $\mathrm{C.m^{-2}}$ at zero temperature (Merz 1953), this yields $\beta \approx 1.3\times10^{-13}$ $\mathrm{m^3J^{-1}}$. Therefore, the value of the coercive field is estimated to be $E_c \approx 4\times10^7$ $\mathrm{Vm^{-1}}$ at zero temperature. It is important to note that the value of β obtained here is not comparable with those provided by Fatuzzo (1976) and Mitsui (1976) due to the difference in the system of units. In fact their free energy coefficients have different dimensions based on the CGS system of units. However, the value of β obtained here is comparable with that of Murgan (2002) who estimated the value of β to be $\approx 1.9\times10^{-13}$ $\mathrm{m^3J^{-1}}$ at room temperature based on a value of P_s = 0.1945 $\mathrm{C.m^{-2}}$ and $C = 1.669\times10^5$ $\boldsymbol{K}$. The small difference between the value of β obtained here and that of Murgan (2002) is due to the difference in the value of the spontaneous polarization P_s and thermodynamic temperature.
The theoretical value of the coercive field value $E_c \approx 4\times10^7$ $\mathrm{Vm^{-1}}$ calculated at zero temperature using the formula $E_c = +\sqrt{4a^3T_c^3/27\varepsilon_0\beta}$ is in good agreement with other theoretical values calculated elsewhere. For example, a theoretical value of $E_c \approx 1.5\times10^7$ $\mathrm{Vm^{-1}}$ for bulk $BaTiO_3$ was mentioned by Mantese (2005). However, the theoretical value of E_c predicted by thermodynamic theory is found to be two orders of magnitude larger than the experimentally observed value (Seitz 1957). For example, an experimental value of $E_c = 3.34\times10^5$ $\mathrm{Vm^{-1}}$ for BatTiO3 at room temperature was mentioned by Feng (2002). Here, we use $E_c = 1.2\times10^5$ $\mathrm{Vm^{-1}}$ for bulk $BaTiO_3$ at zero temperature based on the measurements by Merz (1953) which is more familiar in ferroelectric literature.

6. Numerical procedure

In linear régime, reflectance **R** and transmittance **T** are independent of the electric field input intensity E_0 and the usual results presented in linear optics are **R** and **T** versus the scaled thickness $l = \omega_0 L/c$. However, in nonlinear optics, as seen from equations (17) and (20), **R** and **T** are directly dependent on the electric field incident amplitude e_0, and other material parameters such as temperature and thickness. Nonlinear optics text books usually illustrate the optical bistability by showing **T** versus $|e_0|^2$ for fixed value of thickness L and frequency ω. Therefore, our aim here is to generate graphs of this type within our current formalism. Since there is no incoming wave in medium 3, it is more convenient to integrate equation (12) across the FE medium from the bottom interface at $u = -l = -\omega_0 L/c$ to the top interface at $u = \omega_0 z/c = 0$.
Our numerical strategy is basically similar to the computation presented in chapter three which can be summarized as follows: we assume the polarization at the bottom boundary p_b to take an arbitrary real value $(p_b = p_b^*)$ and evaluate the first derivative dp_b/du from equation (22). The choice p_b to be real rather than complex is justified in the work by Chew (2001). We then integrate equation (12) as an initial value problem from the bottom boundary $u = -l$ to the top boundary $u = 0$. The integration process keep tracks of the polarization and its derivative across the medium up to the top boundary $u = 0$. As a result, for each arbitrary value of p_b at bottom boundary, we obtain the corresponding value of the polarization at top boundary p_t, its complex conjugate p_t^*, its first derivative dp_t/du and its first-derivative complex conjugate dp_t^*/du. For certain input parameters, substituting p_t, p_t^*, dp_t/du, and dp_t^*/du into equation (19), we obtain the corresponding value of electric field incident amplitude e_0. Similarly, the reflectance $\boldsymbol{R} = |r|^2$ is obtained by substituting p_t and p_t^* into equation (17). On the other hand, we evaluate the transmittance $\boldsymbol{T} = |\tau|^2$ at bottom boundary by substituting the polarization at bottom boundary p_b and its complex conjugate p_b^* into equation (20). The integration procedure is then repeated for a large number of arbitrary p_b values and for each time we evaluate e_0, **R**, and **T.**
Similar numerical scheme to integrate a nonlinear dielectric FP resonator is used by Chew (2001) to evaluate the transmittance of dielectric FP resonator. However, Chew (2001) have generated their plots based on a fixed-step 4th order Runge-Kutta solver modified for complex variable. They therefore, had to perform an interpolation and curve fitting to a raw set of points in the $T - |e_0|$ plane to obtain the optical bistability curves. Here, we have found that the explicit Runge-Kutta method with variable-step solver (Dormand 1980) is capable of producing more accurate results and therefore, an interpolation or any curve fitting is not required and the Bistability curves are generated naturally.

7. Effect of mirror reflectivity

To make a physical significance of the mirror parameter η_s that appeares as a result of the existance of partially reflecting mirrors at the interfaces of the Fabry-Perot resonator, it is useful to to find the corresponding mirror reflectivity R_M of each value of η_s. To do so, we use $R_M = \rho_{j+1,j}\rho_{j+1,j}^*$ where $\rho_{j+1,j} = [k_{j+1} - k_j - \eta]/[k_j + k_{j+1} + \eta]$ (Lim 1997) is the elementary reflection coefficient off medium $j+1$ to medium j. k_j and k_{j+1} are the wavenumbers of medium j and medium j+1 respectively. The coeffecient $\rho_{j+1,j}^*$ is the complex conjugate of $\rho_{j+1,j}$ and η accounts for the mirror contribution. In fact, R_M gives the reflectivity of a mirror placed at the interface of a medium in linear regeme. If both media are nonabsorbing dielectric with $k = \omega n/c$, the coefficient $\rho_{j+1,j}$ may be written in terms of refractive index n and a scaled mirror parameter η_s as $\rho_{j+1,j} = [n_{j+1} - n_j - \eta_s]/[n_j + n_{j+1} + \eta_s]$. If a perfect dielectric nondispersive mirror with conductivity $\sigma_M = 0$ is considered, the mirror coefficient η_s reduces to $-i\eta_{s,b}$ and the mirror reflectivity R_M becomes;

$$R_M = \left[\left(n_{j+1} - n_j\right)^2 + \eta_{s,b}^2\right] \Big/ \left[\left(n_{j+1} + n_j\right)^2 + \eta_{s,b}^2\right] \tag{23}$$

For convenience in numerical simulation, it is simpler to consider R_M assuming a range of values between 0 and 1, and then evaluating the corresponding mirror parameter $\eta_{s,b}$ using equation (23).

Fig. 2 shows the mirror parameter $\eta_{s,b}$ versus power reflectivity of the coating mirror R_M based on equation (23). Here, the linear refractive index $n_2 = [\varepsilon(\omega)]^{1/2}$ of the ferroelectric medium calculated using equation (7) is $n_2 \approx 2$ at frequency $f = \omega/\omega_0 = 1.1$. The curve shows that at $\eta_{s,b} = 0$, the reflectivity of the surface is $R \approx 0.11$ and he mirror reflectivity R_M increases gradually with increasing the mirror parameter $\eta_{s,b}$. The corresponding value of R_M is then found for each value of $\eta_{s,b}$ using Fig. 2. To examine the effect of the mirror parameters $\eta_{s,b}$ on the propagation of the polarization wave, we may use equation (12) to plot $|p|$ versus l for different values of $\eta_{s,b}$ (Fig. 3). The solid curve in Fig. 3 shows $|p|$ versus l for $\eta_{s,b} = 0$ (corresponding to $R \approx 0.11$), the dashed curve is for $\eta_{s,b} = 2$ (corresponding to $R_M = 0.38$), the dotted curve for $\eta_{s,b} = 5$ (corresponding to $R_M = 0.76$), and finally the thin-solid curve for $\eta_{s,b} = 10$ (corresponding to $R_M = 0.92$). A comparison between these curves shows a significant increment of the polarization amplitude $|p|$ accompanied by a phase shift which becomes more noticeable with increasing mirror reflectivity R_M. Such increment in the wave amplitude and the corresponding phase change may be due to the constructive interference that gradually builds up as the result of the mirror coating. A highly reflecting mirror plays an important role in improving the bistable performance of a FP resonator particularly it improves its threshold value of bistable operation as will be explained in the upcoming graphs.

As explained in the previous section, the integration of equation (12) as initial value problem together with the boundary conditions allows us to determine the polarization at top and bottom boundary. Further, the electric field incident amplitude is also determined using equation (19). Therefore, we are able to plot the polarization at each boundary as a function of the electric field incident amplitude. To plot the reflectance **R** = $|r|^2$ versus electric field incident amplitude $|e_0|$, both equation (17) and equation (19) are used. Finally, to plot the transmittance **T** = $|\tau|^2$, versus $|e_0|$, both equation (20) and equation (19) are used.

In Figs. 4 we present the optical bistability of a Fabry-Perot resonator coated with an identical pair of partially reflecting dielectric mirrors. The effect of mirror parameter $\eta_{s,b}$ (mirror reflectivity R_M) on the optical bistability is is investigated for various system variables namely, the polarization p, the reflectance **R** and the transmittance **T**. In each graph of Figs. 4 family, the curves are generatd for various mirror parameters ($\eta_{s,b}$ = 0, 0.1, 0.2, 0.5 and 1, which correspond to R_M = 0.11, 0.128, 0.13, 0.15 and 0.2 respectively) while other parameters are fixed at frequency f = 1.1, thickness l = 1.9, $\varepsilon_\infty = 3.84$, resonance $\omega_0 = 1.4 \times 10^{14}$ Hz (evaluated at room temperature). The graphs in general feature typical

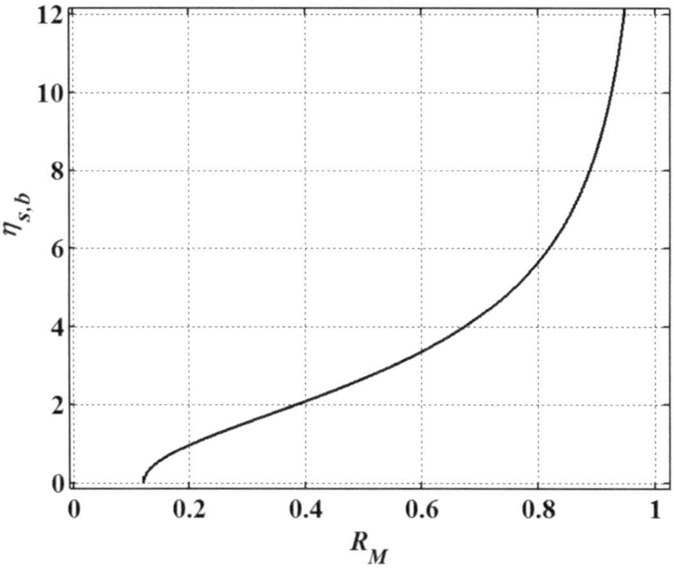

Fig. 2. Scaled mirror parameter $\eta_{s,b} = [\varepsilon_M \delta_M \omega / c]$ versus mirror reflectivity $R_M = \rho_{21}\rho_{21}^*$ placed at single interface between 2 media for scaled frequency f = 1.1, linear refractive index $n_2 = 2$ and $n_1 = 1$.

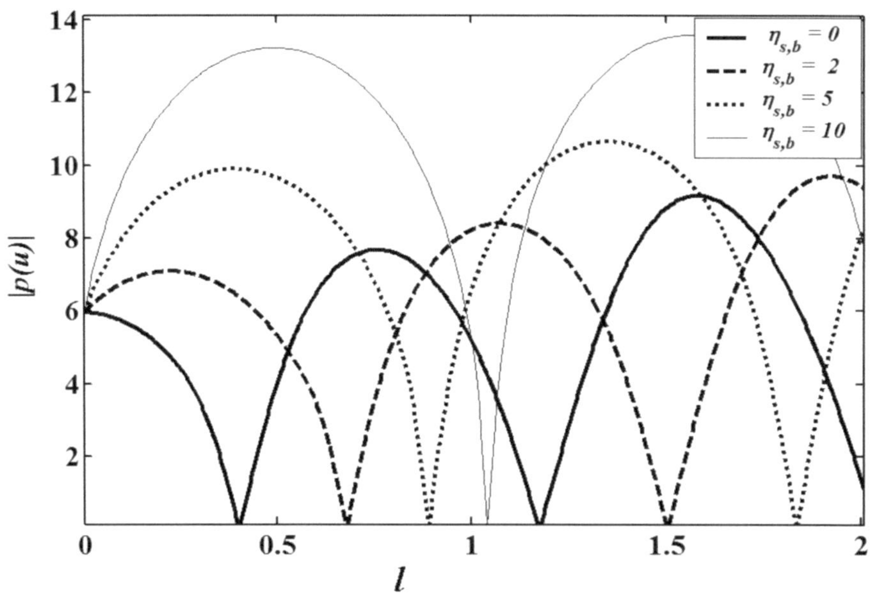

Fig. 3. Scaled polarizations $|p| = |P|/P_0$ versus scaled thickness $l = \omega_0 L/c$ for different mirror parameters $\eta_{s,b}$. Other parameters are $f = \omega/\omega_0$ = 1.1 and ε_∞ = 3.84.

bistability curves. They demonstrate the enhancement of the optical bistability due to the external feedback provided by the coating mirrors. In case $\eta_{s,b} = 0$ (solid curves in Figs. 4) which correspond to a ferroelectric slab, the curves do not show any bistability and the response is essentially linear. In this case, it seems that the intrinsic feedback mechanism is insufficient to generate a bistable behavior. The rest of the curves illustrate that for each value of mirror parameter $\eta_{s,b}$, the bistability in $|p|$, **R**, and **T** is extended over the same range of $|e_0|$. Further, as the mirror reflectivity increases, the bistable behavior becomes more pronounced. For example at $\eta_{s,b} = 0.1$, (the dashed-curves in Figs. 4) the bistability is barely noticeable, it extends over the range $|e_0| \simeq 5204$ to $|e_0| \simeq 5227$. At $\eta_{s,b} = 0.2$, (the dotted-curves in Figs. 4) the bistability is more noticeable and its range is shifted to much lower values of $|e_0|$, where it extends over the range $|e_0| \simeq 5038$ to $|e_0| \simeq 5120$. At $\eta_{s,b} = 0.5$, (the -o- curves in Figs. 4), a wider range of optical bistability is obtained, which begins at $|e_0| \simeq 4720$ and ends at $|e_0| \simeq 4910$. Finally, at $\eta_{s,b} = 1$, (the -□- curves in Figs. 4), the bistability range is shifted to even lower values of $|e_0|$, ranging from 4535 and end at 4725. This certainly suggests that the bistability operation is improved upon increasing the mirror reflectivity. However, from the experimental point of view some important factors should be taken care of, first, the mirror material should be chosen in such a way it does not react chemically with the nonlinear medium. Second, crystal surface should be parallel to the mirror surface which requires careful alignment. Third the operating frequency should not be very close to resonance to avoid highly-absorption rates of the crystal. Unfortunately, we are unable to generate graphs for higher values of $\eta_{s,b}$ due to numerical instability. However, the curves here are sufficient to show the relationship between the optical bistability and mirror reflectivity.

Next is to draw the attention to the relationship between the bistability in the polarization p and reflectance **R** or transmittance **T**. Observation of these graphs shows that the bistability in **R** (Fig. 4(b)) is basically a manifestation of the bistability in $|p_t|$ (Fig. 4(a)) while the bistability in **T** (Fig. 4(d)) is a manifestation to the bistability in $|p_b|$ (Fig. 4(c)). This is explained by equations (17) and (20) where reflectance **R** and transmittance **T** are basically a function of the polarization and other material parameters. Further, the optical bistability in the macroscopic polarizarion is in fact due to the optical bistability in the microscopic polarization where the individual molecules responds nonlinearly to the driving field (Goldstone and Garmire 1984, Ibrahim and Osman 2008). In fact, this is where the advantage of Maxwell-Duffing approach is affirmed over the standard analysis in nonlinear optics, the ability to model both intrinsic and extrinsic optical bistability in one approach.

Finally, curves in Figs. 4 suggest that the threshold value of $|e_0|$ for bistability is significantly reduced by increasing mirror parameter or equivalently the mirror reflectivity. For $\eta_{s,b} = 0.1$, (the dashed-curves in Figs. 4), the threshold value of the optical bistability occurs at $|e_0|_{th} \simeq 5204$, and, for $\eta_{s,b} = 0.2$, (the dotted-curves in Figs. 4) the optical bistability starts at $|e_0|_{th} \simeq 5038$. For $\eta_{s,b} = 0.5$, (the -o- curves in Figs. 4), the optical bistability starts at $|e_0|_{th} \simeq 4720$ and for $\eta_{s,b} = 1$, (the -□- curves in all Figs. 4), the optical bistability begins at $|e_0|_{th} \simeq 4535$. Therefore, a systematic decrease of the threshold value of the driving field $|e_0|_{th}$ required for optical bistability is obtained upon increasing the mirror parameter $\eta_{s,b}$ or the mirror reflectivity.

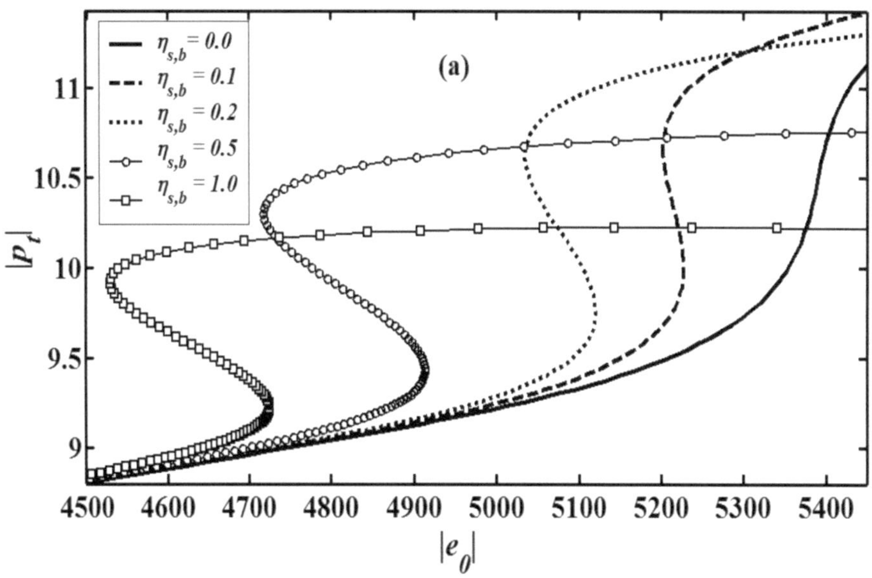

Fig. 4.(a) Polarization $|p_t| = |P_t|/P_s$ at top boundary versus electric field incident amplitude $|e_0| - E_0/E_c$ for different mirror parameters $\eta_{s,b}$ = 0, 0.1, 0.2, 0.5 and 1 corresponds to mirror reflectivity R_M = 0.127, 0.128, 0.13, 0.15 and 0.2. Other parameters are $f = \omega/\omega_0 = 1.1$, $l = \omega_0 L/c = 1.9$, $\varepsilon_\infty = 3.84$ with $t = T/T_c$ evaluated at room temperature.

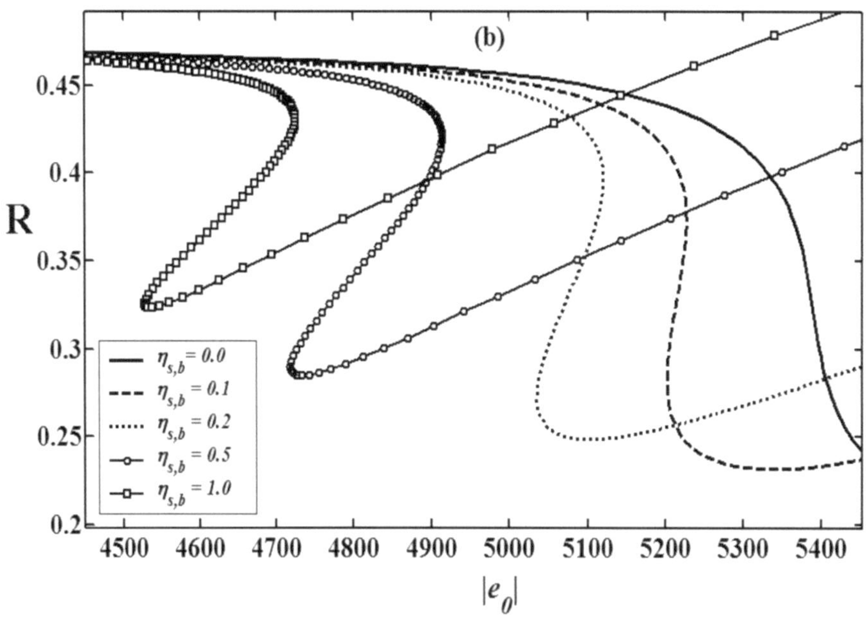

Fig. 4.(b) Reflectance $R = |r|^2$ versus $|e_0|$ for different mirror parameters. Other parameters remain as in Fig. 4(a)

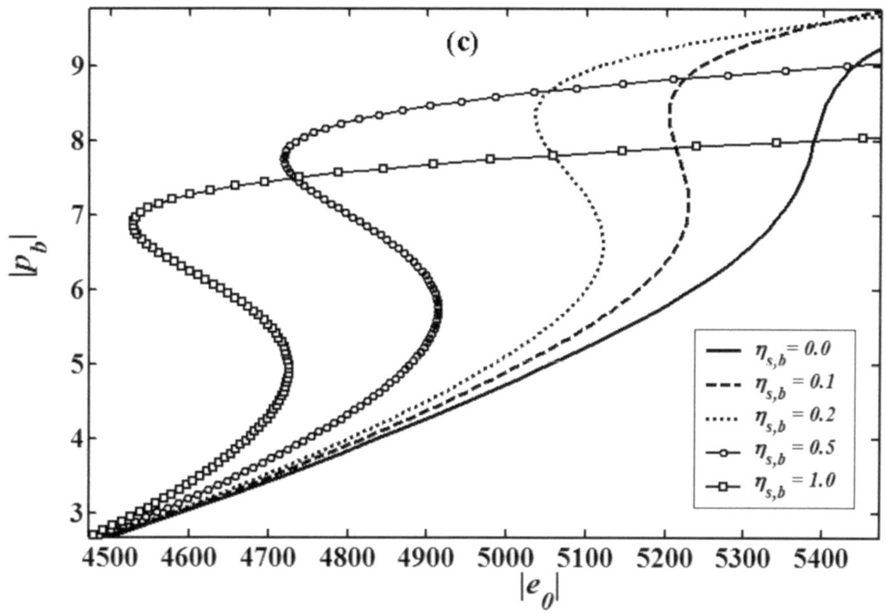

Fig . 4.(c) Polarization $|p_b| = |P_b|/P_s$ at bottom boundary $u = -\omega_0 L/c = -1$ versus electric field incident amplitude $|e_0| = E_0/E_c$ for different mirror parameters $\eta_{s,b}$ as on Fig. 4(a). Other parameters remain as in Fig. 4(a).

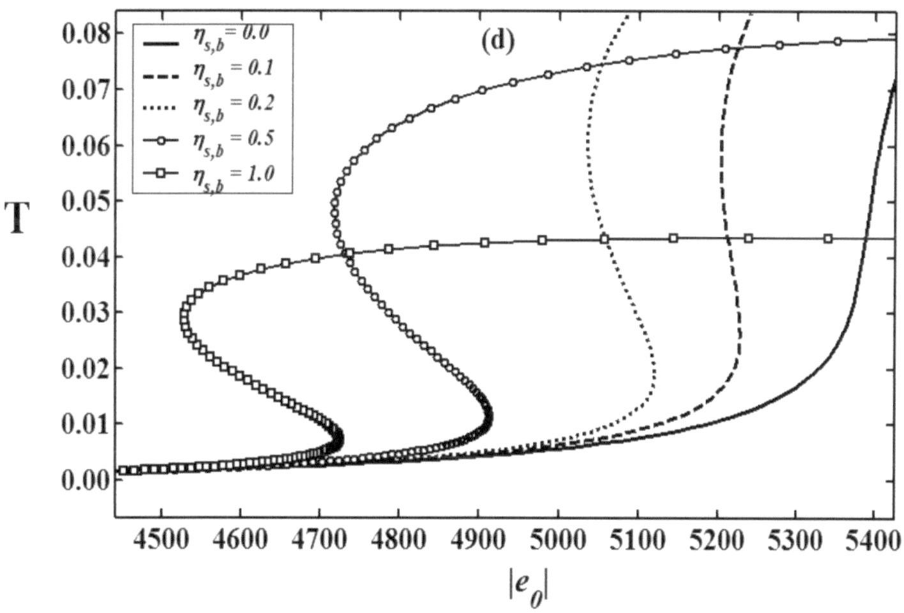

Fig. 4.(d) Transmittance $T = |\tau|^2$ versus electric field incident amplitude $|e_0| = E_0/E_c$ for different mirror parameters as in Fig. 4(a). Other parameters remain as in Fig. 4(a)

An increment in the mirror reflectivity seems to have its effect on both **R** and **T**, the switching "on" and "off" values. Observation of Fig. 4(b) shows that, for $\eta_{s,b} = 0.1$, (the dashed-curve), the reflectance switches between $\boldsymbol{R}_{off} \approx 0.27$ "off state" and $\boldsymbol{R}_{on} \approx 0.45$ "on state". For $\eta_{s,b} = 1$, (the -□- curve), the **R** "off" state increases up to $\boldsymbol{R}_{off} \approx 0.32$ while the "on" state remains at $\boldsymbol{R}_{on} \approx 0.45$. Similar behavior of transmittance **T** is also noticed in Fig. 4(d), while the "off" state remains at nearly $\boldsymbol{T}_{off} \approx 0$, the "on" state decreases gradually with increasing $\eta_{s,b}$. This is an ideal optical switch switches between "1" and "0" as the "on" and "off" state. Therefore, a better bistable operation seems to be a trade-off between mirror reflectivity and other material parameters. In other words, even a highly-reflective mirror may decrease the threshold value of the optical bistability and yet it may decrease the quality of the device as an optical switch.

It is also observed that, the nonlinear response of the system exists only for a certain range of input intensity $|e_0|$. Interestingly, the system's response to the driving field becomes linear again above this range. For example, for $\eta_{s,b} = 1$, the optical bistability starts at $|e_0|_{th} \simeq 4535$, below that threshold value of $|e_0|_{th}$ the system response is linear, and then the system starts a bistable period up to $|e_0|_{final} \simeq 4725$, beyond this value, it responds linearly to the driving field again. We have found that obtaining one bistable period through variation of ..over a wide range is basically due to a choice of relatively thin sample ($l = \omega_0 L/c = 1.9$). Increasing the thickness usually results in multistability as will be explained in the next section. It should be noted that a much higher values of the driving field $|e_0|$ from the laser source is not advisable since it may result in a material breakdown.

8. Effect of frequency

In Figs. 5, the reflectance **R** and transmittance **T** versus electric field incident amplitude $|e_0| = E_0/E_c$ are plotted for different operating frequencies $f = \omega/\omega_0$ at room temperature. Other parameters are fixed at $l = \omega_0 L/c = 3$, $\eta_{s,b} = 0$ and $\varepsilon_\infty = 3.84$. Two important features occur as the result of changing the operating frequency $f = \omega/\omega_0$. First is the change of the threshold value of the electric field incident amplitude required for optical bistability. Second is the change in the switching amplitude (the "on" and "off" state). Observation of Fig. 5(a) shows that, far above the resonance $(f = 3$ or $f \gg 1)$, the threshold value of optical bistability is $|e_0|_{th} \approx 5.87 \times 10^4$. In this case, the reflectance (curve ***i***) switches between $\mathbf{R}_{on} \approx 0.55$ and $\mathbf{R}_{off} \approx 0.08$ while the transmittance (curve ***ii***) switches between $\mathbf{T}_{on} \approx 0.8$ and $\mathbf{T}_{off} \approx 0.15$. Observation of Fig. 5(b) shows that, slightly above the resonance $(f = 1.4$ or $f > 1)$, the threshold value of optical bistability is $|e_0|_{th} \approx 1.85 \times 10^4$. In this case, the reflectance (curve ***i***) switches between $\mathbf{R}_{on} \approx 0.61$ and $\mathbf{R}_{off} \approx 0.05$ while the transmittance (curve ***ii***) switches between $\mathbf{T}_{on} \approx 0.9$ and $\mathbf{T}_{off} \approx 0.1$. Therefore, a comparison between Fig. 5(a) and Fig. 5(b) shows that at $f = 1.4$, a better switching in **T** as well as a lower threshold value is obtained comparing to the case of $f = 3$.

At resonance $(f = 1)$, observation of Fig. 5(c) shows that, the threshold value of the optical bistability is $|e_0|_{th} \approx 5.6 \times 10^3$. Two points are worth noted. First, the bistabile response is possible even at resonance where absorption in the FE material is the highest. Secondly, the threshold value of $|e_0|$ needed to induce bistability is much lower. However, even the optical bistability in reflectance is still noticeable; the transmittance in this case is practically

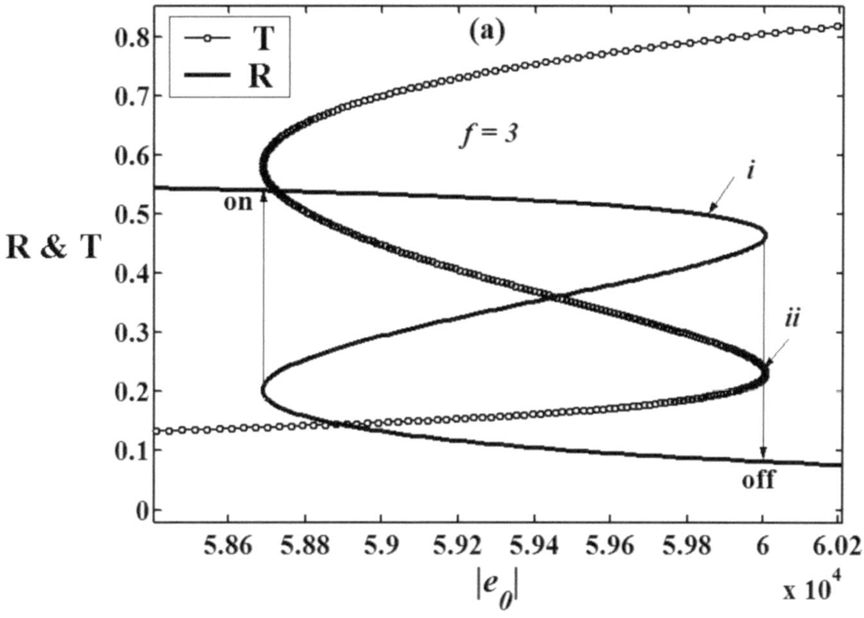

Fig. 5.(a). Reflectance R and transmittance T versus electric field incident amplitude $|e_0| = E_0/E_c$ at frequency f = 3. Other parameters are fixed at thickness $l = \omega_0 L/c = 1$, mirror parameter $\eta_{s,b} = 0$ and $\varepsilon_\infty = 3.84$.

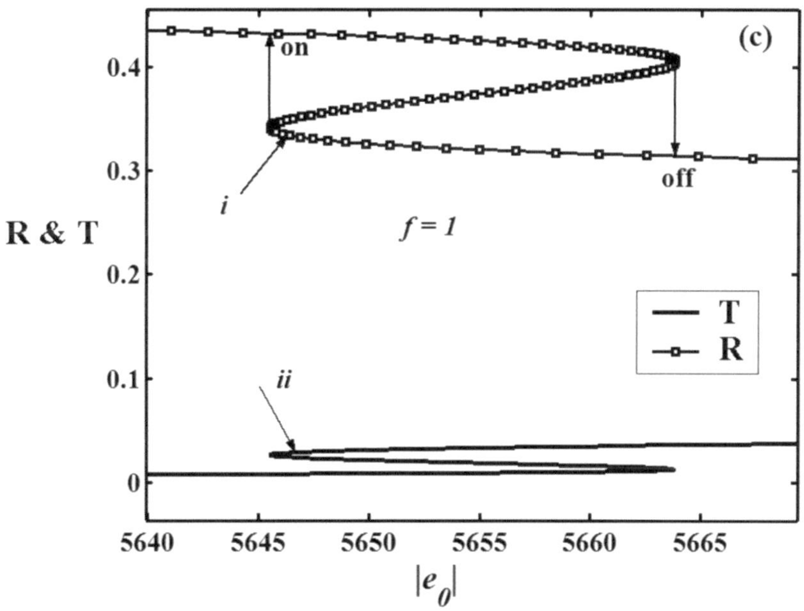

Fig. 5.(b). Reflectance R and transmittance T versus electric field incident amplitude $|e_0| = E_0/E_c$ at frequency f = 1. Other parameters remain as in Fig. 5(a).

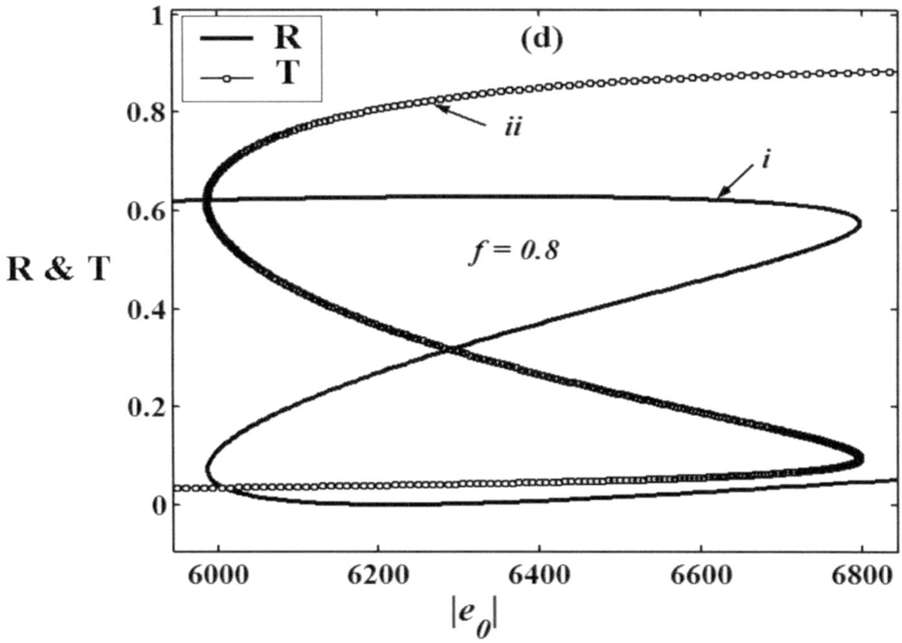

Fig. 5(c). Reflectance *R* and transmittance T versus electric field incident amplitude $|e_0| = E_0/E_c$ at frequency *f* = 0.8. Other parameters remain as in Fig. 5(a)

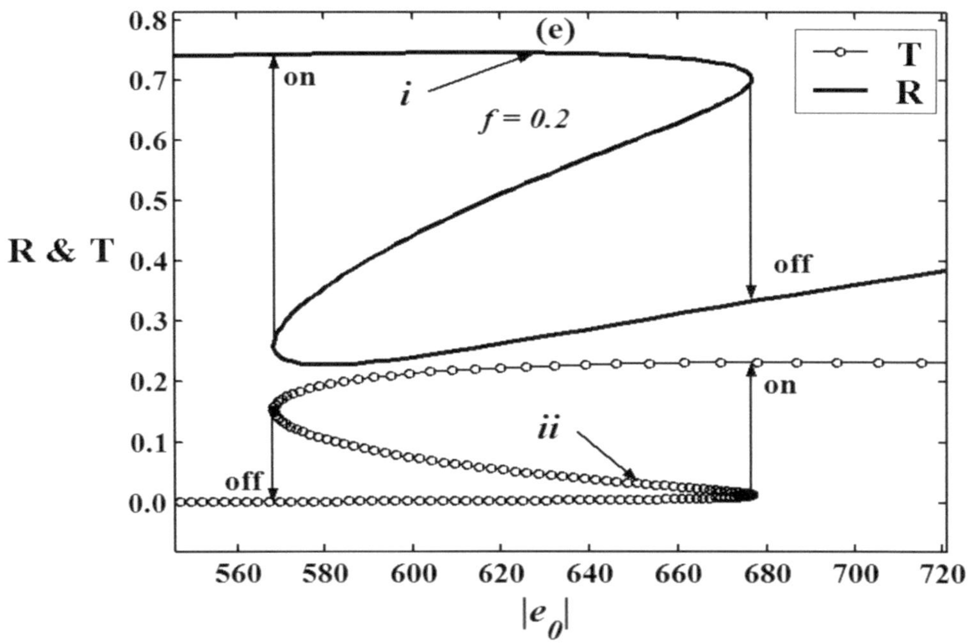

Fig. 5(d). Reflectance *R* and transmittance T versus electric field incident amplitude $|e_0| = E_0/E_c$ at frequency *f* = 0.2. Other parameters remain as in Fig. 5(a)

zero. Observation of Fig. 5(d) shows that, slightly below the resonance $(f = 0.8 \text{ or } f < 1)$, the threshold value of optical bistability is $|e_0|_{th} \approx 6\times10^3$. In this case, the reflectance (curve ***i***) switches between $\mathbf{R}_{on} \approx 0.61$ and $\mathbf{R}_{off} \approx 0.02$ while the transmittance (curve ***ii***) switches between $\mathbf{T}_{on} \approx 0.88$ and $\mathbf{T}_{off} \approx 0.02$. Fig. 5(e) shows that, far below the resonance $(f = 0.2 \text{ or } f \ll 1)$, the threshold value of optical bistability is the lowest $\left(|e_0|_{th} \approx 570\right)$. In this case, the reflectance (curve ***i***) switches between $\mathbf{R}_{on} \approx 0.74$ and $\mathbf{R}_{off} \approx 0.36$ while the transmittance (curve ***ii***) switches between $\mathbf{T}_{on} \approx 0.22$ and $\mathbf{T}_{off} \approx 0$. This means, even the threshold value of bistability is dramatically decreased, yet there is a decrease on the switching contrast between the "on" and "off" states of the switching.
Therefore, in general, the main features of bistability curves above the resonance $(f > 1)$ are found to be similar to those below the resonance $(f < 1)$ and a better bistability is obtained when the operating frequency approaches the resonance of the material from below or above. The enhancment of the optical bistability near resonance can be explained as a result of increasing the magnetude of certain tensor elements of the third-order dielectric susceptibility in this region. On the other hand, the reason for nearly zero transmittance at the exact resonance (curve ***i*** in Fig. 5(c)) may be viewed as a result of rapid depletion of incident optical pump wave or the newly generated signal wave (He 2000). Therefore, the proper choice of the operating frequency is a compromise between the enhancement of the nonlinear susceptibility and the attenuation of the useful optical wave. For this reason, an operating frequency in a quasi-resonance is often employed by tuning the frequency of the incident laser beam to be close but not equal to the resonance of the medium. The threshold value of the optical bistability decreases in general with decreasing *f*. However, from our calculated curves for a larger range of frequencies, some variation in the threshold values are found. For example, the threshold value of $|e_0|$ at $f = 0.8$ is slightly higher than that of $|e_0|$ at $f = 1$ as seen from Fig. 5.7(c) and Fig. 5.7(d). This is may be due to the variation in detuning but the main effect remains.

9. Conclusion

The Maxwell-Duffing analysis has been employed to study the optical bistability of a ferroelectric slab as well as a Fabry-Perot resonator coated with two identical partially-reflecting dielectric mirrors. The nonlinear response of the polarization *P* to an optical driving field *E* using the Landau-Khalatnikov dynamical equation has been modeled. The Landau-Devonshire free energy expression for bulk FE material assumed to exhibit a second-order phase transition has been utilized. Using single frequency approximation and assuming normal incidence, the driving field in the LK equation is substituted into the electromagnetic wave equation to produce a nonlinear polarization equation. For convenience in numerical simulation, all variables are converted into dimensionless form. The resulting nonlinear polarization equation is numerically integrated across the thickness of the FE medium. With the application of the exact nonlinear boundary conditions, expressions for both reflectance *R* and transmittance **T** are derived as a function of the total polarization *p*, electric field incident amplitude $|e_0|$, and other material parameters such as temperature. The behaviors of the polarization at top and bottom interface, reflectance **R**, and transmittance **T** have been plotted versus electric field incident amplitude $|e_0|$. The effects of mirror parameter η_s, and frequency f on the optical bistability have been investigated. The input parameters used in this simulation are based on available experimental data of $BaTiO_3$.

It is found that the system responds linearly to the driving field at relatively low electric field incident amplitude $|e_0|$. After $|e_0|$ exceeds certain threshold value, the response becomes nonlinear. The threshold value of the bistability phenomenon was found to have a function of mirror reflectivity R_M, and operating frequency f. The bi-stability in both reflectance and transmittance has been demonstrated which is a manifestation of the bistability in the polarization itself. The bistability in the macroscopic polarization is proportional to the bistability in the microscopic domain. The current approach is more suitable for ferroelectrics particularly at frequency ranges where the nonlinear response of the material is strong and resonant. The intrinsic optical bistability obtained is in agreement with the experimental results of intrinsic optical bistability obtained recently for $BaTiO_3$ (Ciolek 2006).

This method could be applied to oblique incidence and to multilayers. Apart from the considerable difference between this approach and the standard approach in nonlinear optics, the graphs shown are qualitatively similar to those found in the textbook analysis. It might also be possible to study the behavior of this system without the use of single frequency approximation to examine the characteristics of the full dynamic in time domain which may lead to chaos.

10. Refrences

Biran B., Opt. Commun. 74 183 (1990).

Born M. and Wolf E., Principles of Optics, "*Electromagnetic Theory of vPropagation, Interference and Diffraction of Light*", 6th ed. Pergamon, Oxford, (1980).

Cao W., and Cross L.E., Phys. Rev. B 47 4285 (1993).

Chew K-H, Osman J. and Tilley D. R., Opt. Commun. 191, 393 (2001).

Ciolek R., Osuch K., Pura B., Wierzbicki M., Zagórski A., and Wrzesiński Z., Optical Materials 28, 1341 (2006).

Danckaert J., Thienpont H., Vertennicofff I., Haelterman M., and Mandel P., Opt. Commun. 71 317 (1989).

Dawber M., Rabe K. M., and Scott J. F., Reviews of Modern Physics, 77, 1083 (2005).

Dormand, J. R. and Prince P. J., *J. Comp. Appl. Math.* 6 19 (1980).

Eaton D. F., Science, New Series, Vol. 253, No. 5017, 281 (1991).

Fatuzzo E., and Merz W. J., *Ferroelectricity*. North-Holland Publishing Co, Amsterdam: (1967).

Feng, S.H. and J.S. Chen, *Frontiers of solid state chemistry: proceedings of the International Symposium on Solid State Chemistry in China, Changchun, China,* (2002).

Gibbs H.M., McCall S.L., and Venkatesan T.N.C.: Optical bistability. Optics News, 6 (1977).

Gibbs, H. M., McCall, S. L., Venkatesan T. N. C., Gossard A. C., Passner A., and Wiegmann W., *Appl. Phys. Lett.* 35, 451(1979).

Gibbs H.M. "*Optical bistability: controlling light with light*", Academic Press, Inc., Orlando FL, (1985)

Ginzburg V. L. "Phase Transitions and Critical Phenomena", In *Ferroelectricity: The Fundamentals Collection,* Gonzalo J. A., and Jiménez B., WILEY- VCH Verlag GmbH & Co. KGaA, Weinheim (2005)

Goldstone J. A., and Garmire E., Phys. Rev. Lett. 53 910 (1984)

Gupta S., and Agrawal G. S., J. Opt. Soc. Am. B 4 691 (1987).

Haelterman M., Mandel P., Danckaert J., Thienpont H., Vertennicofff I., Opt. Commun. 74, 238 (1989).

He, G. S., and Liu S. H., "*Physics of Nonlinear Optics*", World Scientific, (2000).

Hehlen M.P., Gudel H.U., Shu Q., Rai J., Rai S., and Rand S.C., Phys. Rev. Lett. 73 1103 (1994).

Hehlen M.P., Kuditcher A., Rand S.C., and Luthi S.R., Phys. Rev. Lett. 82 3050 (1999)

Ibrahim A-B M. A., Tilley D. R., and Osman J., Ferroelectrics, 355 (1) 140 (2007)

Ibrahim, A. B. M. A., and Osman, J., *Euro. Phys. J., B* 63, 193 (2008).

Kanzig W. and Meier R., Helv. Phys. Acta 21, 585 (1949).

Lim S.-C., Osman J., and Tilley D. R., J. Phys.: Condens. Matter 9 8297 (1997).

Lines M. E., and Glass A.M., "*Principles and Applications of Ferroelectrics and Related Materials*", Clarendon Press, Oxford, (1977)

Mantese J. V., and Alpay S. P., "*Graded Ferroelectrics, Transcapacitors and Transponents*"., Springer, New York, (2005).

Marburger J. H., and Felber F.S., Phys. Rev. A. 17 335 (1978).

Merz,W . J., Phys. Rev. 76 1221 (1949)

Merz W. J., Phys. Rev. 91, 513 (1953)

Murgan R., Tilley D. R., Ishibashi Y., Webb J. F., and Osman J., J. Opt. Soc. Am. B 19 2007 (2002).

Murgan R., Razak F., Tilley D. R., Tan T. Y., Osman J., and Halif M. N. A., Computational Material Science., 30 468 (2004).

Mitsui T., Tatsuzaki I., and Nakamura E., "*An Introduction to the Physics of Ferroelectrics*", Gordon and Breach, London, (1976)

Osuch K., Pura B., Petykiewicz J., Wierzbicki M., and Wrzesiński Z., Optical Materials 27, 39 (2004).

Przedmojski J., and Pura B., Ferroelectrics 21, 545 (1978).

Pura B., Jda W., Noniewicz K., and Zagórski A., J. Nonlinear Opt. Phys. Mater. 7 441 (1998)

Rabe K. M., Ahn C. H., and Triscone J-M., "*Physics of ferroelectrics: a modern perspective*", Springer, Berlin, (2007)

Ramesh R. Aggarwal S. and Auciello O., Materials Science and engineering 32 (2001).

Seitz F., and Turnbull D., "*Solid State Physics. Advances in research and application*", vol. 4, Academic Press, New York, (1957).

Shen Y. R., *The Principles of Nonlinear Optics*, Wiley, New York, (1984).

Shi F.W. , Meng X.J., Wang G.S., Sun J.L., Lin T., Ma J.H., Li Y.W., and Chu J.H., Thin Solid Films 496 333 (2006)

Snieder R., "*A guided tour of mathematical methods for the physical sciences*", 2nd ed., Cambridge University Press (2004)

Sutherland R. L, "*Handbook of Nonlinear Optics*", Marcel Dekker Inc, New York: (1996).
Xuan L., Pan S., Chen Z., Wang R., Shi W., and Li C., Appl. Phys. Lett. 73 2896 (1998).
Zhang S., Dong X., and Kojima S.. Jpn. J. Appl. Phys. 36 2994 (1997).
Zhao Q., Liu Y., Shi W., Ren W., Zhang L., and Yao X., Appl. Phys. Lett. 69 458 (1996).

19

Nonlinear Conversion Enhancement for Efficient Piezoelectric Electrical Generators

Daniel Guyomar and Mickaël Lallart
Université de Lyon, INSA-Lyon, LGEF EA 682, F-69621, Villeurbanne FRANCE

1. Introduction

Although primary batteries have initially promoted the development of low-power devices, recent progresses in microelectronics and in ultra-low power system have shown the limit of such a powering solution. In particular, the limited lifespan and complex recycling process of batteries may raise environmental issues as well as maintenance problems for widespread devices. Hence, in order to counteract these drawbacks, recent trends encouraged the research on renewable energy. The possibility of using ambient sources has thus become an important research field (Roundy and Wright, 2004; Krikke, 2005; Ng and Liao, 2005; Paradiso and Starner, 2005; Guyomar *et al.*, 2007a; Lallart *et al.*, 2008a). Such alternative solutions for providing electrical energy to systems may include solar energy (Hamakawa, 2003), thermal energy (Sodano *et al.*, 2006) or mechanical energy. When dealing with small-scale systems, the latter energy source has been of particular interest as vibrations are widely available in many environments (Shearwood and Yates, 1997; Beeby *et al.*, 2007). In addition, the use of piezoelectric transducers for converting mechanical energy into electricity has attracted much attention as such materials offer high energy densities and promising integration potentials, making them a premium choice for the conception of embeddable microgenerators (Anton and Sodano, 2007; Blystad, Halvorsen and Husa, 2008).

However, mechanical energy is still limited in structures, and piezoelectric transducers present moderate coupling coefficients, especially when used in flexural solicitation which is the most common application of such materials when used in energy harvesting applications (Keawboonchuay and Engel, 2003; Richards *et al.*, 2004). Therefore, in order to dispose of efficient devices able to provide a significant amount of energy for powering electronic systems, it is mandatory to enhance the conversion abilities of piezoelectric elements.

To do so, many studies have focused on the material itself, aiming at increasing the piezoelectric activity. In this domain, most of the works performed have consisted in the development of single crystals, which exhibits piezoelectric coefficients d_{31} and g_{31} typically 9 and 4 times higher, respectively, than conventional piezoceramics, leading to performance in terms of energy harvesting 20 times greater (Park and Hackenberger, 2002; Badel *et al.*, 2006a). However, the synthesis procedure for obtaining piezoelectric single crystals is quite complex and not industrializable yet, making them difficult to achieve in large quantities as well as costly. Therefore, the realistic implementation of piezoelectric transducers for energy harvesting purposes necessitates a simpler process. To address this issue, Guyomar *et al.* (2005) proposed a nonlinear treatment for artificially enhancing the conversion abilities of such materials. The principles of this approach, originally applied for vibration damping purposes (Richard et al., 1999; Petit et al., 2004; Qiu, Ji and Zhu, 2009a), is to quickly invert the charges available on the piezomaterial synchronously with the structure motion.

The purpose of this chapter is to expose efficient energy harvesting schemes based on this nonlinear conversion enhancement concept. Several architectures will be presented from the original nonlinear approach, each of them addressing one or several issues for improving the performance of microgenerators (power output, load independency, low voltage systems, broadband excitation performance...), and a comparative analysis between the techniques will also be discussed.

The chapter is organized as follows. Section 2 aims at introducing a simple but realistic model of an electromechanical system that will be used in the following theoretical development, as well as the basic physical principles of the nonlinear approach for improving the conversion abilities of ferroelectric materials. In section 3 the direct application of the nonlinear concept to energy harvesting will be developed. Then section 4 will expose other nonlinear approaches that allow a decoupling of the energy extraction and storage stages, permitting a harvested power independent from the load. Finally, a last architecture based on a bidirectional energy flow and energy injection mechanism will be introduced in section 5, and be demonstrated to offer an "energy resonance" effect thanks to the feedback loop. A particular attention will be placed on the realistic implementation of such microgenerators as well as on systems featuring low voltage output (as piezoelectric elements are particularly interesting for microdevices) in section 6. Because of the similarities between the two conversion effects, the application of the exposed methods to energy harvesting from temperature variation using pyroelectric materials will be discussed in section 7. Finally, section 8 will summarize the obtained results and draw some conclusions about the concepts exposed in this chapter.

2. Modeling & conversion enhancement principles

Before exposing and analyzing the harvesting systems using nonlinear approaches, it is proposed in this section to describe a simple but realistic model developed by Badel *et al.* (2007) of a structure equipped with piezoelectric inserts, along with the physical principles of the nonlinear treatment for enhancing the conversion abilities of piezoelectric materials.

The electromechanical model that will be used in this chapter is based on a simple electromechanically coupled spring-mass-damper system (Figure 1), which however relates quite well the behavior of the system near one of its resonance frequencies. From the Newton's law and piezoelectric constitutive equations, it can be demonstrated under given assumptions[1] that the governing equation of motion and electrical equation are given by (Badel *et al.*, 2007):

$$\left\{ \begin{array}{l} M\ddot{u} + C\dot{u} + K_E u = F - \alpha V \\ I = \alpha \dot{u} - C_0 \dot{V} \end{array} \right. , \tag{1}$$

where u, F, V and I respectively refer to the displacement at a given location of the structure, driving force[2], piezoelectric voltage and current flowing out the piezoelectric element. M,

[1]For the model development, the assumptions are based on plane strain behavior (no stress along z-axis), Euler-Bernoulli hypothesis (plane sections remain plane), and similar dynamic and static deformed shapes (Badel *et al.*, 2007).

[2]For seismic systems, the force may also be expressed as a function of the acceleration. In this case, the applied force is given by $\mu_1 M a$, with a the acceleration, M the dynamic mass and μ_1 a correction factor (Erturk and Inman, 2008).

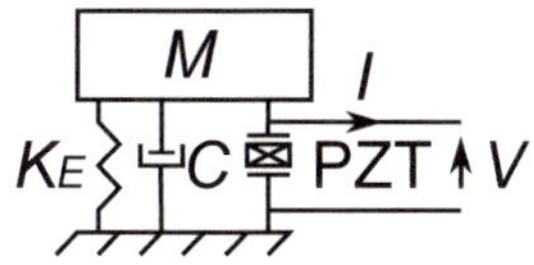

Fig. 1. Single Degree Of Freedom (SDOF) model of an electromechanical structure

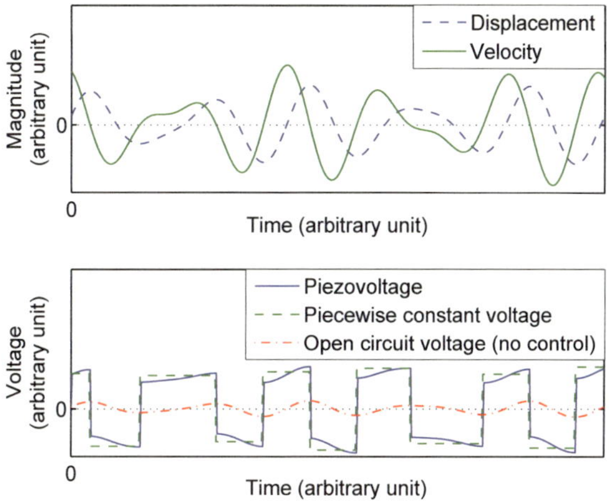

Fig. 2. Waveforms using nonlinear treatment

C and K_E are defined as the dynamic mass, structural damping coefficient and short-circuit stiffness, and α and C_0 are given as the force factor and piezocapacitance. In open circuit condition, it is also possible to define an open-circuit stiffness K_D, whose expression yields:

$$K_D = K_E + \frac{\alpha^2}{C_0}. \quad (2)$$

The energy analysis over a particular time range $[t_0; t_0 + \tau]$ therefore yields:

$$\begin{cases} \int_{t_0}^{t_0+\tau} F\dot{u}dt = \frac{1}{2}M\left[\dot{u}^2\right]_{t_0}^{t_0+\tau} + C\int_{t_0}^{t_0+\tau}\dot{u}^2dt + \frac{1}{2}K_E\left[u^2\right]_{t_0}^{t_0+\tau} + \alpha\int_{t_0}^{t_0+\tau}V\dot{u}dt \\ \alpha\int_{t_0}^{t_0+\tau}V\dot{u}dt = \int_{t_0}^{t_0+\tau}VIdt + \frac{1}{2}C_0\left[V^2\right]_{t_0}^{t_0+\tau} \end{cases}. \quad (3)$$

Hence, it can be seen that from the motion equation that the amount of converted energy is given by the time integral of the product of the voltage by the velocity. Therefore, in order to increase the converted energy, two possibilities may be adopted[3]:

- Increase the voltage magnitude
- Ensure that the voltage is as proportional as possible to the speed (*i.e.*, reduce the time shift between V and $\dot{u}$).

To do so, several approaches are possible, but they have to consume as less energy as possible. In particular, inverting the piezovoltage on its maximum and minimum value allows shaping an additional piecewise constant voltage proportional to the sign of the velocity much larger than the original voltage (Figure 2). Such a process therefore allows benefitting from both effects for improving the conversion.

[3]considering that the velocity remains constant

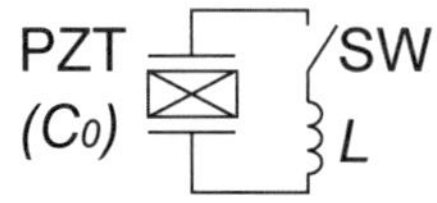

Fig. 3. Implementation of the inversion process

The inversion process can besides be obtained in a very simple way without requiring any external energy. The principles of this process lie on the dielectric properties of piezoelements. Thanks to their capacitive behavior, the voltage is continuous. Hence, after an inversion event, the induced initial condition change is kept on the material, shaping the piecewise function. The inversion process is very simple as well. It consists in connecting the piezoelectric element to an inductor L (Figure 3), which creates an oscillating network. Hence, once the piezoelectric element is connected to the inductance, the voltage starts oscillating around 0, and, if the switching time t_i is chosen so that it equals half the electrical oscillation pseudo-period:

$$t_i = \pi\sqrt{LC_0}, \tag{4}$$

this leads to an inversion of the piezoelectric voltage. This inversion process is however not perfect because of the losses in the circuit, and can be characterized by the inversion coefficient γ giving the ratio of the absolute voltages before and after the inversion, which can also be obtained from the electrical quality factor Q_i of the LC_0 network:

$$\gamma = e^{-\frac{\pi}{2Q_i}} \text{ with } Q_i = \frac{1}{r}\sqrt{\frac{L}{C_0}}, \tag{5}$$

with r the equivalent loss resistance of the circuit.

3. SSH techniques

Now the basic principles of the nonlinear conversion enhancement exposed, this section proposes the direct application of this concept to energy harvesting, leading to the concept of *Synchronized Switch Harvesting on Inductor* (Guyomar *et al.*, 2005; Lefeuvre *et al.*, 2006a; Shu, Lien and Wu, 2007; Liang and Liao, 2009; Qiu *et al.*, 2009b). Considering the standard energy harvesting interface that consists in connecting the piezoelectric element to a smoothing capacitor C_S and load R_L (that represents the connected device) through a diode bridge rectifier (Figure 4(a)), the switching element may be placed in two ways:

- In parallel with the piezoelectric element (Parallel SSHI - Figure 4(b))
- In series with the piezoelectric element and the harvesting stage (Series SSHI - Figure 4(c))

When using the standard interface, it can be demonstrated that the harvested energy under a constant vibration magnitude u_M is given in steady state case by (Guyomar *et al.*, 2005):

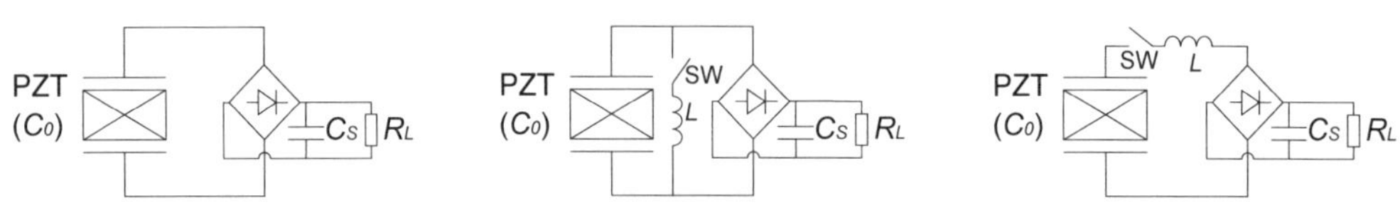

(a) Standard (b) Parallel SSHI (c) Series SSHI

Fig. 4. Energy harvesting interfaces

$$P_{\text{standard}} = \frac{(4\alpha f_0)^2 R_L}{(1+4f_0C_0R_L)^2} u_M{}^2, \tag{6}$$

with f_0 and u_M referring to the vibration frequency and displacement magnitude, respectively. The associated maximal power when using the optimal load yields:

$$P_{\text{standard}}|_{\max} = f_0 \frac{\alpha^2}{C_0} u_M{}^2. \tag{7}$$

However, converting mechanical energy into electrical energy decreases the former, therefore leading to vibration damping effect that limits the effective harvested power. When considering that the system is submitted to a monochromatic driving force with constant magnitude F_M, an energy analysis of the system leads to the expression of the power (Guyomar *et al.*, 2005):

$$P_{\text{standard}} = \frac{(4\alpha f_0)^2 R_L}{(1+4f_0C_0R_L)^2} \left(\frac{F_M}{2\pi C f_0 + \frac{16\alpha^2 f_0 R_L}{\pi + (1+4f_0C_0R_L)^2}} \right)^2, \tag{8}$$

whose maximal value is given by:

$$\begin{cases} P_{\text{standard}}|_{\max} = \frac{k^2 Q_M}{(\pi + k^2 Q_M)^2} \frac{\pi}{2} \frac{F_M{}^2}{C} & \text{for } k^2 Q_M \leq \pi \\ P_{\text{standard}}|_{\max} = \frac{F_M{}^2}{8C} & \text{for } k^2 Q_M \geq \pi \end{cases}, \tag{9}$$

where $k^2 Q_M$ represents the figure of merit given by the product of the mechanical quality factor Q_M:

$$Q_M = \frac{\sqrt{K_D M}}{C}, \tag{10}$$

representing the amount of mechanical energy that can be converted, by the squared coupling coefficient k^2:

$$k^2 = \frac{\alpha^2}{C_0 K_D}, \tag{11}$$

which gives the part of mechanical energy that can effectively be converted into electrical energy.

3.1 Parallel SSHI

The principles of parallel SSHI (Guyomar *et al.*, 2005) consist of inverting the voltage of the piezoelectric element when the velocity cancels. Hence, such an operation leads to three steps in the conversion and harvesting process (Figure 5):

1. Open-circuit phase (step (1) in Figure 5)
2. Harvesting phase (step (2) in Figure 5)
3. Inversion phase (step (3) in Figure 5)

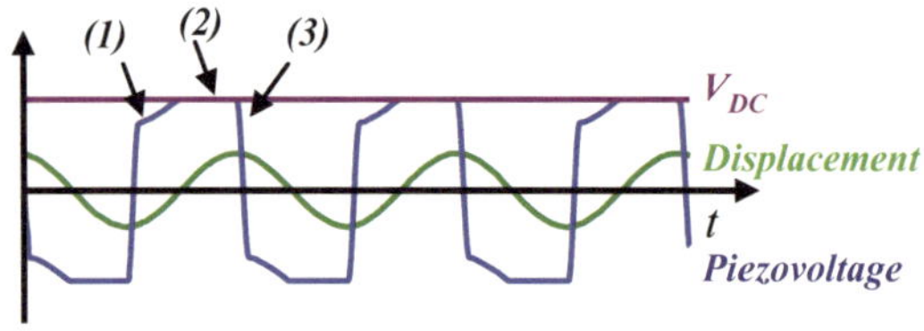

Fig. 5. Parallel SSHI waveforms

The energy harvested using the parallel SSHI approach over a single scavenging cyle may be expressed by:

$$E_{\mathrm{pSSHI}} = \int_{t_1}^{t_1+\tau} V_{DC} I dt, \tag{12}$$

where V_{DC} refers to the rectified voltage (assumed constant as the time constant $R_L C_0$ is far greater than half a vibration period $T/2$) and I the current flowing from the piezoelectric element to the storage stage. t_1 and $t_1 + \tau$ respectively refer to the time when the harvesting process starts (absolute piezovoltage equals to the rectified voltage) and stops (current cancellation, occurring coincidentally with displacement minimum or maximum values). From the electrical equation of Eq. (1), the harvested energy yields:

$$E_{\mathrm{pSSHI}} = \alpha V_{DC} \left(u_M - u_1 \right), \tag{13}$$

with u_1 and u_M the displacement value when the rectifier starts conducting and displacement magnitude, respectively. The value of u_1 may be found by integrating the current equation during the open circuit phase ($I = 0$), and considering that the voltage varies from γV_{DC} (which corresponds to a displacement $-u_M$) to V_{DC} (Figure 5):

$$u_1 = \frac{C_0}{\alpha} \left(1 - \gamma \right) V_{DC} - u_M, \tag{14}$$

leading to the expression of the harvested power as a function of the rectified voltage and displacement magnitude:

$$P_{\mathrm{pSSHI}} = 2 f_0 E_{\mathrm{pSSHI}} = 2 f_0 V_{DC} \left(2 \alpha u_M - C_0 \left(1 - \gamma \right) V_{DC} \right). \tag{15}$$

Noting that the power may also be given by $P = {V_{DC}}^2 / R_L$, the harvested power may also be expressed using the load value:

$$P_{\mathrm{pSSHI}} = \frac{\left(4 f_0 \alpha \right)^2 R_L}{\left(1 + 2 \left(1 - \gamma \right) R_L C_0 f_0 \right)^2} {u_M}^2. \tag{16}$$

Nevertheless, converting mechanical energy into electricity leads to a reduction of the vibrations. Because of this damping effect, less energy is available from the source when the system is driven by a constant force magnitude. In this case, the energy analysis of the equation of motion allows expressing the displacement magnitude in steady state case, assuming an excitation at the resonance frequency:

$$\int_{t_0}^{t_0+T/2} F \dot{u} dt = C \int_{t_0}^{t_0+T/2} \dot{u}^2 dt + \frac{1}{2} C_0 \left(1 - \gamma^2 \right) {V_{DC}}^2 + \frac{T}{2} \frac{{V_{DC}}^2}{R_L}, \tag{17}$$

where the left side member is the provided energy, and the right side members the dissipated energy (through mechanical losses), energy lost in the switching circuit, and harvested energy.

Assuming that the system features relatively high mechanical quality factor ($Q_M > 10$), the velocity and force may be considered in phase at the resonance, yielding the displacement magnitude:

$$u_M\big|_{\text{pSSHI}} = \frac{F_M}{2\pi C f_0 + \frac{16\alpha^2 R_L f_0 (R_L C_0 f_0 (1-\gamma^2)+1)}{\pi(1+2R_L C_0 f_0(1-\gamma))^2}}, \tag{18}$$

The maximal power harvested taking into account the damping effect may also be approximated as a function of the figure of merit $k^2 Q_M$ as (Guyomar *et al.*, 2009):

$$P_{\text{pSSHI}}\Big|_{\max} \approx \frac{k^2 Q_M}{\pi(1-\gamma) + 8k^2 Q_M} \frac{{F_M}^2}{C}. \tag{19}$$

3.2 Series SSHI

The principles of operation of the series SSHI (Taylor *et al.*, 2001; Lefeuvre *et al.*, 2006a) are a little bit different than in the case of the parallel SSHI. Actually in the case of the series SSHI, the harvesting process occurs at the same time than the inversion process (Figure 6), this latter being done with respect to $+V_{DC}$ (switching from positive voltage) or $-V_{DC}$ (switching from negative voltage).

Therefore the energy harvested over a single switching process yields:

$$E_{\text{sSSHI}} = C_0 V_{DC} (V_M + V_m), \tag{20}$$

with V_M and V_m the absolute values of the voltage just before and after the switching process (Figure 6), whose values may be found considering the inversion process (with respect to V_{DC}):

$$V_m + V_{DC} = \gamma (V_M - V_{DC}), \tag{21}$$

as well as the open-circuit stage between two switching events:

$$V_M - V_m = 2\frac{\alpha}{C_0} u_M. \tag{22}$$

Hence, from Eqs. (20), (21) and (22), the harvested power using the series SSHI approach yields:

$$P_{\text{sSSHI}} = 4\frac{1+\gamma}{1-\gamma} (\alpha u_M - C_0 V_{DC}), \tag{23}$$

which can also be expressed as a function of the load:

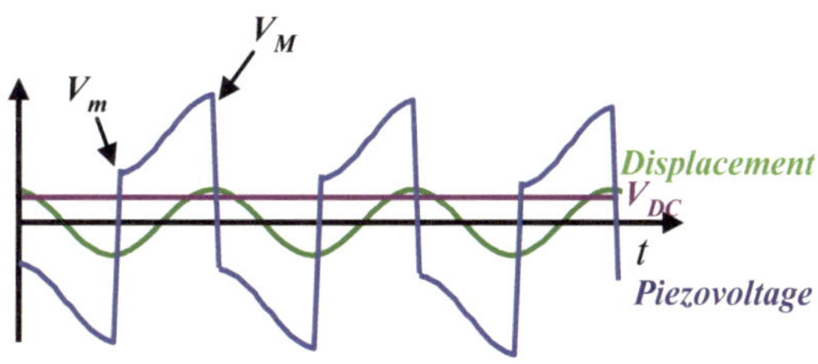

Fig. 6. Series SSHI waveforms

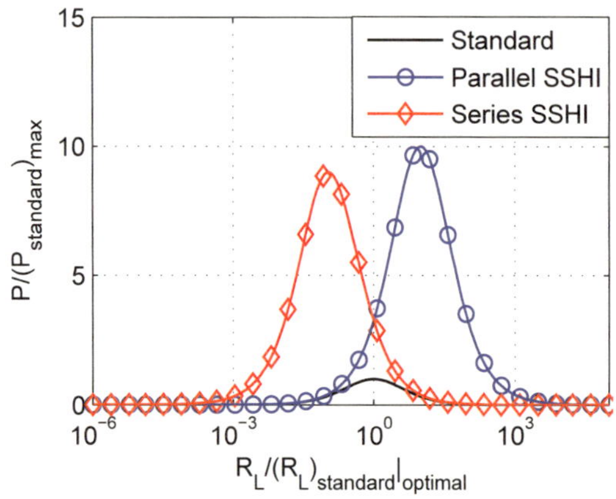

Fig. 7. Normalized harvested power of SSH techniques under constant displacement magnitude and comparison with standard interface ($\gamma = 0.8$)

$$P_{\text{sSSHI}} = \frac{\left(4\left(1+\gamma\right)\alpha f_0\right)^2 R_L}{\left(\left(1-\gamma\right)+4\left(1+\gamma\right)R_L C_0 f_0\right)^2} {u_M}^2. \tag{24}$$

In the same fashion that the standard and parallel SSHI cases, harvesting energy induces vibration damping effect, meaning that less mechanical energy is available for harvesting. From an energy analysis of the system, assuming the structure excited at its resonance frequency by a force with a constant magnitude F_M :

$$\int_{t_0}^{t_0+T/2} F\dot{u}dt = C\int_{t_0}^{t_0+T/2} \dot{u}^2 dt + \frac{1}{2}C_0\left(1-\gamma^2\right)\left(V_M - V_{DC}\right)^2 + \frac{T}{2}\frac{{V_{DC}}^2}{R_L}, \tag{25}$$

it is possible to derive the displacement magnitude taking into account this damping effect:

$$u_M|_{\text{sSSHI}} = \frac{F_M}{2\pi C f_0 + \frac{4\alpha^2(1+\gamma)}{\pi C_0((1-\gamma)+4(1+\gamma)R_L C_0 f_0)}}. \tag{26}$$

It can also be noted that the effect of the series SSHI can also be seen as the semi-active SSDV[4] damping approach (Badel *et al.*, 2006b; Lefeuvre *et al.*, 2006b), but with a negative voltage. It can besides be shown that the maximal harvested power at the resonance when considering the damping effect may be approximated by (Guyomar *et al.*, 2009):

$$P_{\text{sSSHI}}|_{\max} \approx \frac{k^2 Q_M}{2\pi\frac{1-\gamma}{1+\gamma} + 8k^2 Q_M}\frac{{F_M}^2}{C}. \tag{27}$$

3.3 Discussion

The performance of the SSHI techniques, along with the comparison with the standard interface, are depicted in Figure 7, considering that the electromechanical structure features harmonic displacement with a constant amplitude (*i.e.*, no damping effect). In order to make this chart as independent as possible from the device's parameters, it has been normalized along the x-axis according to the optimal load value in the standard case:

[4] *Synchronized Switch Damping on Voltage source*

$$(R_L)_{\text{standard}}|_{\text{optimal}} = \frac{1}{4 f_0 C_0}, \tag{28}$$

and along the y-axis according to the maximal power harvested using the standard interface Eq. (7). Therefore, this figure only depends on the inversion coefficient γ that has been set to 0.8; its typical value being comprised between 0.6 and 0.9.

Figure 7 clearly demonstrates the abilities of the SSHI approaches for greatly increasing the power output abilities of the microgenerator, by a typical factor of 9 for the parallel SSHI and a bit less for the series SSHI (8), thanks to the conversion enhancement offered by the switching process[5]. Actually, it can be demonstrated that the converted energy is actually up to 20 times higher than the converted energy in the standard case at the SSHI optimal loads, but the losses in the inversion circuit leads to an efficiency of 50% between the extraction and harvesting stages (Guyomar *et al.*, 2009).

It can also be noted that the optimal load in the parallel SSHI is higher than the optimal load in the standard case, has the nonlinear treatment leads to an artificial decrease of the piezoelectric capacitance value, while the series SSHI features an optimal load less than in the case of the standard technique, which may be beneficial as the capacitive behavior of piezoelectric elements leads to high optimal loads that may be difficult to interface with electronic components.

However, it has previously been pointed out that harvesting energy from a structure driven by a force of constant amplitude at the resonance frequency leads to a vibration damping effect that actually limits the input energy and thus the harvested energy. Figure 8 depicts the normalized harvested power considering such a damping effect, and shows the performance of the nonlinear approaches for harvesting the same amount of energy than in the standard case with much less piezoelectric materials (represented by a lower value of $k^2 Q_M$). However, for highly coupled, weakly damped structures, there is no significant improvement of the nonlinear approaches when considering steady-state excitation and a power limit is reached:

$$P_{\text{lim}} = \frac{{F_M}^2}{8C}, \tag{29}$$

although the SSHI interfaces may be advantageous considering time-limited excitations (Badel *et al.*, 2005a; Lallart, Inman and Guyomar, 2010a). It can also finally be noted that realistic electromechanical structures usually have a value of the figure of merit $k^2 Q_M$ less than 1, and therefore can significantly benefit from the nonlinear approach.

It can also be noted from Figure 8 that the optimal loads for both the standard and SSHI approaches change as $k^2 Q_M$ increases. Two optimal loads appear for the standard interface after a critical value of $k^2 Q_M$ (π), while the optimal load of the parallel SSHI decreases and increases for the series SSHI to reduce the voltage and limit the damping effect, although it also decreases the energy conversion abilities.

Nevertheless, realistic excitations in most of the cases would be barely a sine, but more likely random (Halvorsen, 2008; Blystad, Halvorsen and Husa, 2010). In this case, it can be demonstrated that a trade-off exists between the number of switching events and the value of the voltage at the switching instants, as the extracted energy is proportional to:

$$E_{\text{extracted}} \propto C_0 \sum_k {V_k}^2 \tag{30}$$

[5]It can be shown that the gains of the nonlinear interfaces are given by $2/(1-\gamma)$ and $(1+\gamma)/(1-\gamma)$ for the parallel and series SSHI techniques, respectively.

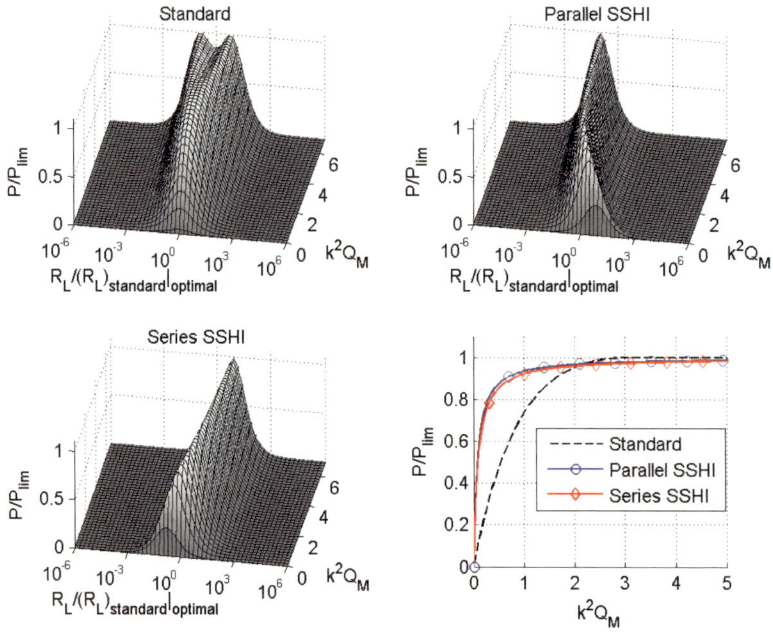

Fig. 8. Normalized harvested power and normalized maximal harvested power of SSH techniques under constant force magnitude and comparison with standard interface ($\gamma = 0.8$)

where V_k denotes the voltage value at the kth instant. This equation shows the trade-off between energy extraction and voltage increase through the cumulative process of the nonlinear technique. Hence, it is possible to improve the SSHI performances in random vibrations by disabling the switch when the voltage or displacement value is less than a user-specified threshold (Guyomar and Badel, 2006; Guyomar, Richard and Mohammadi, 2007b; Lallart *et al.*, 2008b).

4. Charge extraction techniques

The previous section presented the direct application of the nonlinear technique for conversion enhancement to energy harvesting, by connecting the switching element either in parallel (section 3.1) or in series (section 3.2) with the harvesting stage. However, is spite of a great increase of the harvested power, such approaches still suffer from load-dependent power. Hence, the aim of this section is to expose a modified approach still based on nonlinear treatment that not only allows a power output increase (less than the SSHI approaches however), but also a harvested power independent from the load thanks to a decoupling of the extraction stage from the storage, which permits bypassing a supplementary adaptation stage (Ottman *et al.*, 2002; Ottman, Hofmann and Lesieutre, 2003; Han *et al.*, 2004; Lefeuvre *et al.*, 2007a; Lallart and Inman, 2010b) that may dramatically decrease the power because of losses[6].

4.1 SECE technique

The principles of the SECE (*Synchronized Electric Charge Extraction* - Lefeuvre *et al.* (2005)), depicted in Figure 9, consists of extracting all the electrostatic energy available on the

[6]The efficiency of such interface is usually comprised between 70% and 90%.

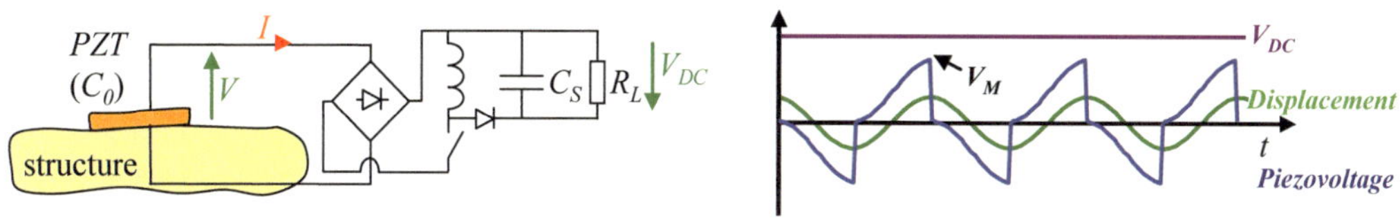

Fig. 9. SECE technique and waveforms

piezoelement when this latter is maximum (otherwise the material is left in open-circuit conditions), which corresponds to minimum and maximum voltages (or equivalently displacement). The extracted energy is then transferred from its electrostatic form into electromagnetic form to an inductance L. After this extraction process, the switch is open and the energy stored in the inductance is transferred to the smoothing capacitor C_S and load R_L. However, because of the losses (mainly in the inductor), a part of the energy is lost. Hence, for one cycle, the energy extracted is given by:

$$E_{\text{SECE}} = \frac{1}{2} C_0 {V_M}^2, \tag{31}$$

with V_M the piezovoltage just before the harvesting process, whose value can be found considering the open-circuit stage such as:

$$V_M = 2\frac{\alpha}{C_0} u_M. \tag{32}$$

Hence, the harvested power by the SECE technique yields:

$$P_{\text{SECE}} = \gamma_C f_0 \frac{\alpha^2}{C_0} {u_M}^2, \tag{33}$$

with γ_C the efficiency of energy transfer and extraction.
In a purely mechanical point of view, the SECE technique is equivalent to a dry friction, and more particularly to the semi-passive SSDS[7] technique (Badel *et al.*, 2006b). Hence, the damping effect induced by the harvesting process leads to the expression of the displacement magnitude u_M at the resonance frequency:

$$u_M|_{\text{SECE}} = \frac{F_M}{2\pi C f_0 + \frac{4}{\pi}\frac{\alpha^2}{C_0}}, \tag{34}$$

and the associated maximal power at the resonance taking into account the damping effect yields:

$$P_{\text{sSSHI}}|_{\max} = \gamma_C \frac{2}{\pi} \frac{k^2 Q_M}{\left(1 + \frac{4}{\pi} k^2 Q_M\right)^2} \frac{{F_M}^2}{C}. \tag{35}$$

4.2 DSSH technique

The main drawback of the SECE technique lies in the fact that the extraction process cannot be controlled; only all the energy can be extracted. Such a process therefore limits the voltage increase process (as no inversion is performed), hence limiting the conversion enhancement and therefore the harvested energy.

[7] *Synchronized Switch Damping on Short-circuit*

To be able to control the trade-off between extracted energy and voltage increase, as well as the trade-off between energy extraction and damping effect (*i.e.*, the balance between mechanical energy and conversion abilities), it is proposed in this section to use a combination of the series SSHI technique and SECE approach. This concept, called DSSH for *Double Synchronized Switch Harvesting* (Figure 10 - Lallart *et al.* (2008c)), lies in extracting a part of the energy (and use the remaining for processing the voltage inversion) on an intermediate capacitance C_{int}, and then transferring all the energy on C_{int} to the inductance, and finally to the storage stage. When using the DSSH technique, it can be demonstrated that the value of transferred energy to the intermediate capacitor yields (Lallart *et al.*, 2008c):

$$E_{\mathrm{DSSH}}|_{int} = 2x\left(\frac{1+\gamma}{2+(1-\gamma)x}\right)^2 \frac{\alpha^2}{C_0} {u_M}^2, \tag{36}$$

with x the ratio of the intermediate capacitance over the piezocapacitance ($x = C_{int}/C_0$), leading to the expression of the harvested power:

$$P_{\mathrm{DSSH}} = 4f_0 x \gamma_C \left(\frac{1+\gamma}{2+(1-\gamma)x}\right)^2 \frac{\alpha^2}{C_0} {u_M}^2. \tag{37}$$

From Eq. (37), it can be shown that an optimal value of x that maximizes the harvested power under constant vibration magnitude u_M exists ($x_{opt} = 2/(1-\gamma)$), leading to the expression of the maximal power when no damping effect is considered:

$$P_{\mathrm{DSSH}}|_{\max}^{u_M} = f_0 \gamma_C \frac{1}{2} \frac{(1+\gamma)^2}{1-\gamma} \frac{\alpha^2}{C_0} {u_M}^2. \tag{38}$$

Another advantage of the DSSH is its ability to control the trade-off between converted energy and damping effect as well. Hence, thanks to its ability to control the amount of extracted energy through the intermediate electrostatic energy tank (intermediate capacitor), the DSSH technique is able to let mechanical energy entering into the system, contrary to fixed system, which, in spite of increasing the conversion abilities of materials, drastically limit the mechanical energy in the system, leading to moderate harvested energy.
In this case, the harvested energy may be expressed as (Lallart *et al.*, 2008c):

$$P_{\mathrm{DSSH}} = \gamma_C \frac{2}{\pi} k^2 Q_M \frac{1+\Gamma}{1-\Gamma} \left(\frac{1}{1+\frac{4}{\pi}\frac{1+\Gamma}{1-\Gamma}k^2 Q_M}\right)^2 \frac{{F_M}^2}{C} \text{ with } \Gamma = -\frac{1-x\gamma}{1+x}. \tag{39}$$

Hence, the trade-off between energy conversion enhancement and input mechanical energy can be tuned through the intermediate capacitor. In particular, an optimal value of x exists that leads to the maximal harvested power at the resonance taking into account the damping effect:

$$\begin{cases} P_{\mathrm{DSSH}}|_{\max} = \gamma_C \frac{2\pi k^2 Q_M (1-\gamma^2)}{(\pi(1-\gamma)+4k^2 Q_M (1+\gamma))^2} \frac{{F_M}^2}{C} & \text{for } k^2 Q_M \le \frac{\pi}{4}\frac{1-\gamma}{1+\gamma} \quad (x_{\mathrm{opt}} = \infty) \\ P_{\mathrm{DSSH}}|_{\max} = \gamma_C \frac{{F_M}^2}{8C} \quad \text{for } k^2 Q_M \le \frac{\pi}{4}\frac{1-\gamma}{1+\gamma} & \left(x_{\mathrm{opt}} = \frac{2\pi}{\pi(1-\gamma)+4k^2 Q_M (1+\gamma)}\right) \end{cases}. \tag{40}$$

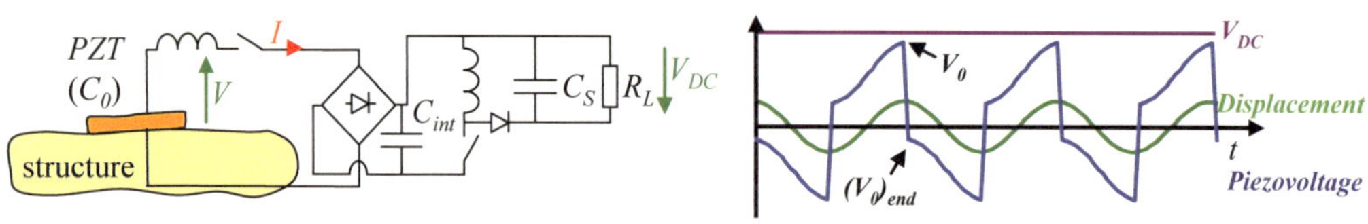

Fig. 10. DSSH technique and waveforms

4.3 Discussion

The performance of the SECE and DSSH techniques considering a monochromatic displacement with a constant amplitude is depicted in Figure 11, which has been normalized in the same way than previously (so that it only depends on the energy transfer efficiency γ_C), with the value of the energy transfer efficiency being given as $\gamma_C = 0.9$. This figure demonstrates that the SECE and DSSH technique allows harvesting 3.5 to 7.5 times more energy than in the standard case, but, in addition to this power output increase, the most remarkable property of these techniques is their independency to the load, which actually leads to performance similar to the SSHI techniques combined with load adaptation interfaces that features typical efficiency of $70 - 90\%$ (Ottman *et al.*, 2002; Ottman, Hofmann and Lesieutre, 2003; Han *et al.*, 2004; Lefeuvre *et al.*, 2007a; Lallart and Inman, 2010b).

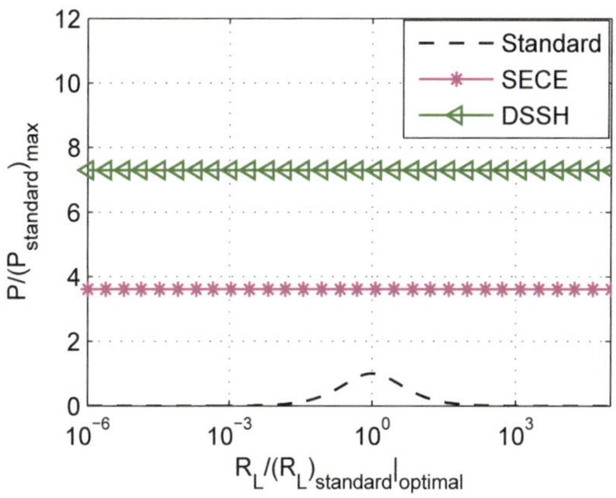

Fig. 11. Normalized harvested power of SECE and DSSH techniques under constant displacement magnitude and comparison with standard interface ($\gamma_C = 0.9$)

When considering the damping effect induced by the energy conversion process, the harvested energy normalized with the power limit (Eq. (29)) is depicted in Figure 12. This chart shows that the SECE is a little bit more efficient than the SSHI techniques for low coupled or highly damped systems (low $k^2 Q_M$), but the power output of this approach is decreasing after reaching an optimal value for large values of $k^2 Q_M$, because the damping effect becomes much larger. This is not the case of the DSSH technique as such a technique allows controlling the trade-off between energy extraction and damping effect through the intermediate capacitance. Such a control also allows the DSSH technique to harvest much more energy for low value of the figure of merit, allowing using up to 10 times less piezoelectric material than the standard interface for realistic values of $k^2 Q_M$. However, although the efficiency of the SECE and DSSH techniques is higher than the SSHI interfaces (Guyomar *et al.*, 2009), the maximal power is less than the power limit because of the energy transfer stage[8].

The SECE and DSSH interface are also very well adapted to multimodal excitation because of the load independency, contrary to the SSHI and standard techniques whose optimal loads depend on the frequency (Lefeuvre *et al.*, 2007b). The SECE technique exhibits relatively

[8]In the case of the standard and SSHI cases, the maximal power would actually be less as well if a load adaptation interface is used to maximize the power.

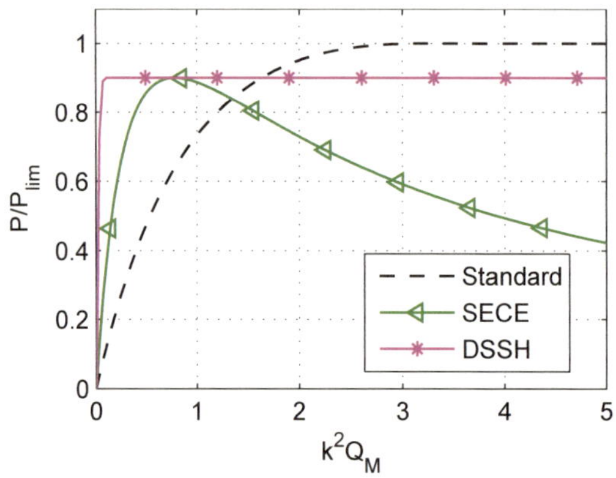

Fig. 12. Normalized harvested power of SECE and DSSH techniques under constant force magnitude and comparison with maximal power of the standard interface ($\gamma_C = 0.9$)

good performances under broadband excitation as well, as this technique relies on voltage cancellation rather than voltage inversion (so that no cumulative process appears).

5. Conversion enhancement and energy harvesting using bidirectional energy transfer and pulsed energy injection

The methods exposed so far considered that once the energy is transferred to the source, it cannot go backward. In this section another concept based on an energy feedback from the storage stage to the source is exposed (Lallart and Guyomar, 2010c). The principles of this method start from the observation that, considering a single energy conversion process, more energy can be converted if an initial energy is given to the system:

$$E_{\text{conv}} = \frac{1}{2}C_0 \left(V_{\text{conv}} + V_{\text{init}}\right)^2 = \frac{1}{2}C_0 {V_{\text{conv}}}^2 + \frac{1}{2}C_0 {V_{\text{init}}}^2 + C_0 V_{\text{conv}} V_{\text{init}}, \tag{41}$$

where V_{conv} is the voltage induced by the conversion process and V_{init} the initial voltage given to the material. In Eq. (41), the first two terms of the right side member respectively correspond to the converted energy without any initial voltage, and initial energy given to the material. Hence, thanks to the quadratic dependence of the energy with the voltage, providing initial energy allows an energy gain given by the cross-product of the two voltage terms, multiplied by the capacitance.
The operations of the energy injection system are as follows (Figure 13):

(1) Energy extraction using SECE technique

(2) Energy injection from the storage stage to the piezoelectric element

(3) Open-circuit

Hence, the energy extracted for a single cycle is given as:

$$E_{\text{extr}} = \frac{1}{2}\gamma_C C_0 {V_M}^2, \tag{42}$$

where V_M is the absolute voltage value when the energy harvesting process is engaged (maximal voltage value) and γ_C the energy transfer efficiency. After this harvesting event, energy is provided from the storage capacitance to the piezoelectric element. In order to reduce the losses, the energy injection is done through an inductor, leading to the value of the piezoelectric voltage after the process:

$$V_{inj} = (1+\gamma)\,V_{DC}, \tag{43}$$

with V_{DC} the value of the rectified voltage. Hence, the energy extracted from the source is given by:

$$E_{source} = (1+\gamma)\,C_0 {V_{DC}}^2. \tag{44}$$

As the piezoelectric element is left in open-circuit condition after the energy injection process, it is therefore possible to derive the value of the voltage V_M as:

$$V_M = (1+\gamma)\,V_{DC} + 2\frac{\alpha}{C_0}u_M, \tag{45}$$

with u_M the displacement magnitude. Hence, the global harvested power using such a technique yields:

$$\begin{aligned} P_{inj} &= 2f_0\,(E_{extr} - E_{source}) \\ &= f_0\left[4\gamma_C\tfrac{\alpha^2}{C_0}{u_M}^2 + 4\gamma_C\alpha\,(1+\gamma)\,u_M V_{DC} + (\gamma_C\,(1+\gamma)-2)\,(1+\gamma)C_0{V_{DC}}^2\right], \end{aligned} \tag{46}$$

which can also be expressed as a function of the load R_L:

$$P_{inj} = 4f_0\gamma_C\left[\frac{(1+\gamma)\sqrt{\gamma_C R_L C_0 f_0} + \sqrt{2(1+\gamma)R_L C_0 f_0 + 1}}{(2-(1+\gamma)\gamma_C)\,(1+\gamma)R_L C_0 f_0 + 1}\right]^2 \frac{\alpha^2}{C_0}{u_M}^2. \tag{47}$$

Considering that the system is excited at its resonance frequency by a driving force of constant amplitude, the damping effect may be taken into account by considering the mechanical effect similar to the one obtained when using the semi-active SSDV damping approach (Badel *et al.*, 2006b; Lefeuvre *et al.*, 2006b), but with a negative voltage, leading to the expression of the displacement magnitude:

$$u_M|_{inj} = \left[\frac{1}{1+\frac{4}{\pi}\left(1+2\frac{\gamma_C R_L C_0 f_0(1+\gamma)+\sqrt{\gamma_C R_L C_0 f_0(2R_L C_0 f_0(1+\gamma)+1)}}{1+R_L C_0 f_0(2-(1+\gamma)\gamma_C)(1+\gamma)}\right)k^2 Q_M}\right]\frac{F_M}{2\pi C f_0}, \tag{48}$$

yielding the harvested power:

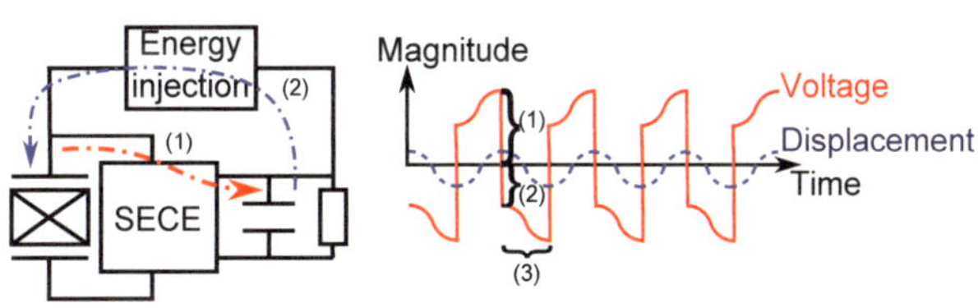

Fig. 13. Bidirectional pulsed energy harvesting principles

$$P_{\text{inj}} = \gamma_C k^2 Q_M \frac{2}{\pi}\frac{F_M{}^2}{C}\left[\frac{(1+\gamma)\sqrt{\gamma_C R_L C_0 f_0}+\sqrt{2(1+\gamma)R_L C_0 f_0+1}}{(2-(1+\gamma)\gamma_C)(1+\gamma)R_L C_0 f_0+1}\right]^2$$
$$\times\left[\frac{1}{1+\frac{4}{\pi}\left(1+2\frac{\gamma_C R_L C_0 f_0(1+\gamma)+\sqrt{\gamma_C R_L C_0 f_0\left(2R_L C_0 f_0(1+\gamma)+1\right)}}{1+R_L C_0 f_0\left(2-(1+\gamma)\gamma_C\right)(1+\gamma)}\right)k^2 Q_M}\right]^2. \tag{49}$$

Figure 14 depicts the harvested power considering constant monochromatic displacement magnitude, also normalized along the x-axis with respect to the optimal load in the standard case and along the y-axis according to the maximal harvested power in the standard case, so that the chart only depends on the energy transfer efficiency γ_C and energy injection coefficient γ (respectively set to 0.9 and 0.8).

This figure clearly demonstrates the performance of the bidirectional pulsed energy extraction and injection in terms of energy harvesting, allowing a gain in terms of power output of 20 using typical components (40 using low-losses devices). Such energy scavenging abilities can be explained by a particular "energy resonance" effect, which comes from the fact that if more power is harvested, this leads to a greater injected energy, which also leads to a higher harvested power according to Eq. (41) and so on, leading to outstanding power output. It can also be noted that for low values of the load, the technique performs as the SECE technique, as almost no energy is injected to the system (the rectified voltage being low), while the power tends to zero for high load values, as higher voltages lead to higher losses.

Another remarkable property of the this technique is the fact that, when considering the damping effect introduced by the scavenging process (Figure 15), the energy extraction/injection concept allows bypassing the output power limit that is common to all the previously exposed energy harvesting approach.

Because of the voltage cancellation process, the performance of the energy extraction/injection injection technique is not significantly compromised in the case of random excitation. However, the load-dependency of the harvested power may alter the output power if the structure features several modes.

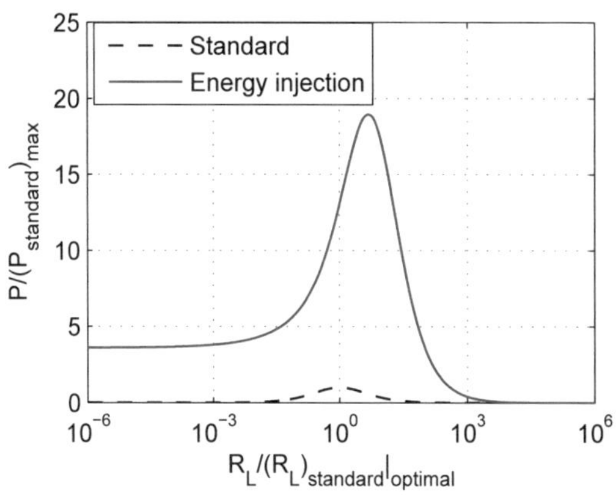

Fig. 14. Normalized harvested power of energy injection technique under constant displacement magnitude and comparison with standard interface ($\gamma = 0.8$, $\gamma_C = 0.9$)

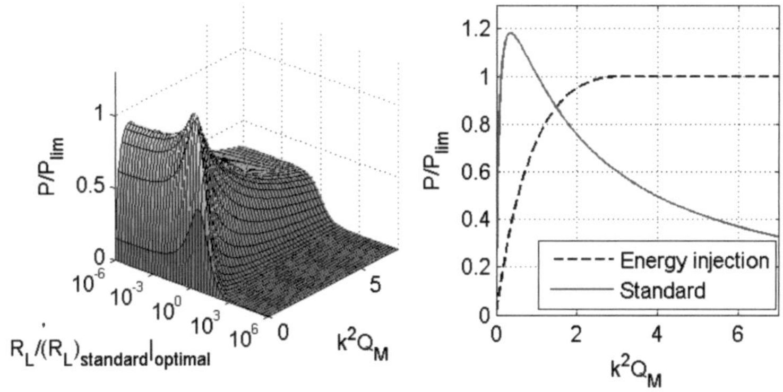

Fig. 15. Normalized harvested power and maximal normalized harvested power of energy injection technique under constant force magnitude and comparison with standard interface ($\gamma = 0.8$, $\gamma_C = 0.9$)

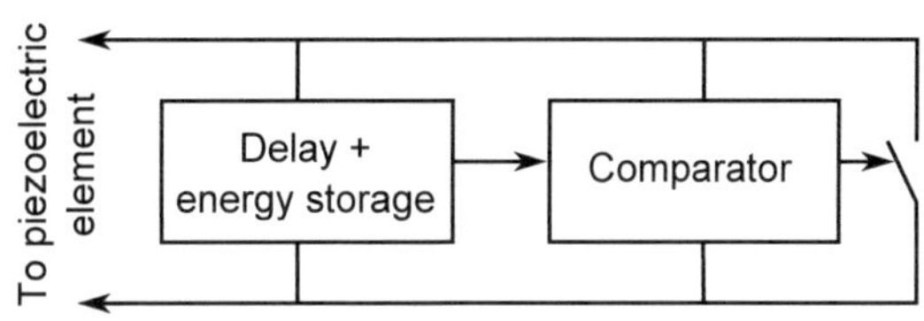

Fig. 16. Principles of the self-powered switch

6. Implementation issues

Through this chapter, it has been demonstrated that using nonlinear treatments for energy harvesting purposes allows a significant increase of the performance of vibration-based microgenerators. However, the nonlinear process may seem to be delicate to implement for realistic applications. Nevertheless, the implementation of the SSHI may be done in an easy and energy-efficient way, based on the detection of maximum values using the delayed version of the voltage (the maximum value is reached when the delayed signal becomes greater then the original piezovoltage) as depicted in Figure 16, generating a pulsed voltage that drives a transistor acting as the digital switching (Richard, Guyomar and Lefeuvre, 2007; Lallart *et al.*, 2008b). Hence, such a process can be made truly self-powered using widely available components and may be easily integrated. It besides consumes very little energy, typically 3% of the electrostatic energy available on the piezoelectric element, hence not compromising the energy harvesting enhancement offered by these techniques. The implementation of the other techniques may also be derived from this concept (Badel, 2005b; Lallart, 2008d; 2010d).
Another concerns about the implementation of piezo-based vibration harvesters is the challenge concerning low-voltage transducers, as piezoelectric elements present their most promising application field in small-scale size. However when dealing with such systems (such as MEMS[9]), the output voltage of the active material is usually very low and cannot bypass the discrete component voltage gap, leading to poor performance in terms of energy generation.

[9]*Micro Electro-Mechanical Systems*

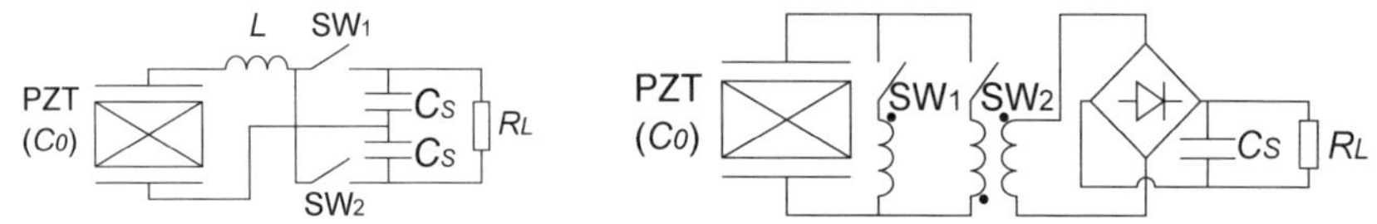

(a) Diodeless series SSHI

(b) SSHI-MR

Fig. 17. Interfaces for low voltage harvesting

In order to counteract this problem, it is possible to take benefit of the nonlinear energy harvesting interfaces (Makihara, Onoda and Miyakawa, 2006; Lallart and Guyomar, 2008e). In particular, the series SSHI approach is the most flexible to be adapted to low-voltage systems. For example, the rectifier bridge may be replaced by the switching elements (Figure 17(a)), allowing the removal of the diodes. Another approach consists of replacing the switching inductance by a transformer (Figure 17(b)), leading to the concept of SSHI-MR[10] (Garbuio *et al.*, 2009), which also presents the advantage of having a higher optimal load and therefore delivers voltage levels that are compatible with electronic systems when the electromechanical structure delivers low voltage levels. Because of the load decoupling offered by the use of the transformer, the SSHI-MR technique may be combined with the parallel energy harvesting system, leading to the concept of hybrid energy harvesting (Lallart, 2008d; 2010d), which allows a decreased sensitivity to load shifts.

7. Application to thermal energy harvesting through pyroelectric effect

While the previous development have been done considering vibration energy harvesting through piezoelectric coupling, it is also possible to apply the exposed approaches to other conversion effects, as the principles of the nonlinear treatment is independent from the energy conversion mechanisms (*e.g.*, electromagnetism[11] - Lallart *et al.* (2008f)). In this section, a particular attention is placed on pyroelectric devices that are able to convert temperature variation into electricity, as these materials behave in a similar fashion than piezoelectric elements. Hence, it is possible to apply the proposed concepts to energy harvesting from temperature time-domain variations using pyroelectric inserts.

Although pyroelectric materials feature low coupling coefficients, the source presents much higher energy than mechanical vibrations. Hence, in terms of energy density, pyroelectric elements present similar energy densities than piezoelectric materials (Table 1), as the low coupling coefficient is compensated by the high input energy levels.

However, contrary to mechanical energy harvesting, thermal devices do no present any resonance effect. Combined with the low coupling coefficient of pyroelectric materials, this leads to the observation that the harvesting process does not induce a significant cooling, and hence does not significantly modify the input energy source.

From the constitutive pyroelectric equations:

$$\left\{ \begin{array}{l} \Delta D = \epsilon_{33}^{\theta} \Delta E + p \Delta\theta \\ \Delta\sigma = p \Delta E + c \frac{\Delta\theta}{\theta_0} \end{array} \right. , \tag{50}$$

where θ, θ_0 and σ respectively refer to the absolute and mean temperatures in Kelvin and entropy of the system, ϵ_{33}^{θ}, p and c represent the permittivity under constant temperature,

[10] *Synchronized Switch Harvesting on Inductor with Magnetic Rectifier*

[11] In this case, the working electrical quantity is the current rather than the voltage.

	Piezoelectricity	Pyroelectricity
Material	Hard ceramic NAVY-III type ($Q\&S\ P1-89$)	PVDF films
Multiphysic coupling coefficient	$e_{33} = -12.79\ C.m^{-1}$	$p = -24.10^{-6}\ C.m^{-2}.K^{-1}$
Permittivity	$\epsilon^{S}{}_{33}/\epsilon_0 = 668$	$\epsilon^{\theta}/\epsilon_0 = 12$
Typical variation of the associated physical quantity	$S_M = 10\ \mu m.m^{-1}$	$\theta_M = 1\ K$
Associated electrostatic energy	$(W_{el})_{\text{piezo}} = 1.4\ \mu J.cm^{-3}$	$(W_{el})_{\text{pyro}} = 2.7\ \mu J.cm^{-3}$

Table 1. Electrostatic energy comparison using piezoelectric or pyroelectric coupling

pyroelectric coefficient and heat capacity, and with Δ the difference operator, it is possible to get the macroscopic model of the pyroelectric coupling (only the equation of the current is given as no significant feedback occurs):

$$I = \alpha\dot{\theta} - C_0\dot{V} \text{ with } \begin{cases} C_0 = \frac{\epsilon^{\theta}_{33} S_0}{l} \\ \alpha = -pS_0 \end{cases}, \tag{51}$$

with S_0 and l the surface and thickness of the material, respectively.
Hence, the output power of each technique would be the same than the previously exposed ones[12], except that the displacement magnitude u_M would be replaced by the temperature variation magnitude θ_M.
Another specificity of thermal energy harvesting from temperature time-domain variations is the low frequency of the system (less than 1 Hz typically), which leads to a decreased value of the inversion coefficient, decreasing the gain of the nonlinear techniques by a typical factor 2 approximately (Guyomar *et al.*, 2009).
This observation, combined with the fact that the system does not feature any resonance effect, shows the advantage of the SECE and DSSH techniques that offers an output power independent from the load, as low frequency variations lead to high optimal load values which can besides change easily due to the non-resonant nature of the device.

8. Conclusion

This chapter exposed the use of nonlinear treatments for energy harvesting enhancement. Thanks to the conversion magnification offered by the switching approach (allowing both a voltage increase and a reduction of the time shift between voltage and velocity), it has been demonstrated that the application of this concept to energy harvesting (SSHI) allows a

[12]Eq. (6) for the standard interface, Eqs. (16) and (24) for the parallel and series SSHI, Eqs. (33) and (37) for the SECE and DSSH approaches, and Eq. (47) for the pulsed energy injection/extraction technique.

significant gain in terms of harvested power (7-8 times greater than the standard interface) or significantly reduce the amount of piezoelectric material required to harvest a given amount of energy.

Then, the principles of a nonlinear pulsed energy extraction have been exposed, showing that such techniques not only still permits an enhancement of the output power (although they may not be as effective as SSHI approaches), but also the harvested energy is independent from the load. In particular, the use of an intermediate energy tank (DSSH) allows controlling the trade-off between energy conversion and damping effect, allowing a great reduction (typically by a factor of 10) of the required amount of piezoelement for the same amount of scavenged energy.

In a third step, the addition of a pulsed energy injection mechanism has been exposed. Thanks to the dependence of the available electrostatic energy with the squared voltage, it has been shown that providing initial energy to a piezoelectric material allows improving its conversion abilities. Applied to energy harvesting, such a process therefore allows an outstanding gain in terms of harvested power, particularly thanks to an "energy resonance" effect created by the feedback loop.

The realistic application of the switching process in a self-powered fashion has also been demonstrated, and it has been shown that such a process can simply be done by comparing the piezoelectric voltage with its delayed version, which can be implemented using widely available and embeddable components, leading to a low-consumption circuit that does not compromise the performance of the nonlinear techniques. The issue of low-voltage energy harvesting has also been discussed, as piezoelectric materials are promising for micro and nano-systems.

Finally, it has been shown that the nonlinear process may be applied to other conversion effects, as the concept is independent from the physical quantities. A particular emphasis has been placed on thermal harvesting from temperature time-domain variations using pyroelectric elements, as, although such elements feature low coupling coefficients, the source presents high energy levels. In this case, it has also been observed that the low-frequency, non-resonant nature of these systems makes the use of load-independent harvesting techniques (*e.g.*, SECE) a premium choice.

As a conclusion, Table 2 proposes a ranking of the techniques exposed in this chapter according to several criteria. Hence, it can be seen that when the system features monochromatic excitation, the use of the energy injection technique is of particular interest

Technique	Harvested energy: Constant displacement magnitude	Harvested energy: Constant force magnitude	Random Excitation	Low-voltage harvesting	Load Independency	Implementation easiness
Standard	☹	☹	☹	☹	☹	☺
Parallel SSHI	☺	😐	☹	☹	☹	😐
Series SSHI (diodeless)	☺	😐	☹	😐	☹	😐
SSHI-MR	☺	😐	☹	☺	☹	😐
Hybrid SSHI	☺	😐	😐	☺	😐	😐
SECE	😐	😐	☺	😐	☺	☹
DSSH	☺	☺	😐	😐	☺	☹
Energy injection	☺☺	☺	😐	😐	😐	☹

Table 2. Comparaison of exposed energy harvesting techniques

if no vibration damping effect appears, although its self-powered implementation is not as simple as the SSHI techniques and not as efficient than the DSSH approach when damping effect appears and if the electromechanical structure features a low value of the figure of merit $k^2 Q_M$. Finally, if the system features random excitations however, the use of load-independent techniques seems more adapted.

9. References

Anton, S. R. & Sodano H. A. (2007). A review of power harvesting using piezoelectric materials (2003-2006). *Smart Mater. Struct.*, Vol. 16(3), R1-R21.

Badel, A.; Guyomar, D.; Lefeuvre, E. & Richard, C. (2005a). Efficiency Enhancement of a Piezoelectric Energy Harvesting Device in Pulsed Operation by Synchronous Charge Inversion. *J. Intell. Mater. Syst. Struct.*, Vol. 16, 889-901.

Badel, A. (2005b). Récupération dénergie et contrôle vibratoire par éléments piézoélectriques suivant une approche non linéaire[13]. Ph.D. Thesis, *in french*.

Badel, A.; Benayad, A.; Lefeuvre, E.; Lebrun, L.; Richard, C. & Guyomar, D. (2006a). Single Crystals and Nonlinear Process for Outstanding Vibration Powered Electrical Generators. *IEEE Trans. on Ultrason., Ferroelect., Freq. Contr.*, Vol. 53, 673-684.

Badel, A.; Sébald, G.; Guyomar, D.; Lallart, M.; Lefeuvre, E.; Richard, C. & Qiu, J. (2006b). Piezoelectric vibration control by synchronized switching on adaptive voltage sources: Towards wideband semi-active damping. *J. Acoust. Soc. Am.*, Vol. 119, No. 5, 2815-2825.

Badel, A.; Lagache, M.; Guyomar, D.; Lefeuvre, E. & Richard, C. (2007). Finite Element and Simple Lumped Modeling for Flexural Nonlinear Semi-passive Damping. *J. Intell. Mater. Syst. Struct.*, Vol. 18, 727-742.

Beeby, S. P.; Torah, R. N.; Tudor, M. J.; Glynne-Jones, P.; ODonnell, T.; Saha, C. R. & Roy, S. (2007). A micro electromagnetic generator for vibration energy harvesting, *J. Micromech. Microeng.*, Vol. 17, 12571265.

Blystad, L.-C. J.; Halvorsen, E. & Husa, S. (2008). Simulation of a MEMS Piezoelectric Energy Harvester Including Power Conditioning and Mechanical Stoppers. *Technical Digest, PowerMEMS 2008*, Sendai, Japan, November 2008, 237-240.

Blystad, L.-C. J.; Halvorsen, E. & Husa, S. (2010). Piezoelectric MEMS energy harvesting systems driven by harmonic and random vibrations. *IEEE Trans. Ultrason., Ferroelect., Freq. Contr.*, Vol. 57(4), 908-919.

Erturk A. & D. J. Inman (2008). Issues in mathematical modeling of piezoelectric energy harvesters. *Smart Mater. Struct. 17*, paper # 065016.

Garbuio, L.; Lallart, M.; Guyomar, D. & Richard, C. (2009). Mechanical Energy Harvester with Ultra-Low Threshold Rectification Based on SSHI Non-Linear Technique, *IEEE Trans. Indus. Elec.*, Vol. 56(4), 048-1056.

Guyomar, D.; Badel, A.; Lefeuvre, E. & Richard, C. (2005). Towards energy harvesting using active materials and conversion improvement by nonlinear processing, *IEEE Trans. Ultrason., Ferroelect., Freq. Contr.*, Vol. 52, 584-595.

Guyomar, D. & Badel, A. (2006). Nonlinear semi-passive multimodal vibration damping: An efficient probabilistic approach. *J. Sound Vib.*, Vol. 294, 249-268.

Guyomar, D.; Jayet, Y.; Petit, L.; Lefeuvre, E.; Monnier, T.; Richard, C. & Lallart, M. (2007a) Synchronized Switch Harvesting applied to Self-Powered Smart Systems :

[13] *Energy harvesting and vibration control using piezoelectric elements and a nonlinear approach*

Piezoactive Microgenerators for Autonomous Wireless Transmitters, *Sens. Act. A: Phys.*, Vol. 138, No. 1, 151-160. doi : 10.1016/j.sna.2007.04.009

Guyomar, D.; Richard, C. & Mohammadi, S. (2007b). Semi-passive random vibration control based on statistics. *J. Sound Vib.*, Vol. 307, 818-833.

Guyomar, D.; Sébald, G.; Pruvost, S.; Lallart, M.; Khodayari, A. & Richard, C. (2009). Energy Harvesting From Ambient Vibrations and Heat, *J. Intell. Mater. Syst. Struct.*, Vol. 20(5), 609-624.

Halvorsen, E. (2008). Energy Harvesters Driven by Broadband Random Vibrations. *J. Microelectromech. Syst.*, Vol. 17(5), 1061-1071.

Hamakawa, Y. (2003). 30 years trajectory of a solar photovoltaic research. In *3rd World Conference on Photovoltaic Energy Conversion*.

Han, J.; Von-Jouanne,A.; Le, T.; Mayaram, K. & Fiez, T. S. (2004). Novel power conditioning circuits for piezoelectric micro power generators. In *Proc. IEEE Appl. Power Electron. Conf. Expo. (APEC)*, vol. 3, 1541-1546.

Keawboonchuay , C. & Engel, T. G. (2003). Electrical power generation characteristics of piezoelectric generator under quasi-static and dynamic stress conditions. *IEEE Trans. Ultrason., Ferroelect., Freq. Contr.*, Vol. 50, 1377-1382.

Krikke, J. (2005). Sunrise for energy harvesting products. *IEEE Pervasive Comput.*, Vol. 4, 4-35.

Lallart, M.; Guyomar, D.; Jayet, Y.; Petit, L.; Lefeuvre, E.; Monnier, T.; Guy, P. & Richard, C. (2008a). Synchronized Switch Harvesting applied to Selfpowered Smart Systems: Piezoactive Microgenerators for Autonomous Wireless Receiver, *Sens. Act. A: Phys.*, Vol. 147, No. 1, 263-272. doi: 10.1016/j.sna.2008. 04.006.

Lallart, M.; Lefeuvre, E.; Richard, C. & Guyomar, D. (2008b). Self-Powered Circuit for Broadband, Multimodal Piezoelectric Vibration Control. *Sens. Act. A: Phys.*, Vol. 143, No. 2, 277-382, 2008. doi : 10.1016/j.sna.2007.11.017

Lallart, M.; Garbuio, L.; Petit, L.; Richard, C. & Guyomar, D. (2008c) Double Synchronized Switch Harvesting (DSSH) : A New Energy Harvesting Scheme for Efficient Energy Extraction, *IEEE Trans. Ultrason., Ferroelect., Freq. Contr.*, Vol. 55, No. 10, 2119-2130.

Lallart, M. (2008d). Amélioration de la conversion électroactive de matériaux piézoélectriques et pyroélectriques pour le contrôle vibratoire et la récupération d'énergie - Application au contrôle de santé structurale auto-alimenté[14]. Ph.D. Thesis, *in french.*

Lallart, M. & Guyomar, D. (2008e). Optimized Self-Powered Switching Circuit for Non-Linear Energy Harvesting with Low Voltage Output, *Smart Mater. Struct.*, Vol. 17, 035030. doi: 10.1088/0964-1726/17/3/035030

Lallart, M.; Magnet, C.; Richard, C.; Lefeuvre, E.; Petit, L.; Guyomar, D. & Bouillault, F. (2008f). New Synchronized Switch Damping Methods Using Dual Transformations. *Sens. Act. A: Phys.*, Vol. 143, No. 2, 302-314. doi : 10.1016/j.sna.2007.12.001

Lallart, M.; Inman, D. J. & Guyomar, D. (2010a - *in press - available online*). Transient Performance of Energy Harvesting Strategies under Constant Force Magnitude Excitation. *J. Intell. Mater. Syst. Struct.*. DOI: 10.1177/1045389X09358334.

Lallart, M. & Inman, D. J. (2010b). Low-Cost Integrable Tuning-Free Converter for Piezoelectric Energy Harvesting Optimization. *IEEE Trans. Power Electron.*, Vol. 25(7), 1811 - 1819.

Lallart, M. & Guyomar, D. (2010c). Piezoelectric conversion and energy harvesting enhancement by initial energy injection. Appl. Phys. Lett., Vol. 97, # 014104.

[14] *Electroactive conversion enhancement of piezoelectric and pyroelectric materials for vibration control and energy harvesting - Application to self-powered Structural Health Monitoring*

Lallart, M. (2010d). Conversion électroactive et application aux systèmes auto-alimentés[15]. Editions Universitaires Européennes, *in french*. ISBN: 978-613-1-50507-2

Lefeuvre, E.; Badel, A.; Richard, C. & Guyomar, D. (2005). Piezoelectric energy harvesting device optimization by synchronous electric charge extraction. *J. Intell. Mat. Syst. Struct.*, Vol. 16, No. 10, 865-876.

Lefeuvre, E.; Badel, A.; Richard, C.; Petit, L. & Guyomar, D. (2006a). A comparison between several vibration-powered piezoelectric generators for standalone systems, *Sens. Act. A: Phys*, Vol. 126, 405-416.

Lefeuvre, E.; Badel, A.; Petit, L.; Richard, C. & Guyomar D. (2006b). Semi-passive Piezoelectric Structural Damping by Synchronized Switching on Voltage Sources. *J. Intell. Mater. Syst. Struct.*, Vol. 17, Nos. 8-9, 653-660.

Lefeuvre, E.; Audigier, D.; Richard, C. & Guyomar, D. (2007a). Buck-boost converter for sensorless power optimization of piezoelectric energy harvester. *IEEE Trans. Power Electron.*, Vol. 22(5), 2018-2025.

Lefeuvre, E.; Badel, A.; Richard, C. & Guyomar, D. (2007b). Energy harvesting using piezoelectric materials: Case of random vibrations. *J. Electrocem.*, Vol. 19(4), 349-355.

Liang, J. R. & Liao, W. H. (2009). An Improved Self-Powered Switching Interface for Piezoelectric Energy Harvesting. In *Proc. of 2009 IEEE International Conference on Information and Automation*, 945-950.

Makihara, K.; Onoda, J. & Miyakawa, T. (2006). Low energy dissipation electric circuit for energy harvesting. *Smart Mater. Struct.*, Vol. 15, 1493-1498.

T. H. Ng, T. H.& Liao, W. H. (2005). Sensitivity Analysis and Energy Harvesting for a Self-Powered Piezoelectric Sensor. *J. Intell. Mat. Syst. Struct.*, Vol. 16(10), 785-797.

Ottman, G. K.; Hofmann, H. F.; Bhatt, A. C. & Lesieutre, G. A. (2002). Adaptive Piezoelectric Energy Harvesting Circuit for Wireless Remote Power Supply. *IEEE Trans. Power Electron.*, vol. 17(5), 669-676.

Ottman, T. S.; Hofmann, H. F. & Lesieutre, G. A. (2003). Optimized piezoelectric energy harvesting circuit using step-down converter in discontinuous conduction mode. *IEEE Trans. Power Electron.*, Vol. 18(2), 696-703.

Paradiso, J. A. & Starner, T. (2005). Energy scavenging for mobile and wireless electronics. *IEEE Pervasive Computing*, Vol. 4, 18-27.

Park, S.-E. & Hackenberger, W. (2002). High performance single crystals, applications and issues. *Current Opinion in Solid State and Material Science*, Vol. 6, 11-18.

Petit, L.; Lefeuvre, E.; Richard, C. & Guyomar, D. (2004). A broadband semi passive piezoelectric technique for structural damping, *Proceedings of SPIE conference on Smart Struct. Mater. 1999: Passive Damping and Isolation*, San Diego, CA, USA, March 2004, Vol. 5386, 414-425. ISBN : 0-8194-5303-X

Qiu, J.; Ji, H. & Zhu, K. (2009a). Semi-active vibration control using piezoelectric actuators in smart structures. *Front. Mech. Eng. China*, Vol. 4(3), 242-251.

Qiu, J.; Jiang, H.; Ji, H. & Zhu, K. (2009b). Comparison between four piezoelectric energy harvesting circuits. *Front. Mech. Eng. China*, Vol. 4(2), 153-159.

Richard, C.; Guyomar, D.; Audigier, D. & Ching, G. (1999). Semi passive damping using continuous switching of a piezoelectric device, *Proceedings of SPIE conference on Smart Struct. Mater. 1999: Passive Damping and Isolation*, Newport Beach, CA, USA, March 1999, Vol. 3672, 104-111. ISBN : 0-8194-3146-X

Richard C.; Guyomar, D. & Lefeuvre, E. (2007). Self-Powered Electronic Breaker With

[15] *Electroactive conversion and application to self-powered systems*

Automatic Switching By Detecting Maxima Or Minima Of Potential Difference Between Its Power Electrodes, *patent # PCT/FR2005/003000*, publication number: WO/2007/063194, 2007.

Richards, C. D.; Anderson, M. J.; Bahr, D. F. & Richards, R. F. (2004). Efficiency of energy conversion for devices containing a piezoelectric component. *J. Micromech. Microeng.*, Vol. 14, 717-721.

Roundy, S. & Wright, P. K. (2004). A piezoelectric vibration based generator for wireless electronics. *Smart Mater. Struct.*, Vol. 13, 1131-1142.

Shearwood C. & Yates, R. B. (1997). Development of an electromagnetic microgenerator. *Electronics Letters*, Vol. 33, 1883-1884.

Shu, Y. C.; Lien, I. C. & Wu, W. J. (2007). An improved analysis of the SSHI interface in piezoelectric energy harvesting. *Smart Mater. Struct.*, Vol. 16, 2253-2264.

Sodano, H. A.; Simmers, G. E.; Dereux, R. & Inman, D. J. (2006). Recharging batteries using energy harvested from thermal gradients. *J. Intell. Mater. Syst. Struct.*, Vol. 18, 4-10.

Taylor, G. W.; Burns, J. R.; Kammann, S. M.; Powers, W. B. & Welsh, T. R. (2001). The Energy Harvesting Eel: A Small Subsurface Ocean/River Power Generator. *IEEE J. Oceanic Eng.*, Vol. 26, 539-547.

20

Quantum Chemical Investigations of Structural Parameters of PVDF-based Organic Ferroelectric Materials

A. Cuán[1], E. Ortiz[1], L. Noreña[1], C. M. Cortés-Romero[1], and Q. Wang[2]
[1]*Basic Science Department and Materials Department, Metropolitan Autonomous University-Azcapotzalco. Av. San Pablo 180, Col. Reynosa Tamaulipas, 02200*
[2]*Department of Materials Science and Engineering. The Pennsylvania State University, University Park PA 16802,*
[1]*México D.F.*
[2]*USA*

1. Introduction

Poly(vinylidene fluoride)-based polymers have been the research focus of many groups due to the discovery of piezoelectricity in this material, by Kawai et al in 1969. PVDF is constituted by sequential [-(CH_2-CF_2)*n*-] chains and known as the first crystalline organic ferroelectric material, which makes it critically important, since there exist only few classes of materials with ferroelectric properties, such as perovskite-type ABO_3 metal oxides ($BaTiO_3$, $PbTiO_3$, or $KNbO_3$) (Sakashita et al, 1987). Drawbacks of the latter material are related to its heavy weight, brittleness and substantial costs of device manufacturing. Thus, the organic ferroelectrics materials arise as suitable alternative to metal oxides because of better prospective technological applications.

PVDF-based polymers have polymorph properties and at least five experimentally characterized crystal phases, namely, the all-trans (Tp) planar zigzag β-phase, TGa and TGp, (where G denotes the gauche form), the α- and δ-phases, and the T3GT3G´(T_3Ga, T_3Gp) γ- and ε-phases (Lovinger, 1982; Broahurst, 1978). Nevertheless, among these phases, only the all-trans conformation (or β-phase) exhibits ferroelectric behavior. The Tp conformation of the all-trans phase, has a highly polarized backbone with the highest spontaneous polarization in a unit crystal cell (Jungnickel, 1999). This spontaneous polarization gives the special ferroelectric properties, leading to a broad potential of application in several new-technology electronics, such as in sensors, transducers, energy storage devices, communications and microphones (Kawai, 1969; Lovinger, 1983; Jungnickel, 1999; Lando et al., 1966; Farmer et al., 1972; Hasegawa et al., 1972; Karasawa & Goddard, 1992; Tashiro et al., 1995; Nalwa, 1995; Lang, 2006; Wang et al., 1988). A drawback, however, of the synthesized β-PVDF pure material, is that the maximum crystallinity reached is 50%, which results in lower polar properties than those of pure Tp phase, if formed.

Some applications currently under development include artificial muscles and harnessing energy from sea waves. The ferroelectric properties of PVDF can be enhanced by the introduction of trifluoroethylene, TrFE, as comonomer. P(VDF-TrFE) exhibits ferroelectric properties at TrFE contents between 50 and 85 mole percent (Lu et al., 2006). At a specific temperature, the Curie temperature, P(VDF-TrFE) copolymers show a conformational and phase transformation, from ferroelectric to paraelectric. The Curie temperature depends on the copolymer composition and at this temperature the PVDF-based materials show some of the highest dielectric constants of any organic polymer, resulting from large crystalline polar domains. However, for energy storage applications, as in capacitors, is more convenient a relatively high dielectric constant at room temperature and a smaller remnant polarization (Lu et al., 2006), that can be accomplished by reducing the crystal domain size through high energy radiation or by the introduction of a third monomer, such as chlorotrifluoroethylene P(VDF-CTFE). Experimental results indicate that the crystalline domain in P(VDF-TrFE)s and its size play an important role in the dielectric response (Zhang et al., Lovinger, 1985, Daudin & Dubus, 1987; Tashiro et al, 1988; Furukawa, 1990; Casalini & Roland, 2001; Gao & Scheinbeim, 2000; Casalini & Roland, 2001).

Both, experimental and theoretical works have been devoted to the study of PVDF-based materials unique piezoelectric, pyroelectric, ferroelectric, electro-acoustic and nonlinear optical properties (Jungnickel, 1999; Lando et al., 1966; Farmer et al., 1972; Hasegawa et al., 1972; Karasawa & Goddard, 1992; Tashiro et al., 1995; Nalwa, 1995; Lang, 2006; Wang et al., 1988). Common dielectric materials may become polarized under an applied electrical field, whereas ferroelectric materials may become spontaneously polarized. Piezoelectric materials can transform a mechanical movement into an electric signal and vice versa. On the other hand, electro-acoustic materials can transform an acoustic wave into an electric signal and vice versa.

The density functional theory (DFT) is a powerful tool for both, elucidation of and understanding the PVDF-based materials structure-property relationships. The relative accuracy of DFT is comparable to that of traditional ab-initio molecular orbital methods. However, the computational requirements of DFT calculations are much less demanding, allowing us to efficiently study the large systems needed for the realistic modeling of PVDF-based materials. Theoretical and experimental studies can greatly aid our atomic-level understanding of ferroelectric properties. Herein, we briefly describe the technical details of DFT methodologies and the factors that influence the accuracy of the results. Examples of DFT applied to PVDF-based materials features, like phase transition and their related electronic properties, are presented, demonstrating the important role DFT plays in the study of these kinds of materials.

Theoretical investigations for the aforementioned materials are presented and discussed within this manuscript. Molecular models commonly employed, methods applied to these materials, and results currently obtained for the energetics and structures corresponding to the different conformations, i.e., the changes in the molecular arrangement associated to T_p, TG_a or TG_p conformations, are also presented. Properties such as the charge polarization and the dipole moment increment produced by enlarging the chain components that actually confer the ferroelectric properties are reviewed as well.

2. Quantum theory as an analytical chemistry tool

The complete characterization of any mechanism, phase stability or phase transitions and its structure-property relationships is a huge task, nevertheless, considering its potential economic and environmental impact in industry, it is clear that deeper knowledge in it would be of great benefit.

The rational optimization and design of a PVDF-based material requires fundamental information concerning the structure and function of these materials under ideal conditions. Quantum mechanics (QM) methods, which we also denote as QCC (Quantum Chemical Calculations) can provide valuable information about the electronic and structural phase transitions for several polymer backbones giving an insight into structure-property relationships, which are not yet completely elucidated (Broahurst et al. 1978; Jungnickel, 1999; Tashiro et al., 1988; Casalini & Roland, 2001; Gao & Scheinbeim, 2000; Casalini & Roland, 2002; Nicholas, 1997; Nicholas et al., 1991; Gómez-Zaravaglia & Fausto, 2004; Teunissen et al., 1994; Sierra, 1993; Evleth et al., 1994; Das & Whittenburg, 1999). Therefore, the first principles-based materials models are helpful for the design of materials with improved properties. In this concern, QCC together with experimental techniques have reached widespread application as a tool for characterization of a reactive system and spectra modeling (Nicholas, 1997).

The experimental spectroscopic results are often confronted with theoretical calculations in which the vibrational frequencies are computed and afterwards analyzing the corresponding mode in order to reproduce the model molecule characterization (Nicholas et al., 1991; Gómez-Zaravaglia & Fausto, 2004). Moreover, it is possible to make a model of a solid crystalline system or instead be treated as a molecule, depending on the properties that we wish to obtain.

2.1 Quantum mechanics methods applied to PVDF-based materials

The QM methods offer the opportunity to study electronic structure at the microscopic level. Within the QCC methods there are different approximations levels, i.e., the Moller-Plesset (MP2) and Configuration Interaction (CI), which are some of the most accurate and involve electronic correlation. The required time for running a calculation is about N^5, where N is the basis set employed for the atomic orbital description (Teunissen, 1994). Other approximations such as the Hartree-Fock (HF) (N^4) and the Density Functional (DFT) (N^3) level of theory require shorter computing time (Sierra, 1993; Evleth et al., 1994) with the possibility of applying them to a larger system size, having less accuracy but similar phenomena description to those of MP2 and CI, (further details are reported in Das & Whittenburg in 1999). The DFT offer some computing time advantages and a quantitative characterization of electronic-structures properties and in order to gain insight in these phenomena it is necessary necessary to build predictive, then the first principles-based materials models. DFT methodology results are helpful for the design of PVDF-based materials.

2.2 Molecular models commonly employed in QCC

Crystalline or semi-crystalline molecular modeling can be of different complexity level, depending on the target of the study. Periodic boundary conditions or finite models are commonly employed in QCC. In this concern, some investigations have been reported with both models as described further on. In addition, QCC contributes to predict the energetic and structures corresponding to the different conformations, as their relationship to charge polarization and to dipole moment enhancement, that in molecular design, can be related to the ferroelectric properties of PVDF-based materials.

Using periodic boundary conditions (Chun-gang et al., 2003) from first-principles, the band structure approach with full-potential linear augmented-plane-wave method (FLAPW)

(Blaha, 1997) can be applied to the study of PVDF and P(VDF-TrFE) and the band gap and symmetries can be correlated in line with experimental photoemission data. As a reminder, P(VDF-TrFE) stands for poly-vinylidene fluoride-triflouoroethylene copolymers. They also found that the interactions between different PVDF chains are rather weak, which reflects clearly the quasi–one-dimensional nature of the system. Concerning the electronic structure of P(TrFE), results indicated that replacing -(CH_2-CF_2)- by -(CHF-CF_2)- does not change major features of the band structure.

Other contribution using periodic conditions has been reported (Haibin Su et al. 2004)). These authors used DFT pseudopotential code SeqQuest (Schultz, 2002; Feibelman, 1987) which makes use of Gaussian basis sets to study static and dynamical mechanical properties of PVDF and its copolymer with trifluoroethylene (TrFE). With this methodology, they found that conformations with T and G bonds are energetically favorable for large interchain separations if compared to all-T structures. These results are in line with the experimental observation that samples irradiated with high-energy electrons favor T_3G nonpolar conformations. With this methodology it is also possible to obtain the crystal cohesive energy (E_{coh}) and energy stabilization of the different conformation or phases.

On the other hand, other research groups, (Nicholas et al., 2006) have performed DFT calculations *via* the ABINIT software package with periodic boundary conditions employing Perdew BurkeeErnzerhof generalized-gradient approximation (GGA) (Perdew, 1996) and pseudopotentials (Rappe, 1990). Together with this strategy they also used the OPIUM code which led to a full reproduction of the structure and NMR and FTIR vibrational frequencies of the α-PVDF phase. With these results, they contributed to the understanding of those mode ambiguities founded experimentally (Makarevich & Nikitin, 1965; Wentink et al., 1961; Cortili & Zerbi, 1967; Kobayashi et al., 1975).

In the case of finite models that are treated as molecules (Das et al., 1999), the DFT methodology with ab-initio methods implemented in the Gaussian 94 software package help in geometric structure, vibrational frequency, dipole moment and singlet–triplet energy separation of CH_2, CHF, CF_2, CCl_2 and CBr_2 studies, using B3LYP, B3P86 and B3PW91 hybrid density functionals. For the case of the methylene group (CH_2), the density functionals predict the equilibrium C–H bond length, the H–C–H bond angle and the vibrational frequency up to an accuracy level comparable to high-level *ab initio* methods. However, they found that only the B3LYP functional produces reasonable singlet–triplet energy separation. Estimates of the singlet–triplet energy separation of CHF and CF_2 indicate B3LYP functional produces better agreement with experimental data than B3P86 or B3PW91.

Another contribution was made by (Zhi-Yin, 2006). They reported a study of the internal rotation, geometry, energy, vibrational spectra, dipole moments and molecular polarizabilities of PVDF in α- and β-chain models employing the density functional theory at B3PW91/6-31G(d) level. They found the effects of chain length and monomer inversion defects on the electric properties and vibrational spectra. They also reported that the average distance between adjacent monomer units in the β-PVDF is 2.567 Å and that the energy difference between the α- and β-chains is about 10 kJ/mol per monomer unit. They found that the dipole moment is also affected by chain curvature and by defect concentration and that the chain length and defects will not significantly affect the polarizability. Those are some of the reported results presented about this topic, as can be seen, depending on the model and methodology, we can obtain different structural and electronic properties relationships. Then, insight has been gained from the QCC literature in which both periodic and finite models have been proposed with considerably good acceptance.

2.3 When do models correctly describe the PVDF-based materials?

There is not a gradual development of quantum chemical models or methodologies applied to PVDF-based materials, which for instance occurred in the case of other macro systems, such as zeolite catalysts (Cuán & Cortés-Romero, 2010). The reason lies on the current development of the software and hardware that, unlike for zeolite-based materials, is not a limitation anymore for PVDF-based materials. Researchers are applying all the possible QM tools to the molecular design of this polymer due to the complexity and importance of this phenomenon.

So far, the crystalline or semi-crystalline molecular modeling of these materials can be of different level of complexity depending on the purpose. Periodic or finite models give valuable information, but the model employed and the correct description will be in terms of some factors, such as the properties to be estimated or reproduced, geometrical description, computational resources (software and hardware) and quality of the theoretical approximation. Nevertheless, in many cases a simple molecular model can give valuable information, as it will be discussed below.

3. Test case of PVDF-based materials using finite models

3.1 Model representation

Five different-length chain molecules were studied for PVDF, H-$(CH_2$-$CF_2)_x$-H, where x=1,2,4,6 for the four different PVDF molecules conformations, namely, I=T_p, II=TG_a, III =TG_p and IV =T_3G, *vide* Figure 1(a), and for the PVDF-based materials PVDF-TrFE and

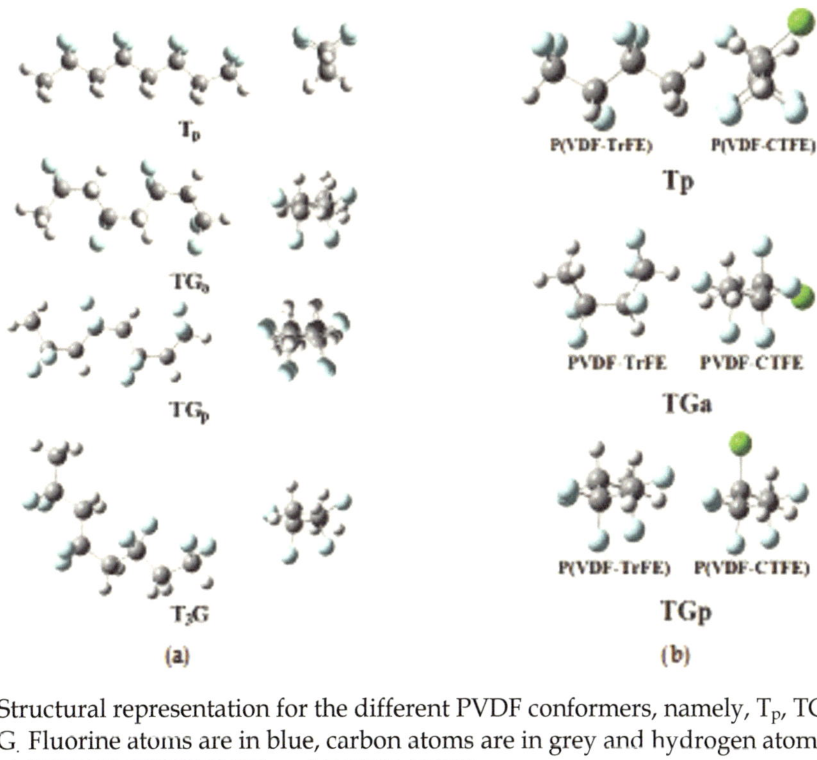

Fig. 1. Structural representation for the different PVDF conformers, namely, T_p, TG_a, TG_p and T_3G. Fluorine atoms are in blue, carbon atoms are in grey and hydrogen atoms are in white. a) PVDF b) P(VDF-TrFE) and P(VDF-CTFE).

PVDF-CTFE, only the three representatives conformations were taken from PVDF and calculated, they are denoted as Tp, TGa and TGp phases, *vide* Figure 1(b). T means all-T, TG indicates TGTG' and T_3G means TTTGTTG', where T indicates trans and G means gauche conformation and the subindexes p and a correspond to polar phases with parallel dipoles and nonpolar phases with antiparallel dipole moments, respectively.

3.2 Stepwise theoretical methodology used

The electronic structure study includes all-electrons within the Kohn-Sham implementation of the Density Functional Theory (DFT). The level of theory used in this work corresponds to the non-local hybrid functional developed by Becke, Lee-Yang-Parr (B3LYP) (Becke, 1993; Lee et al., 1988)whereas the Kohn-Sham orbitals are represented by a triple-ζ numerical with double polarized functions (d,p) plus one diffuse basis set; implemented in the Gaussian 03 code, this methodology was carried out within finite models. The electrostatic potential method was used for the charge calculation (ESP) (Brent, 1990). The Electrostatic Potential (ESP) charge calculation algorithm was chosen because it has no basis set dependence. Geometry optimization calculations were carried out for all the involved systems using the Berny algorithm. The Threshold convergence criterion was 10^{-6} hartrees for the energy, 0.000450 for the Maximum Force and 0.001800 for the Maximum Displacement.

3.3 Active phase formation

In order to understand the differences among the chemical and electrical properties of PVDF with respect to the P(VDF-TrFE) and P(VDF-CTFE) materials, we performed the torsion of the dihedral angle of representative two monomer units, i. e., n_r=4, where n is the total number of carbon atoms in the structure, *vide* Figure 2. Following the dihedral angle torsion of the G -to- T geometry transformation, the potential energy surface (PES) gives the relative energies and structural changes among them. Figure 2 shows that the TG_a, TG_p and T_p are stable geometric conformations for all the materials, because all of them are situated at minimum positions within the PES. We shall bear in mind that the molecular energy computed in this simplified model is far away from that of the real crystal (formed by large domains); nevertheless, computing simulations are consistent with available experimental observations (Wang et al., 2006; Ortiz et al., 2009; Ortiz et al., 2010). According to Figure 2, it is possible to observe that for all the materials, the TG phases are energetically more stable than the T_p phase, and that the PVDF-PES profile shows a symmetric curve distribution, *vide* Figure 2(a); which is mostly due to the symmetric disposition of fluorine substituents. In the P(VDF-TrFE)-PES and the P(VDF-CTFE)-PES this even distribution is lost, because the addition of substituents to the PVDF, such as fluorine and chlorine, *vide* Figure 2(b) and (c). Then, according to Figure 2, to reach the T_p conformation requires an energy supply for all the three cases. *Barrier 2 (Barr2)* indicates the amount of energy required to reach T_p from TG_p; approximately 4 kcal/mol for PVDF, 4.86 kcal/mol for P(VDF-TrFE) and 3.05 kcal/mol for P(VDF-CTFE) (*vide* Table 1). For the PVDF case, the order of the energy difference is in agreement with that reported elsewhere (Wang et al., 2006; Ortiz et al., 2010). On the other hand, *Barrier 3 (Barr3)* indicates the amount of energy required to reach Tp from TGa being around 2.1, 3.5, and 1.0 kcal/mol following the same order mentioned before. Now, *Barrier 1 (Barr1)* represents the amount of energy required to reach TGp from TGa being around 3.9, 4.0, 1.8 kcal/mol, keeping the same ordering. Based on the computed results, the

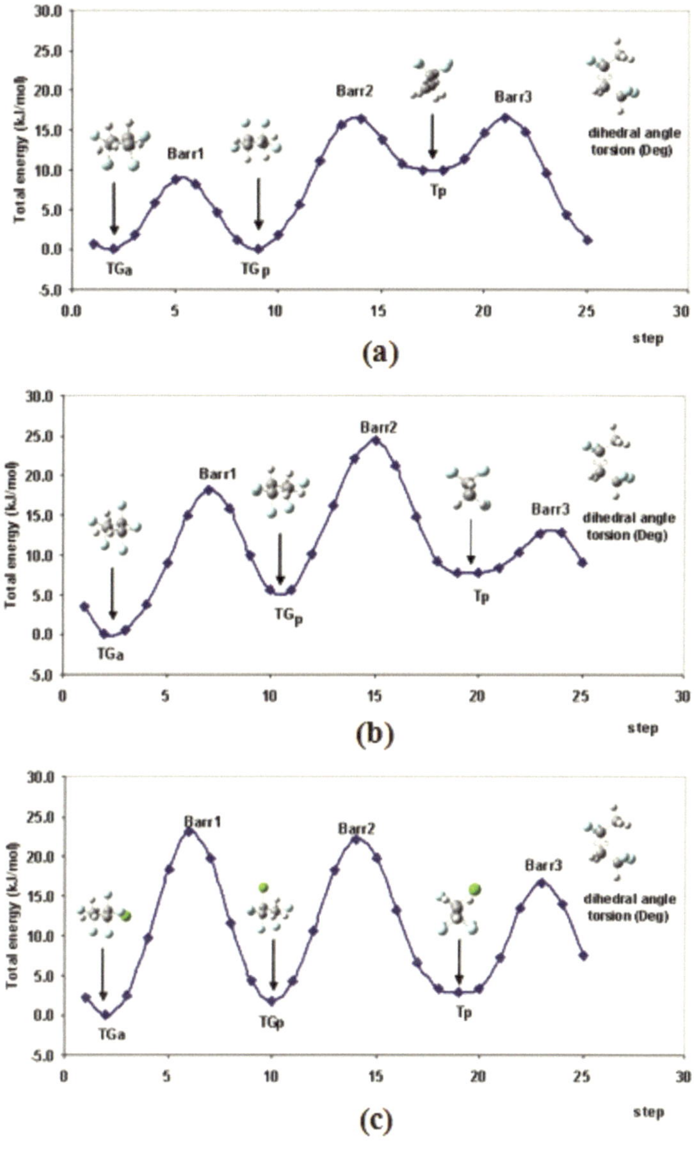

Fig. 2. Graphical representation of the PES for the dihedral angle torsion showed on the right side of the picture. This graph displays the energy associated to the different model structural conformations, TG_a, TG_p and T_p of the different materials: (a) PVDF, (b) P(VDF-TrFE) and (c) P(VDF-CTFE).

energy barrier for converting the TG_a into the TG_p structural conformation is half of the energy required to reach the T_p structural conformation for the PVDF material, as it can be observed in Figure 2, while for the P(VDF-TrFE) increases almost one time and for the P(VDF-CTFE) there is no significant difference. The situation is not the same when Barr2 and Barr3 are compared, here for converting the TGa into the TGp with respect to reach the Tp conformation, for the case of P(VDF-TrFE) the amount of energy diminished three times,

if compared with PVDF, which means that this process would occur more easily. Although for P(VDFCTFE) the energy required increases again one an a half times.
The analysis of the energy differences among the structural phases involves looking at the minima points along the PES for the different materials conformations, as follows. For PVDF, TGa and TGp have a similar energy level, *vide* Figure 2 (a), while the energy difference between them and the Tp conformer is about 2.3 kcal/mol. For P(VDF-TrFE), according to Figure 2 (b), TGa is more stable than TGp by 1.46 kcal/mol; the Tp conformer is placed 1.92 kcal/mol above than TGa and only 0.5 kcal/mol above than TGp. Note that for this material it would be easier to reach the Tp phase from TGa, due to the considerable reduction of the rotational barrier, with respect to the others, *vide* Table 1. Analyzing the graphical representation in Figure 2(c), but now for P(VDF-CTFE), all the conformations, namely, TGa, TGp and Tp have a similar energy level, with less than 1 kcal/mol difference among them. These results mean, that for P(VDF-CTFE), the Tp phase becomes almost as stable as the TGa or TGp phases, although the rotational energy barrier for reaching the Tp conformation is larger than for the other systems; but once the energetic barrier may be overcome, a stable structure is obtained. These structural conformations are related with a corresponding phase in the crystal. Then, the active phase formation could be correlated to the conformational geometry and the rotational dihedral angle during the formation of Tp phase. The rotational barriers found for each different composition affect both the higher mobility of the C-F dipoles and the structural rearrangement of the systems which actually produce the differences in polymers properties and, hence, applications. Somehow, the rotational barriers and its energy level are a function of the fluorine composition and the degree of contamination (chlorine content) which is expected by the presence of this halogen.

System	Phase	ΔE_{Ph-Tr} Kcal/mol (kJmol)	ΔE Kcal/mol (kJmol)		ΔE_{H-L} (eV)	Dipole (D)
PVDF	Tp	2.373 (9.92)	***Barr1***	3.929 (16.44)	9.24	4.68
	TGa	0.000 (0.00)	***Barr2***	3.954 (16.55)	9.57	2.37
	TGp	0.020 (0.08)	***Barr3***	2.095 (8.77)	9.56	2.55
PVDF-TrFE	Tp	1.924 (8.05)	***Barr1***	3.992 (16.70)	8.32	2.54
	TGa	0.000 (0.00)	***Barr2***	4.862 (20.34)	8.43	1.65
	TGp	1.457 (6.10)	***Barr3***	3.501 (14.65)	8.36	2.30
PVDF-CTFE	Tp	0.698 (2.92)	***Barr1***	1.777 (7.44)	9.07	3.65
	TGa	0.000 (0.00)	***Barr2***	3.050 (12.76)	9.43	1.24
	TGp	0.435 (1.82)	***Barr3***	0.608 (2.54)	9.28	3.35

$\boldsymbol{\Delta E_{Ph-Tr}}$=*E(phaseTGa)-E(corresponded – phase), i.e., Tp or TGp*
$\boldsymbol{\Delta E_{H-L}}$= $| E_{HOMO} - E_{LUMO} |$
Barr1= *E(Phase Tp – E(Phase TGa)*
Barr2= *E(Phase Tp – E(Phase TGp)*
Barr3= *E(Phase TGp – E(Phase TGa)*
μ=Debye

Table 1. Phase transition energy (ΔE_{Ph-Tr}), Rotational barrier (ΔE), HOMO-LUMO GAP (ΔE_{H-L}) and Dipole moment (μ).for the representative two monomer units of PVDF, P(VDF-TrFE) and P(VDF-CTFE).

3.4 Physicochemical and electrical properties

PVDF material is the base of the organic ferroelectric material and although β- structure serves well as a simple model for basic understanding of the polar nature of ferroelectric polymer crystal, it is necessary to study the electronic-structure relationship and the behavior of these polymers when the environment changes, i.e., variation in VDF-to-copolymer concentration. We currently used five different-length chain molecules as a model of $H\text{-}(CH_2\text{-}CF_2)_x\text{-}H$, where x = 1, 2, 4, 6 for the four different PVDF conformations, namely, Tp, TGa, TGp and T_3G and compare with PVDF-based materials electronic properties.

3.4.1 PVDF material case

The analysis of the differences among the structural changes obtained along the PES in Figure 2 (a) leads to the values in Table 2 where some electronic properties and their

No. of C Atoms n_r	Length L (Å)	E_{HOMO} (eV)	E_{LUMO} (eV)	ΔE^a (eV)	Dipole (D)
			T_p		
2	3.30	-9.63	0.08	9.72	2.52
4	5.62	-9.24	-0.33	8.91	4.68
6	8.17	-9.08	-0.58	8.50	6.74
8	10.68	-8.99	-0.75	8.25	8.76
12	15.63	-8.93	-0.95	7.98	12.70
			TG_a		
2	3.30	-9.63	0.08	9.72	2.52
4	5.25	-9.57	-0.19	9.38	2.37
6	7.62	-9.47	-0.37	9.11	4.30
8	9.86	-9.41	-0.46	8.95	5.04
12	14.43	-9.30	-0.57	8.72	7.97
			TG_p		
2	3.30	-9.63	0.08	9.72	2.52
4	5.26	-9.56	-0.07	9.49	2.55
6	7.62	-9.48	-0.48	9.01	4.08
8	9.86	-9.42	-0.57	8.84	4.73
12	14.43	-9.31	-0.70	8.605	7.57
			T_3G		
2	3.30	-9.63	0.08	9.72	2.52
4	5.26	-9.57	-0.19	9.38	2.36
6	6.75	-9.38	-0.49	8.89	3.32
8	9.90	-9.19	-0.56	8.632	3.64

[a] $\Delta E = E_{LUMO} - E_{HOMO}$ (eV)

Table 2. Length chains (carbon atom number) for the different PVDF structural conformations and their electronic properties: HOMO energy, LUMO energy, [a] $\Delta E = E_{LUMO} - E_{HOMO}$ and dipole moment.

corresponding parameters are presented. Considering only two monomer units, i. e., n_r=4, where n is the total number of carbon atoms in the structure, the computed results indicate that the dipole moment of the T_p conformation is approximately 40% higher than that of the rest of the structural conformations (TG_a, TG_p and T_3G), *vide* Table 2. The dipole moment of the different conformations is presented in Figure 3. The blue line corresponds to the dipole moment variation depending on the chain length of the T_p conformation. Similarly, the green line corresponds to the TG (a or p) conformation and the orange line corresponds to the T_3G phase. In general, the picture shows that the dipole moment increases as the length of the carbon chain increases. The T_p conformation exhibits the larger increase in the dipole moment. The TG structural conformation shows an intermediate increase whereas the T_3G conformational phase has the lowest one. The dipole moment increase for T_p is about 40% stronger than the TG´s conformations and 55% stronger than the T_3G conformation. This means that the structural arrangement of the T_p phase promotes an increase in the polarity of the system, as already known. The charge polarization can also be obtained (*vide* Figure 4), where the positive and negative ESP charges in the T_p conformation are perfectly ordered at the outside part of the molecule. The TG´s and T_3G conformational systems do not show the same split charge distribution.

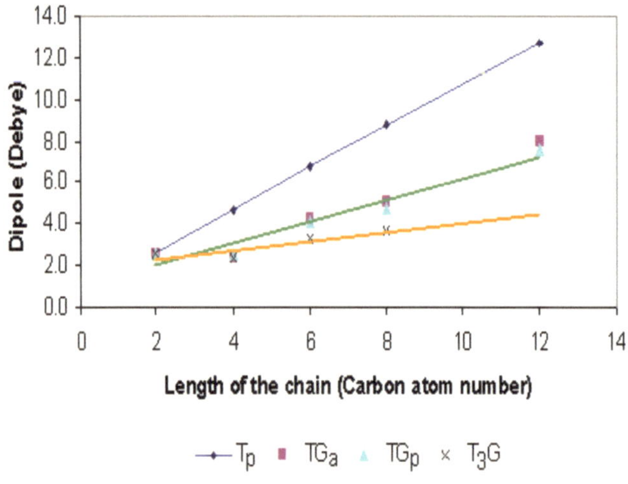

Fig. 3. Dipole moment trend (in Debye) for the different PVDF structural conformations. In blue for T_p, in green for TG_a and TG_p and in orange for T_3G.

Hence, the dipole variation and the charge polarization of the Tp system are characteristic of a ferroelectric material; they together act to form an electric dipole moment even in the absence of an external electrical field, as it is currently demonstrated. This behavior is a result of a change in the phase structure. Moreover, the simple molecular model in Figure 2 shows that the energy barriers among the different structural conformations are not too high to avoid switching among them with a relatively low energy supply.

Figures 5 and 6 show the electronic states distribution of the valence band and the conduction band for the T_p and TG_p conformers, repectivetly. We can observe that the band gap between the valence and the conduction bands decreases as the total number of carbon atoms increases; being shorter in the Tp conformer than in the TG_p conformer. The number of empty molecular orbitals under the Fermi level increases as the length of the chain also increases. Therefore, the |HOMO-LUMO| energy difference also exhibits a decrease, being highly notorious for the T_p conformer, from 9.72 eV to 7.98 eV but they correspond to a simplified model, which tendency is to decrease with the system is enlarged (*vide* Table 2), the experimental value reported is about 6.5 eV (Choi Jaewu et al., 1998). In the T_p conformer there is a noticeable increase of the empty states under the Fermi level. Although, the reduction in the |HOMO-LUMO| band gap is low, whereas the conduction band lies sufficiently low for rendering a negative band gap value, corresponding to a semimetal behavior. However, the spontaneous polarization of the β-phase might be associated to a higher sheet carrier density with respect to the bulk.

Then, we can conclude partially that the changes in the molecular arrangement associated to T_p, TG_a or TG_p and T_3G conformations lead to significant changes in shape and electrical-chemical properties. A larger dipole moment and spatial charge polarization were obtained for the all-trans T_p molecular structure, which can be obtained by accumulative motion of the neighboring groups, through large-scale T-G conformational changes. The dipole moment increase and the charge polarization in the T_p system are characteristics of a ferroelectric material. These contribute to form an electric dipole moment, even in the absence of an external electrical field.

3.4.2 PVDF-based materials

Now, in order to get insight and better understanding of the electronic-structure relationships and changes in behavior, leading towards enhanced ferroelectric properties produced by environment changes, we have also studied P(VDF-TrFE) and P(VDF-CTFE) copolymers and compared the differences or similitudes to the pure PVDF.

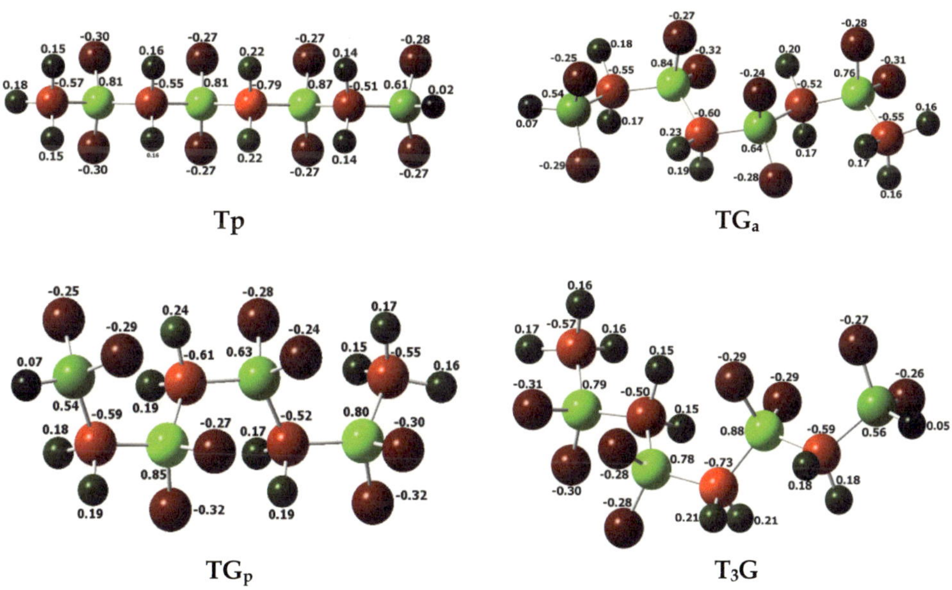

Fig. 4. ESP charge representation for T_p, TG_a, TG_p and T_3G. Negative charge values are in red, positive charge values are in dark green and the larger positive charges are depicted in light green.

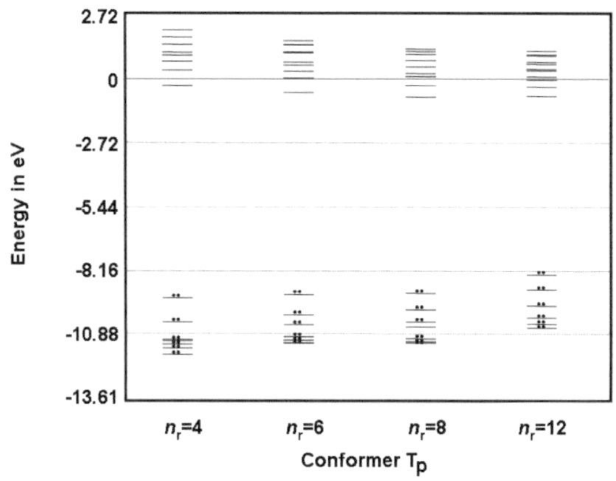

Fig. 5. Electronic states distribution for the valence and conduction bands for the Tp conformer with n_r=4, 6, 8 *and* 12 carbon atoms forming the structure.

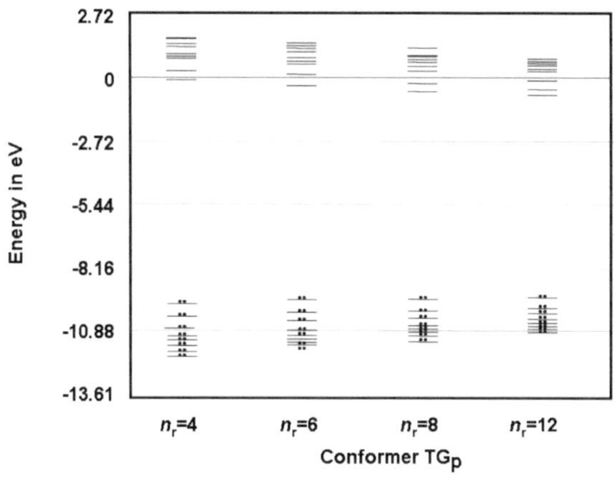

Fig. 6. Electronic states distribution for the valence and conduction bands for the TG_P conformer with n_r=4, 6, 8 *and* 12 carbon atoms forming the structure.

The analysis of the energy differences among the structural changes obtained along the PES in Figure 2 leads to the values reported in Table 1. Considering only two monomer units, i. e., n_r=4, where n is the total number of carbon atoms in the structure, the computed results indicate that the dipole moment of the T_P conformation is always higher than the rest of the structural conformations (TG_a and TG_P). This means that the structural arrangement of the T_P phase promotes an increase in the polarity of the system, as it has been mentioned before. Analyzing the ESP charge distribution for the three cases, charge polarization in PVDF is obtained for the Tp conformation (*vide* Figure 7), where positive and negative ESP charges arrange in the T_P conformation. The quantum mechanics calculations of the ESP charge distribution indicate (Figure 4) that the electrical charges are perfectly ordered at the outside part of the molecule, being one side positive and the other negative. In the same way, for the P(VDF-TrFE) system, although the charge is polarized by fluorine atoms being at one outside part completely negative, the other outside part, is mostly positive with exception of the non-symmetric fluorine atom, a sort of contamination. For the P(VDF-CTFE) case, in which a chorine atom is present in the structure, a charge polarization is also obtained outside the molecule, exhibiting a negative charge where fluorine atoms are present, but on the other side, there is an alternate charge distribution, being negative at fluorine and chlorine atoms positions and positive at hydrogen atoms positions. The TG´s conformational systems do not show a charge polarization or the same split charge distribution (*vide* Figure 4).

So far, the current results for the model's representation give a good description of the system, also in line with previous literature reports. Furthermore, some additional parameters can be presented. On this respect, the |HOMO-LUMO| energy difference was also computed. From Table 1, although the |HOMO-LUMO| energy gap decreases slowly for the Tp conformation (in all the studied systems), in previous results for PVDF (Lu et al., 2006), the |HOMO-LUMO| energy difference also exhibits a decrease, being highly notorious for the T_p conformer in P(VDF-CTFE). It decrease from 9.24 eV in the PVDF to 8.32 eV in the P(VDF-CTFE), whereas for the P(VDF-TrFE) the|HOMO-LUMO|energy difference is between them, about 9.07 eV. The dipole moment decrease for the PVDF-based materials (P(VDF-TrFE) and P(VDF-CTFE)) being the higher dipole moment for the PVDF (*vide* Table 1). Even so, among the different structures, the Tp phases present the highest values in the dipole moment with respect to TG´s phase structures, no matter if PVDF or PVDF-based materials are.

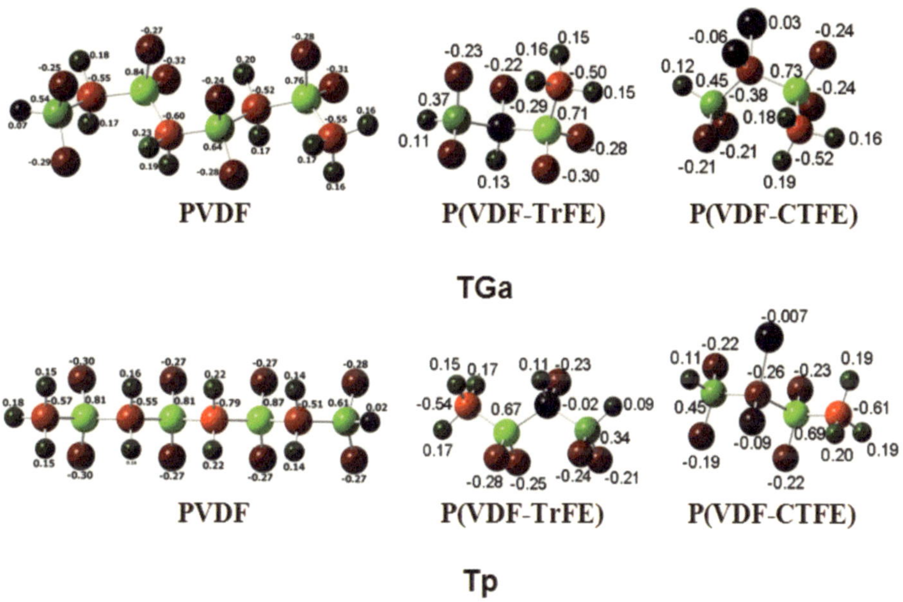

Fig. 7. ESP charge representation for TG_a, and T_p, of the representative model of the different materials, namely, PVDF, P(VDF-TrFE) and P(VDF-CTFE). Negative charge values are in red, positive charge values are in dark green and the larger positive charges are depicted in light green.

When the chain is enlarge for PVDF and P(VDF-CTFE) as a composition of 85 mol % of VDF and 15 mol % of CTFE, as can be seen in the Figure 8, the dipole moment keep similar behavior as in the two monomer units. In this case, for the PVDF in a Tp phase is 12.6 D, for the P(VDF-CTFE) in TGa phase is about 8.95 D and for P(VDF-CTFE) in Tp phase is 11.32 D.

Analyzing the ESP charge distribution for the three last cases, charge polarization in PVDF is obtained for the Tp conformation (*vide* Figure 8a), where positive and negative ESP charges arrange in the T_p conformation. The quantum mechanics calculations of the ESP charge distribution indicate (Figure 8a) that the electrical charges are perfectly ordered at the outside part of the molecule, being one side positive and the other negative. For the P(VDF-CTFE) case, in which a chorine atom is present in the structure, a charge polarization is also obtained outside the molecule, exhibiting a negative charge where fluorine atoms are present, contrary to obtained when only two monomers are taking into account, *vide* Figure7 and 8c. The TGa conformational system for P(VDF-CTFE) as a composition of 85 mol % of VDF and 15 mol % of CTFE, do not show a charge polarization or the same split charge distribution (*vide* Figure 8b), as it is expected.

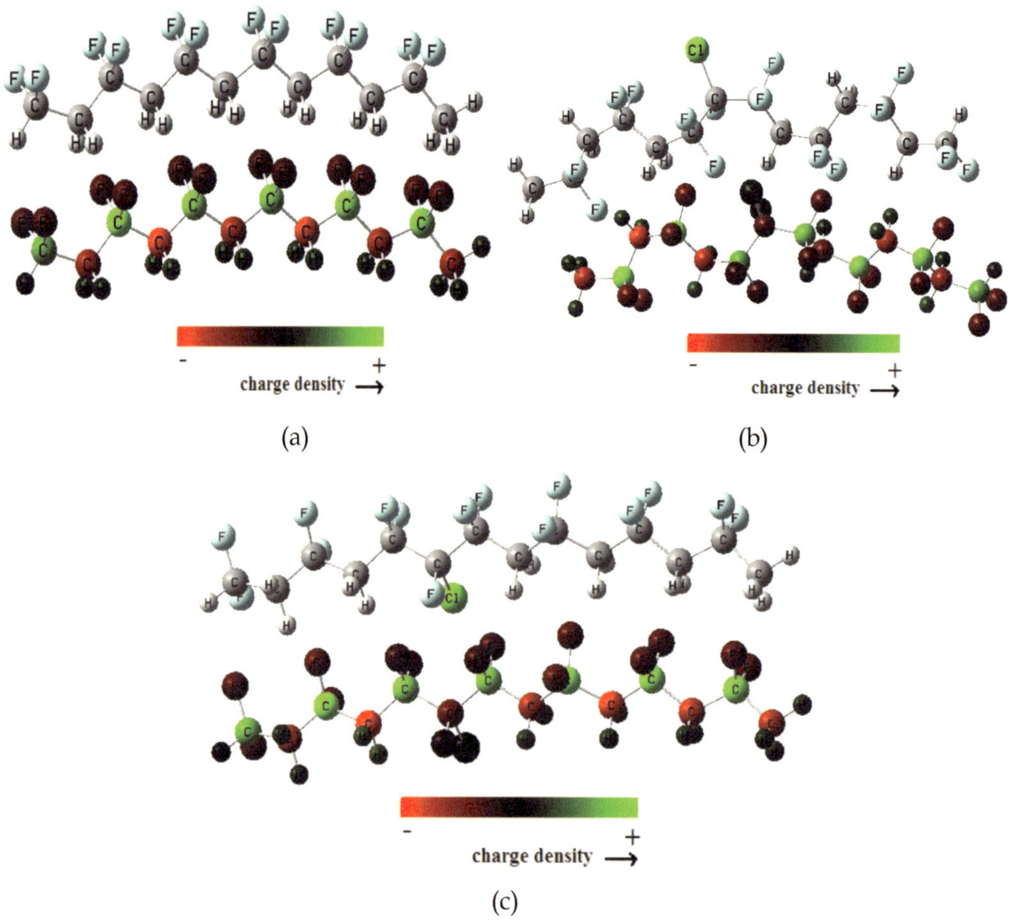

Fig. 8. Geometrical and schematic ESP charge distribution for a representative models for of the different materials, namely, (a) T_P phase of PVDF; (b) TGa phase of P(VDF-CTFE) and (c) Tp phase of P(VDF-CTFE). P(VDF-CTFE) as a composition of 85 mol % of VDF and 15 mol % of CTFE. Negative charge values are in red, positive charge values are in dark green and the larger positive charges are depicted in light green.

For the system which contain a sort of contamination of fluoride and chlorine atom in the PVDF system to generate the P(VDF-CTFE) as a composition of 85 mol % of VDF and 15 mol % of CTFE, composition studied here, it does not has a strongly effect over the charge polarization for the Tp phase, but according to the vibrational frequencies obtained for the both phases Tp and TGa- P(VDF-CTFE), the presence of the chlorine atom in the structure gives not too much flexibility to the chain neighboring to it. In fact, as chlorine atom is heavier than fluorine or hydrogen atoms, it anchors the molecule and restricts the mechanic movement. Maybe this special factor gives different properties to the copolymer with respect to PVDF system.

4. Conclusions

Ferroelectric materials are commonly used in electronic devices taking advantage of, for instance, their piezoelectricity, electricity storage capacity and sensing ability. The actual operability of these materials is based on the charge polarization, which is basically due to the energy associated to electron mobility. The fundamentals of PVDF-based materials have gained more and better understanding in recent years. The key for this advancement, from an experimental point of view, is based on the new methods and techniques that look for the interpretation of structural and composition changes of the polymer through improvement of instrumental sensitivity. Likewise, more efficient computational software and modern hardware allow performing complex theoretical calculations to simulate the local and overall chemical phenomena taking into account as many atoms as necessary to obtain reliable results. Quantum-chemical calculations performed using Density functional Theory yield deep understanding on the electronic-structure relationships from fundamentals with well-suited description of the phase transition phenomena.

Dealing with PVDF-based materials model design, the quantum mechanics calculations of the energetics and structures corresponding to the different structural conformations for the PVDF, P(VDF-TrFE) and P(VDF-CTFE) units show that the T_p conformation is energetically stabilized even with a chlorine substituent. The changes in the molecular arrangement associated to T_p, TG_a or TG_p conformations lead to significant changes in shape and electrical-chemical properties. A larger dipole moment and spatial charge polarization were obtained for the all-trans T_p molecular structure, being more ordered in the PVDF system, which can be obtained by accumulative motion of the neighboring groups through large-scale T-G conformational changes. The molecular model shows that the energy barriers among the different structural conformations are not too high, allowing to switch among them with a relatively low energy supply, more easily for P(VDF-TrFE), according to the computed results. The interphase mobility among the different conformations is an important property for actuators. All these additive properties of the T_p structure model may produce a ferroelectric material when the system size scales up to a polymer crystal.

5. References

ABINIT. The ABINIT programs are a common project of the Universite´ Catholique de Louvain, Corning Inc. and other contributors (see http://www.abinit.org).

ABINIT-P. For a description of the ABINIT project, Gonze X, Beuken JM, Caracas R, Detraux F, Fuchs M, Rignanese GM, et al. Comput Mater Sci 2002;25:478e92; Gonze X,

Rignanese GM, Verstraete M, Beuken JM, Pouillon Y, Caracas R, et al. Z Kristallogr 2005;220:558e62.

A. D. Becke. (1993). Density-functional thermochemistry. III. The role of exact exchange, *J. Chem. Phys.* Vol. 98, pp 5648–5653.

P. Blaha, K.Schwarz and J. Luitz, WIEN97. (1999). A Full Potential Linearized Augmented Plane-Wave Package for Calculating Crystal Properties (Karlheinz Schwarz, Technische Universit¨atWien, Austria); ISBN 3-9501031-0-4.

B. H. Brent, K. M. Merz, Jr., and P. A. Kollman. (1990). Atomic charges derived from semiempirical methods. *J. Comput. Chem. Vol.* 11, pp 431–439.

M. G. Broahurst, G.T. Davis, J.E. McKinney, R.E. Collins. (1978) Piezoelectricity and pyroelectricity in polyvinylidene fluoride—A model. *J Appl Phys* Vol. 49 No. 10, pp 4992-4992.

R. Casalini, C. M. Roland. (2001). Highly electrostrictive poly(vinylidene fluoride-trifluoroethylene) networks. *Appl. Phys. Lett.* Vol. 79, pp 2627.

A Casalini, R.; Roland, C. M. Electromechanical properties of poly(vinylidene fluoride-trifluoroethylene) networks J. Polym. Sci., Part B: Polym. Phys. 2002,40, 1975.

Choi Jaewu, P. A. Dowben, Pebley Shawn, A. V. Bune. S. Ducharme, V. M. Fridkin, S. P. Palto and N. Petukhova. (1998). *Phys. Rev. Lett.*, Vol. 80. pp 1328.

B. Chu, X. Zhou, K. Ren, B. Neese, M. Lin, Q. Wang, F. Bauer, Q. M. Zhang- (2006).A Dielectric Polymer with High Electric Energy Density and Fast Discharge Speed. *Science.* Vol. 313, pp 334-336.

Chun-gang Duan, W. N. Mei, J. R. Hardy, S. Ducharme, Jaewu Choi and P. A. Dowben. (2003). Comparison of the theoretical and experimental band structure of poly(vinylidene fluoride) crystal. *Europhys. Lett*. Vol. 61 No. 1, pp. 81–87.

G. Cortili, G. Zerbi. (1967). Chain conformations of polyvinylidene fluoride as derived from its vibrational spectrum. *Spectrochim Acta*. Vol. 23A pp 285-299.

A. Cuán and C. M. Cortés-Romero. (2010) Theoretical investigations of structural parameters contribution on zeolitic materials and its interaction with model molecules. Molecular Systems: Theory and Modeling, Editor(s) F. Jiménez-Cruz and J. L. García-Gutiérrez, ISBN 978-81-7895-391-5, Under Release.

D. Das, S.L. Whittenburg. (1999). Performance of the hybrid density functionals in the determination of the geometric structure, vibrational frequency and singlet–triplet energy separation of CH2, CHF, CF2, CCl2 and CBr2. *J of Molecular Structure (Theochem)* Vol. 492 pp 175–186.

B. Daudin, M. Dubus. (1987). Effects of electron irradiation and annealing on ferroelectric vinylidene fluoride-trifluoroethylene copolymers. *J. Appl. Phys.* Vol. 62, pp 994.

E. M. Evleth, E. Kassab, and L. R. Sierra (1994). Calculation of the Exchange Mechanism of D2 and CD4 with a Zeolite Model. *J. Phys. Chem. Vol.* 98, pp 1421.

B. L Farmer, A. J Hopfinger. and J. B.Lando. (1972) Effect of trifluoroethylene monomers on molecular conformation of poly (vinylidene fluoride-trifluoroethylene) copolymer. *J. Appl. Phys.* Vol. 43 pp 4293.

P. J. Feibelman. (1987). Force and total-energy calculations for a spatially compact adsorbate on an extended, metallic crystal surface. *Phys. Rev. B*. Vol. 35, pp 2626.

GAUSSIAN (1995) M. J. Frisch, G. W. Trucks, H. B. Schlegel, P. M. W. Gill, B. G. Johnson, M. A. Robb, J. R. Cheeseman, T. Keith, G. A. Petersson, J. A. Montgomery, K. Raghavachari, M. A. A l. Laham, V. G. Zakrzewski, J. V. Ortiz, J. B. Foresman, J.

Cioslowski, B. B. Stefanov, A. Nanayakkara, M. Challacombe, C. Y. Peng, P. Y. Ayala, W. Chen, M. W. Wong, J. L. Andres, E. S. Replogle, R. Gomperts, R. L. Martin, D. J. Fox, J. S. Binkley, D. J. Defrees, J. Baker, J. P. Stewart, M. Head-Gordon, C. Gonzalez, J. A. Pople, Gaussian 94, Revision D.4, Gaussian, Inc., Pittsburgh, PA.

Q. Gao, J. I. Scheinbeim. (2000). Dipolar Intermolecular Interactions, Structural Development, and Electromechanical Properties in Ferroelectric Polymer Blends of Nylon-11 and Poly(vinylidene fluoride). *Macromolecules*. Vol. 33, pp 7564.

A. Gómez-Zaravaglia, and R. Fausto. (2004). Self-Aggregation in Pyrrole: Matrix Isolation, Solid State Infrared Spectroscopy, and DFT Study. *J. Phys. Chem. A*. Vol. 108, pp 6953.

R. Hasegawa, Y. Takahashi, Y. (1972) Chatani and H.Tadokoro, Molecular conformation and packing of poly(vinylidene fluoride). Stability of three crystalline forms and the effect of high pressure *Polym. J.* Vol. 3, pp 591-605.

B. J. Jungnickel. (1999). Poly(vinylidene fluoride) (overview). In: J.C. Salamone, editor. Polymeric materials handbook. New York: CRC Press Inc; pp. 7115-7122.

N. Karasawa and W. A. Goddard III. (1992). Force, fields, structures and properties of poly(viny1idene fluoride) crystals, *Macromolecules. Vol.* 25 pp7268.

H. Kawai. (1969) The piezoelectricity of poly(vinylidene fluoride). *Jpn. J. Appl. Phys.* Vol. 8,pp 975.

M. Kobayashi, K. Tashiro, H. Tadokoro.(1975). Molecular Vibrations of Three Crystal Forms of Poly(vinylidene fluoride). *Macromolecules*. Vol 8, pp 158-171.

J. B. Lando, H. G. Olf and A. Peterlin. (1966) NMR and X-ray determination of the structure of poly(vinylidene fluoride). *J. Polym. Sci*. Vol. 4,pp 941.

S. B. Lang, S. Muensit. (2006). Review of some lesser-known applications of piezoelectric and pyroelectric polymers. *Appl. Phys. A* Vol. 85, pp 125-134.

C. Lee, W. Yang, R.G. Parr. (1988). Development of the Colle-Salvetti correlation-energy formula into a functional of the electron density. *Phys. Rev. B*. Vol. 37,pp 785–789.

A. J. Lovinger. (1982) Annealing of poly(vinylidene Fluoride) and formation of a fifth phase *Macromolecules*. Vol. 15 pp40-44.

A. Lovinger. (1983). Ferroelectric polymers. *J. Science Vol.* 220, pp 1115.

A. J. Lovinger. (1985). Polymorphic transformations in ferroelectric copolymers of vinylidene fluoride induced by electron irradiation. *Macromolecules*. Vol. 18, pp 910-918.

Y. Lu, J. Claude, Q. M. Zhang, Q. Wang. (2006). Microstructures and Dielectric Properties of the Ferroelectric Fluoropolymers Synthesized via Reductive Dechlorination of Poly(vinylidene fluoride-co-chlorotrifluoroethylene)s, *Macromolecules*. Vol. 39, pp 6962-6968.

N. I. Makarevich, V. N. Nikitin. (1965). Study of the structure of the α- and β-forms of polyvinylidenefluoride. *Polym Sci (USSR)*. Vol. 7. Pp 1843-1849.

H. S. Nalwa. (1995). Ferroelectric Polymers: Chemistry, *Physics, and Applications* (Marcel Dekker, New York, Basel, Hong Kong) pp. 895.

J. B. Nicholas, A. J.Hopfinger, I L. Eton, and , F. R. Trow. (1991). Molecular modeling of zeolite structure. 2. Structure and dynamics of silica sodalite and silicate force field *J. Am. Chem. Soc*. Vol. 113, pp 4792

J. B. Nicholas. (1997). Density functional theory studies of zeolite structure, acidity, and reactivity. *Topics in Catalysis*. Vol. 4, pp 157.

J. Nicholas, T. Ramer, K. Marrone, A. Stiso. (2006). Structure and vibrational frequency determination for a-poly(vinylidene fluoride) using density-functional theory. *Polymer*. Vol. 47, pp7160-7165.

OPIUM. For OPIUM pseudopotential generation programs, see http://opium. sourceforge.net.

E. Ortiz, A. Cuán, C. Badillo, C. M. CortésRomero, Q. Wang and L. Noreña. (2009) Electronic properties of poly(vinylidene fluoride): A Density Functional Theory Study. *Molecular Simulation*. Vol. 35: No. 6, pp 477 -482

E. Ortiz, A. Cuán, C. Badillo, C. M. Cortés-Romero, Q. Wang, L. Noreña. (2010). DFT Study of Ferroelectric Propertiesof the Copolymers: Poly(vinylideneflouride-trifluoroethylene) andPoly(vinylidene fluoridechlorotrifluoroethylene) *International Journal of Quantum Chemistry*, Vol. 110, pp 2411–2417

J. P. Perdew, K. Burke, M. Ernzerhof. (1996). Generalized Gradient Approximation Made Simple *Phys Rev Lett*. Vol.77. Pp 3865-3868.

A. M. Rappe, K. M. Rabe, E. Kaxiras, J. D. Joannopoulos. (1990). Optimized pseudopotentials. *Phys Rev B*. Vol. 41,pp 1227-1230.

P. A. Schultz, SeqQuest Electronic Structure Code (Sandia National Laboratories, 2002).

L. R. Sierra, E. Kassab, and E. M. Evleth. (1993). Calculation of hydroxymethylation of a zeolite model *J. Phys. Chem*. Vol. 97, pp 641-646.

H. Su, A. Strachan, and W. A. Goddard III. (2004). Density functional theory and molecular dynamics studies of the energetics and kinetics of electroactive polymers: PVDF and P(VDF-TrFE). Phys Rev B. Vol. 70, pp 64101

K. Tashiro, S.Nishimura, M.Kobayashi. (1988). Thermal contraction and ferroelectric phase transition in vinylidene fluoride-trifluoroethylene copolymers. An effect of tensile stress along the chain axis *Macromolecules*. Vol. 21, pp 2463.

K. Tashiro, Y. Abe, and M. Kobayashi. (1995) Computer simulation of structure and ferroelectric phase transition of vinylidene fluoride copolymers. I. VDF content dependence of the crystal structure. *Ferroelectrics*. Vol. 171, pp 281.

E. H. Teunissen, A. P. J. Jansen, R. A. van Santen, R. Orlando, R. Dovesi. (1994). Adsorption energies of NH3 and NH4+ in zeolites corrected for the long-range electrostatic potential of the crystal. *J. Chem. Phys*. Vol. 101 pp 5865-5874.

T. T. Wang, J. M. Herbert, and A. M. Glass. (1988). The Applications of Ferroelectric Polymers (Blackie, Glasgow).

Z.Y. Wang, H. Q. Fan, K.H. Su, Z.Y. Wen. (2006). Structure and piezoelectric properties of poly(vinylidene fluoride) studied by density functional theory. *Polymer*. Vol. 47, pp 7988-7996.

T. Wentink, L. J. Willworth, J. P. Phaneuf. (1961). Properties of Polyvinylidene Fluoride. Part II. Infrared Transmission of normal and thermally Decomposed Polymer. *J Polym Sci*. Vol. 55, pp 551-562.

Q. M. Zhang, V. Bharti, X. Zhao. (1998). Giant Electrostriction and Relaxor Ferroelectric Behavior in Electron-Irradiated Poly(vinylidene fluoride-trifluoroethylene) Copolymer. *Science* Vol. 280, pp 2102.

21

An Exact Impedance Control of DC Motors Using Casimir Function

Satoru Sakai
Shinshu University
Japan

1. Introduction

This chapter gives a new and exact impedance control of DC motor. From Hogan's original work, the control input for impedance control is torque input since the impedance control is designed for Lagrangian systems. However, in actual situation, there exist dynamics between the torque and control input and this dynamics can be dominant in certain scale. In such situation, if we neglect the dynamics or try to cancel the dynamcis, the standard impedance control can lose the stability or the control performance at least.
To overcome this problem, we need an new impedance control which takes the dynamics into account wihtout canceling any dynamics. In this chapter we give a solution for this problem by focusing on Casimir function which is rarely used in the conventional robotics.
The rest of this chapter is organized as follwows. In Section 2, we give a new model of DC motor with dynamics between the torque and control input. In Section 3, we propose a new impedance control which is based on Casimir function. Casimir function is one of the properties of port-Hamiltonian systems. In Section 4, we confirm the proposed method in numerical simulation and we conclude this chapter in Section 5.

2. Modeling

Let us start from a well-known model of DC motor:

$$\begin{bmatrix} \dot{\theta} \\ \dot{\omega} \\ \dot{i} \end{bmatrix} = \begin{bmatrix} 0 & \frac{1}{J} & 0 \\ 0 & 0 & K \\ 0 & -\frac{K}{L} & -\frac{R}{L} \end{bmatrix} \begin{bmatrix} \theta \\ \omega \\ i \end{bmatrix} + \begin{bmatrix} 0 \\ 0 \\ \frac{1}{L} \end{bmatrix} v \tag{1}$$

where the displacement θ, the velocity ω and the current i are the states , the voltage v is the control input with the torque constant K and the inductance L.
Although the system (1) is a third-order system and thus not mechanical system, the system (1) has a mechanical-like structure, that is, can be modeled as a port-Hamiltonian system van der Schaft (2000), Maschke & van der Schaft (1996) with a Hamilonian $H = (1/2J)p^2 + (1/2)Kr^2$

$$\begin{bmatrix} \dot{q} \\ \dot{p} \\ \dot{r} \end{bmatrix} = \underbrace{\begin{bmatrix} 0 & 1 & 0 \\ -1 & 0 & \bar{K} \\ 0 & -\bar{K} & -\bar{R} \end{bmatrix}}_{A} \nabla_x H + \begin{bmatrix} 0 \\ 0 \\ \bar{v} \end{bmatrix} \tag{2}$$

where $x = \begin{bmatrix} q & p & r \end{bmatrix}^{\mathrm{T}} = \begin{bmatrix} \theta & J\omega & \sqrt{L}i \end{bmatrix}^{\mathrm{T}}$ is the new state, $\bar{v} = v/\sqrt{L}$ is the new input with $\bar{K} = \frac{K}{\sqrt{L}}$ $\bar{R} = \frac{R}{L}$ and

$$y = \nabla_r H$$

is taken as the passive output. It is confirmed that this system is now passive Takegaki & Arimoto (1981) with respect to the Hamiltonian H, that is,

$$\dot{H} \le y^{\mathrm{T}} v$$

holds. This passive property is studied in many systems and used as a structural property to drive robust and nonliear controllers for stabilization Ortega & Garcia-Canseco (2004) Stramigioli et al. (1998), trajectory tracking Fujimoto & Sugie (2001) and motion generation problems Sakai & Stramigioli (2007).

However, in this chapter, we consider a different problem, namely, impedance control problems and we do not focus on the passivity but focus on another structural property:

Lemma 1 Consider the system (1) with zero-input $v \equiv 0$ in the case of no dissipation $R = 0$. Let the skew-symmetric part of the matrix A be $J(x) = J(x)^{\mathrm{T}}$. Then the system has a solluton of the following PDE

$$\nabla_x C(x) J(x) = 0$$

and the solution is characterized as

$$C(x) = \bar{K}q + r.$$

Proof This is confrimed by a direct calculation. (Q.E.D.)

This means that, in the case of no dissipation $R = 0$, not only the Hamiltonian function H but also the Casimir function C are constant

$$\dot{C} = 0 \quad (u \equiv 0)$$

for any the value of the Hamiltonian function H. Then we can express the system (1) by using the Casimir function (with respect to J) explicity.

Lemma 2 (Modeling) Consider the system (1) with zero-input $v \equiv 0$ in the case of no dissipation $R = 0$. Then the coordinate transformation convert the system (1) into

$$\begin{bmatrix} \dot{q} \\ \dot{p} \\ \dot{C} \end{bmatrix} = \begin{bmatrix} 0 & 1 & 0 \\ -1 & 0 & 0 \\ 0 & 0 & 0 \end{bmatrix} \nabla_{x_c} \bar{H} + \begin{bmatrix} 0 \\ 0 \\ \bar{v} \end{bmatrix} \tag{3}$$

with the state $x_c = (q\ p\ C)^{\mathrm{T}}$ and the Hamiltonian function

$$\bar{H} = \frac{p^2}{2J} + \frac{(\bar{K}q)^2}{2} - \bar{K}qC.$$

Proof This is also confirmed by a direct calculation although the old Hamiltonian H is not equal to the new Hamiltonian $\bar{H}$. (Q.E.D.)

3. Exact impedance control

In this section, we give an exact impedance control for DC motor by using Casimir functions.
Proposition 1 (Main result) *Consider the system (1) with the velocity input. Then the following controller*

$$\begin{cases} \dot{\bar{C}} & = & \bar{K}q - (1+k_c)(\bar{K}q + r) \\ v & = & \bar{C}/J_c \end{cases} \tag{4}$$

converts the close-loop system into the mechanical system with the impedance parameters J_c, $k_c > 0$.
Proof First we introduce an artificial Casimir function $\bar{C}$ and via the following dynamic extension

$$\begin{bmatrix} \dot{q} \\ \dot{p} \\ \dot{C} \\ \dot{\bar{C}} \end{bmatrix} = \begin{bmatrix} 0 & 1 & 0 & 0 \\ -1 & 0 & 0 & 0 \\ 0 & 0 & 0 & 0 \\ 0 & 0 & 0 & 0 \end{bmatrix} \nabla_{x_c} \bar{H} + \begin{bmatrix} 0 \\ 0 \\ v \\ \bar{v} \end{bmatrix} \tag{5}$$

where $\bar{v}$ is the input corresponding to the (artificial) Casimir function.

Then the Hamiltonian function $\bar{H}$ is replaced by the following new Hamiltonian function which has a special structure suitable for impedance design with any parameters $k_c > 0$ and $J_c > 0$ as follows:

$$\begin{aligned} \bar{H}_{mec} &= \bar{H} + \bar{K}\frac{C^2}{2} + \frac{\bar{C}^2}{2J_c} + K_c\frac{C^2}{2} \\ &= \frac{p^2}{2J} + \frac{\bar{K}}{2}\left(q - \frac{C}{\bar{K}}\right)^2 + \frac{k_c C^2}{2} + \frac{\bar{C}^2}{2J_c} \end{aligned} \tag{6}$$

due to the definition of the Casimir function. Finally the dynamic controller

$$\begin{cases} v & = & +\nabla_{\bar{C}} H_{mec} \\ \bar{v} & = & -\nabla_{C} H_{mec} \end{cases} \tag{7}$$

converts the system (1) with a dissipation $R \geq 0$ into the the following Hamiltonian system

$$\begin{bmatrix} \dot{q} \\ \dot{p} \\ \dot{C} \\ \dot{\bar{C}} \end{bmatrix} = \begin{bmatrix} 0 & 1 & 0 & 0 \\ -1 & 0 & 0 & 0 \\ 0 & 0 & -R & 1 \\ 0 & 0 & -1 & 0 \end{bmatrix} \nabla_{\bar{x}_c} H_{mec} \tag{8}$$

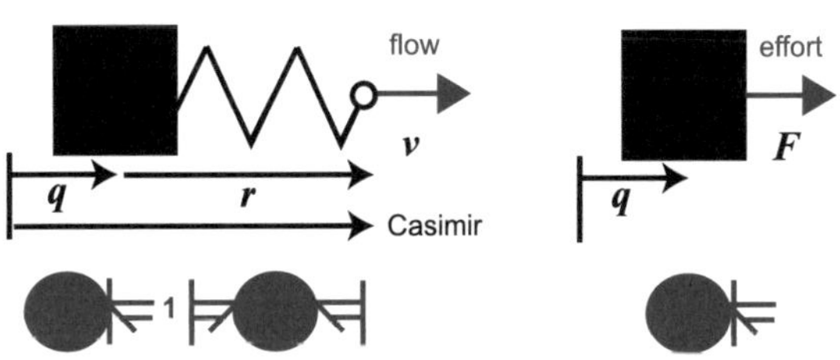

Fig. 1. A port-Hamiltonian system with flow inputs.

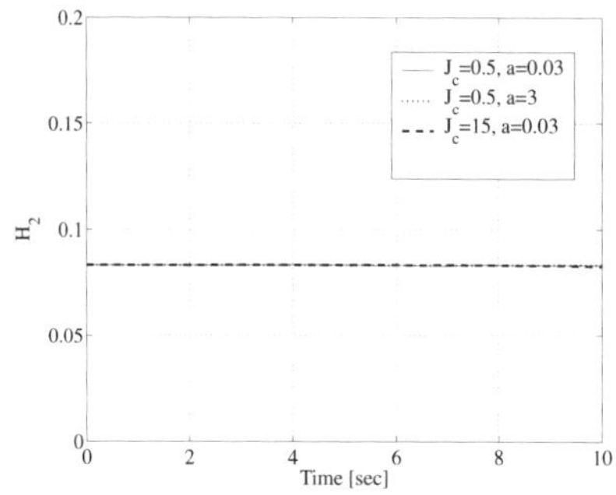

Fig. 2. Time response of H_{mec}

with $\bar{x}_c = (q\ p\ C\ \bar{C})^{\mathrm{T}}$. (Q.E.D.)

The proposed impedance control does not input the torque but the velocity, unlike the conventional impedance control for the mechanical systems. This difference is ullustrated in Fig.1. The spring coefficient k between the real mass and the virtual mass is not design parameter unlike the spring coefficient k_c between the environment and the virtual mass. Note that there is no canceling action in the controller.

4. Numerical simulations

Fig.2 shows the time response of the Hamiltonian function H_{mec} in the case of no dissipation (the Adams method) in the case of the parameters $J = 1.5$ $L = 0.165$ $K = 0.47$, $J_c = 0.5$, $a = 0.03$ and the initial conditions $r(0) = q(0) = 0$ $p(0) = 0.5$. It is confrimed that the value is constant as in the acutal Hamiltonian systems
Figs.3-5 show the time responses of the Casmir function and the state q and p in the case of dissipation $R = 3.2$. The parameters have changed $J_c \to 15$ and $a \to 3$.
In all cases, the nonliner behavior has changed intuitively due to the mechanial strucutre in the closed-loop system. The validity of our methods are confirmed.

5. Conclusions

there exist dynamics between the torque and control input and this dynamics can be dominant in certain scale. In such situation, if we neglect the dynamics or try to cancel the dynamcis, the standard impedance control can lose the stability or the control performance at least.

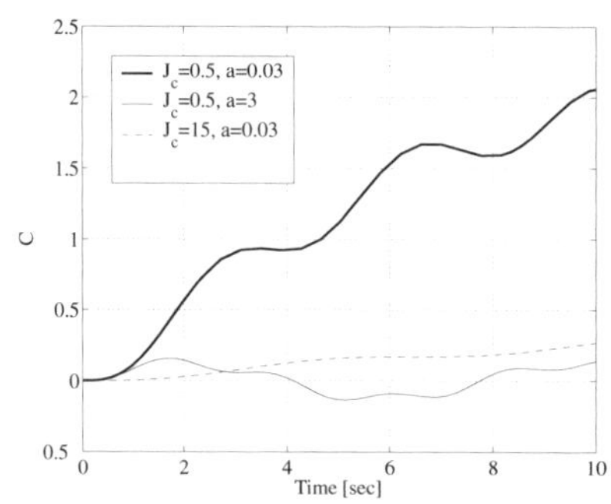

Fig. 3. Time response of $C(t)$

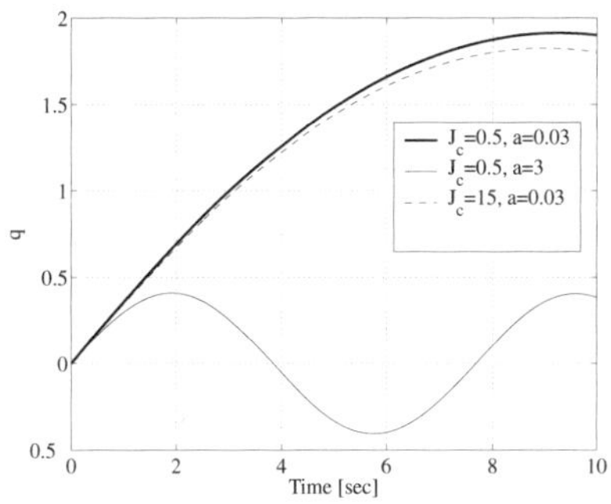

Fig. 4. Time response of $q(t)$

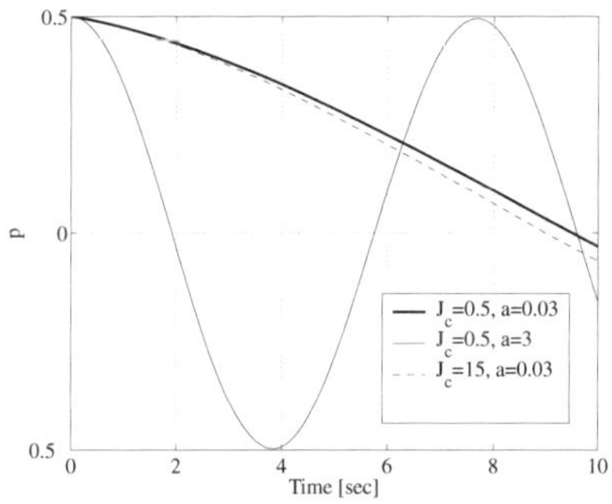

Fig. 5. Time response of $p(t)$

To overcome this problem, we need an new impedance control which takes the dynamics into account wihtout canceling any dynamics. In this chapter we give a solution for this problem by focusing on Casimir function which is rarely used in the conventional robotics.

First we give a new model of DC motor with dynamics between the torque and control input. Second, we propose a new impedance control which is based on Casimir function. Casimir function is one of the properties of port-Hamiltonian systems. Finally, we confirm the proposed method in numerical simulation.

The generalization of the proposed method and appilications to other systems (such as hydraulic systems and mascle-skelton systems) are next works in near future.

6. Acknowledgement

The author would like to show the appreciation to Mr. Ida Shinya for his helps in simulation works.

7. References

Fujimoto, K. & Sugie, T. (2001). Canonical transformation and stabilization of generalized Hamiltonian systems, *Systems & Control Letters* 42(3): 217–227.

Maschke, B. M. J. & van der Schaft, A. J. (1996). Interconnection of systems: the network paradigm, pp. 207–212.

Ortega, R. & Garcia-Canseco, E. (2004). Interconection and damping assignment passivity-based control: A survey, *European Journal of Control* pp. 1–27.

Sakai, S. & Stramigioli, S. (2007). Port-hamiltonian approaches to motion generations for mechanical systems, *Proc. of IEEE Conference on Robotics and Automation*, pp. 69–74.

Stramigioli, S., Maschke, B. M. J. & van der Schaft, A. J. (1998). Passive output feedback and port interconnection, *Proc. 4th IFAC Symp. Nonlinear Control Systems*, pp. 613–618.

Takegaki, M. & Arimoto, S. (1981). A new feedback method for dynamic control of manipulators, *Trans. ASME, J. Dyn. Syst., Meas., Control* 103: 119–125.

van der Schaft, A. J. (2000). *L_2-Gain and Passivity Techniques in Nonlinear Control*, Springer-Verlag, London.

22

Stabilization of Networked Control Systems with Input Saturation

Sung Hyun Kim and PooGyeon Park
Pohang University of Science and Technology
Pohang, Kyungbuk, 790-784
Korea

1. Introduction

The technologies on multi-accessible communication networks have been recently received considerable attention in the area of networked control systems (NCSs) since the multi-accessibility feature leads to flexibility, cost reduction, and distributiveness. However, since the multi-accessibility sometimes causes random network-induced delays that deteriorate the stability and control performance of closed-loop control systems, it is necessary to always handle the delay problem when implementing a feedback control loop closed through multi-accessible communication networks. As a result, numerous investigation and research efforts are underway to deal with the delays (see *e.g.*, Kim *et al.* (2003), Tipsuwan and Chow (2003), Kim *et al.* (2004), Seiler and Sengupta (2005), Yue *et al.* (2005), Yang (2006), and references therein), recent results of which have focused on the problem of attenuating not only the effect of the network-induced delay but also that of unknown disturbances. According to the trend, Kim *et al.* (2004) designed a network-delay-dependent $\mathcal{H}_\infty$ controller for discrete-time systems over communication networks using a deterministic approach. And Seiler and Sengupta (2005) established necessary and sufficient linear matrix inequality (LMI) conditions for the synthesis of the $\mathcal{H}_\infty$ controller for discrete-time systems with Markovian jumping parameters. And Yue *et al.* (2005) designed a robust $\mathcal{H}_\infty$ controller for uncertain continuous-time NCSs with the effects both of the network-induced delay and of data dropout.

In addition to the networked-induced delay and the external disturbance problem, this paper is interested in addressing the input saturation problem since every physical actuator is subject to saturation, and moreover the input saturation deteriorates the stability of the feedback control. Already for the traditional point-to-point communication network, various research results of explicitly handling the input saturation have been published in the system and control literature. Of them, several important results of directly handling the input saturation have recently appeared well in Nguyen and Jabbari (2000), Gomes da Silva *et al.* (2001), Hu and Lin (2001), Hu *et al.* (2002), and references therein. However, to the best of our knowledge, there has been almost no results of considering the input saturation as a constraint in designing the NCSs. Hence, in this paper, we intend to design an $\mathcal{H}_\infty$ control capable of reducing the adverse effects both of the input saturation as well as of the network-induced delay.

The goal of this paper is to design a networked-delay-dependent switching controller for systems with input saturation using the deterministic information from the current *timestamp*. The structural assumption of the networked system follows most of Kim *et al.* (2004). Particulary, we are to employ a reliable transport protocol that guarantees data delivery and supports that transmitted data have their *time-stamp* information. However, taking a step forward in Kim *et al.* (2004), we insert a saturator between the controller and the communication network, and extract built-in memory from the controller for clarity of explanation (see Fig. 1). Based on the modified structure, we propose a

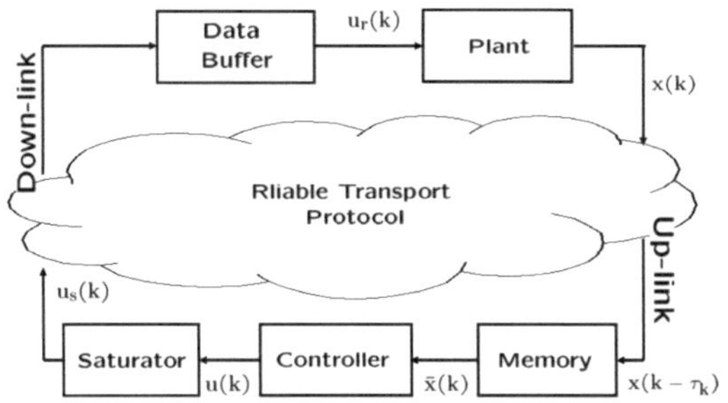

Fig. 1. Networked control system.

systematic method for designing a network-delay-dependent switching controller that achieves the $\mathcal{H}_\infty$ disturbance-rejection performance under an $\mathcal{L}_\infty$ performance representing componentwise input saturation. Methodologically, we first construct a suitable networked control system (NCS) with asymmetric path-delay, and build up the conditions for set invariance, involved in local stabilization, and then incorporate these conditions in the synthesis of a network-delay-dependent $\mathcal{H}_\infty$ controller. In the procedure, we use the polytopic representation method, proposed by Hu and Lin (2001), as one way to deal with the saturation nonlinearity, and employ a quasi-linear parameter-varying (QLPV) structure for the control system as in Wu *et al.* (2005). Then the resultant convex solvability conditions are expressed as a finite number of linear matrix inequalities (LMIs), whose solutions are applied to the construction of the switching controller depending on the previous and current modes characterized by network traffic.

The chapter is organized as follows: Section 2 states target systems and assumptions. Section 3 proposes the condition for set invariance, and explains the procedure of constructing an $\mathcal{H}_\infty$ dynamic state-feedback controller. Section 4 illustrates the performance of the proposed algorithm through an example. Finally, in Section 5, concluding remarks are made.

Notation: Notations in this paper are fairly standard. For $x \in \mathcal{R}^n$, $||x||$ is taken to be the standard Euclidian norm., *i.e.*, $||x|| = (x^T x)^{1/2}$. The Lebesgue space $\mathcal{L}_{2+} = \mathcal{L}_2[0,\infty)$ consists of square-integrable functions on $[0,\infty)$. $\mathcal{L}_\infty$ denotes the space of bounded vector sequences $u(k)$, equipped with the norm $||u||_\infty = \sup_i\{\sup_k |u_i(k)|\}$, and $\mathcal{L}_{\infty,e}$ denotes the space of bounded vector sequences $w(k)$, with the norm $||w||_{\infty,e} = \sup_k\{w^T(k)w(k)\}$. The notation $X \geq Y$ and $X > Y$ means that $X - Y$ is positive semi-definite and positive definite, respectively. Inequalities between vectors mean componentwise inequalities, $\mathrm{svd}(\cdot)$ means a singular value decomposition function, and the saturation function $\mathrm{sat}(u,\bar{u}) : \mathcal{R}^m \to \mathcal{R}^m$ denotes

$$\mathrm{sat}(u,\bar{u}) = [s_1\ s_2\ \cdots\ s_m]^T,\ s_i = \mathrm{sign}(u_i)\cdot\min\{\bar{u}_i, |u_i|\}, \tag{1}$$

where $\mathrm{sign}(\cdot)$ returns the signs of the corresponding argument, and u_i and $\bar{u}_i$ stand for the i-the element of $u \in \mathcal{R}^m$ and $\bar{u} \in \mathcal{R}^m$, respectively. For a matrix $V \in \mathcal{R}^{r\times s}$, $\mathcal{L}(V)$ denotes a linear region as defined in

$$\mathcal{L}(V) \stackrel{\triangle}{=} \{x \in \mathcal{R}^s | -\sigma \leq Vx \leq \sigma\}, \tag{2}$$

where $\sigma \in \mathcal{R}^r$ and $x \in \mathcal{R}^s$. Finally, in symmetric block matrices, $(*)$ is used as an ellipsis for terms that are induced by symmetry.

2. Mathematical representation

Consider the following linear time-invariant (LTI) plant of the form

$$x(k+1) = Ax(k) + B_1 w(k) + B_2 u_r(k), \tag{3}$$
$$z(k) = Cx(k) + D_1 w(k) + D_2 u_r(k), \tag{4}$$

where $x(k) \in \mathcal{R}^n$, $u_r(k) \in \mathcal{R}^m$, $w(k) \in \mathcal{R}^p$ and $z(k) \in \mathcal{R}^q$ denote the state, the input, the disturbance and the performance output, respectively. Throughout this paper, we assume that the disturbance $w(k)$ is unknown but belongs to a known bounded set $\mathcal{W}$ defined as:

$$\mathcal{W} \stackrel{\triangle}{=} \left\{ w \in \mathcal{R}^p \mid ||w(k)||^2 \leq \bar{w},\ \bar{w} \geq 0,\ \forall k \geq 0 \right\}, \tag{5}$$

which implies $w(k) \in \mathcal{L}_{\infty,e}$ for all $k \geq 0$. In addition, we recall some assumptions from Kim *et al.* (2004):

(A1) Network-induced delay is composed of the down-link delay τ_k^d and the up-link delay τ_k^u which are random but bounded as $0 \leq \tau_k^d \leq \bar{\tau}$, $0 \leq \tau_k^u \leq \bar{\tau}$.

(A2) Multiple data can be received at the same time.

(A3) Time-stamp information is appended to the transmitted data between the plant and a controller, which plays an important role in configuring a networked-control system (NCS) for the plant (4).

Fig. 1 shows an NCS subject to our needs, where the difference from that of Kim *et al.* (2004) is that a saturator is inserted between the controller and the communication network, i.e., $u_s(k) = \text{sat}(u(k), \bar{u})$, and a built-in memory is extracted from the controller for clarity of explanation. Except for the difference, the remaining framework follows that of Kim *et al.* (2004), that is, the down-link delay τ_k^d is fixed into its bound value $\bar{\tau}$ through the data buffer, and the up-link delay τ_k^u is confirmed by the real-time information on the up-link delay sequence delivered to the controller.

With the above settings, the system model in the controller-saturator point of view is given as

$$\tilde{x}(k+1) = \tilde{A}\,\tilde{x}(k) + \tilde{B}_1\,w(k) + \tilde{B}_2\,u_s(k), \tag{6}$$
$$z(k) = \tilde{C}\,\tilde{x}(k) + \tilde{D}\,w(k), \tag{7}$$

where $\tilde{x}(k) \triangleq [x^T(k) \mid x^T(k-1) \cdots x^T(k-\bar{\tau}) \mid u_s^T(k-1) \cdots u_s^T(k-\bar{\tau})]^T \in \mathcal{R}^{n+(n+m)\bar{\tau}}$ denotes the augmented state and the matrices are defined as

$$\left[\begin{array}{c|c|c} \tilde{A} & \tilde{B}_1 & \tilde{B}_2 \\ \hline \tilde{C} & \tilde{D} & 0 \end{array}\right] \triangleq \left[\begin{array}{c|cccc|cccc|c|c} A & 0 & \cdots & \cdots & 0 & 0 & \cdots & 0 & B_2 & B_1 & 0 \\ \hline I & 0 & 0 & \cdots & 0 & 0 & \cdots & \cdots & 0 & 0 & 0 \\ 0 & I & 0 & & \vdots & \vdots & & & \vdots & \vdots & \vdots \\ \vdots & \ddots & \ddots & \ddots & \vdots & \vdots & & & \vdots & \vdots & \vdots \\ 0 & \cdots & 0 & I & 0 & 0 & \cdots & \cdots & 0 & 0 & 0 \\ \hline 0 & 0 & \cdots & \cdots & 0 & 0 & 0 & \cdots & 0 & 0 & I \\ 0 & \vdots & & & \vdots & I & 0 & & \vdots & \vdots & 0 \\ \vdots & \vdots & & & \vdots & 0 & I & \ddots & \vdots & \vdots & \vdots \\ \vdots & \vdots & & & \vdots & \vdots & \ddots & \ddots & 0 & \vdots & \vdots \\ 0 & 0 & \cdots & \cdots & 0 & 0 & \cdots & 0 & I & 0 & 0 \\ \hline C & 0 & \cdots & \cdots & 0 & 0 & \cdots & 0 & D_2 & D_1 & 0 \end{array}\right]. \tag{8}$$

And the state available in the controller, *say*, $\bar{x}(k)$, is given as

$$\bar{x}(k) = \tilde{E}_{m(k)}\tilde{x}(k), \tag{9}$$

where

$$\tilde{E}_{m(k)} \triangleq \left[\begin{array}{c|cccc|ccc} \Phi_{k0} & 0 & \cdots & \cdots & 0 & 0 & \cdots & 0 \\ \hline 0 & \Phi_{k1} & 0 & & \vdots & \vdots & & 0 \\ \vdots & 0 & \Phi_{k2} & \ddots & \vdots & \vdots & & \vdots \\ \vdots & \vdots & \ddots & \ddots & 0 & \vdots & & \vdots \\ 0 & 0 & \cdots & 0 & \Phi_{k\bar{\tau}} & 0 & \cdots & 0 \end{array}\right], \tag{10}$$

$$\Phi_{kr} \triangleq \begin{cases} I & \text{if } x(k-r) \text{ is avilable at time } k, \\ 0 & \text{otherwise.} \end{cases} \tag{11}$$

Here, $m(k)$ denotes a mode corresponding to each status of $\tilde{E}_{m(k)}$ with $(2^{\bar{\tau}+1}-1)$ different cases, and the unique number assigned to the mode $m(k)$ can be expressed as $m(k) = (b_0 b_1 \cdots b_{\bar{\tau}})_2$, where $(\cdot)_2$ means the binary representation of $m(k)$, and the r-th bit b_r is set to 1 if the r-delayed state, $x(k-r)$, is available at time k, otherwise, the bit is set to 0. From this binary representaion, we can know that the mode $m(k)$ belongs to a set $\mathcal{M} \triangleq \{m \in \mathcal{R} \mid m = 1, 2, \cdots, 2^{\bar{\tau}+1}-1\}$. Besides, under the assumption (A2), we can uniquely determine a set of transitions, $\mathcal{S}$, only if $\bar{\tau}$ is determined:

$$\mathcal{S} \triangleq \{(m(k), m(k-1)) \mid \text{all possible transition pairs yielding (A2)} \\ \text{for } m(k) \in \mathcal{M},\ m(k-1) \in \mathcal{M},\ \forall k\}. \tag{12}$$

For each case of $\bar{\tau} = 1, 2$, all possible modes and transitions has already been described by Kim *et al.* (2004), where if the initial state set $\{x(0), x(1), \cdots, x(-\bar{\tau})\}$ is given, and the packet loss

does not exist, then there is no necessity for considering all transitions as in Kim *et al.* (2004) since the assumption (A1) ensures $b_{\bar{\tau}} = 1$.
The following lemma presents the polytopic representation method for the input saturation, proposed by Hu and Lin (2001).

Lemma 2.1 *(Hu and Lin (2001)) Let $\mathcal{D}$ be the set of $m \times m$ diagonal matrices whose diagonal elements are either 1 or 0. Suppose that $|v_r| \leq \bar{u}_r$ for all $r = 1, \cdots, m$, where v_r and $\bar{u}_r$ denote the r-th element of $v \in \mathcal{R}^m$ and $\bar{u} \in \mathcal{R}^m$, respectively. Then*

$$\mathrm{sat}(u, \bar{u}) = \sum_{\ell=1}^{2^m} \theta_\ell \left(D_\ell u + D_\ell^- v \right), \; \sum_{\ell=1}^{2^m} \theta_\ell = 1, \; \theta_\ell \geq 0, \tag{13}$$

where D_ℓ denote all elements of $\mathcal{D}$, and $D_\ell^- = I - D_\ell$.

In the following, we present a previous mode (PM)-dependent dynamic quasi-linear parameter varying (QLPV) control law which switches itself depending on its previous and current modes:

$$x_c(k+1) = F_{ji}(\Theta_k) x_c(k) + G_{ji}(\Theta_k) \bar{x}(k), \tag{14}$$

$$u(k) = H_{ji} x_c(k) + J_{ji} \bar{x}(k), \tag{15}$$

$$v(k) = K_{ji} x_c(k) + L_{ji} \bar{x}(k), \tag{16}$$

subject to

$$\left[F_{ji}(\Theta_k) \; G_{ji}(\Theta_k) \right] = \sum_{\ell=1}^{2^m} \theta_\ell(k) \left[F_{ji}^\ell \; G_{ji}^\ell \right], \tag{17}$$

where $v(k)$ is an auxiliary control input, $\Theta_k \in \mathcal{R}^{2m}$ denotes a vector consisting of the time-varying interpolation coefficients $\theta_\ell(k)$ at time k, and the subscripts j and i stand for $m(k)$ and $m(k-1)$, respectively.
Consequently, by Lemma 2.1, the closed-loop system subject to $\hat{x}(k) \in \mathcal{L}([L_{ji}\tilde{E}_j \;\; K_{ji}])$, for all $k \geq 0$, is given as

$$\hat{x}(k+1) = \hat{A}_{ji}(\Theta_k)\hat{x}(k) + \hat{B}w(k), \; w(k) \in \mathcal{W}_\delta, \tag{18}$$

$$z(k) = \hat{C}\hat{x}(k) + \hat{D}w(k), \tag{19}$$

where $\hat{A}_{ji}(\Theta_k) = \sum_{\ell=1}^{2^m} \theta_\ell(k) \hat{A}_{ji}^\ell$, $\hat{B}^T = \left[\tilde{B}_1^T \; 0\right]^T$, $\hat{C} = \left[\tilde{C} \; 0\right]$, $\hat{D} = \tilde{D}$ and

$$\hat{A}_{ji}^\ell = \begin{bmatrix} \tilde{A} + \tilde{B}_2 \left(D_\ell J_{ji} + D_\ell^- L_{ji} \right) \tilde{E}_j & \tilde{B}_2 \left(D_\ell H_{ji} + D_\ell^- K_{ji} \right) \\ G_{ji}^\ell \tilde{E}_j & F_{ji}^\ell \end{bmatrix}. \tag{20}$$

3. Main results

This section is explained in three steps:
- invariant ellipsoid property,
- $\mathcal{H}_\infty$ problem description,
- linear matrix inequality (LMI) formulation.

3.1 Invariant ellipsoid property

Before designing a controller, we shall first derive the conditions for obtaining the ellipsoidal sets $\mathcal{E}(P_i)$ such that, for all $k \geq 0$,

$$\psi(k,\hat{x}(0),w) \in \mathcal{E}(P_i),\ \forall \hat{x}(0) \in \mathcal{E}(P_i),\ i \in \mathcal{M},\ w \in \mathcal{W}, \tag{21}$$

where $\psi(\cdot)$ denotes the state trajectory of the closed-loop system, and $\mathcal{E}(P_i)$ denote PM-dependent ellipsoidal sets defined as

$$\mathcal{E}(P_i) \triangleq \left\{ \hat{x} \in \mathcal{R}^{2(n+(n+m)\bar{\tau})} \mid \hat{x}^T P_i \hat{x} \leq 1,\ P_i > 0 \right\},\ \forall i \in \mathcal{M}. \tag{22}$$

The following lemma presents the conditions for obtaining the ellipsoidal sets $\mathcal{E}(P_i)$ with the property (21).

Lemma 3.1 *Let $\bar{w} \geq 0$ be given. Suppose that there exist $0 \leq \lambda_1 \leq 1$ and $\bar{P}_i > 0$ such that*

$$0 \leq \begin{bmatrix} \lambda_1 P_i & 0 & (*) \\ 0 & (1/\bar{w})(1-\lambda_1)I & (*) \\ \hat{A}_{ji}^{\ell} & \hat{B} & \bar{P}_j \end{bmatrix},\ \forall (j,i) \in \mathcal{S},\ \ell \in [1,2^m], \tag{23}$$

$$\mathcal{E}(P_i) \subset \mathcal{L}(V_{ji}),\ \forall (j,i) \in \mathcal{S}, \tag{24}$$

where $\bar{P}_j \triangleq P_j^{-1}$. Then there exist the ellipsoidal sets $\mathcal{E}(P_i)$ with the property (21).

Proof: The property (21) can be altered as follows: for $k \geq 0$, $(j,i) \in \mathcal{S}$,

$$\hat{x}(k+1) \in \mathcal{E}(P_j) \text{ subject to } \hat{x}(k) \in \mathcal{E}(P_i),\ w^T(k)w(k) \leq \bar{w}. \tag{25}$$

At time k, let $\hat{x}(k) \in \mathcal{E}(P_i)$ and $w^T(k)w(k) \leq \bar{w}$, that is,

$$0 \leq 1 - \hat{x}^T(k)P_i\hat{x}(k) \text{ and } 0 \leq \bar{w} - w^T(k)w(k). \tag{26}$$

Then, by the condition (24), the transition of the state $\hat{x}(k)$ is determined by the closed-loop system (18), and hence $\hat{x}(k+1) \in \mathcal{E}(P_j)$ becomes

$$0 \leq \begin{bmatrix} \hat{x}(k) \\ w(k) \\ 1 \end{bmatrix}^T \begin{bmatrix} -\hat{A}_{ji}^T(\Theta_k)P_j\hat{A}_{ji}(\Theta_k) & (*) & 0 \\ -\hat{B}^T P_j \hat{A}_{ji}(\Theta_k) & -\hat{B}^T P_j \hat{B} & 0 \\ 0 & 0 & 1 \end{bmatrix} \begin{bmatrix} \hat{x}(k) \\ w(k) \\ 1 \end{bmatrix} \forall (j,i) \in \mathcal{S}. \tag{27}$$

To convert, based on (26) and (27), the condition (25) into a matrix inequality. we employ the S-procedure as a constraint-elimination method for the conditions (26), which yields

$$0 \leq \left[\begin{array}{cc|c} -\hat{A}_{ji}^T(\Theta_k)P_j\hat{A}_{ji}(\Theta_k) + \lambda_1 P_i & (*) & 0 \\ -\hat{B}^T P_j \hat{A}_{ji}(\Theta_k) & -\hat{B}P_j\hat{B} + \lambda_2 I & 0 \\ \hline 0 & 0 & 1 - \lambda_1 - \lambda_2 \bar{w} \end{array} \right], \tag{28}$$

where $\lambda_1 \geq 0$ and $\lambda_2 \geq 0$. And, from (28), it is straightforward

$$0 \leq \begin{bmatrix} -\hat{A}_{ji}^T(\Theta_k)P_j\hat{A}_{ji}(\Theta_k) + \lambda_1 P_i & (*) \\ -\hat{B}^T P_j \hat{A}_{ji}(\Theta_k) & -\hat{B}P_j\hat{B} + \lambda_2 I \end{bmatrix},\ \forall (j,i) \in \mathcal{S}, \tag{29}$$

$$0 \leq \lambda_2 \leq (1/\bar{w})(1-\lambda_1). \tag{30}$$

Here, note that $\lambda_2 = (1/\varpi)(1-\lambda_1)$ since, for any given λ_1, the feasibility of P_i increases as λ_2 increases. Thus, with the help of Schur complements, the conditions (29) and (30) becomes

$$0 \leq \begin{bmatrix} \lambda_1 P_i & 0 & (*) \\ 0 & (1/\varpi)(1-\lambda_1)I & (*) \\ \hat{A}_{ji}(\Theta_k) & \hat{B} & P_j^{-1} \end{bmatrix}, \forall (j,i) \in \mathcal{S}. \tag{31}$$

Furthermore, since multiplying (23) by $\theta_\ell(k)$ and summing it from $\ell = 1$ to $\ell = 2^m$ yields (31), it is clear that the condition (31) also holds if the condition (23) holds. ∎

3.2 $\mathcal{H}_\infty$ Problem description

Now let us consider the PM-dependent Lyapunov candidate $V_i(\hat{x}(k))$ given as

$$V_i(\hat{x}(k)) = \hat{x}^T(k) P_i \hat{x}(k), \ \ P_i > 0, \ \forall i \in \mathcal{M}. \tag{32}$$

Then the following two statements are equivalent:

- The closed-loop system (18) is stable with the $\mathcal{H}_\infty$ performance γ.
- There exist P_i such that (23), (24),

$$0 \leq \begin{bmatrix} P_i & 0 & (*) & (*) \\ 0 & \gamma^2 I & (*) & (*) \\ \hat{A}_{ji}(\Theta_k) & \hat{B} & P_j^{-1} & 0 \\ \hat{C} & \hat{D} & 0 & I \end{bmatrix}, \forall (j,i) \in \mathcal{S}. \tag{33}$$

In this equivalence, the condition (33) is directly derived by

$$V_i(\bar{x}(k+1)) - V_i(\bar{x}(k)) + z^T(k)z(k) - \gamma^2 w^T(k) w(k) \leq 0, \ \forall i \in \mathcal{M}, \tag{34}$$

where since the conditions (23) and (24) make the state trajectories remain inside $\mathcal{E}(P_i) \subset \mathcal{L}(V_{ji})$, the transition of the state $\hat{x}(k)$ is always determined by the closed-loop system (18). Consequently, we shall solve the following minimization problem to construct a PM-dependent dynamic QLPV controller which achieves the maximal disturbance rejection capability:

$$\min \gamma \text{ subject to } (23), (24), \text{ and } (33). \tag{35}$$

3.3 LMI formulation

With the help of the replacement method, we shall formulate the conditions in the optimization problem (35) in terms of LMIs. To this end, let us first partition matrices P_i and $\bar{P}_i$ in the form:

$$P_i \triangleq \begin{bmatrix} X_i & Y_i \\ Y_i^T & Z_i \end{bmatrix}, \ \bar{P}_i = P_i^{-1} \triangleq \begin{bmatrix} \bar{X}_i & \bar{Y}_i \\ \bar{Y}_i^T & \bar{Z}_i \end{bmatrix}. \tag{36}$$

Theorem 3.1 *Let $\varpi \geq 0$ be given. Suppose, for a prescribed value $0 \leq \lambda_1 \leq 1$, that there exist matrices X_i, $\bar{X}_i$, $\Psi_{1,ji}^\ell$, $\Psi_{2,ji}$, $\Psi_{3,ji}$, Π_{ji}^ℓ, J_{ji}, L_{ji}, and Γ that are solutions of the following optimization problem:*

$$\gamma^* = \min \gamma \tag{37}$$

subject to, for all $(j,i) \in \mathcal{S}$,

$$0 \leq \begin{bmatrix} \lambda_1 X_i & (*) & (*) & (*) & (*) \\ \lambda_1 I & \lambda_1 \bar{X}_i & (*) & (*) & (*) \\ 0 & 0 & (1/\bar{w})(1-\lambda_1)I & (*) & (*) \\ \Delta_{ji}^{\ell}(1,1) & \Pi_{ji}^{\ell} & X_j \tilde{B}_1 & X_j & (*) \\ \Delta_{ji}^{\ell}(2,1) & \Delta_{ji}^{\ell}(2,2) & \tilde{B}_1 & I & \bar{X}_j \end{bmatrix}, \ \ell \in [1, 2^m], \tag{38}$$

$$0 \leq \begin{bmatrix} \Gamma & L_{ji}\tilde{E}_j & \Psi_{3,ji} \\ (*) & X_i & I \\ (*) & (*) & \bar{X}_i \end{bmatrix}, \ \Gamma_{rr} \leq \bar{u}_r^2, \ r \in [1, m], \tag{39}$$

$$0 \leq \begin{bmatrix} X_i & (*) & (*) & (*) & (*) & (*) \\ I & \bar{X}_i & (*) & (*) & (*) & (*) \\ 0 & 0 & \gamma^2 I & (*) & (*) & (*) \\ \Delta_{ji}^{\ell}(1,1) & \Pi_{ji}^{\ell} & X_j \tilde{B}_1 & X_j & (*) & (*) \\ \Delta_{ji}^{\ell}(2,1) & \Delta_{ji}^{\ell}(2,2) & \tilde{B}_1 & I & \bar{X}_j & (*) \\ \tilde{C} & \tilde{C}\bar{X}_i & \tilde{D} & 0 & 0 & I \end{bmatrix}, \ \ell \in [1, 2^m], \tag{40}$$

where Γ_{rr} denotes the r-th diagonal element of Γ,

$$\Delta_{ji}^{\ell}(1,1) \stackrel{\triangle}{=} X_j \tilde{A} + \Psi_{1,ji}^{\ell} \tilde{E}_j, \tag{41}$$

$$\Delta_{ji}^{\ell}(2,1) \stackrel{\triangle}{=} \tilde{A} + \tilde{B}_2 \left(D_\ell J_{ji} + D_\ell^- L_{ji} \right) \tilde{E}_j, \tag{42}$$

$$\Delta_{ji}^{\ell}(2,2) \stackrel{\triangle}{=} \tilde{A}\bar{X}_i + \tilde{B}_2 \left(D_\ell \Psi_{2,ji} + D_\ell^- \Psi_{3,ji} \right). \tag{43}$$

Then closed-loop system (18) is asymptotically stable in the absence of disturbances, and $||z(k)||_2 \leq \gamma^* ||w(k)||_2$ holds in the presence of disturbances. Moreover, based on the solutions X_i, $\bar{X}_i$, $\Psi_{1,ji}^{\ell}$, $\Psi_{2,ji}$, $\Psi_{3,ji}$, Π_{ji}^{ℓ}, J_{ji} and L_{ji}, the PM-dependent dynamic QLPV control gains $(F_{ji}(\Theta_k), G_{ji}(\Theta_k), H_{ji}, J_{ji})$ can be obtained by the following procedure: :

- **Off-line procedure**
  - *(i) obtain Y_i and Z_i from* $\mathrm{svd}(X_i - \bar{X}_i^{-1}) = Y_i Z_i^{-1} Y_i^T$, *and then obtain $\bar{Y}_i = -\bar{X}_i Y Z_i^{-1}$.*
  - *(ii) reconstruct H_{ji}, K_{ji}, G_{ji}^{ℓ}, and F_{ji}^{ℓ}:*

$$H_{ji} = \left(\Psi_{2,ji} - J_{ji}\tilde{E}_j \bar{X}_i \right) \bar{Y}_i^{-T}, \tag{44}$$

$$K_{ji} = \left(\Psi_{3,ji} - L_{ji}\tilde{E}_j \bar{X}_i \right) \bar{Y}_i^{-T}, \tag{45}$$

$$G_{ji}^{\ell} = Y_j^{-1} \left(\Psi_{1,ji}^{\ell} - X_j \tilde{B}_2 \left(D_\ell J_{ji} + D_\ell^- L_{ji} \right) \right), \tag{46}$$

$$F_{ji}^{\ell} = Y_j^{-1} \left(\Pi_{ji}^{\ell} - X_j \tilde{A} \bar{X}_i - \Psi_{1,ji}^{\ell} \tilde{E}_j \bar{X}_i - X_j \tilde{B}_2 \left(D_\ell H_{ji} + D_\ell^- K_{ji} \right) \bar{Y}_i^T \right) \bar{Y}_i^{-T}. \tag{47}$$

- **On-line procedure**
  - *(i) obtain $u(k)$ and $v(k)$ on-line from (15) and (16), respectively, and then calculate the interpolation coefficient vector Θ_k from (13).*
  - *(ii) update, based on (17), $F_{ji}(\Theta_k)$ and $G_{ji}(\Theta_k)$ at discrete time instance.*

Proof: Before formulating the conditions (23), (24), and (33) in terms of LMIs, let us define matrices W_i as

$$W_i \triangleq \begin{bmatrix} X_i & I \\ Y_i^T & 0 \end{bmatrix}, \forall i \in \mathcal{M}, \tag{48}$$

and, without loss of generality, assume that the matrices Y_i are full rank. And then, to convert the condition (23) into (38), we pre- and post-multiply T_{1ji}^T and $T_{1ji} = \text{blockdiag}(\bar{P}_i W_i, I, W_j)$ on the right-hand side of the inequality (23), respectively, which yields

$$0 \le \begin{bmatrix} \lambda_1 W_i^T \bar{P}_i W_i & 0 & (*) \\ 0 & (1/\delta)(1-\lambda_1) I & \bar{B}^T \\ W_j^T \hat{A}_{ji}^{\ell} \bar{P}_i W_i & W_j^T \hat{B} & W_j^T \bar{P}_j W_j \end{bmatrix}, \tag{49}$$

where

$$W_i^T \bar{P}_i W_i = \begin{bmatrix} X_i & I \\ I & \bar{X}_i \end{bmatrix}, \; W_j^T \hat{B} = \begin{bmatrix} X_j \tilde{B}_1 \\ \tilde{B}_1 \end{bmatrix}, \tag{50}$$

$$W_j^T \hat{A}_{ji}^{\ell} \bar{P}_i W_i = \begin{bmatrix} \Delta_{ji}^{\ell}(1,1) & \Pi_{ji}^{\ell} \\ \Delta_{ji}^{\ell}(2,1) & \Delta_{ji}^{\ell}(2,2) \end{bmatrix}, \tag{51}$$

$$\Delta_{ji}^{\ell}(1,1), \Delta_{ji}^{\ell}(2,1), \text{ and } \Delta_{ji}^{\ell}(2,2) \text{ in (41)–(43)}, \tag{52}$$

$$\Psi_{1,ji}^{\ell} \triangleq X_j \tilde{B}_2 \left(D_\ell J_{ji} + D_\ell^- L_{ji} \right) + Y_j G_{ji}^{\ell}, \tag{53}$$

$$\Psi_{2,ji} \triangleq J_{ji} \tilde{E}_j \bar{X}_i + H_{ji} \bar{Y}_i^T, \tag{54}$$

$$\Psi_{3,ji} \triangleq L_{ji} \tilde{E}_j \bar{X}_i + K_{ji} \bar{Y}_i^T, \tag{55}$$

$$\Pi_{ji}^{\ell} \triangleq X_j \tilde{A} \bar{X}_i + \Psi_{1,ji}^{\ell} \tilde{E}_j \bar{X}_i + X_j \tilde{B}_2 \left(D_\ell H_{ji} + D_\ell^- K_{ji} \right) \bar{Y}_i^T + Y_j F_{ji}^{\ell} \bar{Y}_i^T. \tag{56}$$

Here, from (53)–(56), it follows (44)–(47). Next, to convert the condition $\mathcal{E}(P_i) \subset \mathcal{L}(V_{ji})$, that is,

$$0 \le \begin{bmatrix} \Gamma & V_{ji} \\ (*) & P_i \end{bmatrix}, \; \Gamma_{rr} \le \bar{u}_r^2, \; \forall r \in [1,m], \tag{57}$$

into (39), we pre- and post-multiply T_{2i}^T and $T_{2i} = \text{blockdiag}(I, \bar{P}_i W_i)$ on the right-hand side of (57), respectively, which yields

$$0 \le \begin{bmatrix} \Gamma & V_{ji} \bar{P}_i W_i \\ (*) & W_i^T P_i W_i \end{bmatrix}, \; \Gamma_{rr} \le \bar{u}_r^2, \; \forall r \in [1,m], \tag{58}$$

where $V_{ji} \bar{P}_i W_i = \begin{bmatrix} L_{ji} \tilde{E}_j & \Psi_{3,ji} \end{bmatrix}$. Finally, by pre- and post-multiplying T_{3ji}^T and $T_{3ji} = \text{blockdiag}(\bar{P}_i W_i, I, W_j, I)$ on the right-hand side of the inequality (33), respectively, and then by substituting (50), (51) and $\hat{C} \bar{P}_i W_i = \begin{bmatrix} \tilde{C} & \tilde{C} \bar{X}_i \end{bmatrix}$, we can obtain (40). ∎

Remark 3.1 *Theorem 3.1 can be applied to the case of the delay-free communication network $\bar{\tau} = 0$ by substituting (A, B_1, B_2, C, I) for $(\tilde{A}, \tilde{B}_1, \tilde{B}_2, \tilde{C}, \tilde{E}_j)$,*

Remark 3.2 *The greatest disturbance rejection capability, i.e., the smallest γ^*, can be obtained by tuning the prescribed value λ_1 between 0 and 1.*

4. Numerical example

$\bar{\tau}$	$\bar{u}$	γ^*	$\bar{u}$	γ^*
1	1	0.5005 ($\lambda_1 = 0.91$)	3	0.2910 ($\lambda_1 = 0.94$)
2	1	0.5284 ($\lambda_1 = 0.90$)	3	0.3128 ($\lambda_1 = 0.91$)

Table 1. Minimized $\mathcal{H}_\infty$ γ-performance

To verify the performance of the proposed control algorithm, we consider a classical angular positioning system, adapted from Kwakernaak and Sivan (1972), with the sampling time $0.2s$:

$$\left[\begin{array}{c|c|c} A & B_1 & B_2 \\ \hline C & D_1 & D_2 \end{array}\right] = \left[\begin{array}{cc|c|c} 1.0 & 0.20 & 0.02 & 0.0000 \\ 0.0 & 0.08 & 0.12 & 0.1574 \\ \hline 1.0 & 0.00 & 0.15 & 0.1000 \end{array}\right], \tag{59}$$

where the control problem is to supply the input voltage (u_r volts) to the motor in order to rotate the antenna toward the desired direction. To demonstrate the effectiveness of our

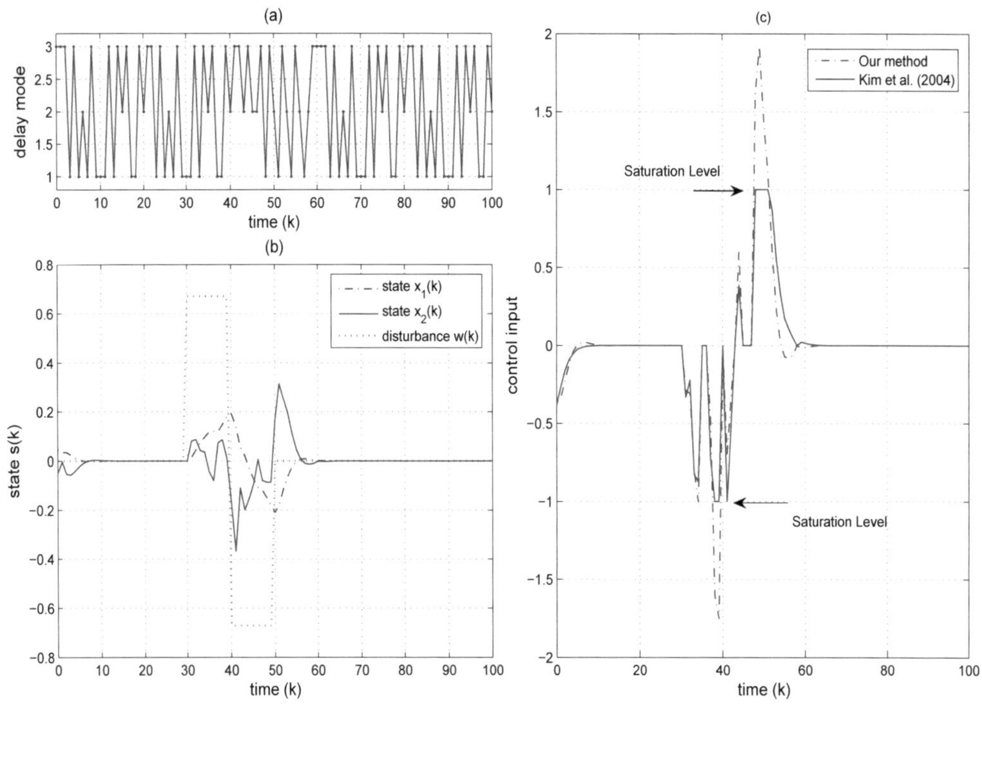

Fig. 2. (a) Delay mode $m(k)$; (b) external disturbance $w(k)$, and state responds $x_1(k)$ and $x_2(k)$; (d) control inputs for our method and Kim *et al.* (2004).

method for the control problem, we assume that the impressed voltage is constrained to be within the limits $-\bar{u}$ to $\bar{u}$ and that a multi-accessible communication network is employed for the transmission of the information between the system and the controller. Based on the setting, for $\bar{\tau} = 1,2$, we first solve the optimization problem in Theorem 3.1 with $\bar{\varpi}^{1/2} = 0.6325$ and $\bar{u} = 1$ (or $\bar{u} = 3$) to obtain the minimized $\mathcal{H}_\infty$ performance γ^*. Table 1 shows the minimized $\mathcal{H}_\infty$ performance γ^* for respective cases, from which we can observe that the disturbance rejection capability increases as the saturation level $\bar{u}$ increases. Next, for the initial conditions $x(0) = [0.45 \;\; -0.5]^T$ and $x(-1) = [0.0 \;\; 0.0]^T$, we simulate the behaviors of the closed-loop systems under the PM-dependent $\mathcal{H}_\infty$ controller corresponding to $\bar{\tau} = 1$ and $\bar{u} = 1$, where the external disturbances $w(k)$ are generated in the form of two-phase pulse with amplitude $\bar{w}^{1/2} = 0.6708$, and the delay sequences $m(k)$ are generated as random integers between 1 and 3. Fig. 2-(b) and (c) show the state and control input profiles, respectively, when $\gamma^* = 0.5084$. Particularly, Fig. 2-(c) depicts the comparison of the control inputs generated by our method and Kim *et al.* (2004), where the dotted line corresponds to the result of Kim *et al.* (2004) with $\gamma = 0.5084$, and the solid line corresponds to our result. As shown in Fig. 2-(c), contrary to Kim *et al.* (2004), our input voltage does never exceed the saturation level of the motor, $\bar{u} = 1$.

5. Concluding remarks

In this paper, we addressed the problem of designing an $\mathcal{H}_\infty$ control for networked control systems (NCSs) with the effects of both the input saturation as well as the network-induced delay. Based on a PM-dependent dynamic QLPV control law, we first found the conditions for set invariance, involved in the local stabilization, and then incorporated these conditions in the synthesis of dynamic state-feedback $\mathcal{H}_\infty$ control. The resultant convex solvability conditions have been expressed as a finite number of LMIs.

6. Acknowledgment

This research was supported by WCU (World Class University) program through the Korea Science and Engineering Foundation funded by the Ministry of Education, Science and Technology (Project No. R31-2008-000-10100-0). This research was supported by the MKE (The Ministry of Knowledge Economy), Korea, under the ITRC (Information Technology Research Center) support program supervised by the NIPA (National IT Industry Promotion Agency) (NIPA-2010-(C1090-1011-0011)). This research was supported by the MKE (The Ministry of Knowledge Economy), Korea, under the ITRC (Information Technology Research Center) support program supervised by the NIPA (National IT Industry Promotion Agency) (NIPA-2010-(C-1090-1021-0006)).

7. References

Fang, H., Lin, Z., & Hu, T. (2004). Analysis of linear systems in the presence of actuator saturation and $\mathcal{L}_2$-disturbances. *Automatica*, 40, pp. 1229–1238.

Gahinet, P., Nemirovski, A., Laub, A. J., & Chilali, M. (1995). LMI Control Tool box: For use with MATLAB. The Mathworks, Natick, MA.

Gomes da Silva, J. M., Jr., & Tarbourich, S. (2001). Local Stabilization of discrete-time linear systems with saturating constrols: an LMI-based approach. *IEEE Trans. Autom. Contr.*, 46(1), pp. 119–125.

Goodwin, G. C., Haimovich, H., Quevedo, D. E., & Welsh, J. S. (2004) A moving horizon approach to networked control system design. *IEEE Trans. Autom. Contr.*, 49(9), pp. 1427–1445.

Hu, T., & Lin, Z. (2001) *Control systems with actuator saturation: Analysis and design*. Vol. xvi (392p). Boston: Birkhäuser.

Nguyen, T., & Jabbari, F. (2000). Output feedback controllers for disturbance attenuation with actuator amplitude and rate saturation. *Automatica*, 36, pp. 1339–1346.

Kim, D.-S., Lee, Y. S., Kwon, W. H., & Park, H. S. (2003) Maximum allowalbe delay bounds of networked control systems. *Control Engineering Practice,* 11, pp. 1301–1313.

Kim, D. K., Park, P.-G., & Ko, J. W. (2004) Output-feedback $\mathcal{H}_\infty$ control of systems over communication networks using a deterministic switching system approach. *Automatica,* 40, pp. 1205–1212.

Kwakernaak, H., & Sivan, R. (1972). Linear Optimal Control Systems. Wiley-Interscience, New York.

Seiler, P., & Sengupta, R. (2005) An $\mathcal{H}_\infty$ approach to networked control. *IEEE Trans. Autom. Contr.,* 50(3), pp. 356–364.

Tipsuwan, Y., & Chow, M.-Y. (2003) Control methodologies in networked control systems. *Control Engineering Practice,* 11, pp. 1099-1111.

Wu, F., Zheng, Q., & Lin, Z. (2005) Disturbance attenuation for linear systems subject to actuator saturation using output feedback. *in Proc. 44th IEEE Conf. Dec. Contr.,* pp. 7546–7551.

Yue, D., Han, Q.-L., & Lam, J. (2005) Network-based robust $\mathcal{H}_\infty$ control of systems with uncertainty. *Automatica,* 41, pp. 999–1007.

Yang, T. C. (2006) Networked conrol system: a brief survey. *IEE Proc.- Control Theory Appl.* 153(4), pp. 403–412.

Zhang, L., Shi, Y., Chen, T., & Huang, B. (2005) A new method for stabilization of networked control systems with random delays. *IEEE Trans. Autom. Contr.,* 50(8), pp. 1177–1181.

23

Robust Sampled-Data Control Design of Uncertain Fuzzy Systems with Discrete and Distributed Delays

Jun Yoneyama, Yuzu Uchida and Makoto Nishigaki
Aoyama Gakuin University
Japan

1. Introduction

Nonlinear time-delay systems appear in many engineering systems and system formulations such as transportation systems, networked control systems, telecommunication systems, chemical processing systems, and power systems. Hence, it is important to analyze and synthesize such time-delay systems. Considerable research on nonlinear time-delay systems has been made via fuzzy system approach in (2), (6), (9), (12), (13) where stability conditions of fuzzy systems with discrete delays have been given in terms of Linear Matrix Inequalities (LMIs). Takagi-Sugeno fuzzy systems, described by a set of if-then rules which gives local linear models of an underlying system, represent a wide class of nonlinear systems. In the last two decade, Takagi-Sugeno fuzzy system has been extensively used for nonlinear control systems since it can universally approximate or exactly describe general nonlinear systems((8)). Theory has been extended to fuzzy systems with distributed delays in (7), (11), (15). Those results are based on continuous-time delay systems. From a practical point of view, sampled-data control is of importance. However, only a few results on sampled-data control for fuzzy system with discrete delays have been given in the literature ((1), (5), (14), (16)). Sampled-data controller design has been made for fuzzy systems with distributed delays in (3) and (4). To the best of our knowledge, no result for fuzzy sampled-data control systems with neutral and distributed delays has appeared yet.

In this paper, we propose a design method for robust sampled-data control of uncertain fuzzy systems with discrete, neutral and distributed delays. A zero-order sampled-data control can be regarded as a delayed control. Hence, a time-varying delay system approach is taken to design a sampled-data controller. We first obtain a stability condition by introducing an appropriate Lyapunov-Krasovskii functional with free weighting matrices, which reduce the conservatism in our stability condition. Then, based on such an LMI condition, we propose a robust sampled-data control design method of fuzzy uncertain systems with discrete, neutral and distributed delays. We also propose a sampled-data observer design method of fuzzy time-delay systems. A similar approach is taken for analysis of a sampled-data observer, and a condition for an existence of an observer is given by another LMI, which is a dual result of stabilizing controller. Finally, we give some illustrative examples to show our design procedures for sampled-data controller and observer.

2. Fuzzy time-delay systems

In this section, we introduce Takagi-Sugeno fuzzy systems with discrete, neutral and distributed delays. Consider the Takagi-Sugeno fuzzy time-delay model, described by the following IF-THEN rule:

$$
\begin{aligned}
&\textbf{IF} \quad \xi_1 \text{ is } M_{i1} \text{ and } \cdots \text{ and } \xi_p \text{ is } M_{ip},\\
&\textbf{THEN} \quad \dot{x}(t) - (A_{ni} + \Delta A_{ni})\dot{x}(t-\gamma) = (A_i + \Delta A_i)x(t) + (A_{di} + \Delta A_{di})x(t-\alpha(t))\\
&\qquad\qquad + (D_i + \Delta D_i)\int_{t-\beta}^{t} x(s)ds + (B_i + \Delta B_i)u(t),\\
&\qquad y(t) = C_i x(t),\ i = 1, \cdots, r
\end{aligned}
$$

where $\alpha(t), \beta$ and γ are time-varying discrete delay, constant distributed delay, and constant neutral delay, respectively. They may be unknown but they satisfy $0 \leq \alpha(t) \leq \alpha_M$, $\dot{\alpha}(t) \leq d < 1$, $0 \leq \beta \leq \beta_M$, $0 \leq \gamma \leq \gamma_M$ where α_M, d, β_M and γ_M are known numbers. $x(t) \in \Re^n$ is the state and $u(t) \in \Re^m$ is the input. The matrices A_i, A_{di}, A_{ni}, B_i and D_i are of appropriate dimensions. r is the number of IF-THEN rule. M_{ij} is a fuzzy set and $\xi_1, \cdots, \xi_p$ are premise variables. We set $\xi = \left[\xi_1, \cdots, \xi_p\right]^T$ and $\xi(t)$ is assumed to be available. The uncertain matrices are of the form

$$
\begin{aligned}
&\left[\begin{array}{ccccc} \Delta A_i(t) & \Delta A_{di}(t) & \Delta A_{ni}(t) & \Delta B_i(t) & \Delta D_i(t) \end{array}\right]\\
&\qquad = H_i F_i(t) \left[\begin{array}{ccccc} E_{1i} & E_{2i} & E_{3i} & E_{bi} & E_{di} \end{array}\right],\ i = 1, \cdots, r
\end{aligned}
$$

where H_i, E_{1i}, E_{2i}, E_{3i}, E_{bi} and E_{di} are known matrices of appropriate dimensions, and each $F_i(t)$ is unknown real time varying matrices satisfying

$$
F_i^T(t)F_i(t) \leq I,\ \ i = 1, \cdots, r.
$$

The system is defined as follows:

$$
\begin{aligned}
\dot{x}(t) \quad - \quad & \sum_{i=1}^{r} \lambda_i(\xi(t))(A_{ni} + \Delta A_{ni})\dot{x}(t-\gamma) = \sum_{i=1}^{r} \lambda_i(\xi(t))\{(A_i + \Delta A_i)x(t)\\
& + (A_{di} + \Delta A_{di})x(t-\alpha(t)) + (D_i + \Delta D_i)\int_{t-\beta}^{t} x(s)ds + (B_i + \Delta B_i)u(t)\}, \qquad (1)\\
y(t) \quad = \quad & \sum_{i=1}^{r} \lambda_i(\xi(t))C_i x(t)
\end{aligned}
$$

where $\lambda_i(\xi) = \frac{\mu_i(\xi)}{\sum_{i=1}^{r}\mu_i(\xi)}$, $\mu_i(\xi) = \prod_{j=1}^{q} M_{ij}(\xi_j)$ and $M_{ij}(\cdot)$ is the grade of the membership function of M_{ij}. We assume $\mu_i(\xi(t)) \geq 0$, $i = 1, \cdots, r$, $\sum_{i=1}^{r}\mu_i(\xi(t)) > 0$ for any $\xi(t)$. Hence $\lambda_i(\xi(t))$ satisfy $\lambda_i(\xi(t)) \geq 0$, $i = 1, \cdots, r$, $\sum_{i=1}^{r}\lambda_i(\xi(t)) = 1$ for any $\xi(t)$. We consider the sampled-data control input. It may be represented as delayed control as follows:

$$
u(t) = u_d(t_k) = u_d(t - (t - t_k)) = u_d(t - h(t)),\ \ t_k \leq t \leq t_{k+1}
$$

where u_d is a zero-order control signal and the time-varying delay $0 \leq h(t) = t - t_k$ is piecewise linear with the derivative $\dot{h}(t) = 1$ for $t \neq t_k$. A sampling time t_k is the time-varying sampling instant satisfying $0 < t_1 < t_2 < \cdots < t_k < \cdots$. Sampling interval $h_k = t_{k+1} - t_k$ may vary but it is bounded. Thus, we assume $h(t) \leq t_{k+1} - t_k = h_k \leq h_M$ for all t_k where h_M is known constant. We consider the following rules for a controller:

$$
\begin{aligned}
&\textbf{IF} \quad \xi_1(t_k) \text{ is } M_{i1} \text{ and } \cdots \text{ and } \xi_p(t_k) \text{ is } M_{ip},\\
&\textbf{THEN} \quad u(t) = K_i x(t_k),\ \ i = 1, \cdots, r
\end{aligned}
$$

where K_i is to be determined. Then, the natural choice of a controller is given by

$$u(t) = \sum_{i=1}^{r} \lambda_i(\xi(t_k))K_i x(t_k). \tag{2}$$

We represent a piecewise control law as a continuous-time one with a time-varying piecewise continuous(continuous from the right) delay $h(t)$. Hence, we look for a state feedback controller of the form

$$u(t) = \sum_{i=1}^{r} \lambda_i(\xi(t_k))K_i x(t - h(t)). \tag{3}$$

that robustly stabilizes the system (1).The system is said to be robustly stable if it is asymptotically stable for all admissible uncertainties. The closed-loop system (1) with (3) becomes

$$\begin{aligned} \dot{x}(t) &- \sum_{i=1}^{r} \lambda_i(\xi(t))(A_{ni} + \Delta A_{ni})\dot{x}(t-\gamma) = \sum_{i=1}^{r}\sum_{j=1}^{r} \lambda_i(\xi(t))\lambda_j(\xi(t_k))\{(A_i + \Delta A_i)x(t) \\ &+ (A_{di} + \Delta A_{di})x(t-\alpha(t)) + (D_i + \Delta D_i)\int_{t-\beta}^{t} x(s)ds + (B_i + \Delta B_i)K_j x(t-h(t))\}. \end{aligned}$$

When we consider a nominal system, we have

$$\begin{aligned} \dot{x}(t) - \sum_{i=1}^{r} \lambda_i(\xi(t))A_{ni}\dot{x}(t-\gamma) &= \sum_{i=1}^{r}\sum_{j=1}^{r} \lambda_i(\xi(t))\lambda_j(\xi(t_k))\{A_i x(t) + A_{di}x(t-\alpha(t)) \\ &\quad + D_i \int_{t-\beta}^{t} x(s)ds + B_i K_j x(t-h(t))\}. \end{aligned} \tag{4}$$

3. Stability analysis

First, we make stability analysis of the nominal closed-loop system (4).

Theorem 3.1 *Given control gain matrices* K_i, $i = 1, \cdots, r$, *the closed-loop system (4) is asymptotically stable if there exist matrices* $P_i > 0$, $R \geq 0$, $X > 0$, $Y_i > 0$, $i = 1,2,3$, $Q_i \geq 0$, $Z_i > 0$, $i = 1,2$, *and*

$$\begin{aligned} N_{ij} &= \begin{bmatrix} N_{1ij}^T & N_{2ij}^T & N_{3ij}^T & N_{4ij}^T & N_{5ij}^T & N_{6ij}^T & N_{7ij}^T & N_{8ij}^T & N_{9ij}^T \end{bmatrix}^T, \\ S_{ij} &= \begin{bmatrix} S_{1ij}^T & S_{2ij}^T & S_{3ij}^T & S_{4ij}^T & S_{5ij}^T & S_{6ij}^T & S_{7ij}^T & S_{8ij}^T & S_{9ij}^T \end{bmatrix}^T, \\ M_{ij} &= \begin{bmatrix} M_{1ij}^T & M_{2ij}^T & M_{3ij}^T & M_{4ij}^T & M_{5ij}^T & M_{6ij}^T & M_{7ij}^T & M_{8ij}^T & M_{9ij}^T \end{bmatrix}^T, \\ V_{ij} &= \begin{bmatrix} V_{1ij}^T & V_{2ij}^T & V_{3ij}^T & V_{4ij}^T & V_{5ij}^T & V_{6ij}^T & V_{7ij}^T & V_{8ij}^T & V_{9ij}^T \end{bmatrix}^T, \\ W_{ij} &= \begin{bmatrix} W_{1ij}^T & W_{2ij}^T & W_{3ij}^T & W_{4ij}^T & W_{5ij}^T & W_{6ij}^T & W_{7ij}^T & W_{8ij}^T & W_{9ij}^T \end{bmatrix}^T, \\ O_{ij} &= \begin{bmatrix} O_{1ij}^T & O_{2ij}^T & O_{3ij}^T & O_{4ij}^T & O_{5ij}^T & O_{6ij}^T & O_{7ij}^T & O_{8ij}^T & O_{9ij}^T \end{bmatrix}^T, \; i,j = 1, \cdots, r, \end{aligned}$$

$$T = \begin{bmatrix} T_1^T & T_2^T & T_3^T & T_4^T & T_5^T & T_6^T & T_7^T & T_8^T & T_9^T \end{bmatrix}^T$$

such that

$$\Phi_{ij} = \begin{bmatrix} \Phi_{11ij} & \Phi_{12ij} \\ \Phi_{12ij}^T & \Phi_{22} \end{bmatrix} < 0, \ i,j = 1,\cdots,r \tag{5}$$

where

$$\begin{aligned}
\Phi_{11ij} &= \Phi_1 + \Phi_{2ij} + \Phi_{2ij}^T + \Phi_{3ij} + \Phi_{3ij}^T, \\
\Phi_1 &= \begin{bmatrix}
\Phi_{111} & 0 & 0 & 0 & 0 & 0 & 0 & P_2 & P_1 \\
0 & 0 & 0 & 0 & 0 & 0 & 0 & 0 & 0 \\
0 & 0 & -R & 0 & 0 & 0 & 0 & 0 & 0 \\
0 & 0 & 0 & -(1-d)Q_1 & 0 & 0 & 0 & 0 & 0 \\
0 & 0 & 0 & 0 & 0 & 0 & 0 & -P_2 & 0 \\
0 & 0 & 0 & 0 & 0 & -Q_2 & 0 & 0 & 0 \\
0 & 0 & 0 & 0 & 0 & 0 & 0 & 0 & 0 \\
P_2 & 0 & 0 & 0 & -P_2 & 0 & 0 & -X - \frac{1}{\beta_M}U & 0 \\
P_1 & 0 & 0 & 0 & 0 & 0 & 0 & 0 & \Phi_{199}
\end{bmatrix}, \\
\Phi_{111} &= Q_1 + R + \beta_M U + \beta_M^2 X, \\
\Phi_{199} &= Q_2 + \alpha_M Y_1 + \beta_M Y_2 + \gamma_M Y_3 + h_M(Z_1 + Z_2), \\
\Phi_{2ij} &= \begin{bmatrix} N_{ij} + M_{ij} + W_{ij} + O_{ij} + V_{ij} & -N_{ij} + S_{ij} & -M_{ij} - S_{ij} & -W_{ij} \end{bmatrix. \\
&\qquad \begin{matrix} -O_{ij} & 0 & -V_{ij} & 0 & 0 \end{matrix} \Big], \\
\Phi_{3ij} &= \begin{bmatrix} -TA_i & -TB_iK_j & 0 & -TA_{di} & 0 & -TA_{ni} & 0 & -TD_i & T \end{bmatrix}, \\
\Phi_{12ij} &= \begin{bmatrix} h_M N_{ij} & h_M S_{ij} & h_M M_{ij} & \alpha_M W_{ij} & \beta_M O_{ij} & \gamma_M V_{ij} \end{bmatrix}, \\
\Phi_{22} &= \mathrm{diag}\begin{bmatrix} -h_M Z_1 & -h_M Z_1 & -h_M Z_2 & -\alpha_M Y_1 & -\beta_M Y_2 & -\gamma_M Y_3 \end{bmatrix}.
\end{aligned}$$

Proof: First, it follows from the Leibniz-Newton formula that the following equations hold for any matrices N_{ij}, S_{ij}, M_{ij}, V_{ij}, W_{ij} and O_{ij}, the forms of which are given in Theorem 3.1.

$$2\sum_{i=1}^{r}\sum_{j=1}^{r}\lambda_i(\xi(t))\lambda_j(\xi(t_k))\zeta^T(t)N_{ij}\left[x(t) - x(t-h(t)) - \int_{t-h(t)}^{t}\dot{x}(s)ds\right] = 0, \tag{6}$$

$$2\sum_{i=1}^{r}\sum_{j=1}^{r}\lambda_i(\xi(t))\lambda_j(\xi(t_k))\zeta^T(t)S_{ij}\left[x(t-h(t)) - x(t-h_M) - \int_{t-h_M}^{t-h(t)}\dot{x}(s)ds\right] = 0, \tag{7}$$

$$2\sum_{i=1}^{r}\sum_{j=1}^{r}\lambda_i(\xi(t))\lambda_j(\xi(t_k))\zeta^T(t)M_{ij}\left[x(t) - x(t-h_M) - \int_{t-h_M}^{t}\dot{x}(s)ds\right] = 0, \tag{8}$$

$$2\sum_{i=1}^{r}\sum_{j=1}^{r}\lambda_i(\xi(t))\lambda_j(\xi(t_k))\zeta^T(t)V_{ij}\left[x(t) - x(t-\gamma) - \int_{t-\gamma}^{t}\dot{x}(s)ds\right] = 0, \tag{9}$$

$$2\sum_{i=1}^{r}\sum_{j=1}^{r}\lambda_i(\xi(t))\lambda_j(\xi(t_k))\zeta^T(t)W_{ij}\left[x(t) - x(t-\alpha(t)) - \int_{t-\alpha(t)}^{t}\dot{x}(s)ds\right] = 0, \tag{10}$$

$$2\sum_{i=1}^{r}\sum_{j=1}^{r}\lambda_i(\xi(t))\lambda_j(\xi(t_k))\zeta^T(t)O_{ij}\left[x(t) - x(t-\beta) - \int_{t-\beta}^{t}\dot{x}(s)ds\right] = 0 \tag{11}$$

where

$$\zeta(t) = \begin{bmatrix} x^T(t) & x^T(t-h(t)) & x^T(t-h_M) & x^T(t-\alpha(t)) & x^T(t-\beta) \\ & & \dot{x}(t-\gamma) & x(t-\gamma) & \int_{t-\beta}^{t} x^T(s)ds \quad \dot{x}^T(t) \end{bmatrix}^T.$$

It is also clear from the closed-loop system (4) that the following is true for any matrix T.

$$2\sum_{i=1}^{r}\sum_{j=1}^{r}\lambda_i(\xi(t))\lambda_j(\xi(t_k))\zeta^T(t)T\big[\dot{x}(t) - A_i x(t)$$

$$-A_{di}x(t-\alpha(t)) - A_{ni}\dot{x}(t-\gamma) - D_i\int_{t-\beta}^{t} x(s)ds - B_iK_jx(t-h(t))\big] = 0. \qquad (12)$$

Now, we consider the following Lyapunov-Krasovskii functional:

$$V(x_t) = V_1(x) + V_2(x_t) + V_3(x_t) + V_4(x_t)$$

where $x_t = x(t+\theta)$, $-\max(h_M, \alpha_M, \beta_M) \le \theta \le 0$,

$$V_1(x) = x^T(t)P_1x(t) + \left[\int_{t-\beta}^{t} x(s)ds\right]^T P_2 \int_{t-\beta}^{t} x(s)ds,$$

$$V_2(x_t) = \int_{t-\alpha(t)}^{t} x^T(s)Q_1x(s)ds + \int_{t-\gamma}^{t} \dot{x}^T(s)Q_2\dot{x}(s)ds + \int_{t-h_M}^{t} x^T(s)Rx(s)ds,$$

$$V_3(x_t) = \int_{-\beta}^{0}\int_{t+\theta}^{t} x^T(s)Ux(s)dsd\theta + \int_{-\alpha_M}^{0}\int_{t+\theta}^{t} \dot{x}^T(s)Y_1\dot{x}(s)dsd\theta$$

$$+\int_{-\beta}^{0}\int_{t+\theta}^{t} \dot{x}^T(s)Y_2\dot{x}(s)dsd\theta + \int_{-\gamma}^{0}\int_{t+\theta}^{t} \dot{x}^T(s)Y_3\dot{x}(s)dsd\theta$$

$$+\int_{-h_M}^{0}\int_{t+\theta}^{t} \dot{x}^T(s)(Z_1+Z_2)\dot{x}(s)dsd\theta,$$

$$V_4(x_t) = \int_{t-\beta}^{t}\left[\int_{\theta}^{t} x^T(s)ds\right]X\left[\int_{\theta}^{t} x(s)ds\right]d\theta + \int_{0}^{\beta}\int_{t-\theta}^{t}(s-t+\theta)x^T(s)Xx(s)dsd\theta,$$

and $P_i > 0$, $R \ge 0$, $U > 0$, $X > 0$, $Y_i > 0$, $i = 1,2,3$, $Q_i \ge 0$, $Z_i > 0$, $i = 1,2$ are to be determined. We take the derivative of $V(x_t)$ with respect to t along the solution of the system (4) and add the left-hand-sides of (6)-(12):

$$
\begin{aligned}
\dot{V}(x_t) \leq\ & 2\dot{x}^T(t)P_1x(t) + 2x^T(t)P_2\int_{t-\beta}^{t} x(s)ds - 2x^T(t-\beta)P_2\int_{t-\beta}^{t} x(s)ds \\
& + x^T(t)(Q_1 + R + \beta_M U + \beta_M^2 X)x(t) - (1-d)x^T(t-\alpha(t))Q_1x(t-\alpha(t)) \\
& - \dot{x}^T(t-\gamma)Q_2\dot{x}(t-\gamma) - x^T(t-h_M)Rx(t-h_M) \\
& - \left[\int_{t-\beta}^{t} x(s)ds\right]^T \frac{1}{\beta_M}U\left[\int_{t-\beta}^{t} x(s)ds\right] \\
& + \dot{x}^T(t)[Q_2 + \alpha_M Y_1 + \beta_M Y_2 + \gamma_M Y_3 + h_M(Z_1+Z_2)]\dot{x}(t) \\
& - \int_{t-\alpha_M}^{t} x^T(s)Y_1x(s)ds - \int_{t-\beta}^{t} \dot{x}^T(s)Y_2\dot{x}(s)ds - \int_{t-\gamma}^{t} \dot{x}^T(s)Y_3\dot{x}(s)ds \\
& - \int_{t-h(t)}^{t} \dot{x}^T(s)Z_1\dot{x}(s)ds - \int_{t-h_M}^{t-h(t)} \dot{x}^T(s)Z_1\dot{x}(s)ds \\
& - \int_{t-h_M}^{t} \dot{x}^T(s)Z_2\dot{x}(s)ds - \left[\int_{t-\beta}^{t} x^T(s)ds\right]X\left[\int_{t-\beta}^{t} x(s)ds\right] \\
& + 2\sum_{i=1}^{r}\sum_{j=1}^{r}\lambda_i(\xi(t))\lambda_j(\xi(t_k))\zeta^T(t)N_{ij}\left[x(t) - x(t-h(t)) - \int_{t-h(t)}^{t}\dot{x}(s)ds\right] \\
& + 2\sum_{i=1}^{r}\sum_{j=1}^{r}\lambda_i(\xi(t))\lambda_j(\xi(t_k))\zeta^T(t)S_{ij}\left[x(t-h(t)) - x(t-h_M) - \int_{t-h_M}^{t-h(t)}\dot{x}(s)ds\right] \\
& + 2\sum_{i=1}^{r}\sum_{j=1}^{r}\lambda_i(\xi(t))\lambda_j(\xi(t_k))\zeta^T(t)M_{ij}\left[x(t) - x(t-h_M) - \int_{t-h_M}^{t}\dot{x}(s)ds\right] \\
& + 2\sum_{i=1}^{r}\sum_{j=1}^{r}\lambda_i(\xi(t))\lambda_j(\xi(t_k))\zeta^T(t)V_{ij}\left[x(t) - x(t-\gamma) - \int_{t-\gamma}^{t}\dot{x}(s)ds\right] \\
& + 2\sum_{i=1}^{r}\sum_{j=1}^{r}\lambda_i(\xi(t))\lambda_j(\xi(t_k))\zeta^T(t)W_{ij}\left[x(t) - x(t-\alpha(t)) - \int_{t-\alpha(t)}^{t}\dot{x}(s)ds\right] \\
& + 2\sum_{i=1}^{r}\sum_{j=1}^{r}\lambda_i(\xi(t))\lambda_j(\xi(t_k))\zeta^T(t)O_{ij}\left[x(t) - x(t-\beta) - \int_{t-\beta}^{t}\dot{x}(s)ds\right] \\
& + 2\sum_{i=1}^{r}\sum_{j=1}^{r}\lambda_i(\xi(t))\lambda_j(\xi(t_k))\zeta^T(t)T\big[\dot{x}(t) - A_ix(t) - A_{di}x(t-\alpha(t)) \\
& - A_{ni}\dot{x}(t-\gamma) - D_i\textstyle\int_{t-\beta}^{t} x(s)ds - B_iK_jx(t-h(t))\big] \\
\leq\ & \sum_{i=1}^{r}\sum_{j=1}^{r}\lambda_i(\xi(t))\lambda_j(\xi(t_k))\Big\{\zeta^T(t)\Psi_{ij}\zeta(t) \\
& - \int_{t-h(t)}^{t}\left[\zeta^T(t)N_{ij} + \dot{x}^T(s)Z_1\right]Z_1^{-1}\left[N_{ij}^T\zeta(t) + Z_1\dot{x}(s)\right]ds \\
& - \int_{t-h_M}^{t-h(t)}\left[\zeta^T(t)S_{ij} + \dot{x}^T(s)Z_1\right]Z_1^{-1}\left[S_{ij}^T\zeta(t) + Z_1\dot{x}(s)\right]ds \\
& - \int_{t-h_M}^{t}\left[\zeta^T(t)M_{ij} + \dot{x}^T(s)Z_2\right]Z_2^{-1}\left[M_{ij}^T\zeta(t) + Z_2\dot{x}(s)\right]ds \\
& - \int_{t-\alpha(t)}^{t}\left[\zeta^T(t)W_{ij} + \dot{x}^T(s)Y_1\right]Y_1^{-1}\left[W_{ij}^T\zeta(t) + Y_1\dot{x}(s)\right]ds \\
& - \int_{t-\beta}^{t}\left[\zeta^T(t)O_{ij} + \dot{x}^T(s)Y_2\right]Y_2^{-1}\left[O_{ij}^T\zeta(t) + Y_2\dot{x}(s)\right]ds
\end{aligned}
$$

$$- \int_{t-\gamma}^{t} \left[\zeta^T(t)V_{ij} + \dot{x}^T(s)Y_3\right] Y_3^{-1} \left[V_{ij}^T\zeta(t) + Y_3\dot{x}(s)\right] ds \Big\} \tag{13}$$

where

$$\begin{aligned}\Psi_{ij} &= \Phi_{11ij} + h_M(N_{ij}Z_1^{-1}N_{ij}^T + S_{ij}Z_1^{-1}S_{ij}^T + M_{ij}Z_2^{-1}M_{ij}^T) + \alpha_M W_{ij}Y_1^{-1}W_{ij}^T \\ &\quad +\beta_M O_{ij}Y_2^{-1}O_{ij}^T + \gamma_M V_{ij}Y_3^{-1}V_{ij}^T.\end{aligned}$$

Now, if (5) is satisfied, then by Schur complement formula we have

$$\Psi_{ij} < 0, \ i,j = 1,\cdots,r. \tag{14}$$

If (14) holds, we have $\sum_{i=1}^{r}\sum_{j=1}^{r}\lambda_i(\xi(t))\lambda_j(\xi(t_k))\zeta^T(t)\Psi_{ij}\zeta(t) < 0$, which implies that $\dot{V}(x_t) < 0$ because $Y_i > 0$, $Z_i > 0$, $i = 1,2$ and the last five terms in (13) are all less than 0. This proves that conditions (5) suffice to show the asymptotic stability of the system (4).

4. Sampled-data control design

In this section, we seek a design method of a sampled-data control for fuzzy time-delay systems based on Theorem 3.1. Unfortunately, however, Theorem 3.1 does not give feasible LMI conditions for obtaining state feedback control gain matrices K_i. To this end, we take an appropriate congruence transformation to obtain feasible LMI conditions and a design method of a sampled-data state feedback controller.

Theorem 4.1 *Given scalars ρ_i, $i = 1,\cdots,9$, the sampled-data controller (2) asymptotically stabilizes the nominal system (4) if there exist matrices $\bar{P}_i > 0$, $\bar{R} \geq 0$, $\bar{U} > 0$, $\bar{X} > 0$, $\bar{Y}_i > 0$, $i = 1,2,3$, $\bar{Q}_i \geq 0$, $\bar{Z}_i > 0$, $i = 1,2$, L, G_j, $j = 1,\cdots,r$,*

$$\begin{aligned}
\bar{N}_{ij} &= \begin{bmatrix} \bar{N}_{1ij}^T & \bar{N}_{2ij}^T & \bar{N}_{3ij}^T & \bar{N}_{4ij}^T & \bar{N}_{5ij}^T & \bar{N}_{6ij}^T & \bar{N}_{7ij}^T & \bar{N}_{8ij}^T & \bar{N}_{9ij}^T \end{bmatrix}^T, \\
\bar{S}_{ij} &= \begin{bmatrix} \bar{S}_{1ij}^T & \bar{S}_{2ij}^T & \bar{S}_{3ij}^T & \bar{S}_{4ij}^T & \bar{S}_{5ij}^T & \bar{S}_{6ij}^T & \bar{S}_{7ij}^T & \bar{S}_{8ij}^T & \bar{S}_{9ij}^T \end{bmatrix}^T, \\
\bar{M}_{ij} &= \begin{bmatrix} \bar{M}_{1ij}^T & \bar{M}_{2ij}^T & \bar{M}_{3ij}^T & \bar{M}_{4ij}^T & \bar{M}_{5ij}^T & \bar{M}_{6ij}^T & \bar{M}_{7ij}^T & \bar{M}_{8ij}^T & \bar{M}_{9ij}^T \end{bmatrix}^T, \\
\bar{V}_{ij} &= \begin{bmatrix} \bar{V}_{1ij}^T & \bar{V}_{2ij}^T & \bar{V}_{3ij}^T & \bar{V}_{4ij}^T & \bar{V}_{5ij}^T & \bar{V}_{6ij}^T & \bar{V}_{7ij}^T & \bar{V}_{8ij}^T & \bar{V}_{9ij}^T \end{bmatrix}^T, \\
\bar{W}_{ij} &= \begin{bmatrix} \bar{W}_{1ij}^T & \bar{W}_{2ij}^T & \bar{W}_{3ij}^T & \bar{W}_{4ij}^T & \bar{W}_{5ij}^T & \bar{W}_{6ij}^T & \bar{W}_{7ij}^T & \bar{W}_{8ij}^T & \bar{W}_{9ij}^T \end{bmatrix}^T, \\
\bar{O}_{ij} &= \begin{bmatrix} \bar{O}_{1ij}^T & \bar{O}_{2ij}^T & \bar{O}_{3ij}^T & \bar{O}_{4ij}^T & \bar{O}_{5ij}^T & \bar{O}_{6ij}^T & \bar{O}_{7ij}^T & \bar{O}_{8ij}^T & \bar{O}_{9ij}^T \end{bmatrix}^T, \ i,j = 1,\cdots,r
\end{aligned}$$

such that

$$\Theta_{ij} = \begin{bmatrix} \Theta_{11ij} & \Theta_{12ij} \\ \Theta_{12ij}^T & \Theta_{22} \end{bmatrix} < 0, \ i,j = 1,\cdots,r \tag{15}$$

where

$$
\begin{aligned}
\Theta_{11ij} &= \Theta_1 + \Theta_{2ij} + \Theta_{2ij}^T + \Theta_{3ij} + \Theta_{3ij}^T, \\
\Theta_1 &= \begin{bmatrix}
\Theta_{111} & 0 & 0 & 0 & 0 & 0 & 0 & \bar{P}_2 & \bar{P}_1 \\
0 & 0 & 0 & 0 & 0 & 0 & 0 & 0 & 0 \\
0 & 0 & -\bar{R} & 0 & 0 & 0 & 0 & 0 & 0 \\
0 & 0 & 0 & -(1-d)\bar{Q}_1 & 0 & 0 & 0 & 0 & 0 \\
0 & 0 & 0 & 0 & 0 & 0 & 0 & -\bar{P}_2 & 0 \\
0 & 0 & 0 & 0 & 0 & -Q_2 & 0 & 0 & 0 \\
0 & 0 & 0 & 0 & 0 & 0 & 0 & 0 & 0 \\
\bar{P}_2 & 0 & 0 & 0 & -\bar{P}_2 & 0 & 0 & -\bar{X} - \frac{1}{\beta_M}\bar{U} & 0 \\
\bar{P}_1 & 0 & 0 & 0 & 0 & 0 & 0 & 0 & \Theta_{199}
\end{bmatrix}, \\
\Theta_{111} &= \bar{Q}_1 + \bar{R} + \beta_M \bar{U} + \beta_M^2 \bar{X}, \\
\Theta_{199} &= Q_2 + \alpha_M \bar{Y}_1 + \beta_M \bar{Y}_2 + \gamma_M \bar{Y}_3 + h_M(\bar{Z}_1 + \bar{Z}_2), \\
\Theta_{2ij} &= [\ \bar{N}_{ij} + \bar{M}_{ij} + \bar{W}_{ij} + \bar{O}_{ij} + \bar{V}_{ij} \quad -\bar{N}_{ij} + \bar{S}_{ij} \quad -\bar{M}_{ij} - \bar{S}_{ij} \quad -\bar{W}_{ij} \\
&\qquad\qquad -\bar{O}_{ij} \quad 0 \quad -\bar{V}_{ij} \quad 0 \quad 0\ ],
\end{aligned}
$$

$$
\begin{aligned}
\Theta_{3ij} &= -\begin{bmatrix}
\rho_1 A_i L^T & \rho_1 B_i G_j & 0 & \rho_1 A_{di} L^T & 0 & \rho_1 A_{ni} L^T & 0 & \rho_1 D_i L^T & -\rho_1 L^T \\
\rho_2 A_i L^T & \rho_2 B_i G_j & 0 & \rho_2 A_{di} L^T & 0 & \rho_2 A_{ni} L^T & 0 & \rho_2 D_i L^T & -\rho_2 L^T \\
\rho_3 A_i L^T & \rho_3 B_i G_j & 0 & \rho_3 A_{di} L^T & 0 & \rho_3 A_{ni} L^T & 0 & \rho_3 D_i L^T & -\rho_3 L^T \\
\rho_4 A_i L^T & \rho_4 B_i G_j & 0 & \rho_4 A_{di} L^T & 0 & \rho_4 A_{ni} L^T & 0 & \rho_4 D_i L^T & -\rho_4 L^T \\
\rho_5 A_i L^T & \rho_5 B_i G_j & 0 & \rho_5 A_{di} L^T & 0 & \rho_5 A_{ni} L^T & 0 & \rho_5 D_i L^T & -\rho_5 L^T \\
\rho_6 A_i L^T & \rho_6 B_i G_j & 0 & \rho_6 A_{di} L^T & 0 & \rho_6 A_{ni} L^T & 0 & \rho_6 D_i L^T & -\rho_6 L^T \\
\rho_7 A_i L^T & \rho_7 B_i G_j & 0 & \rho_7 A_{di} L^T & 0 & \rho_7 A_{ni} L^T & 0 & \rho_7 D_i L^T & -\rho_7 L^T \\
\rho_8 A_i L^T & \rho_8 B_i G_j & 0 & \rho_8 A_{di} L^T & 0 & \rho_8 A_{ni} L^T & 0 & \rho_8 D_i L^T & -\rho_8 L^T \\
\rho_9 A_i L^T & \rho_9 B_i G_j & 0 & \rho_9 A_{di} L^T & 0 & \rho_9 A_{ni} L^T & 0 & \rho_9 D_i L^T & -\rho_9 L^T
\end{bmatrix}, \\
\Theta_{12ij} &= [\ h_M \bar{N}_{ij} \quad h_M \bar{S}_{ij} \quad h_M \bar{M}_{ij} \quad \alpha_M \bar{W}_{ij} \quad \beta_M \bar{O}_{ij} \quad \gamma_M \bar{V}_{ij}\ ], \\
\Theta_{22} &= \mathrm{diag}[\ -h_M \bar{Z}_1 \quad -h_M \bar{Z}_1 \quad -h_M \bar{Z}_2 \quad -\alpha_M \bar{Y}_1 \quad -\beta_M \bar{Y}_2 \quad -\gamma_M \bar{Y}_3\ ].
\end{aligned}
$$

In this case, state feedback control gains in (2) are given by

$$K_i = G_i L^{-T},\ i = 1, \cdots, r. \tag{16}$$

Proof: We let $T_i = \rho_i \bar{L}$, $i = 1, \cdots, 9$ where each ρ_i is given and $\bar{L}$ is defined later, and substitute them into (5). If (5) holds, it follows that (9,9)-block of Φ_{11ij} must be negative definite. It follows that $T_9 + T_9^T = \rho_9(\bar{L} + \bar{L}^T) < 0$, which implies that $\bar{L}$ is nonsingular. Then, we define $\bar{L} = L^{-1}$ and calculate $\Theta_{ij} = \Sigma \Phi_{ij} \Sigma^T$ with $\Sigma = \mathrm{diag}[\, L\, L\, L\, L\, L\, L\, L\, L\, L\, L\, L\, L\, L\, L\, L\,]$. Defining $\bar{P}_i = LP_iL^T$, $\bar{R} = LRL^T$, $\bar{X} = LXL^T$, $\bar{Y}_i = LY_iL^T$, $i = 1,2,3$, $\bar{Q}_i = LQ_1L^T$, $\bar{Z}_i = LZ_iL^T$, $i = 1,2$, $\bar{N}_{kij} = LN_{kij}L^T$, $\bar{S}_{kij} = LS_{kij}L^T$, $\bar{M}_{kij} = LM_{kij}L^T$, $\bar{V}_{kij} = LV_{kij}L^T$, $\bar{W}_{kij} = LW_{kij}L^T$, $\bar{O}_{kij} = LO_{kij}L^T$, $i, j = 1, \cdots, r, k = 1, \cdots, 9$, we obtain Θ_{ij} in (15) where we let $G_j = K_jL^T$. If the conditions (15) hold, state feedback control gain matrix K_i is obviously given by (16).

We extend the result to the case of the uncertain system (1). The following lemma is necessary to prove Theorem 4.3.

Lemma 4.2 *((10)) Given matrices $Q = Q^T$, H, E and $R = R^T > 0$ of appropriate dimensions*

$$Q + HF(t)E + E^T F^T(t) H^T < 0$$

for all $F(t)$ satisfying $F^T(t)F(t) \leq R$ if and only if there exists a scalar $\varepsilon > 0$ such that

$$Q + \frac{1}{\varepsilon} HH^T + \varepsilon E^T RE < 0.$$

Theorem 4.3 *Given scalars ρ_i, $i = 1, \cdots, 9$, the sampled-data controller (2) robustly stabilizes the uncertain system (1) if there exist matrices $\bar{P}_i > 0$, $\bar{Q}_i \geq 0$, $\bar{Z}_i > 0$, $i = 1,2$, $\bar{R} \geq 0$, $\bar{U} > 0$, $\bar{X} > 0$, $\bar{Y}_i > 0$, $i = 1,2,3$, L, G_j, $j = 1, \cdots, r$, and*

$$\begin{aligned}
\bar{N}_{ij} &= \begin{bmatrix} \bar{N}_{1ij}^T & \bar{N}_{2ij}^T & \bar{N}_{3ij}^T & \bar{N}_{4ij}^T & \bar{N}_{5ij}^T & \bar{N}_{6ij}^T & \bar{N}_{7ij}^T & \bar{N}_{8ij}^T & \bar{N}_{9ij}^T \end{bmatrix}^T, \\
\bar{S}_{ij} &= \begin{bmatrix} \bar{S}_{1ij}^T & \bar{S}_{2ij}^T & \bar{S}_{3ij}^T & \bar{S}_{4ij}^T & \bar{S}_{5ij}^T & \bar{S}_{6ij}^T & \bar{S}_{7ij}^T & \bar{S}_{8ij}^T & \bar{S}_{9ij}^T \end{bmatrix}^T, \\
\bar{M}_{ij} &= \begin{bmatrix} \bar{M}_{1ij}^T & \bar{M}_{2ij}^T & \bar{M}_{3ij}^T & \bar{M}_{4ij}^T & \bar{M}_{5ij}^T & \bar{M}_{6ij}^T & \bar{M}_{7ij}^T & \bar{M}_{8ij}^T & \bar{M}_{9ij}^T \end{bmatrix}^T, \\
\bar{V}_{ij} &= \begin{bmatrix} \bar{V}_{1ij}^T & \bar{V}_{2ij}^T & \bar{V}_{3ij}^T & \bar{V}_{4ij}^T & \bar{V}_{5ij}^T & \bar{V}_{6ij}^T & \bar{V}_{7ij}^T & \bar{V}_{8ij}^T & \bar{V}_{9ij}^T \end{bmatrix}^T, \\
\bar{W}_{ij} &= \begin{bmatrix} \bar{W}_{1ij}^T & \bar{W}_{2ij}^T & \bar{W}_{3ij}^T & \bar{W}_{4ij}^T & \bar{W}_{5ij}^T & \bar{W}_{6ij}^T & \bar{W}_{7ij}^T & \bar{W}_{8ij}^T & \bar{W}_{9ij}^T \end{bmatrix}^T, \\
\bar{O}_{ij} &= \begin{bmatrix} \bar{O}_{1ij}^T & \bar{O}_{2ij}^T & \bar{O}_{3ij}^T & \bar{O}_{4ij}^T & \bar{O}_{5ij}^T & \bar{O}_{6ij}^T & \bar{O}_{7ij}^T & \bar{O}_{8ij}^T & \bar{O}_{9ij}^T \end{bmatrix}^T, \quad i,j = 1, \cdots, r
\end{aligned}$$

and scalars $\varepsilon_{ij} > 0$, $i,j = 1, \cdots, r$ such that

$$\begin{bmatrix} \Theta_{ij} + \varepsilon_i \bar{H}_i^T \bar{H}_i & \bar{E}_{ij}^T \\ \bar{E}_{ij} & -\varepsilon_{ij} I \end{bmatrix} < 0, \ i,j = 1, \cdots, r \tag{17}$$

where Θ_{ij} is given in Theorem 4.1, and

$$\begin{aligned}
\bar{H}_i &= -[\ \rho_1 H_i^T \ \ \rho_2 H_i^T \ \ \rho_3 H_i^T \ \ \rho_4 H_i^T \ \ \rho_5 H_i^T \ \ \rho_6 H_i^T \ \ \rho_7 H_i^T \ \ \rho_8 H_i^T \\
&\qquad\qquad \rho_9 H_i^T \ \ 0 \ \ 0 \ \ 0 \ \ 0 \ \ 0 \ \ 0 \ ]^T, \\
\bar{E}_{ij} &= [\ E_{1i} L^T \ \ E_{bi} G_j \ \ 0 \ \ E_{2i} L^T \ \ 0 \ \ E_{3i} L^T \ \ 0 \ \ E_{di} L^T \ \ 0 \ \ 0 \ \ 0 \ \ 0 \ \ 0 \ \ 0 \ \ 0 \ ].
\end{aligned}$$

In this case, state feedback control gains in (2) are given by (16).

Proof: Replacing A_i, A_{di}, A_{ni}, B_i, D_i by $A_i + H_i F_i(t) E_{1i}$, $A_{di} + H_i F_i(t) E_{2i}$, $A_{ni} + H_i F_i(t) E_{3i}$, $B_i + H_i F_i(t) E_{bi}$, $D_i + H_i F_i(t) E_{di}$, we have

$$\Theta_{ij} + \bar{H}_i F_i(t) \bar{E}_{ij} + (\bar{H}_i F_i(t) \bar{E}_{ij})^T < 0, \ i,j = 1, \cdots, r.$$

It follows from Lemma 4.2 that the above LMIs hold if and only if there exist $\varepsilon_{ij} > 0$ such that

$$\Theta_{ij} + \varepsilon_{ij} \bar{H}_i \bar{H}_i + \frac{1}{\varepsilon_{ij}} \bar{E}_{ij} \bar{E}_{ij} < 0, \ i,j = 1, \cdots, r.$$

Applying the Schur complement formula, we have (17).

5. Numerical examples for controller design

Let us design robust sampled-data controllers for the system (1) with the following matrices.

$$A_1 = \begin{bmatrix} 0 & 0 \\ 0 & 1 \end{bmatrix}, A_2 = \begin{bmatrix} 0 & 0 \\ 0 & 1.5 \end{bmatrix}, A_{d1} = \begin{bmatrix} -1 & -1 \\ 0 & -0.9 \end{bmatrix}, A_{d2} = \begin{bmatrix} -1 & -1 \\ 0 & -1.4 \end{bmatrix},$$
$$A_{n1} = \begin{bmatrix} 0.1 & 0 \\ 0 & 0.2 \end{bmatrix}, A_{n2} = \begin{bmatrix} 0.1 & 0 \\ 0 & 0.3 \end{bmatrix}, B_1 = \begin{bmatrix} 0.1 \\ 1 \end{bmatrix}, B_2 = \begin{bmatrix} 0.1 \\ 1.2 \end{bmatrix},$$
$$D_1 = \begin{bmatrix} 0.1 & 0 \\ 0 & 0.3 \end{bmatrix}, D_2 = \begin{bmatrix} 0.2 & 0 \\ 0 & 0.4 \end{bmatrix}, H_1 = H_2 = 0.2I, E_{11} = E_{12} = 0.2I,$$
$$E_{21} = E_{22} = 0.2I, E_{31} = E_{32} = 0.1I, E_{b1} = E_{b2} = \begin{bmatrix} 0 \\ 0.1 \end{bmatrix}, E_{d1} = E_{d2} = 0.1I.$$

The grades are given by $\lambda_1(x_1) = \sin^2(x_1)$ and $\lambda_2(x_1) = 1 - \lambda_1(x_1)$. The maximum upper bound of the sampling time $h_M = 0.1$ and $d = 0.5$ are assumed. First, we let $\beta_M = \gamma_M = 0.1$. Theorem 4.3 with $\rho_1 = 5.46$, $\rho_2 = -0.01$, $\rho_3 = -2.19$, $\rho_4 = 0.60$, $\rho_5 = -0.01$, $\rho_6 = -0.01$, $\rho_7 = 0.50$, $\rho_8 = 0.10$, $\rho_9 = 1.96$ guarantees the existence of the sampled-data controller for the maximum upper bound of the time-delay $\alpha_M = 0.42$. In this case, control gains in (2) are given by

$$K_1 = \begin{bmatrix} -0.1800 & -0.9934 \end{bmatrix}, K_2 = \begin{bmatrix} -0.1808 & -0.9942 \end{bmatrix}.$$

Next, we let $\alpha_M = \gamma_M = 0.1$. Theorem 4.3 with $\rho_1 = 5.74$, $\rho_2 = 0.50$, $\rho_3 = -2.19$, $\rho_4 = -0.60$, $\rho_5 = -0.01$, $\rho_6 = -0.42$, $\rho_7 = -0.50$, $\rho_8 = 0.16$, $\rho_9 = 1.96$ gives a robust sampled-data controller for the maximum upper bound $\beta_M = 3.43$. In this case, control gains in (2) are given by

$$K_1 = \begin{bmatrix} 0.1794 & -2.6198 \end{bmatrix}, K_2 = \begin{bmatrix} 0.1795 & -2.6194 \end{bmatrix}.$$

Finally, we let $\alpha_M = \beta_M = 0.1$. Theorem 4.3 with $\rho_1 = 4.74$, $\rho_2 = -0.01$, $\rho_3 = -2.19$, $\rho_4 = -0.60$, $\rho_5 = -0.01$, $\rho_6 = 0.01$, $\rho_7 = -0.50$, $\rho_8 = 0.07$, $\rho_9 = 1.96$ gives a robust sampled-data controller for the maximum upper bound $\gamma_M = 2.90$. In this case, control gains in (2) are given by

$$K_1 = \begin{bmatrix} -0.0265 & -0.7535 \end{bmatrix}, K_2 = \begin{bmatrix} -0.0260 & -0.7515 \end{bmatrix}.$$

6. Application to observer design

In this section, using the results in the previous sections we consider an observer design for the system (1), which estimates the state variables of the system using sampled-data measurement outputs. Here, we assume that the system does not contain any uncertain parameters so that it is given by

$$\dot{x}(t) - \sum_{i=1}^{r} \lambda_i(\xi(t)) A_{ni} \dot{x}(t-\gamma) = \sum_{i=1}^{r} \lambda_i(\xi(t)) \{ A_i x(t) + A_{di} x(t-\alpha(t)) + D_i \int_{t-\beta}^{t} x(s)ds + B_i u(t) \}, \quad (18)$$

$$y(t) = \sum_{i=1}^{r} \lambda_i(\xi(t)) C_i x(t) \quad (19)$$

where all the time delays are assumed to be measurable.
The sampled-data measurement output may be represented as delayed measurement as follows:

$$y(t) = y_d(t_k) = y_d(t-(t-t_k)) = y_d(t-h(t)),\ t_k \le t \le t_{k+1}$$

where y_d is a zero-order measurement signal and the time-varying delay $0 \le h(t) = t - t_k$ is piecewise linear with the derivative $\dot{h}(t) = 1$ for $t \neq t_k$ as before. We consider the following rules for a system to estimate the state variables:

IF $\xi_1(t_k)$ is M_{i1} and $\cdots$ and $\xi_p(t_k)$ is M_{ip},

THEN $\dot{\hat{x}}(t) - A_{ni}\dot{\hat{x}}(t-\gamma) = A_i\hat{x}(t) + A_{di}\hat{x}(t-\alpha(t)) + D_i\int_{t-\beta}^{t}\hat{x}(s)ds + B_iu(t) + \bar{K}(y(t_k) - C_i\hat{x}(t_k)),\ i = 1,\cdots,r$

where $\hat{x}$ is the estimated state and $\bar{K} = \sum_{j=1}^{r}\lambda_j(\xi(t_k))\bar{K}_j$ is an observer gain to be determined. Then, the overall system is given by

$$\begin{aligned}\dot{\hat{x}}(t) - \sum_{i=1}^{r}\lambda_i(\xi(t))A_{ni}\dot{\hat{x}}(t-\gamma) &= \sum_{i=1}^{r}\sum_{j=1}^{r}\lambda_i(\xi(t))\lambda_j(\xi(t_k))\{A_i\hat{x}(t) + A_{di}\hat{x}(t-\alpha(t)) \\ &\quad + D_i\int_{t-\beta}^{t}\hat{x}(s)ds + B_iu(t) + \bar{K}_j(y(t_k) - C_i\hat{x}(t_k))\}.\end{aligned} \tag{20}$$

where we see the measurement output as

$$y(t) = \sum_{i=1}^{r}\lambda_i(\xi(t_k))C_ix(t-h(t)).$$

It follows from (18), (19) and (20) that the error $e(t) = x(t) - \hat{x}(t)$ satisfies

$$\begin{aligned}\dot{e}(t) - \sum_{i=1}^{r}\lambda_i(\xi(t))A_{ni}\dot{e}(t-\gamma) &= \sum_{i=1}^{r}\sum_{j=1}^{r}\lambda_i(\xi(t))\lambda_j(\xi(t_k))\{A_ie(t) + A_{di}e(t-\alpha(t)) \\ &\quad + D_i\int_{t-\beta}^{t}e(s)ds - \bar{K}_jC_ie(t_k)\}.\end{aligned} \tag{21}$$

We shall find conditions for (21) to be asymptotically stable. In this case, (20) becomes an observer for the system (18) and (19).
The following theorem gives conditions for the error system (21) to be asymptotically stable.

Theorem 6.1 *Given control gain matrices $\bar{K}_i$, $i = 1, \cdots, r$, the error system (21) is asymptotically stable if there exist matrices $P_i > 0$, $R \geq 0$, $X > 0$, $Y_i > 0$, $i = 1,2,3$, $Q_i \geq 0$, $Z_i > 0$, $i = 1,2$, and*

$$
\begin{aligned}
N_{ij} &= \begin{bmatrix} N_{1ij}^T & N_{2ij}^T & N_{3ij}^T & N_{4ij}^T & N_{5ij}^T & N_{6ij}^T & N_{7ij}^T & N_{8ij}^T & N_{9ij}^T \end{bmatrix}^T, \\
S_{ij} &= \begin{bmatrix} S_{1ij}^T & S_{2ij}^T & S_{3ij}^T & S_{4ij}^T & S_{5ij}^T & S_{6ij}^T & S_{7ij}^T & S_{8ij}^T & S_{9ij}^T \end{bmatrix}^T, \\
M_{ij} &= \begin{bmatrix} M_{1ij}^T & M_{2ij}^T & M_{3ij}^T & M_{4ij}^T & M_{5ij}^T & M_{6ij}^T & M_{7ij}^T & M_{8ij}^T & M_{9ij}^T \end{bmatrix}^T, \\
V_{ij} &= \begin{bmatrix} V_{1ij}^T & V_{2ij}^T & V_{3ij}^T & V_{4ij}^T & V_{5ij}^T & V_{6ij}^T & V_{7ij}^T & V_{8ij}^T & V_{9ij}^T \end{bmatrix}^T, \\
W_{ij} &= \begin{bmatrix} W_{1ij}^T & W_{2ij}^T & W_{3ij}^T & W_{4ij}^T & W_{5ij}^T & W_{6ij}^T & W_{7ij}^T & W_{8ij}^T & W_{9ij}^T \end{bmatrix}^T, \\
O_{ij} &= \begin{bmatrix} O_{1ij}^T & O_{2ij}^T & O_{3ij}^T & O_{4ij}^T & O_{5ij}^T & O_{6ij}^T & O_{7ij}^T & O_{8ij}^T & O_{9ij}^T \end{bmatrix}^T, \; i,j = 1, \cdots, r,
\end{aligned}
$$

$$
T = \begin{bmatrix} T_1^T & T_2^T & T_3^T & T_4^T & T_5^T & T_6^T & T_7^T & T_8^T & T_9^T \end{bmatrix}^T
$$

such that

$$
\Phi_{ij} = \begin{bmatrix} \Phi_{11ij} & \Phi_{12ij} \\ \Phi_{12ij}^T & \Phi_{22} \end{bmatrix} < 0, \; i,j = 1, \cdots, r \tag{22}
$$

where

$$
\begin{aligned}
\Phi_{11ij} &= \Phi_1 + \Phi_{2ij} + \Phi_{2ij}^T + \Phi_{3ij} + \Phi_{3ij}^T, \\
\Phi_1 &= \begin{bmatrix}
\Phi_{111} & 0 & 0 & 0 & 0 & 0 & 0 & P_2 & P_1 \\
0 & 0 & 0 & 0 & 0 & 0 & 0 & 0 & 0 \\
0 & 0 & -R & 0 & 0 & 0 & 0 & 0 & 0 \\
0 & 0 & 0 & -(1-d)Q_1 & 0 & 0 & 0 & 0 & 0 \\
0 & 0 & 0 & 0 & 0 & 0 & 0 & -P_2 & 0 \\
0 & 0 & 0 & 0 & 0 & -Q_2 & 0 & 0 & 0 \\
0 & 0 & 0 & 0 & 0 & 0 & 0 & 0 & 0 \\
P_2 & 0 & 0 & 0 & -P_2 & 0 & 0 & -X - \frac{1}{\beta_M}U & 0 \\
P_1 & 0 & 0 & 0 & 0 & 0 & 0 & 0 & \Phi_{199}
\end{bmatrix}, \\
\Phi_{111} &= Q_1 + R + \beta_M U + \beta_M^2 X, \\
\Phi_{199} &= Q_2 + \alpha_M Y_1 + \beta_M Y_2 + \gamma_M Y_3 + h_M(Z_1 + Z_2), \\
\Phi_{2ij} &= \begin{bmatrix} N_{ij} + M_{ij} + W_{ij} + O_{ij} + V_{ij} & -N_{ij} + S_{ij} & -M_{ij} - S_{ij} & -W_{ij} \end{bmatrix. \\
&\qquad \begin{matrix} -O_{ij} & 0 & -V_{ij} & 0 & 0 \end{matrix} \Big], \\
\Phi_{3ij} &= \begin{bmatrix} -TA_i & T\bar{K}_j C_i & 0 & -TA_{di} & 0 & -TA_{ni} & 0 & -TD_i & T \end{bmatrix}, \\
\Phi_{12ij} &= \begin{bmatrix} h_M N_{ij} & h_M S_{ij} & h_M M_{ij} & \alpha_M W_{ij} & \beta_M O_{ij} & \gamma_M V_{ij} \end{bmatrix}, \\
\Phi_{22} &= \mathrm{diag}\begin{bmatrix} -h_M Z_1 & -h_M Z_1 & -h_M Z_2 & -\alpha_M Y_1 & -\beta_M Y_2 & -\gamma_M Y_3 \end{bmatrix}.
\end{aligned}
$$

Proof: Proof is similar to that of Theorem 3.1. We first note that the following is true for any matrix T.

$$2\sum_{i=1}^{r}\sum_{j=1}^{r}\lambda_i(\xi(t))\lambda_j(\xi(t_k))\bar{\zeta}^T(t)T\big[\dot{e}(t) - A_i e(t)$$
$$- A_{di}e(t-\alpha(t)) - A_{ni}\dot{e}(t-\gamma) - D_i\int_{t-\beta}^{t} e(s)ds - \bar{K}_j C_i e(t-h(t))\big] = 0 \tag{23}$$

where

$$\bar{\zeta}(t) = \big[\; e^T(t) \quad e^T(t-h(t)) \quad e^T(t-h_M) \quad e^T(t-\alpha(t)) \quad e^T(t-\beta)$$
$$\dot{e}(t-\gamma) \quad e(t-\gamma) \quad \int_{t-\beta}^{t} e^T(s)ds \quad \dot{e}^T(t) \;\big]^T.$$

Now, we take the derivative of $V(e_t)$, which is defined as $V(x_t)$ with replacing x_t by e_t, with respect to t along the solution of the error system (21) and add the left-hand-sides of (6)-(11) with replacing x by e and (23):

$$\begin{aligned}
\dot{V}(e_t) \;\le\; & \sum_{i=1}^{r}\sum_{j=1}^{r}\lambda_i(\xi(t))\lambda_j(\xi(t_k))\Big\{\bar{\zeta}^T(t)\Xi_{ij}\bar{\zeta}(t) \\
& - \int_{t-h(t)}^{t}\big[\bar{\zeta}^T(t)N_{ij} + \dot{e}^T(s)Z_1\big] Z_1^{-1}\big[N_{ij}^T\bar{\zeta}(t) + Z_1\dot{e}(s)\big]ds \\
& - \int_{t-h_M}^{t-h(t)}\big[\bar{\zeta}^T(t)S_{ij} + \dot{e}^T(s)Z_1\big] Z_1^{-1}\big[S_{ij}^T\bar{\zeta}(t) + Z_1\dot{x}(s)\big]ds \\
& - \int_{t-h_M}^{t}\big[\bar{\zeta}^T(t)M_{ij} + \dot{x}^T(s)Z_2\big] Z_2^{-1}\big[M_{ij}^T\bar{\zeta}(t) + Z_2\dot{e}(s)\big]ds \\
& - \int_{t-\alpha(t)}^{t}\big[\bar{\zeta}^T(t)W_{ij} + \dot{e}^T(s)Y_1\big] Y_1^{-1}\big[W_{ij}^T\bar{\zeta}(t) + Y_1\dot{e}(s)\big]ds \\
& - \int_{t-\beta}^{t}\big[\bar{\zeta}^T(t)O_{ij} + \dot{e}^T(s)Y_2\big] Y_2^{-1}\big[O_{ij}^T\bar{\zeta}(t) + Y_2\dot{e}(s)\big]ds \\
& - \int_{t-\gamma}^{t}\big[\bar{\zeta}^T(t)V_{ij} + \dot{e}^T(s)Y_3\big] Y_3^{-1}\big[V_{ij}^T\bar{\zeta}(t) + Y_3\dot{e}(s)\big]ds\Big\}
\end{aligned} \tag{24}$$

where

$$\begin{aligned}
\Xi_{ij} \;=\; & \Theta_{11ij} + h_M(N_{ij}Z_1^{-1}N_{ij}^T + S_{ij}Z_1^{-1}S_{ij}^T + M_{ij}Z_2^{-1}M_{ij}^T) + \alpha_M W_{ij}Y_1^{-1}W_{ij}^T \\
& + \beta_M O_{ij}Y_2^{-1}O_{ij}^T + \gamma_M V_{ij}Y_3^{-1}V_{ij}^T.
\end{aligned}$$

Now, if (22) is satisfied, then by Schur complement formula we have

$$\Theta_{ij} < 0, \; i,j = 1,\cdots,r. \tag{25}$$

If (25) holds, we have $\sum_{i=1}^{r}\sum_{j=1}^{r}\lambda_i(\xi(t))\lambda_j(\xi(t_k))\bar{\zeta}^T(t)\Psi_{ij}\bar{\zeta}(t) < 0$, which implies that $\dot{V}(x_t) < 0$ because $Y_i > 0$, $Z_i > 0$, $i = 1,2$ and the last five terms in (24) are all less than 0. This proves that conditions (22) suffice to show the asymptotic stability of the system (21).
Theorem 6.1 still does not propose an observer design method. Hence, we give the following result.

Theorem 6.2 *Given scalars ρ_i, $i = 1, \cdots, 9$, (20) becomes an observer for the nominal system (18) and (19) if there exist matrices $P_i > 0$, $R \geq 0$, $U > 0$, $X > 0$, $Y_i > 0$, $i = 1,2,3$, $Q_i \geq 0$, $Z_i > 0$, $i = 1,2$, L, G_j, $j = 1, \cdots, r$,*

$$
\begin{aligned}
N_{ij} &= \begin{bmatrix} N_{1ij}^T & N_{2ij}^T & N_{3ij}^T & N_{4ij}^T & N_{5ij}^T & N_{6ij}^T & N_{7ij}^T & N_{8ij}^T & N_{9ij}^T \end{bmatrix}^T, \\
S_{ij} &= \begin{bmatrix} S_{1ij}^T & S_{2ij}^T & S_{3ij}^T & S_{4ij}^T & S_{5ij}^T & S_{6ij}^T & S_{7ij}^T & S_{8ij}^T & S_{9ij}^T \end{bmatrix}^T, \\
M_{ij} &= \begin{bmatrix} M_{1ij}^T & M_{2ij}^T & M_{3ij}^T & M_{4ij}^T & M_{5ij}^T & M_{6ij}^T & M_{7ij}^T & M_{8ij}^T & M_{9ij}^T \end{bmatrix}^T, \\
V_{ij} &= \begin{bmatrix} V_{1ij}^T & V_{2ij}^T & V_{3ij}^T & V_{4ij}^T & V_{5ij}^T & V_{6ij}^T & V_{7ij}^T & V_{8ij}^T & V_{9ij}^T \end{bmatrix}^T, \\
W_{ij} &= \begin{bmatrix} W_{1ij}^T & W_{2ij}^T & W_{3ij}^T & W_{4ij}^T & W_{5ij}^T & W_{6ij}^T & W_{7ij}^T & W_{8ij}^T & W_{9ij}^T \end{bmatrix}^T, \\
O_{ij} &= \begin{bmatrix} O_{1ij}^T & O_{2ij}^T & O_{3ij}^T & O_{4ij}^T & O_{5ij}^T & O_{6ij}^T & O_{7ij}^T & O_{8ij}^T & O_{9ij}^T \end{bmatrix}^T, \quad i,j = 1, \cdots, r
\end{aligned}
$$

such that

$$
\Theta_{ij} = \begin{bmatrix} \Theta_{11ij} & \Theta_{12ij} \\ \Theta_{12ij}^T & \Theta_{22} \end{bmatrix} < 0, \; i,j = 1, \cdots, r \tag{26}
$$

where

$$
\begin{aligned}
\Theta_{11ij} &= \Theta_1 + \Theta_{2ij} + \Theta_{2ij}^T + \Theta_{3ij} + \Theta_{3ij}^T, \\
\Theta_1 &= \begin{bmatrix}
\Theta_{111} & 0 & 0 & 0 & 0 & 0 & 0 & P_2 & P_1 \\
0 & 0 & 0 & 0 & 0 & 0 & 0 & 0 & 0 \\
0 & 0 & -R & 0 & 0 & 0 & 0 & 0 & 0 \\
0 & 0 & 0 & -(1-d)Q_1 & 0 & 0 & 0 & 0 & 0 \\
0 & 0 & 0 & 0 & 0 & 0 & 0 & -P_2 & 0 \\
0 & 0 & 0 & 0 & 0 & -Q_2 & 0 & 0 & 0 \\
0 & 0 & 0 & 0 & 0 & 0 & 0 & 0 & 0 \\
P_2 & 0 & 0 & 0 & -P_2 & 0 & 0 & -X - \frac{1}{\beta_M}U & 0 \\
P_1 & 0 & 0 & 0 & 0 & 0 & 0 & 0 & \Theta_{199}
\end{bmatrix}, \\
\Theta_{111} &= Q_1 + R + \beta_M U + \beta_M^2 X, \\
\Theta_{199} &= Q_2 + \alpha_M Y_1 + \beta_M Y_2 + \gamma_M Y_3 + h_M (Z_1 + Z_2), \\
\Theta_{2ij} &= \begin{bmatrix} N_{ij} + M_{ij} + W_{ij} + O_{ij} + V_{ij} & -N_{ij} + S_{ij} & -M_{ij} - S_{ij} & -W_{ij} \end{bmatrix. \\
&\qquad \begin{matrix} -O_{ij} & 0 & -V_{ij} & 0 & 0 \end{matrix} \Big],
\end{aligned}
$$

$$\Theta_{3ij} = -\begin{bmatrix} \rho_1 LA_i & -\rho_1 G_j C_i & 0 & \rho_1 LA_{di} & 0 & \rho_1 LA_{ni} & 0 & \rho_1 LD_i & -\rho_1 L \\ \rho_2 LA_i & -\rho_2 G_j C_i & 0 & \rho_2 LA_{di} & 0 & \rho_2 LA_{ni} & 0 & \rho_2 LD_i & -\rho_2 L \\ \rho_3 LA_i & -\rho_3 G_j C_i & 0 & \rho_3 LA_{di} & 0 & \rho_3 LA_{ni} & 0 & \rho_3 LD_i & -\rho_3 L \\ \rho_4 LA_i & -\rho_4 G_j C_i & 0 & \rho_4 LA_{di} & 0 & \rho_4 LA_{ni} & 0 & \rho_4 LD_i & -\rho_4 L \\ \rho_5 LA_i & -\rho_5 G_j C_i & 0 & \rho_5 LA_{di} & 0 & \rho_5 LA_{ni} & 0 & \rho_5 LD_i & -\rho_5 L \\ \rho_6 LA_i & -\rho_6 G_j C_i & 0 & \rho_6 LA_{di} & 0 & \rho_6 LA_{ni} & 0 & \rho_6 LD_i & -\rho_6 L \\ \rho_7 LA_i & -\rho_7 G_j C_i & 0 & \rho_7 LA_{di} & 0 & \rho_7 LA_{ni} & 0 & \rho_7 LD_i & -\rho_7 L \\ \rho_8 LA_i & -\rho_8 G_j C_i & 0 & \rho_8 LA_{di} & 0 & \rho_8 LA_{ni} & 0 & \rho_8 LD_i & -\rho_8 L \\ \rho_9 LA_i & -\rho_9 G_j C_i & 0 & \rho_9 LA_{di} & 0 & \rho_9 LA_{ni} & 0 & \rho_9 LD_i & -\rho_9 L \end{bmatrix},$$

$$\Theta_{12ij} = \begin{bmatrix} h_M N_{ij} & h_M S_{ij} & h_M M_{ij} & \alpha_M W_{ij} & \beta_M O_{ij} & \gamma_M V_{ij} \end{bmatrix},$$

$$\Theta_{22} = \text{diag}\begin{bmatrix} -h_M Z_1 & -h_M Z_1 & -h_M Z_2 & -\alpha_M Y_1 & -\beta_M Y_2 & -\gamma_M Y_3 \end{bmatrix}.$$

In this case, observer gains in (20) are given by

$$\bar{K}_i = L^{-1} G_i, \ i = 1, \cdots, r. \tag{27}$$

Proof: We let $T_i = \rho_i \bar{L}$, $i = 1, \cdots, 9$ where each ρ_i is given and $\bar{L}$ is defined later, and substitute them into (26). If (26) holds, it follows that (9,9)-block of Θ_{11ij} must be negative definite. It follows that $T_9 + T_9^T = \rho_9(\bar{L} + \bar{L}^T) < 0$, which implies that $\bar{L}$ is nonsingular. Then, we calculate $\Theta_{ij} = \Sigma\Phi_{ij}\Sigma^T$ with $\Sigma = \text{diag}[\, L\ L\ L\ L\ L\ L\ L\ L\ L\ L\ L\ L\ L\ L\ L \,]$, and obtain Θ_{ij} in (26) where we let $G_j = L\bar{K}_j$. If the conditions (26) hold, observer gain matrix $\bar{K}_i$ is obviously given by (27).

7. Numerical examples for observer design

Let us design sampled-data observers for the system (18) and (19) with the following matrices.

$$A_1 = \begin{bmatrix} 0 & 0 \\ 0 & 1 \end{bmatrix}, A_2 = \begin{bmatrix} 0 & 0 \\ 0 & 1.5 \end{bmatrix}, A_{d1} = \begin{bmatrix} -1 & -1 \\ 0 & -0.9 \end{bmatrix}, A_{d2} = \begin{bmatrix} -1 & -1 \\ 0 & -1.4 \end{bmatrix},$$
$$A_{n1} = \begin{bmatrix} 0.1 & 0 \\ 0 & 0.2 \end{bmatrix}, A_{n2} = \begin{bmatrix} 0.1 & 0 \\ 0 & 0.3 \end{bmatrix}, B_1 = \begin{bmatrix} 0.1 \\ 1 \end{bmatrix}, B_2 = \begin{bmatrix} 0.1 \\ 1.2 \end{bmatrix},$$
$$C_1 = \begin{bmatrix} 0.2 & 0.5 \end{bmatrix}, C_2 = \begin{bmatrix} 0.3 & 1.2 \end{bmatrix}, D_1 = \begin{bmatrix} 0.1 & 0 \\ 0 & 0.3 \end{bmatrix}, D_2 = \begin{bmatrix} 0.2 & 0 \\ 0 & 0.4 \end{bmatrix}.$$

The grades are given by $\lambda_1(x_1) = \frac{1}{1+e^{-x_1}}$ and $\lambda_2(x_1) = 1 - \lambda_1(x_1)$. The maximum upper bound of the sampling time $h_M = 0.1$ and $d = 0.4$ are assumed. We let $\alpha_M = \beta_M = 0.2$. Theorem 6.2 with $\rho_1 = 4.76$, $\rho_2 = -0.03$, $\rho_3 = -2.19$, $\rho_4 = -0.60$, $\rho_5 = -0.01$, $\rho_6 = 0.01$, $\rho_7 = -0.50$, $\rho_8 = 0.09$, $\rho_9 = 1.96$ guarantees the existence of the sampled-data observer for the maximum upper bound of the time-delay $\gamma_M = 2.41$. In this case, observer gains in (2) are given by

$$\bar{K}_1 = \begin{bmatrix} -1.1239 \\ 0.7925 \end{bmatrix}, \bar{K}_2 = \begin{bmatrix} -1.1254 \\ 0.7916 \end{bmatrix}.$$

8. Conclusion

In this paper, robust sampled-data control and observer design for uncertain fuzzy systems with discrete, neutral and distributed delays has been considered. Less conservative robust stability conditions were obtained as LMI conditions via time-varying delay system approach. Then, a controller design method was proposed via LMI conditions. As a dual result, an observer design method was also given. Finally, some examples were given to illustrate our design approach.

9. References

[1] Gao, H. & Chen, T. (2007). Stabilization of nonlinear systems under variable sampling: a fuzzy control approach, *IEEE Transactions on Fuzzy Systems*, Vol.15, 972-983.

[2] Guan, X.P. & Chen, C.L. (2004). Delay-dependent guaranteed cost control for T-S fuzzy systems with time-delay, *IEEE Transactions on Fuzzy Systems*, Vol.12, 236-249.

[3] Ishihara, T. & Yoneyama, J. (2009). H_∞ sampled-data control for fuzzy systems with discrete and distributed Delays, 2, *International Review of Automatic Control*, Vol.6, 654-660.

[4] Ishihara, T. & Yoneyama, J. (2010). Robust sampled-data control for uncertain fuzzy systems with discrete and distributed delays, *Advances in Fuzzy Sets and Systems*, Vol.5, 135-151.

[5] Lam, H.K. & Leung, F.H.F. (2007). Sampled-data fuzzy controller for time-delay nonlinear systems: fuzzy-model-based LMI approach, *IEEE Transactions on Fuzzy Systems, Man, and Cybernetics*, Part B, Vol.37, 617-629.

[6] Li, C.;Wang, H. & Liao, X. (2004). Delay-dependent robust stability of uncertain fuzzy systems with time-varying delays, *IEE Proceedings on Control Theory Applications*, Vol.151, 417-421.

[7] Li, Y.;Xu, S.;Zhang, B. & Chu, Y. (2008). Robust stabilization and H_∞ control for uncertain fuzzy neutral systems with mixed time delays, *Fuzzy Sets and Systems*, Vol.159, 2730-2748.

[8] Takagi, T. & Sugeno, M. (1985). Fuzzy identification of systems and its applications to modeling and control, *IEEE Transactions on Systems, Man, Cybernetics*, Vol.15, 116-132.

[9] Tian, E. & Pang, C. (2006). Delay-dependent stability analysis and synthesis of uncertain T-S fuzzy systems with time-varying delay, *Fuzzy Sets and Systems*, Vol. 157, 544-559.

[10] Xie, L. (1996). Output feedback H_∞ control of systems with parameter uncertainty, *International Journal of Control*, Vol.63, 741-59.

[11] Yoneyama, J. (2006). Robust stability and stabilization for fuzzy systems with discrete and distributed delays, *Proceedings of IEEE International Symposium on Intelligent Systems*, pp.2372-2377, Munich, Germany, October 2006.

[12] Yoneyama, J. (2007). New robust stability conditions and design of robust stabilizing controllers for Takagi-Sugeno fuzzy time-delay systems, *IEEE Transactions on Fuzzy Systems*, Vol.15, 828-839.

[13] Yoneyama, J. (2007). New delay-dependent approach to robust stability and stabilization for Takagi-Sugeno fuzzy time-delay systems, *Fuzzy Sets and Systems*, Vol.158, 2225-2237.

[14] Yoneyama, J. (2008). Robust stabilization of fuzzy systems under sampled-data control, *Proceedings of IEEE International Conference on Fuzzy Systems*, pp.1236-1241, Hong Kong, June 2008.

[15] Yoneyama, J. (2008). Robust stability and stabilizing controller design of fuzzy systems with discrete and distributed delays, *Information Sciences*, Vol.178, 1935-1947.

[16] Yoneyama, J. (2008). H_∞ sampled-data control of fuzzy systems with input delay, *Journal of Computers*, Vol.3, 25-31.

24

Numerical Solution of a System of Polynomial Parametric form Fuzzy Linear Equations

Majid Amirfakhrian
Department of Mathematics, Islamic Azad University, Central Tehran Branch (IAUCTB)
Iran

1. Introduction

Since fuzzy logic was introduced by Lotfi Zadeh in 1965 (41), it has had many successful applications in all fields that one can imagine. The reason is that many real-world applications problems are involved the systems in which at least some parameters are represented by fuzzy numbers rather than crisp numbers and linguistic labels such as small and large are also associated with the fuzzy sets. On the other hand a system of fuzzy linear equations may appear in a wide variety of problems in various areas such as mathematics, statistics, physics, engineering and social sciences.

The objective of this chapter is to introduce a method to find a good approximate solution to a system of fuzzy linear equations, and first we need to be familiar with some notations on fuzzy numbers in this chapter, however it is assumed that the reader is relatively familiar with the elementary fuzzy logic concepts.

In (14), Chong-Xin and Ming represented a fuzzy number $\tilde{u}$ by an ordered pair of functions $(\underline{u}(r), \overline{u}(r))$, $0 \le r \le 1$, which satisfies some requirements.

In papers and books the authors mainly have used linear membership functions as spreads, because they are conceptually the simplest, have a clear interpretation and play a crucial role in many areas of fuzzy applications, and almost every works on this field of study have been done on triangular or trapezoidal fuzzy numbers, but polynomial form fuzzy numbers are simple and have a clear interpretation too, and in order to obtain a richer class of fuzzy numbers we use polynomials of the degree higher than one, as the spreads of membership functions of fuzzy numbers. Thus in (2) we introduced a type of fuzzy numbers in which both left spread function $\underline{u}(r)$ and right spread function $\overline{u}(r)$ are polynomials of degree at most m. We named this type of fuzzy numbers, m-degree polynomial-form fuzzy numbers.

The main aim of introducing this type of fuzzy numbers is that in many applications of fuzzy logic and fuzzy mathematics we need (or it is better) to work with the same fuzzy numbers.

It has been shown in (2) that a fuzzy number $\tilde{u}$ with continuous left and right spread functions can be approximated by a fuzzy number with m-degree polynomial-form, where choosing m depends on the shape of left and right spread functions L and R, and the derivation order of them.

Some applications of this approximation in the case $m = 1$ (trapezoidal fuzzy numbers) are given in (3), and some properties of this approximation operator are recently given in (10).

There are many other literatures which authors tried to approximate a fuzzy number by a simpler one (2; 3; 4; 18; 25; 27; 28; 29; 30; 31; 32; 42). Also there are some distances defined by authors to compare fuzzy numbers (35; 37).

Obviously, if we use a defuzzification rule which replaces a fuzzy set by a single number, we generally loose too many important information. Also, an interval approximation is considered for fuzzy numbers in (25), where a fuzzy computation problem is converted into interval arithmetic problem. But, in this case, we loose the fuzzy central concept. Even in some works such as (4; 31; 32; 42), authors solve an optimization problem to obtain the nearest triangular or trapezoidal fuzzy number which is related to an arbitrary fuzzy number, however in these cases there is not any guarantee to have the same modal value (or interval). But by parametric polynomial approximation we are able to approximate many fuzzy numbers in a good manner.
The problem of finding the nearest parametric approximation of a fuzzy number with respect to the average Euclidean distance is completely solved in (13). Ban point out the wrongs and inadvertence in some recent papers, then correct the results in (12). A parametric fuzzy approximation method based on the decision maker's strategy as an extension of trapezoidal approximation of a fuzzy number offered in (34). An improvement of the nearest trapezoidal approximation operator preserving the expected interval, which is proposed by Grzegorzewski and MrXowka is studied in (39). There are some trapezoidal approximation operators introduced in (26; 29; 38)
But the main aim of this chapter is to give a method to solve linear system of fuzzy equations. There are three categories of a linear system of fuzzy equations

$$\mathbf{Ax} = \mathbf{b}.$$

- In the first category, the coefficient matrix arrays are crisp numbers, the right-hand side column is an arbitrary fuzzy vector and the unknowns are fuzzy numbers.
- In the second category, the coefficient matrix arrays are fuzzy numbers, the right-hand side column is an arbitrary fuzzy vector and the unknowns are crisp numbers.
- In the third category, all the coefficient matrix arrays, the right-hand side arrays and the unknowns, are fuzzy numbers.

There have been few papers of fuzzy linear equations with crisp unknowns, and in this chapter we propose a method for solving an $n \times n$ linear system from second category based on (9). However there are many reported studies in which researchers tried to solve a system of linear fuzzy equations numerically (1; 5; 6; 7; 11; 15; 21; 22). In most of these studies the problem is considered for a system with fuzzy unknowns and crisp coefficients. Some new work has been done on fully fuzzy linear systems of equations. Some works are done on fully fuzzy linear system of equations (16; 17). Some works have been done on rectangular $m \times n$ system of equations (11; 24; 43) and some works have been done on blocked matrices (33; 36). A minimal solution for dual fuzzy linear system is given in (8). A class of methods is considered in (40).
The structure of this chapter is as follows: First we represent *source distance* and $m-$*source distance*; and we present the nearest approximation of a fuzzy number in polynomial parametric form, introduced in (2) and present some properties of it where recently published in (10). Next we present a method for solving a fuzzy system of linear equations with m-degree polynomial parametric-form fuzzy coefficients and crisp unknowns by a least squares method. Then we introduce the extension of this method to solve a general fuzzy system of linear equations with LR fuzzy coefficients and crisp unknowns.
One of the advantages of the proposed method is that by this method not only one can find a good approximation of the solution of such fuzzy systems with polynomial-form fuzzy arrays

but even a system of fuzzy linear equations with general type of fuzzy number arrays can be solved by this method.
Finally some applied examples are given at the end of this chapter .

2. Preliminaries

Let $\mathcal{F}(\mathbb{R})$ be the set of all normal and convex fuzzy numbers on the real line (44).

Definition 2.1 *(19), A generalized LR fuzzy number $\tilde{u}$ with the membership function $\mu_{\tilde{u}}(x), x \in \mathbb{R}$ can be defined as*

$$\mu_{\tilde{u}}(x) = \begin{cases} l_{\tilde{u}}(x) & , \quad a \leq x \leq b, \\ 1 & , \quad b \leq x \leq c, \\ r_{\tilde{u}}(x) & , \quad c \leq x \leq d, \\ 0 & , \quad otherwise, \end{cases} \tag{2.1}$$

where $l_{\tilde{u}}(x)$ is the left membership function that is an increasing function on $[a,b]$ and $r_{\tilde{u}}(x)$ is the right membership function that is a decreasing function on $[c,d]$. Furthermore we want to have $l_{\tilde{u}}(a) = r_{\tilde{u}}(d) = 0$ and $l_{\tilde{u}}(b) = r_{\tilde{u}}(c) = 1$. In addition, if $l_{\tilde{u}}(x)$ and $r_{\tilde{u}}(x)$ are linear, then $\tilde{u}$ is a trapezoidal fuzzy number which is denoted by (a,b,c,d). If $b = c$, we denoted it by (a,c,d), which is a triangular fuzzy number.
For $0 < \alpha \leq 1$; α-cut of a fuzzy number $\tilde{u}$ is defined by (20),

$$[\tilde{u}]^{\alpha} = \{t \in \mathbb{R} \mid \mu_{\tilde{u}}(t) \geq \alpha\}. \tag{2.2}$$

Definition 2.2 *(35), A continuous function $s : [0,1] \longrightarrow [0,1]$ with the following properties is a regular reducing function :*

1. *$s(r)$ is increasing.*
2. *$s(0) = 0$,*
3. *$s(1) = 1$,*
4. *$\int_0^1 s(r)dr = \frac{1}{2}$.*

In (14), Chong-Xin and Ming represented a fuzzy number $\tilde{u}$ by an ordered pair of functions $(\underline{u}(r), \overline{u}(r))$:
The *parametric form* of a fuzzy number is shown by $\tilde{v} = (\underline{v}(r), \overline{v}(r))$, where functions $\underline{v}(r)$ and $\overline{v}(r)$; $0 \leq r \leq 1$ satisfy the following requirements:

1. $\underline{v}(r)$ is monotonically increasing left continuous function.
2. $\overline{v}(r)$ is monotonically decreasing left continuous function.
3. $\underline{v}(r) \leq \overline{v}(r)$, $0 \leq r \leq 1$.
4. $\overline{v}(r) = \underline{v}(r) = 0$ for $r < 0$ or $r > 1$.

Definition 2.3 *(35), The* value *and* ambiguity *of a fuzzy number $\tilde{u}$ are defined by*

$$\boldsymbol{val}(\tilde{u}) := \int_0^1 s(r)[\overline{u}(r) + \underline{u}(r)]dr, \tag{2.3}$$

and

$$\boldsymbol{amb}(\tilde{u}) := \int_0^1 s(r)[\overline{u}(r) - \underline{u}(r)]dr, \tag{2.4}$$

respectively.

Definition 2.4 *(2; 9) We say a fuzzy number $\tilde{u}$ has $m-$degree polynomial-form, if there exist two polynomials $p_m(r)$ and $q_m(r)$, of degree at most m; such that $\tilde{u} = (p_m(r), q_m(r))$.*

Let $\mathcal{PF}_m(\mathbb{R})$ be the set of all $m-$degree polynomial-form fuzzy numbers.

Definition 2.5 *(2), Let for positive integer k, we have $\underline{u}, \overline{u} \in C^k[0,1]$. We define k-validity and k-unworthiness of a fuzzy number $\tilde{u}$ by*

$$\boldsymbol{val}_k(\tilde{u}) := \int_0^1 s(r)[\overline{u}^{(k)}(r) + \underline{u}^{(k)}(r)]dr, \tag{2.5}$$

and

$$\boldsymbol{amb}_k(\tilde{u}) := \int_0^1 s(r)[\overline{u}^{(k)}(r) - \underline{u}^{(k)}(r)]dr, \tag{2.6}$$

respectively.

0-validity and 0-unworthiness of a fuzzy number are its value and ambiguity, respectively. i.e.

$$\mathbf{val}_0 := \mathbf{val}, \quad \mathbf{amb}_0 := \mathbf{amb}.$$

Proposition 2.1 *For any nonnegative integer k, the k-validity and k-unworthiness have the following properties:*

1. $\boldsymbol{val}_k(\tilde{u} \pm \tilde{v}) = \boldsymbol{val}_k(\tilde{u}) \pm \boldsymbol{val}_k(\tilde{v})$,
2. $\boldsymbol{amb}_k(\tilde{u} \pm \tilde{v}) = \boldsymbol{amb}_k(\tilde{u}) \pm \boldsymbol{amb}_k(\tilde{v})$.

Definition 2.6 *Let for positive integer m, we have $\underline{u}, \overline{u} \in C^{m-1}[0,1]$. For $k = 0,1,\ldots,m-1$, we define*

$$d_k(\tilde{u},\tilde{v}) := |\boldsymbol{val}_k(\tilde{u} - \tilde{v})| + |\boldsymbol{amb}_k(\tilde{u} - \tilde{v})|. \tag{2.7}$$

Definition 2.7 *Let X and Y be two non-empty subsets of a metric space (M,d). We define their Hausdorff distance $d_H(X,Y)$ by*

$$d_H(X,Y) = \max\{\sup_{x\in X}\inf_{y\in Y} d(x,y), \sup_{y\in Y}\inf_{x\in X} d(x,y)\}. \tag{2.8}$$

3. Source distance and m-source distance

Definition 3.1 *(2), Let $\mathcal{AF}(\mathbb{R})$ be a subset of $\mathcal{F}(\mathbb{R})$. $\tilde{v}^* \in \mathcal{AF}(\mathbb{R})$, is the nearest approximation of an arbitrary fuzzy number $\tilde{u} \in \mathcal{F}(\mathbb{R})$ out of $\mathcal{AF}(\mathbb{R})$, with respect to a meter d if and only if*

$$d(\tilde{u},\tilde{v}^*) = \min_{\tilde{v}\in\mathcal{AF}(\mathbb{R})} d(\tilde{u},\tilde{v}). \tag{3.1}$$

Definition 3.2 *(2), For $\tilde{u},\tilde{v} \in \mathcal{F}(\mathbb{R})$, we define **source distance** of $\tilde{u}$ and $\tilde{v}$ by*

$$D(\tilde{u},\tilde{v}) := \frac{1}{2}\{d_0(\tilde{u},\tilde{v}) + d_H([\tilde{u}]^1, [\tilde{v}]^1)\}, \tag{3.2}$$

where d_H is the Hausdorff metric.

In (2) we proved that the source distance, D, is a metric on the set of all trapezoidal fuzzy numbers.

Definition 3.3 *For a nonnegative integer j, we define j^{th}-source number by*

$$I_j := \int_0^1 r^j s(r)dr. \tag{3.3}$$

Lemma 3.1 *Let I_j be the jth-source number, then $\{I_j\}$ is a positive decreasing sequence, where $I_0 = \frac{1}{2}$. i.e. $0 \leq \ldots < I_2 < I_1 < I_0 = \frac{1}{2}$.*

Proof: By using the definition of jth-source number and regular reducing function we have $I_0 = \frac{1}{2}$. By Mean Value Theorem for integrals for any integer $j \geq 1$, there exists a $\psi \in (0,1)$, such that

$$I_j = \psi \int_0^1 r^{j-1} s(r)dr < I_{j-1}.$$

□

Definition 3.4 *For $\tilde{u},\tilde{v} \in \mathcal{F}(\mathbb{R})$, we define m-**source distance** of $\tilde{u}$ and $\tilde{v}$ by*

$$D_m(\tilde{u},\tilde{v}) := \frac{1}{2} d_H([\tilde{u}]^1,[\tilde{v}]^1) + \sum_{k=0}^{m-1}{}' d_k(\tilde{u},\tilde{v}).$$

where $\sum_{k=0}^{m-1}{}'$ is defined by

$$\sum_{i=0}^{m-1}{}' a_i = \frac{1}{2} a_0 + \sum_{i=1}^{m-1} a_i.$$

Theorem 3.2 *For $\tilde{u},\tilde{v},\tilde{w},\tilde{z} \in \mathcal{F}(\mathbb{R})$, the distance, D_m, satisfies the following properties:*

1. $D_m(\tilde{u},\tilde{u}) = 0$,
2. $D_m(\tilde{u},\tilde{v}) = D_m(\tilde{v},\tilde{u})$,
3. $D_m(\tilde{u},\tilde{v}) \leq D_m(\tilde{u},\tilde{w}) + D_m(\tilde{w},\tilde{v})$,
4. $D_m(k\tilde{u},k\tilde{v}) = |k| D_m(\tilde{u},\tilde{v})$ *for* $k \in \mathbb{R}$,
5. $D_m(\tilde{u}+\tilde{w},\tilde{v}+\tilde{z}) \leq D_m(\tilde{u},\tilde{v}) + D_m(\tilde{w},\tilde{z})$.

Example 3.1 *For two crisp real numbers a and b we have*

$$D_m(a,b) = |a-b|.$$

Proposition 3.3 *The fuzzy number $\tilde{v}^*$ is a nearest approximation of $\tilde{u}$ out of $\mathcal{PF}_m(\mathbb{R})$ if and only if*

$$D_m(\tilde{u},\tilde{v}^*) = \min_{\tilde{v} \in \mathcal{PF}_m(\mathbb{R})} D_m(\tilde{u},\tilde{v}). \tag{3.4}$$

We denote the set of all the nearest approximations of a fuzzy number $\tilde{u}$, out of $\mathcal{PF}_m(\mathbb{R})$, by $N^*_{\mathcal{PF}_m(\mathbb{R})_m}(\tilde{u})$.

Theorem 3.4 *Let $\tilde{u}$ be a fuzzy number. If for positive integer m we have $\underline{u},\overline{u} \in C^{m-1}[0,1]$, then $\tilde{v}$ is the nearest approximation of $\tilde{u}$ out of $\mathcal{PF}_m(\mathbb{R})$ if and only if $D_m(\tilde{u},\tilde{v}) = 0$.*

Proof: See (2). □

Theorem 3.5 *Let m be a positive integer and suppose that $\tilde{u} \in \mathcal{F}(\mathbb{R})$ be an arbitrary fuzzy number. $\tilde{v} \in \mathcal{PF}_m(\mathbb{R})$ is the nearest approximation of $\tilde{u}$ out of $\mathcal{PF}_m(\mathbb{R})$, if and only if for $k = 0,\dots,m-1$; $\tilde{v}$ and $\tilde{u}$ have the same k-validity and k-unworthiness and furthermore $[\tilde{u}]^1 = [\tilde{v}]^1$.*

Proof: See (2). □

Corollary 3.6 *Let $\tilde{u}$ be a generalized LR fuzzy number. If for positive integer m we have $\underline{u},\overline{u} \in C^{m-1}[0,1]$, then the nearest approximation of $\tilde{u}$ out of $\mathcal{PF}_m(\mathbb{R})$ exists.*

Theorem 3.7 *(10) The nearest approximation of a m-degree polynomial-form fuzzy number, out of $\mathcal{PF}_m(\mathbb{R})$, is itself.*

Proof: Let $\tilde{v} \in \mathcal{PF}_m(\mathbb{R})$ be the nearest approximation of $\tilde{u} \in \mathcal{PF}_m(\mathbb{R})$. Also let

$$\underline{v}(r) = \underline{d}_m r^m + \ldots + \underline{d}_1 r + \underline{d}_0,$$

and

$$\underline{u}(r) = \underline{c}_m r^m + \ldots + \underline{c}_1 r + \underline{c}_0,$$

By defining $w(r) = \underline{v}(r) - \underline{u}(r) = \sum_{j=0}^{m} a_j r^j$, where $a_j = \underline{c}_j - \underline{d}_j$ for $j = 0,1,\dots,m$; we have $w(1) = 0$ and

$$\int_0^1 w^{(k)}(r)s(r)dr = 0, \qquad k = 0,1,\dots,m-1.$$

Therefore we have a homogenous system of linear equations with the following nonsingular coefficients matrix

$$\begin{pmatrix} 1 & 1 & 1 & 1 & \cdots & 1 \\ I_0 & I_1 & I_2 & I_3 & \cdots & I_m \\ 0 & I_0 & 2I_1 & 3I_2 & \cdots & mI_{m-1} \\ 0 & 0 & 2I_0 & 3!I_1 & \cdots & m(m-1)I_{m-2} \\ \vdots & \vdots & \vdots & \vdots & & \vdots \\ 0 & 0 & 0 & 0 & \cdots & m!I_1 \end{pmatrix}$$

Therefore $w \equiv 0$. i.e. $\underline{u} \equiv \underline{v}$. In a similar way $\overline{u} \equiv \overline{v}$. Thus $\tilde{u} = \tilde{v}$. □

Corollary 3.8 *$D_m(.,.)$ is a meter on $\mathcal{PF}_m(\mathbb{R})$.*

Lemma 3.9 *(10) Let $\tilde{u}$ be a fuzzy number. For all $m \geq 1$ if $N^*_{PF_m}(\tilde{u})$ is not empty, then we have*

$$|N^*_{\mathcal{PF}_m(\mathbb{R})_m}(\tilde{u})| = 1.$$

Proof: Let $\tilde{u}_1^*$ and $\tilde{u}_2^*$ be two nearest approximations of a fuzzy number $\tilde{u}$ out of $\mathcal{PF}_m(\mathbb{R})$. Thus $D_m(\tilde{u}_1^*,\tilde{u}) = D_m(\tilde{u}_1^*,\tilde{u}) = 0$, and we have

$$D_m(\tilde{u}_1^*,\tilde{u}_2^*) \leq D_m(\tilde{u}_1^*,\tilde{u}) + D_m(\tilde{u},\tilde{u}_2^*) = 0.$$

From lemma 3.7 we have $\tilde{u}_1^* = \tilde{u}_2^*$. □

Corollary 3.10 *Let $\tilde{u}$ be a l-degree polynomial-form fuzzy number where $l \leq m$. If $\tilde{v}^*$ be the nearest approximations of $\tilde{u}$ out of $\mathcal{PF}_m(\mathbb{R})$, then $\tilde{v}^* = \tilde{u}$.*

Lemma 3.11 *(10) Let $\tilde{u}^*$ and $\tilde{v}^*$ be the nearest approximations of two fuzzy numbers $\tilde{u}$ and $\tilde{v}$, respectively. Then we have*

$$D_m(\tilde{u}^*, \tilde{v}^*) = D_m(\tilde{u}, \tilde{v})$$

Proof: We have

$$D_m(\tilde{u}^*, \tilde{v}^*) \leq D_m(\tilde{u}^*, \tilde{u}) + D_m(\tilde{u}, \tilde{v}) + D_m(\tilde{v}, \tilde{v}^*) = D_m(\tilde{u}, \tilde{v}).$$

In a similar way $D_m(\tilde{u}, \tilde{v}) \leq D_m(\tilde{u}^*, \tilde{v}^*)$. □

Lemma 3.12 *(10) Let $\tilde{v}$ and $\tilde{u}$ be two fuzzy numbers where $\underline{u}, \underline{v}, \overline{u}, \overline{v} \in C^{m-1}[0,1]$. If $D_m(\tilde{u}, \tilde{v}) = 0$, then there are two sequences of points $\{\delta_{i,k}\}_{k=0}^{m-1}$, $i = 1,2$ such that for $k = 0, 1, \cdots, m-1$,*

$$\underline{u}^{(k)}(\delta_{1,k}) = \underline{v}^{(k)}(\delta_{1,k}),$$

and

$$\overline{u}^{(k)}(\delta_{2,k}) = \overline{v}^{(k)}(\delta_{2,k}).$$

Proof: Let for two fuzzy numbers $\tilde{v}$ and $\tilde{u}$ we have $D_m(\tilde{u}, \tilde{v}) = 0$. Thus for $k = 0, 1, \ldots, m-1$, we have

$$\begin{aligned} &[\tilde{u}]^1 = [\tilde{v}]^1, \\ &\mathbf{val}_k(\tilde{v}) = \mathbf{val}_k(\tilde{u}), \qquad k = 0, \ldots, m-1, \\ &\mathbf{amb}_k(\tilde{v}) = \mathbf{amb}_k(\tilde{u}), \qquad k = 0, \ldots, m-1, \end{aligned}$$

Therefore

$$\int_0^1 s(r)[\overline{u}^{(k)}(r) - \overline{v}^{(k)}(r)]dr = \pm \int_0^1 s(r)[\underline{u}^{(k)}(r) - \underline{v}^{(k)}(r)]dr, \qquad k = 0, \ldots, m-1,$$

Thus

$$\begin{aligned} &\int_0^1 s(r)[\overline{u}^{(k)}(r) - \overline{v}^{(k)}(r)]dr = 0, \qquad k = 0, \ldots, m-1, \\ &\int_0^1 s(r)[\underline{u}^{(k)}(r) - \underline{v}^{(k)}(r)]dr = 0, \qquad k = 0, \ldots, m-1. \end{aligned}$$

Thus by Mean Value Theorem for integrals, for any $k = 0, 1, \ldots, m-1$, there are two numbers $\delta_{1,k}$ and $\delta_{2,k}$ such that

$$\begin{aligned} &\overline{u}^{(k)}(\delta_{1,k}) = \overline{v}^{(k)}(\delta_{1,k}), \\ &\underline{u}^{(k)}(\delta_{2,k}) = \underline{v}^{(k)}(\delta_{2,k}). \end{aligned}$$

□

4. Properties of $m-$source distance

Some properties of the approximation operators are presented by Grzegorzewski and Mrówka (28). In this section we consider some properties of the approximation operator suggested in Section 3.

Let $T_m : \mathcal{F}(\mathbb{R}) \longrightarrow \mathcal{PF}_m(\mathbb{R})$ be the approximation operator which produces the nearest approximation fuzzy number out of $\mathcal{PF}_m(\mathbb{R})$ to a given original fuzzy number using Theorem 3.5. Almost all of the theorems of this section are taken out from (10).

Theorem 4.1 *The nearest approximation operator is 1-cut invariance.*

Proof: It is a necessary condition for this approximation that

$$[T_m(\tilde{u})]^1 = [\tilde{u}]^1.$$

□

Theorem 4.2 *The nearest approximation operator is invariant to translations. i.e. for each $\tilde{u} \in \mathcal{F}(\mathbb{R})$ and $a \in \mathbb{R}$, we have $T_m(\tilde{u}+a) = T_m(\tilde{u}) + a$.*

Proof: Let a be a real number. Then $\underline{u+a} = \underline{u} + a$ and $\overline{u+a} = \overline{u} + a$. Therefore

$$\mathbf{val}(T_m(\tilde{u}+a)) = \mathbf{val}(\tilde{u}+a) = \mathbf{val}(\tilde{u}) + a = \mathbf{val}(T_m(\tilde{u})) + a,$$

$$\mathbf{amb}(T_m(\tilde{u}+a)) = \mathbf{amb}(\tilde{u}+a) = \mathbf{amb}(\tilde{u}) = \mathbf{amb}(T_m(\tilde{u})) = \mathbf{amb}(T_m(\tilde{u})+a),$$

Also for $k = 1,2,\ldots,m-1$, we have

$$\mathbf{val}_k((T_m(\tilde{u})+a)) = \mathbf{val}_k((\tilde{u}+a)) = \mathbf{val}_k(\tilde{u}) = \mathbf{val}_k(T_m(\tilde{u})),$$

and

$$\mathbf{amb}_k((T_m(\tilde{u})+a)) = \mathbf{amb}_k((\tilde{u}+a)) = \mathbf{amb}_k(\tilde{u}) = \mathbf{amb}_k(T_m(\tilde{u})).$$

Thus

$$D_m(T_m(\tilde{u}+a), T_m(\tilde{u})+a) = 0.$$

Since both $T_m(\tilde{u}+a)$ and $T_m(\tilde{u})+a$ have m-degree polynomial-form, then from Lemma 3.9 we have

$$T_m(\tilde{u}+a) = T_m(\tilde{u}) + a.$$

□

Theorem 4.3 *The nearest approximation operator is scale invariant. i.e. for each $\tilde{u} \in \mathcal{F}(\mathbb{R})$ and $\lambda \in \mathbb{R}$, we have $T_m(\lambda\tilde{u}) = \lambda T_m(\tilde{u})$.*

Proof: Let $\lambda \neq 0$ be a real number. Thus for $k = 0,1,\ldots,m-1$, we have

$$\mathbf{val}_k(\lambda\tilde{u}) = \lambda\mathbf{val}_k(\tilde{u}) \quad , \quad \mathbf{amb}_k(\lambda\tilde{u}) = \lambda\ \mathbf{amb}_k(\tilde{u}).$$

and

$$D_m(T_m(\lambda\tilde{u}), \lambda T_m(\tilde{u})) = 0.$$

Therefore from Lemma 3.9 we have

$$T_m(\lambda\tilde{u}) = \lambda T_m(\tilde{u}).$$

□

Theorem 4.4 *The nearest approximation operator fulfills the nearness criterion with respect to $m-$source metric D_m defined in Definition 3.4, on the set of all $m-$degree polynomial-form fuzzy numbers.*

Proof: By Lemma 3.3, we have

$$D_m(\tilde{u}, T_m(\tilde{u})) = \min_{\tilde{v} \in \mathcal{PF}_m(\mathbb{R})} D_m(\tilde{u}, \tilde{v}),$$

therefore

$$D_m(\tilde{u}, T_m(\tilde{u})) \leq D_m(\tilde{u}, \tilde{v}), \quad \forall \tilde{v} \in \mathcal{PF}_m(\mathbb{R}).$$

□

Theorem 4.5 *The nearest approximation operator is continuous.*

Proof: An approximation operator T is continuous if for any $\tilde{u}, \tilde{v} \in \mathcal{F}(\mathbb{R})$ we have

$$\forall \epsilon > 0, \exists \delta > 0, D_m(\tilde{u}, \tilde{v}) < \delta \Longrightarrow D_m(T(\tilde{u}), T(\tilde{v})) < \epsilon.$$

Let $D_m(\tilde{u}, \tilde{v}) < \delta$. By Theorem 3.2 we have

$$D_m(T_m(\tilde{u}), T_m(\tilde{v})) \leq D_m(T_m(\tilde{u}), \tilde{u}) + D_m(\tilde{u}, \tilde{v}) + D_m(\tilde{v}, T_m(\tilde{v}))$$

and by Theorem 3.4 we have $D_m(T_m(\tilde{u}), \tilde{u}) = D_m(\tilde{v}, T_m(\tilde{v})) = 0$. Thus

$$D_m(T_m(\tilde{u}), T_m(\tilde{v})) \leq D_m(\tilde{u}, \tilde{v}) < \delta.$$

Therefore its suffices to take $\delta \leq \epsilon$. □

Theorem 4.6 *The nearest trapezoidal approximation operator (case m=1) is monotonic on any set of fuzzy numbers with equal cores.*

Proof: See (2). □

Theorem 4.7 *The nearest approximation operator is order invariant with respect to value function.*

Proof: The proof is trivial, because $\mathbf{val}(T_m(\tilde{u})) = \mathbf{val}(\tilde{u})$ and $\mathbf{val}(T_m(\tilde{v})) = \mathbf{val}(\tilde{v})$. □

Theorem 4.8 *The nearest approximation operator does not change the distance of fuzzy numbers by m-source distance. i.e.*

$$D_m(T_m(\tilde{u}), T_m(\tilde{v})) = T_m(D_m(\tilde{u}, \tilde{v})) = D_m(\tilde{u}, \tilde{v}).$$

Proof: The proof is trivial by Lemma 3.11.

□

By the following Theorem we show that the nearest approximation operator is a linear operator:

Theorem 4.9 *The nearest approximation operator is a linear operator on the set of all fuzzy numbers. i.e. for a real number λ and two fuzzy numbers $\tilde{u}$ and $\tilde{v}$ we have*

$$T_m(\lambda \tilde{u} + \tilde{v}) = \lambda T_m(\tilde{u}) + T_m(\tilde{v}).$$

Proof: From Lemma 3.2 we have

$$\begin{aligned} D_m(\lambda\tilde{u}+\tilde{v},\lambda T_m(\tilde{u})+T_m(\tilde{v})) &\leq D_m(\lambda\tilde{u},\lambda T_m(\tilde{u}))+D_m(\tilde{v},T_m(\tilde{v})) \\ &= D_m(\lambda\tilde{u},\lambda T_m(\tilde{u})) \\ &= |\lambda|D_m(\tilde{u},T_m(\tilde{u}))=0. \end{aligned}$$

Also we have

$$D_m(\lambda\tilde{u}+\tilde{v},T_m(\lambda\tilde{u}+\tilde{v}))=0,$$

therefore

$$D_m(T_m(\lambda\tilde{u}+\tilde{v}),\lambda T_m(\tilde{u})+T_m(\tilde{v}))=0.$$

Since both fuzzy numbers $T_m(\lambda\tilde{u}+\tilde{v})$ and $\lambda T_m(\tilde{u})+T_m(\tilde{v})$ belong to $\mathcal{PF}_m(\mathbb{R})$, we have

$$T_m(\lambda\tilde{u}+\tilde{v})=\lambda T_m(\tilde{u})+T_m(\tilde{v}).$$

□

5. Linear system of fuzzy equations

In this section we introduce a method for solving fuzzy linear system of equations, and this method is taken from (9).
Let $\tilde{\mathbf{A}}$ and $\tilde{\mathbf{b}}$ be a matrix and a vector with fuzzy number arrays, respectively. A system of fuzzy linear equations with crisp variables is a system with the following form:

$$\tilde{\mathbf{A}}\mathbf{x}=\tilde{\mathbf{b}}. \tag{5.1}$$

Equivalently we have

$$\begin{cases} \tilde{a}_{11}x_1+\tilde{a}_{12}x_2+\ldots+\tilde{a}_{1n}x_n &= \tilde{b}_1 \\ \tilde{a}_{21}x_1+\tilde{a}_{22}x_2+\ldots+\tilde{a}_{2n}x_n &= \tilde{b}_2 \\ &\vdots \\ \tilde{a}_{n1}x_1+\tilde{a}_{n2}x_2+\ldots+\tilde{a}_{nn}x_n &= \tilde{b}_n \end{cases} \tag{5.2}$$

where for$i,j=1,\cdots,n$; $\tilde{a}_{ij}$'s and $\tilde{b}_i$'s are fuzzy numbers.

Lemma 5.1

Consider two systems of linear equations as follows

$$\tilde{\mathbf{A}}\mathbf{x}=\tilde{\mathbf{b}}. \tag{5.3}$$

and

$$[\tilde{\mathbf{A}}]^1\mathbf{x}=[\tilde{\mathbf{b}}]^1. \tag{5.4}$$

If system (5.3) has a solution then the system (5.4) has a solution. Furthermore if $\mathbf{x}^*$ be the solution of (5.3) then $\mathbf{x}^*$ is the solution of (5.4) too.
Proof: It is straightforward. □

Corollary 5.2 *If system (5.4) hasn't any solution then the system (5.3) hasn't any solution too.*

6. Solution of fuzzy linear equations

In this Section we try to solve a system of fuzzy linear equations with $m-$degree polynomial-form fuzzy number coefficients. Let for positive integer m, all $\tilde{a}_{ij}$'s and $\tilde{b}_i$'s are fuzzy numbers with $m-$degree polynomial-form, for $i,j=1,\cdots,n$. Our equations now are as follows:

$$\sum_{j=1}^{n}\tilde{a}_{ij}x_j=\tilde{b}_i\ ,\quad i=1,\ldots,n. \tag{6.1}$$

where for $i,j=1,\ldots,n$; $\tilde{a}_{ij},\tilde{b}_i\in\mathcal{PF}_m(\mathbb{R})$, and $x_j\in\mathbb{R}$. We can consider (6.1) as

$$\sum_{j=1}^{n}(\underline{a_{ij}}(r),\overline{a_{ij}}(r))x_j=(\underline{b_i}(r),\overline{b_i}(r))\ ,\quad i=1,\ldots,n.$$

It means that

$$\sum_{x_j\geq 0}\underline{a_{ij}}(r)x_j+\sum_{x_j<0}\overline{a_{ij}}(r)x_j=\underline{b_i}(r),\quad i=1,\ldots,n, \tag{6.2}$$

and

$$\sum_{x_j\geq 0}\overline{a_{ij}}(r)x_j+\sum_{x_j<0}\underline{a_{ij}}(r)x_j=\overline{b_i}(r),\quad i=1,\ldots,n. \tag{6.3}$$

Let $x_j=x'_j-x''_j$, where

$$x'_j=\begin{cases} x_j, & x_j\geq 0,\\ 0, & x_j<0,\end{cases}$$

and

$$x''_j=\begin{cases} 0, & x_j\geq 0,\\ -x_j, & x_j<0,\end{cases}$$

thus $x'_j,x''_j\geq 0$, and we have

$$\sum_{j=1}^{n}\underline{a_{ij}}(r)x'_j-\sum_{j=1}^{n}\overline{a_{ij}}(r)x''_j=\underline{b_i}(r),\quad i=1,\ldots,n, \tag{6.4}$$

and

$$\sum_{j=1}^{n}\overline{a_{ij}}(r)x'_j-\sum_{j=1}^{n}\underline{a_{ij}}(r)x''_j=\overline{b_i}(r),\quad i=1,\ldots,n. \tag{6.5}$$

Since $\tilde{a}_{ij}$'s and $\tilde{b}_i$'s are fuzzy numbers with $m-$degree polynomial-form, there exist coefficients c_{ijk}, d_{ijk}, e_{ik} and f_{ik} such that

$$\begin{cases} \underline{a_{ij}}(r)=\sum_{k=0}^{m}c_{ijk}r^k, & i,j=1,\ldots,n,\\ \overline{a_{ij}}(r)=\sum_{k=0}^{m}d_{ijk}r^k, & i,j=1,\ldots,n,\\ \underline{b_i}(r)=\sum_{k=0}^{m}e_{ik}r^k, & i=1,\ldots,n,\\ \overline{b_i}(r)=\sum_{k=0}^{m}f_{ik}r^k, & i=1,\ldots,n.\end{cases} \tag{6.6}$$

By substituting (6.6) into the equations (6.4) and (6.5), we have

$$\sum_{j=1}^{n}\sum_{k=0}^{m}c_{ijk}r^kx'_j-\sum_{j=1}^{n}\sum_{k=0}^{m}d_{ijk}r^kx''_j=\sum_{k=0}^{m}e_{ik}r^k,\quad i=1,\ldots,n,$$

and

$$\sum_{j=1}^{n}\sum_{k=0}^{m} d_{ijk}r^k x'_j - \sum_{j=1}^{n}\sum_{k=0}^{m} c_{ijk}r^k x''_j = \sum_{k=0}^{m} f_{ik}r^k, \quad i = 1,\ldots,n.$$

Consequently for all $i = 1,2,\ldots,n$, and $k = 0,1,\ldots,m$, we have

$$\sum_{j=1}^{n} c_{ijk}x'_j - \sum_{j=1}^{n} d_{ijk}x''_j = e_{ik}, \tag{6.7}$$

and

$$\sum_{j=1}^{n} d_{ijk}x'_j - \sum_{j=1}^{n} c_{ijk}x''_j = f_{ik}. \tag{6.8}$$

Considering

$$\mathbf{C} = \begin{pmatrix} c_{110} & \cdots & c_{1n0} \\ \vdots & & \vdots \\ c_{11m} & \cdots & c_{1nm} \\ \vdots & & \vdots \\ c_{n10} & \cdots & c_{nn0} \\ \vdots & & \vdots \\ c_{n1m} & \cdots & c_{nnm} \end{pmatrix}, \quad \mathbf{D} = \begin{pmatrix} d_{110} & \cdots & d_{1n0} \\ \vdots & & \vdots \\ d_{11m} & \cdots & d_{1nm} \\ \vdots & & \vdots \\ d_{n10} & \cdots & d_{nn0} \\ \vdots & & \vdots \\ d_{n1m} & \cdots & d_{nnm} \end{pmatrix}$$

and

$$\mathbf{e} = \begin{pmatrix} e_{10} \\ \vdots \\ e_{1m} \\ \vdots \\ e_{n0} \\ \vdots \\ e_{nm} \end{pmatrix}, \quad \mathbf{f} = \begin{pmatrix} f_{10} \\ \vdots \\ f_{1m} \\ \vdots \\ f_{n0} \\ \vdots \\ f_{nm} \end{pmatrix}, \quad \mathbf{x}' = \begin{pmatrix} x'_1 \\ \vdots \\ x'_n \end{pmatrix}, \quad \mathbf{x}'' = \begin{pmatrix} x''_1 \\ \vdots \\ x''_n \end{pmatrix}, \quad \mathbf{y} = \begin{pmatrix} \mathbf{x}' \\ \mathbf{x}'' \end{pmatrix}$$

one should solve the following system of linear equations,

$$\mathbf{Sy} = \mathbf{d}, \tag{6.9}$$

where

$$S = \begin{pmatrix} \mathbf{C} & -\mathbf{D} \\ \mathbf{D} & -\mathbf{C} \end{pmatrix}, \quad \mathbf{x} = \begin{pmatrix} \mathbf{x}' \\ \mathbf{x}'' \end{pmatrix}, \quad \mathbf{d} = \begin{pmatrix} \mathbf{e} \\ \mathbf{f} \end{pmatrix}$$

and we have $\mathbf{x} = \mathbf{x}' - \mathbf{x}''$.

S is a $2n(m+1) \times 2n$ matrix, and since $m > 0$, we try to solve the following least squares problem:

$$\min_{\mathbf{y}\in\mathbb{R}^{2n}} \|\mathbf{Sy} - \mathbf{d}\|_2. \tag{6.10}$$

It is well known that the solution of (6.10) is equal to the solution of its corresponding normal system of equations as follows:

$$\mathbf{S}^T\mathbf{S}\mathbf{y} = \mathbf{S}^T\mathbf{d}. \tag{6.11}$$

Lemma 6.1

The system (6.11) always has a solution even when $\mathbf{S}^T\mathbf{S}$ is singular.
Proof: See (23).

□

Let $\mathbf{x}^*$ and $\overline{\mathbf{x}}$ be the solution of systems (5.3) and (5.4), respectively and let $\tilde{\mathbf{x}}$ and $\widehat{\mathbf{x}}$ be the obtained solutions corresponding to the systems (6.9) and (6.11), respectively. From the last lemma it is known that $\widehat{\mathbf{x}}$ always exists.

Lemma 6.2

Let all $\tilde{a}_{ij}$'s and $\tilde{b}_i$'s have $m-$degree polynomial-form. If $\mathbf{x}^*$ exists then $\overline{\mathbf{x}}$ and $\tilde{\mathbf{x}}$ exist and $\mathbf{x}^* = \overline{\mathbf{x}} = \tilde{\mathbf{x}} = \widehat{\mathbf{x}}$.
Proof: It is straightforward.

□

$\widehat{\mathbf{x}}$ is the least squares solution of the system $\tilde{\mathbf{A}}\mathbf{x} = \tilde{\mathbf{b}}$, also if it's corresponding vector $\mathbf{y}$ satisfies in the system $\mathbf{S}\mathbf{y} = \mathbf{d}$ then $\widehat{\mathbf{x}}$ is the exact solution of the system $\tilde{\mathbf{A}}\mathbf{x} = \tilde{\mathbf{b}}$.
If $\mathbf{S}^T\mathbf{S}$ is nonsingular then the solutions of two systems (6.9) and (6.11) are the same and $\mathbf{y}^* = (\mathbf{S}^T\mathbf{S})^{-1}\mathbf{S}^T\mathbf{d}$.
By a simple computations it can be shown that

$$\mathbf{S}^T\mathbf{S} = \begin{pmatrix} \mathbf{L} & -\mathbf{M} \\ -\mathbf{M} & \mathbf{L} \end{pmatrix},$$

where

$$\mathbf{L} = \mathbf{C}^T\mathbf{C} + \mathbf{D}^T\mathbf{D} \quad , \quad \mathbf{M} = \mathbf{C}^T\mathbf{D} + \mathbf{D}^T\mathbf{C}.$$

Thus

$$\mathbf{S}^T\mathbf{S} \longrightarrow \begin{pmatrix} \mathbf{L} - M & \mathbf{L} - \mathbf{M} \\ -\mathbf{M} & \mathbf{L} \end{pmatrix} \longrightarrow \begin{pmatrix} \mathbf{L} - \mathbf{M} & 0 \\ -\mathbf{M} & \mathbf{L} + \mathbf{M} \end{pmatrix},$$

therefore

$$|\mathbf{S}^T\mathbf{S}| = |\mathbf{L} - \mathbf{M}| \cdot |\mathbf{L} + \mathbf{M}|.$$

Also we have

$$\mathbf{L} - \mathbf{M} = (\mathbf{C} - \mathbf{D})^T(\mathbf{C} - \mathbf{D}) \quad , \quad \mathbf{L} + \mathbf{M} = (\mathbf{C} + \mathbf{D})^T(\mathbf{C} + \mathbf{D}).$$

If $\mathbf{C} - \mathbf{D}$ be a full rank matrix then $\mathbf{L} - \mathbf{M}$ is nonsingular and if $\mathbf{C} + \mathbf{D}$ be a full rank matrix then $\mathbf{L} + \mathbf{M}$ is nonsingular. i.e. if $rank(\mathbf{C} \pm \mathbf{D}) = n$ then $\mathbf{S}^T\mathbf{S}$ is nonsingular.

7. Numerical solution of fuzzy linear equations

In this section we consider a system of fuzzy linear equations as

$$\tilde{\mathbf{A}}\mathbf{x} = \tilde{\mathbf{b}}, \tag{7.1}$$

where for $i, j = 1, \cdots, n$; $\tilde{a}_{ij}$'s and $\tilde{b}_i$'s, are from LR type fuzzy numbers.

For the purpose of solving (7.1), we consider the following system of equations

$$\tilde{\mathbf{A}}^{(m)}\mathbf{x} = \tilde{\mathbf{b}}^{(m)}, \tag{7.2}$$

where for $i,j = 1,\cdots,n$; the quantities $\tilde{a}_{ij}^{(m)}$'s and $\tilde{b}_i^{(m)}$'s are the nearest approximations of the given numbers $\tilde{a}_{ij}$'s and $\tilde{b}_i$'s out of $\mathcal{PF}_m(\mathbb{R})$, respectively. Existence of the nearest approximations of the given fuzzy numbers is known from Lemma 3.6.
From Lemma 6.1 we know that this system has a least squares solution. Using the process in the Section 6 we obtain a least squares solution $\mathbf{x}^{(m)}$ for this system. We name this solution, the *m-degree nearest least squares* ($m-$**DNLS**) solution of the system (5.3). Thus we have

$$\tilde{\mathbf{A}}^{(m)}\mathbf{x}^{(m)} = \tilde{\mathbf{b}}^{(m)}, \tag{7.3}$$

Choosing m depends on the shape of left and right spread functions L and R, and the derivation order of them.

8. Numerical examples

In this Section we present some numerical examples which have been solved on both $m-$degree polynomial-form fuzzy number coefficients and general LR fuzzy number coefficients.

Example 8.1 *Consider the following 2 by 2 system of equations with $m = 1$:*

$$\begin{cases} (-1+2r, 4-2r)x_1 + (-2+3r, 3-2r)x_2 = (-8+13r, 17-10r), \\ (1+r, 4-r)x_1 + (2r, 5-2r)x_2 = (2+8r, 23-8r), \end{cases}$$

For this system of equations by solving the least squares system we have $x_1 = 2$ and $x_2 = 3$.

Example 8.2 *Consider the following 2 by 2 system of equations with $m = 1$:*

$$\begin{cases} (-1+r, 3-r)x_1 + (1+2r, 4-r)x_2 = (-12+11r, 17-8r), \\ (-1+2r, 3-2r)x_1 + (3r, 6-2r)x_2 = (-15+19r, 23-16r), \end{cases}$$

For this system of equations by solving the least squares system we have $x_1 = -5$ and $x_2 = 3$.

Example 8.3 *Consider the following 2 by 2 system of equations with $m = 2$:*

$$\begin{cases} \tilde{a}_{11}x_1 + \tilde{a}_{12}x_2 = \tilde{b}_1, \\ \tilde{a}_{21}x_1 + \tilde{a}_{22}x_2 = \tilde{b}_2, \end{cases}$$

where

$$\begin{aligned} \tilde{a}_{11} &= (3r + r^2, 7 - 3r + 2r^2) \\ \tilde{a}_{12} &= (2r + r^2, 4 - 2r + 2r^2) \\ \tilde{a}_{21} &= (1 + 2r + r^2, 8 - 3r + r^2) \\ \tilde{a}_{22} &= (1 + 2r + r^2, 6 - 3r + 2r^2) \\ \tilde{b}_1 &= (48.45r + 17.1r^2, 111.15 - 48.45r + 34.2r^2) \\ \tilde{b}_2 &= (17.1 + 34.2r + 17.1r^2, 131.1 - 51.3r + 19.95r^2), \end{aligned}$$

For this system of equations by solving the least squares system we have $x_1 = 14.85$ and $x_2 = 2.25$.

Example 8.4 *Consider the following 2 by 2 system of equations*

$$\begin{cases} \tilde{a}_{11}x_1 + \tilde{a}_{12}x_2 = \tilde{b}_1, \\ \tilde{a}_{21}x_1 + \tilde{a}_{22}x_2 = \tilde{b}_2, \end{cases}$$

where

$$\begin{cases} \tilde{a}_{11} = \tilde{a}_{22} = (2 + \ln\{(e-1)r+1\}, 4 - \ln\{(e-1)r+1\}), \\ \tilde{a}_{12} = \tilde{a}_{21} = (3 + \ln(2r+1), 3 + 2\ln 3 - \ln(2r+1)), \\ \tilde{b}_1 = (13 + \ln\{((e-1)r+1)^2(2r+1)^3\}, 17 + 6\ln 3 - \ln\{((e-1)r+1)^2(2r+1)^3\}), \\ \tilde{b}_2 = (12 + \ln\{((e-1)r+1)^3(2r+1)^2\}, 18 + 4\ln 3 - \ln\{((e-1)r+1)^3(2r+1)^2\}). \end{cases}$$

By choosing $m = 3$, we have

$$\begin{cases} \underline{a_{11}} = \underline{a_{22}} = 2.01846 + 1.48660r - 0.64225r^2 + 0.137185r^3, \\ \overline{a_{11}} = \overline{a_{22}} = 3.98154 - 1.48660r + 0.64225r^2 - 0.13718r^3, \\ \underline{a_{12}} = \underline{a_{21}} = 3.02605 + 1.66763r - 0.75820r^2 + 0.16313r^3, \\ \overline{a_{12}} = \overline{a_{21}} = 5.17117 - 1.66763r + 0.75820r^2 - 0.16313r^3, \end{cases}$$

$$\begin{cases} \underline{b_1} = 13.11509 + 7.97609r - 3.55909r^2 + 0.76375r^3, \\ \overline{b_1} = 23.47658 - 7.97609r + 3.55909r^2 - 0.76375r^3, \\ \underline{b_2} = 12.107499 + 7.79506r - 3.44314r^2 + 0.73781r^3, \\ \overline{b_2} = 22.28695 - 7.79506r + 3.44314r^2 - 0.73781r^3, \end{cases}$$

and the 3−DNLS solution of this system is $x_1 = 2.000000$ and $x_2 = 3.000000$.

Example 8.5 *Consider the following 2 by 2 system of equations*

$$\begin{cases} \tilde{a}_{11}x_1 + \tilde{a}_{12}x_2 = \tilde{b}_1, \\ \tilde{a}_{21}x_1 + \tilde{a}_{22}x_2 = \tilde{b}_2, \end{cases}$$

where

$$\begin{cases} \tilde{a}_{11} = \tilde{a}_{21} = (\ln(r+1), \ln(4-2r)), \\ \tilde{a}_{12} = (\ln(2r+1), \ln(4-r)), \\ \tilde{a}_{22} = (\ln(r+1), \ln(5-3r)), \\ \tilde{b}_1 = (\ln\{(r+1)^2(2r+1)\}, \ln\{(4-2r)^2(4-r)\}), \\ \tilde{b}_2 = (3\ln(r+1), \ln\{(4-2r)^2(5-3r)\}). \end{cases}$$

the 1−DNLS solution of this system is $x_1 = 2.022850$ and $x_2 = 0.06491$, meanwhile the 2−DNLS solution of this system is $x_1 = 2 + 0.0 \times 10^{-6}$ and $x_2 = 1 + 0.0 \times 10^{-6}$, also the 3−DNLS solution of this system is $x_1 = 2 + 0.0 \times 10^{-13}$ and $x_2 = 1 + 0.0 \times 10^{-13}$.

9. Conclusion

In this chapter a new method was proposed to solve a system of linear equations. If all coefficients are polynomial form fuzzy numbers then the system may have an exact solution but otherwise we can approximate fuzzy coefficients with m-degree polynomial-form fuzzy numbers and find an approximated solution of the system. Choosing m depends on the shape of left and right spread functions L and R, and the derivation order of them. The presented method can be applied on any system of equations with LR fuzzy number coefficients.

10. References

[1] Abbasbandy, S. & Alavi, M. (2005), A method for solving fuzzy linear system, *Iranian Journal of Fuzzy Systems*, Vol. 2, pp. 37–43.

[2] Abbasbandy, S. & Amirfakhrian, M. (2006). The nearest approximation of a fuzzy quantity in parametric form, *Applied Mathematics and Computations*, Vol. 172, pp. 624–632.

[3] Abbasbandy, S. & Amirfakhrian, M. (2006). The nearest trapezoidal form of a generalized left right fuzzy number, *International Journal of Approximate Reasoning* Vol. 43, pp. 166-178.

[4] Abbasbandy, S. & Asady B., (2004). The nearest trapezoidal fuzzy number to a fuzzy quantity, *Applied Mathematics and Computations* Vol. 156, pp. 381-386.

[5] Abbasbandy, S.; Ezzati, R. & Jafarian, A. (2006). LU decomposition method for solving fuzzy system of linear equations, *Applied Mathematics and Computations* Vol. 172 pp. 633–643.

[6] Abbasbandy, S.; Jafarian, A. & Ezzati, R. (2005). Conjugate gradient method for fuzzy symmetric positive definite system of linear equations,*Applied Mathematics and Computations* Vol. 171, pp. 1184–1191.

[7] Abbasbandy, S. & Jafarian, A. (2006). Steepest descent method for system of fuzzy linear equations, *Applied Mathematics and Computations* Vol. 175, pp. 823–833.

[8] Abbasbandy, S.; Otadi M. & Mosleh, M. (2008). Minimal solution of general dual fuzzy linear systems, *Chaos, Solitons and Fractals*, Vol. 37, pp. 1113-1124.

[9] Amirfakhrian, M. (2007). Numerical solution of a fuzzy system of linear equations with polynomial parametric form, *International Journal of Computer Mathematics*, Vol. 84, pp. 1089-1097.

[10] Amirfakhrian, M. (2010). Properties of parametric form approximation operator of fuzzy numbers, *Analele Stiintifice ale Universitatii Ovidius Constanta*, Vol. 18, pp. 23–34.

[11] Asadi, B.; Abbasbandy, S. & Alavi, M. (2005). Fuzzy general linear systems, *Applied Mathematics and Computations* Vol. 169, pp. 34–40.

[12] Ban, A. (2008). Approximation of fuzzy numbers by trapezoidal fuzzy numbers preserving the expected interval, Fuzzy Sets and Systems, Vol. 159, pp. 1327-1344.

[13] Ban, A. I. (2009) Triangular and parametric approximations of fuzzy numbers-inadvertences and corrections, *Fuzzy Sets and Systems*, , Vol. 160, pp. 3048-3058.

[14] Chong-Xin, W. & Ming, M. (1991). Embedding problem of of fuzzy number space: Part I, *Fuzzy Sets and Syatems*, Vol. 44, pp. 33–38.

[15] Dehghan, M. & Hashemi, B. (2006). Iterative solution of fuzzy linear systems, *Applied Mathematics and Computations*, Vol. 175, pp. 645–674.

[16] Dehghan, M. & Hashemi, B. (2006). Solution of the fully fuzzy linear systems using the decomposition procedure, *Applied Mathematics and Computations*, Vol. 182 pp. 1568–1580.

[17] Dehghan, M.; Hashemi, B. & Ghatee, M. (2006). Computational methods for solving fully fuzzy linear systems, *Applied Mathematics and Computations*, Vol. 179, pp. 328-343.

[18] Delgado, M.; Vila M. A. & Voxman, W. (1998). On a canonical representation of fuzzy numbers, *Fuzzy Sets and Systems* Vol. 93, pp. 125-135.

[19] Dubois, D. & Prade, H. (1978), Operations on fuzzy numbers, *Intenational Journal of Systems Sci.* Vol. 9, pp. 613–626.

[20] Dubois, D. & Prade, H. (1980). Fuzzy Sets and Systems: Theory and Application, Academic Press, New York.

[21] Friedman, M.; Ming, M. & Kandel, A. (1998). Fuzzy linear systems, *Fuzzy Sets and Systems* Vol. 96, pp. 201–209.
[22] Friedman, M.; Ming, M. & Kandel, A. (2000). Duality in Fuzzy linear systems, *Fuzzy Sets and Systems* Vol. 109, pp. 55–58.
[23] Gill, P. E.; Murray, W. & Wright, M. H. (1991). Numerical Linear Algebra and Optimization, Addison-Wesley Publishing Company.
[24] Ghanbari, R. & Mahdavi-Amiri, N. (2010). New solutions of *LR* fuzzy linear systems using ranking functions and *ABS* algorithms, *Applied Mathematical Modelling*, Vol. 34, pp. 3363-3375.
[25] Grzegorzewski, P. (2002). Nearest interval approximation of a fuzzy number, *Fuzzy Sets and Systems* Vol. 130, pp. 321-330.
[26] Grzegorzewski, P. (2008). Trapezoidal approximations of fuzzy numbers preserving the expected interval - Algorithms and properties, *Fuzzy Sets and Systems*, Vol. 159, pp. 1354-1364.
[27] Grzegorzewski, P. & Mrówka, E. (2003). Trapezoidal approximations of fuzzy numbers, In: Fuzzy Sets and Systems IFSA 2003, T. Bilgic, B. De Baets, O. Kay nak (Eds.), Lecture Notes in Artificial Intelligence 2715, Springer, pp. 237-244.
[28] Grzegorzewski, P. & Mrówka, E. (2005). Trapezoidal approximations of fuzzy numbers, *Fuzzy Sets and Systems* Vol. 153, pp. 115-135.
[29] Grzegorzewski, P. & Mrówka, E. (2007). Trapezoidal approximations of fuzzy numbers-revisited, *Fuzzy Sets and Systems*, Vol. 158, pp. 757-768.
[30] Liua, X. & Lin, H. (2007). Parameterized approximation of fuzzy number with minimum variance weighting functions, *Mathematica and Computer Modelling*, Vol. 46, pp. 1398-1409.
[31] Ma, M.; Kandel, A. & Friedman, M. (2000). A new approach for defuzzification, *Fuzzy Sets and Systems* Vol. 111, pp. 351-356.
[32] Ma, M.; Kandel, A. & Friedman, M. (2002). Correction to "A new approach for defuzzification", *Fuzzy Sets and Systems* Vol. 128, pp. 133-134.
[33] Miao, S. X.; Zheng, B. & Wang, K. (2008). Block SOR method for fuzzy linear systems, *Journal of Applied Mathematics and Computations*, Vol. 26, pp. 201-218.
[34] Nasibov, Efendi N. & Peker, S. (2008). On the nearest parametric approximation of a fuzzy number, Fuzzy Sets and Systems, Vol. 159, pp. 1365-1375.
[35] Voxman, W. (1998). Some remarks on distance between fuzzy numbers, *Fuzzy Sets and Systems* Vol. 100, pp. 353–365.
[36] Wang, K. & Zheng, B. (2007). Block iterative methods for fuzzy linear systems, *Journal of Applied Mathematics and Computating*, Vol. 25, pp. 119-136.
[37] Yao, J.S. & Wu, K. (2000). Ranking fuzzy numbers based on decomposition principle and signed distance, *Fuzzy Sets and Systems* Vol. 116, pp. 275-288.
[38] Yeh, C. T. (2007). A note on trapezoidal approximations of fuzzy numbers *Fuzzy Sets and Systems*, Vol. 158, pp. 747-754.
[39] Yeh, C. T. (2008). Trapezoidal and triangular approximations preserving the expected interval, *Fuzzy Sets and Systems*, Vol. 159, pp. 1345-1353.
[40] Yin, J. F. & Wang, K. (2009). Splitting iterative methods for fuzzy system of linear equations, *Computational Mathematics and Modeling*, Vol. 20, pp. 326-335.
[41] Zadeh, L. A. (1965). Fuzzy Sets, *Information and Control*, Vol. 8, pp. 338-353.
[42] Zeng, W. & Li, H. (2007). Weighted triangular approximation of fuzzy numbers, *International Journal of Approximate Reasoning*, Vol. 46, pp. 137-150.

[43] Zheng, B. & Wang, K. (2006). General fuzzy linear systems, *Applied Mathematics and Computations*, Vol. 181, pp. 1276-1286.

[44] Zimmermann, H. J. (1991). Fuzzy Set Theory and Its Applications, 2nd Edition, Kluwer Academic, Boston.